Additional Features

- **Activities**, which help you learn and review key structures and important processes.
- **"A Step Further"** topics, which expand on the material covered in the textbook.
- **Flashcards** for each chapter offer a convenient way to learn and review the many new terms introduced in the textbook.
- A complete **glossary** for quick access to definitions.

QR Codes

Using your smartphone or tablet, scan the QR codes that appear throughout the textbook (or type in the Web address shown) to instantly access animations, videos, and visual summaries on the Companion Website. (QR code reader app required. These are available for free from your device's app store.)

To see the video
Nature and Nurture,
go to

2e.mindsmachine.com/1.1

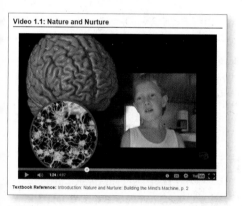

Video 1.1: Nature and Nurture

► 1:24 / 4:07

Textbook Reference: Introduction: Nature and Nurture: Building the Mind's Machine, p. 2

Videos

Videos present real-world examples of some of the key concepts and conditions discussed in the textbook.

chapter
1

An Introduction to Brain and Behavior

Nature and Nurture: Building the Mind's Machine

We humans have a long history of using contemporary technology as a metaphor for the mysterious workings of the brain. Scholars of old, influenced by inventions such as aqueducts, plumbing, and ornamental fountains, proposed that behavior was the result of liquid "animal spirits" jetting around within the body. Later thinkers emphasized the possible importance of wires, switches, and relays: the technology of their day. Today it is commonplace to see the brain described as a computer, with "hardware" and "software." Perhaps tomorrow's neuroscientists will describe the brain in terms of holograms, quantum devices, or a technology that has yet to be imagined.

Modern research aims to describe the complicated machine within each of our heads, but we also want to know how the operation of the brain produces the *mind*—the perceptions, emotions, thoughts, self-awareness, and other cognitive processes that inform our behavior. During the twentieth century, a lot of ink was spilled over the "nature-nurture" con-

troversy, with scholars arguing passionately about the extent to which mental characteristics and abilities are the result of learning experiences versus innate, "hardwired" genetic programs.

The two perspectives have often been presented as mutually incompatible alternatives, but thanks to more-powerful techniques, we have come to realize that there is nothing controversial about nature versus nurture: they are two sides of the same coin. Consider the case of rat pups who have inattentive mothers. As adults, the formerly neglected pups show elevated stress hormone responses to stressors that have little effect on rats that were not neglected as pups (T. Y. Zhang and Meaney, 2010). How does this lasting reactivity develop—through experience or as a result of innate biological factors? Both, it turns out. As we'll see later in this chapter, and throughout the book, the mind and its machine are shaped by a precise combination of genes and experience, inextricably tied together.

To see the video
Nature and Nurture,
go to

2e.mindsmachine.com/1.1

D0721748

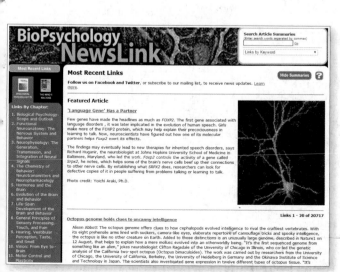

BIOLOGICAL PSYCHOLOGY NewsLink
2e.mindsmachine.com/news

This invaluable online resource helps you make connections between the science of biological psychology and your daily life, and keeps you apprised of the latest developments in the field. The site is updated 3–4 times per week, and contains thousands of news stories organized both by keyword and by textbook chapter.

Find us on Facebook (Biopsychology).

COMPANION WEBSITE Resources (2e.mindsmachine.com)

The Mind's Machine Companion Website includes animations, videos, and activities for each chapter (listed below) as well as an interactive version of each chapter's visual summary. Animations and Videos are referenced throughout the book with QR codes and direct Web addresses. Activities are referenced in each chapter's summary.

Animations & Videos

Activities

The Mind's Machine
Foundations of Brain and Behavior

SECOND EDITION

The Mind's Machine

Foundations of Brain and Behavior

INTERNATIONAL SECOND EDITION

Neil V. Watson
Simon Fraser University

S. Marc Breedlove
Michigan State University

This version of the text has been adapted and
customized. Not for sale in the USA or Canada.

 SINAUER ASSOCIATES

NEW YORK   OXFORD
OXFORD UNIVERSITY PRESS

Oxford University Press is a department of the University of Oxford. It furthers
the University's objective of excellence in research, scholarship, and education
by publishing worldwide.

© 2018 Oxford University Press
Sinauer Associates is an imprint of Oxford University Press.

Published in the United States of America by Oxford University Press
198 Madison Avenue, New York, NY 10016, United States of America

ISBN 9781605357393

Printing number: 9 8 7 6 5 4 3 2 1

Printed in the United States of America

Brief Contents

Table of Contents

CHAPTER 3
Neurophysiology: The Generation, Transmission, and Integration of Neural Signals **50**

CHAPTER 4
The Chemistry of Behavior: Neurotransmitters and Neuropharmacology **78**

CHAPTER 5
The Sensorimotor System 110

CHAPTER 6
Hearing, Balance, Taste, and Smell 144

CHAPTER 7
Vision: From Eye to Brain 174

CHAPTER 8
Hormones and Sex   208

CHAPTER 9
Homeostasis: Active Regulation of the Internal Environment   254

CHAPTER 10
Biological Rhythms and Sleep   278

CHAPTER 11
Emotions, Aggression, and Stress   310

CHAPTER 12
Psychopathology: The Biology of Behavioral Disorders   338

CHAPTER 13
Memory, Learning, and Development   368

Preface

It's getting difficult to browse the internet, open a newspaper, or flip through a magazine without encountering reports of astonishing discoveries about the brain's structure, function, and—alas—malfunction. The main reason for all this coverage is that the subject matter is intrinsically fascinating; who has not pondered their own consciousness, marveled at their many sensory experiences, or wondered how a small and lumpy organ can process so much information? But another reason neuroscience is in the news so frequently is simply that it has become one of the most active branches of science. The pace of discoveries about brain and behavior has increased at an exponential rate over the last few decades.

Every new edition of one of our books requires substantial updating because so much is happening all the time. (It's exciting, but boy, do we read a lot of reports and articles!) In fact, by far the hardest part of our job as authors lies in deciding which discoveries to include and which to (reluctantly) leave out: As the Red Queen remarked to Alice in Wonderland, "it takes all the running you can do, to keep in the same place." Our website (2e.mindsmachine.com/news) boasts a collection of more than 20,000 news stories, all drawn from the mainstream media, that relate to the topics covered in the book. You can follow updates on the website, via email, or Facebook (www.facebook.com/biopsychology).

While we are sampling from this almost boundless scientific smorgasbord, we have to watch our weight. Our goal for *The Mind's Machine*, Second Edition is to introduce you to the *basics* of behavioral neuroscience in a way that focuses on the foundational topics in the field—with a generous sprinkling of the newest and most fascinating discoveries—and leaves you with an appetite for more. Whether you are beginning a program of study centered on the brain and behavior, or are just adding some breadth to your education, you will find that behavioral neuroscience now permeates all aspects of modern psychology, along with related life sciences like physiology, biology, and the health sciences. But that's not all. The tools and techniques of behavioral neuroscience are also creating new ways of looking at questions in many nontraditional areas, such as economics, the performing arts, anthropology, sociology, computer science, and engineering. Researchers are beginning to probe mental processes that seemed impenetrable only a decade or two ago: the neural bases of decision making, love and attachment, memory and learning, consciousness, and much of what we call the mind. A few examples of formerly mysterious questions that are being answered with cutting-edge research include:

- Do prenatal events influence the probability that a child will develop a heterosexual or homosexual orientation?
- Does the brain make new neurons throughout life, in numbers large enough to make a functional difference?
- Can we improve memory performance with some drugs, and use other drugs to erase unwanted, traumatic memories?
- What happens in the brain as we develop trust in another person?
- Does strong liking for sweet foods involve the same brain mechanisms as addiction to drugs?
- How can we share so many genes with chimpanzees and other primates, and yet be so different from them?
- How can recent discoveries about the neural control of appetite help us to curb the obesity epidemic?
- Does a gene that predisposes for Alzheimer's disease in old age actually improve cognitive functioning earlier in life?

Understanding the research probing these sorts of questions requires some familiarity with the physiology of behavior and experience. Our aim in *The Mind's Machine*, Second Edition is to provide a foundation that places these and other important problems in a unified scientific context.

We've found that students enrolled in our courses have diverse academic backgrounds and personal interests. In this book, we've tried to avoid making too many assumptions about our readers, and have focused on providing both behavioral and biological perspectives on major topics. If you've had some high-school level biology you should have no trouble with most of the material in the book.

For those readers who have more experience in science—or who want more detail—we have peppered the chapters with embedded links to more advanced material located on our website. These links, called A Step Further, are just one of several novel features we have included to aid your learning. Throughout the book you will find QR codes (small square bar codes) that will link your smartphone to animated versions of many figures, video clips, and more. (You'll need to download an app to read the codes; a variety of free or inexpensive Code Readers are available for the major smartphones like iPhone, Android, and Blackberry.) Or you may use your computer to go to the web address provided below every QR code.

Each chapter also features a segment called Researchers at Work, which illustrates the nuts and bolts of experimentation through real-world examples, and a new segment called Signs & Symptoms that relates a real-world clinical issue relevant to the chapter topic. Every few pages, you will find a feature called How's it Going?, with self-test questions that will help you to gauge your progress. And every chapter ends with a Visual Summary, an innovative combination of the main points and figures from the chapter, which you can also view in an interactive format on the companion website. We encourage you to explore the website for the book (2e.mindsmachine. com), which contains a free comprehensive set of study questions. This website is a powerful companion to the textbook that enhances the learning experience with a variety of multi-media resources.

The chapter lineup in this edition of *The Mind's Machine* encompasses several major themes. In the opening chapters, we trace the origins of behavioral neuroscience and introduce you to the structure of the brain, both as seen by the naked eye and as revealed by the microscope. We discuss how the cells of the brain use electrical signals to process information, and how they transmit that information to other cells within larger circuits. Along the way we'll look at the ways in which drugs affect nerve cells in order to change behavior, as well as some of the remarkable technology that lets us study the activity of the conscious brain as it perceives and thinks.

In the middle part of the book we look at the neural systems that underlie fundamental capabilities like feeling, moving, seeing, smelling, and hearing. We'll also consider biological and behavioral aspects of "mission-critical" functions such as feeding, sleeping, and sexual behavior. And we'll look at how the endocrine system acts as an interface between the brain and the rest of the body, as well as the reverse—ways in which the environment and behavior alter hormones and thus alter brain activity.

In the latter part of the book we turn to some of the high-level emotional and cognitive processes that color our lives and define us as individuals. We'll survey the systems that allow us to learn and remember information and skills, and the brain systems dedicated to language and spatial cognition. Research on processes of attention has made great progress in recent years, and we'll also consider consciousness and decision-making from a neuroscientific perspective. Finally, we'll review some of the consequences of brain dysfunction, ranging from psychopathology to behavioral manifestations of brain damage, and some of the innovative strategies being developed to counter these problems.

As you make your way through the book, you'll learn that one of the outstanding features of the brain is its ability to remodel. Every new experience, every piece of information that you learn, every skill that you master, causes changes in the brain that can alter your future behavior. The changes may involve physical alterations in the connections between cells, or in the chemicals they use to communicate, or even the addition of whole new cells and circuits. It's a property that we neuroscientists refer to as "plasticity." And it's something that we aim to exploit—if we've done our job properly, *The Mind's Machine*, Second Edition should cause lots of changes in *your* brain. We hope you enjoy the process.

Neil V. Watson   S. Marc Breedlove

2e.mindsmachine.com/watson   2e.mindsmachine.com/breedlove

We welcome feedback on any aspect of
The Mind's Machine, Second Edition.
Simply drop us a line at mindsmachine@sinauer.com.

Acknowledgments

This book bears the strong imprint of our late colleagues and coauthors Arnold Leiman (1932–2000) and Mark Rosenzweig (1922–2009). Arnie and Mark prepared the earliest editions of our more advanced text, *Biological Psychology*, and many illustrations and concepts in *this* book originated in their minds' machines.

In writing this book we benefited from the help of many highly skilled people. These include members of the staff of Sinauer Associates: Syd Carroll, Editor; Kathaleen Emerson, Production Editor; Christopher Small, Production Manager; Joanne Delphia, Production Specialist; Jason Dirks, Media and Supplements Editor, and Mara Silver, Production Editor for Media and Supplements. Copy Editor Lou Doucette skillfully edited the text, and Photo Researcher David McIntyre found many of the photographs. Mike Demaray, Craig Durant, and colleagues at Dragonfly Media Group transformed our rough sketches and wish list into the handsome and dynamic art program of this text.

Our greatest source of inspiration (and critical feedback) has undoubtedly been the thousands of students to whom we have had the privilege of introducing the mysteries and delights of behavioral neuroscience, over the course of the last couple of decades. We have benefited from wisdom generously contributed by a legion of academic colleagues, whose advice and critical reviews have enormously improved our books. In particular we are grateful to: Duane Albrecht, Anne E. Powell Anderson, Michael Antle, Benoit Bacon, Scott Baron, Mark S. Blumberg, William Boggan, Eliot A. Brenowitz, Chris Brill, Peter C. Brunjes, Rebecca D. Burwell, Aryn Bush, Catherine P. Cramer, Betty Deckard, Brian Derrick, Karen De Valois, Russell De Valois, Tiffany Donaldson, Rena Durr, Thomas Fischer, Julia Fisher, Loretta M. Flanagan-Cato, Francis W. Flynn, Lauren Fowler, Michael Foy, Kara Gabriel, John D. E. Gabrieli, Jack Gallant, Kimberley P. Good, Diane C. Gooding, Janet M. Gray, James Gross, Ervin Hafter, Mary E. Harrington, Ron Harris, Christian Hart, Chris Hayashi, Wendy Heller, Mark Hollins, Dave Holtzman, Rick Howe, Richard Ivry, Lucia Jacobs, Janice Juraska, Dacher Keltner, Raymond E. Kesner, Mike Kisley, Keith R. Kluender, Leah A. Krubitzer, Joseph E. LeDoux, Diane Lee, Robert Lennartz, Michael A. Leon, Simon LeVay, Jeannie Loeb, Stephen G. Lomber, Jeffrey Love, Donna Maney, Stephen A. Maren, Joe L. Martinez, Jr., John J. McDonald, Robert J. McDonald, James L. McGaugh, Robert L. Meisel, Ralph E. Mistlberger, Jeffrey S. Mogil, Randy J. Nelson, Chris Newland, Miguel Nicolelis, Michelle Niculescu, Lee Osterhout, Linda Perrotti, James Pfaus, Eleni Pinnow, Helene S. Porte, George V. Rebec, Thomas Ritz, Scott R. Robinson, David A. Rosenbaum, Lawrence Ryan, Martin F. Sarter, Jeffrey D. Schall, Stan Schein, Frederick Seil, Dale R. Sengelaub, Victor Shamas, Matthew Shapiro, Arthur Shimamura, Rachel Shoup, Rae Silver, Cheryl L. Sisk, Laura Smale, Robert L. Spencer, Jeffrey Stowell, Steven K. Sutton, Harald K. Taukulis, Jessica Thompson, Sandra Trafalis, Lucy J. Troup, Meg Upchurch, Franco J. Vaccarino, David R. Vago, Cyma Van Petten, Charles J. Vierck, Robert Wickesberg, Christoph Wiedenmayer, Walter Wilczynski, S. Mark Williams, Richard D. Wright, Mark C. Zrull, and Irving Zucker.

A dedicated group of proofreaders pored over the text under severe time constraints; thanks to Will Vickerman and Maria Watson for catching errors we repeatedly missed. The following external reviewers read and critiqued the draft chapters of the Second Edition of *The Mind's Machine*; their contributions and corrections really honed the final product. Any errors that remain are thus entirely our fault.

John Agnew, *University of Colorado, Boulder*

Evangelia Chrysikou, *University of Kansas*

Heidi Day, *University of Colorado*

Steven I. Dworkin, *Western Illinois University*

Joyce A. Furfaro, *Pennsylvania State University*

Eric W. Gobel, *University of Illinois at Chicago*

Gary Greenberg, *Wichita State University*, Emeritus

Karin Hu, *City College of San Francisco*

Karen Jennings, *Keene State College*

Erin Keen-Rinehart, *Susquehanna University*

Ralph Mistlberger, *Simon Fraser University*

Daniel Montoya, *Fayetteville State University*

Antonio A. Nunez, *Michigan State University*

Kathleen Page, *Bucknell University*

John Pellitteri, *Mt. San Antonio College*

Joseph Porter, *Virginia Commonwealth University*

Stephen Sammut, *Franciscan University*

Steve St. John, *Rollins College*

Patrick Steffen, *Brigham Young University*

Bruce Svare, *State University of New York at Albany*

Beth Wee, *Tulane University*

Carmen Westerberg, *Texas State University*

Finally, we would like to thank all our colleagues whose ideas and discoveries make behavioral neuroscience so much fun.

Media and Supplements

to accompany
The Mind's Machine
Second Edition

For the Student

Companion Website
2e.mindsmachine.com

The Mind's Machine, Second Edition Companion Website contains a wide range of study and review resources to help students master the material presented in the textbook and to help engage them in the subject with fascinating examples. Access to the site is free and requires no passcode. (Instructor registration is required in order for students to access the online quizzes.) Tightly integrated with the text, with content corresponding to every major heading in the book, this online resource greatly enhances the learning experience. Key resources are linked throughout the textbook via QR codes and direct Web addresses, making access to animations and videos easy from any smartphone, tablet, or computer.

The Companion Website includes:

- Chapter outlines
- Extensive study questions
- Animations, videos, and activities
- Online, interactive versions of the visual summaries
- Online quizzes (multiple choice and essay)
- Flashcards
- "A Step Further," additional coverage of selected topics
- Complete glossary

Biological Psychology NewsLink
2e.mindsmachine.com/news

This invaluable online resource helps students make connections between the science of biological psychology and their daily lives, and keeps them apprised of the latest developments in the field. The site includes links to thousands of news stories, all organized both by keyword and by textbook chapter. The site is updated 3–4 times per week. Find us on Facebook (facebook.com/Biopsychology).

For the Instructor

Instructor's Resource Library

(Available to qualified adopters)
The Mind's Machine Instructor's Resource Library includes a variety of resources to aid in planning the course, developing lectures, and assessing students.

The Instructor's Resource Library includes:

- *Figures and Tables*: All of the illustrations, photos, and tables from the textbook are provided as both high-resolution and low-resolution JPEGs, all optimized for use in presentation software (such as PowerPoint).
- *PowerPoint Resources*: Two PowerPoint presentations are provided for each chapter of the textbook:
  - All figures, photos, and tables
  - A complete lecture outline, including selected figures
- *Animations*: These detailed animations help enliven lectures and illustrate dynamic processes
- *Videos*: A collection of video segments that illustrate interesting concepts and phenomena
- *Instructor's Manual and Test Bank* in Word format (details below)
- *Computerized Test Bank*: The entire Test Bank is provided in Diploma format (software included), making it easy to quickly assemble exams using any combination of publisher-provided and custom questions. Includes the Companion Website quiz questions.

Instructor's Manual & Test Bank

(Included in the Instructor's Resource Library)
The Mind's Machine, Second Edition Instructor's Manual & Test Bank includes useful resources for planning the course, building lectures, and creating assessments. For each chapter of the textbook, the Instructor's Manual & Test Bank includes the following:

- Chapter overview
- Complete chapter outline
- Detailed key concepts
- Additional references for lecture and course development
- Key terms
- Comprehensive test questions, including multiple choice, fill-in-the-blank, matching, short answer, and essay questions. The Companion Website quiz questions are also included.

Online Quizzing

The Companion Website includes online quizzes that can be assigned by instructors or (at the instructor's discretion) used as self-review exercises by students. Instructors can create custom quizzes with any combination of their own questions and publisher-provided questions. Results of the quizzes are stored in the online gradebook. (Instructors must register in order for their students to be able to access the online quizzes.)

Course Management System Support

Using the Computerized Test Bank provided in the Instructor's Resource Library, instructors can easily create and export quizzes and exams (or the entire test bank) for import into many common course management system, including Blackboard, Moodle, and Desire2Learn.

Value Options

eBook

The Mind's Machine is available as an eBook, in several different formats, including VitalSource CourseSmart, Yuzu, and BryteWave. The eBook can be purchased as either a 180-day rental or a permanent (non-expiring) subscription. All major mobile devices are supported. For details on the eBook platforms offered, please visit www.sinauer.com/ebooks.

Looseleaf Textbook

(ISBN 978-1-60535-444-6)
The Mind's Machine is also available in a three-hole punched, looseleaf format. Students can take just the sections they need to class and can easily integrate instructor material with the text.

The Mind's Machine
Foundations of Brain and Behavior

SECOND EDITION

An Introduction to Brain and Behavior

Nature and Nurture: Building the Mind's Machine

We humans have a long history of using contemporary technology as a metaphor for the mysterious workings of the brain. Scholars of old, influenced by inventions such as aqueducts, plumbing, and ornamental fountains, proposed that behavior was the result of liquid "animal spirits" jetting around within the body. Later thinkers emphasized the possible importance of wires, switches, and relays: the technology of their day. Today it is commonplace to see the brain described as a computer, with "hardware" and "software." Perhaps tomorrow's neuroscientists will describe the brain in terms of holograms, quantum devices, or a technology that has yet to be imagined.

Modern research aims to describe the complicated machine within each of our heads, but we also want to know how the operation of the brain produces the *mind*—the perceptions, emotions, thoughts, self-awareness, and other cognitive processes that inform our behavior. During the twentieth century, a lot of ink was spilled over the "nature-nurture" con-

troversy, with scholars arguing passionately about the extent to which mental characteristics and abilities are the result of learning experiences versus innate, "hardwired" genetic programs.

The two perspectives have often been presented as mutually incompatible alternatives, but thanks to more-powerful techniques, we have come to realize that there is nothing controversial about nature versus nurture: they are two sides of the same coin. Consider the case of rat pups who have inattentive mothers. As adults, the formerly neglected pups show elevated stress hormone responses to stressors that have little effect on rats that were not neglected as pups (T. Y. Zhang and Meaney, 2010). How does this lasting reactivity develop—through experience or as a result of innate biological factors? Both, it turns out. As we'll see later in this chapter, and throughout the book, the mind and its machine are shaped by a precise combination of genes and experience, inextricably tied together.

To see the video
Nature and Nurture,
go to

2e.mindsmachine.com/1.1

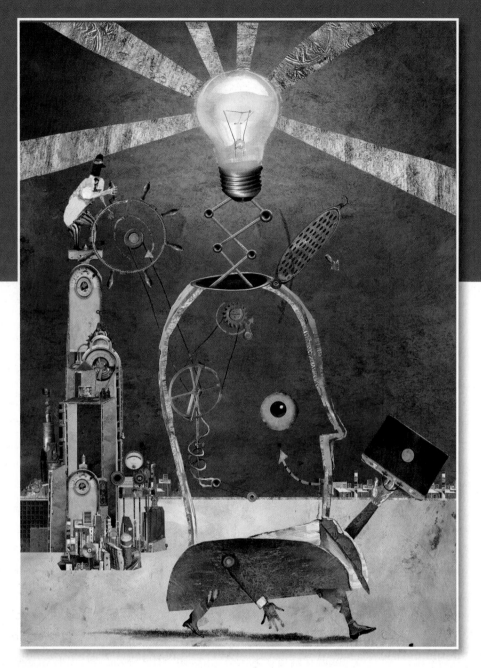

There are now almost 7.5 billion of us, and while we can debate whether to fear or celebrate that number, there is no doubt that each of those billions of human brains will at times contemplate its own existence and meaning. How does the operation of a three-pound organ generate our sense of self, express our unique personalities, record information, and guide our actions? Evolution has shaped our bodies and brains so that we closely resemble one another, yet our brains remain malleable throughout life, continually remolded by our environments, experiences, and interactions with other people. So, through a remarkable intersection of genetic heritage and environmental influences, 7.5 billion unique individuals have been formed, and we literally change each other's minds on a daily basis.

Our goal in this book is to introduce you to some of the many ways in which the structures and actions of the brain produce mind and behavior, and also the reverse: some of the many ways in which experiences change the brain. We hope to kindle in you the same interest and excitement that we experienced as students (and still feel today) when reading and thinking about the biology of behavior. Let's start by considering the aims and scope of the science called *biological psychology* or *behavioral neuroscience*.

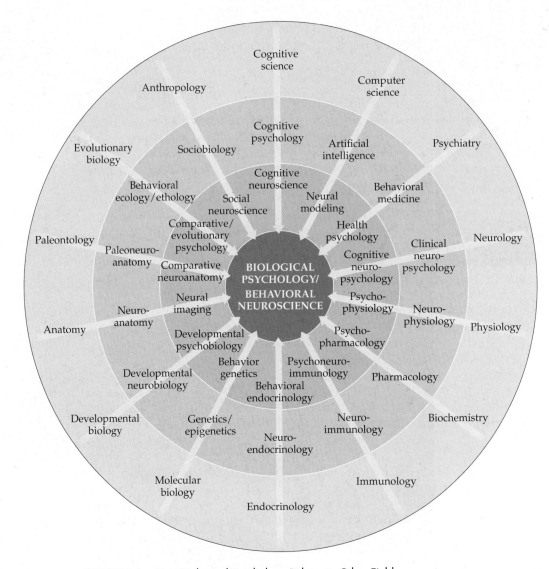

FIGURE 1.1 How Biological Psychology Relates to Other Fields

What's in a Name?

neuroscience   The scientific study of the nervous system.

biological psychology   Also called *behavioral neuroscience*, *brain and behavior*, and *physiological psychology*. The study of the biological bases of psychological processes and behavior.

The general field of **neuroscience**—the scientific study of the nervous system—is divided into many subdisciplines because the topic is so vast. The first scholars to study the relationships between brain and behavior called themselves philosophers, because it was philosophy that established the scientific method as our best tool for finding new knowledge. Philosophers had long been concerned with the sources of human behavior, so **biological psychology**, the field that relates behavior to bodily processes, naturally evolved from those beginnings. The names *behavioral neuroscience*, *brain and behavior*, and *physiological psychology* are all synonyms for *biological psychology*, but whichever name is used, the main goal of this field is to understand the brain structures and functions that respond to experiences and generate behavior.

Researchers with dramatically varied backgrounds—psychologists, biologists, physiologists, engineers, neurologists, psychiatrists, and many others—together make up the field of biological psychology. It is a field that spans both academia and industry, with focus that ranges from pure research on basic processes to entirely applied work directly translating findings into goods and services (Hitt, 2007). The diverse branches of science that overlap with biological psychology are mapped in **FIGURE 1.1**.

To see the
Brain Explorer,
go to

2e.mindsmachine.com/1.2

The Science of Brain and Behavior Spans Past, Present, and Future

An early textbook famously opened with the observation that, as a science, "psychology has a long past but only a short history" (Ebbinghaus, 1908). That's certainly an apt description of biological psychology. The modern era of biological psychology—characterized by objective experimentation and use of the scientific method to test hypotheses—has a formal history of only 100 years or so. But curiosity about the genesis of behavior reaches much further into the past, shaped by religious ideas, folk knowledge, and ancient observations about the biology of humans and nonhuman animals. Where does behavior come from?

The behavioral role of the brain was uncertain to early scholars

The elaborate preparation of tombs and careful mummification of important people in ancient Egypt (especially about 1500–1000 BCE) reflected the belief that the dead would enter an afterlife that entailed both struggle and—for the adequately equipped individual—great reward. So, in addition to embalming the body with special salts and oils, the usual practice was to preserve four important organs in alabaster jars in the tomb: liver, lungs, stomach, and intestines. The heart, being especially esteemed, was preserved in its place within the body. The brain, however, was plucked out and unceremoniously discarded; apparently it was considered to be of no particular value in the afterlife.

There is little or no mention of the brain in the Quran, and it is likewise never mentioned in either the Old Testament or New Testament of the Bible, but the heart is mentioned hundreds of times, along with several references each to the liver, the stomach, and the bowels as the seats of passion, courage, and pity, respectively. Aristotle (about 350 BCE), the most prominent scientist of ancient Greece, likewise considered mental capacities to be properties of the heart. When we call people *kindhearted*, *openhearted*, *fainthearted*, *hardhearted*, or *heartless*, and when we speak of learning *by heart*, we are using language echoing this ancient notion. Aristotle thought the brain was little more than a cooling system for hot blood from the heart. But Aristotle's near contemporary, the great Greek physician Hippocrates (about 400 BCE), already suspected that Aristotle's view was—ahem—wrongheaded, and he instead ascribed emotion, perception, and thought to the functioning of the brain.

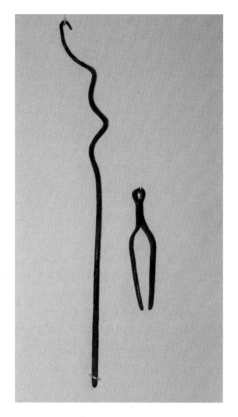

Brain Removal Kit It seems the ancient Egyptians had little regard for the brain. During the mummification process, embalmers used specialized tools, like these examples in the British Museum, to first break up the small bones behind the nose and then extract the brain through the opening. Unlike other major organs, the brain was discarded, and the cranium was stuffed with linen or straw. (Photograph by Neil Watson.)

A Brain on the Ceiling of the Sistine Chapel? Between 1508 and 1512, Michelangelo painted the Sistine Chapel in the Vatican. In one panel of Michelangelo's masterpiece, God is depicted reaching out to bestow the gift of life upon humanity, through Adam. But neuroscientists have noted that the oddly shaped drapery behind God, and the arrangement of his attendants, closely resembles the human brain (Meshberger, 1990); compare it with the sagittal view in Figure 2.16. A keen student of anatomy, Michelangelo probably knew perfectly well what a dissected human brain looks like. So, was Michelangelo having some fun, making a subtle commentary about the origins of human behavior? We probably will never know. But we can all agree that our uniquely human qualities—language, reason, emotion, and the rest—are products of the brain.

FIGURE 1.2 Leonardo da Vinci's Changing View of the Brain

(A)

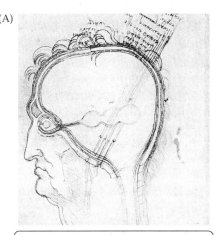

In an early sketch, Leonardo simply copied old drawings that bore little resemblance to the actual structure of the brain, showing the fluid-filled ventricles as a balloon connected to the eye.

(B)

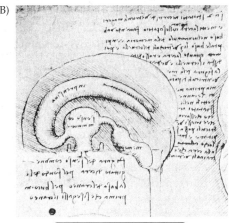

Leonardo's later drawings, made from direct observations, were much more anatomically correct.

dualism The notion, promoted by René Descartes, that the mind has an immaterial aspect that is distinct from the material body and brain.

By the second century CE, this brain-centered view of mental processes had become more entrenched, appearing in the writings of the Greco-Roman physician Galen (the "Father of Medicine"). Galen's experiences in treating head injuries of gladiators lead him to propose that behavior results from the movement of "animal spirits" from the brain through nerves to the body, but his understanding of the relevant anatomy was poor. Not until much later were techniques developed for making highly detailed anatomical studies of the fine structure of the brain.

Skillfully applying newly developed innovations in drawing technique, Renaissance painter and scientist Leonardo da Vinci (1452–1519) produced exquisite neuroanatomical illustrations of nerves and brain structures (**FIGURE 1.2**). Religious dogma dominated Renaissance science—just ask Galileo—with the result that scientific writing from that era often presents the brain as a mysterious and intricate gift from God. But perhaps some thinkers of the day secretly held a more secular view of neuroscience; for example, it has been observed that the depiction of God on the ceiling of the Sistine Chapel, painted by Michelangelo (1475–1564) (see the photo on the previous page), bears a striking resemblance to a midline view of the human brain (Meshberger, 1990). It is believed that Michelangelo was conducting dissections of cadavers at about the time that the painting was created.

In any event, weighing religious notions of the soul against increasingly mechanistic views of the brain became a major preoccupation for later scholars. Among his many contributions to math and science, René Descartes (1596–1650) tried to explain how the control of behavior might resemble the workings of a machine, proposing the concept of spinal reflexes and a neural pathway for them (**FIGURE 1.3**). But Descartes also argued (perhaps in order to deflect criticism) that free will and moral choice could not arise from a mere machine. So Descartes asserted that humans, at least, had a nonmaterial soul as well as a material body and that the soul governed behavior through a point of contact (possibly the pineal gland) in the brain. This notion of **dualism** spread widely and left other thinkers with the task of trying to explain how a nonmaterial soul could exert influence over a material body and brain. Today, biological psychologists reject dualism in favor of the much simpler view that the workings of the mind can be understood as purely physical processes taking place in the material brain.

Thanks in large part to systematic studies of the relation between various disorders and damage to regions of the human

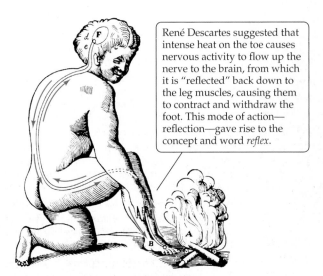

René Descartes suggested that intense heat on the toe causes nervous activity to flow up the nerve to the brain, from which it is "reflected" back down to the leg muscles, causing them to contract and withdraw the foot. This mode of action—reflection—gave rise to the concept and word *reflex*.

FIGURE 1.3 An Early Account of Reflexes

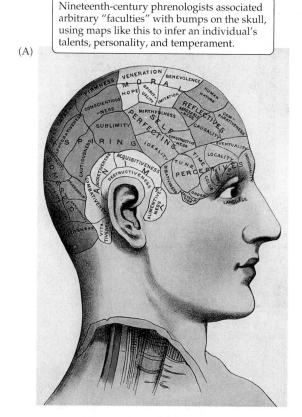

Nineteenth-century phrenologists associated arbitrary "faculties" with bumps on the skull, using maps like this to infer an individual's talents, personality, and temperament.

(A)

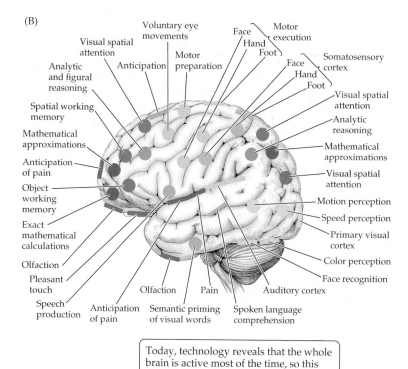

(B)

Today, technology reveals that the whole brain is active most of the time, so this map simply reflects the *peaks* of brain activation during various behaviors.

FIGURE 1.4 Old and New Phrenology (Part B after Nichols and Newsome, 1999.)

brain that were conducted by the English physician Thomas Willis (1621–1675), the notion that the brain coordinates and controls behavior eventually became widely accepted (Zimmer, 2004). A pseudoscientific fad of the early 1800s called **phrenology** (**FIGURE 1.4A**) capitalized on the emerging idea that specific behaviors, feelings, and personality traits were controlled by corresponding specific regions of the brain. Although phrenology was plainly wrong in several fundamental ways—for example, phrenologists believed they could "read" a person's character by feeling the bumps on that person's head—the field helped establish the concept of **localization of function**, which asserts that different brain regions specialize in specific behaviors.

Later researchers found that damage to specific regions of the brain causes predictable impairments in people; for example, Paul Broca (1824–1880) noted that damage to a particular region of the left side of the brain reliably causes problems with speech production (see Chapter 15). Neuroscientists today accept that the localization of function within the brain is more or less true. Although the whole brain is active most of the time, when we are performing particular tasks, certain brain regions become *even more* activated, and different tasks activate different brain regions. So modern functional maps of the human brain track the locations where these *peaks* of activation occur (**FIGURE 1.4B**). A parallel concern that harks back to the phrenologists has been the importance of brain size to intellectual function. After a couple of centuries of study, the evidence indicates that while the overall size of your brain matters, it matters a lot less than you might expect (**FIGURE 1.5**).

In 1890, William James's book *Principles of Psychology* signaled the beginnings of a modern approach to biological psychology. In James's work, psychological ideas such as consciousness and other aspects of human experience came to be seen as properties of the nervous system. A true biological psychology began to emerge from this approach.

phrenology The belief that bumps on the skull reflect enlargements of brain regions responsible for certain behavioral faculties.

localization of function The concept that different brain regions specialize in specific behaviors.

(A)

Whether a bigger brain indicates greater intelligence is an old question. Investigators in the nineteenth century measured the volumes of skulls of various groups and estimated intelligence on the basis of people's occupations, teachers' guesstimates, and other doubtful criteria.

(B)

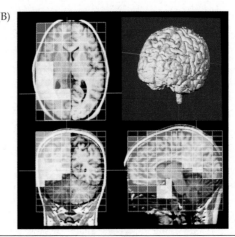

IQ tests improved the objectivity of intelligence measurement. When carefully controlled, MRI studies like this one do find a correlation between brain size and IQ scores (Andreasen et al., 1993), and IQ seems to correlate better with the volume of the front of the brain than the back (Colom et al., 2009). So it seems that bigger is better, but the correlations are of only modest size, meaning that other factors must also contribute to IQ scores (Stanovich, 2009).

FIGURE 1.5   Does Size Matter? (Part A from the Bettmann Archive; B courtesy of Nancy Andreasen.)

conserved In the context of evolution, referring to a trait that is passed on from a common ancestor to two or more descendant species.

Advances in experimental methodology propel modern biological psychology

By the beginning of the twentieth century, researchers were applying newly developed tools to study mental processes that had previously seemed unknowable. Rapid progress was made in developing techniques for measuring learning and memory in humans and animals, and Russian physiologist Ivan P. Pavlov (1849–1936) made his landmark discoveries of classical conditioning in animals—Nobel Prize–winning work that influences scientists to this day.

This rapid progress prompted a parallel interest in understanding the neural basis of learning, marked by one of the first true biological psychology research programs: the "search for the engram" by Karl Lashley (1890–1958). Although he would not accomplish his goal of linking a specific brain region to the formation of a specific long-term memory (an "engram"), Lashley gave us the idea (now well established) that memory is not localized to only one region of the brain.

One of Lashley's students, Donald O. Hebb (1904–1985), had a particularly profound impact on biological psychology. Hebb showed that cognitive processing could be accomplished by networks of active neurons, molded by repeated activation patterns into functional circuits. His hypothesis about how neurons strengthen their connections as a consequence of experiences led to the idea of the *Hebbian synapse*, a type of plastic (changeable) connection between neurons that remains a hot topic in neuroscience, as discussed in Chapter 13.

Present-day biological psychologists may draw on several different theoretical perspectives. Here are some of the major ones:

1. *Systematic description of behavior* Until we describe what we want to study, we cannot accomplish much. Depending on our goals, we may describe behavior in terms of detailed acts or processes, or in terms of results or functions. To be useful for scientific study, a description must be precise, using accurately defined terms and units.

2. *The evolution of brain and behavior* Darwin's theory of evolution through natural selection is central to all modern biology and psychology. Biological psychologists employ evolutionary theory in two ways: by evaluating *similarities* among species due to shared ancestry, and by looking for species-specific *differences* in behavior and biology that have evolved as adaptations to different environments (**BOX 1.1**). We will discuss many examples of both perspectives in this book.

BOX 1.1   We Are All Alike, and We Are All Different

Each person has some characteristics shared by...

all animals...

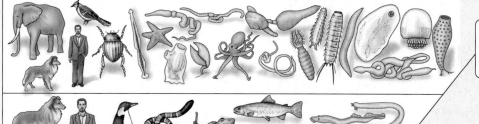

> All animals use DNA to store genetic information.

all vertebrates...

> All vertebrates have a backbone and spinal cord.

all mammals...

> All mammals suckle their young.

all primates...

> All primates have a relatively large, complex brain, and many primates have a hand with an opposable thumb.

all humans (people)...

> All humans use symbolic language to communicate with each other.

some people...

> Some people like to eat beets (no one knows why).

no other person.

> No two people, even identical twins, are alike in each and every way, as individual experiences leave their unique stamp on every brain.

How do similarities and differences among people and animals fit into biological psychology? Each person is in some ways like all other people, in some ways like some other people, and in some ways like no other person. As the figure shows, we can extend this observation to the much broader range of animal life. In some ways each person is like all other animals (e.g., needing to ingest complex organic nutrients), in some ways like all other vertebrates (e.g., having a spinal column), in some ways like all other mammals (e.g., nursing our young), and in some ways like many other primates (e.g., having a relatively large, complex brain and a hand with an opposable thumb). The electrical messages used by nerve cells (see Chapter 3) are essentially the same in a jellyfish, a cockroach, and a human being, and many species employ identical hormones. Species share these **conserved** characteristics because the features first arose in a shared ancestor. But mere similarity of a feature between species does not guarantee that the feature came from a common ancestral species. Our eyes resemble those of octopuses, but certain key differences reveal that their eyes and our eyes evolved separately.

Whether knowledge gained about a process in another species applies to humans depends on whether we are like that species in regard to that process. The fundamental research on the mechanisms of inheritance in the bacterium *Escherichia coli* proved so widely applicable that some molecular biologists proclaimed, "What is true of *E. coli* is true of the elephant." To a remarkable extent, that statement is true, but there are also some important differences in the genetic mechanisms of *E. coli* and mammals.

With respect to each biological property, researchers must determine how animals are identical and how they are different. When we seek animal models for studying human behavior or biological processes, we must ask the following question: Does the proposed animal model really have some things in common with the process at work in humans? We will see many cases in which it does.

Even within the same species, however, individuals differ from one another: cat from cat, blue jay from blue jay, and person from person. Biological psychology seeks to understand individual differences as well as similarities. Therefore, the way in which each person is able to process information and store the memories of these experiences is another part of our story.

(A) Prevalence of neurological disorders   (B) Prevalance of psychiatric disorders

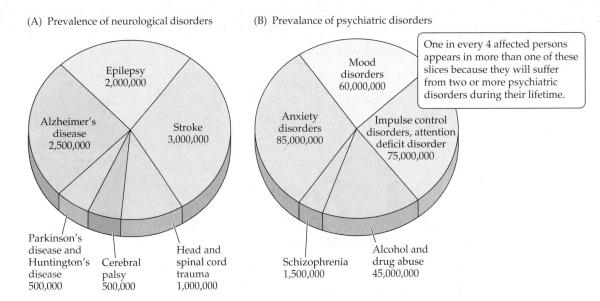

FIGURE 1.6   The Toll of Brain Disorders (Data for part A from Hirtz et al., 2007; for B, from Kessler et al., 2005.)

3. *Life-span development of the brain and behavior* **Ontogeny** is the process by which an individual changes in the course of its lifetime—that is, grows up and grows old. Observing the way a particular behavior changes during ontogeny may give us clues to its functions and mechanisms. For example, we know that learning ability in monkeys increases over several years of development. Therefore, we can speculate that prolonged maturation of brain circuits is required for complex learning tasks. Studying the development of reproductive capacity and of differences in behavior between the sexes, along with changes in body structures and processes, enables us to throw light on body mechanisms underlying sexual behaviors.

4. *The biological mechanisms of behavior* To learn about the mechanisms of an individual's behavior, we study how his or her present body works, separate from evolutionary or developmental concerns. To understand the underlying mechanisms of behavior, we must regard the organism (with all due respect) as a "machine," made up of billions of nerve cells, or **neurons**. In a sense, the mechanistic questions are the "how" questions of biological psychology, in contrast to the "why" questions that derive from the evolutionary and developmental perspectives. So in the case of learning and memory, for example, we might try to understand how a sequence of electrical and biochemical processes allows us to store information in our brains, and how a different process retrieves it.

The future of biological psychology is in interdisciplinary discovery and knowledge translation

There is reason to hope that some of the fundamental discoveries of biological psychology will be translated into a greater understanding of brain disorders and the development of effective treatments. At least one person in five around the world currently suffers from a neurological or psychiatric illness. These disorders vary in severity, from illnesses that cause significant but manageable changes in the quality of everyday life, to devastating conditions that completely disable. **FIGURE 1.6** illustrates the estimated numbers of U.S. residents afflicted by some of the main neurological and psychiatric disorders. The toll of these disorders is enormous: just the cost for treatment of dementia (severely disordered thinking) alone exceeds the costs of treating cancer and heart disease combined.

ontogeny   The process by which an individual changes in the course of its lifetime—that is, grows up and grows old.

neuron   Also called *nerve cell*. The basic unit of the nervous system.

(A)

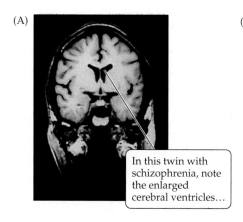

In this twin with schizophrenia, note the enlarged cerebral ventricles…

(B)

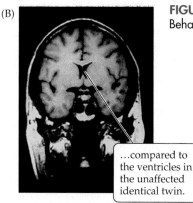

…compared to the ventricles in the unaffected identical twin.

FIGURE 1.7 Identical Twins but Nonidentical Brains and Behavior (Images courtesy of E. Fuller Torrey.)

As the quest to understand and relieve these diseases gathers speed, some of the historical distinction between clinical and laboratory approaches has begun to fade away. For example, when clinicians encounter a pair of twins, one of whom has schizophrenia while the other seems healthy, the discovery of structural differences in their brains (**FIGURE 1.7**) immediately raises questions for laboratory scientists: Did the structural differences arise before the symptoms of schizophrenia, or the other way around? Were the brain differences present at birth, or did they arise during puberty? Does medication that reduces symptoms affect brain structure? (We'll consider schizophrenia again in Chapter 12.)

The twenty-first century has brought an explosion in new research areas and perspectives in biological psychology. In addition to the rapid progress that is occurring in the more established topic areas of biological psychology, which we survey in this book, some emerging research areas are attracting intense interest. In the paragraphs that follow, we'll take a brief look at a handful of these.

NEUROPLASTICITY   When you think about it, the only explanation for our ability to learn skills and form memories is that the brain physically changes in some way to encode and store that information. In this book we'll see many examples of how behavior and experiences alter the physical brain—a phenomenon called **neuroplasticity** (from the Greek *plassein*, "to mold or form") or *neural plasticity*—but much remains to be discovered about mechanisms of neuroplasticity.

We know that experiences can alter the size of brain regions and interconnections between neurons, and cellular changes have been discovered that *could* be mechanisms for storing memories; scientists are actively testing that idea. We also know that certain experiences and physiological states can modify the rate at which new neurons are born in the adult brain, but again, the functional significance of this **adult neurogenesis** (see Chapter 13) remains to be determined. Perhaps nothing distinguishes biological psychology from other neurosciences more clearly than a fascination with neuroplasticity and the role of experience.

SOCIAL NEUROSCIENCE   Because of neuroplasticity, even simple interactions with other people can remodel our brains. Indeed, the whole point of coming to a lecture hall is to have the instructor use words and figures to alter your brain so that you can retrieve that information in the future (in other words, she is teaching you something). Most aspects of our social behavior are learned—from the language we speak to the clothes we wear and the kinds of foods we eat, as well as our identity in belonging to larger groups (clubs, teams, schools, nationalities, and so on).

Social neuroscience is an emerging discipline that uses the tools of neuroscience to discover how biological and social factors continually interact and affect each other as behavior unfolds. For example, the amount of testosterone in a male's circulation affects his dominance behavior and aggression, expressed in social settings ranging from friendly games to overt physical aggression (see Chapter 11). But the *outcome* of

neuroplasticity Also called *neural plasticity*. The ability of the nervous system to change in response to experience or the environment.

adult neurogenesis The creation of new neurons in the brain of an adult.

social neuroscience A field of study that uses the tools of neuroscience to discover both the biological bases of social behavior and the effects of social circumstances on brain activity.

evolutionary psychology   A field of study devoted to asking how natural selection has shaped behavior in humans and other animals.

epigenetics   The study of factors that affect gene expression without making any changes in the nucleotide sequence of the genes themselves.

gene expression   The turning on or off of specific genes.

neuroeconomics   The study of brain mechanisms at work during economic decision making.

those contests can cause changes in relative testosterone levels—winners show more testosterone, and losers have less—so testosterone concentration in the blood at any particular moment is determined (in part) by the male's recent history of dominant-submissive social experience. The modified testosterone level, in turn, helps determine the male's dominance and aggression in the future. So it goes.

EVOLUTIONARY PSYCHOLOGY   Zoologists have long viewed animal behaviors as adaptations that evolved to solve specific ecological pressures, such as the need to find food and avoid predators. More recently, speculations about how natural selection might have shaped our own behavior, including specific cognitive abilities, have given rise to a lively and controversial field called **evolutionary psychology** (Barkow et al., 1992; Buss, 2013). For example, it has been argued (G. F. Miller, 2000) that sexual selection was crucial for evolution of the human brain. If early ancestors of modern humans came to favor mates who sang, made jokes, or produced artistic works, an "arms race" might have ensued as the ever more discriminating brains of one sex demanded ever more impressive performances from the brains of the other sex. Did humor, song, and art originate from the drive to be sexy? And does sexual selection account for the large size of the human brain?

Although evolutionary psychology excels at generating intriguing hypothetical accounts of the evolution of behaviors, the challenge for the future is to come up with ways to test and potentially disprove these hypotheses.

EPIGENETICS   Nearly all of the cells in your body have a complete copy of your *genome* (that is, a copy of all your genes), but each cell *uses* only a small subset of those genes at any one time. **Epigenetics** is a young field focusing on factors that have a lasting effect on patterns of **gene expression**—the turning on or off of specific genes—without changing the structure of the genes themselves. In some cases the acquired alteration in gene expression is passed down through generations, from parent to child, despite the absence of genetic modifications.

At the beginning of the chapter, we briefly discussed rats that will show heightened stress reactivity throughout their lives if neglected by their mothers while they are pups, and we asked whether this phenomenon was more attributable to "nature" or to "nurture." The answer is: neither. Or perhaps both. It is an epigenetic phenomenon, in which the maternal neglect causes lasting inactivation of a gene (a process called *methylation*) in the pup's brain that causes the pup to be hyperresponsive to stress for the rest of its life. So, early experience produces a permanent change in the way in which genes are expressed in the neglected rat, thus altering its behavior in adulthood—nurture *and* nature.

This same gene is also more likely to be methylated in the postmortem brains of humans who have committed suicide, *but only if the victim was subjected to childhood abuse* (McGowan et al., 2009). So, methylation of the gene in abused children may make them more susceptible to stress as adults and put them at risk for suicide—a powerful demonstration of epigenetic influences on behavior. (For additional examples of social influences on the structure of the brain, see **A STEP FURTHER 1.1**, on the website.)

NEUROECONOMICS   A growing number of neuroscience labs focus on the neural bases of decision making. Involving perspectives ranging from philosophy to social psychology to traditional experimental psychology and economics, **neuroeconomics** exploits new brain-imaging technologies to identify brain regions that are especially active when decisions are being made while playing games, managing resources, making strategic choices, and so on. Naturally, this research has some shorter-term benefits relating to product marketing—what makes us want to buy something? But researchers mostly hope that, over the longer term, we will develop a more complete understanding of the massive brain networks that are active while we choose among various alternatives and decide what to do next, and of the ways in which we perceive and express our free will (as well as if, indeed, we actually *have* free will!).

Early indications are that we possess brain mechanisms dedicated to neuroeconomic evaluations, assessing the relative value of the each choice available and then sifting through the evaluated choices in order to make a conscious decision (Kable and Glimcher, 2009; Rustichini and Padoa-Schioppa, 2015), along with a system that inhibits impulsive decision making (Muhlert and Lawrence, 2015). There is good reason to expect exciting new discoveries about how these neural systems interact with other cognitive systems to produce our conscious feeling of self.

THE TRULY FINAL FRONTIER: CONSCIOUSNESS   Ultimately, many of the traditional and emerging topics in biological psychology converge on the problem of **consciousness**: the personal, private awareness of our emotions, intentions, thoughts, and movements and of the sensations that impinge upon us. How is it possible that you are aware of the words on this page, the room you're occupying, the goals you have in life? Scientists have laid some of the groundwork for conceptualizing consciousness as a property of the brain, and for establishing it as an area of scientific inquiry (Zeman, 2002), but the devil is in the details. Later in the book we will consider experiments that demonstrate properties of consciousness, including the role of arousal systems of the brain, and evidence that synchronized activity within large cortical networks gives rise to conscious introspection (Raichle, 2015). And some of these experiments lead us to interesting—possibly disturbing—conclusions that our consciousness may track the operation of our brains much less accurately than we perceive (as discussed in Chapter 14).

However it is brought about, any satisfying account of consciousness should be able to explain, for example, why a certain pattern of activity in your brain causes you to experience the sensation of blue when looking at the sky. And a really good theory should tell us how we could cause you to experience the sky as yellow, just by changing your brain activity for you (not, of course, by simply wearing colored goggles, but rather through a change in the deeper *conscious* experience of the world around you). However, we are nowhere near understanding consciousness this clearly. We generally have no idea what the actual *inner experience* of a human or animal is—only what an individual's behavior tells us about it. So even if you tell me that the sky is blue to you, I can't tell whether blue *feels* the same in your mind as it does in mine. Until we can get a more complete grasp of these complex issues of brain and behavior, we are in no position to know whether complicated machines like computers are, or might one day be, conscious.

"How Beautifully Blue the Sky" We would all agree that this sky is the color we call "blue." But in Chapter 14 we'll ask whether everyone who sees that sky has the same experience of the color.

consciousness The state of awareness of one's own existence, thoughts, emotions, and experiences.

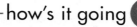

how's it going ❓

1. Define *biological psychology*. Name some fields that are closely allied to biological psychology.

2. What do we mean when we say that biological psychology has a long past but only a short history? Review the prehistory of biological psychology and the gradual process of elimination that linked the brain and behavior.

3. Discuss the concept of localization of function.

4. Describe the four major theoretical perspectives employed by modern-day biological psychologists that we discussed in the chapter: behavioral description, evolution, development, and biological mechanisms.

5. Comment on the prevalence of psychiatric and neurological disorders in contemporary society, and discuss their economic and emotional impact.

6. Where is biological psychology headed? Discuss some of the probable hot topics for future biological psychologists.

Note: Near the end of each section of text, we stop to ask you some questions so you can assess how well you've absorbed the main points. These questions are intended as a self-check, so if you're having problems answering them, you may find it helpful to review the preceding section before moving on.

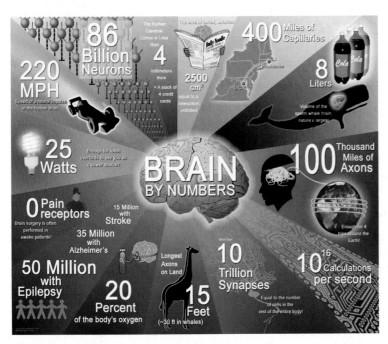

FIGURE 1.8 Your Brain, by the Numbers

Careful Research Design Is Crucial for Brain Research

Because it is both fantastically complex (**FIGURE 1.8**) and somewhat inaccessible, the brain poses special challenges when it comes to formulating research questions and designing experiments to answer those questions. For example, it is difficult to physically manipulate the structure and activity of the brain with pinpoint accuracy, so technological advances have been responsible for the modern explosion in neuroscience research. And because we can't ethically manipulate the brains of human research participants (other than through transient and noninvasive means), we often must rely on alternatives such as purely observational research, research in which behavioral means are used to alter brain activity, or research in which we study neural function in lab animals in order to gain insights about our own nervous systems.

Three kinds of studies probe brain-behavior relationships

Biological psychologists use three general types of studies for research on the biological bases of behavior. In an experiment employing **somatic intervention** (**FIGURE 1.9A**), we alter a structure or function of the brain or body to see how this alteration changes behavior. In this sort of experiment, the physical alteration is an **independent variable** (a general term used to describe the manipulated aspect of any experiment), and the behavioral effect is the **dependent variable** (a general term used to describe the measured consequence of an experimental manipulation). Some examples of somatic intervention experiments include (1) administering a hormone to some animals, but not others, and comparing the later sexual behavior of both groups; (2) electrically stimulating a specific brain region and measuring alterations in movement; and (3) destroying a specific region in the brain and observing subsequent changes in sleep patterns. In each case, the behavioral measurements follow the bodily intervention; furthermore, in each case the behavioral measurements are compared with those of a **control group**. In a **within-participants experiment**, the control group is simply the same individuals, tested before the somatic intervention occurs. In a **between-participants experiment**, the experimental group of individuals is compared with a different group of individuals who are treated identically in every way except that they don't receive the somatic intervention.

The approach opposite to somatic intervention is **behavioral intervention** (**FIGURE 1.9B**). In this approach the scientist alters or controls the behavior of an organism and looks for resulting changes in body structure or function. Here, behavior is the independent variable, and change in the body is the dependent variable. A few examples include (1) allowing adults of each sex to interact, and measuring subsequent changes in sex hormones; (2) having a person perform a cognitive task while in a brain scanner, and measuring changes in activity in specific regions of the brain; and (3) training an animal to fear a previously neutral stimulus, and observing electrical changes in the brain that may encode the newly learned association. As with somatic intervention, these experimental approaches may employ either within-group or between-groups designs.

The third type of study is **correlation** (**FIGURE 1.9C**), which measures how closely changes in one variable are associated with changes in another variable. Two examples of correlational studies include (1) observing the extent to which memory ability is associated with the size of a certain brain structure and (2) noting that increases in a certain hormone are accompanied by increases in aggressive behavior. Note that while this type of study tells us if the measured variables are associated in some way, it can't tell us which causes the other. We can't tell, for example, whether the hormones cause the aggression or aggression increases the hormones. But even though it can't

somatic intervention An approach to finding relations between body variables and behavioral variables that involves manipulating body structure or function and looking for resultant changes in behavior.

independent variable The factor that is manipulated by an experimenter.

dependent variable The factor that an experimenter measures to monitor a change in response to changes in an independent variable.

control group In research, a group of individuals that are identical to those in an experimental (or test) group in every way except that they do not receive the experimental treatment or manipulation. The experimental group is then compared with the control group to assess the effect of the treatment.

within-participants experiment An experiment in which the same set of individuals is compared before and after an experimental manipulation. The experimental group thus serves as its own control group.

between-participants experiment An experiment in which an experimental group of individuals is compared with a control group of individuals that have been treated identically in every way except that they haven't received the experimental manipulation.

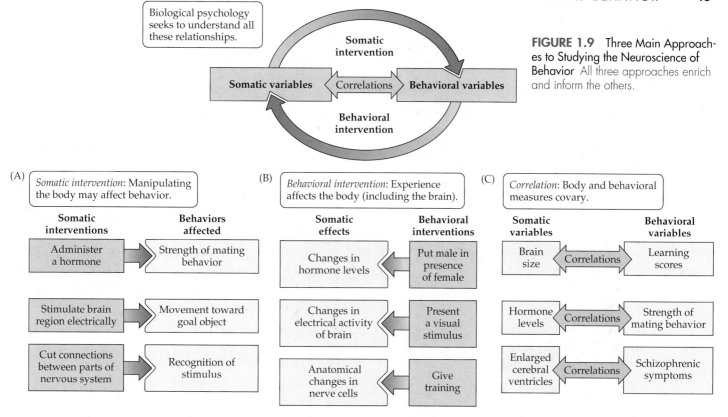

FIGURE 1.9 Three Main Approaches to Studying the Neuroscience of Behavior All three approaches enrich and inform the others.

establish **causality**, correlational research can help researchers identify which things are linked, and thus it helps us to develop hypotheses that can be tested experimentally using behavioral and somatic interventions.

As Figure 1.9 illustrates, biological psychology research is guided by the understanding that relations between brain and body are reciprocal: each affects the other in an ongoing cycle of bodily and behavioral interactions. We will see examples of this reciprocal relationship throughout the book.

Biological psychologists use several levels of analysis

Another consideration that researchers must weigh in designing experiments is the level of complexity at which to work. Even the most complex behavior could, *in theory*, be understood at the level of cellular activity or even lower, at the level of biochemistry and molecular interactions. This idea, that we can understand complex systems by dissecting their simpler constituent parts, is known as **reductionism**. But we wouldn't get very far if we set out to explain, say, the use of grammar, in terms of chemical reactions; the behavior is so complex that an explanation at the molecular level would involve a vast amount of data. So instead, the reductionist approach aims to identify **levels of analysis** that are *just simple enough* that they allow us to make rapid progress on the more complex phenomena under study. Finding explanations for behavior often requires several levels of biological analysis, ranging from social interactions, to brain systems, to circuits and single nerve cells and their even simpler, molecular constituents.

Naturally, in all fields different problems are carried to different levels of analysis, and fruitful work is often being done simultaneously by different workers at several levels. For example, in their research on visual perception, some cognitive psychologists carefully analyze behavior. They try to determine how the eyes move while looking at a visual pattern, or how the contrast among parts of the pattern determines its visibility. Meanwhile, other biological psychologists study the differences in visual abilities among species and try to determine the adaptive significance of these differences. For example, how is the presence (or absence) of color vision related to the lifestyle of a species? At the same time, other investigators trace out brain structures and networks involved in different visual tasks. Still other scientists try to understand the electrical and chemical events that occur in the brain during vision (**FIGURE 1.10**).

behavioral intervention An approach to finding relations between body variables and behavioral variables that involves intervening in the behavior of an organism and looking for resultant changes in body structure or function.

correlation The tendency of two measures to vary in concert, such that a change in one measure is matched by a change in the other.

causality The relation of cause and effect, such that we can conclude that an experimental manipulation has specifically caused an observed result.

reductionism The scientific strategy of breaking a system down into increasingly smaller parts in order to understand it.

levels of analysis The scope of experimental approaches. A scientist may try to understand behavior by monitoring molecules, nerve cells, brain regions, or social environments or using some combination of these levels of analysis.

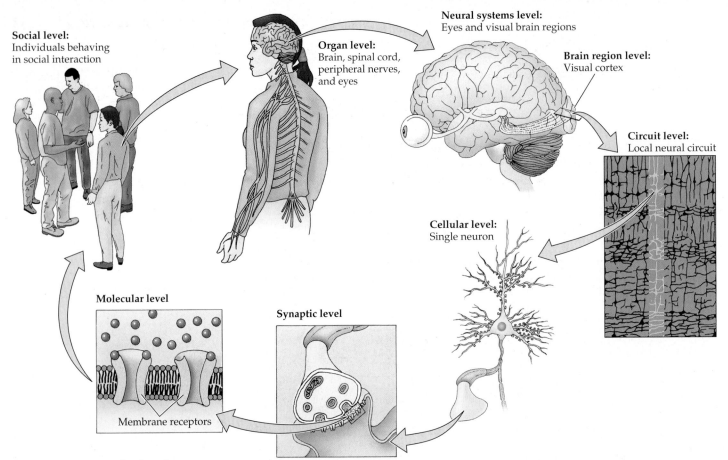

Social level:
Individuals behaving in social interaction

Organ level:
Brain, spinal cord, peripheral nerves, and eyes

Neural systems level:
Eyes and visual brain regions

Brain region level:
Visual cortex

Circuit level:
Local neural circuit

Cellular level:
Single neuron

Synaptic level

Molecular level

Membrane receptors

FIGURE 1.10 Levels of Analysis in Biological Psychology

Animal research is an essential part of life sciences research, including biological psychology

Because we will draw on animal research throughout this book, we should comment on some of the ethical issues of experimentation on animals. Human beings' involvement and concern with other species predates recorded history; early humans had to study animal behavior and physiology in order to escape some species and hunt others. To study the biological bases of behavior inevitably requires research on animals of other species, as well as on human beings. Psychology students usually underestimate the contributions of animal research to psychology because the most widely used introductory psychology textbooks often present major findings from animal research as if they were obtained with human participants (Domjan and Purdy, 1995).

A vocal minority of people believe that research with animals, even if it does lead to lasting benefits, is unethical. Others, like Peter Singer in his influential book *Animal Liberation* (1975), argue that animal research is acceptable only when it produces immediate and measurable benefits. The problem with this perspective is that we have no way of predicting which experiments will lead to a breakthrough. The whole point of studying the unknown is that it is *unknown*; there is a long history of chance observation, based on the steady accumulation of basic knowledge, leading to unexpected benefits.

Of course, researchers have an obligation to minimize the discomfort of their animal subjects, and ironically enough, animal research has provided us with drugs and techniques that make most research painless for lab animals, while also leading to improved veterinary care for our animal companions (Sunstein and Nussbaum, 2004). Researchers are also subject to continual ethical oversight, and they are bound by animal protection legislation. They also adhere to nationally mandated animal care policies, formally adopted by virtually all university administrations, that emphasize the use of as few animals as possible without jeopardizing research integrity, as well as the use of the simplest species that can answer the questions under study.

As human beings with the full range of emotions and empathetic feelings toward animals, we all wish there was an alternative to the use of animals in research. But if we want to understand how the nervous system works, we have to actually study it, in detail. The life sciences would slow to a crawl without the basic knowledge that we derive from studying animals.

---how's it going ?

1. What are the three general forms of research studies in biological psychology? What is the issue of *causality*? How do the three research perspectives inform and shape one another?

2. Define *independent variable*, *dependent variable*, *control group*, *within-participants experiment*, and *between-participants experiment*.

3. What is the general principle behind reductionism? How does this influence the level of analysis at which a researcher works. And for that matter, what is meant by *level of analysis*?

4. Consider both sides of the debate over animal research, weighing the pros and cons of the "for" and "against" positions. How do you think animal use should be regulated?

Looking Forward: A Glimpse inside the Mind's Machine

Our mission in this book is to acquaint you with the broad topic of biological psychology, from historical underpinnings to cutting-edge investigations of the most complex aspects of our intellect. Along the way, we are going to touch on some of the most interesting questions in modern neuroscience, including these:

• How does the nervous system capture, process, and represent information about the environment? For example, sometimes brain damage causes a person to lose the ability to identify other people's faces; what does that tell us about how the brain works during face recognition?

• What brain sites and activities underlie feelings and emotional expression? Are particular parts of the brain active in romantic love, for example (**FIGURE 1.11A**)?

• Why are different brain regions active during different language tasks (**FIGURE 1.11B**)?

• Some people suffer damage to the brain and afterward seem alarmingly unconcerned about dangerous situations and unable to judge the emotions of other people. What parts of the brain are damaged to cause such changes?

(A)
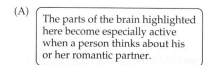
The parts of the brain highlighted here become especially active when a person thinks about his or her romantic partner.

(B)
Different brain regions are activated when people perform four different language tasks.

FIGURE 1.11 "Tell Me Where Is Fancy Bred?" (Part A from Bartels and Zeki, 2000; B courtesy of Marcus Raichle.)

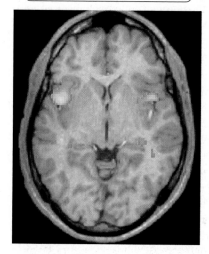

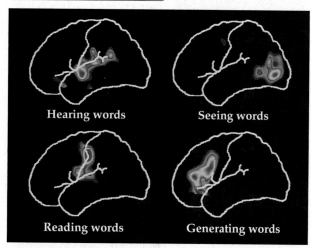

Hearing words   Seeing words

Reading words   Generating words

TABLE 1.1 Neuromythology: Facts or Fables?

Some human nerve cells are more than 3 feet long.	True
Nerve impulses travel at the speed of light.	False
More people die each year from the use of legal drugs than illegal drugs.	True
Our bodies make chemicals that are similar in structure to heroin and marijuana.	True
Testosterone is made only by males, and estrogen is made only by females.	False
Once our brains are developed, we can never grow new nerve cells.	False
Some people are incapable of feeling pain.	True
We have five senses.	False
Different parts of the tongue are specialized to recognize certain tastes.	False
Dogs are color-blind.	False
Each side of the brain controls the muscles on the opposite side of the body.	True
There are no anatomical differences between men's and women's brains.	False
Some people are "born gay."	Uncertain
During sleep the brain is relatively inactive.	Not always
Sleepwalkers are acting out dreams.	False
In some animals, half the brain can be asleep while the other half is awake.	True
The left side of the face is more emotionally expressive than the right side.	True
All cultural groups recognize the same facial expressions for various emotions.	Uncertain
Scientists are not sure why antidepressant drugs work.	True
Some people are incapable of producing any new memories.	True
Each memory is stored in its own brain cells.	False (probably)
Our memory contains accurate accounts of past experiences.	False (most of the time)
A stimulating environment can change the structure of an animal's brain.	True
We can take in a whole visual scene in just a single glance.	False
My brain decides what I will do next, before my conscious self is aware of the decision.	Uncertain
People are "right-brained" or "left-brained."	False
A child can have half of the brain removed and still develop normal intelligence.	True

A Bright Idea The symbolic lightbulb coming on over someone's head, representing a sudden insight or idea, is an especially apt metaphor for the functioning of the brain, given the lightbulb's rapid action and ability to penetrate the gloom (both literal and metaphorical). This particular lightbulb, in a fire station in Livermore, California, is the world's longest-burning bulb (you can check in on it, live, at www.centennialbulb.org/cam.htm). It has been continually lit for more than 1 million hours (about 115 years)—since the dawn of the scientific discipline of biological psychology.

- How does the brain manage to change during learning, and how are memories retrieved?
- Some brain regions are different in heterosexual versus homosexual men. What do those differences tell us about the development of human sexual orientation?

The relationship between the brain and behavior is very mysterious because it is difficult to understand how a physical device, the brain, could be responsible for our subjective experiences of fear, love, and awe. Perhaps it is the "everyday miracle" aspect of the topic that has generated so much folk wisdom—and unfounded mythology—about the brain. For example, the oft-repeated claim that we normally use only 10% of the brain is total nonsense; brain scans show that the entire brain is active most of the time while we go about normal daily activities. There are lots of other examples of commonplace beliefs about the brain. We've compiled a few of these in **TABLE 1.1**. Some of them are unfounded; others are true but may sound improbable. No doubt you can think of others.

Of course, it wasn't that long ago that the idea of making light from electricity seemed far-fetched. With each passing year, technological developments and the progress of thousands of neuroscientists in labs around the world provide a clearer view of what is happening when a lightbulb goes on in the mind and someone has a clever idea. Our hope for this book is that it will turn on a few lightbulbs for you too.

Recommended Reading

Decety, J., and Cacioppo, J. T. (2011). *The Oxford Handbook of Social Neuroscience.* New York, NY: Oxford University Press.

Doidge, N. (2007). *The Brain That Changes Itself.* New York, NY: Penguin.

Finger, S. (2001). *Origins of Neuroscience.* New York, NY: Oxford University Press.

Koch, C. (2012). *Consciousness: Confessions of a Romantic Reductionist.* Cambridge, MA: MIT Press.

Mandler, G. (2007). *A History of Modern Experimental Psychology: From James and Wundt to Cognitive Science.* Cambridge, MA: MIT Press, 2007.

2e.mindsmachine.com/vs1

VISUAL SUMMARY

1

You should be able to relate each summary to the adjacent illustration, including structures and processes. Go to the online version of this summary (scan the QR code above) for links to figures, animations, and activities that will help you consolidate the material.

1 **Biological psychology** is a branch of **neuroscience** that focuses on the biological bases of behavior. It is closely related to many other neuroscience disciplines. Review **Figure 1.1, Animation 1.2**

2 Although humans have wondered about the control of behavior for thousands of years, only comparatively recently has a mechanistic view of the brain taken hold. Review **Figure 1.3**

3 The concept of **localization of function**, which originated in **phrenology**—despite obvious flaws with the phrenologists' methodology—was an important milestone for biological psychology. Today we know that the part of the brain that shows a *peak* of activity varies in a predictable way depending on what task we're doing. Review **Figure 1.4**

4 Localization of cognitive functions remains a major focus of biological psychology. With modern imaging technology and a more carefully validated understanding of cognitive abilities, a detailed view of the organization of the brain is emerging. Review **Figure 1.4**

5 The prevalence of neurological and psychiatric disorders exacts a very high emotional and economic toll. Review **Figure 1.6**

6 Although genes can have a major impact on brain function, it is clear that experience physically alters the brain, and that genetically identical people will not necessarily suffer from the same brain disorders. Review **Figure 1.7**

7 Biological psychologists balance three general research perspectives— **correlation, somatic intervention,** and **behavioral intervention**— in designing their research. Review **Figure 1.9**

8 Research in biological psychology is conducted at **levels of analysis** ranging from molecular events to the functioning of the entire brain and complex social situations. Review **Figure 1.10**

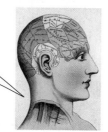

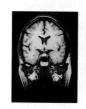

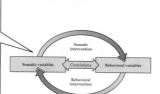

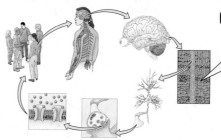

Go to **2e.mindsmachine.com** for study questions, quizzes, flashcards, and other resources.

Cells and Structures
The Anatomy of the Nervous System

Electrical Storm

To see the video
Inside the Brain,
go to

2e.mindsmachine.com/2.1

Sam had been complaining about feeling a bit odd all day; he thought perhaps he was coming down with a bug. But when he collapsed unconscious to the floor of the lunchroom at work and began twitching and jerking, showing the classic signs of a seizure, it was clear that he had a much bigger problem than the flu. By the time Sam arrived at the hospital, the seizure had stopped, and although he was confused and slow to respond to commands, he didn't seem to be in distress. But Sam was in his thirties and had never had a seizure before, or any other illness. In the emergency room, Sam smiled at Dr. Cheng, the attending neurologist, and offered to shake her hand. Alarmed by what she observed, Dr. Cheng ordered immediate brain scans: Sam could offer only half a

smile, because only the left side of his face was working, and he was unable to grip Dr. Cheng's hand at all.

Although we now know quite a bit about the neural organization of basic functions, the control of complex cognition remains a tantalizing mystery. However, the advent of sophisticated brain-imaging technology has invigorated the search for answers to fundamental questions about brain organization: Does each brain region control a specific behavior? Conversely, can every behavior be linked to a particular brain region? Or do some regions act as general purpose processors? Is everybody's brain organized in the same way? Can an understanding of the pathways between brain and body provide clues to what's happening to Sam?

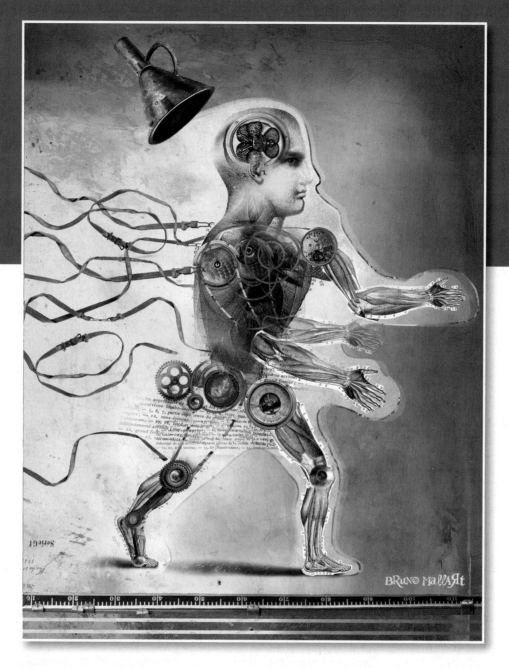

Almost everything about us—our thoughts, feelings, and behavior, however serious or silly—is the product of a knobbly three-pound organ that, despite its
unremarkable appearance, is the most complicated object in the known universe. In
this chapter we'll have a look at the structure of the brain from several different perspectives: the brain's cellular composition, its major anatomical divisions, and its appearance in computerized brain imaging. In later chapters we will build on this information as we learn how cells within the brain communicate through electrical (see
Chapter 3) and chemical (see Chapter 4) signals.

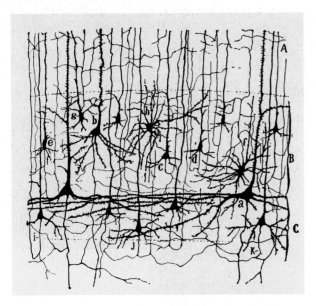

Nineteenth-Century Drawings of Neurons Santiago Ramón y Cajal perfected methods for visualizing the detailed structure of individual neurons. Ramón y Cajal and his collaborators proposed that neurons are discrete cells that communicate via tiny contacts, which were later named *synapses*.

neuron   Also called *nerve cell*. The basic unit of the nervous system, each composed of receptive extensions called dendrites, an integrating cell body, a conducting axon, and a transmitting axon terminal.

glial cells   Also called *glia*. Nonneuronal brain cells that provide structural, nutritional, and other types of support to the brain.

synapse   The cellular location at which information is transmitted from a neuron to another cell.

input zone   The part of a neuron that receives information from other neurons or from specialized sensory structures.

To view the
Brain Explorer,
go to

2e.mindsmachine.com/2.2

The Nervous System Is Composed of Cells

All of your organs and muscles are in communication with the nervous system, which, like all other living tissue, is made up of cells. The most important of these are the **neurons** (or nerve cells), arranged into the neural circuits that underlie the simplest and the most complex of behaviors. Each neuron receives inputs from many other neurons, integrates those inputs, and then distributes the processed information to other neurons. Your brain contains nearly 100 billion of these tiny cellular computers (Herculano-Houzel, 2012), working together to process vast amounts of information with apparent ease. A similarly huge number of **glial cells** (sometimes called glia) are found in the human brain, mostly providing a variety of support functions but also participating in information processing. Because neurons are larger and produce readily measured electrical signals, we know much more about them than about glial cells.

An important early controversy in neuroscience concerned the functional independence of individual neurons: Was each neuron a discrete component? Or were the cells of the nervous system fused together into larger functional units, like continuous circuits? Through painstaking study of the fine details of individual neurons, the celebrated Spanish anatomist Santiago Ramón y Cajal (1852–1934) was able to show that although neurons come very close together, they are not quite *continuous* with one another. Ramón y Cajal and his contemporaries showed that (1) the neurons and other cells of the brain are structurally, metabolically, and functionally independent; and (2) information is transmitted from neuron to neuron across tiny gaps, later named **synapses**—a notion that came to be known as the *neuron doctrine*.

It has been estimated that the brain may have as many as 10^{15} synapses. This is a remarkably large number: if you gathered that many grains of sand, each a millimeter in diameter, they would fill a cube wider (and deeper and taller) than an American football field—well over a million cubic yards! These vast networks of connections are responsible for all of humanity's achievements.

The neuron has four principal divisions

Like most human cells, a neuron contains DNA inside a cell nucleus, as well as a wide assortment of organelles. As in other cells of the body, some organelles (the mitochondria) produce energy, while others (the ribosomes) manufacture proteins under the control of DNA in the nucleus. But the neuron also has some unique, highly specialized features that allow it to collect inputs from multiple sources, integrate this information, and distribute the processed information to other cells. These information-processing features, illustrated in **FIGURE 2.1**, can be viewed as belonging to four functional zones:

1. **Input zone**   At cellular extensions called **dendrites** (from the Greek *dendron*, "tree"), neurons receive information from other neurons. Dendrites may be elaborately branched, to accommodate synapses from many other neurons.

2. **Integration zone**   In addition to receiving additional synaptic inputs, the neuron's **cell body** (or *soma*, plural *somata*) integrates (combines) the information that has been received to determine whether or not to send a signal of its own.

3. **Conduction zone**   A single extension, the **axon** (or *nerve fiber*), conducts the neuron's output information, in the form of electrical impulses, away from the cell body.

FIGURE 2.1 The Major Parts of the Neuron

Information from other neurons is passed to the dendrites and cell body via synapses. Some neurons receive only a few synaptic inputs; other receive thousands.

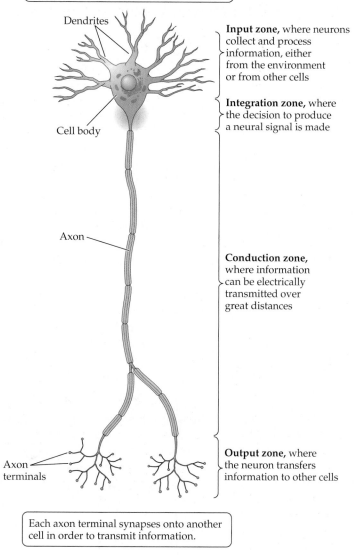

Dendrites

Input zone, where neurons collect and process information, either from the environment or from other cells

Cell body

Integration zone, where the decision to produce a neural signal is made

Axon

Conduction zone, where information can be electrically transmitted over great distances

Axon terminals

Output zone, where the neuron transfers information to other cells

Each axon terminal synapses onto another cell in order to transmit information.

4. **Output zone** Specialized swellings at the ends of the axon, called **axon terminals** (or *synaptic boutons*), transmit the neuron's signals across synapses to other cells.

There are hundreds of different forms of neurons, their shapes reflecting the functions that they are specialized to perform. For example, **motoneurons** (also called *motor neurons*) are large and have long axons reaching out to synapse on muscles, causing muscular contractions. As their name implies, **sensory neurons** are specialized to gather sensory information, and they have very diverse shapes, depending on whether they detect light or sound or touch and so on. Most of the neurons in the brain are **interneurons**, which analyze information gathered from one set of neurons and communicate with others. The axons of interneurons may measure only a few micrometers (μm; one micrometer is a millionth of a meter), while motoneurons and sensory neurons may have axons a meter or more in length, conveying information to and from the most distant parts of the body. In general, larger neurons tend to have more-complex inputs and outputs, cover greater distances, and/or convey information more rapidly than smaller neurons. The relative sizes of some of the neural structures that we will be discussing throughout the book are illustrated in **FIGURE 2.2**.

dendrite An extension of the cell body that receives information from other neurons.

integration zone The part of a neuron that initiates neural electrical activity.

cell body Also called *soma*. The region of a neuron that is defined by the presence of the cell nucleus.

conduction zone The part of a neuron—typically the axon—over which the action potential is actively propagated.

axon Also called *nerve fiber*. A single extension from the nerve cell that carries action potentials from the cell body toward the axon terminals.

output zone The part of a neuron at which the cell sends information to another cell.

axon terminal Also called *synaptic bouton*. The end of an axon or axon collateral, which forms a synapse with a neuron or other target cell.

motoneuron Also called *motor neuron*. A neuron that transmits neural messages to muscles (or glands).

sensory neuron A nerve cell that is directly affected by changes in the environment, such as light, odor, or touch.

interneuron A nerve cell that is neither a sensory neuron nor a motoneuron; interneurons receive input from and send output to other neurons.

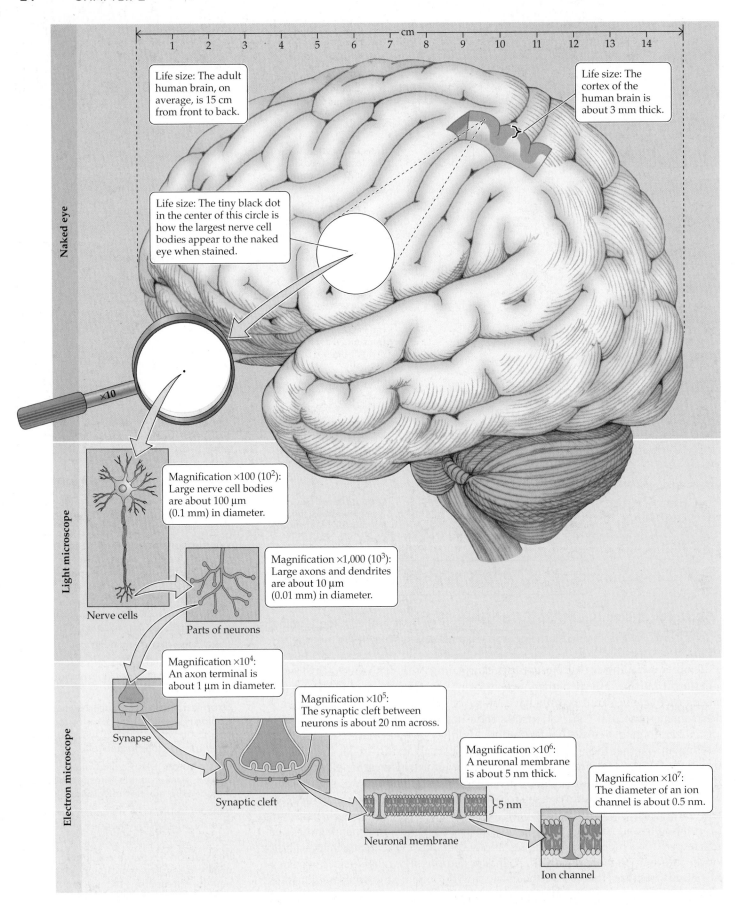

cm 1 2 3 4 5 6 7 8 9 10 11 12 13 14

Naked eye

Life size: The adult human brain, on average, is 15 cm from front to back.

Life size: The cortex of the human brain is about 3 mm thick.

Life size: The tiny black dot in the center of this circle is how the largest nerve cell bodies appear to the naked eye when stained.

×10

Light microscope

Magnification ×100 (10^2): Large nerve cell bodies are about 100 µm (0.1 mm) in diameter.

Nerve cells

Parts of neurons

Magnification ×1,000 (10^3): Large axons and dendrites are about 10 µm (0.01 mm) in diameter.

Electron microscope

Magnification ×10^4: An axon terminal is about 1 µm in diameter.

Synapse

Magnification ×10^5: The synaptic cleft between neurons is about 20 nm across.

Synaptic cleft

Magnification ×10^6: A neuronal membrane is about 5 nm thick.

5 nm

Neuronal membrane

Magnification ×10^7: The diameter of an ion channel is about 0.5 nm.

Ion channel

FIGURE 2.2 Sizes of Some Neural Structures and the Units of Measure and Magnification Used in Studying Them

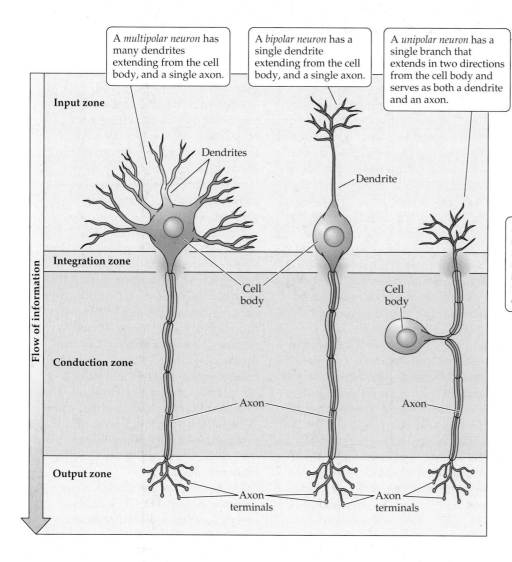

A *multipolar neuron* has many dendrites extending from the cell body, and a single axon.

A *bipolar neuron* has a single dendrite extending from the cell body, and a single axon.

A *unipolar neuron* has a single branch that extends in two directions from the cell body and serves as both a dendrite and an axon.

FIGURE 2.3 Neurons Are Classified into Three Principal Types

All neurons have the same four functional zones—*input, integration, conduction,* and *output*—although they are organized in different ways.

Input zone

Dendrites

Dendrite

Integration zone

Cell body

Cell body

Conduction zone

Flow of information

Axon

Axon

Output zone

Axon terminals

Axon terminals

In addition to size, anatomists classify neurons according to three general shapes, each specialized for a particular kind of information processing (**FIGURE 2.3**):

1. **Multipolar neurons** have many dendrites and a single axon. They are the most common type of neuron.
2. **Bipolar neurons** have a single dendrite at one end of the cell and a single axon at the other end. Bipolar neurons are especially common in sensory systems, such as vision.
3. **Unipolar neurons** (also called *monopolar neurons*) have a single extension (or process), usually thought of as an axon, that branches in two directions after leaving the cell body. One end is the input zone with branches like dendrites; the other, the output zone. Unipolar neurons transmit touch information from the body into the spinal cord.

In all three types of neurons, the dendrites comprise the input zone. In multipolar and bipolar neurons, the cell body also receives synaptic inputs, so it is also part of the input zone. Some of the techniques used to visualize neurons are discussed in **BOX 2.1**.

Information is received through synapses

A neuron's dendrites reflect the complexity of the inputs that are received. Some simple neurons have just a couple of short dendritic branches, while others have huge and complex dendritic trees covered in many thousands of synaptic contacts from other neurons. At each synapse, information is transmitted from an axon terminal of a

multipolar neuron A nerve cell that has many dendrites and a single axon.

bipolar neuron A nerve cell that has a single dendrite at one end and a single axon at the other end.

unipolar neuron Also called *monopolar neuron*. A nerve cell with a single branch that leaves the cell body and then extends in two directions; one end is the input zone, and the other end is the output zone.

BOX 2.1 Visualizing the Cellular Structure of the Brain

In the mid-1800s, dyes used to color fabrics provided a breakthrough in anatomical studies. Preserved nerve cells treated with these dyes suddenly become vivid, and hidden parts become evident. Different dyes have special affinities for different parts of the cell, such as membranes, the cell body, or the sheaths surrounding axons.

Two traditional cell stains allow detailed study of the size and shape of neurons. **Golgi stains** fill the whole cell, including details such as dendritic spines (**FIGURE A**). For reasons that remain a mystery, this technique stains only a small number of cells, each of which stands out in dramatic contrast to adjacent unstained cells. This property makes Golgi staining useful for identifying the type and shape of cells in a region. In contrast, **Nissl stains** outline *all* cell bodies because the dyes are attracted to RNA distributed within the cell (**FIGURE B**). Nissl stains allow us to measure cell body size and the density of cells in particular regions.

In a procedure known as **autoradiography**, cells are manipulated into taking photographs of themselves. For example, in order to identify the parts of the brain

that are affected by a newly discovered drug, experimenters might bathe thin sections of brain tissue in a solution with a radioactively labeled form of the drug. Time is allowed for the radioactive drug to reach its target, and then the brain sections are rinsed and placed on slides covered with photographic emulsion. Radioactivity emitted by the labeled drug in the tissue "exposes" the emulsion—like light striking film—producing a collection of fine, dark grains wherever the drug has accumulated.

Another method for labeling cells that have an attribute in common—termed **immunocytochemistry (ICC)**—capitalizes on the affinity of antibodies for specific proteins. Brain slices are exposed to antibodies that are selective for a particular protein of interest (it is possible to make antibodies for almost any protein). After time is allowed for the antibodies to attach to molecules of the target protein, unattached antibodies are rinsed off and chemical treatments make the antibodies visible. The process reveals only those cells that were making that particular protein (**FIGURE C**). A related procedure, called **in situ hybridization**, goes a step further and, using radioactively labeled lengths

of nucleic acid (RNA or DNA), identifies neurons in which a gene of interest has been turned on.

When neurons become more active, they tend to express **immediate early genes (IEGs)**, such as *c-fos*. By using ICC to label the IEG product, researchers can identify brain regions that were active during particular behaviors being performed by the animal shortly before it was sacrificed (**FIGURE D**).

Some research questions are most concerned with connections between brain regions, so methods for labeling axon pathways have also been developed. The old way to do this was simply to damage neurons of interest and then look for their degenerating axons. Newer procedures accomplish the same goal via injection of radioactively labeled amino acids into a collection of cell bodies. These radioactive molecules are taken up by the cell, incorporated into proteins, and transported to the tips of the axons (a process termed *anterograde labeling*), where they are then visualized through autoradiography, described above.

Alternatively, the cells of origin of a particular set of axons can be identified using a tract tracer such as **horseradish peroxidase (HRP)**, an enzyme found in the roots of horseradish. HRP is taken up into the axon at the terminals and transported back to the cell body, where it is readily visualized (a process known as *retrograde labeling*) (**FIGURE E**). All along the way, visible reaction products are formed—akin to footprints along a pathway. (Figure A courtesy of Timothy DeVoogd; C from Daniele and MacDermott, 2009; D from Sunn et al., 2002; E courtesy of Dale Sengelaub.)

(A) Golgi stain

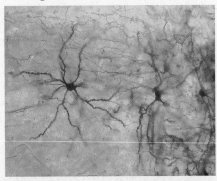

(B) Nissl stain

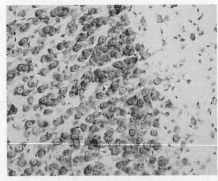

(C) Immunocytochemistry

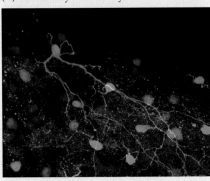

(D) Expression of *c-fos* in activated cells

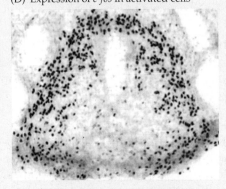

(E) HRP-filled motoneuron

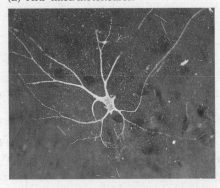

FIGURE 2.4 Synapses

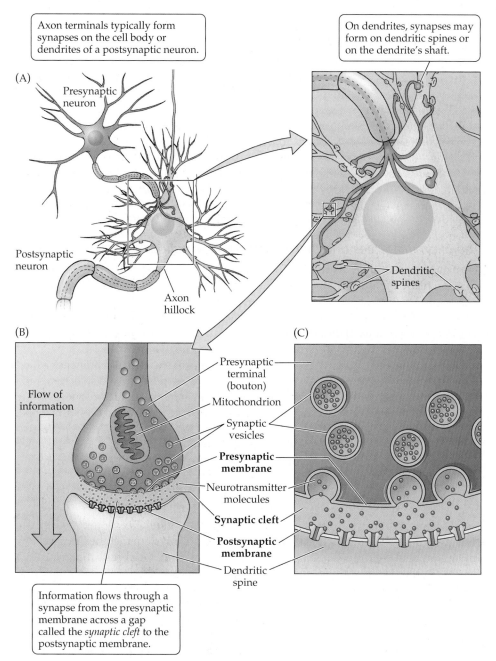

Axon terminals typically form synapses on the cell body or dendrites of a postsynaptic neuron.

(A)

Presynaptic neuron

Postsynaptic neuron

Axon hillock

On dendrites, synapses may form on dendritic spines or on the dendrite's shaft.

Dendritic spines

(B)

Flow of information

Presynaptic terminal (bouton)

Mitochondrion

Synaptic vesicles

Presynaptic membrane

Neurotransmitter molecules

Synaptic cleft

Postsynaptic membrane

Dendritic spine

(C)

Information flows through a synapse from the presynaptic membrane across a gap called the *synaptic cleft* to the postsynaptic membrane.

presynaptic Located on the "transmitting" side of a synapse.

postsynaptic Referring to the region of a synapse that receives and responds to neurotransmitter.

presynaptic membrane The specialized membrane on the axon terminal of a nerve cell that transmits information by releasing neurotransmitter.

postsynaptic membrane The specialized membrane on the surface of a neuron that receives information by responding to neurotransmitter from a presynaptic neuron.

synaptic cleft The space between the presynaptic and postsynaptic neurons at a synapse.

synaptic vesicle A small, spherical structure that contains molecules of neurotransmitter.

neurotransmitter Also called *synaptic transmitter*, *chemical transmitter*, or simply *transmitter*. The chemical released from the presynaptic axon terminal that serves as the basis of communication between neurons.

presynaptic neuron to the **postsynaptic** neuron (**FIGURE 2.4A**). A synapse is typically composed of the following elements (**FIGURE 2.4B**):

1. The specialized **presynaptic membrane** of the axon terminal of the presynaptic neuron

2. The specialized **postsynaptic membrane** on the dendrite or cell body of the postsynaptic neuron

3. The **synaptic cleft**, a gap that separates the presynaptic and postsynaptic membranes, which is tiny, measuring only 20–40 nanometers (nm; one nanometer is a billionth of a meter)

The presynaptic axon terminals contain many microscopic hollow spheres called **synaptic vesicles**. Each synaptic vesicle contains molecules of **neurotransmitter**, the special chemical with which a presynaptic neuron communicates with postsynaptic cells. To signal the postsynaptic cell, the presynaptic neuron fuses many synaptic vesicles to its presynaptic membrane, releasing their contents into the synaptic cleft

neurotransmitter receptor Also called simply *receptor*. A specialized protein, often embedded in the cell membrane, that selectively senses and reacts to molecules of a corresponding neurotransmitter or hormone.

neuroplasticity Also called *neural plasticity*. The ability of the nervous system to change in response to experience or the environment.

axon hillock The cone-shaped area on the cell body from which the axon originates.

innervate To provide neural input to.

axon collateral A branch of an axon.

axonal transport The transportation of materials from the neuronal cell body toward the axon terminals, and from the axon terminals back toward the cell body.

oligodendrocyte A type of glial cell that forms myelin in the central nervous system.

Schwann cell A type of glial cell that forms myelin in the peripheral nervous system.

myelin The fatty insulation around an axon, formed by glial cells. This sheath boosts the speed at which nerve impulses are conducted.

node of Ranvier A gap between successive segments of the myelin sheath where the axon membrane is exposed.

astrocyte A star-shaped glial cell with numerous processes (extensions) that run in all directions.

(see Figure 2.4B). After crossing the cleft, the released neurotransmitter interacts with matching **neurotransmitter receptors**, special protein molecules that stud the postsynaptic membrane. These receptors capture and react to molecules of the neurotransmitter, altering the level of excitation of the postsynaptic neuron and thus affecting the likelihood that the postsynaptic neuron will in turn release its own neurotransmitter from its axon terminals. Molecules of neurotransmitter generally do not enter the postsynaptic neuron; they simply bind to the receptors momentarily, and then disengage.

The configuration of synapses on a neuron's dendrites and cell body is constantly changing—synapses come and go, and dendrites change their shapes—in response to new patterns of synaptic activity and the formation of new neural circuits. We use the general term **neuroplasticity** to refer to this capacity for continual remodeling of the connections between neurons. We will take a much more detailed look at neurotransmission in Chapters 3 and 4.

The axon integrates and then transmits information

Most neurons feature a distinctive cone-shaped enlargement on the cell body called an **axon hillock** ("little hill"), from which the neuron's axon projects. The hillock has unique properties that allow it to gather and integrate incoming information from the synapses on the dendrites and cell body, converting those inputs into a code of electrical impulses that we'll describe in detail in Chapter 3. These electrical signals race down the axon toward the targets that the neuron is said to **innervate**.

The axon itself is a narrow tube that may divide near the end into branches called **axon collaterals**. Various important substances, such as enzymes and structural proteins, are conveyed inside the axon from the cell body, where they are produced, to the axon terminals, where they are used. This **axonal transport** works in both directions, also carrying used materials back to the cell body for recycling. Thus the axon has two quite different functions: the rapid transmission of electrical signals along the outer membrane, and the much slower transportation of substances within the axon, to and from the axon terminals.

Glial cells protect and assist neurons

As the name glial cell suggests (from the Greek glia, "glue"), early neuroscientists thought of glial cells as mostly filler, holding the nervous system together. But we now know that glial cells are much more important than that. Glial cells directly affect neuronal processes by providing neurons with raw materials, chemical signals, and specialized structural components. These diverse functions are accomplished by four basic classes of glial cells (**FIGURE 2.5**).

Two of the four types of glia—**oligodendrocytes** and **Schwann cells**—wrap around successive segments of axons to insulate them with a fatty substance called **myelin**. These myelin sheaths give an axon the appearance of a string of elongated slender beads. Between adjacent beads, small uninsulated patches of axonal membrane, called **nodes of Ranvier**, remain exposed (**FIGURE 2.5A**). Within the brain and spinal cord, myelination is provided by the oligodendrocytes, each cell typically supplying myelin beads to several nearby axons (also illustrated in Figure 2.5A). In the rest of the body, it is Schwann cells that do the ensheathing, with each Schwann cell wrapping itself around a segment of one axon to provide a single bead of myelin. But whether it is provided by oligodendrocytes or by Schwann cells, myelination has the same result: a large increase in the speed with which electrical signals pass down the axon, jumping from one node of Ranvier to the next. (In Chapter 3 we discuss the diverse abnormalities that arise when the myelin insulation is compromised in the disease multiple sclerosis [MS].)

The other two types of glial cells—astrocytes and microglial cells—perform more diverse functions in the brain. **Astrocytes** (from the Greek astron, "star") weave around and between neurons with tentacle-like extensions (**FIGURE 2.5B**). Some astrocytes stretch between neurons and fine blood vessels, controlling local blood flow to increase the amount of blood reaching more-active brain regions (Schummers et al.,

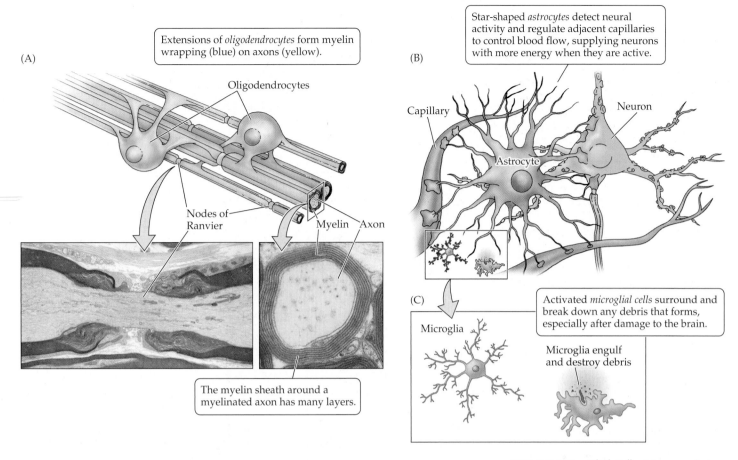

(A)

Extensions of *oligodendrocytes* form myelin wrapping (blue) on axons (yellow).

Oligodendrocytes

Nodes of Ranvier

Myelin   Axon

The myelin sheath around a myelinated axon has many layers.

(B)

Star-shaped *astrocytes* detect neural activity and regulate adjacent capillaries to control blood flow, supplying neurons with more energy when they are active.

Capillary

Neuron

Astrocyte

(C)

Microglia

Activated *microglial cells* surround and break down any debris that forms, especially after damage to the brain.

Microglia engulf and destroy debris

FIGURE 2.5 Glial Cells (Micrograph A *left* courtesy of Mark Ellisman and the National Center for Microscopy and Imaging Research; *right* from Peters et al., 1991.)

2008). Astrocytes help to form the tough outer membranes that swaddle the brain, and also secrete chemicals that modulate neural activity and the formation of synapses (Mauch et al., 2001; R. D. Fields and Stevens-Graham, 2002). In contrast, **microglial cells** (or microglia) are tiny and mobile (**FIGURE 2.5C**). Their primary job appears to be to contain and clean up sites of injury (Davalos et al., 2005), but research is revealing unexpected additional roles for microglia, such as the maintenance of synapses and contributions to pain perception (Graeber, 2010; S. Beggs et al., 2012).

Although glial cells perform many beneficial functions, they can also cause problems. For one thing, because they continue to divide in adulthood (unlike neurons), glial cells can give rise to deadly brain tumors. In addition, some glial cells, especially astrocytes, respond to brain injury by changing size—that is, by swelling. This **edema** damages neurons and is responsible for many symptoms of brain injuries.

Supported and influenced by glial cells, and sharing information through synapses, neurons form the vast ensembles of information-processing circuits that give the brain its visible form. Powerful anatomical and genetic initiatives, such as the Allen Institute brain mapping project (www.brain-map.org), are now creating detailed maps of the cellular compositions of the major divisions of the brain. These major divisions are our next topic.

microglial cells Also called *microglia*. Extremely small motile glial cells that remove cellular debris from injured or dead cells.

edema The swelling of tissue in response to injury.

how's it going ?

1. What are the four "zones" common to all neurons, and what are their functions?
2. Compare and contrast axonal signal transmission and axonal transport.
3. Describe the three main components of the synapse. What are some of the specialized structures found on each side of the synapse?
4. What are the names and general functions of the four types of glial cells?
5. What special properties does myelin have?

(A)

Here, a modern view of the central nervous system (CNS) is superimposed on the peripheral nervous system as drawn by Andreas Vesalius (1514–1564). The peripheral nervous system connects the body to the CNS.

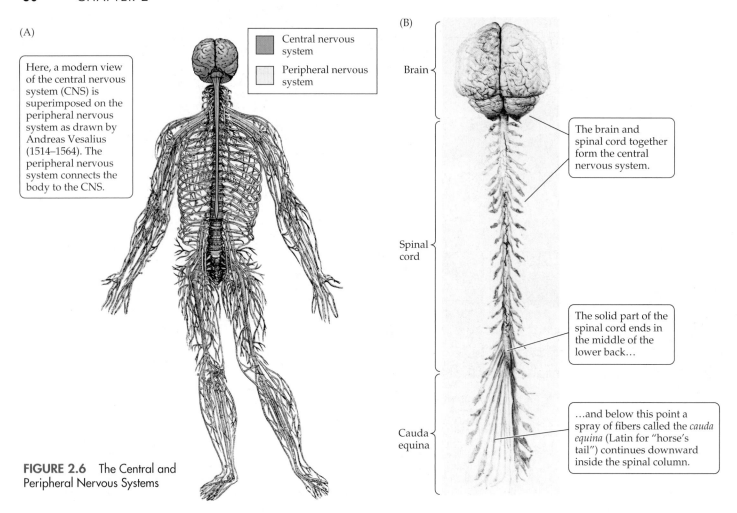

Central nervous system

Peripheral nervous system

(B)

Brain

The brain and spinal cord together form the central nervous system.

Spinal cord

The solid part of the spinal cord ends in the middle of the lower back…

Cauda equina

…and below this point a spray of fibers called the *cauda equina* (Latin for "horse's tail") continues downward inside the spinal column.

FIGURE 2.6   The Central and Peripheral Nervous Systems

gross neuroanatomy   Anatomical features of the nervous system that are apparent to the naked eye.

central nervous system (CNS)   The portion of the nervous system that includes the brain and the spinal cord.

peripheral nervous system   The portion of the nervous system that includes all the nerves and neurons outside the brain and spinal cord.

nerve   A collection of axons bundled together outside of the central nervous system.

motor nerve   A nerve that transmits information from the central nervous system to the muscles and glands.

sensory nerve   A nerve that conveys information from the body to the central nervous system.

somatic nervous system   A part of the peripheral nervous system that supplies neural connections mostly to the skeletal muscles and sensory systems of the body. It consists of cranial nerves and spinal nerves.

The Nervous System Extends throughout the Body

In this section we'll describe the **gross neuroanatomy** of the nervous system—the components that are visible to the unaided eye (in this context gross means "large," not "yucky"). The gross view of the entire human nervous system presented in **FIGURE 2.6** reveals the basic division between the **central nervous system** (**CNS**; consisting of the brain and spinal cord) and the **peripheral nervous system** (everything else).

The peripheral nervous system has two divisions

The peripheral nervous system consists of **nerves**—collections of axons bundled together—that extend throughout the body. Some nerves, called **motor nerves**, transmit information from the spinal cord and brain to muscles and glands; others, called **sensory nerves**, convey information from the body to the CNS. The various nerves of the body are divided into two distinct systems:

1. The **somatic nervous system**, which consists of nerves that interconnect the brain and the major muscles and sensory systems of the body

2. The **autonomic nervous system**, which consists of the nerves that connect to the viscera (internal organs)

THE SOMATIC NERVOUS SYSTEM   Taking its name from the Latin word for "body"— soma—the somatic nervous system is the main pathway through which the brain controls movement and receives sensory information from the body and from the sensory organs of the head. The nerves that make up the somatic nervous system form two anatomical groups: the cranial nerves and the spinal nerves.

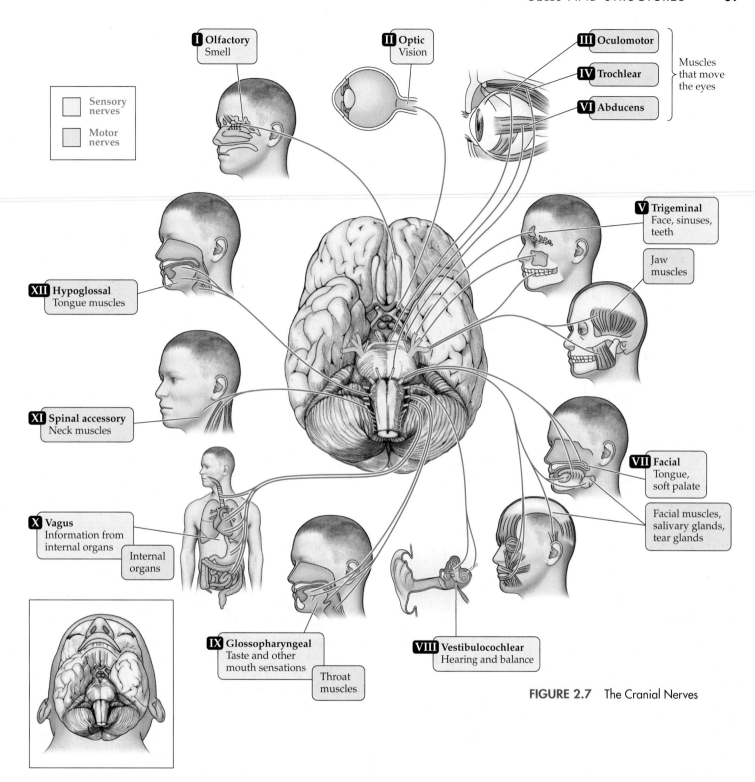

I **Olfactory**
Smell

II **Optic**
Vision

III **Oculomotor**

IV **Trochlear**

VI **Abducens**

Muscles
that move
the eyes

Sensory
nerves

Motor
nerves

V **Trigeminal**
Face, sinuses,
teeth

Jaw
muscles

XII **Hypoglossal**
Tongue muscles

XI **Spinal accessory**
Neck muscles

VII **Facial**
Tongue,
soft palate

Facial muscles,
salivary glands,
tear glands

X **Vagus**
Information from
internal organs

Internal
organs

IX **Glossopharyngeal**
Taste and other
mouth sensations

Throat
muscles

VIII **Vestibulocochlear**
Hearing and balance

FIGURE 2.7   The Cranial Nerves

We each have 12 pairs (left and right) of **cranial nerves** that arise from the brain and innervate the head, neck, and visceral organs directly, without ever joining the spinal cord. As you can see in **FIGURE 2.7**, some of these nerves are exclusively sensory: the olfactory (I) nerves transmit information about smell, the optic (II) nerves carry visual information from the eyes, and the vestibulocochlear (VIII) nerves convey auditory and balance information. Five pairs of cranial nerves are exclusively motor pathways from the brain: the oculomotor (III), trochlear (IV), and abducens (VI) nerves innervate muscles to move the eye; the spinal accessory (XI) nerves control neck muscles; and the hypoglossal (XII) nerves control the tongue. The remaining cranial nerves have

autonomic nervous system   A part of the peripheral nervous system that provides the main neural connections to glands and to smooth muscles of internal organs.

cranial nerve   A nerve that is connected directly to the brain.

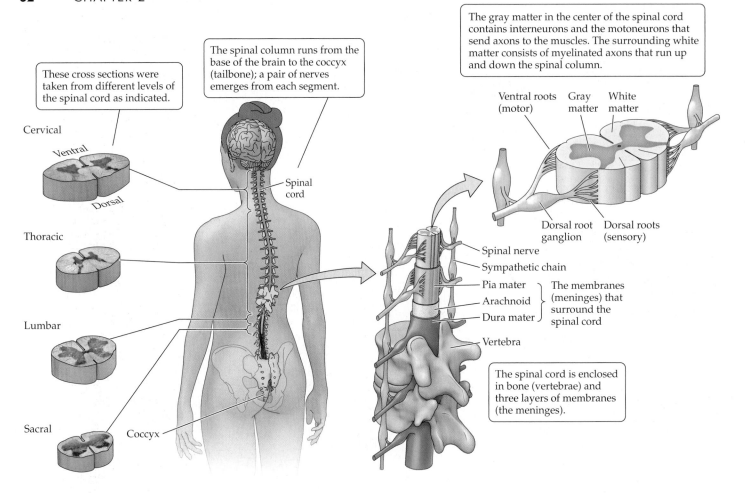

These cross sections were taken from different levels of the spinal cord as indicated.

The spinal column runs from the base of the brain to the coccyx (tailbone); a pair of nerves emerges from each segment.

The gray matter in the center of the spinal cord contains interneurons and the motoneurons that send axons to the muscles. The surrounding white matter consists of myelinated axons that run up and down the spinal column.

Cervical
Ventral
Dorsal
Thoracic
Lumbar
Sacral
Coccyx

Spinal cord

Ventral roots (motor)   Gray matter   White matter

Dorsal root ganglion   Dorsal roots (sensory)

Spinal nerve
Sympathetic chain
Pia mater
Arachnoid   The membranes (meninges) that surround the spinal cord
Dura mater
Vertebra

The spinal cord is enclosed in bone (vertebrae) and three layers of membranes (the meninges).

FIGURE 2.8   The Spinal Cord and Spinal Nerves

spinal nerve   A nerve that emerges from the spinal cord.

cervical   Referring to the topmost eight segments of the spinal cord, in the neck region.

thoracic   Referring to the 12 spinal segments below the cervical (neck) portion of the spinal cord, corresponding to the chest.

lumbar   Referring to the five spinal segments that make up the upper part of the lower back.

sacral   Referring to the five spinal segments that make up the lower part of the lower back.

coccygeal   Referring to the lowest spinal vertebra (the coccyx, also known as the "tailbone").

both sensory and motor functions. The trigeminal (V) nerves, for example, transmit facial sensation through some axons but control the chewing muscles through other axons. The facial (VII) nerves control facial muscles and receive some taste sensation, and the glossopharyngeal (IX) nerves receive additional taste sensations and sensations from the throat, and control the muscles there. The vagus (X) nerve extends far from the head, running to the heart, liver, and intestines. Its long, convoluted route is the reason for its name, which is Latin for "wandering."

An additional 31 pairs of **spinal nerves**—again, one member of each pair serves each side of the body—are connected to the spinal cord through regularly spaced openings along both sides of the backbone (**FIGURE 2.8**). Each nerve is made up of a group of motor fibers, projecting from the ventral (front) part of the spinal cord to the organs and muscles, and a group of sensory fibers that enter the dorsal (rear) part of the spinal cord. Spinal nerves are named according to the segments of the spinal cord to which they are connected. There are 8 **cervical** (neck), 12 **thoracic** (trunk), 5 **lumbar** (lower back), 5 **sacral** (pelvic), and 1 **coccygeal** (bottom) spinal segments. The name of each spinal nerve reflects the position of the spinal cord segment to which it is connected; for example, the nerve connected to the twelfth thoracic segment is called T12, the nerve connected to the seventh cervical segment is called C7, and so on. After leaving the spinal cord, axons from the spinal nerves spread out in the body and may join with axons from different spinal nerves to form the various peripheral nerves.

THE AUTONOMIC NERVOUS SYSTEM   Although it is "autonomous" in the sense that we have little conscious, voluntary control over its actions, the autonomic nervous system is the brain's main system for controlling the organs of the body. These control functions are performed by two major divisions—the sympathetic and parasympathetic nervous systems—that act more or less in opposition (**FIGURE 2.9**).

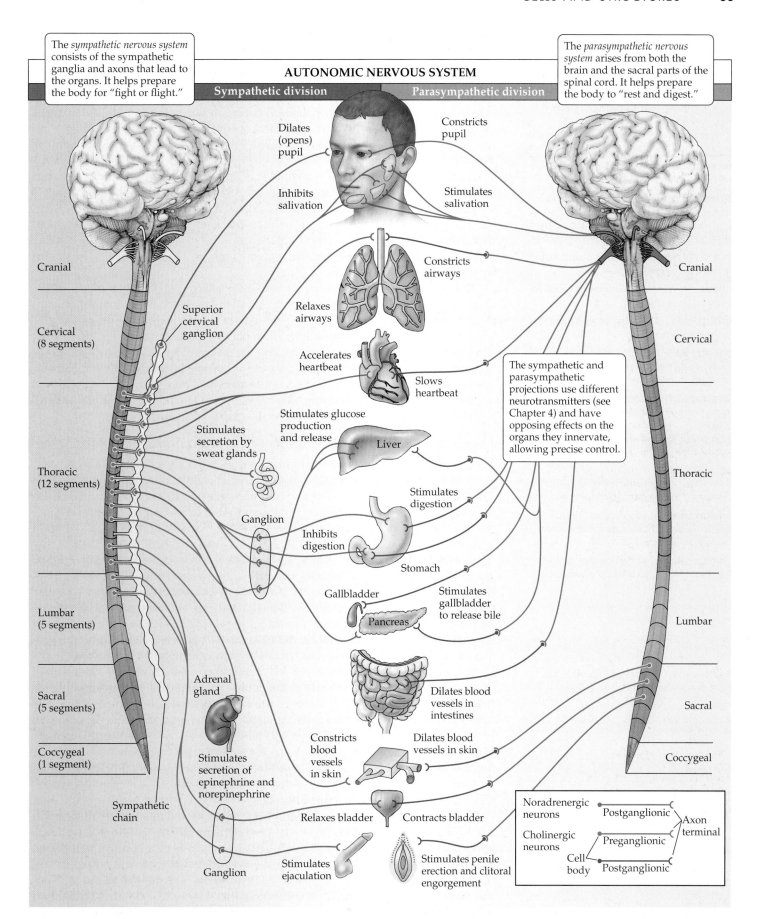

The *sympathetic nervous system* consists of the sympathetic ganglia and axons that lead to the organs. It helps prepare the body for "fight or flight."

AUTONOMIC NERVOUS SYSTEM

Sympathetic division

Parasympathetic division

The *parasympathetic nervous system* arises from both the brain and the sacral parts of the spinal cord. It helps prepare the body to "rest and digest."

Dilates (opens) pupil

Constricts pupil

Inhibits salivation

Stimulates salivation

Cranial

Cranial

Cervical (8 segments)

Cervical

Superior cervical ganglion

Constricts airways

Relaxes airways

Accelerates heartbeat

Slows heartbeat

The sympathetic and parasympathetic projections use different neurotransmitters (see Chapter 4) and have opposing effects on the organs they innervate, allowing precise control.

Stimulates glucose production and release

Stimulates secretion by sweat glands

Liver

Thoracic (12 segments)

Thoracic

Stimulates digestion

Ganglion

Inhibits digestion

Stomach

Gallbladder

Stimulates gallbladder to release bile

Pancreas

Lumbar (5 segments)

Lumbar

Dilates blood vessels in intestines

Sacral (5 segments)

Adrenal gland

Constricts blood vessels in skin

Dilates blood vessels in skin

Sacral

Coccygeal (1 segment)

Stimulates secretion of epinephrine and norepinephrine

Coccygeal

Sympathetic chain

Relaxes bladder

Contracts bladder

Ganglion

Stimulates ejaculation

Stimulates penile erection and clitoral engorgement

Noradrenergic neurons

Cholinergic neurons

Postganglionic

Preganglionic

Cell body

Postganglionic

Axon terminal

FIGURE 2.9 The Autonomic Nervous System

sympathetic nervous system   The part of the autonomic nervous system that acts as the "fight or flight" system, generally activating the body for action.

parasympathetic nervous system   The part of the autonomic nervous system that generally prepares the body to relax and recuperate.

cerebral hemisphere   One of the two halves—right or left—of the forebrain.

cerebral cortex   Also called simply *cortex*. The outer covering of the cerebral hemispheres, which consists largely of nerve cell bodies and their branches.

gyrus   A ridged or raised portion of a convoluted brain surface.

sulcus   A crevice or valley of a convoluted brain surface.

frontal lobe   The most anterior portion of the cerebral cortex.

parietal lobe   The large region of cortex lying between the frontal and occipital lobes in each cerebral hemisphere.

temporal lobe   The large lateral region of cortex in each cerebral hemisphere. It is continuous with the parietal lobe posteriorly and separated from the frontal lobe by the Sylvian fissure.

occipital lobe   A large region of cortex that covers much of the posterior part of each cerebral hemisphere.

Sylvian fissure   Also called *lateral sulcus*. A deep fissure that demarcates the temporal lobe.

central sulcus   A fissure that divides the frontal lobe from the parietal lobe.

Axons of the **sympathetic nervous system** exit from the middle parts of the spinal cord, travel a short distance, and then innervate the sympathetic ganglia (small clusters of neurons found outside the CNS), which run in two chains along the spinal cord, one on each side (see Figure 2.9 left). Axons from the sympathetic ganglia then course throughout the body, innervating all the major organ systems. In general, sympathetic innervation prepares the body for immediate action: blood pressure increases, the pupils of the eyes widen, and the heart quickens. This set of reactions is sometimes called simply the "fight or flight" response.

In contrast to the effects of sympathetic activity, the **parasympathetic nervous system** generally helps the body to relax, recuperate, and prepare for future action—sometimes called the "rest and digest" response. Arising above and below the sympathetic projections, in the brain and sacral spinal cord (hence the name parasympathetic, from the Greek para, "around"), parasympathetic axons travel a longer distance before terminating in parasympathetic ganglia, usually located close to the organs they serve (see Figure 2.9 right).

The sympathetic and parasympathetic systems have very different effects on individual organs because the two opposing systems release different neurotransmitters. The balance between the two systems determines the state of the internal organs at any given moment. So, for example, when parasympathetic activity predominates, heart rate slows, blood pressure drops, and digestive processes are activated. As the brain causes the balance of autonomic activity to become predominantly sympathetic, opposite effects are seen: increased heart rate and blood pressure, inhibited digestion, and so on. This tension between parasympathetic and sympathetic activity ensures that the individual is appropriately prepared for current circumstances.

The central nervous system consists of the brain and spinal cord

The spinal cord funnels sensory information from the body up to the brain and conveys the brain's motor commands out to the body. The spinal cord also contains circuits that perform local processing and control simple units of behavior, such as reflexes. We will discuss other aspects of the spinal cord in later chapters, but for now let's focus on the anatomy of the executive portion of the CNS: the brain.

THE OUTER SURFACE OF THE BRAIN   On average, the human brain weighs only 1,400 grams (about 3 pounds), accounting for just 2% of the average body weight. Put your two fists together and you get a sense of the size of the two **cerebral hemispheres**—smaller than most people expect. But what the brain lacks in size and weight it makes up for in intricacy. Anatomists use standard terminology to help identify structures, locations, and directions in the brain, as described in **BOX 2.2**. (It's a bit of a chore, but learning the anatomical lingo now will make later discussions of brain structures much easier to follow.)

One obvious feature of the brain is its lumpy, convoluted surface—the result of elaborate folding of a thick sheet of tissue, mostly the dendrites, cell bodies, and axonal projections of neurons, called the **cerebral cortex** (or sometimes just cortex). The resultant ridges of tissue, called **gyri** (singular gyrus), are separated from each other by crevices called **sulci** (singular sulcus). By being crumpled up in this way, much more of the cortex can be crammed into the confines of the skull, and about two-thirds of the cerebral cortex is hidden in the depths of these folds. The pattern of folding is not random; in fact, it is similar enough between brains that we can name the various gyri and sulci and group them together into lobes.

Neuroscientiush rely on a combination of landmarks and functions to distinguish among four major cortical regions of the cerebral hemispheres: the **frontal**, **parietal**, **temporal**, and **occipital lobes** (**FIGURE 2.10**). In some cases the boundaries between adjacent lobes are very clear; for example, the **Sylvian fissure** (or lateral sulcus) divides the temporal lobe from other regions of the hemisphere. The **central sulcus** provides a distinct landmark dividing the frontal and parietal lobes. The physical boundaries between the occipital lobe and the temporal and parietal lobes are less obvious, but the lobes are quite different with regard to the functions they perform.

BOX 2.2   Three Customary Orientations for Viewing the Brain and Body

Because the nervous system is a three-dimensional structure, two-dimensional illustrations and diagrams cannot represent it completely. The brain is usually cut in one of three main planes to obtain a two-dimensional section from this three-dimensional object. Although it takes time and practice to master the lingo, it is useful to know the terminology that applies to these sections, as illustrated in the figure.

The plane that bisects the body into right and left halves is called the **sagittal plane**. The plane that divides the body into a front (anterior) and a back (posterior) part is called the **coronal plane** (also known as the *frontal* or *transverse plane*). The third main plane, which divides the brain into upper and lower parts, is called the **horizontal plane**.

In addition, several directional terms are used. **Medial** means "toward the middle" and is contrasted with **lateral**, "toward the side." Relative to one location, a second location is **ipsilateral** if it

is on the same side of the body and **contralateral** if on the opposite side of the body. These terms are all relative, as are the terms **superior** ("above") and **inferior** ("below"); for example, the eye is lateral to the nose but medial to the ear, and the mouth is inferior to the nose but superior to the chin. The term **basal** simply means "toward the base" or "toward the bottom" of a structure.

The head end of the body, and therefore the front of the brain, is referred to as **anterior** or **rostral**; the tail end of the body, and the back of the head, is described as **posterior** or **caudal** (from the Latin *cauda*, "tail"). **Proximal** means "near the center," and **distal** means "toward the periphery" or "toward the end of a limb." We call an axon, tract,

or nerve **afferent** if it carries information into a region that we're interested in, and **efferent** if it carries information away from the region of interest (a handy way to remember this is that *e*fferents *e*xit but *a*fferents *a*rrive, relative to the region of interest).

Dorsal means "toward or at the back," and **ventral** means "toward or at the belly." In four-legged animals, such as cats or rats, *dorsal* refers to both the back of the body and the top of the head and brain. For consistency in comparing brains among species, this term is also used to refer to the top of the brain of a human or of a chimpanzee, even though in such two-legged animals the top of the brain is not at the back of the body. Similarly, *ventral* is understood to designate the bottom of the brain of a two-legged as well as of a four-legged animal. (Photographs courtesy of S. Mark Williams and Dale Purves, Duke University Medical Center.)

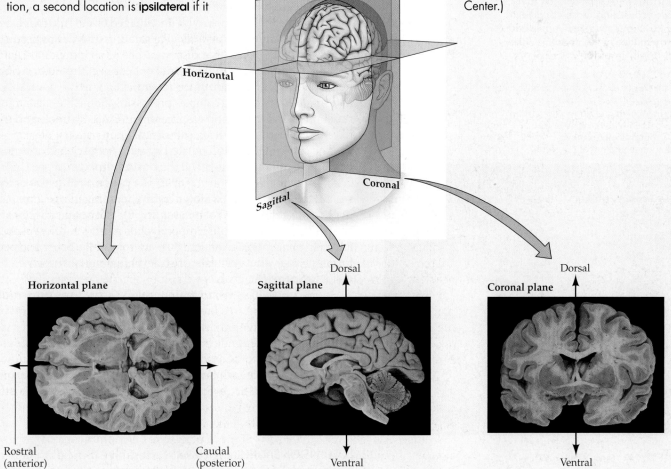

Horizontal

Sagittal

Coronal

Horizontal plane

Rostral (anterior)   Caudal (posterior)

Sagittal plane

Dorsal

Ventral

Coronal plane

Dorsal

Ventral

FIGURE 2.10 The Human Brain Has Four Distinct Lobes

To see the video
Brain Development,
go to

2e.mindsmachine.com/2.3

corpus callosum The main band of axons that connects the two cerebral hemispheres.

postcentral gyrus The strip of parietal cortex, just behind the central sulcus, that receives somatosensory information from the entire body.

precentral gyrus The strip of frontal cortex, just in front of the central sulcus, that is crucial for motor control.

gray matter Areas of the brain that are dominated by cell bodies and are devoid of myelin. Gray matter mostly receives and processes information.

white matter A light-colored layer of tissue, consisting mostly of myelin-sheathed axons, that lies underneath the gray matter of the cortex. White matter mostly transmits information.

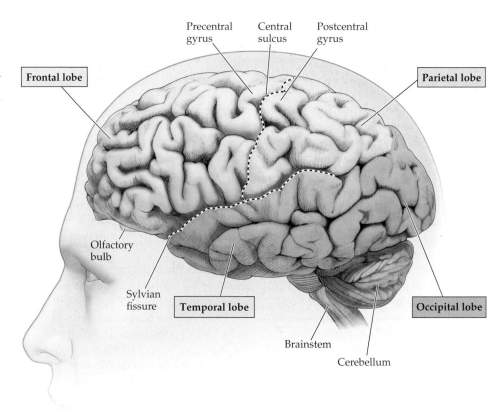

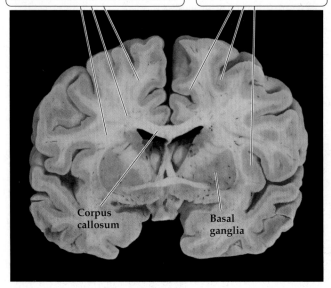

The lighter-colored interior is *white matter*, packed with the fatty myelin that surrounds axons sending information in and out of the cortex.

Gray matter consists of cell bodies that form the outer layers of the cortex and nuclei within the brain.

FIGURE 2.11 Gray Matter, White Matter (Photograph courtesy of S. Mark Williams and Dale Purves, Duke University Medical Center.)

The cortex is the seat of complex cognition. Depending on the specific regions affected, cortical damage can cause symptoms ranging from impairments of movement or body sensation; through speech errors, memory problems, and personality changes; to many kinds of visual impairments. In people with undamaged brains, the four lobes of the cortex are continually communicating and collaborating in order to produce the seamless control of complex behavior that distinguishes us as individuals. Furthermore, hundreds of millions of axons connect the left and right hemispheres via the **corpus callosum**, allowing the brain to act as a single entity during complex processing. Some life-sustaining functions—heart rate and respiration, reflexes, balance, and the like—are governed by lower, subcortical brain regions.

The sense of touch is mediated by a strip of parietal cortex just behind the central sulcus called the **postcentral gyrus** (see Figure 2.10). In front of the central sulcus, the **precentral gyrus** of the frontal lobe is crucial for motor control, organized like a map of the body (Penfield and Rasmussen, 1950). The occipital lobes are crucial for vision, and the temporal lobes receive auditory inputs and help in memory formation. But each lobe of the brain also performs a wide variety of other high-level functions. These will be major topics in later chapters.

Most people have heard brain tissue referred to as **gray matter**. When you cut into a brain, you see that the outer layers of the cortex have a darker grayish shade (**FIGURE 2.11**). This is because they contain a preponderance of neuronal cell bodies and dendrites. In contrast, the underlying **white matter** gets its snowy appearance from the whitish fatty myelin that insulates the axons of many neurons. So, a simple view is that gray matter mostly receives and processes information, while white matter mostly transmits information.

SUBDIVISIONS WITHIN THE BRAIN It can be difficult to understand some of the anatomical distinctions applied to the adult human brain. For example, the part of the brain closest to the back

FIGURE 2.12 Divisions of the Human Nervous System in the Embryo and the Adult

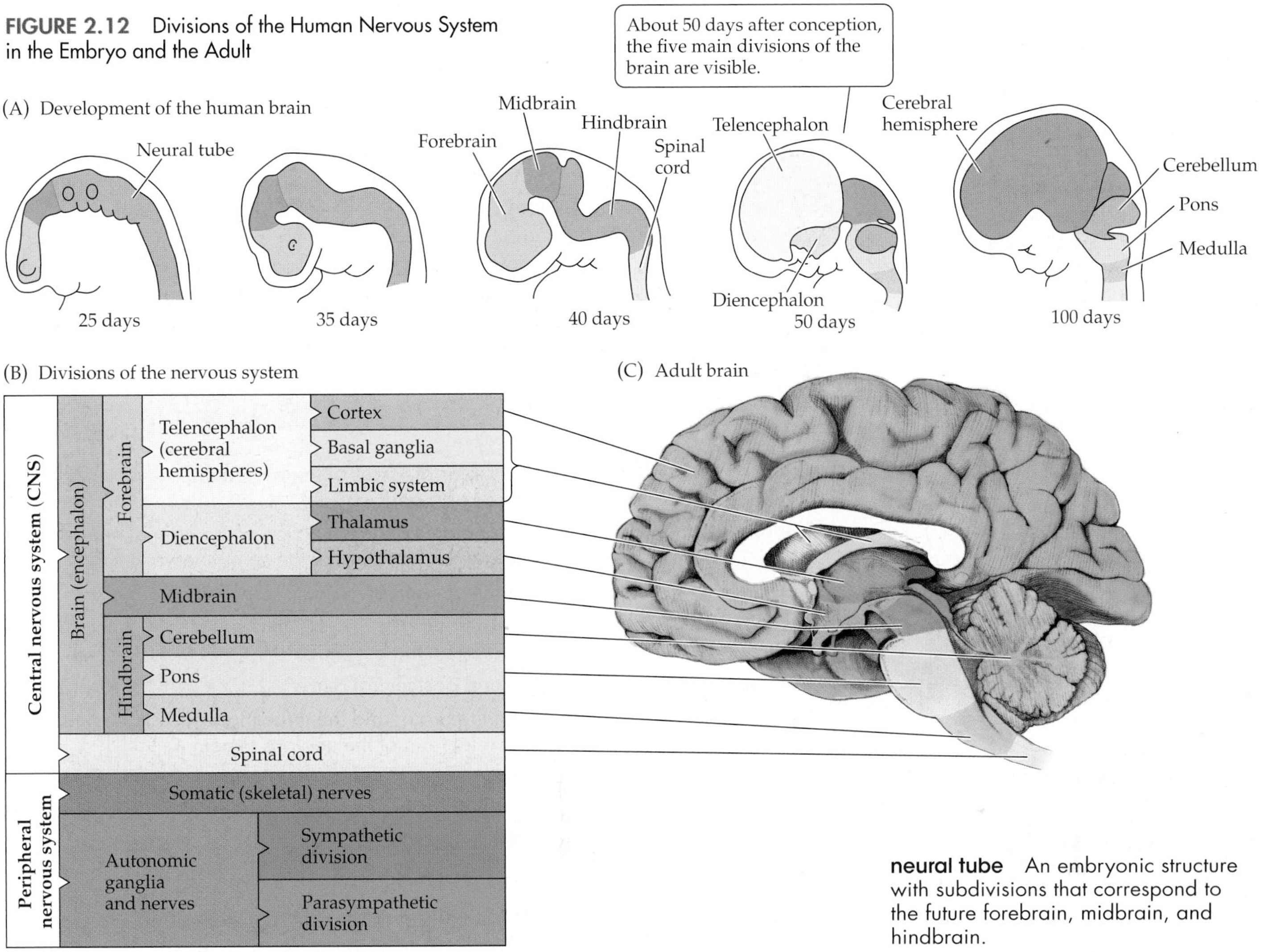

(A) Development of the human brain

About 50 days after conception, the five main divisions of the brain are visible.

25 days   35 days   40 days   50 days   100 days

Neural tube

Midbrain
Forebrain   Hindbrain
Spinal cord

Telencephalon   Cerebral hemisphere

Cerebellum
Pons
Medulla

Diencephalon

(B) Divisions of the nervous system

Central nervous system (CNS)

Brain (encephalon)

Forebrain
- Telencephalon (cerebral hemispheres)
  - Cortex
  - Basal ganglia
  - Limbic system
- Diencephalon
  - Thalamus
  - Hypothalamus

Midbrain

Hindbrain
- Cerebellum
- Pons
- Medulla

Spinal cord

Peripheral nervous system
- Somatic (skeletal) nerves
- Autonomic ganglia and nerves
  - Sympathetic division
  - Parasympathetic division

(C) Adult brain

neural tube An embryonic structure with subdivisions that correspond to the future forebrain, midbrain, and hindbrain.

forebrain The frontal division of the neural tube, containing the cerebral hemispheres, the thalamus, and the hypothalamus.

midbrain The middle division of the brain.

hindbrain The rear division of the brain, which in the mature vertebrate contains the cerebellum, pons, and medulla.

telencephalon The anterior part of the fetal forebrain, which will become the cerebral hemispheres in the adult brain.

diencephalon The posterior part of the fetal forebrain, which will become the thalamus and hypothalamus in the adult brain.

brainstem The region of the brain that consists of the midbrain, the pons, and the medulla.

nucleus Here, a collection of neuronal cell bodies within the central nervous system (e.g., the caudate nucleus).

tract A bundle of axons found within the central nervous system.

of the head is anatomically identified as part of the forebrain. Why? The key to understanding this confusing terminology is to consider how the gross anatomy of the brain develops early in life.

In a very young embryo of any vertebrate, the CNS looks like a tube. The walls of this **neural tube** are made of cells, and the interior is filled with fluid. A few weeks after conception, the human neural tube begins to show three separate swellings at the head end (**FIGURE 2.12A**): the **forebrain**, the **midbrain**, and the **hindbrain**. By about 50 days, the fetal forebrain features two clear subdivisions. At the very front is the **telencephalon** (from the Greek encephalon, "brain"), which will become the cerebral hemispheres (consisting of cortex plus some deeper structures). The other part of the forebrain is the **diencephalon**, which will go on to become the thalamus and the hypothalamus, two of the many subcortical structures of the forebrain.

Similarly, the hindbrain further develops into several large structures: the cerebellum, pons, and medulla. The term **brainstem** usually refers to the midbrain, pons, and medulla combined (some scientists include the diencephalon too). **FIGURE 2.12B** and **C** shows the positions of these structures and their relative sizes in the adult human brain. Even when the brain achieves its adult form, it is still a fluid-filled tube, but a tube of very complicated shape.

The main sections of the brain can be subdivided in turn. We can work our way from the largest, most general divisions of the nervous system on the left of the schematic in Figure 2.12B to more-specific ones on the right.

Within and between the major brain regions are collections of neurons called **nuclei** (singular nucleus) and bundles of axons called **tracts**. Recall that outside the CNS,

pyramidal cell   A type of large nerve cell that has a roughly pyramid-shaped cell body and is found in the cerebral cortex.

cortical column   One of the vertical columns that constitute the basic organization of the cerebral cortex.

basal ganglia   A group of forebrain nuclei, including the caudate nucleus, globus pallidus, and putamen, found deep within the cerebral hemispheres.

collections of neurons are called ganglia, and bundles of axons are called nerves. Unfortunately, the same word nucleus can mean either "a collection of neurons in the CNS" or "the spherical DNA-containing organelle within a single cell." You must rely on the context to understand which meaning is intended. Because brain tracts and nuclei are the same in different individuals, and often the same in different species, they have names too (many, many names).

You are probably more interested in the functions of all these parts of the brain than in their names, but as we noted earlier, each region serves more than one function, and our knowledge of the functional organization of the brain is continually being updated with new research findings. So, with that caution in mind, we'll briefly survey the functions of specific brain structures next, leaving the detailed discussion for later chapters.

how's it going ?

1. Name and briefly describe the major divisions of the peripheral nervous system. What general function does each part perform?
2. Briefly sketch and describe the anatomical organization of the cranial nerves. How many nerves are there? Now do the same for the spinal nerves.
3. Give some examples of how each division of the autonomic nervous system affects organs of the body.
4. Why does the cortex look so lumpy on the outside?
5. What is special about the pre- and postcentral gyri?
6. Review the fetal development of the brain, and the three major divisions of the brain that arise from the earlier fetal form of the nervous system.

The Brain Is Described in Terms of Both Structure and Function

Vertebrates are bilaterally symmetrical: our bodies have mirror-image left and right sides. The brain is no exception, and almost all the structures of the brain also come in twos. One important principle of the vertebrate brain is that each side of the brain controls the opposite (or contralateral; see Box 2.2) side of the body: the right side of the brain controls and receives sensory information from the left side of the body, while the left side of the brain monitors and controls the right side of the body. In Chapter 15 we'll learn about how the two cerebral hemispheres interact, but for now let's review the various components of the brain and their functions.

The cerebral cortex performs complex cognitive processing

Neuroscientists are only just beginning to understand how the structures and functions of the cerebral cortex accomplish the feats of human cognition. If the human cortex were unfolded, it would occupy an area of about 2,000 square centimeters (315 square inches)—more than 3 times the area of this book's front cover. How are all those millions of cells arranged?

Cortical neurons make up six distinct layers, as shown in **FIGURE 2.13**. Each cortical layer has a unique appearance because it consists of either a band of similar neurons, or a particular pattern of dendrites or axons. For example, the outermost layer, layer I, is

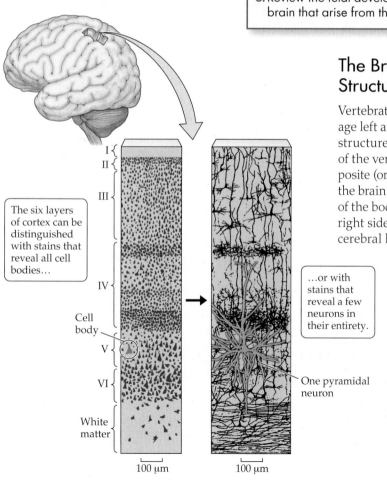

The six layers of cortex can be distinguished with stains that reveal all cell bodies...

...or with stains that reveal a few neurons in their entirety.

One pyramidal neuron

Cell body

I
II
III
IV
V
VI
White matter

100 μm   100 μm

FIGURE 2.13   Layers of the Cerebral Cortex

distinct because it has few cell bodies; layers V and VI stand out because of their many neurons with large cell bodies. The most prominent kind of neuron in the cerebral cortex—the **pyramidal cell**—usually has its pyramid-shaped cell body in layer III or V.

In some regions of the cerebral cortex, neurons are organized into regular columns, perpendicular to the layers, that seem to serve as information-processing units (Horton and Adams, 2005). These **cortical columns** extend through the entire thickness of the cortex, from the white matter to the surface. Within each column, most of the synaptic interconnections of neurons are vertical, although there are some horizontal connections as well (Mountcastle, 1979).

Important nuclei are hidden beneath the cerebral cortex

Buried within the cerebral hemispheres are several large gray matter structures, richly connected to each other and to other brain regions, and contributing to a wide variety of behaviors. One prominent cluster—the **basal ganglia**, consisting primarily of the **caudate nucleus**, the **putamen**, and the **globus pallidus** (**FIGURE 2.14A**)—plays a critical role in the control of movement, which we will discuss in Chapter 5.

Overlapping and curled around the basal ganglia, the **limbic system** is a loosely defined, widespread network of structures (identified in **FIGURE 2.14B**) that are involved in emotion and learning. The **amygdala** consists of several subdivisions with quite diverse functions, including emotional regulation (see Chapter 11) and the perception of odor (see Chapter 6). The **hippocampus** and **fornix** are important for learning and memory (discussed in Chapter 13). Other components of the limbic system include a strip of cortex in each hemisphere called the **cingulate gyrus**, which is implicated in many cognitive functions, including the direction of attention (see Chapter 14), as well as the **olfactory bulb**, which is involved in the sense of smell. Limbic structures near the base of the brain, especially the hypothalamus, help to govern highly motivated behaviors like sex and aggression, and to regulate the hormonal systems of the body.

Toward the medial (middle) and basal (bottom) aspects of the forebrain are found the **thalamus** and the **hypothalamus** (the latter means simply "under thalamus"). You can see the thalamus in **FIGURE 2.15** and both the hypothalamus and thalamus in Figures 2.14B and 2.16A. The thalamus is the brain's traffic cop, directing virtually

caudate nucleus   One of the basal ganglia. It has a long extension or tail.

putamen   One of the basal ganglia.

globus pallidus   One of the basal ganglia.

limbic system   A loosely defined, widespread group of brain nuclei that innervate each other and form a network.

amygdala   A group of nuclei in the medial anterior part of the temporal lobe.

hippocampus   A medial temporal lobe structure that is important for learning and memory.

fornix   A fiber tract that extends from the hippocampus to the mammillary body.

cingulate gyrus   A strip of cortex, found in the frontal and parietal midline, that is part of the limbic system and is implicated in many cognitive functions.

olfactory bulb   An anterior projection of the brain that terminates in the upper nasal passages and, through small openings in the skull, provides receptors for smell.

thalamus   The brain regions that surround the third ventricle.

hypothalamus   Part of the diencephalon, lying ventral to the thalamus.

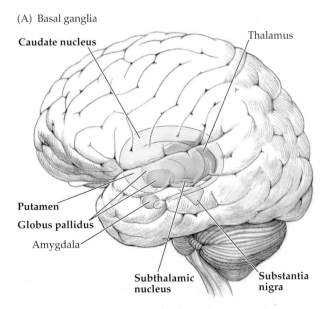

(A) Basal ganglia

Caudate nucleus
Thalamus
Putamen
Globus pallidus
Amygdala
Subthalamic nucleus
Substantia nigra

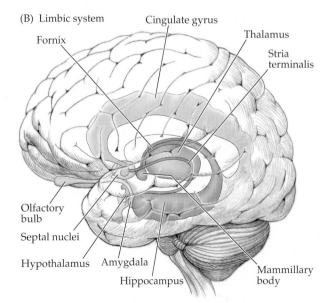

(B) Limbic system

Cingulate gyrus
Fornix
Thalamus
Stria terminalis
Olfactory bulb
Septal nuclei
Hypothalamus   Amygdala
Hippocampus
Mammillary body

FIGURE 2.14   Two Important Brain Systems

(A) Lateral view showing planes of section

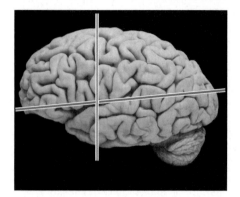

(B) Horizontal section

Basal ganglia · Thalamus · Gray matter (cortex) · White matter

Frontal poles

Occipital poles

Third ventricle

Posterior horn of lateral ventricle

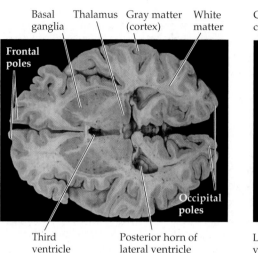

(C) Coronal (transverse) section

Corpus callosum

Basal ganglia
Caudate nucleus · Putamen

Lateral ventricle · Amygdala · Temporal lobe

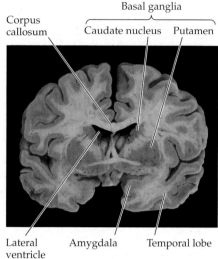

FIGURE 2.15   Inside the Brain (Photographs courtesy of S. Mark Williams and Dale Purves, Duke University Medical Center.)

tectum   The dorsal portion of the midbrain consisting of the inferior and superior colliculi.

superior colliculi   Paired gray matter structures of the dorsal midbrain that process visual information.

inferior colliculi   Paired gray matter structures of the dorsal midbrain that process auditory information.

tegmentum   The main body of the midbrain, containing the substantia nigra, periaqueductal gray, part of the reticular formation, and multiple fiber tracts.

substantia nigra   A brainstem structure that innervates the basal ganglia and is a major source of dopaminergic projections.

periaqueductal gray   A midbrain region involved in pain perception.

reticular formation   An extensive region of the brainstem, extending from the medulla through the thalamus, that is involved in sleep and arousal.

cerebellum   A structure located at the back of the brain, dorsal to the pons, that is involved in the central regulation of movement, and in some forms of learning.

all incoming sensory information to the appropriate regions of the cortex for further processing, and receiving instructions back from the cortex about which sensory information is to be transmitted. The small but mighty hypothalamus has a much different role: it is packed with discrete nuclei involved in many vital functions, such as hunger, thirst, temperature regulation, sex, and many more. Furthermore, because the hypothalamus also controls the pituitary gland, it serves as the brain's main interface with the hormonal systems of the body. We'll encounter the hypothalamus again in several later chapters.

The midbrain has sensory and motor systems

Compared with the forebrain and hindbrain, the midbrain doesn't encompass a lot of tissue, but that doesn't mean its components are unimportant. The top part of the midbrain, called the **tectum** (from the Latin for "roof," because it's atop the midbrain), features two pairs of bumps—one pair in each hemisphere—with specific roles in sensory processing. The more rostral bumps are called the **superior colliculi** (singular colliculus), and they have specific roles in visual processing. The more caudal bumps, called the **inferior colliculi** (see Figure 2.16A), process information about sound.

The main body of the midbrain is called the **tegmentum**, and it also contains several important structures. The **substantia nigra** is in many ways a part of the basal ganglia, and loss of its neurons (which normally release the neurotransmitter dopamine within the forebrain) leads to Parkinson's disease, discussed in Chapter 5. The **periaqueductal gray** is a midbrain structure implicated in the perception of pain (discussed in Chapter 4). The **reticular formation** (reticular means "netlike") is a loose collection of neurons that are important in a variety of behaviors, including sleep and arousal (see Chapter 10). Multiple large tracts of nerve fibers in and out of the midbrain connect the brain to the rest of the body.

The brainstem controls vital body functions

The midsagittal and basal views of the brain in **FIGURE 2.16** show the hemispheres of the **cerebellum**, which is tucked up under the posterior cortex and attached to the dorsal brainstem. Like the cerebral cortex, the cerebellum is highly convoluted, but it is made up of a simpler three-layered tissue instead of the six layers found in the cerebral

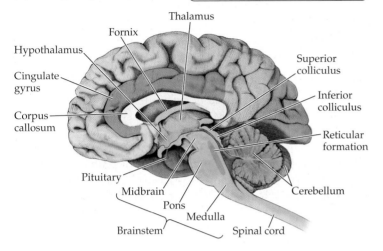

The four lobes of the cerebral cortex are color coded here as in Figure 2.10. In addition, the cingulate gyrus and brainstem are shaded red and yellow, respectively.

(A) Midsagittal (midline) view

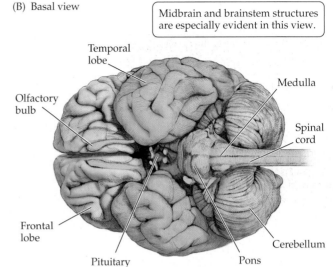

(B) Basal view

Midbrain and brainstem structures are especially evident in this view.

FIGURE 2.16 Midline and Basal Structures of the Brain

cortex. The cerebellum has long been known to be crucial for motor coordination and control, but we now know that it also participates in certain aspects of cognition, including learning. The adjacent **pons** (from the Latin word for "bridge") contains many nerve fibers and important motor control and sensory nuclei; it is the point of origin for several cranial nerves. The reticular formation, which we first saw in the midbrain, stretches down through the pons and ends in the medulla.

The **medulla** marks the transition from the brain to the spinal cord. In addition to conveying all of the major motor and sensory fibers to and from the body, the medulla contains nuclei that drive such essential processes as respiration and heart rate, so brainstem injuries are often lethal. And like other parts of the brainstem, the medulla gives rise to several cranial nerves.

how's it going ❓

1. How are the cells of the cerebral cortex organized?
2. Name the major components of the basal ganglia and the limbic system. What behaviors especially rely on these systems?
3. What functions are served by the thalamus and hypothalamus?
4. Name and describe the general functions of the major components of the midbrain and hindbrain.
5. Why are injuries to the medulla often fatal?

Specialized Support Systems Protect and Nourish the Brain

The brain is relatively soft and easily damaged. It also needs a steady and substantial supply of fuel to maintain normal functioning, and thus keep us alive. Fortunately, the brain is equipped with systems that protect and cushion the brain and that provide a continual source of energy, nutrients, and important chemicals.

The brain floats within layers of membranes

Within the bony skull and vertebrae, the brain and spinal cord are swaddled by three protective membranes called **meninges** (see Figure 2.8). Between a tough outer sheet called the **dura mater** (in Latin, literally "tough mother") and the delicate **pia mater** ("tender mother") that adheres tightly to the surface of the brain, a webby substance

pons The portion of the brainstem that connects the midbrain to the medulla.

medulla The posterior part of the hindbrain, continuous with the spinal cord.

meninges The three protective membranes—dura mater, pia mater, and arachnoid—that surround the brain and spinal cord.

dura mater The outermost of the three meninges that surround the brain and spinal cord.

pia mater The innermost of the three meninges that surround the brain and spinal cord.

arachnoid The thin covering (one of the three meninges) of the brain that lies between the dura mater and the pia mater.

cerebrospinal fluid (CSF) The fluid that fills the cerebral ventricles.

meningitis An acute inflammation of the meninges, usually caused by a viral or bacterial infection.

meningioma A noninvasive tumor of the meninges.

ventricular system A system of fluid-filled cavities inside the brain.

lateral ventricle A complex C-shaped lateral portion of the ventricular system within each hemisphere of the brain.

choroid plexus A specialized membrane lining the ventricles that produces cerebrospinal fluid by filtering blood.

third ventricle The midline ventricle that conducts cerebrospinal fluid from the lateral ventricles to the fourth ventricle.

fourth ventricle The passageway within the pons that receives cerebrospinal fluid from the third ventricle and releases it to surround the brain and spinal cord.

cerebral arteries The three pairs of large arteries within the skull that supply blood to the cerebral cortex.

called the **arachnoid** ("spiderweb-like") suspends the brain in a bath of a watery liquid called **cerebrospinal fluid (CSF)**. The meninges can become inflamed by infections, and because the inflammation ends up squeezing the brain, the resultant **meningitis** is a medical emergency. Sometimes the meninges form into quite large tumors; these **meningiomas** are usually classified as benign, because most do not invade the brain tissue, but in an enclosed space like the cranium, any mass that takes up space is far from harmless.

The brain relies on two fluids for survival

The brain essentially floats in cerebrospinal fluid, cushioning it from minor blows to the head. But CSF is also a source of important materials, such as nutrients and signaling chemicals. Inside the brain is a series of chambers called the cerebral ventricles, which are filled with CSF (**FIGURE 2.17**). These chambers comprise the **ventricular system**. Each hemisphere of the brain contains a **lateral ventricle** extending into all four lobes of the hemisphere. The lateral ventricles are lined with a specialized membrane called the **choroid plexus**, which produces CSF by filtering blood. The CSF flows from the lateral ventricles into a midline **third ventricle** (so named because it follows the two lateral ventricles) and continues down a narrow passage (the *cerebral aqueduct*) to the **fourth ventricle**, which lies between the cerebellum and the pons. Just below the cerebellum, three small openings allow CSF to exit the ventricular system and circulate over the outer surface of the brain and spinal cord. The CSF is absorbed back into the circulatory system through large veins beneath the top of the skull.

The second crucial fluid for the brain is, of course, blood. Without a continual supply of a high volume of oxygen- and nutrient-rich blood, the tissue of the brain would swiftly die. That's because brain tissue is unusually needy: it accounts for only 2% of the average human body but consumes more than 20% of the body's energy. So the brain is critically dependent on a set of large blood vessels. Blood arrives in the brain via two pairs of arteries: the carotid arteries in the neck and the vertebral arteries that ascend to either side of the spinal column. Inside the skull these arteries give rise to a set of three pairs of **cerebral arteries** that supply the cortex, plus a number of smaller vessels that penetrate and supply other regions of

(A) Cerebral ventricles of the brain

The positions of the cerebral ventricles are shown here within an adult brain.

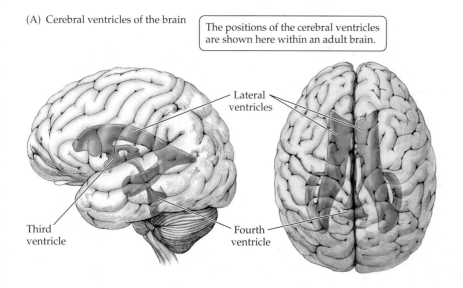

Lateral ventricles

Third ventricle

Fourth ventricle

(B) A closer view

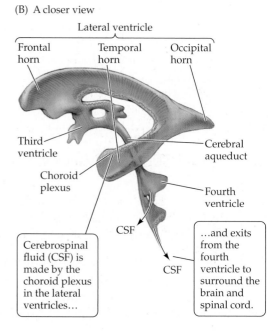

Lateral ventricle

Frontal horn

Temporal horn

Occipital horn

Third ventricle

Choroid plexus

Cerebral aqueduct

Fourth ventricle

CSF

Cerebrospinal fluid (CSF) is made by the choroid plexus in the lateral ventricles...

...and exits from the fourth ventricle to surround the brain and spinal cord.

CSF

FIGURE 2.17 The Cerebral Ventricles

the brain. Fine vessels and capillaries branching off from the arteries deliver nutrients and other substances to brain cells and remove waste products. In contrast to capillaries in the rest of the body, capillaries in the brain are highly resistant to the passage of large molecules across their walls and into neighboring neurons. This **blood-brain barrier** probably evolved to help protect the brain from infections and blood-borne toxins, but it also makes the delivery of drugs to the brain more difficult. You can learn more about the brain's elaborate vascular system in **A STEP FURTHER 2.1**, on the website.

blood-brain barrier   The mechanisms that make the movement of substances from blood vessels into brain cells more difficult than exchanges in other body organs, thus affording the brain greater protection from exposure to some substances found in the blood.

stroke   Damage to a region of brain tissue that results from the blockage or rupture of vessels that supply blood to that region.

transient ischemic attack (TIA) A temporary blood restriction to part of the brain that causes strokelike symptoms that quickly resolve, serving as a warning of elevated stroke risk.

how's it going ❓

1. Name the three meninges, and describe how they're organized. Identify one special characteristic of each.
2. What is CSF? What function does it serve, where does it come from, and where does it go?
3. Describe the ventricular system of the brain.
4. What is the blood-brain barrier?

SIGNS & SYMPTOMS

Stroke

The general term **stroke** applies to a situation in which a clot, a narrowing, or a rupture interrupts the supply of blood to a particular brain region, causing the affected region to stop functioning or die (**FIGURE 2.18**). Although the exact effects of stroke depend on the region of the brain that is affected, the five most common warning signs are sudden numbness or weakness, altered vision, dizziness, severe headache, and confusion or difficulty speaking. Effective treatments are available to help restore blood flow and minimize the long-term effects of a stroke, but only if the victim is treated immediately. Some people experience temporary strokelike symptoms lasting for a few minutes. Caused by a brief interruption of blood supply to some part of the brain, this **transient ischemic attack**, or **TIA**, (from the Greek *ischemia*, "interrupted blood") is a serious warning sign that a major stroke may be imminent, and it should be treated as a medical emergency.

FIGURE 2.18   Stroke

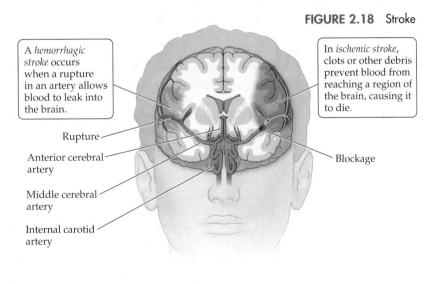

A *hemorrhagic stroke* occurs when a rupture in an artery allows blood to leak into the brain.

In *ischemic stroke*, clots or other debris prevent blood from reaching a region of the brain, causing it to die.

Rupture

Anterior cerebral artery

Middle cerebral artery

Internal carotid artery

Blockage

Brain-Imaging Techniques Reveal the Structure and Function of the Human Brain

Although techniques for diagnosing brain injury have been around for decades—using X-rays of the head and its blood vessels, for example—more recent technological developments have allowed researchers to study the brains of healthy humans too. Some of these techniques are used primarily to identify brain structure; others focus more on tracking changes in brain activity as behavior occurs.

CT uses X-rays to reveal brain structure

In **computerized axial tomography** (**CAT** or **CT**), X-ray energy is used to generate images. In a CT scanner, an X-ray source is moved by steps in an arc around the head. At each point, detectors on the opposite side of the head measure the amount of X-ray radiation that is absorbed; this value is proportional to the density of the tissue through which the X-rays passed. When this process is repeated from many angles and the results are mathematically combined, an anatomical map of the brain based on tissue density can be generated by computer (**FIGURE 2.19A**). CT scans are medium-resolution images, useful for visualizing problems such as strokes, tumors, or cortical shrinkage.

Sam, whom we met at the beginning of the chapter, had developed a type of tumor called a meningioma that was pressing on and deforming the motor cortex on the left side of his brain. The pressure caused abnormal firing—a seizure—that spread to adjacent areas, resulting in unconsciousness. The tumor subsequently impaired the functioning of regions of the motor cortex responsible for voluntary control of the muscles of Sam's right arm and the right side of his face, producing his alarming symptoms. Fortunately, his emergency CT scan was able to pinpoint the meningioma within Sam's skull, and later that evening a surgeon removed the tumor in pieces without damaging nearby tissue. Sam experienced immediate improvement and has been healthy ever since.

MRI maps density to deduce brain structure with high detail

Magnetic resonance imaging (**MRI**) provides higher-resolution images than CT, and because MRI uses magnetic fields and radio waves instead of X-rays, MRI also has fewer damaging effects than CT. When an MRI image of the brain is made, the person's head is first placed in an extremely powerful magnet that causes all the protons in the brain's tissues and fluids to line up in parallel, instead of in their usual random orientations (protons are found in the nuclei of atoms; in body tissues, most protons are found within water molecules). Next, the protons are knocked over by a powerful pulse of radio waves. When this pulse is turned off, the protons relax back to their original configuration, emitting radio waves as they go. Detectors surrounding the head measure those radio waves, which differ for tissues of varying densities. This density-based information is then used by a computer to create a detailed cross-sectional view of the brain (**FIGURE 2.19B**). Thanks to their high resolution, MRI images can reveal subtle changes in the brain, such as the local loss of myelin that is characteristic of some diseases.

Functional MRI uses local changes in metabolism to identify active brain regions

Functional MRI (fMRI), which generates images of the brain's activity rather than details of its structure, has revolutionized cognitive neuroscience. Offering both reasonable speed (temporal resolution) and good sharpness (spatial resolution), fMRI uses rapidly oscillating magnetic fields to detect small changes in brain metabolism, particularly moment-to-moment oxygen use and blood flow in the most active regions of the brain. Scientists can use fMRI data to create what are known as "difference images" of the specific activity of different parts of the brain while people engage in various experimental tasks. When combined with conventional anatomical MRI, the detailed

computerized axial tomography (CAT or CT) A noninvasive technique for examining brain structure through computer analysis of X-ray absorption at several positions around the head.

magnetic resonance imaging (MRI) A noninvasive brain imaging technology that uses magnetism and radio-frequency energy to create images of the gross structure of the living brain.

functional MRI (fMRI) Magnetic resonance imaging that detects changes in blood flow and therefore identifies regions of the brain that are particularly active during a given task.

A *Computerized tomography* (CT) helps us spot problems such as strokes or tumors, like the one evident here.

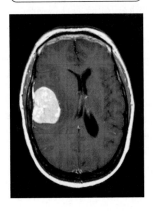

B *Magnetic resonance imaging* (MRI) shows great detail, enabling us to see fine structure and recognize subtle changes in the brain.

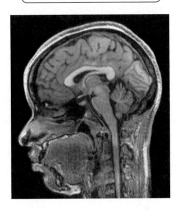

C *Functional MRI (fMRI)* detects small changes in brain metabolism. Changes in brain activity associated with visual or auditory stimuli are highlighted in this example.

Anterior 3-D view

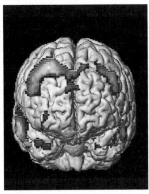

Lateral 3-D view of right hemisphere

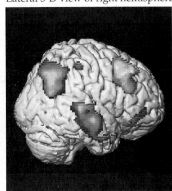

D *Positron emission tomography* (PET) provides a portrait of the brain's activity. Here, PET shows that metabolic activity is diminished in the brain of a person with Alzheimer's disease.

Normal (horizontal view)

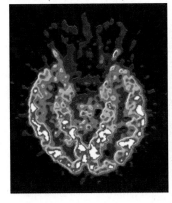

Patient with Alzheimer's disease

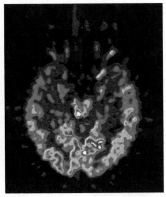

E This patient is entering an MRI scanner.

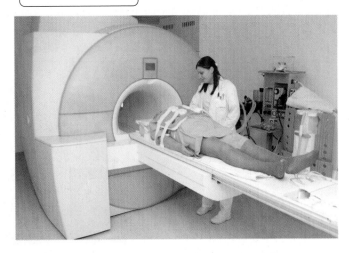

FIGURE 2.19 Visualizing the Living Human Brain (Images in part C courtesy of Semir Zeki.)

To see the animation **Visualizing the Living Human Brain**, go to

2e.mindsmachine.com/2.4

activity maps provided by fMRI reveal how networks of brain structures collaborate on complex cognitive processes (**FIGURE 2.19C** and **E**).

PET tracks radioactive substances to produce images of brain activity

Like fMRI, **positron emission tomography** (**PET**) provides a means to visualize the brain's activity during behavioral tasks. Short-lived radioactive chemicals are injected into the bloodstream, and radiation detectors encircling the head map the destination of these chemicals in the brain. A particularly effective strategy is to inject radioactively labeled glucose ("blood sugar") while the person is engaged in a cognitive task of interest to the researcher. Because the radioactive glucose is selectively taken up and used by the most active parts of the brain, a moment-to-moment color-coded portrait of brain activity can be created (**FIGURE 2.19D**) (P. E. Roland, 1993). Although PET can't match the detailed resolution of fMRI, it tends to be faster and thus better able to track quick changes in brain activity. Next we'll discuss how precise experimental methods allow researchers to identify brain regions that contribute to specific functions.

positron emission tomography (PET) A brain imaging technology that tracks the metabolism of injected radioactive substances in the brain, in order to map brain activity.

researchers at work

Subtractive analysis isolates specific brain activity

Modern brain imaging provides dramatic pictures showing the particular brain regions that are activated during specific cognitive processes; there are many such images in this book. But if you do a PET scan of a healthy person, you find that almost all of the brain is active at any given moment (showing that the old notion that "we use only 10% of our brain" is nonsense). How do researchers obtain these highly specific images of brain activity?

In order to associate specific brain regions with particular cognitive operations, researchers developed a sort of algebraic technique, in which activity during one behavioral condition is subtracted from activity during a different condition. So, for example, the data from a control PET scan made while a person was gazing at a blank wall might be subtracted from a PET scan collected while that person studied a complex visual stimulus. Averaged over enough trials, the specific regions that are almost always active during the processing task become apparent, even though, on casual inspection, a single experimental scan might not look much different from a single control scan (**FIGURE 2.20**). It is important to keep in mind that although functional brain images seem unambiguous and easy to label, they are subject to a variety of procedural and experimental limitations (Racine et al., 2005).

FIGURE 2.20   Isolating Specific Brain Activity

■ **Hypothesis**

Brain regions engaged in a specific behavior can be isolated by algebraic means, subtracting resting scans from scans during activity.

■ **Test**

Participants are scanned twice—once while looking at a blank screen, and once while looking at test stimuli. The control scan is then subtracted from the test scan.

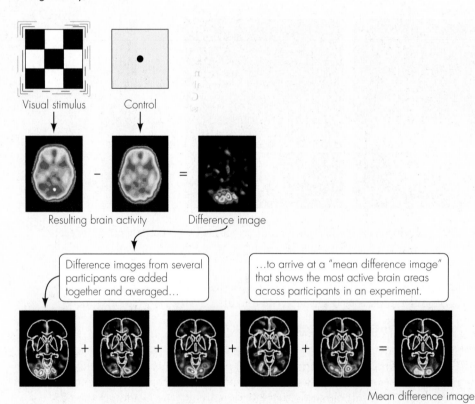

Visual stimulus   Control

Resulting brain activity   Difference image

Difference images from several participants are added together and averaged…

…to arrive at a "mean difference image" that shows the most active brain areas across participants in an experiment.

■ **Result**

By repeating the process and averaging the results across multiple participants, a stable "difference image" showing activation of just a few brain regions is formed.

Mean difference image

■ **Conclusion**

The activated brain regions in the difference image—in this case, in the occipital cortex—are selectively involved in the particular cognitive processing required by the stimulus.

Magnetism can be used to study the brain

It is a simple matter to pass magnetic fields into the brain. However, it is technically more challenging to project magnetic fields in a highly focused and precise manner. In **transcranial magnetic stimulation** **(TMS)** (**FIGURE 2.21**), focal magnetic currents are used to briefly stimulate the cortex of alert people directly, without any lasting physical alterations or surgery (Y. Noguchi et al., 2003). Using TMS allows experimenters to activate a discrete area of the brain while simultaneously tracking any resulting changes in behavior.

Not only can magnets stimulate neurons, but neurons also act as tiny electromagnets themselves! In **magnetoencephalography** **(MEG)**, a large array of ultrasensitive detectors measures the minuscule magnetic fields produced by the electrical activity of cortical neurons. This information is used to construct real-time maps of brain activity during ongoing cognitive processing (**FIGURE 2.22**). Because MEG can track quick, moment-by-moment changes in brain activity, it is excellent for studying rapidly shifting patterns of brain activity in cortical circuits (F. H. Lin et al., 2004). In the next chapter, we'll learn about electroencephalography (EEG), which measures electrical activity in the brain.

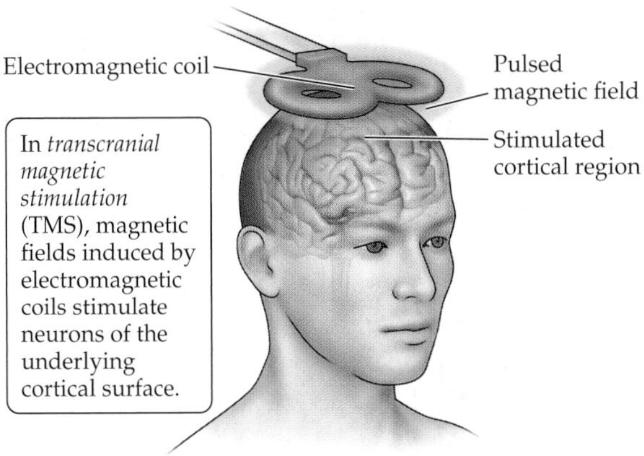

Electromagnetic coil

Pulsed magnetic field

Stimulated cortical region

In *transcranial magnetic stimulation* (TMS), magnetic fields induced by electromagnetic coils stimulate neurons of the underlying cortical surface.

FIGURE 2.21   Transcranial Magnetic Stimulation

transcranial magnetic stimulation (TMS)   A noninvasive technique for examining brain function that applies strong magnetic fields to stimulate cortical neurons, in order to identify discrete areas of the brain that are particularly active during specific behaviors.

magnetoencephalography (MEG) A noninvasive brain-imaging technology that creates maps of brain activity during cognitive tasks by measuring tiny magnetic fields produced by active neurons.

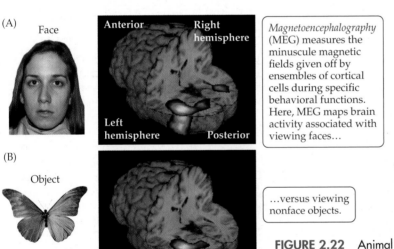

(A) Face

Anterior   Right hemisphere

Left hemisphere   Posterior

Magnetoencephalography (MEG) measures the minuscule magnetic fields given off by ensembles of cortical cells during specific behavioral functions. Here, MEG maps brain activity associated with viewing faces…

(B) Object

…versus viewing nonface objects.

FIGURE 2.22   Animal Magnetism (MEG images courtesy of Drs. Mario Liotti and Anthony Herdman, Simon Fraser University.)

how's it going ❓

1. Compare and contrast the main methods for producing still images of the structure of the brain. What do you think are some of the advantages and disadvantages of each method?

2. Compare and contrast the main functional-imaging technologies used for visualizing the activity of brain regions. What do you think are some of the advantages and disadvantages of each method?

3. Describe the process that neuroscientists can use to isolate brain activity associated with a specific behavior, as visualized by functional-imaging techniques.

Recommended Reading

Blumenfeld, H. (2010). *Neuroanatomy through Clinical Cases* (2nd ed.). Sunderland, MA: Sinauer.

Cabeza, R., and Kingstone, A. (2006). *Handbook of Functional Neuroimaging of Cognition* (2nd ed.). Cambridge, MA: MIT Press.

Huettel, S. A., Song, A. W., and McCarthy, G. (2014). *Functional Magnetic Resonance Imaging* (3rd ed.). Sunderland, MA: Sinauer.

Schoonover, C. (2010). *Portraits of the Mind: Visualizing the Brain from Antiquity to the 21st Century*. New York, NY: Abrams.

2e.mindsmachine.com/vs2

VISUAL SUMMARY
2

You should be able to relate each summary to the adjacent illustration, including structures and processes. Go to the online version of this summary (scan the QR code above) for links to figures, animations, and activities that will help you consolidate the material.

1 Neurons (*nerve cells*) are the basic units of the nervous system. The typical neuron has four main parts: (1) **dendrites**, which receive information; (2) the **cell body** (*soma*), which integrates the information; (3) an **axon**, which carries impulses from the neuron; and (4) **axon terminals**, which transmit the neuron's signals to other cells. Neurons almost universally feature an **input zone**, an **integration zone**, a **conduction zone**, and an **output zone**. Review Figure 2.1, Animation 2.2

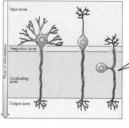

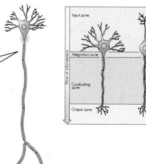

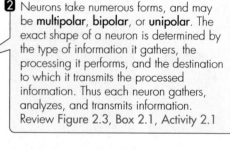

2 Neurons take numerous forms, and may be **multipolar**, **bipolar**, or **unipolar**. The exact shape of a neuron is determined by the type of information it gathers, the processing it performs, and the destination to which it transmits the processed information. Thus each neuron gathers, analyzes, and transmits information. Review Figure 2.3, Box 2.1, Activity 2.1

3 Neurons make functional contacts with other cells at specialized junctions called **synapses**. By changing their shape or function in response to experiences, synapses exhibit **neuroplasticity**. At most synapses a chemical **neurotransmitter** released from the **presynaptic membrane** diffuses across the **synaptic cleft** and binds to special **neurotransmitter receptor** molecules in the **postsynaptic membrane**. Review Figure 2.4

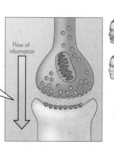

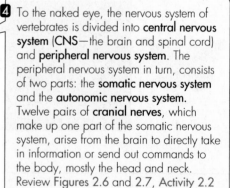

4 To the naked eye, the nervous system of vertebrates is divided into **central nervous system** (CNS—the brain and spinal cord) and **peripheral nervous system**. The peripheral nervous system in turn, consists of two parts: the **somatic nervous system** and the **autonomic nervous system**. Twelve pairs of **cranial nerves**, which make up one part of the somatic nervous system, arise from the brain to directly take in information or send out commands to the body, mostly the head and neck. Review Figures 2.6 and 2.7, Activity 2.2

5 A second major part of the somatic nervous system consists of the 31 pairs of **spinal nerves**, spaced through the **cervical, thoracic, lumbar, sacral,** and **coccygeal** segments of the spinal cord. Review Figure 2.8, Activity 2.3

6 The autonomic nervous system consists of the **sympathetic nervous system** (which tends to ready the body for immediate action) and the **parasympathetic nervous system** (which tends to have an effect opposite to that of the sympathetic system). We cannot consciously control autonomic activity. Review Figure 2.9, Activity 2.4

7 The human brain is dominated by the **cerebral hemispheres**, which include the **frontal, parietal, temporal,** and **occipital lobes**. The outermost parts of the cerebral hemispheres are known as **cerebral cortex**, or simply *cortex*. Review Figure 2.10, Activity 2.5

8 The major divisions of the brain are established during fetal development. These include the **forebrain** (cortex and embedded structures), the **thalamus** and **hypothalamus** (together called the **diencephalon**), and the **brainstem** components (**midbrain, pons,** and **medulla**), as well as the **cerebellum** atop the pons. Review Figure 2.12, Box 2.2, Video 2.3, Activity 2.6

9 The **cerebral cortex** is an extensive sheet of folded tissue. The six-layered cerebral cortex is responsible for higher-order functions such as vision, language, and memory. Review Figure 2.13

10 Important subcortical systems include the **basal ganglia**, which regulate movement; the **limbic system**, which controls emotional behaviors; and the **cerebellum**, which aids motor control. Review Figure 2.14, Activities 2.7–2.10

11 Modern brain-imaging technology provides us with a variety of ways to visualize the living brain. Structural imaging technologies like **computerized axial tomography (CT or CAT)** and **magnetic resonance imaging (MRI)** provide high-resolution images of the structure of the brain. Functional imaging technologies, such as **positron emission tomography (PET)**, **functional MRI (fMRI)**, and **magnetoencephalography (MEG)**, provide maps of brain activity during behavior. Review Figures 2.19 and 2.20, Animation 2.4

12 The brain and spinal cord, surrounded and protected by the three **meninges** (**dura mater, pia mater,** and **arachnoid**), float in **cerebrospinal fluid (CSF)**, which is produced in the **lateral ventricles** and exits the ventricles to surround the brain. Review Figure 2.17, Activity 2.11

13 The brain requires a constant supply of blood via the **cerebral arteries**, to fuel the activity of neurons. An interruption in blood flow, caused by a blockage of hemorrhage, is called a **stroke**. The effect or stroke is determined by its size and location in the brain. Review Figure 2.18

14 Algebraic operations on scans under different experimental conditions can help pinpoint peaks of activity during various mental activities in functional brain imaging like PET and fMRI. Review Figure 2.20

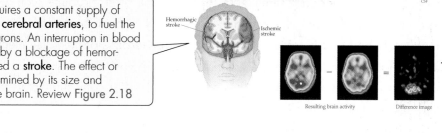

Go to **2e.mindsmachine.com** for study questions, quizzes, flashcards, and other resources.

chapter

3

Neurophysiology
The Generation, Transmission, and Integration of Neural Signals

Stimulating Conversation

Perhaps the most dramatic demonstration in the history of neuroscience occurred in 1964 when Yale Professor José Delgado strolled into an arena in Spain to face an enraged bull trained to attack humans. Armed only with a remote control, Delgado watched the massive bull paw the earth, lower its head, and charge right at him. Just before the bull reached him, Delgado pressed a button on the remote control that caused a wire, called an *electrode*, in the bull's brain to deliver a tiny trickle of electricity. The bull stopped cold. When Delgado electrically stimulated another part of the bull's brain, the animal turned to the right and calmly trotted away. Repeated stimulations rendered the bull docile for several minutes. Other animals responded differently to brain stimulation, depending on the brain region targeted. One bull produced a single "moo" for every button press—a hundred times in a row.

Delgado also electrically stimulated electrodes in the brains of people, in an attempt to pinpoint the cause of a neurological disorder. Depending on which part of the brain was stimulated, patients might suddenly become anxious or angry (Delgado, 1969). Maybe the creepiest response elicited by brain stimulation was in women who suddenly became romantically interested in the man interviewing them. Yet as soon as the electrical stimulation of their brains stopped, the women returned to their usual reserved behavior.

Why does a tiny bit of electrical stimulation in the brain produce such dramatic changes in mood and behavior? The reason is that neurons normally use electrical signals to sum up vast amounts of information. When Delgado electrically stimulated the brain, he was triggering those normal electrical signals in a very abnormal way, with sometimes dramatic results. To understand how even a tiny electrical charge to the brain can so dramatically affect the mind, we need to understand how electrical signaling works in the brain.

To see the video
**Electrical Stimulation
of the Brain,**
go to

2e.mindsmachine.com/3.1

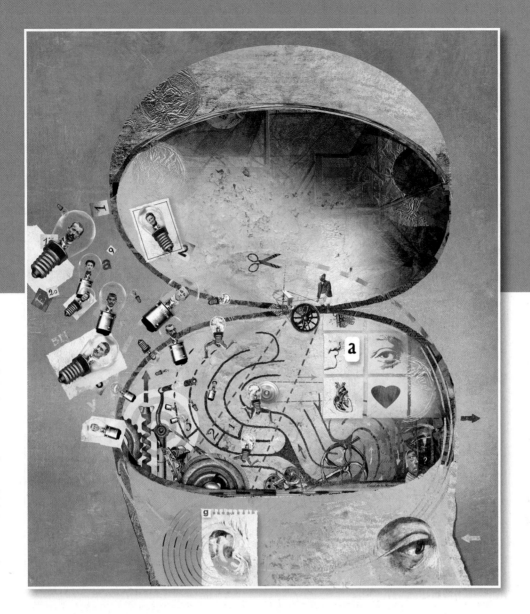

Neurophysiology is the study of the specialized life processes that allow neurons to use chemical and electrical processes to sum up vast amounts of information, and then pass information on to other neurons. In this chapter we'll study the *electrical* processes at work *within* a neuron; in the next chapter we'll look at the *chemical* signals that pass *between* neurons. We'll see that brain function is an alternating series of electrical signals within neurons and of chemical signals between neurons.

For example, a doctor may use a small rubber mallet to strike just below your knee and watch your leg kick upward in what is known as the *knee-jerk reflex*. Simple as it appears, a lot happens during this test. First, sensory neurons in the muscle detect the hammer tap and send a rapid electrical signal along their axons to your spinal cord. That rapid electrical signal along the axon from knee to spinal cord is called an *action potential*, which is a major topic in this chapter. When the action potential reaches the axon terminals, it releases a chemical, called a *neurotransmitter,* to stimulate spinal motoneurons. In response to the neurotransmitter, the motoneurons send action potentials down their own axons to release yet another neurotransmitter onto muscles. In response to that neurotransmitter, the muscles contract, kicking your foot into the air. Problems in electrical or chemical signaling in this circuit might cause the kick to be stronger or weaker, faster or slower, than it should be.

Hold It Right There! Dr. José Delgado stops a bull in the middle of a charge, using the remote control in his hand.

To view the
Brain Explorer,
go to

2e.mindsmachine.com/3.2

neurophysiology   The study of the life processes of neurons.

ion   An atom or molecule that has acquired an electrical charge by gaining or losing one or more electrons.

anion   A negatively charged ion, such as a protein or a chloride ion.

cation   A positively charged ion, such as a potassium or sodium ion.

intracellular fluid   Also called *cytoplasm*. The watery solution found within cells.

extracellular fluid   The fluid in the spaces between cells (interstitial fluid).

cell membrane   The lipid bilayer that ensheathes a cell.

microelectrode   An especially small electrode used to record electrical potentials inside living cells.

resting potential   The difference in electrical potential across the membrane of a nerve cell at rest.

millivolt (mV)   A thousandth of a volt.

ion channel   A pore in the cell membrane that permits the passage of certain ions through the membrane when the channels are open.

So this "simple" behavior involves several rounds of signaling: first electrical (along sensory neuron axons), then chemical (sensory neurons to motoneurons), then electrical again (along motoneuron axons), and finally chemical again (motoneurons to muscle). This is the classic pattern of neural function: information flows *within* a neuron via electrical signals (action potentials) and passes *between* neurons through chemical signals (neurotransmitters). This sequence also reflects the organization of this chapter. First we explain how neurons produce action potentials and send them along their axons. Then we describe how the action potential causes axon terminals to release neurotransmitter into the synapse. Next we discuss how the neurotransmitter affects the electrical state of the neuron on the other side of the synapse. And we conclude the chapter by discussing how electrical probing of the brain revealed that the brain's surface reflects a map of the body. All this will prepare us for the next chapter, where we will learn more details about the chemical signals between neurons.

Electrical Signals Are the Vocabulary of the Nervous System

All living cells are more negative on the inside than on the outside, so we say they are *polarized*, meaning there is a difference in electrical charge between the inside and outside of the cell. Long ago, nerve cells began to exploit this electrical property to keep track of information, and works in much the same way in the neurons of animals as different as human beings and jellyfish. These neural signals underlie all our thoughts and actions, from composing music or solving a mathematical problem to feeling an itch on the skin and swatting a mosquito. You'll understand this system more easily if we first review the physical forces at work and then discuss some details of why nerve cells are electrically polarized, how neuronal polarity is influenced by other cells, and how a change of polarity in one part of a neuron can spread throughout the cell.

A balance of electrochemical forces produces the resting potential of neurons

Let's start by considering a neuron at rest, neither perturbed by other neurons nor producing its own signals. Of the many **ions** (electrically charged molecules) that a neuron contains, a majority are **anions** ("ANN-eye-ons"; negatively charged ions), especially large protein anions that cannot exit the cell; the rest are **cations** ("CAT-eye-ons"; positively charged ions—it may help you to remember that the letter *t*, which occurs in the word *cation*, is shaped a bit like a plus sign "+"). All of these ions are dissolved in the **intracellular fluid** inside the cell and the **extracellular fluid** surrounding the **cell membrane**.

If we insert a fine **microelectrode** into the interior of a neuron and place an electrode in the extracellular fluid and take a reading (as in **FIGURE 3.1**), we find that the neuron is more negative on the inside than on the outside. Specifically, a neuron at rest exhibits a characteristic **resting potential** (an electrical difference across the membrane) of about −50 to −80 thousandths of a volt, or **millivolts** (**mV**) (the negative sign indicates that the cell's interior is more negative than the outside). To understand how this negative membrane potential comes about, we have to consider some special properties of the cell membrane, as well as two forces that drive ions across it.

The cell membrane is a double layer of fatty molecules studded with many sorts of specialized proteins. One important type of membrane-spanning protein is the **ion channel**, a tubelike pore that allows ions of a specific type to pass through the

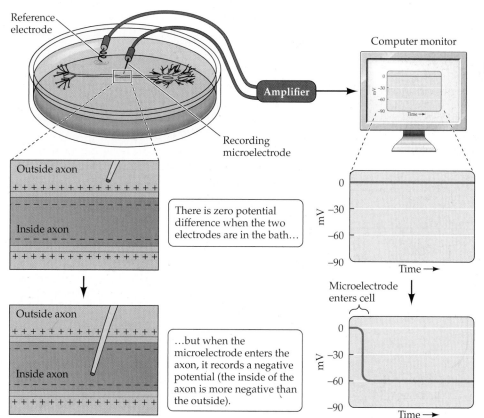

FIGURE 3.1 Measuring the Resting Potential

Reference electrode

Recording microelectrode

Amplifier

Computer monitor

Outside axon

+ + + + + + + + + + + + + +

Inside axon

+ + + + + + + + + + + + + +

There is zero potential difference when the two electrodes are in the bath...

Outside axon

+ + + + + + + + + + + + + +

Inside axon

+ + + + + + + + + + + + + +

...but when the microelectrode enters the axon, it records a negative potential (the inside of the axon is more negative than the outside).

Microelectrode enters cell

membrane (**FIGURE 3.2**). As we'll see later, some types of ion channels are *gated*: they can open and close rapidly in response to various influences. But some ion channels stay open all the time, and the cell membrane of a neuron contains many such channels that selectively allow **potassium ions (K⁺)** to cross the membrane, but not **sodium ions (Na⁺)**. Because it is studded with these K⁺ channels, we say that the cell membrane of a neuron exhibits **selective permeability**, allowing some things to pass through, but not others. The membrane allows K⁺ ions, but not Na⁺ ions, to enter or exit the cell fairly freely.

potassium ion (K⁺) A potassium atom that carries a positive charge.

sodium ion (Na⁺) A sodium atom that carries a positive charge.

selective permeability The property of a membrane that allows some substances to pass through, but not others.

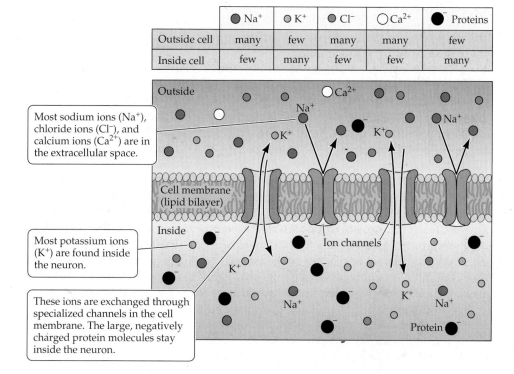

	● Na⁺	● K⁺	● Cl⁻	○ Ca²⁺	● Proteins
Outside cell	many	few	many	many	few
Inside cell	few	many	few	few	many

Most sodium ions (Na⁺), chloride ions (Cl⁻), and calcium ions (Ca²⁺) are in the extracellular space.

Most potassium ions (K⁺) are found inside the neuron.

These ions are exchanged through specialized channels in the cell membrane. The large, negatively charged protein molecules stay inside the neuron.

Outside

Cell membrane (lipid bilayer)

Inside

Ion channels

FIGURE 3.2 The Distribution of Ions Inside and Outside of a Neuron

(A) Diffusion

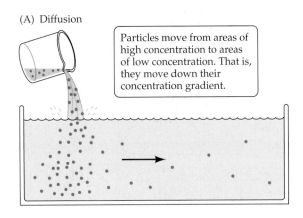

Particles move from areas of high concentration to areas of low concentration. That is, they move down their concentration gradient.

(B) Diffusion through semipermeable membranes

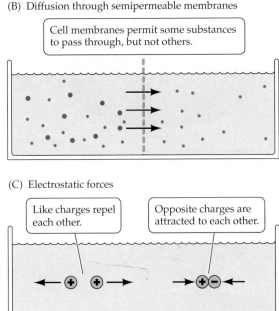

Cell membranes permit some substances to pass through, but not others.

(C) Electrostatic forces

Like charges repel each other.

Opposite charges are attracted to each other.

FIGURE 3.3 Ionic Forces Underlying Electrical Signaling in Neurons

diffusion The spontaneous spread of molecules from an area of high concentration to an area of low concentration until a uniform concentration is achieved.

electrostatic pressure The propensity of charged molecules or ions to move toward areas with the opposite charge.

sodium-potassium pump The energetically expensive mechanism that pushes sodium ions out of a cell, and potassium ions in.

equilibrium potential The point at which the movement of ions across the cell membrane is balanced, as the electrostatic pressure pulling ions in one direction is offset by the diffusion force pushing them in the opposite direction.

The resting potential of the neuron reflects a balancing act between two opposing forces that drive K^+ ions in and out of the neuron. The first of these is **diffusion** (**FIGURE 3.3A**), which is the force that causes molecules of a substance to spread from regions of high concentration to regions of low concentration. For example, when placed in a glass of water, the molecules in a drop of food coloring will tend to spread from the drop out into the rest of the glass, where they are less concentrated. So we say that molecules tend to move down their concentration gradient until they are evenly distributed. If a selectively permeable membrane divides the fluid, particles that can pass through the membrane, such as K^+, will diffuse across until they are equally concentrated on both sides. Other ions, unable to cross the membrane, will remain concentrated on one side (**FIGURE 3.3B**).

The second force at work is **electrostatic pressure**, which arises from the distribution of electrical charges rather than the distribution of molecules. Charged particles exert electrical force on one another: like charges repel, and opposite charges attract (**FIGURE 3.3C**). Positively charged cations like K^+ are thus attracted to the negatively charged interior of the cell; conversely, anions are repelled by the cell interior and so tend to exit to the extracellular fluid.

Now let's consider the situation across a neuron's cell membrane. Much of the energy consumed by a neuron goes into operating specialized membrane proteins called **sodium-potassium pumps** that pump three Na^+ ions out of the cell for every two K^+ ions pumped in (**FIGURE 3.4A**). This action results in a buildup of K^+ ions inside the cell (and Na^+ outside the cell), but as we explained earlier, the membrane is selectively permeable to K^+ ions (but not Na^+ ions). Therefore, K^+ ions can leave the interior, moving down their concentration gradient and causing a net buildup of negative charges inside the cell (**FIGURE 3.4B**). As negative charge builds up inside the cell, it begins to exert electrostatic pressure to pull positively charged K^+ ions back inside. Eventually the opposing forces exerted by the K^+ concentration gradient and by electrostatic pressure reach the **equilibrium potential**, the electrical charge that exactly balances the concentration gradient: any further movement of K^+ ions into the cell (drawn by electrostatic attraction) is matched by the flow of K^+ ions out of the cell (moving down their concentration gradient). This point corresponds to the cell's resting potential of about −60 mV (values may range between −50 and −80 mV), as **FIGURE 3.4C** depicts.

The resting potential of a neuron provides a baseline level of polarization found in all cells. But unlike most other cells, neurons routinely undergo a brief but radical *change* in polarization, sending an electrical signal from one end of the neuron to the other, as we'll discuss next.

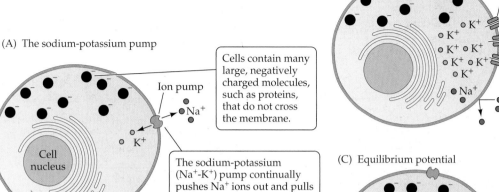

(A) The sodium-potassium pump

Cells contain many large, negatively charged molecules, such as proteins, that do not cross the membrane.

The sodium-potassium (Na⁺-K⁺) pump continually pushes Na⁺ ions out and pulls K⁺ ions in. This ion pump requires considerable energy.

(B) Membrane permeability to potasssium (K⁺) ions

K⁺ ions pass back out again through channels down their concentration gradient. The departure of K⁺ ions leaves the inside of the cell more negative than the outside. Na⁺ ions cannot pass back inside.

(C) Equilibrium potential

When enough K⁺ ions have departed to bring the membrane potential to –60 mV or so, the electrical attraction pulling K⁺ in is exactly balanced by the concentration gradient pushing K⁺ out. This is the K⁺ equilibrium potential, approximately the cell's resting potential.

FIGURE 3.4   The Ionic Basis of the Resting Potential

To see the animation **The Resting Membrane Potential**, go to

2e.mindsmachine.com/3.3

how's it going ?

1. What two physical forces determine a neuron's resting potential?
2. What does it mean if K⁺ ions are at equilibrium? How does that state relate to a neuron's resting potential?
3. What does the sodium-potassium pump do, and how does its action relate to the resting potential?

A threshold amount of depolarization triggers an action potential

Action potentials are very brief but large changes in neuronal polarization that arise in the initial segment of the axon, just after the **axon hillock** (the specialized membrane located where the axon emerges from the cell body; see Figure 2.4A), and then move rapidly down the axon. The information that a neuron sends to other cells is encoded in patterns of these action potentials, so we need to understand their properties—where they come from, how they race down the axon, and how they send information across synapses to other cells. Let's turn first to the creation of the action potential.

Two concepts are central to understanding how action potentials are triggered. **Hyperpolarization** is an increase in membrane potential (that is, the neuron becomes *even more negative* on the inside, relative to the outside). So if the neuron already has a resting potential of, say, –60 mV, hyperpolarization makes it even *farther from zero*, maybe –70 mV. **Depolarization** is the reverse, referring to a decrease in membrane potential. The depolarization of a neuron from a resting potential of –60 mV to, say, –50 mV makes the inside of the neuron more like the outside. In other words, depolarization of a neuron brings its membrane potential *closer to zero*.

Let's use an apparatus to apply hyperpolarizing and depolarizing stimuli to a neuron, via electrodes. (Later we'll talk about how synapses produce similar hyperpolarizations and depolarizations.) Applying a *hyperpolarizing* stimulus to the membrane

axon hillock   The cone-shaped area on the cell body from which the axon originates.

hyperpolarization   An increase in membrane potential (the interior of the neuron becomes even more negative).

depolarization   A decrease in membrane potential (the interior of the neuron becomes less negative).

To see the animation **The Action Potential**, go to

2e.mindsmachine.com/3.4

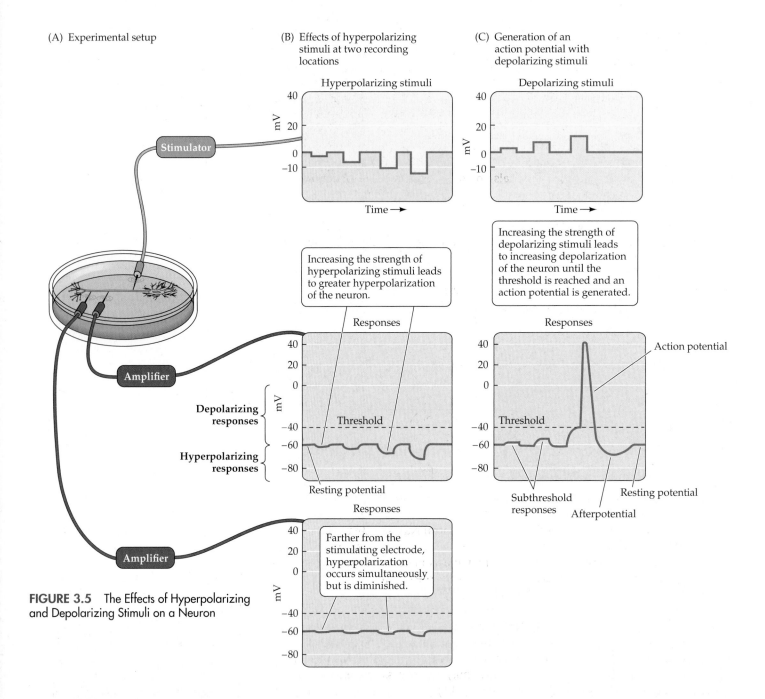

(A) Experimental setup

(B) Effects of hyperpolarizing stimuli at two recording locations

Hyperpolarizing stimuli

(C) Generation of an action potential with depolarizing stimuli

Depolarizing stimuli

Increasing the strength of hyperpolarizing stimuli leads to greater hyperpolarization of the neuron.

Increasing the strength of depolarizing stimuli leads to increasing depolarization of the neuron until the threshold is reached and an action potential is generated.

Responses

Depolarizing responses

Threshold

Hyperpolarizing responses

Resting potential

Responses

Action potential

Threshold

Subthreshold responses

Afterpotential

Resting potential

Responses

Farther from the stimulating electrode, hyperpolarization occurs simultaneously but is diminished.

FIGURE 3.5 The Effects of Hyperpolarizing and Depolarizing Stimuli on a Neuron

local potential An electrical potential that is initiated by stimulation at a specific site, which is a graded response that spreads passively across the cell membrane, decreasing in strength with time and distance.

produces an immediate response that passively mirrors the stimulus pulse (**FIGURE 3.5A** and **B**). The greater the stimulus, the greater the response, so the neuron's change in potential is called a *graded response*.

If we measured the membrane response at locations farther and farther away from the stimulus location, we would see another way in which the membrane response seems passive. Like the ripples spreading from a pebble dropped in a pond, these graded **local potentials** across the membrane get smaller as they spread away from the point of stimulation (see Figure 3.5B *bottom*).

Up to a point, the application of *depolarizing* pulses to the membrane follows the same pattern as for hyperpolarizing stimuli, producing local, graded responses. However, the situation changes suddenly if the stimulus depolarizes the axon hillock to −40 mV or so (the exact value varies slightly among neurons). At this point, known as the **threshold**, a sudden and brief (0.5- to 2.0-millisecond) response—the **action potential**, sometimes referred to as a *spike* because of its shape—is provoked (**FIGURE 3.5C**). An action potential is a rapid reversal of the membrane potential that momentarily makes the inside of the membrane *positive* with respect to the outside. Unlike the passive graded potentials that we have been discussing, the action potential is actively reproduced (or *propagated*) down the axon, through mechanisms that we'll discuss shortly.

Applying strong stimuli to produce depolarizations that far exceed the neuron's threshold reveals another important property of action potentials: larger depolarizations do not produce larger action potentials. In other words, the size (or *amplitude*) of the action potential is independent of stimulus size. This characteristic is referred to as the **all-or-none property** of the action potential: either it fires at its full amplitude, or it doesn't fire at all. It turns out that information is encoded by changes in the *number* of action potentials rather than in their amplitude. With stronger stimuli, *more* action potentials are produced, but the size of each action potential remains the same.

A closer look at the form of the action potential shows that the return to baseline membrane potential is not simple. Many axons exhibit small potential changes immediately following the spike; these changes are called **afterpotentials** (see Figure 3.5C), and they are also related to the movement of ions in and out of the cell, which we take up next.

Ionic mechanisms underlie the action potential

What events explain the action potential? The action potential is created by the sudden movement of Na^+ ions into the axon (Hodgkin and Katz, 1949). At its peak, the action potential approaches the equilibrium potential for Na^+: about +40 mV. At this point, the concentration gradient pushing Na^+ ions *into* the cell is exactly balanced by the positive charge pushing them *out*. The action potential thus involves a rapid shift in membrane properties, switching suddenly from the potassium-dependent resting state to a primarily sodium-dependent active state, and then swiftly returning to the resting state. This shift is accomplished through the actions of a very special kind of ion channel: the **voltage-gated Na^+ channel**. Like other ion channels, this channel is a tubular, membrane-spanning protein, but its central Na^+-selective pore is *gated*. The gate is ordinarily closed. But if we electrically stimulate the neuron, or if synapses affect the neuron in ways we'll describe later, then the axon may be depolarized. If the axon is depolarized enough to reach threshold levels, the channel's shape changes, opening the "gate" to allow Na^+ ions through for a short while.

Consider what happens when a patch of axonal membrane depolarizes. As long as the depolarization is below threshold, Na^+ channels remain closed. But when the depolarization reaches threshold, a few Na^+ channels open at first, allowing a few ions to start entering the neuron. The positive charges of those ions depolarize the membrane even further, opening still more Na^+ channels. Thus, the process accelerates until the barriers are removed and Na^+ ions rush in (**FIGURE 3.6**).

The voltage-gated Na^+ channels stay open for a little less than a millisecond, and then they automatically close again. By this time, the membrane potential has shot up to about +40 mV. Positive charges inside the nerve cell start to push K^+ ions out, aided by the opening of additional voltage-gated K^+ channels that let lots of K^+ ions rush out quickly, restoring the resting potential.

Applying very strong stimuli reveals another important property of axonal membranes. As we bombard the axon with ever-stronger stimuli, an upper limit to the frequency of action potentials becomes apparent at about 1,200 spikes per second. (Many neurons have even slower maximum rates of response.) Similarly, applying pairs of stimuli that are spaced closer and closer together reveals a related phenomenon: beyond a certain point, only the first stimulus is able to elicit an action potential. The axonal membrane is said to be **refractory** (unresponsive) to the second stimulus.

threshold The stimulus intensity that is just adequate to trigger an action potential in an axon.

action potential Also called *spike*. A rapid reversal of the membrane potential that momentarily makes the inside of the membrane positive with respect to the outside.

all-or-none property Referring to the fact that the size (amplitude) of the action potential is independent of the size of the stimulus.

afterpotential The positive or negative change in membrane potential that may follow an action potential.

voltage-gated Na^+ channel A Na^+-selective channel that opens or closes in response to changes in the voltage of the local membrane potential. It mediates the action potential.

refractory Temporarily unresponsive or inactivated.

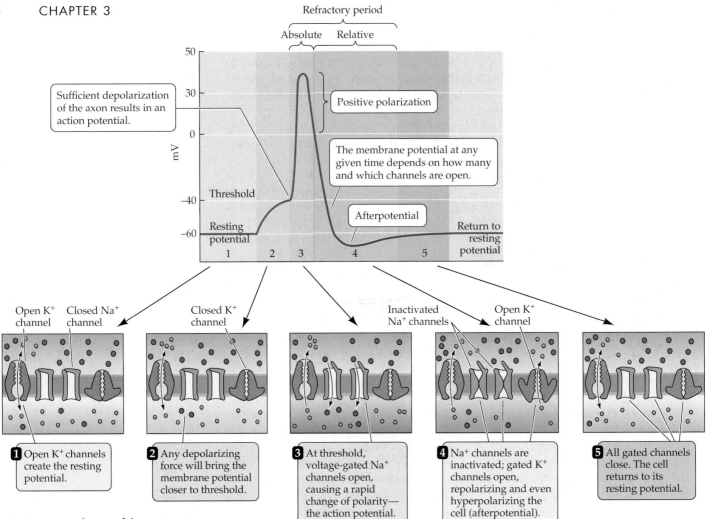

Open K+   Closed Na+   Closed K+   Inactivated   Open K+
channel   channel   channel   Na+ channels   channel

1 Open K+ channels create the resting potential.

2 Any depolarizing force will bring the membrane potential closer to threshold.

3 At threshold, voltage-gated Na+ channels open, causing a rapid change of polarity—the action potential.

4 Na+ channels are inactivated; gated K+ channels open, repolarizing and even hyperpolarizing the cell (afterpotential).

5 All gated channels close. The cell returns to its resting potential.

FIGURE 3.6 Mediation of the Action Potential by Voltage-Gated Sodium Channels

Refractoriness has two phases: During the **absolute refractory phase**, a brief period immediately following the production of an action potential, no amount of stimulation can induce another action potential, because the voltage-gated Na+ channels can't respond (in Figure 3.6, see the brackets above the graph, as well as step 4). The absolute phase is followed by a period of reduced sensitivity, the **relative refractory phase**, during which only strong stimulation, well beyond threshold, can produce another action potential. The neuron is *relatively* refractory because K+ ions are still flowing out, so the cell is temporarily hyperpolarized after firing an action potential (see Figure 3.6, step 4). The overall duration of the refractory phase is what determines a neuron's maximal rate of firing.

You might wonder whether the repeated inrush of Na+ ions would allow them to build up, affecting the cell's resting potential. In fact, relatively few Na+ ions need to enter to change the membrane potential, and the K+ ions quickly restore the resting potential. In the long run, the sodium-potassium pump enforces the concentrations of ions that maintain the resting potential.

This tiny protein molecule, the voltage-gated Na+ channel, is really quite a complicated machine. It monitors the axon's polarity, and at threshold the channel changes its shape to open the pore, shutting down again just a millisecond later. The channel then "remembers" that it was recently open and refuses to open again for a short time. These properties of the voltage-gated Na+ channel are responsible for the characteristics of the action potential.

In general, the transmission of action potentials is limited to axons. Cell bodies and dendrites usually have few voltage-gated Na+ channels, so they do not conduct action potentials. The ion channels on the cell body and dendrites are stimulated chemically

absolute refractory phase A brief period of complete insensitivity to stimuli.

relative refractory phase A period of reduced sensitivity during which only strong stimulation produces an action potential.

at synapses, as we'll discuss later in this chapter. Because the axon has many such channels, an action potential that occurs at the origin of the axon regenerates itself down the length of the axon, as we discuss next.

how's it going

1. What does it mean if a neuron is depolarized or hyperpolarized, and which action brings the cell closer to threshold?
2. Describe how the polarity of a neuron changes during the phases of an action potential.
3. How does the flow of ions account for that sequence of changes in electrical potential?
4. What mechanisms underlie the two phases of the refractory period?

Action potentials are actively propagated along the axon

Now that we've explored how voltage-gated channels underlie action potentials, we can turn to the question of how action potentials spread down the axon—another function for which voltage-gated channels are crucial. Consider an experimental setup like the one pictured in **FIGURE 3.7**: recording electrodes are positioned along the length of the axon, allowing us to record an action potential at various points on the axon. Recordings like this show that an action potential begun at the axon hillock spreads in a sort of chain reaction down the length of the axon.

How does the action potential travel? It is important to understand that the action potential is *regenerated* along the length of the axon. Remember, the action potential is a spike of depolarizing electrical activity (with a peak of about +40 mV), so it strongly depolarizes the next adjacent axon segment. Because this adjacent axon segment is similarly covered with voltage-gated Na⁺ channels, the depolarization immediately

To see the animation
Action Potential Propagation,
go to

2e.mindsmachine.com/3.5

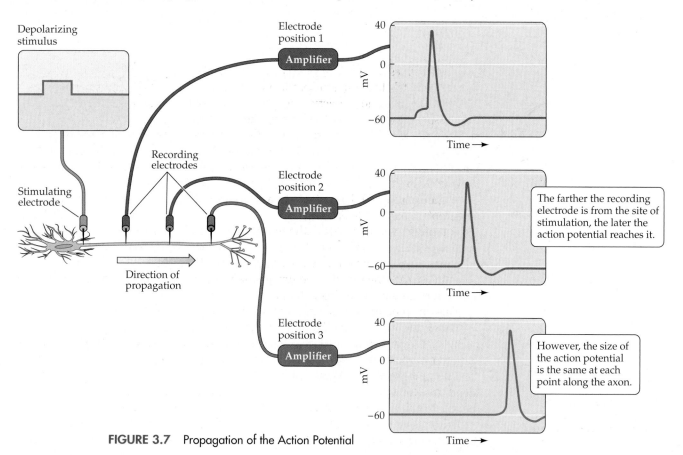

FIGURE 3.7 Propagation of the Action Potential

FIGURE 3.8 Conduction along Unmyelinated versus Myelinated Axons

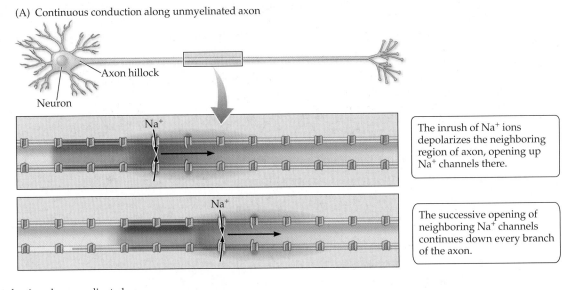

(A) Continuous conduction along unmyelinated axon

Axon hillock

Neuron

Na⁺

The inrush of Na⁺ ions depolarizes the neighboring region of axon, opening up Na⁺ channels there.

Na⁺

The successive opening of neighboring Na⁺ channels continues down every branch of the axon.

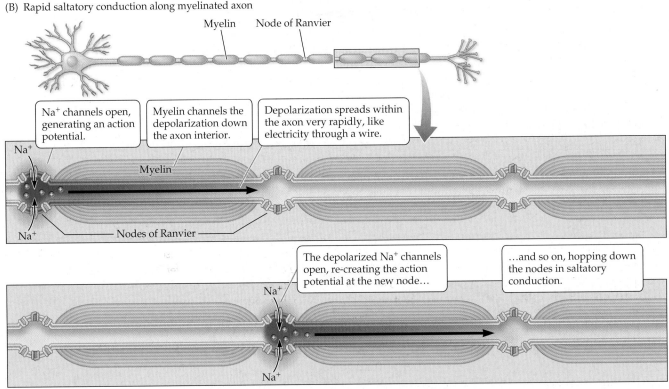

(B) Rapid saltatory conduction along myelinated axon

Myelin   Node of Ranvier

Na⁺ channels open, generating an action potential.

Myelin channels the depolarization down the axon interior.

Depolarization spreads within the axon very rapidly, like electricity through a wire.

Na⁺

Myelin

Na⁺

Nodes of Ranvier

The depolarized Na⁺ channels open, re-creating the action potential at the new node…

…and so on, hopping down the nodes in saltatory conduction.

Na⁺

Na⁺

creates a new action potential, which in turn depolarizes the *next* patch of membrane, which generates yet another action potential, and so on all down the length of the axon (**FIGURE 3.8A**). An analogy is the spread of fire along a row of closely spaced match heads in a matchbook. When one match is lit, its heat is enough to ignite the next match, and so on along the row. Voltage-gated Na⁺ channels open when the axon is depolarized to threshold. In turn, the influx of Na⁺ ions—the movement of positive charges into the axon—depolarizes the adjacent segment of axonal membrane and therefore opens new gates for the movement of Na⁺ ions.

The axon normally conducts action potentials in only one direction—from the axon hillock toward the axon terminals—because, as it progresses along the axon, the action potential leaves in its wake a stretch of refractory membrane (see Figure 3.8A). The action potential does not spread back over the axon hillock and the cell body and dendrites, because the membranes there have too few voltage-gated Na⁺ channels to be able to produce an action potential.

If we record the speed of action potentials along axons that differ in diameter, we see that **conduction velocity** varies with the diameter of the axon. Larger axons allow the

conduction velocity The speed at which an action potential is propagated along the length of an axon.

BOX 3.1 | How Is an Axon Like a Toilet?

It might help you to remember basic facts about action potentials if you consider how much they resemble a flushing toilet. For example, if you gently push the lever on a toilet, nothing much happens. As you gradually increase the force you apply to the lever, you will eventually find the threshold—the amount of force that is just enough to trigger a flush (step 1 of the figure). Likewise, the neuron has a *threshold*—the amount of depolarization that is just enough to trigger an action potential at the start of an axon (see Figure 3.5). Once you're past the threshold, it doesn't matter how hard you pushed the toilet lever; the flush will always be the same. Similarly, once the neuron is pushed past threshold, the action potential will be the same. This is the *all-or-nothing* property of action potentials (step 2).

After a toilet has flushed, it takes a while (about a minute) before it can flush again (step 3). Likewise, after a neuron fires, it takes a while (about a millisecond) before it can fire again. This is the neuron's *refractory period*. Why do neurons have a refractory period for action potentials? As you may remember, the voltage-gated sodium channels always slam shut for a while after they've opened, no matter what the membrane potential is. Until that time is up, the sodium channels won't open again. If the situation is urgent, we can flush our toilet about 60 times per hour or fire our neurons about 1,000 times per second. (If things are *really* urgent, we might do both.)

1 A toilet flush is similar to an action potential.

2 *All-or-none property* Pushing the toilet lever harder does not produce a bigger flush. Pushing a neuron past threshold does not increase the size of the action potential.

3 *Refractory phase* Until the tank is full, the toilet will not flush again. Until the Na^+ channels recover, the neuron cannot produce another action potential.

4 *Direction* Like water in a properly operating toilet, an action potential always travels in one direction only.

Also notice that when a properly working toilet flushes, the water always goes in the same direction (no one wants a toilet that sometimes flushes backward!) (step 4). Likewise, the neuron's action potential goes in only one direction down the axon, from the end attached to the cell body to the axon terminals.

Of course, neurons are different from toilets in many ways. The outflow of a toilet goes only to the single sewer line leaving a house, but an action potential may flow down many axon branches, to communicate with hundreds of other neurons. Voltage-gated sodium channels on the axon branches ensure that the action potential is just as large in each branch, so it's not diminished by spreading out among branches. A toilet has only one lever, but each neuron has hundreds or thousands of synapses, and by producing different sorts of local, graded potentials, some synapses make the neuron more likely to reach threshold, while others make it less likely.

depolarization to spread faster through the interior. In mammals, the conduction velocity in large fibers may be as fast as 150 meters per second. (We will discuss axon diameter and conduction velocity again in Chapter 5.) Although not as fast as the speed of light (as was once believed), neural conduction can nevertheless be very fast: over 300 miles per hour. This relatively high rate of conduction ensures rapid sensory and motor processing.

The highest conduction velocities require more than just large axons. **Myelin** sheathing also greatly speeds conduction. As we described in Chapter 2, the myelin sheath is provided by glial cells. This sheath surrounding the axon is interrupted by **nodes of Ranvier**, small gaps spaced about every millimeter along the axon (see Figure 2.5A). Because the myelin insulation resists the flow of ions across the membrane, the action potential "jumps" from node to node. This process is called **saltatory conduction** (from the Latin *saltare*, "to jump") (**FIGURE 3.8B**). The evolution of rapid saltatory conduction in vertebrates has given them a major behavioral advantage over the invertebrates, whose axons are unmyelinated and thus slower in conduction. **Multiple sclerosis (MS)** is a disease in which myelin is compromised, with highly variable effects on brain function. We describe MS in more detail in Signs & Symptoms at the end of this chapter.

Many of the characteristics of action potentials have been whimsically likened to the action of a toilet, as **BOX 3.1** explores.

myelin The fatty insulation around an axon, formed by glial cells. This sheath boosts the speed at which action potentials are conducted.

node of Ranvier A gap between successive segments of the myelin sheath where the axon membrane is exposed.

saltatory conduction The form of conduction that is characteristic of myelinated axons, in which the action potential jumps from one node of Ranvier to the next.

multiple sclerosis (MS) Literally "many scars." A disorder characterized by the widespread degeneration of myelin.

1. How is the action potential propagated along the axon?
2. What factor causes saltatory conduction, and why does it speed propagation of the action potential?
3. Why do action potentials move only *away* from the cell body?

Synapses cause local changes in the postsynaptic membrane potential

At the beginning of the chapter, we told you that when the action potential reaches the end of an axon, it causes the axon to release a chemical, called a **neurotransmitter** (or *transmitter*) into the synapse. We will discuss the many different types of transmitters in detail in Chapter 4. For now, what you need to know is that when an axon releases neurotransmitter molecules into a synapse, they briefly alter the membrane potential of the other cell. Because information is moving from the axon to the target cell on the other side of the synapse, we say the axon is from the **presynaptic** cell, and the target neuron on the other side of the synapse is the **postsynaptic** cell.

The brief changes in the membrane potential of the postsynaptic cell in response to neurotransmitter are called, naturally enough, **postsynaptic potentials**. A given neuron, receiving synapses from hundreds of other cells, is subject to hundreds or thousands of postsynaptic potentials. When added together, this massive array of local potentials determines whether the axon hillock's membrane potential will reach threshold and therefore trigger an action potential. The nervous system employs electrical synapses too (see **A STEP FURTHER 3.1**, on the website), but the vast majority of synapses use neurotransmitters to produce postsynaptic potentials.

We can study postsynaptic potentials with a setup like that shown in **FIGURE 3.9**. This setup allows us to compare the effects of activity of excitatory versus inhibitory

neurotransmitter   Also called simply *transmitter, synaptic transmitter,* or *chemical transmitter*. The chemical released from the presynaptic axon terminal that serves as the basis of communication between neurons.

presynaptic   Located on the "transmitting" side of a synapse.

postsynaptic   Referring to the region of a synapse that receives and responds to neurotransmitter.

postsynaptic potential   A local potential that is initiated by stimulation at a synapse, which can vary in amplitude, and spreads passively across the cell membrane, decreasing in strength with time and distance.

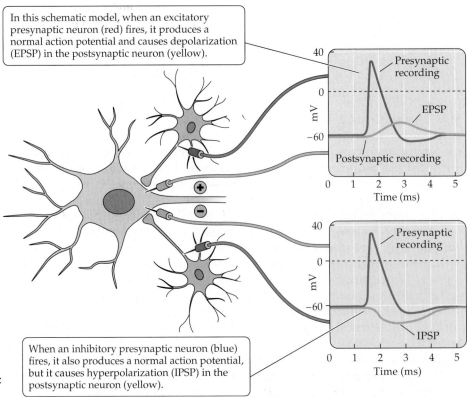

In this schematic model, when an excitatory presynaptic neuron (red) fires, it produces a normal action potential and causes depolarization (EPSP) in the postsynaptic neuron (yellow).

When an inhibitory presynaptic neuron (blue) fires, it also produces a normal action potential, but it causes hyperpolarization (IPSP) in the postsynaptic neuron (yellow).

FIGURE 3.9   Recording Postsynaptic Potentials

TABLE 3.1 Comparing Axons and Dendrites

	Property			
	Size	Number per neuron	Information flow	Voltage changes
Axon	Thin, uniform	One (but may have branches)	Away from cell body	All-or-none
Dendrite	Thick, variable	Many	Into cell body	Graded variable

synapses on the local membrane potential of a postsynaptic cell. The responses of the presynaptic and postsynaptic cells are shown on similar graphs in Figure 3.9 for easy comparison of their timing. It is important to remember that excitatory and inhibitory neurons get their names from their *actions on postsynaptic neurons*, not from their effects on behavior.

Stimulation of the excitatory presynaptic neuron (red in Figure 3.9) causes it to produce an all-or-none action potential that spreads to the end of the axon, releasing transmitter. After a brief delay, the postsynaptic cell (yellow) displays a small local depolarization, as Na^+ channels open to let the positive ions in. This postsynaptic membrane depolarization is known as an **excitatory postsynaptic potential (EPSP)** because it pushes the postsynaptic cell a little closer to the threshold for an action potential.

The action potential of the inhibitory presynaptic neuron (blue in Figure 3.9) looks exactly like that of the excitatory presynaptic neuron; all neurons use the same kind of action potential, as Figure 3.9 shows. But the effect on the postsynaptic side is quite different. When the inhibitory presynaptic neuron is activated, the postsynaptic membrane potential becomes even more negative, or hyperpolarized. This hyperpolarization moves the cell membrane potential away from threshold—it decreases the probability that the neuron will fire an action potential—so it is called an **inhibitory postsynaptic potential (IPSP)**.

Usually IPSPs result from the opening of channels that permit **chloride ions (Cl^-)** to enter the cell. Because Cl^- ions are much more concentrated outside the cell than inside (see Figure 3.2), they rush into the cell, making its membrane potential more negative. What determines whether a synapse excites or inhibits the postsynaptic cell? One factor is the particular neurotransmitter released by the presynaptic cell. Some transmitters typically generate an EPSP in the postsynaptic cells; others typically generate an IPSP. So in the end, whether a neuron fires an action potential at any given moment is decided by the balance between the number of excitatory and the number of inhibitory signals that it is receiving, and it receives many signals of both types at all times.

Now that you know more about the parts of neurons and how they communicate, we summarize the differences between axons (which send information via action potentials) and dendrites (which receive information from synapses) in **TABLE 3.1**.

Spatial summation and temporal summation integrate synaptic inputs

Synaptic transmission is an impressive process, but complex behavior requires more than the simple arrival of signals across synapses. Neurons must also be able to integrate the messages they receive. In other words, they perform *information processing*—by using a sort of neural algebra, in which each nerve cell adds and subtracts the many inputs it receives from other neurons. As we'll see next, this is possible because of the characteristics of synaptic inputs, the way in which the neuron integrates the postsynaptic potentials, and the trigger mechanism that determines whether a neuron will fire an action potential.

We've seen that postsynaptic potentials are caused by transmitter chemicals that can be either depolarizing (excitatory) or hyperpolarizing (inhibitory). From their points of origin on the dendrites and cell body, these graded EPSPs and IPSPs spread

excitatory postsynaptic potential (EPSP) A depolarizing potential in the postsynaptic neuron that is normally caused by synaptic excitation. EPSPs increase the probability that the postsynaptic neuron will fire an action potential.

inhibitory postsynaptic potential (IPSP) A hyperpolarizing potential in the postsynaptic neuron. IPSPs decrease the probability that the postsynaptic neuron will fire an action potential.

chloride ion (Cl^-) A chlorine atom that carries a negative charge.

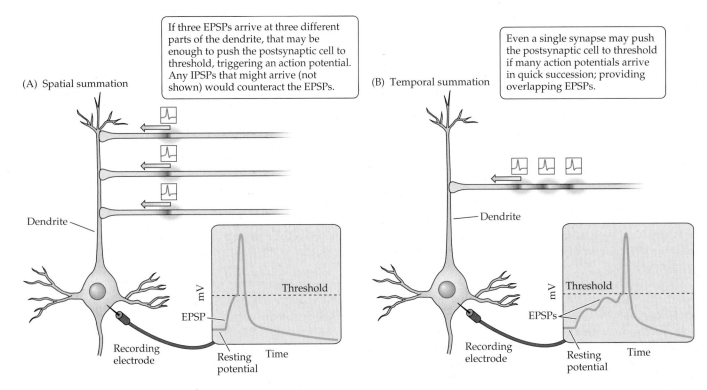

FIGURE 3.10 Spatial versus Temporal Summation

To see the animation
Spatial Summation,
go to

2e.mindsmachine.com/3.6

spatial summation The summation of postsynaptic potentials that reach the axon hillock from different locations across the cell body. If this summation reaches threshold, an action potential is triggered.

temporal summation The summation of postsynaptic potentials that reach the axon hillock at different times. The closer in time the potentials occur, the more complete the summation is.

passively over the postsynaptic neuron, decreasing in strength over time and distance. Whether the postsynaptic neuron will fire depends on whether a depolarization exceeding threshold reaches the axon hillock, triggering an action potential. If many EPSPs are received, the axon may reach threshold and fire. But if both EPSPs and IPSPs arrive at the axon hillock, they partially cancel each other. Thus, the net effect is the *difference* between the two: the neuron subtracts the IPSPs from the EPSPs. Simple arithmetic, right?

Well, yes, summed EPSPs and IPSPs do tend to cancel each other out. But because postsynaptic potentials spread passively and dissipate as they cross the cell membrane, the resulting sum is also influenced by *distance*. For example, EPSPs from synapses close to the axon hillock will produce a larger sum there than will EPSPs from farther away. The summation of potentials originating from different physical locations across the cell body is called **spatial summation**. Only if the overall sum of *all* the potentials—both EPSPs and IPSPs—is sufficient to depolarize the cell to threshold at the axon hillock is an action potential triggered (**FIGURE 3.10A**). Usually it takes excitatory messages from many presynaptic neurons to cause a postsynaptic neuron to fire an action potential.

Postsynaptic effects that are not absolutely simultaneous can also be summed, because the postsynaptic potentials last a few milliseconds before fading away. The closer they are in time, the greater is the overlap and the more complete is the summation, which in this case is called **temporal summation**. Temporal summation is easily understood if you imagine a neuron with only one input. If EPSPs arrive one right after the other, they sum and the postsynaptic cell eventually reaches threshold and produces an action potential (**FIGURE 3.10B**). But these graded potentials fade quickly, so if too much time passes between successive EPSPs, they will never sum and no action potentials will be triggered. **TABLE 3.2** summarizes the many properties of action potentials, EPSPs, and IPSPs, noting the similarities and differences among the three kinds of neural potentials.

It should now be clear that although action potentials are all-or-none phenomena, the postsynaptic effect they produce is graded in size and determined by the processing of many inputs occurring close together in time. The membrane potential at the axon hillock thus reflects the moment-to-moment integration of all the neuron's inputs, which the axon hillock encodes into action potentials.

TABLE 3.2 Characteristics of Electrical Signals of Nerve Cells

Type of signal	Signaling role	Typical duration (ms)	Amplitude	Character	Mode of propagation	Ion channel opening	Channel sensitive to:
Action potential	Conduction along an axon	1–2	Overshooting, 100 mV	All-or-none, digital	Actively propagated, regenerative	First Na+, then K+, in different channels	Voltage (depolarization)
Excitatory postsynaptic potential (EPSP)	Transmission between neurons	10–100	Depolarizing, from less than 1 to more than 20 mV	Graded, analog	Local, passive spread	Na+–K+	Chemical (neurotransmitter)
Inhibitory postsynaptic potential (IPSP)	Transmission between neurons	10–100	Hyperpolarizing, from less than 1 to about 15 mV	Graded, analog	Local, passive spread	Cl−–K+	Chemical (neurotransmitter)

Dendrites add to the story of neuronal integration. A vast number of synaptic inputs, arrayed across the dendrites and cell body, can induce postsynaptic potentials. So dendrites expand the receptive surface of the neuron and increase the amount of input the neuron can handle. All other things being equal, the farther out on a dendrite a potential occurs, the less effect it should have at the axon, because the potential decreases in size as it passively spreads. When the potential arises at a dendritic *spine* (see Figure 2.4), its effect is even smaller because it has to spread down the shaft of the spine. Thus, information arriving at various parts of the neuron is *weighted*, in terms of the distance to the axon hillock and the path resistance along the way.

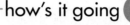how's it going

1. What are EPSPs and IPSPs?
2. Compare and contrast spatial summation versus temporal summation.
3. Discuss the electrical properties of a neuron that allow it to process information.
4. Where does information enter a neuron, and how does a neuron send information to other cells?

Synaptic Transmission Requires a Sequence of Events

The following overview summarizes the events that take place during chemical synaptic transmission:

1. The action potential arrives at the presynaptic axon terminal.
2. Voltage-gated calcium channels in the membrane of the axon terminal open, and calcium ions (Ca^{2+}) enter the axon terminal.
3. Ca^{2+} causes synaptic vesicles filled with neurotransmitter to fuse with the presynaptic membrane and rupture, releasing the transmitter molecules into the synaptic cleft.
4. Some transmitter molecules bind to special receptor molecules in the postsynaptic membrane, leading—directly or indirectly—to the opening of ion channels in the postsynaptic membrane. The resulting flow of ions creates a local EPSP or IPSP in the postsynaptic neuron.
5. The IPSPs and EPSPs in the postsynaptic cell spread toward the axon hillock. (If the sum of all the EPSPs and IPSPs ultimately depolarizes the axon hillock enough to reach threshold, an action potential will arise.)
6. Synaptic transmission is rapidly stopped, so the message is brief and accurately reflects the activity of the presynaptic cell.

To see the animation
Synaptic Transmission,
go to

2e.mindsmachine.com/3.7

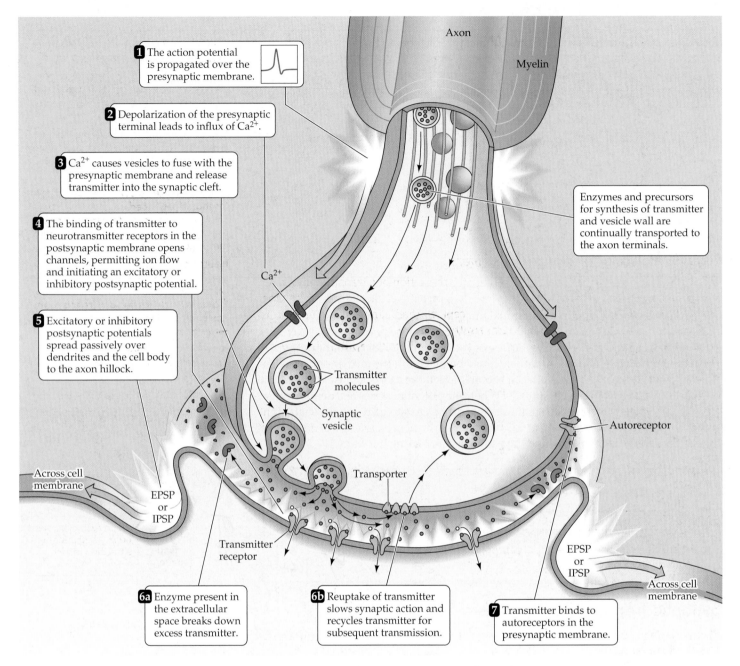

1 The action potential is propagated over the presynaptic membrane.

2 Depolarization of the presynaptic terminal leads to influx of Ca^{2+}.

3 Ca^{2+} causes vesicles to fuse with the presynaptic membrane and release transmitter into the synaptic cleft.

4 The binding of transmitter to neurotransmitter receptors in the postsynaptic membrane opens channels, permitting ion flow and initiating an excitatory or inhibitory postsynaptic potential.

5 Excitatory or inhibitory postsynaptic potentials spread passively over dendrites and the cell body to the axon hillock.

Axon

Myelin

Enzymes and precursors for synthesis of transmitter and vesicle wall are continually transported to the axon terminals.

Ca^{2+}

Transmitter molecules

Synaptic vesicle

Autoreceptor

Across cell membrane

EPSP or IPSP

Transporter

Transmitter receptor

EPSP or IPSP

Across cell membrane

6a Enzyme present in the extracellular space breaks down excess transmitter.

6b Reuptake of transmitter slows synaptic action and recycles transmitter for subsequent transmission.

7 Transmitter binds to autoreceptors in the presynaptic membrane.

FIGURE 3.11   Steps in Transmission at a Chemical Synapse

7. Synaptic transmitter may also activate presynaptic receptors, resulting in a decrease in transmitter release.

Let's look at these seven steps (**FIGURE 3.11**) in a little more detail.

Action potentials cause the release of transmitter molecules into the synaptic cleft

When an action potential reaches a presynaptic terminal, it causes hundreds of **synaptic vesicles** near the presynaptic membrane to fuse with the membrane and discharge their contents—molecules of neurotransmitter—into the **synaptic cleft** (the space between the presynaptic and postsynaptic membranes). The key event

synaptic vesicle   A small, spherical structure that contains molecules of neurotransmitter.

synaptic cleft   The space between the presynaptic and postsynaptic neurons at a synapse. This gap measures about 20–40 nm.

in this process is an influx of **calcium ions (Ca²⁺)**, rather than K⁺ or Na⁺, into the axon terminal. These ions enter through voltage-gated Ca²⁺ channels opening in response to the arrival of an action potential. **Synaptic delay** is the time needed for Ca²⁺ to enter the terminal, for the transmitter to diffuse across the synaptic cleft, and for transmitter molecules to interact with their receptors before the postsynaptic cell responds.

The presynaptic terminal normally produces and stores enough transmitter to ensure that it is ready for activity. Intense activity of the neuron reduces the number of available vesicles, but soon more vesicles are produced to replace those that were discharged. Neurons differ in their ability to keep pace with a rapid rate of incoming action potentials. Furthermore, the rate of making the transmitter is regulated by enzymes that are manufactured in the neuronal cell body and transported actively down the axons to the terminals.

Receptor molecules recognize transmitters

The action of a key in a lock is a good analogy for the action of a transmitter on a receptor protein. Just as a particular key can open a door, a molecule of the correct shape, called a **ligand** (see Chapter 4), can fit into a receptor protein and activate or block it. So, for example, at synapses where the transmitter is **acetylcholine (ACh)**, the ACh fits into areas called *ligand-binding sites* in **neurotransmitter receptor** molecules located in the postsynaptic membrane (**FIGURE 3.12**).

The nature of the postsynaptic receptors at a synapse determines the action of the transmitter (see Chapter 4). For example, ACh can function as either an inhibitory or an excitatory neurotransmitter, at different synapses. At excitatory synapses, binding of ACh to one type of receptor opens channels for Na⁺ and K⁺ ions. At inhibitory synapses, ACh may act on another type of receptor to open channels that allow Cl⁻ ions to enter, thereby hyperpolarizing the membrane (that is, making it more negative and so less likely to create an action potential).

The lock-and-key analogy is strengthened by the observation that various chemicals can fit onto receptor proteins and block the entrance of the key. Some of the preparations used in this research sound like the ingredients for a witches' brew. As an example, consider some potent poisons that block ACh receptors: curare and bungarotoxin. **Curare** is an arrowhead poison used by native South Americans. Extracted from a plant, it greatly increases the efficiency of hunting: if the hunter hits any part of the prey, the arrow's poison soon blocks ACh receptors on muscles, paralyzing the animal. **Bungarotoxin**, another blocker of ACh receptors, is found in the venom of the many-banded krait (*Bungarus multicinctus*), a snake native to China and southeast Asia.

The chemical nicotine, found in tobacco products, mimics the action of ACh at some synapses, increasing alertness and heart rate. Molecules such as nicotine that act like transmitters at a receptor are called

calcium ion (Ca²⁺) A calcium atom that carries a double positive charge.

synaptic delay The brief delay between the arrival of an action potential at the axon terminal and the creation of a postsynaptic potential.

ligand A substance that binds to receptor molecules, such as a neurotransmitter or drug that binds postsynaptic receptors.

acetylcholine (ACh) A neurotransmitter that is produced and released by parasympathetic postganglionic neurons, by motoneurons, and by neurons throughout the brain.

neurotransmitter receptor Also called simply *receptor*. A specialized protein, often embedded in the cell membrane, that selectively senses and reacts to molecules of a corresponding neurotransmitter or hormone.

curare A neurotoxin that causes paralysis by blocking acetylcholine receptors in muscle.

bungarotoxin A neurotoxin, isolated from the venom of the many-banded krait, that selectively blocks acetylcholine receptors.

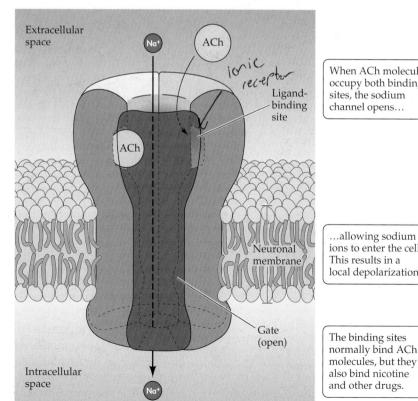

When ACh molecules occupy both binding sites, the sodium channel opens…

…allowing sodium ions to enter the cell. This results in a local depolarization.

The binding sites normally bind ACh molecules, but they also bind nicotine and other drugs.

FIGURE 3.12 A Nicotinic Acetylcholine Receptor

agonist   A substance that mimics or potentiates the actions of a transmitter or other signaling molecule.

antagonist   A substance that blocks or attenuates the actions of a transmitter or other signaling molecule.

cholinergic   Referring to cells that use acetylcholine as their synaptic transmitter.

degradation   The chemical breakdown of a neurotransmitter into inactive metabolites.

acetylcholinesterase (AChE)   An enzyme that inactivates the transmitter acetylcholine.

reuptake   The process by which released synaptic transmitter molecules are taken up and reused by the presynaptic neuron, thus stopping synaptic activity.

transporter   A specialized membrane component that returns transmitter molecules to the presynaptic neuron for reuse.

agonists (from the Greek *agon*, "contest" or "struggle") of that transmitter. Conversely, molecules that interfere with or prevent the action of a transmitter, like curare, are called **antagonists**.

Just as there are master keys that fit many different locks, there are submaster keys that fit a certain group of locks, as well as keys that fit only a single lock. Similarly, each chemical transmitter binds to several different receptor molecules. ACh acts on at least four subtypes of **cholinergic** receptors. Nicotinic cholinergic receptors are found at synapses on muscles and in autonomic ganglia; it is the blockade of these receptors that causes paralysis brought on by curare and bungarotoxin. Most nicotinic sites are excitatory, but there are also inhibitory nicotinic synapses. The many types of receptors for each transmitter have evolved to enable a variety of actions in the nervous system.

The nicotinic ACh receptor resembles a lopsided dumbbell with a tube running down its central axis (see Figure 3.12). The handle of the dumbbell spans the cell membrane, with two sites on the outside that fit ACh molecules (Karlin, 2002). For the channel to open, both of the ACh-binding sites must be occupied. Receptors for some of the synaptic transmitter molecules that we will consider in later chapters, such as gamma-aminobutyric acid (GABA), glycine, and glutamate, are similar.

The coordination of different transmitter systems of the brain is incredibly complex. Each subtype of neurotransmitter receptor has a unique pattern of distribution within the brain. Different receptor systems become active at different times in fetal life. The number of any given type of receptor remains plastic in adulthood: not only are there seasonal variations, but many kinds of receptors show a regular daily variation of 50% or more in number, affecting the sensitivity of cells to that particular transmitter. Similarly, the numbers of some receptors have been found to vary with the use of drugs. We'll learn more about these properties of neurotransmitter receptors in Chapter 4.

The action of synaptic transmitters is stopped rapidly

When a chemical transmitter such as ACh is released into the synaptic cleft, its postsynaptic action is not only prompt but usually very brief as well. It is important that each activation of the synapse be brief in order to maximize how much information can be transmitted. Think of it this way: the worst doorbell in the world is one that, when the button is pushed, rings forever. Such a doorbell would be able to transmit only one piece of information, and only once. But a doorbell that could ring as fast as a thousand times per minute would be able to send a lot of information—Morse code maybe. Likewise, a synapse can signal over a thousand times per second, potentially sending a lot of information (but not by Morse code).

Two processes bring transmitter effects to a prompt halt:

1. *Degradation* Transmitter molecules can be rapidly broken down and thus inactivated by special enzymes—a process known as **degradation** (step 6a in Figure 3.11). For example, the enzyme that inactivates ACh is **acetylcholinesterase (AChE)**. AChE breaks down ACh very rapidly into products that are recycled (at least in part) to make more ACh in the axon terminal.

2. *Reuptake* Alternatively, transmitter molecules may be swiftly cleared from the synaptic cleft by being absorbed back into the axon terminal that released them—a process known as **reuptake** (step 6b in Figure 3.11). Norepinephrine, dopamine, and serotonin are examples of transmitters whose activity is terminated mainly by reuptake. In these cases, special receptors for the transmitter, called **transporters**, are located on the presynaptic axon terminal and bring the transmitter back inside. Once taken up into the presynaptic terminal, transmitter molecules may be repackaged into newly formed synaptic vesicles to await re-release, conserving the resources that would be needed to make new transmitter molecules. Malfunction of reuptake mechanisms is suspected to cause some kinds of mental illness, such as depression (see Chapter 12).

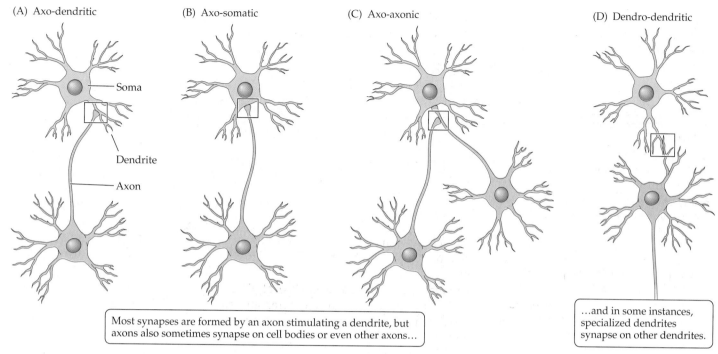

Most synapses are formed by an axon stimulating a dendrite, but axons also sometimes synapse on cell bodies or even other axons…

…and in some instances, specialized dendrites synapse on other dendrites.

FIGURE 3.13   Different Types of Synaptic Connections

Neural circuits underlie reflexes

For simplicity, so far we have focused on the classic, **axo-dendritic synapses** (from axon to dendrite) and **axo-somatic synapses** (from axon to cell body, or soma). But many nonclassic forms of chemical synapses exist in the nervous system. As the name implies, **axo-axonic synapses** form on axons, often near the axon terminal, allowing the presynaptic neuron to strongly facilitate or inhibit the activity of the postsynaptic axon. Similarly, neurons may form **dendro-dendritic contacts**, allowing coordination of their activities (**FIGURE 3.13**).

Now that we know more about the electrical signaling that takes place *within* each neuron and the neurotransmitter signaling that goes on *between* neurons, we can revisit the **knee-jerk reflex** that we discussed at the start of the chapter (**FIGURE 3.14**). Note that this reflex is extremely fast: only about 40 milliseconds elapse between the hammer tap and the start of the kick. Several factors account for this speed: (1) both the sensory and the motor axons involved are myelinated and of large diameter, so they conduct rapidly; (2) the sensory cells synapse directly on the motor neurons; and (3) both the central synapse and the neuromuscular junction are fast synapses. We discuss other aspects of neural circuits in **A STEP FURTHER 3.2**, on the website.

We offered this reflex at the start of the chapter as an example of all neural processing—electrical signaling within each neuron alternating with chemical signaling between neurons. The next chapter will explain how drugs can interfere with the chemical signaling between neurons. For the final part of this chapter, let's see how scientists exploit the electrical signaling within neurons to learn more about brain function.

axo-dendritic synapse   A synapse at which a presynaptic axon terminal synapses onto a dendrite of the postsynaptic neuron, either via a dendritic spine or directly onto the dendrite itself.

axo-somatic synapse   A synapse at which a presynaptic axon terminal synapses onto the cell body (soma) of the postsynaptic neuron.

axo-axonic synapse   A synapse at which a presynaptic axon terminal synapses onto the axon terminal of another neuron.

dendro-dendritic synapse   A synapse at which a synaptic connection forms between the dendrites of two neurons.

knee-jerk reflex   A variant of the stretch reflex in which stretching of the tendon beneath the knee leads to an upward kick of the leg.

how's it going ❓

1. Recount the seven steps in synaptic transmission, including processes that end the signal.
2. What ion must enter the axon terminal to trigger neurotransmitter release?
3. What are drug agonists and antagonists?
4. Describe how information is processed within neurons by electrical signals, yet communicated to other neurons by chemical signals.

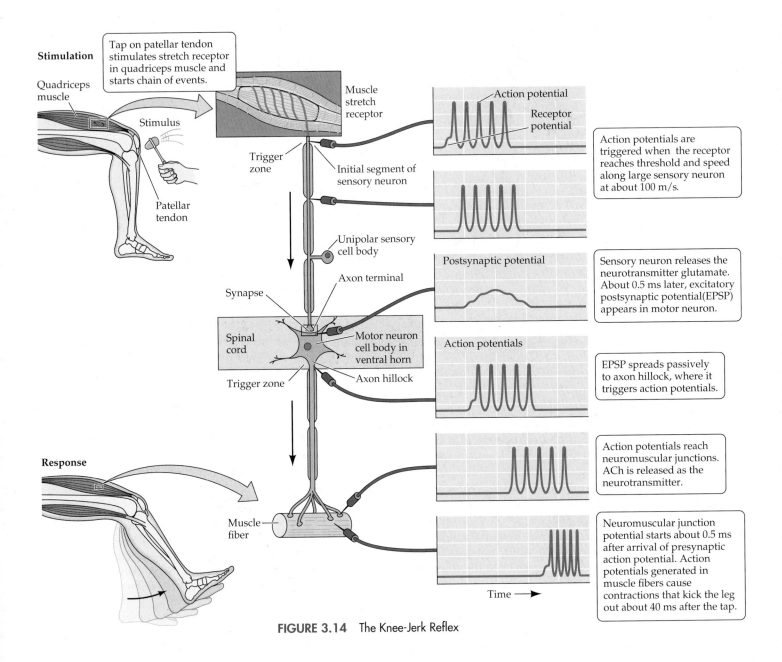

Stimulation

Tap on patellar tendon stimulates stretch receptor in quadriceps muscle and starts chain of events.

Quadriceps muscle

Stimulus

Patellar tendon

Muscle stretch receptor

Trigger zone

Initial segment of sensory neuron

Unipolar sensory cell body

Synapse

Axon terminal

Spinal cord

Motor neuron cell body in ventral horn

Trigger zone

Axon hillock

Response

Muscle fiber

Action potential
Receptor potential

Action potentials are triggered when the receptor reaches threshold and speed along large sensory neuron at about 100 m/s.

Postsynaptic potential

Sensory neuron releases the neurotransmitter glutamate. About 0.5 ms later, excitatory postsynaptic potential(EPSP) appears in motor neuron.

Action potentials

EPSP spreads passively to axon hillock, where it triggers action potentials.

Action potentials reach neuromuscular junctions. ACh is released as the neurotransmitter.

Time →

Neuromuscular junction potential starts about 0.5 ms after arrival of presynaptic action potential. Action potentials generated in muscle fibers cause contractions that kick the leg out about 40 ms after the tap.

FIGURE 3.14 The Knee-Jerk Reflex

EEGs Measure Gross Electrical Activity of the Human Brain

electroencephalogram (EEG)
A recording of gross electrical activity of the brain via large electrodes placed on the scalp.

event-related potential (ERP) Also called *evoked potential*. Averaged EEG recordings measuring brain responses to repeated presentations of a stimulus. Components of the ERP tend to be reliable because the background noise of the cortex has been averaged out.

The electrical activity of millions of cells working together combines to produce electrical potentials large enough that we can detect them with electrodes applied to the surface of the scalp. Recordings of these spontaneous brain potentials (or *brain waves*), called **electroencephalograms** (**EEGs**) (**FIGURE 3.15A**), can provide useful information about the activity of brain regions during behavioral processes. As we will see in Chapter 10, EEG recordings can distinguish whether a person is asleep or awake. In many countries, EEG activity determines whether someone is legally dead.

   Event-related potentials (**ERPs**) are EEG responses to a single stimulus, such as a flash of light or a loud sound. Typically, many ERP responses to the same stimulus are averaged to obtain a reliable estimate of brain activity (**FIGURE 3.15B**). ERPs have very distinctive characteristics of wave shape and time delay (or *latency*) that reflect the type

(A) Multichannel EEG recording

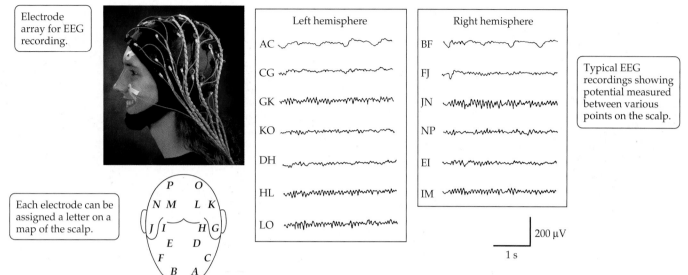

Electrode array for EEG recording.

Each electrode can be assigned a letter on a map of the scalp.

Left hemisphere

AC
CG
GK
KO
DH
HL
LO

Right hemisphere

BF
FJ
JN
NP
EI
IM

Typical EEG recordings showing potential measured between various points on the scalp.

200 μV

1 s

FIGURE 3.15   Gross Potentials of the Human Nervous System

(B) Event-related potentials (average of many stimulus presentations)

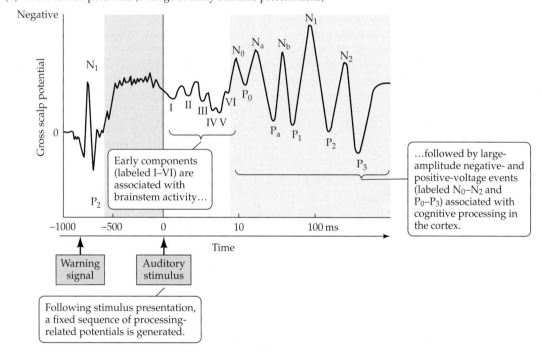

Negative

Gross scalp potential

N_1

N_0 N_a N_b N_1

N_2

P_0

I   II   III   VI

IV V

P_a P_1

P_2

P_3

P_2

0

Early components (labeled I–VI) are associated with brainstem activity…

…followed by large-amplitude negative- and positive-voltage events (labeled N_0–N_2 and P_0–P_3) associated with cognitive processing in the cortex.

−1000   −500   0   10   100 ms

Time

Warning signal   Auditory stimulus

Following stimulus presentation, a fixed sequence of processing-related potentials is generated.

of stimulus, the state of the participant, and the site of recording (Luck, 2005). ERPs can also be used to detect hearing problems in babies, evident as reduced or absent auditory ERPs in response to sounds. In Chapter 14 we'll learn how ERPs are used to study subtler psychological processes, such as attention.

EEG recordings can also provide vital information for diagnosing seizure disorders, as we discuss next.

Electrical storms in the brain can cause seizures

Since the dawn of civilization, people have pondered the causes of **epilepsy**, a disorder in which **seizures** lasting for a few seconds or minutes may produce dramatic behavioral changes such as alterations or loss of consciousness and rhythmic convulsions of the

epilepsy   A brain disorder marked by major, sudden changes in the electro-physiological state of the brain that are referred to as seizures.

seizure   A wave of abnormally synchronous electrical activity in the brain.

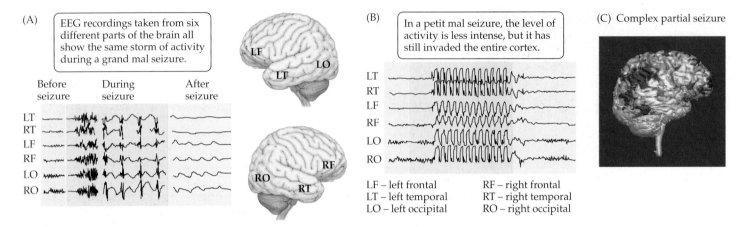

(A) EEG recordings taken from six different parts of the brain all show the same storm of activity during a grand mal seizure.

	Before seizure	During seizure	After seizure
LT			
RT			
LF			
RF			
LO			
RO			

(B) In a petit mal seizure, the level of activity is less intense, but it has still invaded the entire cortex.

LT
RT
LF
RF
LO
RO

LF – left frontal   RF – right frontal
LT – left temporal   RT – right temporal
LO – left occipital   RO – right occipital

(C) Complex partial seizure

FIGURE 3.16   Seizure Disorders

grand mal seizure   A type of generalized epileptic seizure in which nerve cells fire in high-frequency bursts, usually accompanied by involuntary rhythmic contractions of the body.

petit mal seizure   Also called *absence attack*. A seizure that is characterized by a spike-and-wave EEG and often involves a loss of awareness and inability to recall events surrounding the seizure.

complex partial seizure   In epilepsy, a type of seizure that doesn't involve the entire brain and therefore can cause a wide variety of symptoms.

aura   In epilepsy, the unusual sensations or premonition that may precede the beginning of a seizure.

body. Worldwide, about 30 million people suffer from epilepsy, which we now know to be a disorder of electrical potentials in the brain.

In the normal, active brain, electrical activity tends to be desynchronized; that is, different brain regions carry on their functions more or less independently. In contrast, during a seizure there is widespread synchronization of electrical activity: broad stretches of the brain start firing in simultaneous waves, which are evident in the EEGs as an abnormal "spike-and-wave" pattern of brain activity. Many abnormalities of the brain, such as trauma, injury, or metabolic problems, can predispose brain tissue to produce such synchronized activity, which, once begun in one brain region, may readily spread to others.

There are several major categories of seizure disorders. The most dramatic, with loss of consciousness and rhythmic convulsions, are called **grand mal seizures** and are accompanied by abnormal EEG activity all over the brain (**FIGURE 3.16A**). In the more subtle **petit mal seizures**, the characteristic spike-and-wave EEG activity is evident for 5–15 seconds at a time (**FIGURE 3.16B**), sometimes occurring many times per day. The person is unaware of the environment during these periods and later cannot recall events that occurred during the episodes. Behaviorally, people experiencing petit mal seizures show no unusual muscle activity; they just stop what they're doing and stare into space.

Complex partial seizures do not involve the entire brain and thus can produce a wide variety of symptoms, often preceded by an unusual sensation, or **aura**. In one example, a woman felt an unusual sensation in the abdomen, a sense of foreboding, and tingling in both hands before the seizure spread. At the height of the episode, she was unresponsive and rocked her body back and forth while speaking nonsensically, twisting her left arm, and looking toward the right. Of course, her seemingly random set of behavioral symptoms actually reflected the functions of the particular brain regions affected by the seizure (**FIGURE 3.16C**); others experiencing seizures would produce a completely different set of behaviors. In some individuals, complex partial seizures may be provoked by stimuli like loud noises or flashing lights.

Many seizure disorders can be effectively controlled with the aid of antiepileptic drugs. Although these drugs have a wide variety of neural targets, they tend to selectively reduce the excitability of neurons (Rogawski and Löscher, 2004). But some cases of epilepsy do not respond to medication, and if the seizures are severe, or happen very often, they may be life threatening. Individuals suffering from such severe epilepsy may resort to the drastic step of having parts of the brain removed, as we'll see next.

Surgical probing of the brain revealed a map of the body

Usually the electrical activity causing seizures begins in one part of the brain and then spreads to others. So in the twentieth century, neurosurgeons began taking dramatic measures to help people with severe epilepsy that did not respond to medication: surgical removal of the part of the brain where the seizures begin. The trick, of course, is to remove the part of the brain where the seizures begin, and *only* that part. Otherwise the patient might take the risks of surgery and still suffer from seizures, or might suffer impairment of a vital function, such as verbal or memory skills, if healthy tissue is removed. One way to locate the origin of the seizures is to compare EEG readings from different places on the skull (see Figure 3.16). But this approach gives only a rough idea of where the seizures

begin, and it is problematic because the recording must be made when a seizure is actually taking place.

To improve the success rate of such surgeries, Canadian neurosurgeon Wilder Penfield developed a procedure that, 85 years later, is still compelling (Foerster and Penfield, 1930). Using only local anesthesia to deaden the pain of cutting the scalp and opening up the skull, Penfield had patients remain awake and alert as he exposed the brain. Then he used electrodes to provide a tiny electrical stimulation to the surface of the cortex, asking the patient to report the results. One strategy for people whose epileptic seizures were preceded by an aura was to try to find the point where stimulation recreated the aura.

continued

FIGURE 3.17   Mapping the Human Brain

■ **Hypothesis**
Sensory information from the body arrives in an organized fashion in the cortex.

■ **Experiment**
Electrically stimulate the surface of the cortex in alert patients, carefully recording the patient's experience with stimulation at each site. Compare these maps in various patients.

■ **Result**
Each side of the brain receives sensory information from the opposite side of the body, organized along the postcentral gyrus of the parietal lobe. Across the central sulcus, in the precentral gyrus, cortical regions control movement of that same part of the body, so that sensory and motor regions are aligned.

■ **Conclusion**
The maps of sensory cortex and motor cortex are remarkably consistent from one person to another.

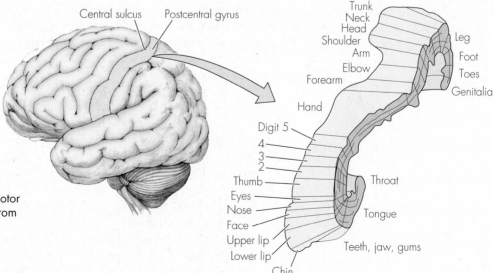

In one famous case of a woman whose seizures were preceded by the smell of burnt toast, Penfield was able to find a spot where stimulation caused her to smell burnt toast and, presuming that region was the origin of the seizures, surgically removed it. (The brain itself has no pain receptors, so cutting the cortex didn't hurt.) Using this refined technique, Penfield was able to cure about half of his patients, and seizures were reduced in another 25%. José Delgado's stimulation of patients' brains, which we discussed at the start of the chapter, also sought the origin of seizures. Delgado altered the technique by implanting several electrodes temporarily, so the patient could walk around while doctors electrically stimulated different brain regions and observed the results. In Chapters 5 and 12 we'll learn that today electrodes are sometimes implanted in the brain as a treatment for other disorders.

In his pioneering work, Penfield did more than help his patients. He also made major discoveries about the organization of the human cortex (**FIGURE 3.17**). By carefully recording the effects of stimulating different regions of the brain, he found that stimulation of occipital cortex often caused the patient to "see" flashes of light. Stimulating another region might cause the person's thumb to tingle, while stimulation elsewhere might cause the patient's leg to move. Through these studies, Penfield confirmed that each side of the cortex receives information from, and sends commands to, the opposite side of the body. He found that stimulating the postcentral gyrus of the parietal cortex caused patients to experience sensations on various parts of the body in a way that was consistent from one person to another (see Figure 5.9). Just across the central sulcus from each site, in the precentral gyrus,

stimulations caused that same part of the body to move (see Figure 5.22). These "maps" of how the various parts of the body are laid out on the cortex (Jasper and Penfield, 1954) have been reproduced in countless textbooks, providing the basis of what is called the *homunculus*, the "little man" drawn on the surface of the cortex to depict Penfield's maps, as we will discuss further in Chapter 5.

These groundbreaking observations taught us that brain function is organized in a map that reproduces body parts. We also learned that the map is distorted, in the sense that parts of the body that are especially sensitive to touch, such as the lips or fingers, are monitored by a relatively large area of cortex compared with, say, the backs of the legs.

Penfield's studies also stimulated a rich store of speculation about the relationship between the workings of the brain and the mind. In a small minority of patients, electrical stimulation in some sites would sometimes elicit a memory of, for example, sitting on the porch step and hearing a relative's voice, or hearing a snatch of music. As we saw at the start of this chapter, other researchers would find that electrical stimulation of the brain could make people think they loved their examiner or could make an angry, murderous bull peaceful and calm. In another case, electrical stimulation of one part of her brain caused a young woman to find whatever was happening around her to be humorous (Fried et al., 1998).

These startling observations—showing that electrical stimulation of the brain triggers mental processes—remain a cornerstone of neuroscience, and a tantalizing demonstration that our mind is a result of physical processes at work in the machine we call the brain.

SIGNS & SYMPTOMS

Multiple Sclerosis

Most cases of multiple sclerosis (MS) are due to the body's immune system generating antibodies that attack one or more molecules in myelin. If the myelin is damaged enough, then saltatory conduction of axon potentials is disrupted, throwing off the brain's timing in coordinating behavior and interpreting sensory input (**FIGURE 3.18**). MS can present with a bewildering variety of motor and/or sensory symptoms, depending on where, exactly, myelin damage occurs. For many people, the first symptoms they notice are blurred vision or poor color perception. For others, it is unexpected tingling sensations, or a difficulty coordinating their walking, with the feeling of stiff legs. Eventually, almost all MS patients experience fatigue. The various symptoms wax and wane for reasons no one understands. While MS symptoms generally worsen over time, the rate varies a great deal across patients, making it impossible to predict how it might progress.

continued

SIGNS & SYMPTOMS *continued*

There is currently no cure for MS, but there are medicines to manage the symptoms, mostly by interfering with the immune system to curtail myelin damage. It's also important to get physical therapy to learn how to maintain normal activity despite the symptoms. Smoking increases the risk of MS, and it is more common in women than in men. Symptoms tend to be reduced during pregnancy when estrogen levels are high, so there is growing interest in whether hormone treatment might offer an effective treatment (Wisdom et al., 2013).

FIGURE 3.18 MS Impairs Axonal Conduction

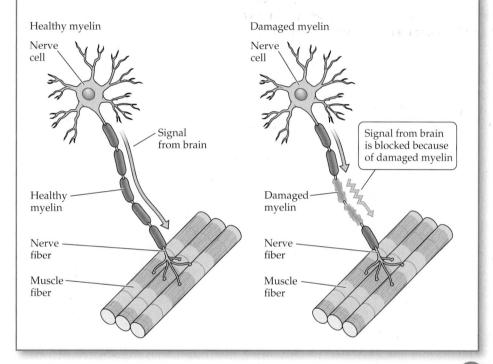

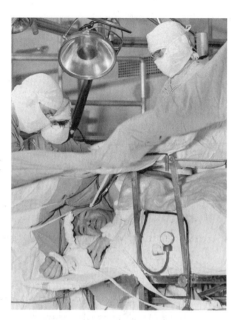

Brain Stimulation Surgeon Wilder Penfield electrically stimulating the surface of the exposed brain in an awake patient.

how's it going ❓

1. What are EEGs and ERPs? How have these techniques been useful?
2. What are the two main categories of epileptic seizures, and what are the main characteristics of each type?
3. Describe Penfield's surgical procedure and what it revealed about organization of the brain.
4. What is the underlying cause of multiple sclerosis, and what are some symptoms of the disease?

Recommended Reading

Kandel, E. R., Schwartz, J. H., Jessell, T. M., Siegelbaum, S. A., and Hudspeth, A. J. (2013). *Principles of Neural Science* (5th ed.). New York, NY: McGraw-Hill.

LeDoux, J. (2002). *Synaptic Self: How Our Brains Become Who We Are.* New York, NY: Viking.

Luck, S. J. (2005). *An Introduction to the Event-Related Potential Technique.* Cambridge, MA: MIT Press.

Nicholls, J. G., Martin, A. R., Fuchs, P. A., Brown, D. A., Diamond, M. E., and Weisblat, D. A. (2012). *From Neuron to Brain* (5th ed.). Sunderland, MA: Sinauer.

Purves, D., Augustine, G. J., Fitzpatrick, D., Hall, W. C., et al. (2012). *Neuroscience* (5th ed.). Sunderland, MA: Sinauer.

Valenstein, E. S. (2005). *The War of the Soups and the Sparks: The Discovery of Neurotransmitters and the Dispute over How Neurons Communicate.* New York, NY: Columbia University Press.

VISUAL SUMMARY

2e.mindsmachine.com/vs3

3

You should be able to relate each summary to the adjacent illustration, including structures and processes. Go to the online version of this summary (scan the QR code above) for links to figures, animations, and activities that will help you consolidate the material.

1 Chemical signals transmit information between neurons; electrical signals transmit information within a neuron. The **resting potential** is a small electrical potential across the neuron's membrane. Review **Figure 3.1, Animation 3.2**

2 Different concentrations of **ions** inside and outside the neuron—especially **potassium ions (K+)**, to which the resting membrane is **selectively permeable**—account for the **resting potential**. At the K+ equilibrium potential, the **electrostatic pressure** pulling K+ ions into the neuron is balanced by the concentration gradient pushing them out. Review **Figures 3.2–3.4, Activity 3.1, Animation 3.3**

3 Reducing the resting potential (**depolarization**) of the axon until it reaches a **threshold** value opens **voltage-gated sodium (Na+) channels**, making the membrane completely permeable to Na+. The **sodium ions (Na+)** rush in, and the axon becomes briefly more positive inside than outside. This event is called an **action potential**. Review **Figure 3.5, Animation 3.4**

4 Following the action potential, the resting potential is quickly restored by the influx of K+ ions. The **sodium-potassium pumps** maintain the resting potential in the long run, counteracting the influx of Na+ ions during action potentials. Review **Figure 3.6**

5 The action potential strongly depolarizes the adjacent patch of axonal membrane, causing it to generate its own action potential, propagating down the axon. **Saltatory conduction** of the action potential along the **nodes of Ranvier** between **myelin** sheaths speeds propagation down the axon. Review **Figures 3.7 and 3.8, Box 3.1, Animation 3.5**

6 Like all other **local potentials**, **postsynaptic potentials** spread very rapidly but are not regenerated, so they diminish as they spread passively along dendrites and the cell body. **Excitatory postsynaptic potentials (EPSPs)** are **depolarizing** (they decrease the resting potential) and increase the likelihood that the neuron will fire an action potential. **Inhibitory postsynaptic potentials (IPSPs)** are **hyperpolarizing**, decreasing the likelihood that the neuron will fire. Review **Figure 3.9**

7 Neurons process information by integrating the postsynaptic potentials through both **spatial summation** (summing potentials from different locations) and **temporal summation** (summing potentials across time). Review **Figure 3.10, Animation 3.6**

8 Action potentials are initiated at the **axon hillock** when the excess of EPSPs over IPSPs reaches threshold. During the action potential, the neuron cannot be excited by a second stimulus; it is **absolutely refractory**. For a few milliseconds afterward, the neuron is **relatively refractory**, requiring a stronger stimulation than usual in order to fire. Review **Figure 3.6**

9 Synaptic transmission occurs when a chemical neurotransmitter diffuses across the **synaptic cleft** and binds to **neurotransmitter receptors** in the postsynaptic membrane. Review **Figures 3.11 and 3.12, Animation 3.7**

10 Summing electrical activity over millions of nerve cells as detected by electrodes on the scalp, **electroencephalograms (EEGs)** can reveal rapid changes in brain function—for example, in response to a brief, controlled stimulus that evokes an **event-related potential (ERP)**. They can also reveal a **seizure** in people with **epilepsy**. Review **Figures 3.15–3.17**

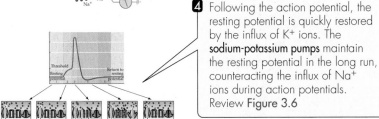
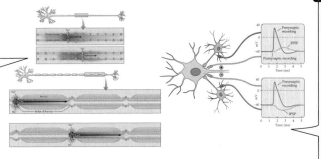
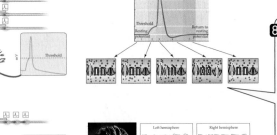
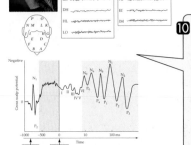
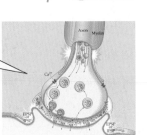

Go to **2e.mindsmachine.com** for study questions, quizzes, flashcards, and other resources.

chapter

4

The Chemistry of Behavior
Neurotransmitters and Neuropharmacology

A Dream of Soups and Sparks

As the twentieth century began, scientists knew that neurons were important for brain function, but there was a big controversy about how neurons communicated. What happened at those newly discovered synapses between one neuron and another? Did sparks of electricity pass from cell to cell? Or was some unknown chemical substance involved? Some scientists, nicknamed "sparks," favored the idea that electrical signals crossed synapses; other scientists, the "soups," thought neurons released a chemical that flowed across synapses. No one knew how to distinguish between these two possibilities.

Otto Loewi was so consumed with the question of neural communication that he even dreamed about it. One night he awoke suddenly, remembering a dream in which one experiment could provide a definitive answer to whether the "soups" or the "sparks" were right. He made a few notes and went back to sleep, only to discover the next day that he couldn't make any sense of the previous night's scribblings. So when he had the same dream again the following night, he got up and went straight to the lab to do the experiment while it was still fresh in his mind. The result was a discovery that would revolutionize the study of the brain.

To see the video
Synaptic Transmission,
go to

2e.mindsmachine.com/4.1

T hroughout the ages, people have used **exogenous** substances (substances from outside the body) to try to change the functioning of their bodies and brains. Our ancestors sipped, swallowed, and smoked their way to euphoria, calmness, pain relief, and hallucination. They discovered deadly poisons in frogs, miraculous antibiotics in mold, powerful painkillers in poppies, and all the rest of a vast catalog of helpful and harmful substances. By studying the physiological actions of these substances, modern scientists have been able to unlock many mysteries of brain function.

The preceding chapters showed us that the brain is an electrochemical system. Today we know that, in general, each neuron *electrically* processes information received through many synapses, and then releases a *chemical* to pass the result of that information processing to the next cell. Specifically, a presynaptic neuron releases an **endogenous** substance (a substance from inside the body), a chemical called a *neurotransmitter*. The neurotransmitter then communicates with the postsynaptic cell. As you might have guessed, most drugs that affect behavior do so by affecting this chemical communication process at millions, or even billions, of synapses.

exogenous Arising from outside the body.

endogenous Produced inside the body.

presynaptic Located on the "transmitting" side of a synapse.

synapse The cellular location at which information is transmitted from a neuron to another cell.

Let's start the chapter by reviewing synaptic transmission. Then we'll dig deeper into the chemical anatomy of the brain, looking at the major families of neurotransmitters, especially how each exerts its distinctive influence over unique networks in the brain. This survey will set the stage for discussing how drugs and toxins alter brain function and thereby modify behavior. We'll wrap up the chapter with a discussion of substance abuse and dependency.

Electrical Signals Are Turned into Chemical Signals at Synapses

As we learned in Chapter 3, electrical signals called *action potentials* move down axons until they reach the *axon terminal*, which forms the **presynaptic** side of a **synapse**. **FIGURE 4.1** recaps the events at a typical synapse. When action potentials reach the terminal, they cause voltage-gated calcium (Ca^{2+}) channels to open, allowing Ca^{2+} ions to enter the terminal. This calcium induces synaptic vesicles to fuse with the presynaptic

FIGURE 4.1 Synapses Convert Electrical Signals into Chemical Signals

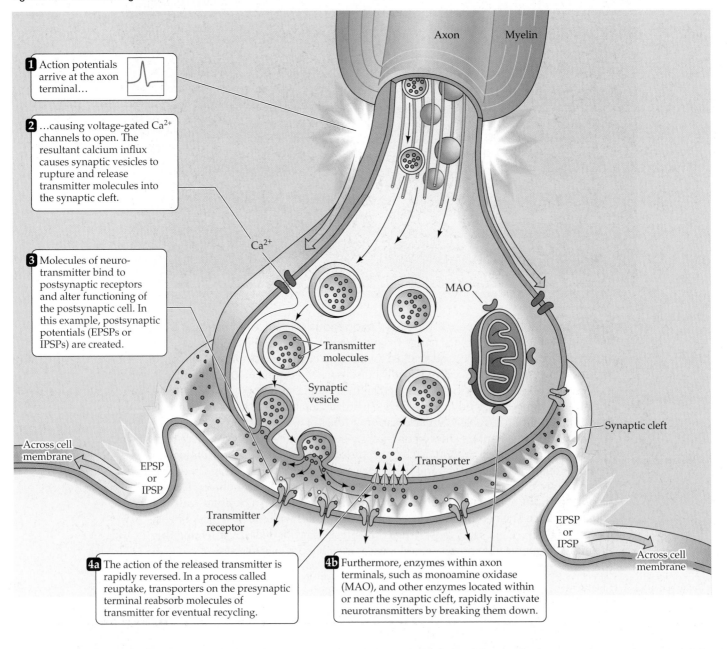

1 Action potentials arrive at the axon terminal…

2 …causing voltage-gated Ca^{2+} channels to open. The resultant calcium influx causes synaptic vesicles to rupture and release transmitter molecules into the synaptic cleft.

3 Molecules of neurotransmitter bind to postsynaptic receptors and alter functioning of the postsynaptic cell. In this example, postsynaptic potentials (EPSPs or IPSPs) are created.

Axon   Myelin

Ca^{2+}

MAO

Transmitter molecules

Synaptic vesicle

Synaptic cleft

Across cell membrane

EPSP or IPSP

Transmitter receptor

Transporter

EPSP or IPSP

Across cell membrane

4a The action of the released transmitter is rapidly reversed. In a process called reuptake, transporters on the presynaptic terminal reabsorb molecules of transmitter for eventual recycling.

4b Furthermore, enzymes within axon terminals, such as monoamine oxidase (MAO), and other enzymes located within or near the synaptic cleft, rapidly inactivate neurotransmitters by breaking them down.

membrane, spilling molecules of **neurotransmitter** (also called just *transmitter*) into the synaptic cleft—a process known as **exocytosis**. After diffusing across the synaptic cleft, the neurotransmitter molecules bind to **neurotransmitter receptors** that span the membrane of the **postsynaptic** cell, which is receiving the chemical signal.

The binding of neurotransmitter molecules to receptors can affect the postsynaptic cell in several different ways, which we'll discuss shortly. Eventually the neurotransmitter molecules are either (1) broken down by enzymes into simpler chemicals or (2) brought back into the presynaptic terminal in a process called **reuptake**. Reuptake of transmitters relies on special proteins called **transporters**, complex structures that bind molecules of transmitter and conduct them back inside the presynaptic terminal. Once inside, the transmitter can be broken down or repackaged in vesicles for reuse.

Let's look at what happens when the neurotransmitter binds to the postsynaptic receptors.

Receptor proteins recognize transmitters and their mimics

Neurotransmitter receptors are very picky about the substances that they will respond to. For this reason, the action of transmitters on receptors is often likened to a key opening a lock. In Chapter 3 we saw that receptors vary enormously in the chemical "keys" they respond to, and in the effects they produce. Nevertheless, the various neurotransmitter receptors can all be categorized as belonging to one of two general kinds: ionotropic receptors or metabotropic receptors.

Ionotropic receptors are just fancy ion channels; when bound by transmitter molecules, they quickly change their shape to open or close the ion channel (**FIGURE 4.2**). The opening of the channel allows more (or fewer) of the channel's favorite ions to flow into or out of the postsynaptic neuron, thus changing the local membrane potential. If that change in the postsynaptic membrane potential is a depolarization, bringing the cell closer to its threshold, then it will be more likely to produce an action potential. In that case we say the synapse has an excitatory effect. Conversely, if ions passing

neurotransmitter   Also called simply *transmitter*. A signaling chemical, released by a presynaptic neuron, that diffuses across the synaptic cleft to alter the functioning of the postsynaptic neuron.

exocytosis   A cellular process that results in the release of a substance into the extracellular space.

neurotransmitter receptor   Also called simply *receptor*. A specialized protein that is embedded in the cell membrane, allowing it to selectively sense and react to molecules of the corresponding neurotransmitter.

postsynaptic   Located on the "receiving" side of a synapse.

reuptake   The reabsorption of molecules of neurotransmitter by the neurons that released them, thereby ending the signaling activity of the transmitter molecules.

transporter   A specialized membrane component that returns transmitter molecules to the presynaptic neuron for reuse.

ionotropic receptor   Also called *ligand-gated ion channel*. A receptor protein containing an ion channel that opens when the receptor is bound by an agonist.

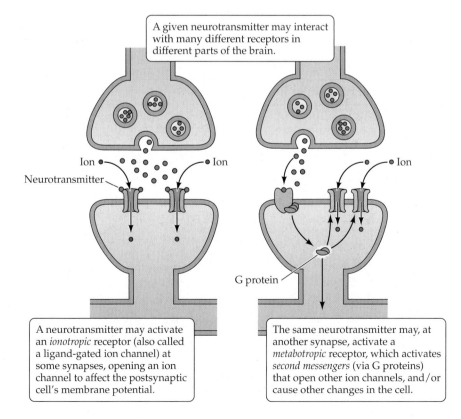

A given neurotransmitter may interact with many different receptors in different parts of the brain.

A neurotransmitter may activate an *ionotropic* receptor (also called a ligand-gated ion channel) at some synapses, opening an ion channel to affect the postsynaptic cell's membrane potential.

The same neurotransmitter may, at another synapse, activate a *metabotropic* receptor, which activates *second messengers* (via G proteins) that open other ion channels, and/or cause other changes in the cell.

To view the
Brain Explorer,
go to

2e.mindsmachine.com/4.2

FIGURE 4.2   The Versatility of Neurotransmitters

metabotropic receptor A receptor protein that does not contain ion channels but may, when activated, use a second messenger system to open nearby ion channels or to produce other cellular effects.

amino acid neurotransmitter A neurotransmitter that is itself an amino acid. Examples include GABA, glycine, and glutamate.

peptide neurotransmitter Also called *neuropeptide*. A neurotransmitter consisting of a short chain of amino acids.

through the ionotropic receptor hyperpolarize the postsynaptic cell, then it will be *less* likely to produce an action potential, and we say the synapse is inhibitory.

But not all receptors work by opening an ion channel. Receptors belonging to the other major category—the **metabotropic receptors**—don't contain ion channels. Instead, they link across the cell membrane to complicated chemical machinery inside the postsynaptic neuron (see Figure 4.2). When activated, metabotropic receptors alter the inner workings of the postsynaptic cell, using chemicals called *second messengers* to cause changes in excitability or other, slower but larger-scale responses. For example, the metabotropic receptor may kick off a chain of chemical reactions that will affect gene expression (the use of genes to produce proteins). Such changes in gene expression can have many lasting effects, such as changes in the excitability or connections of the cell, or the production of more receptors and signaling chemicals.

Together, the two major categories of transmitter receptors allow not only rapid responses where timing is crucial, but also more complex, integrative, slower behaviors such as emotional responses, social behavior, and so on. The specific response of a postsynaptic neuron to molecules of transmitter is therefore determined by the particular types of receptors present on the postsynaptic membrane.

Many neurotransmitters have been identified

Over the years, neuroscientists have agreed on some basic principles for deciding whether a brain chemical qualifies as a classical neurotransmitter. We can conclude that a candidate substance is a transmitter if it meets the following qualifications:

- It can be synthesized by presynaptic neurons and stored in axon terminals.
- It is released when action potentials reach the terminals.
- It is recognized by specific receptors located on the postsynaptic membrane.
- It causes changes in the postsynaptic cell.
- Blocking its release interferes with the ability of the presynaptic cell to affect the postsynaptic cell.

According to these criteria, the brain contains many different neurotransmitters. **TABLE 4.1** summarizes the major categories of transmitters that we will consider in the next portion of this chapter. **Amino acid neurotransmitters** and **peptide neurotransmitters** (or *neuropeptides*), as their names suggest, are based on single amino

TABLE 4.1 Some Synaptic Transmitters and Families of Transmitters

Family and subfamily	Transmitter(s)
AMINO ACIDS	Gamma-aminobutyric acid (GABA), glutamate, glycine, histamine
AMINES	
Quaternary amines	Acetylcholine (ACh)
Monoamines	*Catecholamines*: Norepinephrine (NE), epinephrine (adrenaline), dopamine (DA)
	Indoleamines: Serotonin (5-hydroxytryptamine [5-HT]), melatonin
NEUROPEPTIDES	
Opioid peptides	*Enkephalins*: Met-enkephalin, leu-enkephalin
	Endorphins: Beta-endorphin
	Dynorphins: Dynorphin A
Other neuropeptides	Oxytocin, substance P, cholecystokinin (CCK), vasopressin, neuropeptide Y (NPY), hypothalamic-releasing hormones
GASES	Nitric oxide, carbon monoxide

acid molecules or on short chains of amino acids (called *peptides*), respectively. The family of **amine neurotransmitters** includes some of the best-known classical transmitters, such as acetylcholine, dopamine, and serotonin. Each year the list of probable neurotransmitters grows, and the search for new transmitters has occasionally yielded surprises like the **gas neurotransmitters**, soluble gases that diffuse between neurons to alter ongoing processes (and defy several of the criteria for transmitters that we just listed!). Considering the rate at which these substances are being discovered and characterized, it would not be surprising if there turned out to be several hundred different neurotransmitters at work in synapses throughout the central nervous system. But for now, let's content ourselves with a look at a few of the best-known neurotransmitter systems in the brain. We'll see that each neurotransmitter is distributed in a unique pattern across brain regions. (To learn more about how neurons synthesize neurotransmitters, see **A STEP FURTHER 4.1**, on the website.)

amine neurotransmitter   A neurotransmitter based on modifications of a single amino acid nucleus. Examples include acetylcholine, serotonin, and dopamine.

gas neurotransmitter   A neurotransmitter that is a soluble gas. Examples include nitric oxide and carbon monoxide.

acetylcholine (ACh)   A neurotransmitter that is produced and released by the autonomic nervous system, by motoneurons, and by neurons throughout the brain.

how's it going ❓

1. Distinguish between endogenous and exogenous substances, and give a few examples of each.
2. Review the sequence of events that occurs when an action potential arrives at the axon terminal and causes a release of neurotransmitter. Use the following terms in your answer: *exocytosis, receptors, ionotropic, metabotropic, reuptake.*
3. Identify the criteria that are used to establish whether a substance in the brain can be considered a neurotransmitter. Briefly discuss why each one is important.
4. Name and briefly describe each of the major categories of neurotransmitters.

researchers at work

The first transmitter to be discovered was acetylcholine

Our chapter began with the classic story of Otto Loewi's dream of an experiment that would reveal the first neurotransmitter. After forgetting the details of the dream the first night, Loewi was quick to get out of bed when the dream returned the following night. He went straight to the lab to conduct the experiment depicted in **FIGURE 4.3**. Loewi's first step was to electrically stimulate the vagus nerve in a frog, which he knew would cause its heart to slow down. The question was, why did the heart slow down? Had electricity jumped from the vagus nerve to the heart as the "sparks" believed? Or were the "soups" right? Had the nerve released a chemical to slow the heart?

Loewi's critical, dream-inspired experiment was to collect the fluid that surrounded the slowing heart. Then he applied that fluid to the heart of another frog. If the first heart had been slowed by electrical signals from the vagus, then the fluid should have no effect on the second heart. But if activation of the vagus had caused it to release a chemical, then the fluid from the first heart should alter the beating of the second. In fact, the transferred fluid caused the second heart to slow its rate, providing Loewi with conclusive evidence of chemical neurotransmission (and a nice shiny Nobel Prize, in 1936). The neurotransmitter was later chemically identified as **acetylcholine (ACh** for short). The "soups" were vindicated.

Otto Loewi Following the Nazi annexation of Austria in 1938, Loewi, then a professor at the University of Graz, was forced to flee to the United States, where he remained for the rest of his life. He was photographed in the summer of 1955 by his associate, Dr. Morris Rockstein, at the Woods Hole Marine Biological Laboratory. (Photograph courtesy of Dr. Edward D. Rockstein.)

continued

researchers at work *continued* —————

FIGURE 4.3   "Soups" versus "Sparks": The First Neurotransmitter

■ **Question**

Do neurons release a chemical to communicate with other cells, or is the communication based on electrical signals?

■ **Experiment**

Stimulate the vagus nerve to slow the heart. Collect fluid from around the slowed heart and apply the fluid to a second heart.

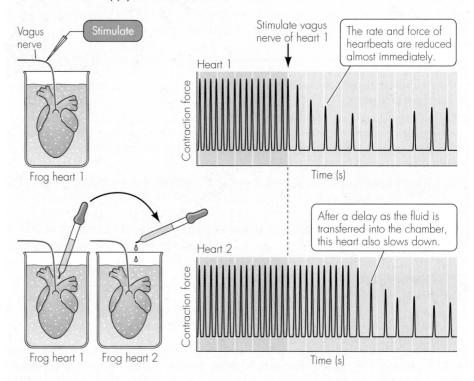

Vagus nerve

Stimulate

Frog heart 1

Frog heart 1   Frog heart 2

Stimulate vagus nerve of heart 1

Heart 1

Contraction force

Time (s)

The rate and force of heartbeats are reduced almost immediately.

After a delay as the fluid is transferred into the chamber, this heart also slows down.

Heart 2

Contraction force

Time (s)

■ **Result**

The second heart also slowed.

■ **Conclusion**

The vagus nerve uses a chemical neurotransmitter, not a direct electrical connection, to communicate to cells of the heart and cause it to slow down.

Neurotransmitter Systems Form a Complex Array in the Brain

It is possible to stain brain tissue in such a way that only the neurons making a particular neurotransmitter end up being labeled (see Box 2.1). Studies of brain sections stained in this way have shown that transmitters are found in complex networks of neurons that extend throughout the brain. In **FIGURE 4.4**, this complicated anatomy is depicted for some of the major amine transmitters. The point to remember is that each of these amine neurotransmitters is carried by a different set of axons, and those axons project to different brain regions. Each "flavor" of neurotransmitter is talking to a distinct set of brain targets, and there may be overlap as two different transmitters arrive at the same target. How those targets respond depends on which neurotransmitter

FIGURE 4.4 Neurotransmitter Pathways in the Brain

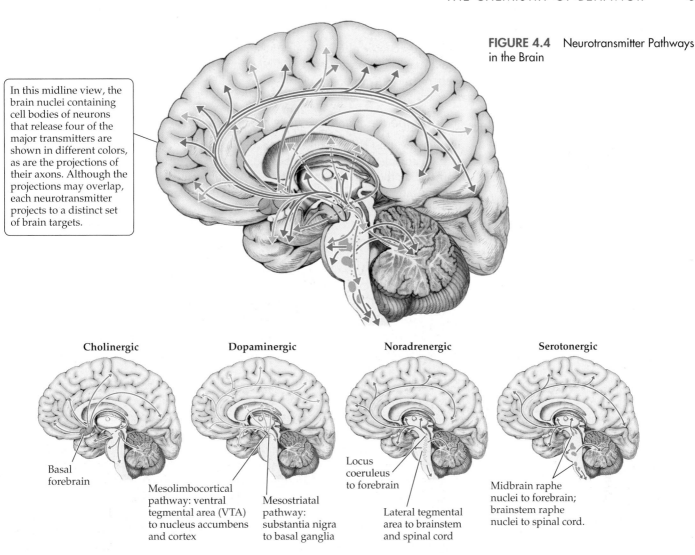

In this midline view, the brain nuclei containing cell bodies of neurons that release four of the major transmitters are shown in different colors, as are the projections of their axons. Although the projections may overlap, each neurotransmitter projects to a distinct set of brain targets.

Cholinergic

Basal forebrain

Dopaminergic

Mesolimbocortical pathway: ventral tegmental area (VTA) to nucleus accumbens and cortex

Mesostriatal pathway: substantia nigra to basal ganglia

Noradrenergic

Locus coeruleus to forebrain

Lateral tegmental area to brainstem and spinal cord

Serotonergic

Midbrain raphe nuclei to forebrain; brainstem raphe nuclei to spinal cord.

is being released *and* which kind of receptors for that transmitter the target neurons possess. Despite its appearance, Figure 4.4 is a simplification; there are many more transmitters at work than the few we have shown here, and they are arranged in much more complicated networks.

The most abundant excitatory and inhibitory neurotransmitters are amino acids

The most common transmitters in the brain are amino acids, and of these, the two best studied are **glutamate**, the most widespread excitatory transmitter in the brain, and **gamma-aminobutyric acid (GABA)**, the most widespread inhibitory transmitter. Both have wide-ranging effects at synapses throughout the central nervous system, and from an evolutionary perspective they are among the most ancient transmitter substances.

Glutamate interacts with several subtypes of receptors (see Table 4.2). Activation of AMPA receptors, the most plentiful receptors in the brain, has rapid excitatory effects. NMDA receptors are another subtype of glutamate receptors, with unique characteristics that suggest they play a central role in memory formation (discussed in Chapter 13, and on the website in **A STEP FURTHER 4.2**).

Among the subtypes of receptors for GABA, the $GABA_A$ receptors have received decades of special scrutiny because of their relationship to anxiety relief. $GABA_A$ receptors are ionotropic; when activated, they allow more Cl^- ions to flow into the postsynaptic cell, thus inhibiting that cell's activity. Drugs that mimic this action of

To see the animation
Neurotransmitter Pathways,
go to

2e.mindsmachine.com/4.3

glutamate An amino acid transmitter, the most common excitatory transmitter.

gamma-aminobutyric acid (GABA) A widely distributed amino acid transmitter, the main inhibitory transmitter in the mammalian nervous system.

GABA$_A$ tend to be effective calming agents because they produce a widespread decrease in neural activity. Drugs called *benzodiazepines*—examples include Valium (diazepam) and Ativan (lorazepam)—are widely used to reduce anxiety, as well as to aid muscle relaxation, sleep induction, and the like.

Four amine neurotransmitters modulate brain activity

Some of the most thoroughly studied transmitters are *amines* (amines are a large family of nitrogen-containing compounds related to ammonia, often derived from an amino acid); they have been implicated in many categories of behavior and pathology. Because they are involved in a wide range of disorders—Parkinson's disease, schizophrenia, Alzheimer's disease, and depression, to name just a few—amine neurotransmitter mechanisms are a major target for drug development. Here we look at four important amine transmitters: acetylcholine, dopamine, serotonin, and norepinephrine.

ACETYLCHOLINE   We now know that acetylcholine (ACh) plays a major role in transmission in the forebrain. Many **cholinergic** (ACh-containing) neurons are found in nuclei within the **basal forebrain**. These cholinergic cells project widely in the brain, to sites such as the cerebral cortex, amygdala, and hippocampus (see Figure 4.4). Widespread loss of cholinergic neurons is associated with Alzheimer's disease, and experimental disruption of cholinergic pathways in rats interferes with learning and memory.

DOPAMINE   Of the nearly 100 billion neurons in the human brain, only about a million synthesize **dopamine** (**DA**), but they are critically important for many aspects of behavior. Figure 4.4 shows the path of the major **dopaminergic** projections. One of these projections, called the *mesostriatal pathway,* originates in and around the **substantia nigra** of the midbrain. From there, axons project to parts of the basal ganglia. There aren't all that many neurons in the system—hundreds of thousands—but keep in mind that a single axon can divide to supply thousands of synapses. Those synapses play a crucial role in motor control. When people lose a significant number of mesostriatal dopaminergic neurons, from either exposure to poisons or just old age, they develop the profound movement problems of Parkinson's disease (described in Chapter 5), including tremors.

Another dopaminergic projection, called the *mesolimbocortical pathway,* also originates in the midbrain, in a region called the **ventral tegmental area** (**VTA**) (see Figure 4.4). From there, the pathway projects to various locations in the limbic system (see Chapter 2) and cortex. The mesolimbocortical system appears to be especially important for the processing of reward; it's probably where feelings of pleasure arise. Thus, it makes sense that the mesolimbocortical dopamine system is important for learning that is shaped by positive reinforcement (which usually involves a reward; see Chapter 13), especially via the D2 dopamine receptor subtype. Abnormalities in the mesolimbocortical pathway are associated with some of the symptoms of schizophrenia, as we discuss in Chapter 12. At the end of this chapter we'll look at the role of this pathway in addictive behaviors.

SEROTONIN   Dopaminergic neurons may be scarce, but there are even fewer **serotonergic** neurons in the human brain—just 200,000 or so. Nevertheless, wide expanses of the brain are innervated by serotonergic fibers, originating from neurons sprinkled along the midline of the midbrain and brainstem in the **raphe nuclei** (*raphe* is pronounced "rafay") (see Figure 4.4).

Serotonin (**5-HT**, short for its chemical name, 5-hydroxytryptamine) participates in the control of all sorts of behaviors: mood, vision, sexual behavior, anxiety, sleep, and many other functions. As we'll see a little later, drugs that either directly mimic serotonin or, like Prozac, cause serotonin to accumulate are often effective for relieving depression and anxiety. The precise behavioral actions of serotonergic drugs depend on which of the many 5-HT receptor subtypes are affected (see Table 4.2) (Gorzalka et al., 1990; Miczek et al., 2002).

cholinergic   Referring to cells that use acetylcholine as their synaptic transmitter.

basal forebrain   A region, ventral to the basal ganglia, that is the major source of acetylcholine in the brain.

dopamine (DA)   A monoamine transmitter found in the midbrain—especially the substantia nigra—and in the basal forebrain.

dopaminergic   Referring to cells that use dopamine as their synaptic transmitter.

substantia nigra   A brainstem structure that innervates the basal ganglia and is a major source of dopaminergic projections.

ventral tegmental area (VTA)   A portion of the midbrain that projects dopaminergic fibers to the nucleus accumbens.

serotonergic   Referring to cells that use serotonin as their synaptic transmitter.

raphe nuclei   A string of nuclei in the midline of the midbrain and brainstem that contain most of the serotonergic neurons of the brain.

serotonin (5-HT)   A synaptic transmitter that is produced in the raphe nuclei and is active in structures throughout the cerebral hemispheres.

NOREPINEPHRINE As Figure 4.4 shows, many of the brain's **noradrenergic** neurons (we use this term because **norepinephrine [NE]** is also known as *noradrenaline*) have their cell bodies in two regions of the brainstem and midbrain: the **locus coeruleus** ("blue spot") and the **lateral tegmental area**. Noradrenergic axons from these regions project broadly throughout the cerebrum, including the cerebral cortex, limbic system, and thalamic nuclei. They participate in the control of behaviors ranging from alertness to mood to sexual behavior (and many more).

Many peptides function as neurotransmitters

Peptides are very important signaling chemicals in both the brain and in the other organs of the body. Some examples of peptides that act as neurotransmitters include:

- The **opioid peptides** (peptides that mimic the actions of opiate drugs like morphine) include met-enkephalin, leu-enkephalin, beta-endorphin, and dynorphin A. Like morphine, these peptides act in the brain to reduce our perception of pain.

- A variety of peptides originally discovered in the periphery, and especially in the organs of the gut (which explains some of their names), are also made by neurons in the spinal cord and brain. These peptides may act as synaptic transmitters, and they are often synthesized alongside and released with classical transmitters (this is known as *co-localization*). Examples include substance P, cholecystokinin (CCK), neurotensin, neuropeptide Y (NPY), and vasoactive intestinal polypeptide (VIP).

- A variety of peptide hormones, such as oxytocin and vasopressin, are produced by the hypothalamus and pituitary. Peptides are involved in an astonishing range of functions; vasopressin, for one, was first known for its role in urine production, but it also contributes to such high-level processes as memory and pair-bonding.

Some neurotransmitters are gases

Neurons sometimes use certain gas molecules to communicate information; the best studied of these is nitric oxide (not to be confused with "laughing gas," which is ni*trous* oxide). Carbon monoxide also serves as a transmitter in some cells. Although we call them *gas neurotransmitters*, these substances are different from traditional neurotransmitters in at least three important ways:

1. Gas transmitter is produced outside axon terminals, especially in the dendrites, and is not held in vesicles; the substance simply diffuses out of the neuron as it is produced.

2. No receptors in the membrane of the target cell are involved. Instead, the gas transmitter diffuses into the target cell to trigger second messengers inside.

3. Most important, these gases can function as **retrograde transmitters**: by diffusing from the postsynaptic neuron back to the presynaptic neuron, a gas transmitter conveys information that is used to physically change the synapse. This process may be crucial for memory formation (see Chapter 13).

noradrenergic Referring to cells using norepinephrine (noradrenaline) as a transmitter.

norepinephrine (NE) Also called *noradrenaline*. A neurotransmitter produced and released by sympathetic postganglionic neurons to accelerate organ activity.

locus coeruleus A small nucleus in the brainstem whose neurons produce norepinephrine and modulate large areas of the forebrain.

lateral tegmental area A brainstem region that provides some of the norepinephrine-containing projections of the brain.

opioid peptide A type of endogenous peptide that mimics the effects of morphine in binding to opioid receptors and producing marked analgesia and reward.

retrograde transmitter A neurotransmitter that diffuses from the postsynaptic neuron back to the presynaptic neuron.

how's it going ❓

1. Describe how the first neurotransmitter was discovered. Why do you think we selected this discovery for this chapter's "Researchers at Work" feature?

2. What are the main excitatory and inhibitory amino acid transmitters of the brain, and what importance do they have?

3. Name and describe the anatomical organization of the four major "classical" amine neurotransmitters. What are some of the functions in which each transmitter has been implicated?

4. What is a peptide? Where do peptides come from—give a few examples— and what are some functions they perform?

5. Discuss the ways in which the gas transmitters resemble and differ from traditional amine transmitters.

Drugs Fit Like Keys into Molecular Locks

Any substance that binds to a receptor is termed a **ligand**. The natural ligands for receptors are molecules of neurotransmitters, of course. But as you may have guessed, many (but not all) drugs that affect the brain are also receptor ligands.

In everyday English, we use the term *drug* in different ways. One common meaning is "a medicine used in the treatment of a disease" (as in *prescription drug* or *over-the-counter drug*). Many *psychoactive drugs*—compounds that alter the function of the brain and thereby affect conscious experiences—fall into this category, and they may be useful in psychiatric settings. Some psychoactive drugs are used recreationally, with varying degrees of risk to the user; these are sometimes referred to as *drugs of abuse*. Some psychoactive drugs affect the brain by altering enzyme action or modifying other internal cellular processes, but as you may have anticipated, most drugs of interest in neuroscience are receptor ligands.

Drugs that mimic or potentiate the actions of a transmitter are called **agonists**. A substance that mimics the normal action of a neurotransmitter on its receptors (**FIGURE 4.5A**) by binding to the receptors and activating them (**FIGURE 4.5B**) is thus a *receptor agonist*. Similarly, drugs that reduce the normal actions of a neurotransmitter system are called **antagonists**. Drugs classified as *receptor antagonists* bind to receptors but do *not* activate them—instead, they block the receptors from being activated by their normal neurotransmitter (**FIGURE 4.5C**). For that reason, drugs with this action are sometimes called receptor *blockers*.

Some drugs—caffeine, opium, nicotine, and cocaine are just a few examples—originally evolved in plants, often as a defense against being eaten. Other modern drugs are synthetic (human-made) and tuned to target specific transmitter systems. For example, benzodiazepine antianxiety drugs like Ativan enhance GABA neurotransmission; classic antipsychotics like haloperidol (trade name Haldol) block dopamine receptors; selective serotonin reuptake inhibitors like fluoxetine (Prozac) are antidepressants. So to understand how drugs work, we must perform analyses at many levels—from molecules to anatomical systems to behavioral effects and experiences.

Earlier we explained that a given neurotransmitter interacts with a variety of different *subtypes* of receptors. To give you a sense of just how diverse the list of

ligand   A substance that binds to receptor molecules, such as a neurotransmitter or drug that binds postsynaptic receptors.

agonist   A substance that mimics or potentiates the actions of a transmitter or other signaling molecule.

antagonist   A substance that blocks or attenuates the actions of a transmitter or other signaling molecule.

To see the animation
Agonists and Antagonists,
go to

2e.mindsmachine.com/4.4

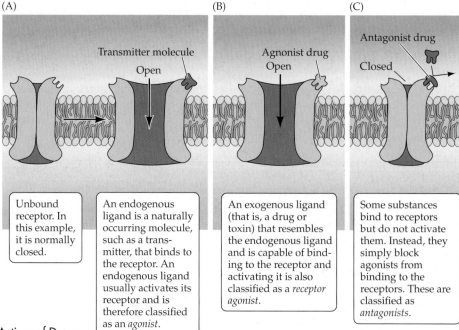

FIGURE 4.5   The Agonistic and Antagonistic Actions of Drugs

transmitter receptor subtypes can be, we've compiled the subtypes for a few transmitters in **TABLE 4.2**. For example, there are more than a dozen different subtypes of serotonin receptors. Some are inhibitory, some excitatory, some ionotropic, some metabotropic; in fact, the only thing they all really share is that they are normally stimulated by serotonin. They even differ in their anatomical distribution within the brain. This division of transmitter receptors into multiple subtypes presents us with an opportunity because, although the natural transmitter will act on *all* its receptor subtypes, we humans can craftily design drugs that single out just one or a few receptor subtypes. Selectively activating or blocking specific subtypes of receptors can produce diverse effects, some of which are beneficial. For example, treating someone with large doses of the neurotransmitter serotonin would necessarily activate *all* of her different subtypes of serotonin receptors, producing a confusing welter of different effects. But drugs that selectively block $5\text{-}HT_3$ receptors while ignoring other subtypes of serotonin receptors produce a powerful and specific anti-nausea effect that brings relief to people undergoing cancer chemotherapy.

TABLE 4.2 The Bewildering Multiplicity of Transmitter Receptor Subtypes

Transmitter	Known receptor subtypes	Function
Glutamate	AMPA, kainate, and NMDA receptors (ionotropic) mGluR's (*metabotropic glutamate receptors*)	Glutamate is the most abundant of all neurotransmitters and the most important excitatory transmitter. Glutamate receptors are crucial for excitatory signals, and NMDA receptors are especially implicated in learning and memory.
Gamma-aminobutyric acid (GABA)	$GABA_A$ (ionotropic)	GABA receptors mediate most of the brain's inhibitory activity, balancing the excitatory actions of glutamate. $GABA_A$ receptors are inhibitory in many brain regions, reducing excitability and preventing seizure activity.
	$GABA_B$ (metabotropic)	$GABA_B$ receptors are also inhibitory, by a different mechanism.
Acetylcholine (ACh)	Muscarinic receptors (metabotropic)	Both are involved in cholinergic transmission in the cortex.
	Nicotinic receptors (ionotropic)	Nicotinic receptors are crucial for muscle contraction.
Dopamine (DA)	D_1 through D_5 receptors (all metabotropic)	Found throughout the forebrain
	D_6 and D_7 probable	Involved in complex behaviors, including motor function, reward, higher cognition
Norepinephrine (NE)	α_1, α_2, β_1, and β_2 receptors (all metabotropic)	Multiple effects in visceral organs; important part of sympathetic nervous system and "fight or flight" responses. In the brain, NE transmission provides an alerting and arousing function.
Serotonin	$5\text{-}HT_1$ receptor family (5 members) $5\text{-}HT_2$ receptor family (3 members) $5\text{-}HT_3$ through $5\text{-}HT_7$ receptors All but one subtype ($5\text{-}HT_3$) are metabotropic.	Different subtypes differ in their distribution in the brain. May be involved in mood, sleep, and higher cognition $5\text{-}HT_3$ receptors are particularly involved in nausea.
Miscellaneous peptides	Many specific receptors for peptides such as opiates (delta, kappa, and mu receptors), cholecystokinin (CCK), neurotensin, neuropeptide (NPY), and dozens more (metabotropic)	Peptide transmitters have many different functions, depending on their anatomical localization. Some important examples include the control of feeding, sexual behaviors, and social functions.

binding affinity   Also called simply *affinity*. The propensity of molecules of a drug (or other ligand) to bind to receptors.

efficacy   Also called *intrinsic activity*. The extent to which a drug activates a response when it binds to a receptor.

dose-response curve (DRC)   A formal graph of a drug's effects (on the *y*-axis) versus the dose given (on the *x*-axis).

The effects of a drug depend on its dose

The tuning of drug molecules to receptor subtypes is not absolutely specific. In reality, a particular drug will generally bind strongly to one kind of receptor, more weakly to a few other types, and not at all to many others. This chemical attraction is known as **binding affinity** (or simply *affinity*). At low doses, when relatively few drug molecules are in circulation, drugs will preferentially bind to their highest-affinity receptors. At higher doses, enough molecules of the drug are available that they can bind to both the highest-affinity receptors and lower-affinity receptors. It's interesting to note that neurotransmitter molecules are low-affinity ligands: they bind only comparatively weakly to their receptors, so they can rapidly detach a moment later, allowing the synapse to reset in preparation for the next presynaptic signal.

Once it is bound to a receptor, a drug molecule has a certain propensity to *activate* the receptor; this propensity is referred to as its **efficacy** (or *intrinsic activity*). As you might guess, agonists have high efficacy, and antagonists have low or no efficacy (see Figure 4.5). So it is a combination of affinity and efficacy—where it binds and what it does—that determines the overall action of a drug. For example, the classic antipsychotic drugs tend to have high affinity and low efficacy at the D_2 subtype of dopamine receptors; in other words, they are D_2 blockers. It's a topic we'll revisit later in this chapter and in Chapter 12.

Administering larger doses of a drug ultimately increases the proportion of receptors that are bound and affected by the drug. Within certain limits, this increase in receptor binding also increases the response to the drug; in other words, greater doses tend to produce greater effects. When plotted as a graph, the relationship between drug doses and observed effects is called a **dose-response curve** (**DRC**), typically taking the sigmoidal shape shown in **FIGURE 4.6A**. Careful analysis of DRCs reveals many aspects of a drug's activity, such as useful and safe dose ranges (**FIGURE 4.6B**), and it is one of the main tools for understanding the functional relationships between drugs and their targets.

Drug doses are administered in many different ways

Drug molecules aren't magic bullets; they don't somehow know where to go to find particular receptor molecules. Instead, drug molecules just spread widely throughout

FIGURE 4.6   The Dose-Response Curve

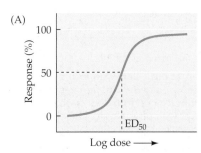

(A)

The basic dose-response curve (DRC) plots increasing drug doses (usually on a logarithmic scale) against increasing strength of the response being studied. The dose at which the drug shows half of its maximal effect is termed the *effective dose 50%* (ED_{50}). Drugs with high *potency*—determined by affinity and efficacy in concert—have a lower ED_{50}.

The *therapeutic index* refers to the separation between useful doses of the drug and dangerous doses. This is determined by comparing the ED_{50} dose of the drug with the dose at which 50% of the animals either show symptoms of toxicity (*toxic dose 50%*; TD_{50}) or outright die (*lethal dose 50%*; LD_{50}). In this example, a greater difference between ED_{50} and LD_{50} is observed for the anxiolytic drug lorazepam (Ativan) than for the older anxiolytic phenobarbital, indicating that lorazepam is the safer drug. Many deaths—accidental and not—have resulted from phenobarbital overdose.

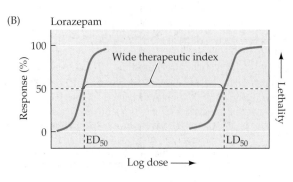

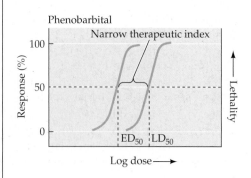

TABLE 4.3 The Relationship between Routes of Administration and Effects of Drugs

Route of administration	Examples and mechanisms	Typical speed of effects
INGESTION Tablets and capsules Syrups Infusions and teas Suppositories	Many sorts of drugs and remedies; depends on absorption by the gut, which is somewhat slower than most other routes and affected by digestive factors such as acidity in the stomach and the presence of food.	Slow to moderate
INHALATION Smoking Nasal absorption (snorting) Inhaled gases, powders, and sprays	Nicotine, cocaine, organic solvents such as airplane glue and gasoline, and other drugs of abuse; also used for a variety of prescription drugs and hormone treatments. Inhalation methods take advantage of the rich vascularization of the nose and lungs to convey drugs directly into the bloodstream.	Moderate to fast
PERIPHERAL INJECTION Subcutaneous Intramuscular Intraperitoneal (abdominal) Intravenous	Many drugs; subcutaneous (under the skin) injections tend to have the slowest effects because they must diffuse into nearby tissue in order to reach the bloodstream; intravenous injections have very rapid effects because the drug is placed directly into circulation.	Moderate to fast
CENTRAL INJECTION Intracerebroventricular (into ventricular system) Intrathecal (into the cerebrospinal fluid of the spine) Epidural (under the dura mater) Intracerebral (directly into a brain region)	Central methods involve injection directly into the central nervous system; used in order to circumvent the blood-brain barrier, to rule out peripheral effects, or to directly affect a discrete brain location.	Fast to very fast

the body, binding to their favorite receptors when they happen to encounter them. This binding triggers a chain of cellular events, but it is usually temporary, and when the drug (or transmitter) breaks away from the receptor, the receptor resumes its unbound shape and functioning.

The amount of a drug that gets to the brain, and how fast it gets there, depends in part on the drug's route of administration. Some routes, such as smoking or intravenous injection, rapidly ramp up the amount of drug that is **bioavailable** (free to act on the target tissue, and thus not in use elsewhere or in the process of being eliminated). With other routes, such as ingestion (swallowing), the concentration of drug builds up more slowly over longer periods of time. The duration of a drug effect also depends on how the drug is metabolized and excreted from the body—via the kidneys, liver, lungs, or other routes. In some cases, the metabolites of drugs are themselves active; this **biotransformation** of drugs can produce substances with beneficial or harmful actions. The factors that affect the movement of a drug into, through, and out of the body are collectively referred to as **pharmacokinetics**.

Humans have devised a wide variety of techniques for introducing substances into the body; these are summarized in **TABLE 4.3**. In Chapter 2 we described how tight junctions between the cells of the walls of blood vessels create a **blood-brain barrier** that inhibits the movement of larger molecules out of the bloodstream and into the brain. This barrier poses a major challenge for neuropharmacology because many drugs that might be useful are too large to cross the blood-brain barrier into the brain. To a limited extent this problem can be circumvented by administering the drugs directly into the brain, but that is a drastic step. Alternatively, some drugs can take advantage of active transport systems that normally move nutrients out of the bloodstream and into the brain.

bioavailable Referring to a substance, usually a drug, that is present in the body in a form that is able to interact with physiological mechanisms.

biotransformation The process in which enzymes convert a drug into a metabolite that is itself active, possibly in ways that are substantially different from the actions of the original substance.

pharmacokinetics Collective name for all the factors that affect the movement of a drug into, through, and out of the body.

blood-brain barrier The protective property of cerebral blood vessels that impedes the movement of some harmful substances from the blood stream into the brain.

drug tolerance Also called simply *tolerance*. A condition in which, with repeated exposure to a drug, an individual becomes less responsive to a constant dose.

metabolic tolerance The form of drug tolerance that arises when repeated exposure to the drug causes the metabolic machinery of the body to become more efficient at clearing the drug.

functional tolerance The form of drug tolerance that arises when repeated exposure to the drug causes receptors to be up-regulated or down-regulated.

down-regulation A compensatory decrease in receptor availability at the synapses of a neuron.

up-regulation A compensatory increase in receptor availability at the synapses of a neuron.

cross-tolerance A condition in which the development of tolerance for one drug causes an individual to develop tolerance for another drug.

Repeated treatments can reduce the effectiveness of drugs

Our bodies are well equipped to maintain a constant internal environment, optimized for cellular activities. Many body systems may change to counteract challenges such as drug treatments. For example, following exposure to a drug, physiological systems may change in order to counteract the drug's effects. This response, called **drug tolerance** (or just *tolerance*), is evident when a drug's effectiveness diminishes over repeated treatments. Consequently, successively larger and larger doses of drug are needed to cause the same effect.

Drug tolerance can develop in several different ways. Some drugs provoke **metabolic tolerance**, in which the body (especially metabolic organs, such as the liver with its specialized enzymes) becomes more effective at eliminating the drug from the bloodstream before it can have an effect. Alternatively, the target tissue may change its sensitivity to the drug—a phenomenon called **functional tolerance**. One important way in which a cell develops functional tolerance is by changing how many receptors it has on its surface. So, for example, after repeated doses of an *agonist* drug, neurons may **down-regulate** their receptors (decrease the number of available receptors to which the drug can bind), thereby becoming less sensitive and countering the drug effect. If the drug is an *antagonist*, target neurons may instead **up-regulate** (increase the number of receptors).

Tolerance to a particular drug often generalizes to other drugs of the same chemical class; this effect is termed **cross-tolerance**. For example, people who have developed tolerance to heroin tend to exhibit a degree of tolerance to all the other drugs in the opiate category, including codeine, morphine, and methadone. This is because all those drugs act on the same family of receptors.

how's it going ❓

1. Briefly explain agonist and antagonist actions of a ligand, with respect to effects on receptors.
2. What are receptor subtypes? What is their significance for drug development?
3. Briefly explain how dose-response curves are calculated, and why they are useful to pharmacologists. Distinguish between a drug's binding affinity and its efficacy.
4. Provide a review of the different ways in which drugs can be administered, in particular noting some considerations that are taken into account when deciding on a route of administration. How does repeated exposure to a drug alter its effects? (*Hint*: Use the words *tolerance* and *regulate* in your answer.)

Drugs Affect Each Stage of Neural Conduction and Synaptic Transmission

As the saying goes, it takes two to tango. Synaptic transmission involves a complicated choreography of the two participating neurons, and drugs that affect the brain and behavior may act on either side of the synapse. Let's consider these two sites of action in turn.

Some drugs alter presynaptic processes

One of the ways that a drug may change synaptic transmission is by altering the presynaptic neuron, changing the system that converts an electrical signal (an action potential) into a chemical signal (secretion of neurotransmitter). As **FIGURE 4.7** illustrates, the most common presynaptic drug effects can be grouped into three main categories: effects on transmitter *production*, effects on transmitter *release*, and effects on transmitter *clearance*.

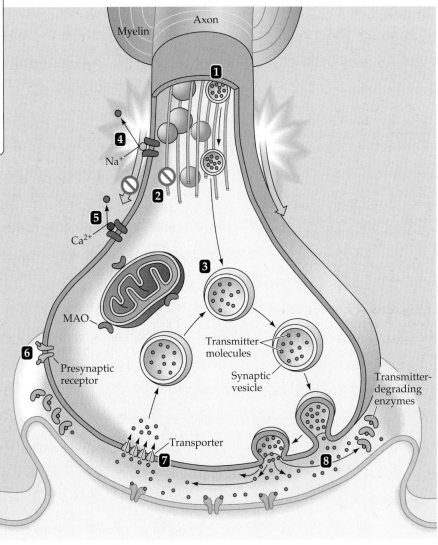

A Effects on Transmitter Production

1 **Inhibition of transmitter synthesis**
Para-chlorophenylalanine inhibits tryptophan hydroxylase, preventing synthesis of serotonin from its metabolic precursor.

2 **Blockade of axonal transport**
Colchicine impairs maintenance of microtubules and blocks axonal transport.

3 **Interference with the storage of transmitters**
Reserpine blocks the packaging of transmitter molecules within vesicles, thereby allowing the transmitter to be broken down by enzymes.

B Effects on Transmitter Release

4 **Prevention of synaptic transmission**
Tetrodotoxin, found in puffer fish, blocks voltage-gated Na⁺ channels and prevents nerve conduction.

5 **Alteration of synaptic transmitter release**
Calcium channel blockers (e.g., verapamil) inhibit release of transmitters. Amphetamine stimulates release of catecholamine transmitters. Black widow spider venom causes overrelease, and thus depletion, of ACh.

6 **Modulation of transmitter release by presynaptic receptors**
Caffeine competes with adenosine for presynaptic receptors, and thus prevents adenosine from inhibiting transmitter release.

C Effects on Transmitter Clearance

7 **Inactivation of transmitter reuptake**
Cocaine and amphetamine inhibit reuptake mechanisms, thus prolonging synaptic activity. The antidepressant drug Prozac is a selective serotonin reuptake inhibitor (SSRI).

8 **Blockade of transmitter degradation**
Some drugs (e.g., monoamine oxidase inhibitors) inhibit enzymes that normally break down neurotransmitter molecules in the axon terminal or in the synaptic cleft. As a result, transmitter remains active longer and to greater effect.

FIGURE 4.7 Drug Effects on Presynaptic Mechanisms, with Examples

TRANSMITTER PRODUCTION   In order for the presynaptic neuron to produce neurotransmitter, a steady supply of raw materials and enzymes must arrive at the axon terminals and carry out the needed reactions. Drugs are available that alter this process in various ways (**FIGURE 4.7A**). For example, drugs may inhibit enzymes that synthesize a transmitter, resulting in depletion of the transmitter. Alternatively, drugs that block axonal transport prevent raw materials from reaching the axon terminals in the first place, which could also cause the presynaptic terminal to run out of neurotransmitter. In both cases, affected presynaptic neurons are prevented from having their usual effects on postsynaptic neurons, with sometimes profound effects on behavior. A third class of drug (e.g., reserpine) doesn't prevent the *production* of transmitter but instead interferes with the cell's ability to *store* the transmitter in synaptic vesicles for later release. The effect on behavior may be complicated, depending on how much transmitter is able to reach the postsynaptic cell.

TRANSMITTER RELEASE   As we saw in Chapter 3, transmitter is released when action potentials arrive at the axon terminal and trigger an inflow of calcium ions. But a number of drugs and toxins can block those action potentials from ever arriving. For example, compounds that block sodium channels (like the toxin that makes puffer fish a dangerous delicacy, called *tetrodotoxin*) prevent axons from firing action potentials,

autoreceptor A receptor for a synaptic transmitter that is located in the presynaptic membrane and tells the axon terminal how much transmitter has been released.

shutting down synaptic transmission with deadly results. And drugs called *calcium channel blockers* do exactly as their name suggests, blocking the calcium influx that normally drives the release of transmitter into the synapse (**FIGURE 4.7B**). The active ingredient in Botox—botulinum toxin—specifically blocks ACh release from axon terminals near the injection site. The resulting local paralysis of underlying muscles reduces wrinkling of the overlying skin, but it may also interfere with the ability to produce normal facial expressions.

A different way to alter transmitter release is to modify the systems that the neuron normally uses to monitor and regulate its own transmitter release. For example, presynaptic neurons often use **autoreceptors** to monitor how much transmitter they have released; it's a kind of feedback system. Drugs that stimulate these receptors provide a false feedback signal, prompting the presynaptic cell to release less transmitter. Drugs that instead *block* autoreceptors prevent the presynaptic neuron from receiving its normal feedback, tricking the cell into releasing more transmitter than usual.

TRANSMITTER CLEARANCE   After action potentials have arrived at the axon terminals and prompted a release of transmitter substance, the transmitter is rapidly cleared from the synapse by several processes. Obviously, getting rid of the used transmitter is an important step, because until it is gone, new releases of transmitter from the presynaptic side won't be able to have much extra effect. However, researchers think that under certain circumstances, neurons may be *too* good at clearing the used transmitter and that a significant lack of transmitter in certain synapses may contribute to disorders such as depression. As we'll see shortly, some important psychiatric drugs, called *reuptake inhibitors*, work by blocking the presynaptic system that normally reabsorbs transmitter molecules after their release; this blocking action allows transmitter molecules to accumulate in the synaptic cleft and have a bigger effect on the postsynaptic cell. Other drugs achieve a similar result by blocking the enzymes that normally break up molecules of neurotransmitter into inactive metabolites, again allowing the transmitter to accumulate, having a greater effect on the postsynaptic cell (**FIGURE 4.7C**).

Some drugs alter postsynaptic processes

An alternate way for drugs to change synaptic transmission is by altering the postsynaptic systems that respond to the released neurotransmitter. This change may be accomplished either by direct actions on the transmitter receptors of the postsynaptic membrane, or by indirect tinkering with other cellular processes within the postsynaptic cell, as illustrated in **FIGURE 4.8**.

TRANSMITTER RECEPTOR–SELECTIVE DRUGS   As we discussed earlier in the chapter, selective receptor antagonists bind directly to postsynaptic receptors and block them from being activated by their neurotransmitter (**FIGURE 4.8A**). The results may be immediate and dramatic. Curare, for example, blocks the nicotinic ACh receptors found on muscles, resulting in immediate paralysis of all skeletal muscles, including those used for breathing (which is why curare is an effective arrow poison).

Selective receptor *agonists* bind to specific receptors and activate them, mimicking the natural neurotransmitter at those receptors. These drugs are often very potent, with effects that vary depending on the particular types of receptors activated. LSD is an example, producing bizarre visual experiences through strong stimulation of a serotonin receptor subtype found in visual cortex.

PROCESSES INSIDE POSTSYNAPTIC NEURONS   When they bind to their matching receptors on postsynaptic membranes, neurotransmitters can stimulate a variety of changes within the postsynaptic cell, such as the activation of second messengers, the activation of genes, and the production of various proteins. These intracellular processes present additional targets for drug action (**FIGURE 4.8B**). For example, some drugs induce the postsynaptic cell to up-regulate its receptors, thus changing the sensitivity of the synapse. Other drugs cause a down-regulation in receptor density. Some drugs, like lithium

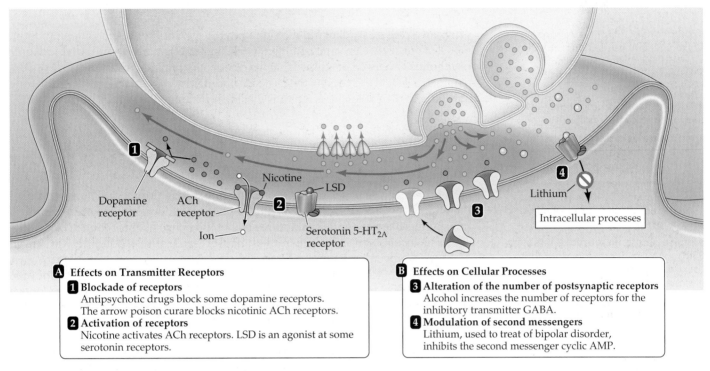

A Effects on Transmitter Receptors
1 Blockade of receptors
Antipsychotic drugs block some dopamine receptors.
The arrow poison curare blocks nicotinic ACh receptors.
2 Activation of receptors
Nicotine activates ACh receptors. LSD is an agonist at some
serotonin receptors.

B Effects on Cellular Processes
3 Alteration of the number of postsynaptic receptors
Alcohol increases the number of receptors for the
inhibitory transmitter GABA.
4 Modulation of second messengers
Lithium, used to treat of bipolar disorder,
inhibits the second messenger cyclic AMP.

FIGURE 4.8 Drug Effects on Postsynaptic Mechanisms

chloride, directly alter second-messenger systems, with widespread effects in the brain. Future research will probably focus on drugs to selectively activate, alter, or block targeted genes within the DNA of neurons. These *genomic* effects could produce profound long-term changes in the structure and function of neurons.

Drugs That Affect the Brain Can Be Divided into Functional Classes

Classifying drugs on the basis of their cellular actions is useful for understanding the physiology of drug effects, but many people are more concerned with the behavioral and therapeutic aspects of these compounds. In the sections that follow, we will briefly review some of the major categories of psychoactive drugs, based on how they affect behavior.

Psychoactive drugs may relieve severe symptoms

Mental disorders have bedeviled people throughout history. No doubt many ancient beliefs about miraculous visions, possession, sorcery, and so on had their origins in the symptoms of severe mental disturbances (the topic of Chapter 12). Sadly, because of the lack of effective treatments and the strong stigma attached to mental disorders, the tendency was to separate the afflicted from the rest of society, locking them away in dismal institutions. But during the mid-twentieth century the development of potent drug therapies revolutionized psychiatry.

The Antipsychotic Revolution The introduction of antipsychotic drugs relieved the suffering of millions of patients who had previously required hospitalization in psychiatric institutions like this one. Antipsychotics dramatically curb the striking hallucinations and delusions that are symptomatic of schizophrenia.

ANTIPSYCHOTICS   Prior to 1950, almost half of all hospital beds in the United States were occupied by psychiatric patients (Menninger, 1948), and, owing to its debilitating nature, a high proportion of these were people suffering from the delusions and hallucinations of schizophrenia. This grim situation was suddenly and dramatically improved by the development of a family of drugs called **antipsychotics** (or *neuroleptics*). The first of these drugs, chlorpromazine (Thorazine), and successors like haloperidol (Haldol) and loxapine (Loxitane), all share one crucial feature: they act as selective antagonists of dopamine D_2 receptors in the brain. These

antipsychotics   Also called *neuroleptics*. Any of a class of antipsychotic drugs that alleviate symptoms of schizophrenia, typically by blocking dopamine receptors.

atypical antipsychotic An antipsychotic drug that has actions other than or in addition to the dopamine D_2 receptor antagonism that characterizes the typical antipsychotics.

antidepressant A drug that relieves the symptoms of depression.

monoamine oxidase (MAO) An enzyme that breaks down monoamine transmitters, thereby inactivating them.

tricyclic antidepressant An antidepressant that acts by increasing the synaptic accumulation of serotonin and norepinephrine.

selective serotonin reuptake inhibitor (SSRI) An antidepressant drug that blocks the reuptake of transmitter at serotonergic synapses.

depressant A drug that reduces the excitability of neurons.

barbiturate An early anxiolytic drug and sleep aid that has depressant activity in the nervous system.

anxiolytic A drug that is used to combat anxiety.

benzodiazepine Any of a class of antianxiety drug that are agonists of $GABA_A$ receptors in the central nervous system. One example is diazepam (Valium).

opium An extract of the opium poppy, *Papaver somniferum*. Drugs based on opium are potent painkillers.

morphine An opiate compound derived from the poppy flower.

The Source of Opium and Morphine The opium poppy has a distinctive flower and seedpod. The bitter flavor and brain actions of opium may provide the poppy plant a defense against being eaten.

drugs are so good at relieving the symptoms of schizophrenia that a dopaminergic model of the disease became dominant (see Chapter 12). More recently, **atypical antipsychotics** have been developed that have both dopaminergic and additional, nondopaminergic actions, especially the blockade of certain serotonin receptors. These drugs may be helpful in relieving symptoms that are resistant to the typical antipsychotics.

ANTIDEPRESSANTS   Disturbances of mood called *affective disorders* are among the most common of all psychiatric complaints (World Health Organization, 2001). In contrast to the antipsychotic drugs, which reduce synaptic activity by blocking receptors, effective **antidepressant** drugs act to *increase* synaptic transmission. Some of the earliest antidepressants were the **monoamine oxidase (MAO)** inhibitors, which, as their name suggests, block the enzyme responsible for breaking down monoamine transmitters like dopamine, serotonin, and norepinephrine. This action allows transmitter molecules to accumulate in the synapses (see Figure 4.7, step 8), with an associated improvement in mood. Later generations of antidepressants also increase synaptic transmitter availability, but they focus on specific transmitters: the **tricyclic antidepressants** block the reuptake of serotonin and norepinephrine, while **selective serotonin reuptake inhibitors (SSRIs)** like fluoxetine (Prozac) and citalopram (Celexa) are so named because they act specifically at serotonergic synapses.

ANXIOLYTICS   Severe anxiety, in the form of panic attacks, phobias (specific irrational fears), and generalized anxiety, can spiral out of control and become disabling; many millions of people suffer from anxiety disorders (see Chapter 12). Anything that reduces or *depresses* the excitability of neurons tends to counter these states, which explains some of the historical popularity of **depressants** like alcohol and opium. Unfortunately, these substances are burdened with strong potential for intoxication and addiction, so they are not suitable for therapeutic use. **Barbiturate** drugs, such as phenobarbital, were originally developed to reduce anxiety, promote sleep, and avoid epileptic seizures. They are still used occasionally for those purposes, but they are also addictive and easy to overdose on, often fatally, as illustrated in Figure 4.6B.

Over the last 50 years, the most heavily prescribed **anxiolytics** (antianxiety drugs) have been the **benzodiazepines**; compared with the barbiturates, benzodiazepines are both more specific and safer (as illustrated for Ativan [lorazepam] in Figure 4.6B), although they still carry some risk of addiction. Members of this class of drug, such as Valium (diazepam) and Ativan, bind to specific sites on $GABA_A$ receptors and enhance the activity of GABA (Walters et al., 2000). Because $GABA_A$ receptors are inhibitory, benzodiazepines help GABA to produce larger inhibitory postsynaptic potentials than GABA would produce alone. The net effect is a reduction in the excitability of neurons. The hunt for new antianxiety agents—both exogenous and endogenous—is an area of intense research efforts. Hormones that interact with GABA receptors, as well as drugs that subtly alter serotonergic neurotransmission, are examples of these novel anxiolytics. Antidepressant drugs are often effective anxiolytics too.

OPIATES   **Opium**, extracted from poppy flower seedpods, has been used by humans since at least the Stone Age. **Morphine**, the major active substance in opium, is a very effective **analgesic** (painkiller) that has brought relief from severe pain to many millions of people (see Chapter 5). Unfortunately, because it produces powerful feelings of euphoria, morphine also has a strong potential for addiction, as do close relatives like **heroin** (diacetylmorphine) and opiate painkillers like OxyContin (oxycodone) and Vicodin (hydrocodone and acetaminophen).

Opiates like morphine, heroin, and codeine bind to specific receptors—**opioid receptors**—that are concentrated in various regions of the brain. The profusion of opioid receptors in an area called the **periaqueductal gray** (**FIGURE 4.9**) is especially important, because it is here that opiates exert their painkilling effects (see Chapter 5).

As we mentioned earlier in the chapter, we now know that the brain makes its own morphinelike compounds, called **endogenous opioids**. Researchers have identified three major families of these potent peptides: the **enkephalins**, from the Greek *en*, "in," and *kephale*, "head" (Hughes et al., 1975); the **endorphins**, a contraction of *endogenous morphine*; and the **dynorphins**, short for *dynamic endorphins*, in recognition of their potency and speed of action (see Table 4.1). There are also three main kinds of opioid receptors—delta (δ), kappa (κ), and mu (μ)—all of which are metabotropic receptors (see Table 4.2). Powerful drugs that block opioid receptors—naloxone (Narcan) is an example—can rapidly reverse the effects of opiates and rescue people from overdose. Opiate antagonists also block the rewarding aspects of drugs like heroin, so they can be helpful for treating addiction, as we discuss at the end of this chapter.

Psychoactive drugs can affect conscious experience

Whether to experience pleasurable sensations, to artificially increase vigor and wakefulness, or simply to satisfy curiosity, people have a long history of chemically altering their perceptions of themselves and the world. Unfortunately, the habitual use of drugs to alter consciousness is often costly to the user and to society. Governments attempt to minimize these costs by controlling (or preventing) the production and distribution of designated drugs, but the division of drugs into licit and illicit categories is largely a matter of historical accident. Some classes of drugs—the opiates, for example—span both categories, being both useful medicines and harmful drugs of abuse. And some substances, like tobacco, are legal only because they have been cultivated for centuries and are backed by powerful economic interests. In terms of illness, death, lost productivity, and sheer human misery, some of the legal drugs may be the worst offenders.

TOBACCO Tobacco is native to the Americas, where European explorers first encountered smoking; these explorers brought tobacco back to Europe with them. Tobacco use became much more widespread following technological innovations that made it easier to smoke, in the form of cigarettes (W. Bennett, 1983). Delivered to the large surface of the lungs, the **nicotine** from cigarettes enters the blood and brain much more rapidly than does nicotine from other tobacco products. Nicotine acts as a **stimulant**, increasing heart rate, blood pressure, digestive action, and alertness. In the short run, these effects make tobacco use pleasurable. But these alterations of body function, quite apart from the effects of tobacco tar on the lungs, make prolonged tobacco use very unhealthful. Smoking and nicotine exposure in adolescence has a lasting impact on attention and cognitive development, likely as a consequence of impairments of glutamatergic synapses in the prefrontal cortex (Counotte et al., 2011).

The *nicotinic* ACh receptors didn't get their name by coincidence; it is through these receptors that the nicotine from tobacco exerts its effects in the body. Nicotinic receptors drive the contraction of skeletal muscles, and the activation of various visceral organs, but they are also found in high concentrations in the brain, including the cortex. This is one way in which nicotine enhances some aspects of cognitive performance. Nicotine also stimulates the ventral tegmental area to exert its rewarding/addicting effects (Maskos et al., 2005). (We will discuss the ventral tegmental area in more detail when we discuss positive reward models later in the chapter.)

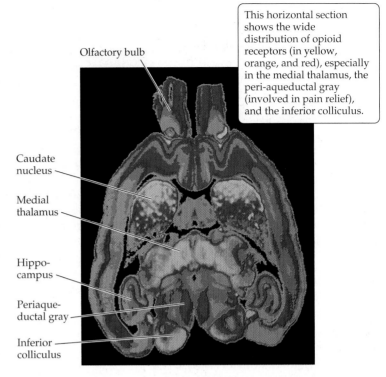

This horizontal section shows the wide distribution of opioid receptors (in yellow, orange, and red), especially in the medial thalamus, the peri-aqueductal gray (involved in pain relief), and the inferior colliculus.

Olfactory bulb

Caudate nucleus

Medial thalamus

Hippocampus

Periaqueductal gray

Inferior colliculus

FIGURE 4.9 The Distribution of Opioid Receptors in the Rat Brain (Courtesy of Miles Herkenham, National Institute of Mental Health.)

analgesic Having painkilling properties.

heroin Diacetylmorphine, an artificially modified, very potent form of morphine.

opioid receptor A receptor that responds to endogenous opioids and/or exogenous opiates.

periaqueductal gray A midbrain region involved in pain perception.

endogenous opioid Any of a class of opium-like peptide transmitters that have been referred to as the body's own narcotics. The three kinds are enkephalins, endorphins, and dynorphins.

enkephalin One of the three kinds of endogenous opioids.

endorphin One of the three kinds of endogenous opioids.

dynorphin One of the three kinds of endogenous opioids.

nicotine A compound found in plants, including tobacco, that acts as an agonist on a large class of cholinergic receptors.

stimulant A drug that enhances the excitability of neurons.

FIGURE 4.10 Abnormal Brain Development in Fetal Alcohol Syndrome

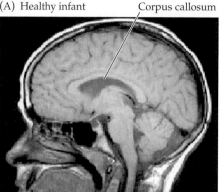

(A) Healthy infant   Corpus callosum   (B) Infant with fetal alcohol syndrome

Compared to MRI images of a healthy infant's brain…

…the MRI of an infant affected by fetal alcohol syndrome—caused by heavy consumption of alcohol by the mother during pregnancy—shows extensive abnormality, including reduced gray matter, complete absence of the corpus callosum, abnormal organization of the brain, and characteristic deformities of the head and face.

fetal alcohol syndrome   A disorder, including intellectual disability and characteristic facial abnormalities, that affects children exposed to too much alcohol (through maternal ingestion) during fetal development.

delta-9-tetrahydrocannabinol (THC)   The major active ingredient in marijuana.

Relaxation   Historically, the production and distribution of marijuana have been illegal activities, but in many jurisdictions these laws are currently being relaxed. Legal access, via licensed marijuana shops like this one, is expected to reduce criminal activity associated with black market distribution, and will benefit users who find that marijuana has medical benefits. However, possible health risks remain a concern for adolescents and long-term heavy users.

ALCOHOL   The most widely consumed psychoactive drug, alcohol, is easily produced by the fermentation of fruit or grains. Taken in moderation (perhaps one or two drinks per day at most), alcohol is harmless or even beneficial to the health of adults (Leroi et al., 2002; Mukamal et al., 2003), but *excessive* alcohol consumption is very damaging and linked to more than 60 disease processes.

Alcohol has a biphasic effect on the nervous system: at first it acts as a stimulant, and then it has a more prolonged depressant phase. (Remember, the word *depressant* relates to a depression or inhibition of *neural activity*, not an effect on mood.) Like the anxiety-reducing benzodiazepines we discussed earlier, alcohol acts via GABA receptors to reduce postsynaptic excitation, resulting in the social disinhibition, poor motor control, and sensory disturbances that we call *drunkenness*. Alcohol additionally activates dopamine-mediated reward systems of the brain, accounting for some of the pleasurable aspects of drinking.

Chronic abuse of alcohol damages or destroys nerve cells in many regions of the brain. Alcohol abuse by expectant mothers can cause grievous permanent damage to the developing fetus, termed **fetal alcohol syndrome**, which is characterized by facial deformities and stunted brain growth, sometimes including the absence of the corpus callosum that normally connects the two hemispheres of the brain (**FIGURE 4.10**). In adults, the frontal lobes are especially affected by chronic alcohol use (Kril et al., 1997). Happily, some of the anatomical changes associated with chronic alcoholism may be reversible with abstinence. In humans suffering from alcoholism, MRI studies show an increase in the volume of cortical gray matter and an associated reduction in ventricular volume within weeks of giving up alcohol (**FIGURE 4.11**) (Pfefferbaum et al., 1995).

Even in the absence of clear-cut alcoholism, periodic binge drinking (generally, five or more drinks on a single occasion) may cause brain damage. After only 4 days of bingeing on alcohol, rats exhibit neural degeneration in several brain regions. Alcohol bingeing also significantly reduces the rate of neurogenesis—the formation of new neurons—in the adult hippocampus (Nixon and Crews, 2002).

MARIJUANA   Smoking marijuana (or *cannabis*) or related preparations, such as hashish, brings a rush of dozens of active ingredients into the bloodstream and brain. Chief among them is the compound **delta-9-tetrahydrocannabinol** (**THC**) (Gaoni and Mechoulam, 1964). Marijuana use usually produces pleasant relaxation and mood

(A)

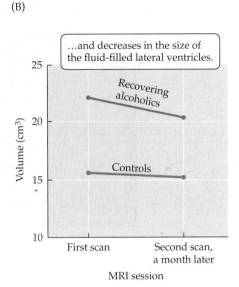

MRIs of recovering alcoholics show that after they've abstained from alcohol for 30 days, there are pronounced increases in cortical gray matter...

(B)

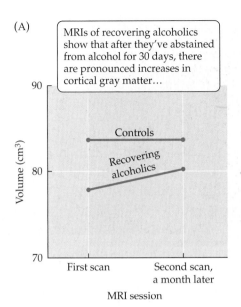

...and decreases in the size of the fluid-filled lateral ventricles.

FIGURE 4.11 The Effects of Alcoholism on the Structure of the Brain (After Pfefferbaum et al., 1995.)

alteration, although the drug can occasionally cause stimulation and paranoia instead. Casual use of marijuana seems to be mostly harmless, but as with other substances, heavy use can be harmful. For example, heavy use may be associated with respiratory problems, addiction, cognitive decline, and psychiatric disorders (Meier et al., 2012). A particular concern is that adolescent use of marijuana may increase the risk of later developing schizophrenia (M. J. Smith et al., 2014), although it remains to be determined whether the marijuana use *causes* the illness, or conversely whether adolescents who are prepsychotic are more drawn to marijuana use.

As with opiates and benzodiazepines, researchers found that the brain contains specific cannabinoid receptors that mediate the effects of compounds like THC. Cannabinoid receptors are found in the substantia nigra, the hippocampus, the cerebellar cortex, and the cerebral cortex (**FIGURE 4.12**) (Devane et al., 1988). Later research revealed that the brain makes several THC-like endogenous ligands for these receptors. The most studied of these **endocannabinoids** is **anandamide** (from the Sanskrit *ananda*, "bliss") (Devane et al., 1992), which produces some of the most familiar physiological and psychological effects of marijuana use, such as mood improvement, pain relief, lowered blood pressure, relief from nausea, improvements in the eye disease glaucoma, and so on. Cannabinoids are thus targets of an intense research effort aimed at developing drugs with some of the specific beneficial effects of marijuana. The documented use of marijuana for recreational and medicinal purposes spans over 6,000 years, but from the early twentieth century until recently it has remained illegal in most jurisdictions. Several U.S. states have now legalized the sale and use of marijuana for recreational purposes, and medical marijuana use is now "decriminalized" in many U.S. states, Canada, and other countries. It seems likely that the relaxation of marijuana laws will continue.

STIMULANTS Proper functioning of the nervous system involves a fine balance between excitatory and inhibitory influences. Stimulants are drugs that tip the balance toward the excitatory side; they therefore have an alerting, activating effect. Some stimulants act directly by increasing excitatory synaptic potentials. Others act by blocking normal inhibitory processes. For example, the adenosine

endocannabinoid An endogenous ligand of cannabinoid receptors, thus an analog of marijuana that is produced by the brain.

anandamide An endogenous substance that binds the cannabinoid receptor molecule.

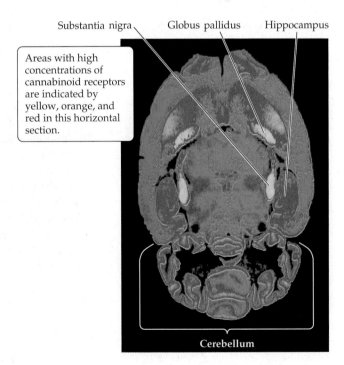

Areas with high concentrations of cannabinoid receptors are indicated by yellow, orange, and red in this horizontal section.

Substantia nigra   Globus pallidus   Hippocampus

Cerebellum

FIGURE 4.12 The Distribution of Cannabinoid Receptors in the Rat Brain (Courtesy of Miles Herkenham, National Institute of Mental Health.)

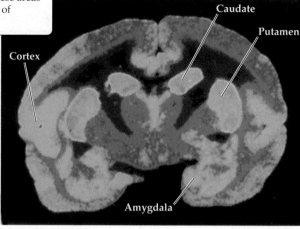

In this coronal image, brain regions with high degrees of cocaine binding are shown in orange and yellow. Cocaine acts in these areas to cause an accumulation of monoamine transmitters.

FIGURE 4.13 Cocaine Action in the Monkey Brain (Courtesy of Bertha K. Madras.)

cocaine A drug of abuse, derived from the coca plant, that acts by enhancing catecholamine neurotransmission.

amphetamine A molecule that resembles the structure of the catecholamine transmitters and enhances their activity.

receptors that are found on some presynaptic terminals are activated by adenosine that is coreleased with the neuron's transmitter; this prompts the neuron to reduce its release of transmitter. By blocking these adenosine receptors, caffeine prevents this inhibitory effect of adenosine, resulting in more transmitter release and, consequently, increased brain activity.

Some stimulants have a much more powerful effect than caffeine or tobacco, and they are among the most addictive drugs. Leaves of the South American coca shrub—either chewed or brewed as a tea—increase endurance, alleviate hunger, and promote a sense of well-being without causing many problems. But processing and concentrating an extract from this plant produces a much more potent and dangerous compound: **cocaine**. Although cocaine was initially used as a food additive (e.g., in *Coca*-Cola) and as a local anesthetic, people soon discovered that the rapid hit of cocaine resulting from snorting or smoking it (see Table 4.3) has a stimulant effect that is powerful and pleasurable. Unfortunately, it is also highly addictive. Furthermore, heavy cocaine use raises the risk of serious side effects like stroke, psychosis, loss of gray matter, and severe mood disturbances (Franklin et al., 2002). Cocaine acts by blocking the reuptake of monoamine transmitters from synapses. This action causes monoamines to accumulate in synapses throughout much of the brain (**FIGURE 4.13**), therefore boosting their effects.

The synthetic stimulant **amphetamine** (and its relatives, like methamphetamine, or "meth") has a mode of action that superficially resembles that of cocaine, inducing an accumulation of the synaptic monoamine transmitters norepinephrine and dopamine. However, the mechanics of amphetamine's actions, involving two steps, are quite different from those of cocaine. First, amphetamine acts within axon terminals to cause a larger-than-normal release of neurotransmitter when the synapse is activated. Second, amphetamine then interferes with the clearance of the released transmitter by blocking its reuptake. The result is that monoaminergic synapses become unnaturally potent, having strong effects on behavior.

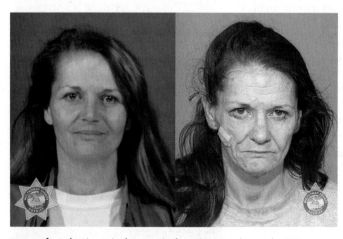

Faces of Meth These before and after photos, taken only 2½ years apart, are a stark testament to the heavy toll that accompanies chronic methamphetamine abuse. Meth causes multiple severe problems such as motor disorders, cognitive impairment, psychosis, and rapid changes in appearance due to accelerated tooth decay ("meth mouth"), skin pathology, and excessive weight loss. (Mug shots courtesy of the Multnomah County Sheriff's Office and the Faces of Meth™ program.)

Over the short term, amphetamine causes increased vigor and stamina, wakefulness, decreased appetite, and feelings of euphoria. For these reasons, amphetamine has historically been used in military applications and other settings where intense sustained effort is required. However, the quality of the work being performed may suffer, and the costs of amphetamine use soon outweigh the benefits. Addiction and tolerance to amphetamine and methamphetamine develop rapidly, requiring

These images are portraits produced by a professional artist just after taking LSD (leftmost drawing) and then at three successive time points as the drug took effect. The model for all four drawings is the same man (the researcher, in fact).

FIGURE 4.14 Changes in Visual Perception after Taking LSD

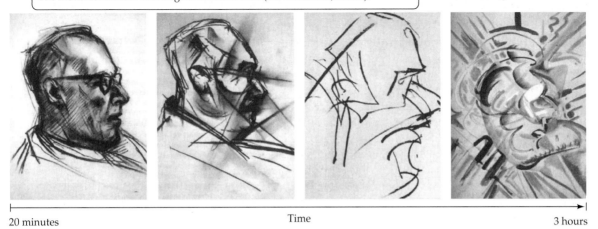

20 minutes   Time   3 hours

ever-larger doses that lead to sleeplessness, severe weight loss, and general deterioration of mental and physical condition.

Prolonged use of amphetamine or methamphetamine may lead to symptoms that closely resemble those of schizophrenia: compulsive, agitated behavior and irrational suspiciousness. In their pursuit of more drug, users may neglect their diet and basic hygiene, aging rapidly. Users also experience a variety of peripheral effects, like high blood pressure, tremor, dizziness, sweating, rapid breathing, and nausea. And worst of all, people who chronically abuse amphetamine often display symptoms of brain damage—such as memory and attention deficits, psychosis, and symptoms of Parkinson's disease—long after they quit using the drug (Ernst et al., 2000). As we'll discuss a little later in the chapter, increased activation of the mesolimbocortical dopaminergic reward system of the brain appears to be crucial for the rewarding aspects of drug use.

HALLUCINOGENS Humans have long prized **hallucinogens**, substances that produce powerful sensory alterations, often believing the resultant experiences to have deep spiritual or psychological meaning. Dozens of such hallucinogens are found in nature—psilocybin and muscarine (from "magic" mushrooms), mescaline (from the peyote plant), and bufotenine (from toads) are a few examples. But the term *hallucinogen* is really a misnomer because, whereas a hallucination is a novel perception that takes place in the absence of sensory stimulation (hearing voices, or seeing something that isn't there), the drugs in this category mostly alter or distort *existing* perceptions (mainly visual in nature). Users may see fantastic images (**FIGURE 4.14**), often with intense colors, but often they are aware that these strangely altered perceptions are not real events.

Hallucinogenic agents are diverse in their neural actions. Whereas muscarine affects the ACh system, mescaline acts via noradrenergic and serotonergic systems. The herb *Salvia divinorum* is unusual among hallucinogens because it acts on the opioid kappa receptor. But research with LSD and related drugs suggests that perhaps the most important shared neural action of hallucinogens is the stimulation of serotonin receptors. Discovered by Albert Hofmann in the 1940s, **LSD** (lysergic acid diethylamide, or more simply *acid*) structurally resembles serotonin. Even in tiny doses, LSD strongly activates serotonin 5-HT$_{2A}$ receptors that are found in especially heavy concentrations in the visual cortex. Other hallucinogens, such as mescaline and psilocybin, share this action. Research with psilocybin has also demonstrated disinhibition of emotion-processing regions in the limbic system (Carhart-Harris et al., 2012), perhaps accounting for some of the drug's emotional, mystical qualities. In addition to its impressive effects on visual perception, LSD produces mood changes and feelings of creativity that have led to resurgent interest in the possible application of hallucinogens to treat various psychiatric disorders, including depression and obsessive-compulsive disorder (Moreno et al., 2006).

hallucinogen A drug that alters sensory perception and produces peculiar experiences.

LSD Also called *acid*. Lysergic acid diethylamide, a hallucinogenic drug.

The Father of LSD Albert Hofmann discovered LSD by accidentally taking some in 1943, and he devoted the rest of his career to studying it. A prohibited drug in most jurisdictions, LSD is sometimes distributed on colorful blotter paper. This example, picturing Hofmann and the LSD molecule, is made up of 1,036 individual doses, or "hits."

MDMA Also called *Ecstasy*. 3,4-Methylenedioxymethamphetamine, a drug of abuse.

Like LSD, the hallucinogenic amphetamine derivative **MDMA** (3,4-methylenedioxymethamphetamine, or *Ecstasy*) stimulates visual cortical 5-HT$_{2A}$ receptors, but it also changes the levels of dopamine and certain hormones, such as prolactin. Exactly how these activities account for the subjective effects of MDMA—positive emotions, empathy, euphoria, a sense of well-being, and colorful visual phenomena—remains uncertain.

The major hallucinogens seem to have low addiction potential, and LSD has relatively few negative side effects (although some users report long-lasting visual changes). Long-term Ecstasy use may cause problems with mood and cognitive performance (Sumnall and Cole, 2005; Parrott, 2013) and long-lasting changes in patterns of brain activation, even at low doses (de Win et al., 2008). However, short-term MDMA treatment is also being investigated as a possible treatment for persistent post-traumatic stress disorder (Mithoefer et al., 2013).

---how's it going ❓

1. Compare and contrast the three major categories of presynaptic effects of psychoactive drugs. Give examples of each kind of action. (*Hint*: The words *production*, *release*, and *clearance* will be important for your discussion.)

2. Compare and contrast the main postsynaptic actions of psychotropic drugs, with examples. Be sure to distinguish between actions at receptors versus actions within the postsynaptic neuron.

3. At least four general categories of psychoactive drugs are used to relieve disorders. Describe these categories, atnd give some examples of each class of drugs. Be sure to discuss the mode of action of the drugs you cite.

4. Identify and discuss the major categories of drugs that people use to alter their consciousness. In what ways are the major categories similar, and in what ways do they differ? What are some of the threats to health that these compounds present?

Drug Abuse Is Pervasive

Substance abuse and addiction have become a social problem that afflicts many millions of people and disrupts the lives of their families, friends, and associates. Just one example reveals the extent of the problem: in the United States each year, more men and women die of smoking-related lung cancer than of colon, breast, and prostate cancers *combined*. In addition to the personal impact of so much illness and early death, there are dire social costs: huge expenses for medical and social services; millions of hours lost in the workplace; elevated rates of crime associated with illicit drugs; and scores of children who are damaged by their parents' substance abuse behavior, in the uterine environment as well as in the childhood home. Males are much more likely than females to abuse drugs and have their lives disrupted by drug use. It is unclear whether this sex difference is related to biological differences between the sexes, or to differences in social influences on males versus females.

For medical purposes, addiction is defined as "substance use disorder" (SUD) in the *Diagnostic and Statistical Manual of Mental Disorders*, Fifth Edition (*DSM-5*; American Psychiatric Association, 2013). SUD can take multiple forms, and it varies in severity from mild to severe. The *DSM-5* criteria for a diagnosis of SUD appear in **TABLE 4.4**. Almost everyone will meet some of the criteria some of the time; a diagnosis of SUD requires more-sustained problems and a pattern of use that interferes with normal daily functioning. In the discussion that follows, we will look at some of the most prevalent perspectives on addiction: what it is, where it comes from, how it can be treated. Some of these models stem from social forces; others are more deeply rooted in scientific observations and theories. But for any model of drug abuse, the challenge is to come up with a single account that can explain the addicting power

"The Needle and the Damage Done"
Addiction has a powerful grip, inducing people to go to sometimes extreme lengths to obtain larger and more frequent doses.

TABLE 4.4 *DSM-5* Criteria for Substance Use Disorder

A *mild* substance use disorder is diagnosed if 3 of the following criteria are met. People meeting 4 or 5 criteria are classified as having *moderate* substance use disorder, and *severe* substance use disorder is diagnosed in cases where 6 or more of the criteria are met.

1. Taking the substance in larger amounts or for longer than you meant to
2. Wanting to cut down or stop using the substance but not managing to
3. Spending a lot of time getting, using, or recovering from use of the substance
4. Cravings and urges to use the substance
5. Not managing to do what you should at work, home, or school because of substance use
6. Continuing to use, even when it causes problems in relationships
7. Giving up important social, occupational, or recreational activities because of substance use
8. Using the substance again and again, even when it puts you in danger
9. Continuing to use, even when you know you have a physical or psychological problem that could have been caused or made worse by the substance
10. Needing more of the substance to get the effect that you want (tolerance)
11. Development of withdrawal symptoms, which can be relieved by taking more of the substance

Source: American Psychiatric Association, 2013.

of substances as diverse as, for example, cocaine (a stimulant), heroin (an analgesic and euphoriant), and alcohol (largely a sedative). We will focus primarily on addiction to cocaine, the opiate drugs (such as morphine and heroin), nicotine, and alcohol because these substances have been studied the most thoroughly. According to the 2010 National Survey of Drug Use and Health, some 22.1 million people in the United States alone suffer from substance-related disorders (Substance Abuse and Mental Health Services Administration, 2011). Worldwide, the number is probably in the hundreds of millions.

Several perspectives help us understand drug abuse

Any comprehensive model of drug abuse has to answer several difficult questions: What social and environmental factors cause someone to start abusing a substance? What factors cause the person to continue abusing? What physiological mechanisms make a substance rewarding? What is addiction, physiologically and behaviorally, and why is it so hard to quit? Four major models attempt to answer at least some of these questions:

1. The *moral model* takes the perspective that addiction results from weakness of character and a lack of self-control. Proponents may apply exhortation, peer pressure, and/or religious intervention in order to curb abusive practices, but usually with limited success. The temperance movement that commenced in the early 1800s did seem to reduce alcohol consumption, but the evidence suggests that despite multibillion dollar expenditures, more-modern moral campaigns (such as the Just Say No and D.A.R.E. campaigns) have had little effect on rates of drug addiction (West and O'Neal, 2004; Vincus et al., 2010).

2. The *disease model* takes the view that the person who abuses drugs requires medical treatment rather than moral exhortation or punishment. The problem is that substance abuse is not like any other disease we know about. We usually reserve the term *disease* for cases involving a physical abnormality, and no such condition has been found in the case of drug addiction (although some people are genetically more susceptible to addiction than others). Nevertheless, this model continues to appeal to many, and much research is focused on looking for pathological states that create addiction after initial exposure to a drug.

A drug's rewarding properties and addictive potential are reflected in the number of lever presses performed to receive a dose. Lab animals will press the lever thousands of times to receive a single small dose of highly addictive compounds like cocaine and methamphetamine.

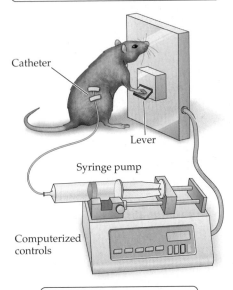

Catheter

Lever

Syringe pump

Computerized controls

A syringe pump is controlled by a computer, delivering a small dose of drug after a certain number of lever presses.

FIGURE 4.15 Experimental Setup for Drug Self-Administration

withdrawal symptom An uncomfortable symptom that arises when a person stops taking a drug that he or she has used frequently, especially at high doses.

dysphoria Unpleasant feelings; the opposite of euphoria.

nucleus accumbens A region of the forebrain that receives dopaminergic innervation from the ventral tegmental area, often associated with reward and pleasurable sensations.

insula A region of cortex lying below the surface, within the lateral sulcus, of the frontal, temporal, and parietal lobes.

3. The *physical dependence model* argues that people keep taking drugs in order to avoid unpleasant **withdrawal symptoms**. The specific withdrawal symptoms depend on the drug, but they are often the opposite of the effects produced by the drug itself. For example, withdrawal from morphine causes irritability, a racing heart, and waves of goose bumps (that's where the term *cold turkey* comes from—the skin looks like the skin of a plucked turkey). And of course, the opposite of the euphoria caused by many drugs is **dysphoria**: strongly negative feelings that can be relieved only by administration of the withdrawn drug. So the model does a good job of explaining why addicts will go to great lengths to obtain the drug they are addicted to, but it has an important shortcoming: the model is mute on how the addiction becomes established in the first place. Why do some people, but not all, start to abuse a drug before physical dependence (tolerance) has developed? And how is it that some people can become addicted to some drugs even in the absence of clear physical withdrawal symptoms? For example, cocaine withdrawal is not accompanied by the shaking and vomiting that are seen during heroin withdrawal, yet cocaine seems to be at least as addictive as heroin.

4. The *positive reward model* proposes that people get started with drug abuse, and become addicted, because the abused drug provides powerful reinforcement. Using an apparatus that allows animals to administer drugs to themselves (**FIGURE 4.15**), researchers have collected plenty of evidence in support of this view. Laboratory animals will quickly learn to press a lever repeatedly in order to receive a small dose of an addictive drug like cocaine or morphine (T. Thompson and Schuster, 1964; McKim, 1991). We can infer that the more lever presses animals will perform for a single dose, or the smaller the dose that will support the lever-pressing behavior, the more rewarding and addictive the drug must be. For example, it turns out that animals will happily self-administer doses of morphine that are so low that no signs of physical dependence ever develop (Schuster, 1970). Animals will also furiously press a lever to self-administer tiny doses of cocaine and other stimulants (Pickens and Thompson, 1968; Koob, 1995; Tanda et al., 2000). In fact, cocaine supports some of the highest rates of lever pressing ever recorded.

Experiments using drug self-administration suggest that, by itself, the physical dependence model is inadequate to explain drug addiction, although physical dependence and tolerance may contribute to drug hunger. The more comprehensive view of drug self-administration interprets it as a behavior controlled by a powerful pattern of positive and negative rewards (a variant of operant conditioning theory; see Chapter 13), without the need to implicate a disease process.

Many addictive drugs cause the release of dopamine in the **nucleus accumbens**, just as occurs with more conventional rewards, such as food, sex, and gambling (Di Chiara et al., 1999; Reuter et al., 2005; D'Ardenne et al., 2008). As we mentioned previously, dopamine released from axons originating in the ventral tegmental area (VTA), part of the mesolimbocortical dopaminergic pathway illustrated in Figure 4.4, has been widely implicated in the perception of reward (**FIGURE 4.16**). If the dopaminergic pathway from the VTA to the nucleus accumbens serves as a reward system for a wide variety of experiences, then the addictive power of drugs may come from their artificial stimulation of this pathway. When the drug hijacks this system, providing an unnaturally powerful reward, the user learns to associate the drug-taking behavior with that pleasure and begins seeking out drugs more and more until life's other pleasures fade into the background. If natural activities like conversation, food, and even sex no longer provide appreciable reward, addicts may seek drugs as the only source of pleasure available to them.

Intriguingly, people suffering damage to a brain region tucked within the frontal cortex, called the **insula** (Latin for "island"), are reportedly able to effortlessly quit smoking (Naqvi et al., 2007), indicating that this brain region is also involved

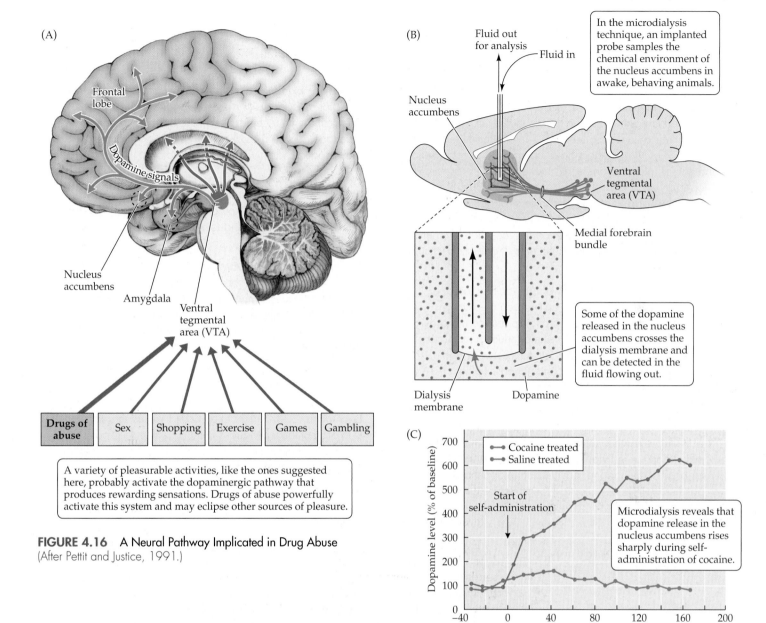

FIGURE 4.16 A Neural Pathway Implicated in Drug Abuse
(After Pettit and Justice, 1991.)

in addiction. The reciprocal connections between the VTA and the insula (Oades and Halliday, 1987) suggest that these two regions may normally interact to mediate addiction.

Not everyone who uses an addictive drug becomes addicted. For example, very, very few hospitalized patients treated with opiates for pain relief go on to abuse opiates after leaving the hospital (Brownlee and Schrof, 1997). However, prescription painkillers are highly effective at activating the dopamine reward system, so the use of these drugs outside of medical contexts carries a high risk of addiction. The individual and environmental factors that account for differential susceptibility are the subject of active investigation (Glantz and Pickens, 1992; Karch, 2006). Some of the major risk factors include biological factors (being male; heritable tendencies to addiction), poor family life, personality factors (poor emotional control), and environmental factors (living in a neighborhood with high rates of addiction). Simply returning to a neighborhood where drugs were previously used can trigger drug craving in an addict (Ciccocioppo et al., 2004).

SIGNS & SYMPTOMS

Medical Interventions for Substance Abuse

Some people can overcome their dependence on substances by themselves. For example, the great majority of ex-smokers, and about half of ex-alcoholics, appear to have quit on their own (S. Cohen et al., 1989; Institute of Medicine, 1990). Many others have benefited from counseling and social interventions such as the 12-step program developed by Alcoholics Anonymous in the 1930s. However, overcoming addiction may require stronger measures in some cases, especially for the most powerfully addictive substances. An intensive research effort has identified a variety of medicines that can help lessen the grip of addiction through the following strategies:

- *Lessening the discomfort of withdrawal.* Benzodiazepines and other sedatives, anti-nausea medications, and drugs that promote sleep all help reduce withdrawal symptoms. *Other medications* help reduce uncomfortable cravings for the abused substance; for example, acamprosate (trade name Campral) eases alcohol-associated withdrawal symptoms.

- *Providing an alternative to the addictive drug.* Agonist or partial agonist analogs of the addictive drug weakly activate the same mechanisms as the addictive drug, to help wean the individual. For example, the opioid receptor agonist methadone reduces heroin appetite; nicotine patches work in a similar fashion to reduce cravings for cigarettes.

- *Directly blocking the actions of the addictive drug.* Specific receptor antagonists can prevent an abused drug from interacting with receptors (e.g., the opiate antagonist naloxone [Narcan] blocks heroin's actions), but they also may produce harsh withdrawal symptoms.

- *Altering metabolism of the addictive drug.* Changing the breakdown of a drug can reduce or reverse its rewarding properties. Disulfiram (Antabuse) changes alcohol metabolism such that a nausea-inducing metabolite (acetaldehyde) accumulates.

- *Blocking the brain's reward circuitry.* When a person takes drugs (e.g., dopamine receptor blockers) that blunt the activity of the mesolimbocortical dopamine reward system, the addictive drugs lose their pleasurable qualities (but at the cost of a general loss of pleasurable feelings called *anhedonia*).

- *Immunization to render the drug ineffective.* Vaccines against such drugs as cocaine, heroin, and nicotine have been developed and are being tested (Maurer and Bachmann, 2007; Hicks et al., 2011). Here the strategy is to prompt the individual's immune system to produce antibodies that remove the targeted drugs from circulation before they ever reach the brain (**FIGURE 4.17**).

No single approach appears to be uniformly effective, and rates of relapse remain high. Research breakthroughs are therefore badly needed.

FIGURE 4.17 The Needle and the Damage Undone (After Hicks et al., 2011.)

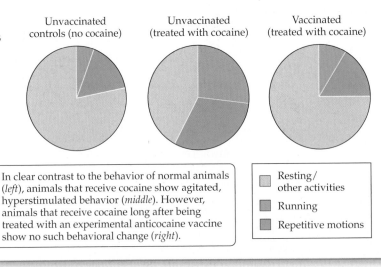

In clear contrast to the behavior of normal animals (*left*), animals that receive cocaine show agitated, hyperstimulated behavior (*middle*). However, animals that receive cocaine long after being treated with an experimental anticocaine vaccine show no such behavioral change (*right*).

───────────────────────────how's it going ❓

1. Define *substance abuse*. How prevalent is drug abuse in the population?

2. Summarize the major models of drug abuse and addiction, highlighting the strengths and shortcomings of each perspective.

3. Describe an experimental setup for measuring the rewarding properties of a drug.

4. Provide a survey of the anatomical system that mediates reward. What happens when this system is activated? What are some triggers that can activate the system, and how does activity of the reward system relate to drug addiction?

5. Provide a thorough overview of medical approaches and interventions in substance abuse.

Recommended Reading

Advokat, C. D., Comaty, J. E., and Julien, R. M. (2014). *Julien's Primer of Drug Action* (13th ed.). New York, NY: Worth.

Grilly, D. M., and Salamone, J. (2011). *Drugs, Brain and Behavior* (6th ed.). Boston, MA: Allyn & Bacon.

Karch, S. B., and Drummer, O. (2008). *Karch's Pathology of Drug Abuse* (4th ed.). Boca Raton, FL: CRC Press.

Meyer, J. S., and Quenzer, L. F. (2013). *Psychopharmacology: Drugs, the Brain, and Behavior* (2nd ed.). Sunderland, MA: Sinauer.

Nestler, E., Hyman, S., and Malenka, R. (2014). *Molecular Neuropharmacology* (3rd ed.). New York, NY: McGraw-Hill.

Schatzberg, A. F., and Nemeroff, C. B. (Eds.). (2013). *Essentials of Clinical Psychopharmacology*. Arlington, VA: American Psychiatric Publishing.

Thombs, D. L. (2006). *Introduction to Addictive Behaviors* (3rd ed.). New York, NY: Guilford Press.

2e.mindsmachine.com/vs4

VISUAL SUMMARY

4

You should be able to relate each summary to the adjacent illustration, including structures and processes. Go to the online version of this summary (scan the QR code above) for links to figures, animations, and activities that will help you consolidate the material.

1 The function of the classical **synapse** is to communicate information from a **presynaptic** neuron to a **postsynaptic** cell. It does so by converting an electrical signal—the action potential—into secretion of a **neurotransmitter** that crosses to the postsynaptic cell and alters that cell's functioning. Review **Figure 4.1**, **Animation 4.2**

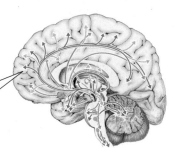

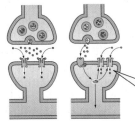

2 Neurotransmitters exert their effects via **neurotransmitter receptors**; these are also the site of action for many psychoactive drugs. Most transmitters have several different subtypes of receptors, which may be individually targeted by drugs. A given neurotransmitter may normally bind several different subtypes of receptors. Review **Figure 4.2**, **Table 4.2**

3 The major categories of neurotransmitters are **amine**, **amino acid**, **peptide**, and soluble **gas neurotransmitters**. Neurotransmitter systems form complex, overlapping patterns of projections throughout the brain. Review **Figures 4.3** and **4.4**, **Table 4.1**, **Animation 4.3**, **Activity 4.1**

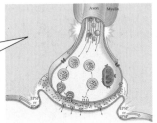

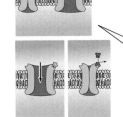

4 Drugs classified as **agonists** activate transmitter pathways, and **antagonists** block transmitter pathways. Repeated exposure to drugs may cause a compensatory decrease (**down-regulation**) or increase (**up-regulation**) in the number of receptors. Changes in receptor density are one mechanism of **drug tolerance**. Review **Figure 4.5**, **Animation 4.4**

5 The **dose-response curve (DRC)** quantifies the relationship between doses of a drug and its physiological effects, and it can be used to deduce characteristics such as drug potency and safety. Review **Figure 4.6**

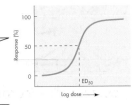

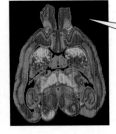

6 Psychoactive drugs have three main presynaptic actions: some drugs alter transmitter synthesis, others alter transmitter release, and some block the clearance of transmitter after it has been released. Review **Figure 4.7**

7 Many psychoactive drugs have postsynaptic effects, especially activation or blockade of postsynaptic receptors. Other drugs affect metabolic processes within the postsynaptic neuron, such as alterations in second-messenger systems, or changes in the production of crucial proteins (like transmitter receptors). Most antipsychotic medications block postsynaptic **dopamine (DA)** receptors, but some also block **serotonin (5-HT)** receptors. Review **Figure 4.8**

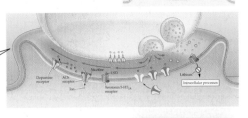

9 **Alcohol** acts on **gamma-aminobutyric acid (GABA)** receptors to produce some of its effects. In moderation, alcohol has beneficial effects; but in higher doses it is very harmful, damaging neurons in many areas of the brain. Review **Figures 4.10** and **4.11**

8 Opiates such as **morphine** are potent painkillers; the brain also makes its own **endogenous opioids** with a variety of effects. Opiates act on specific **opioid receptors**, especially in a major pain pathway that includes the **periaqueductal gray**. Review **Figure 4.9**

10 The active ingredient in marijuana— **delta-9-tetrahydro-cannabinol**, or THC—acts on cannabinoid receptors to produce its effects. An **endocannabinoid**, **anandamide** acts at cannabinoid receptors to affect mood, pain sensitivity, blood pressure, and other functions. Review **Figure 4.12**

11 Some stimulants, such as **nicotine**, imitate an excitatory synaptic transmitter. Others, such as **amphetamine**, cause the release of excitatory synaptic transmitters and block the reuptake of transmitters. **Cocaine** causes the release of transmitters, especially dopamine, in wide regions of the brain. Review **Figure 4.13**

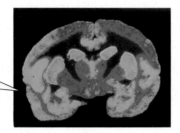

12 Some drugs are called **hallucinogens** because they alter sensory perception and produce peculiar experiences. Although hallucinogens vary in their actions, many share activity at serotonin receptors in the visual cortex, perhaps explaining their effects on visual perception. Review **Figure 4.14**

13 Substance abuse and dependence (addiction) are being studied intensively, and several explanatory models have been proposed. The positive reward model, based on the observation that animals will work very hard to self-administer highly addictive drugs, has received the most support from research. Review **Figure 4.15, Table 4.4**

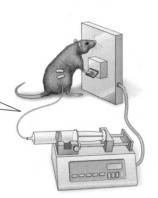

14 A dopamine-based neural pathway from the **ventral tegmental area** to the **nucleus accumbens** appears to be a system for experiencing pleasure and reward. Researchers believe that this reward system plays an important role in the formation of addictions. Some experimental treatments for addiction involve blocking the reward signal from drugs. Review **Figures 4.16** and **4.17**

Go to **2e.mindsmachine.com** for study questions, quizzes, flashcards, and other resources.

chapter

5

The Sensorimotor System

What You See Is What You Get

Ian Waterman had a perfectly ordinary life until he caught a viral infection at age 19. For reasons no one understands, the infection targeted a very specific set of nerves sending information from his body to his brain. Ian can still feel pain or deep pressure, as well as warm and cool surfaces on his skin, but he has no sensation of light touch below his neck. What's more, although Ian can still move all of his muscles, he receives no information about muscle activity or body position (Cole, 1995). You might think this deficiency wouldn't cause any problem, because you've probably never thought much about your "body sense"; it's not even one of the five senses that people talk about, is it?

In fact, however, the loss of this information was devastating. Ian couldn't walk across a room without falling down, and he couldn't walk up or down stairs. The few other people suffering a loss like this have spent the rest of their lives in wheelchairs. But Ian was a young and determined person, so he started teaching himself how to walk using another source of feedback about his body: his vision. Now, as long as the lights are on, Ian can carefully watch his moving body to judge which motor commands to send out to keep walking. If the lights go out, however, he collapses, and he has learned that in that circumstance he just has to lie where he is until the lights come on again. He has so finely honed this ability to guide movements with vision that if asked to point repeatedly to the same location in the air, he does so more accurately than control participants do. Still, it's a mental drain to have to watch and attend constantly to his body just to do everyday tasks.

Today Ian has a good job and an active, independent life, but he is always vigilant. Lying in bed, he has to be very careful to remain calm, tethering his limbs with the covers to prevent them from flailing about. And the lights are always on at Ian's house.

To see the video
Sensory Systems,
go to

2e.mindsmachine.com/5.1

Every species, including our own, is surrounded by forms of environmental energy that may signal life-or-death events. Molecules in the air are sensed as odors—of food, or mates, or smoke. Vibrations traveling through air are perceived as sounds, ranging from infant cries to the roar of a predator (or a waterfall). Light particles reflected from surfaces present a visual representation of the world. And because species differ in the environmental features they must sense for survival, evolution has endowed each species with its own unique set of capabilities. Bats are specially equipped to detect their own ultrasonic cries, which we humans are unable to hear. Some snakes have infrared-sensing organs in their faces that allow them to "see" heat sources (like a warm, tasty mouse) in the dark. Some of the impressive array of sensory modalities that animals may possess are listed in **TABLE 5.1** on the next page.

TABLE 5.1 Classification of Sensory Systems

System type	Modality	Sensed stimuli
Mechanical	Touch	Contact with body surface
	Hearing	Sound vibrations in air or water
	Vestibular	Head movement and orientation
	Joint	Position and movement
	Muscle	Tension
Photic	Vision	Photons, from light sources or reflected from surface
Thermal	Cold	Decrease in skin temperature
	Warmth	Increase in skin temperature
Chemical	Smell	Odorant chemicals in air
	Taste	Substances in contact with the tongue or other taste receptor
	Vomeronasal	Pheromones in air or water
Electrical	Electroreception	Differences in density of electrical currents
Magnetic	Magnetoreception	Magnetic fields for orientation

Sensory systems rely on specialized receptor cells to detect specific energies, as well as brain systems that receive input from these receptors. We open this chapter by considering basic principles of sensory processing, using the sense of touch to illustrate some of the major concepts. In Part II of the chapter, we take a closer look at an unpleasant but crucial sense: pain. And in Part III of the chapter, we turn our attention to the integration of sensory inputs and motor control mechanisms: the streamlined system that allows us to interact with our environment.

PART I
Sensory Processing and the Somatosensory System

All animals have sensory organs containing **receptor cells** that sense some forms of energy—called **stimuli**—but not others. So in a way, receptor cells act as filters, ignoring the environmental background and converting the key stimuli into the language of the nervous system: electrical signals. Information from sensory receptors floods the brain in an unending barrage of action potentials traveling along millions of axons, and our brains must make sense of it all. Of course, different kinds of energy—light, sound, touch, and so on—need different sensory organs to convert them into neural activity, just as taking a photograph requires a camera, not a microphone. There is tremendous diversity in sensory organs across the animal kingdom; for example, the eye is just one type of sensory organ, yet it is found in a dazzling array of sizes, shapes, and forms, reflecting the varying survival needs of different animals. Likewise, the specific auditory abilities of species reflect their unique ecological pressures (**FIGURE 5.1**).

receptor cell A specialized cell that responds to a particular energy or substance in the internal or external environment and converts this energy into a change in the electrical potential across its membrane.

stimulus A physical event that triggers a sensory response.

To view the Brain Explorer, go to

2e.mindsmachine.com/5.2

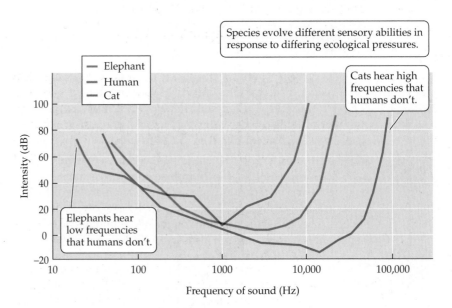

FIGURE 5.1 Do You Hear What I Hear?

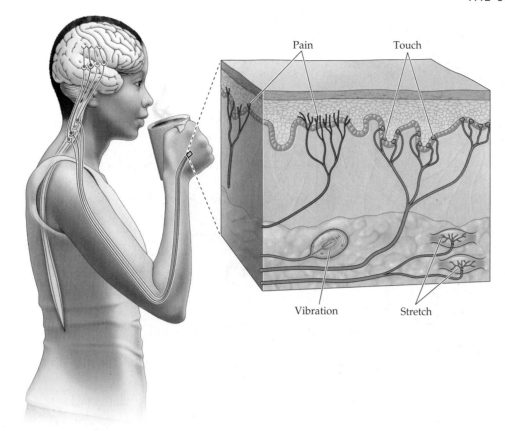

Pain   Touch

Vibration   Stretch

FIGURE 5.2   Labeled Lines

Although the end product of sensory receptors—action potentials—is the same for all the different sensory modalities, the brain recognizes the modalities as separate and distinct because the action potentials for each sense are carried in separate nerve tracts. This is the concept of **labeled lines**: particular neurons that are, right from the outset, labeled for distinctive sensory experiences. Action potentials in one line signal a sound, activity in another line signals a smell, and activity in other lines signals touch. And there are labeled lines within general sensory categories too; for example, we can distinguish different types of touch because our skin contains a variety of receptors and uses some lines to signal light touch, others to signal vibration, and yet other lines to signal stretching of the skin (**FIGURE 5.2**).

Receptor Cells Convert Sensory Signals into Electrical Activity

The structure of a receptor cell determines the particular kind of energy or chemical to which it will respond. And although a wide variety of cellular mechanisms are used to detect different stimuli, the outcome is always the same: an electrical change in the receptor, called a **generator potential**, that resembles the excitatory postsynaptic potentials we discussed in Chapter 3. Converting the signal in this way—from environmental stimuli into action potentials that our brain can understand—is called **sensory transduction**.

Our skin contains a rich array of receptors that transduce different forms of energy to provide our sense of touch. But touch is not just touch. Careful studies of skin sensations reveal qualitatively different sensory experiences: pressure, vibration, tickle, "pins and needles," and more-complex dimensions, such as smoothness or wetness—all recorded by the receptors in the skin (**FIGURE 5.3**).

A skin receptor that provides a clear example of the process of sensory transduction is the **Pacinian** (or *lamellated*) **corpuscle** (Loewenstein, 1971), a tiny onionlike structure embedded in the innermost layer of the skin that selectively responds to vibration and pressure. Acting as a filter, the corpuscle allows only vibrations of more

labeled lines   The concept that each nerve input to the brain reports only a particular type of information.

generator potential   A local change in the resting potential of a receptor cell in response to stimuli, which may initiate an action potential.

sensory transduction   The process in which a receptor cell converts the energy in a stimulus into a change in the electrical potential across its membrane.

Pacinian corpuscle   Also called *lamellated corpuscle*. A skin receptor cell type that detects vibration and pressure.

FIGURE 5.3 Receptors in Skin

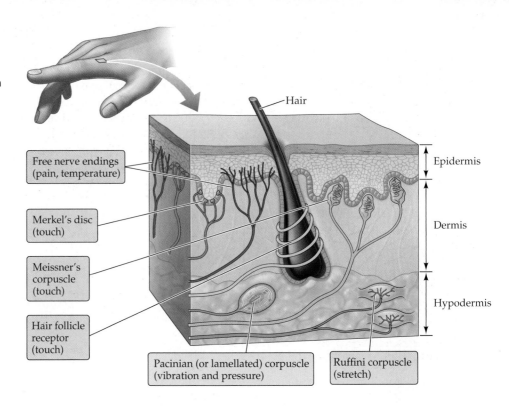

Hair

Free nerve endings (pain, temperature)

Merkel's disc (touch)

Meissner's corpuscle (touch)

Hair follicle receptor (touch)

Epidermis

Dermis

Hypodermis

Pacinian (or lamellated) corpuscle (vibration and pressure)

Ruffini corpuscle (stretch)

threshold   Here, the stimulus intensity that is just adequate to trigger an action potential in a sensory cell.

Meissner's corpuscle   Also called *tactile corpuscle*. A skin receptor cell type that detects light touch, responding especially to changes in stimuli.

than about 200 cycles per second to stimulate the sensory nerve ending inside it; this type of stimulation is what's created when we feel a texture against our skin (see Figure 5.3). By stretching the membrane of the sensory nerve ending, stimuli cause mechanically gated sodium channels to pop open, creating a graded generator potential (**FIGURE 5.4**). The amplitude (size) of this generator potential is directly proportional to the strength of the stimulus that was received. If the generator potential exceeds the firing **threshold**, action potentials are generated that travel via sensory nerves to the spinal cord.

Other dimensions of the sense of touch are mediated by their own unique sensory receptors. In contrast to the texture sensitivity of Pacinian corpuscles, **Meissner's**

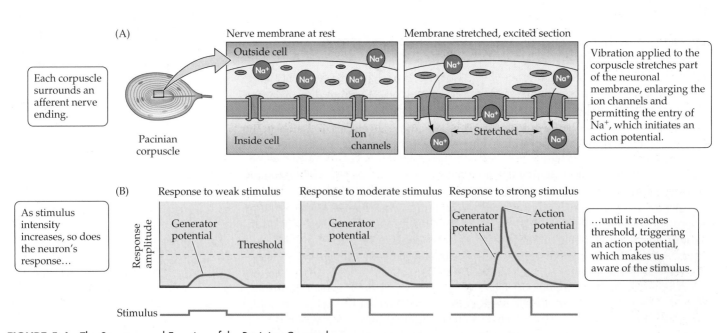

(A)

Each corpuscle surrounds an afferent nerve ending.

Pacinian corpuscle

Nerve membrane at rest

Outside cell

Na⁺

Inside cell

Ion channels

Membrane stretched, excited section

Na⁺

←Stretched→

Vibration applied to the corpuscle stretches part of the neuronal membrane, enlarging the ion channels and permitting the entry of Na⁺, which initiates an action potential.

(B)

As stimulus intensity increases, so does the neuron's response...

Response amplitude

Stimulus

Response to weak stimulus

Generator potential

Threshold

Response to moderate stimulus

Generator potential

Response to strong stimulus

Generator potential

Action potential

...until it reaches threshold, triggering an action potential, which makes us aware of the stimulus.

FIGURE 5.4 The Structure and Function of the Pacinian Corpuscle

corpuscles (also known as *tactile corpuscles*) and **Merkel's discs** mediate most of our ability to perceive the forms of objects we touch. While Merkel's discs are especially responsive to edges and to isolated *points* on a surface, the more numerous Meissner's corpuscles seem to respond to *changes* in stimuli, allowing them to detect localized movement between the skin and a surface (Heidenreich et al., 2011). **Ruffini corpuscles**, which are only sparsely distributed in the skin (Pare et al., 2003), detect *stretching* of patches of the skin when we move fingers or limbs (Johansson and Flanagan, 2009). Finally, pain, heat, and cold stimuli are detected by **free nerve endings** in the skin (see Figure 5.3), which we'll return to a little later in the chapter. All of these sensory receptors are found in their highest concentrations in regions of the skin where our sense of touch is finest (fingertips, tongue, and lips).

Merkel's disc   A skin receptor cell type that detects light touch, responding especially to edges and isolated points on a surface.

Ruffini corpuscle   A skin receptor cell type that detects stretching of the skin.

free nerve ending   An axon that terminates in the skin and has no specialized cell associated with it. Free nerve endings detect pain and/or changes in temperature.

somatosensory system   A set of specialized receptors and neural mechanisms responsible for body sensations such as touch and pain.

how's it going ❓

1. Discuss the relationship between the ecology of a species and its sensory capabilities.
2. What are labeled lines? What do they transmit?
3. Give a general explanation of sensory transduction. What is a generator potential?
4. Identify and describe four sensory receptors found in the skin.

Sensory Information Processing Is Selective and Analytical

Many people assume that the sensory systems simply capture an accurate snapshot of stimulation and transmit it to the brain—in other words, that the sensory systems provide an uncolored window on the world. But neuroscientists realize that the sensory organs and pathways convey only limited—*even distorted*—information to the brain. A good deal of selection and analysis takes place along sensory pathways, before the information ever reaches the brain. So the brain ultimately receives a highly filtered representation of the external world, in which stimuli that are critical for survival are strongly emphasized at the expense of less important stimuli. This processing and filtering is seen in several aspects of sensory transduction, including stimulus coding and processing across receptive fields, as well as in adaptation and active suppression by the brain, which we discuss next.

Sensory events are encoded as streams of action potentials

We've already seen that the nervous system uses labeled lines to identify the *type* of stimulus that is encountered. But how do sensory neurons tell the brain about the *intensity* or *location* of a stimulus? Because the action potentials produced by a sensory neuron always have the same size and duration, the other characteristics of a sensory stimulus must be *encoded* in the number and frequency of the action potentials, the rhythm in which clusters of action potentials occur, and so on.

We can respond to amazingly small differences in stimulus intensity, over a wide range of intensities. Although a single sensory receptor neuron could simply encode the intensity of a stimulus in the frequency of action potentials that the cell produces, only a very limited range of intensities could be represented this way, because neurons can fire only so fast (up to maybe 1,200 action potentials per second, and probably less in most nerves). Some sensory systems solve this problem by employing multiple sensory receptor cells, each specializing in just one part of the overall range of intensities, to cover the whole range. As the strength of a stimulus increases, additional sensory neurons sensitive to the higher intensities are "recruited"; thus, intensity of a stimulus can be represented by the number and thresholds of activated cells.

The position of a stimulus, either outside or inside the body, is likewise an important piece of information. Some sensory systems—the **somatosensory** ("body sensation") **system**, for example—reveal this information by the position of receptors on the sensory surface. Thanks to labeled lines that uniquely convey spatial information, we can directly encode which patch of skin that darn mosquito is biting, in order to know exactly where to aim the slap. Similarly, in the visual system an object's spatial location determines which receptors in the eye are stimulated. In bilateral receptor systems—the two eyes, two ears, and two nostrils—differences in stimulation of the left and right receptors are encoded, providing the brain with additional cues to the location of the stimulus (this type of processing is discussed in more detail in Chapter 6).

Neurons at all levels of the visual and the touch pathways—from the surface sheet of receptors all the way up to the cerebral cortex—are arranged in an orderly, maplike manner. The map at each level is not exact, but it does reflect both spatial positions and receptor density. More cells are allocated to the spatial representation of sensitive, densely innervated sites, like the lips, than to sites that are less sensitive, such as the skin of the back. Each cell in the sensory map thus preferentially responds to a particular type of stimulus occurring in a particular place, as we'll see next.

Sensory neurons respond to stimuli falling in their receptive fields

The **receptive field** of a sensory neuron consists of a region of space in which a stimulus will alter that neuron's firing rate. To determine this receptive field, investigators record the neuron's electrical responses to a variety of stimuli to see what makes the activity of the cell change from its resting rate. For example, which patch of skin must we stimulate to change the activity of one particular touch receptor? Experiments show that these somatosensory receptive fields are shaped like doughnuts, with either an excitatory center and an inhibitory surround, or an inhibitory center and an excitatory surround (**FIGURE 5.5**). The somatosensory receptive fields make it easier to detect edges on the objects we feel. Receptive fields differ in size and shape, and in the quality of stimulation that activates them. For example, some neurons respond preferentially to light touch, while others fire most rapidly in response to painful stimuli, and still others respond to cooling.

To see the animation
Somatosensory Receptive Fields,
go to

2e.mindsmachine.com/5.3

receptive field   The stimulus region and features that affect the activity of a cell in a sensory system.

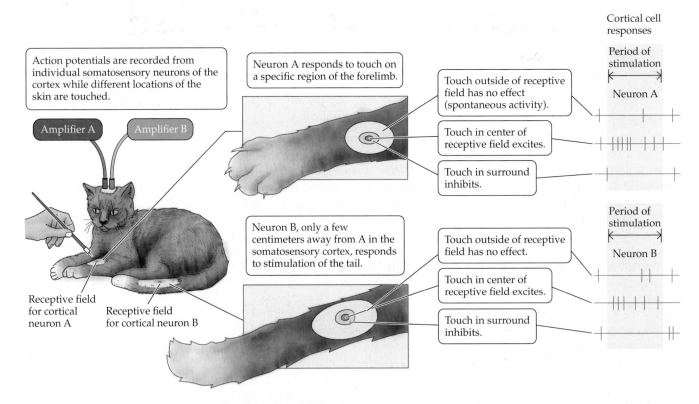

FIGURE 5.5   Identifying Somatosensory Receptive Fields

Experiments tracing sensory information along the pathway from the receptor cell to the brain show that neurons at every level will respond to particular stimuli, so each of these cells has its own receptive field. But as each successive neuron performs additional processing, the receptive fields change considerably, as we will see later in this chapter and in later chapters.

Receptors may show adaptation to unchanging stimuli

Sensory adaptation is the progressive decrease in a receptor's response to sustained stimulation (**FIGURE 5.6**). This process allows us to ignore unimportant events. By not noticing the touch of our clothes on our skin, the buzz of overhead lights, and other stimuli that are unchanging, our sensory systems avoid overload and can remain vigilant for critical events. Neuroscientists distinguish between **phasic receptors**, which display this sort of adaptation; and **tonic receptors**, which show little or no adaptation and thus can signal the duration of a stimulus. (As each of us knows all too well, pain sensors are often tonic receptors, maintaining a high level of activity to help us avoid further injury.)

The process of adaptation illustrates the principle we stated earlier: that sensory systems often shift *away from accurate portrayal* of the external world. In some mechanical receptors, such as the Pacinian corpuscle described earlier, adaptation develops from the elasticity of the receptor cell itself. When the corpuscle (which is a separate, accessory structure) is removed, the uncovered sensory nerve fiber does not adapt, but continues discharging action potentials in response to a constant stimulus.

Sometimes we need receptors to be quiet

We've already noted that survival depends more on sensitivity to important *changes* than on exact reporting of stimuli. To maintain such sensitivity, we need to suppress unneeded or unimportant sensory activity. As we just discussed, adaptation is one way in which sensory activity is controlled, and we are equipped with two additional suppression systems.

One way to suppress sensory activity is simply to physically prevent the stimuli from reaching the sensors. Closing the eyelids provides this function in the visual system; in the auditory system, tiny middle ear muscles reduce the intensity of sounds that reach the inner ear. A second kind of suppression of sensory inputs is entirely neural in nature. In many sensory and pain pathways, reciprocal neural connections descend from the brain to synapse on lower sensory levels, where they can then actively inhibit activity in the ascending sensory axons. This **central modulation of sensory information**, whereby the brain actively controls the information it receives, is a feature of many sensory and pain pathways.

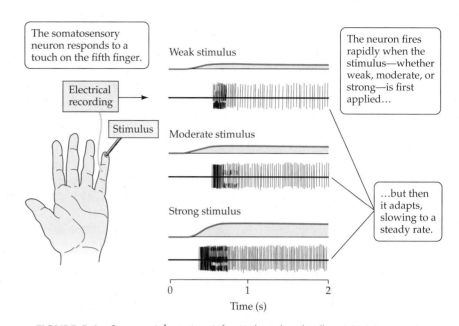

FIGURE 5.6 Sensory Adaptation (After Knibestol and Valbo, 1970.)

sensory adaptation The progressive loss of receptor response as stimulation is maintained.

phasic receptor A receptor in which the frequency of action potentials drops rapidly as stimulation is maintained.

tonic receptor A receptor in which the frequency of action potentials declines slowly or not at all as stimulation is maintained.

central modulation of sensory information The process in which higher brain centers, such as the cortex and thalamus, suppress some sources of sensory information and amplify others.

how's it going ?

1. In general terms, explain how a sensory event is encoded in action potentials in sensory fibers.
2. Why do some receptor cells respond only to strong stimuli?
3. Describe receptive fields, and how scientists detect them.
4. Name and briefly describe a couple of processes that change a sensory neuron's response to stimuli.

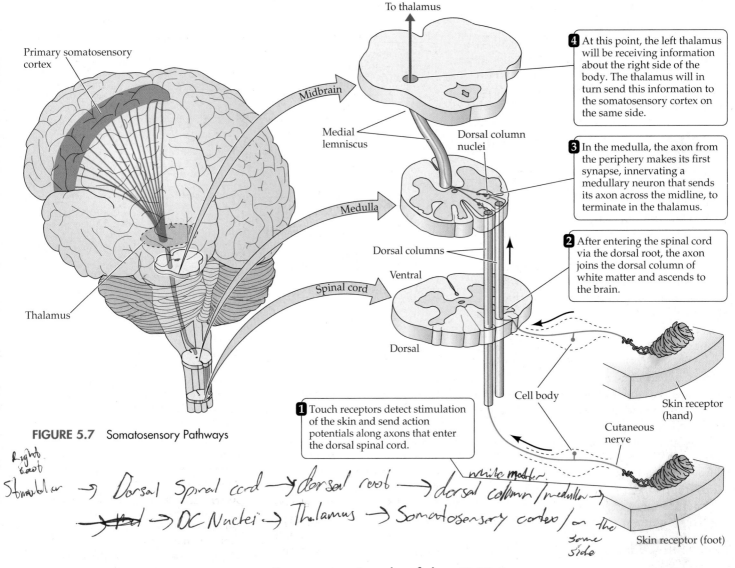

FIGURE 5.7 Somatosensory Pathways

The figure contains the following labels and callouts:

Primary somatosensory cortex

Thalamus

To thalamus

Midbrain

Medial lemniscus

Medulla

Spinal cord

Dorsal column nuclei

Dorsal columns

Ventral

Dorsal

Cell body

Skin receptor (hand)

Cutaneous nerve

Skin receptor (foot)

4 At this point, the left thalamus will be receiving information about the right side of the body. The thalamus will in turn send this information to the somatosensory cortex on the same side.

3 In the medulla, the axon from the periphery makes its first synapse, innervating a medullary neuron that sends its axon across the midline, to terminate in the thalamus.

2 After entering the spinal cord via the dorsal root, the axon joins the dorsal column of white matter and ascends to the brain.

1 Touch receptors detect stimulation of the skin and send action potentials along axons that enter the dorsal spinal cord.

(Handwritten annotation below figure):
Right root
Stimulation → Dorsal Spinal cord → dorsal root → dorsal column /medulla → ~~fed~~ → DC Nuclei → Thalamus → Somatosensory cortex /on the same side.
white matter.

Successive Levels of the CNS Process Sensory Information

dorsal column system A somatosensory system that delivers most touch stimuli via the dorsal columns of spinal white matter to the brain.

dermatome A strip of skin innervated by a particular spinal nerve.

Sensory information travels from the sensory surface to the highest levels of the brain, and each sensory system—such as touch, vision, or hearing—has its own distinctive pathway from the periphery to successively higher levels of the spinal cord and/or brain. For example, the somatosensory touch receptors that we've been discussing send their axons—eventually bundled into sensory nerves—from the skin to the dorsal (rear) part of the spinal cord. On entering the cord, the somatosensory projections ascend as part of the spinal cord's **dorsal column system**, a large wedge of white matter in the dorsal spinal cord (**FIGURE 5.7**). These axons go all the way up to the brainstem, where they synapse onto neurons that project across the midline to the opposite side and then go to the thalamus. From there, the incoming sensory information is directed to cortex. At all levels, the inputs are organized according to a somatosensory map in which the body surface is divided into discrete bands. Each band, called a **dermatome** (from the Greek *derma*, "skin," and *tome*, "part" or "segment"), is the strip of skin that is innervated by a particular spinal nerve. This maplike organization of sensory inputs is a feature of several sensory systems, including touch, vision, and hearing.

Each station in a sensory pathway accomplishes a basic aspect of information processing. For example, painful stimulation of the finger leads to reflexive withdrawal of the hand, which is mediated by spinal circuits before we even feel any pain. At the brainstem level, other circuits turn the head toward the source of pain.

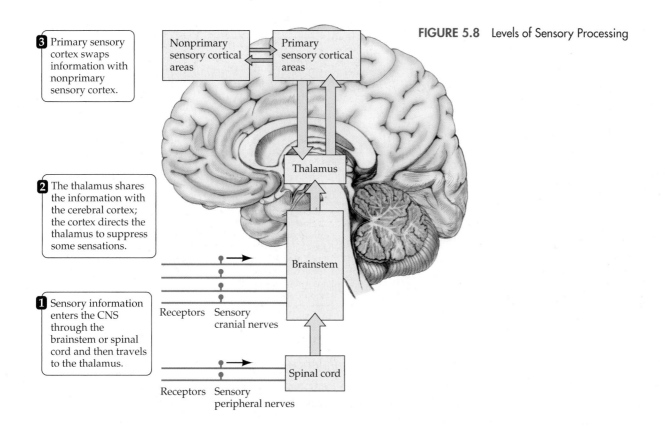

3 Primary sensory cortex swaps information with nonprimary sensory cortex.

2 The thalamus shares the information with the cerebral cortex; the cortex directs the thalamus to suppress some sensations.

1 Sensory information enters the CNS through the brainstem or spinal cord and then travels to the thalamus.

FIGURE 5.8 Levels of Sensory Processing

Eventually, sensory pathways reach the cerebral cortex, where the most complex aspects of sensory processing take place, perhaps consciously identifying the source of the pain (darn, another sliver!) and planning a response (where did I leave those tweezers?). For most senses, information reaches the **thalamus** before being relayed to the cortex (**FIGURE 5.8**). Information about each sensory modality is sent to a separate division of the thalamus. One way for the brain to suppress particular stimuli is for the cortex to direct the thalamus to emphasize some sensory information and suppress other information.

Sensory cortex is highly organized

Researchers have identified a region designated as **primary sensory cortex** for each sensory modality—primary somatosensory cortex, primary auditory cortex, and so on—that is generally the initial destination of sensory inputs to the cortex. However, other cortical regions may receive and process the same information, often in collaboration with the primary sensory cortex; sensibly enough (pardon the pun), we call these regions **nonprimary sensory cortex** (see Figure 5.8). Each cortical sensory region processes different aspects of our perceptual experiences.

Primary somatosensory cortex (also called *somatosensory 1* or *S1*) of each hemisphere is located in the postcentral gyrus, the long strip of tissue that lies just posterior to the central sulcus dividing the parietal lobe from the frontal lobe (**FIGURE 5.9A**). S1 receives touch information from the opposite side of the body. The cells in S1 are arranged as a map of the body (**FIGURE 5.9B**), but it is a very unusual map: it is distorted so that the size of each region on the map is proportional to the density of sensory receptors found in that region of the skin. Parts of the body where we are especially sensitive to touch (like the hand and fingers) have large representations in S1 compared with less sensitive areas (like the shoulder). This proportional mapping is illustrated in the strange-looking character in **FIGURE 5.9C**, called a *sensory homunculus*, in whom the size of each body part reflects the proportion of S1 devoted to that part. We discuss other aspects of cortical organization in **A STEP FURTHER 5.1**, on the website.

thalamus The brain regions at the top of the brainstem that trade information with the cortex.

primary sensory cortex For a given sensory modality, the region of cortex that receives most of the information about that modality from the thalamus (or, in the case of olfaction, directly from the secondary sensory neurons).

nonprimary sensory cortex Also called *secondary sensory cortex*. For a given sensory modality, the cortical regions receiving direct projections from primary sensory cortex for that modality.

primary somatosensory cortex Also called *somatosensory 1* or *S1*. Primarily the postcentral gyrus of the parietal lobe, where sensory inputs from the body surface are mapped.

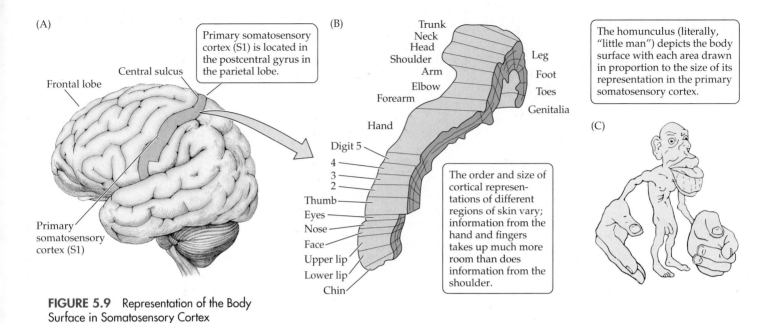

(A)

Frontal lobe

Central sulcus

Primary somatosensory cortex (S1) is located in the postcentral gyrus in the parietal lobe.

Primary somatosensory cortex (S1)

(B)

Trunk
Neck
Head
Shoulder
Arm
Elbow
Forearm

Leg
Foot
Toes
Genitalia

Hand

Digit 5
4
3
2
Thumb
Eyes
Nose
Face
Upper lip
Lower lip
Chin

The order and size of cortical representations of different regions of skin vary; information from the hand and fingers takes up much more room than does information from the shoulder.

The homunculus (literally, "little man") depicts the body surface with each area drawn in proportion to the size of its representation in the primary somatosensory cortex.

(C)

FIGURE 5.9 Representation of the Body Surface in Somatosensory Cortex

Sensory brain regions influence one another and change over time

Often the use of one sensory system influences perception from another sensory system. For example, humans detect a visual signal more accurately if it is accompanied by a sound from the same part of space (McDonald et al., 2000).

Many sensory areas in the brain—called *association areas*—process a mixture of inputs from different modalities. Some "visual" cells, for instance, also respond to auditory or touch stimuli. The convergence of information from different sensory systems on these **polymodal neurons** allows different sensory systems to interact (B. E. Stein and Stanford, 2008). And for a few people, a stimulus in one sensory modality may evoke an additional perception in another sensory modality, as when seeing a number evokes a color, or music literally becomes a matter of taste, where each note has both a sound and a flavor (Beeli et al., 2005). This condition is known as **synesthesia**. For more information and an example of synesthesia, see **A STEP FURTHER 5.2**, on the website.

At one time, most researchers thought that sensory regions of cortex were fixed early in life. Now, however, we know that cortical maps are highly plastic, changing considerably as a result of experience (D. T. Blake et al., 2006). For example, professional musicians who play stringed instruments have expanded cortical representations of their left fingers, presumably because they use these fingers to depress the strings for precisely the right note (Elbert et al., 1995; Münte et al., 2002). Brain imaging also reveals cortical reorganization in people who lose a hand in adulthood (**FIGURE 5.10**). One man received a transplanted hand (from an accident victim) 35 years after losing his own. Despite the length of time that had passed, his brain reorganized in just a few months to receive sensation from the hand in the appropriate part of S1 (Frey et al., 2008). Some changes in cortical maps occur after weeks or months of use or disuse; they may arise from the production of new synapses and dendrites (Florence et al., 1998; Hickmott and Steen, 2005) or from the loss of others.

polymodal neuron A neuron upon which information from more than one sensory system converges.

synesthesia A condition in which stimuli in one modality evoke the involuntary experience of an additional sensation in another modality.

how's it going ?

1. Name the main somatosensory pathway to the brain, describe its organization, and name its main components.

2. Where is the primary somatosensory cortex located? How is it organized?

3. Discuss interactions between sensory modalities—for example, effects of auditory inputs on visual perception.

(A)

Normally, the hand region of S1 lies between the regions representing the upper arm and the face.

(B)

After the loss of one hand, the cortical regions representing the upper arm and face expand, taking over the cortical region previously representing the missing hand.

FIGURE 5.10 Plasticity in Somatosensory Cortex (After T. T. Yang et al., 1994.)

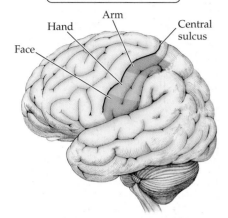

Face
Hand
Arm
Central sulcus

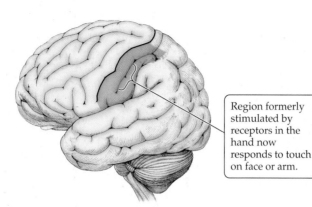

Region formerly stimulated by receptors in the hand now responds to touch on face or arm.

PART II
Pain: The Body's Emergency Signaling System

One important aspect of body sensation is at best a mixed blessing. The International Association for the Study of Pain defines **pain** as "an unpleasant sensory and emotional experience associated with actual or potential tissue damage, or described in terms of such damage." Pain forcefully guides our behavior in several ways that minimize the risk to our bodies (Dennis and Melzack, 1983). Immediate, short-lasting pain causes us to withdraw from the source, often reflexively, thus preventing further damage. Longer-lasting pain encourages behaviors, such as sleep, inactivity, grooming, feeding, and drinking, that promote recuperation. And the pain-related social communication—grimacing, groaning, shrieking, and the rest of the miserable line-up—provides a warning to kin and elicits caregiving behaviors from them, including grooming, defending, and feeding.

pain The discomfort normally associated with tissue damage.

Human Pain Varies in Several Dimensions

Learning, experience, emotion, and culture all affect our perception of pain in striking ways, and these factors may strongly influence people's descriptions of pain, ranging from an apparent absence of pain in badly injured soldiers and athletes, to the anguish of a child with a paper cut. A widely used quantitative measure of pain perception—the McGill Pain Questionnaire (Melzack, 1984)—asks people to select words that tap three different dimensions of pain:

1. The *sensory-discriminative* dimension (e.g., throbbing, gnawing, shooting)
2. The *motivational-affective* (emotional) dimension (e.g., tiring, sickening, fearful)
3. An overall *cognitive-evaluative* dimension (e.g., no pain, mild, excruciating)

Researchers found that people use different constellations of descriptors in various forms of pain: tooth pain is described differently from arthritic pain, which in turn is described differently from menstrual pain. This more detailed analysis provides better information for the diagnosis and treatment of illness.

A Discrete Pain Pathway Projects from Body to Brain

Most tissues of the body (but not all) contain receptors specialized for detecting painful stimuli. These receptors are particularly well studied in the skin; in this section we discuss some features of these receptors, along with the peripheral and CNS pathways that mediate pain.

THE HUMAN PINCUSHION WHO INCURS CONSTANT RISKS OF BLOOD POISONING.

Doesn't That Hurt? Although it might seem like a blessing, people with congenital insensitivity to pain, like the "Human Pincushion" pictured here (Dearborn, 1932), tend to die young as a consequence of repeated body injuries. We'll meet a little girl growing up with the hazards of insensitivity to pain at the end of this chapter. (Photograph by Culver Pictures, Inc.)

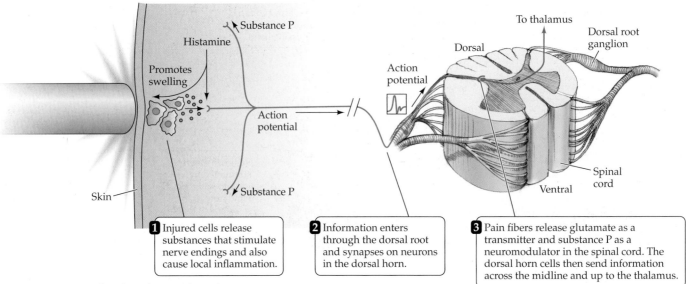

1 Injured cells release substances that stimulate nerve endings and also cause local inflammation.

2 Information enters through the dorsal root and synapses on neurons in the dorsal horn.

3 Pain fibers release glutamate as a transmitter and substance P as a neuromodulator in the spinal cord. The dorsal horn cells then send information across the midline and up to the thalamus.

FIGURE 5.11 Peripheral Mediation of Pain

nociceptor A receptor that responds to stimuli that produce tissue damage or pose the threat of damage.

transient receptor potential 2 (TRP2) A receptor, found in some free nerve endings, that opens its channel in response to rising temperatures.

A delta (Aδ) fiber A moderately large, myelinated, and therefore fast-conducting, axon, usually transmitting pain information.

C fiber A small, unmyelinated axon that conducts pain information slowly and adapts slowly.

Peripheral receptors get the initial message

When tissue is injured, the affected cells release chemicals that activate nearby pain receptors (called **nociceptors**) on free nerve endings, as well as causing inflammation (**FIGURE 5.11**). Many different substances in injured tissue—serotonin, histamine, and various enzymes and peptides, to name just a few—can stimulate these nociceptors. Different nociceptors respond to various stimuli, such as pain and/or changes in temperature.

Studies of capsaicin, the chemical that makes chili peppers spicy hot, helped reveal the receptor that signals sudden increases in temperature (this action is the reason spicy food seems to *burn*) (Caterina et al., 1997). This receptor, with the not-so-spicy name *transient receptor potential vanilloid type 1* (*TRPV1*; or just *vanilloid receptor 1*), belongs to a larger family of proteins called *transient receptor potential (TRP) ion channels*. Mice lacking the gene for TRPV1 still respond to mechanosensory pain, but not to mild heat or capsaicin (Caterina et al., 2000).

TRPV1's normal job is to report a rise in temperature to warn us of danger, so chili peppers cleverly evolved capsaicin to ward off mammalian predators—by falsely signaling burning heat. A related receptor, **transient receptor potential 2 (TRP2)**, detects even higher temperatures than does TRPV1, but it does *not* respond to capsaicin. TRP2 receptors are found on **A delta (Aδ) fibers**, which are large-diameter, myelinated axons. Because of the relatively large axon diameter and myelination, action potentials in these fibers reach the spinal cord very quickly. In contrast, the nerve fibers that possess TRPV1 receptors consist of thin, unmyelinated fibers called **C fibers**. So, when you burn your hand on that hot pan, the initial sharp pain you feel is conducted by the fat A delta fibers activated by their TRP2 receptors, and the long-lasting dull ache that follows arises from slower C fibers and their TRPV1 receptors. Other members of the TRP family of receptors detect coolness as well as constituents of spices like oregano, cloves, garlic, and wasabi (Jordt et al., 2004; Bautista et al., 2007; Salazar et al., 2008), but their relation to pain receptors remains a delicious mystery (sorry).

Identification of the nociceptor that detects *physical* damage is tricky because so many different chemicals are released by damaged tissue. Which one is the key signal? Through careful study of the family of a Pakistani boy who died in tragic circumstances—performing dangerous pranks because he could feel no pain—scientists isolated a mutation in a gene (called *SCN9A*) that appears to be responsible for pain insensitivity in other members of the boy's family. The gene encodes a sodium channel expressed in many free nerve endings (Cox et al., 2006). Researchers

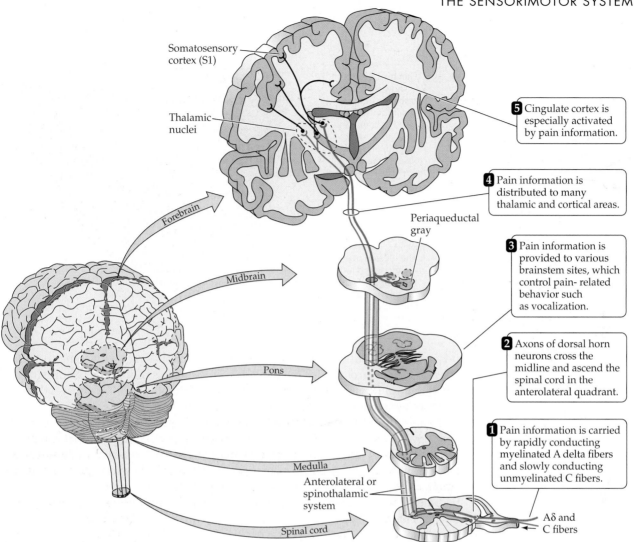

FIGURE 5.12 Ascending Pain Pathways in the CNS Pain sensation travels from its origin to the brain via the spinothalamic system, crossing the midline in the spinal cord.

are hopeful that this sodium channel may provide a new target for developing high-potency pain medication.

Special neural pathways carry pain information to the brain

Nerve fibers carrying information about pain and temperature send their axons to enter the dorsal horns of the spinal cord, where they synapse onto spinal neurons that project across the midline to the opposite side and then up toward the thalamus of the brain (via several brainstem sites), forming the **anterolateral** (or *spinothalamic*) **system** (**FIGURE 5.12**). This projection is distinct from the somatosensory system that we discussed earlier (the dorsal column system; see Figure 5.7), but like that system, each hemisphere receives its inputs from the contralateral side of the body. Within the spinal cord, the arriving pain fibers release the excitatory transmitter glutamate along with a peptide, **substance P**, that selectively boosts pain signals and remodels pain pathway neurons (Mantyh et al., 1997). Mice lacking substance P cannot feel intense pain, but they still feel mild pain (Cao et al., 1998; De Felipe et al., 1998).

Sometimes pain persists long after the injury that started it has healed. This **neuropathic pain** is a disagreeable example of neuplasticity, where neurons continue to directly signal pain, and indeed *amplify* the pain signal, in the absence of any tissue damage (Woolf and Salter, 2000). In one example of neuropathic pain called *phantom limb pain*, patients experience great pain that seems to come from an amputated limb.

anterolateral system   Also called *spinothalamic system*. A somatosensory system that carries most of the pain information from the body to the brain.

substance P   A peptide transmitter that is involved in pain transmission.

neuropathic pain   Pain that persists long after the injury that started it has healed.

FIGURE 5.13 Using a Visual Illusion to Relieve Phantom Limb Pain (After Ramachandran and Rogers-Ramachandran, 2000.)

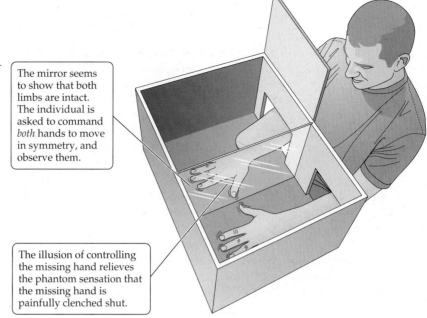

The mirror seems to show that both limbs are intact. The individual is asked to command *both* hands to move in symmetry, and observe them.

The illusion of controlling the missing hand relieves the phantom sensation that the missing hand is painfully clenched shut.

It is notoriously difficult to treat. One approach that has met with some success involves using a mirror to trick the brain into believing it is controlling the missing limb (**FIGURE 5.13**) (Ramachandran and Rogers-Ramachandran, 2000); apparently, visual feedback (even if false) allows the brain to recalibrate the pain signal.

Pain information is eventually integrated in the **cingulate cortex**. The extent of activation in the cingulate (as well as in somatosensory) cortex correlates with how much discomfort different people report in response to the same mildly painful stimulus (Coghill et al., 2003). Different subregions of the cingulate cortex seem to mediate emotional versus sensory aspects of pain (Vogt, 2005); one part of the cingulate cortex becomes active even when we just empathize with a loved one experiencing pain (T. Singer et al., 2004).

how's it going ?

1. Define *pain*. Why should it be viewed as a positive adaptation?
2. Provide a general explanation of the way pain receptors work. How do pain receptors differ from touch receptors?
3. Name and distinguish between the two types of pain fibers that carry pain information from the periphery to the spinal cord.
4. Sketch the pain pathways from the periphery to the cortex.

Pain Control Can Be Difficult

Throughout history, suffering humans have sought remedies to reduce their experience of pain. It's not easy; even cutting nerves may provide only temporary relief, until the pain system finds a way to restore its signal to the brain. A dominant model of pain transmission, called the *gate control theory*, hypothesizes that spinal "gates"—modulation sites at which pain can be facilitated or blocked—control the signal that gets through to the brain (Melzack and Wall, 1965). If this theory is right, effective pain relief may depend on finding ways to keep the gates closed, cutting off the pain signal. Popular strategies for **analgesia** (absence of pain; from the Greek *an*, "not," and *algesis*, "feeling of pain") fall into four general categories, which we'll discuss next.

Analgesic drugs are highly effective

The opiates (opium-related drugs, like morphine) have been known for centuries to relieve pain sensations. Along with the endogenous opioids (which are brain-derived painkillers such as the **endorphins**), opiate drugs bind to specific receptors in the brain

cingulate cortex Also called *cingulum*. A region of medial cerebral cortex that lies dorsal to the corpus callosum.

analgesia Absence of or reduction in pain.

endorphin One of three kinds of endogenous opioids.

to reduce pain (see Chapter 4). Researchers have found that this action is especially pronounced in the brainstem region called the *periaqueductal gray* (see Figure 5.12); one possibility is that the brainstem system activates the pain-gating mechanism of the spinal cord via descending projections, thereby blocking the transmission of pain signals. Similar benefits can be obtained by (carefully!) injecting opiates directly into the spinal cord; this is called an *epidural* or *intrathecal* injection (Landau and Levy, 1993).

Although people sometimes do become addicted to painkillers, that is usually not true of people who are using them to treat severe pain; in fact, the danger of addiction from the use of morphine to relieve surgical pain has been vastly exaggerated (Melzack, 1990) and is estimated to be no more than 0.04% (Brownlee and Schrof, 1997). A greater concern seems to be the undertreatment of severe pain, which can lead to the development of chronic pain.

Of course, there are other painkilling drugs, but none are as effective as the opiates. Over-the-counter medications like aspirin and acetaminophen (Tylenol) act via non-opiate mechanisms (especially the cyclooxygenase enzymes COX-1 and COX-2) to reduce pain and inflammation. Marijuana reduces pain by stimulating endogenous cannabinoid receptors (CB_1 receptors) in the spinal cord and in the brain (Agarwal et al., 2007; Pernía-Andrade et al., 2009).

Electrical stimulation can sometimes relieve pain

In **transcutaneous electrical nerve stimulation** (**TENS**), mild electrical stimulation is applied to nerves around the injury sites to relieve pain. The exact mechanism of this pain relief is not clear, but one possibility is that TENS closes the spinal "gate" for pain that Melzack and Wall (1965) described. Recall, for example, the last time you stubbed your toe. In addition to expelling a string of expletives, you may have vigorously rubbed the injured area, bringing a little relief. TENS is a more efficient way of stimulating those adjacent nerves, and it may bring dramatic relief lasting for hours (Vance et al., 2014). We know that TENS acts at least in part by releasing endogenous opioids, because administration of **naloxone**, an opioid antagonist, partially blocks this analgesic action (Gonçalves et al., 2014).

Placebos effectively control pain in some people, but not all

In some people, simply believing that they are receiving a proven treatment can effectively relieve pain. In a striking example of this **placebo effect**, participants who had just had their wisdom teeth extracted were given morphine or a placebo (J. D. Levine et al., 1978). Fully a third of patients receiving the placebo experienced pain relief. But when the placebo was coadministered with a drug that blocks opioid receptors (naloxone), the participants did not experience the benefits of the placebo effect. This latter finding strongly implies that placebos work by activating the brain's endogenous opioid system. In fact, functional brain imaging indicates that opioids and placebos activate the same brain regions (Petrovic et al., 2002; D. J. Scott et al., 2008). For reasons unknown, some people consistently experience relief from placebos while others do not (**FIGURE 5.14**).

Activation of endogenous opioids relieves pain

Although the ancient pain-relieving technique **acupuncture** remains very popular, only a minority of people using acupuncture achieve lasting relief from chronic pain. In those people for whom acupuncture is effective, a release of endorphins may be an important part of the process, since treatment with naloxone blocks acupuncture's effectiveness (N. M. Tang et al., 1997). Acupuncture thus resembles placebos in this regard. Although many rules govern needle placement in acupuncture, systematic research indicates that the placement of the needles actually has little to do with its effects on pain (Linde et al., 2009). The *expectation* that the needles will relieve pain appears to be the important factor, presumably inducing a release of endogenous opioids.

transcutaneous electrical nerve stimulation (TENS) The delivery of electrical pulses through electrodes attached to the skin, which excite nerves that supply the region to which pain is referred.

naloxone A potent antagonist of opiates that is often administered to people who have taken drug overdoses. It binds to receptors for endogenous opioids.

placebo effect Relief of a symptom, such as pain, that results following a treatment that is known to be ineffective or inert.

acupuncture The insertion of needles at designated points on the skin to alleviate pain or neurological malfunction.

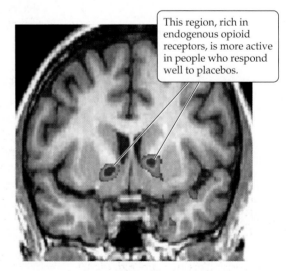

This region, rich in endogenous opioid receptors, is more active in people who respond well to placebos.

FIGURE 5.14 Placebos Affect Opioid Systems in the Brain (From Wager et al., 2007, courtesy of Jon-Kar Zubieta.)

TABLE 5.2 Types of Pain Relief

Type	Mechanism
PSYCHOGENIC	
Placebo	May activate endorphin-mediated pain control system
Hypnosis	Alters brain's perception of pain
Stress	Both opioid and non-opioid mechanisms
Cognitive (learning, coping strategies)	May activate endorphin-mediated pain control system
PHARMACOLOGICAL	
Opiates	Bind to opioid receptors in periaqueductal gray and spinal cord
Spinal block	Drugs block pain signals in spinal cord
Anti-inflammatory drugs	Block chemical inflammatory signals at the site of injury (see Figure 5.11)
Cannabinoids	Act in spinal cord and on nociceptor endings
STIMULATION	
TENS/mechanical	Tactile or electrical stimulation of large fibers blocks or alters pain signal to brain
Acupuncture	Activation of endogenous opioids and/or placebo-like effect; possible modulating effect on activity of peripheral pain pathways
Central gray	Electrical stimulation activates endorphin-mediated pain-control systems, blocking pain signal in spinal cord

Likewise, stressful life events can produce significant analgesia; for example, tales abound of gravely wounded soldiers who feel no pain for some time after their injuries occur (Bowman, 1997). Research in animals indicates that stress activates both an opioid-dependent form of analgesia, which can be blocked by naloxone, and another, non-opioid analgesia system that has not yet been characterized (but may rely on endocannabinoids) (A. G. Hohmann et al., 2005).

Pain relief remains a major challenge for neuroscience research. Chronic pain can have dramatic effects on the brain: for example, the prefrontal cortex in people with chronic back pain shrinks much faster than normal, as if the patients are rapidly aging (Apkarian et al., 2004). The wide range of pain relief strategies (summarized in **TABLE 5.2**), some of which reflect desperation in the face of great anguish, testifies to the elusive nature of pain. As we learn more about how the brain controls pain, we can hope for better, safer analgesics in the future.

how's it going ❓

1. What is the most effective pharmacological method of pain control? How and where do these drugs work in the brain?
2. How is TENS thought to work to control pain?
3. Compare and contrast placebos and acupuncture for pain. Discuss the possibility that they act on the same neural system.

PART III
Movement and the Motor System

Our apparently effortless adult motor abilities—such as reaching out and picking up an object, walking across the room, sipping a cup of coffee—require complex muscular systems with constant feedback from the body. Ian, whom we met at the beginning of the chapter, knows this all too well. Our survey of motor control starts with a discussion of a theoretical framework for studying motor behavior, followed by a tour of the anatomy and pathology of movement.

Behavior Requires Movements That Are Precisely Programmed and Monitored

When you think about it, *all* behavior must involve **movements**—contractions of muscles that provide our sole means of interacting with the world around us (with only rare, technological exception, as we'll see later). Centuries of research focused on the organization of motor behavior as both an engineering problem (how is movement programmed?) and a physiological system (how is movement produced?). Early discoveries suggested that **reflexes**—simple, unvarying, and unlearned responses to sensory stimuli such as touch, pressure, and pain—might be the basic units of behavior. It was thought that more-complex behaviors, or **acts**, such as getting dressed, walking, or speaking a sentence, might result from simply connecting together different reflexes, the sensation from one reflex triggering the next.

The flaws of this perspective soon became apparent: for most acts we have a *plan* in which several units (arm movements, leg movements, speech sounds) are placed in a larger pattern (the intended complete act), and they are not always produced in the same (or even the correct) order. So, researchers realized that acts require a **motor plan** (or *motor program*), a complex set of commands to muscles that is established *before* an act occurs. Feedback from movements informs and fine-tunes the motor program as the execution is unfolding, but the basic sequence of movements is planned. Examples of behaviors that exhibit this kind of internal plan range from highly skilled acts, such as piano playing, to the simple escape behaviors of animals such as crayfish.

Researchers can track the simple movements that make up an act by recording the electrical activity of muscles as they contract—a technique called **electromyography (EMG)**—and the moment-to-moment positions of the body. The EMG recordings in **FIGURE 5.15** show that a person pulling a lever will adjust his legs just before moving his arm—an example of motor planning. Motor plans resemble engineering concepts that are applied to the operation of machines. In designing machines,

movement A single relocation of a body part, usually resulting from a brief muscle contraction. It is less complex than an act.

reflex A simple, highly stereotyped, and unlearned response to a particular stimulus (e.g., an eye blink in response to a puff of air).

act Complex behavior, as distinct from a simple movement.

motor plan Also called *motor program*. A plan for a series of muscular contractions, established in the nervous system prior to its execution.

electromyography (EMG) The electrical recording of muscle activity.

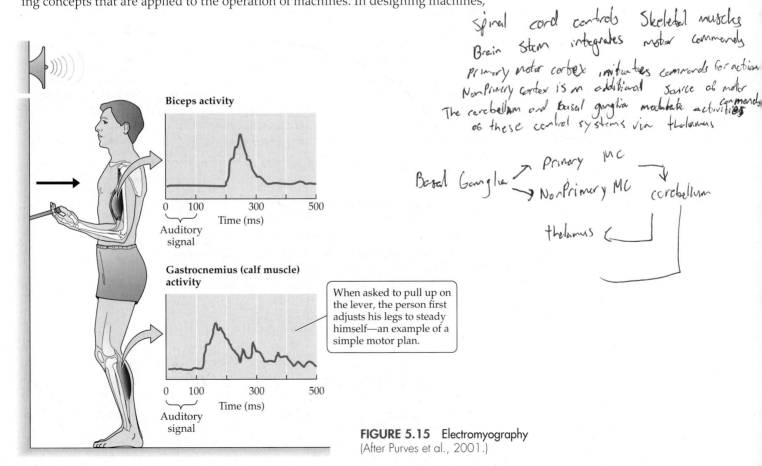

Biceps activity

0   100   300   500
Time (ms)
Auditory
signal

Gastrocnemius (calf muscle) activity

When asked to pull up on the lever, the person first adjusts his legs to steady himself—an example of a simple motor plan.

0   100   300   500
Time (ms)
Auditory
signal

FIGURE 5.15 Electromyography (After Purves et al., 2001.)

engineers commonly have two goals: (1) accuracy, to prevent or minimize error; and (2) speed, to complete a task quickly and efficiently. Improvements in one goal usually come at some cost to the other goal; in other words, there is a trade-off between speed and accuracy, and this trade-off is also apparent in motor planning by the nervous system.

The neuromuscular system consists of the muscles of the body plus a collection of brain mechanisms and nerves that prepare and execute motor plans and obtain feedback information from the sensory system for use in error correction. The system is organized according to a distinct hierarchy:

1. The *skeletal system* and the muscles attached to it determine which movements are possible.

2. The *spinal cord* controls skeletal muscles in response to motor commands from the brain or, in the case of simple reflexes, in direct response to sensory inputs.

3. The *brainstem* integrates motor commands from higher levels of the brain and transmits them to the spinal cord. It also relays sensory information about the body from the spinal cord to the forebrain.

4. Some of the main commands for action are initiated in the *primary motor cortex*.

5. Areas adjacent to the primary motor cortex, *nonprimary motor cortex*, provide an additional source of motor commands, acting indirectly via primary motor cortex and through direct connections to lower levels of the motor hierarchy. At the very top of the movement hierarchy is the prefrontal cortex, which is crucial to the conscious formulation of behavioral plans.

6. Other brain regions—the *cerebellum* and *basal ganglia,* via the *thalamus*—modulate the activities of the other parts of the control system.

Through the remainder of the chapter we'll look at the elements of this hierarchy, as outlined in **FIGURE 5.16**, in a bit more detail.

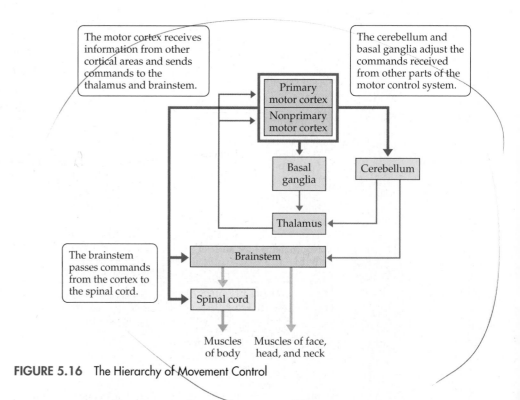

FIGURE 5.16 The Hierarchy of Movement Control

how's it going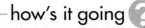

1. Distinguish among reflexes, movements, and acts.
2. Discuss the importance of sensory feedback for the control of movements. How are speed and accuracy related, in the context of movement control?
3. What is a motor plan?
4. Identify the six major levels of the motor control hierarchy.

A Complex Neural System Controls Muscles to Create Behavior

So much processing power is devoted to movement that the range of possible behaviors and their speed of execution can be astonishing. One of the primary factors in determining the range of movements of a species is the type and arrangement of its muscles around joints, so we begin there and then work our way up through the central nervous system.

Muscles and the skeleton work together to move the body

Our skeleton, like those of other species with bones, is articulated with joints that vary in their planes of movement—ranging from "universal" joints, like the hip or shoulder, to joints that act more like hinges and move mostly in one direction, such as the elbow or knee. Around a joint, different muscles, connected to the bones by tendons, are arranged in a reciprocal fashion such that when one muscle group contracts, it stretches the other group; that is, the muscles are **antagonists**. Some groups of muscles, called **synergists**, may work together to move a limb in one direction. A simple example of muscle action around a joint is shown in **FIGURE 5.17**.

antagonist   A muscle that counteracts the effect of another muscle.

synergist   A muscle that acts together with another muscle.

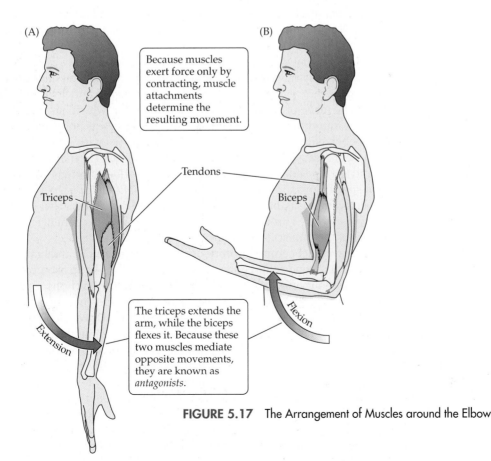

Because muscles exert force only by contracting, muscle attachments determine the resulting movement.

Tendons

Triceps

Biceps

The triceps extends the arm, while the biceps flexes it. Because these two muscles mediate opposite movements, they are known as *antagonists*.

Extension

Flexion

FIGURE 5.17   The Arrangement of Muscles around the Elbow

FIGURE 5.18 The Innervation of Muscle

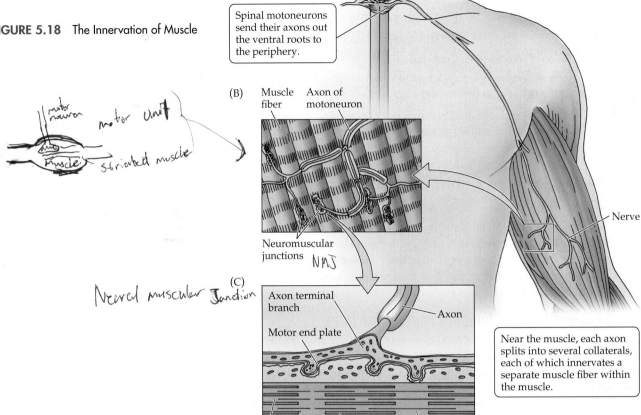

(A) Spinal motoneurons send their axons out the ventral roots to the periphery.

(B) Muscle fiber · Axon of motoneuron

Neuromuscular junctions *NMJ*

Nerve

Near the muscle, each axon splits into several collaterals, each of which innervates a separate muscle fiber within the muscle.

Neural muscular Junction

(C) Axon terminal branch

Motor end plate

Axon

Muscle fiber · Myosin · Actin

motor neuron · motor unit · muscle · striated muscle

---

skeletal muscle   A muscle that is used for movement of the skeleton, typically under our conscious control.

striate muscle   A type of muscle that has a striped appearance. It is generally under voluntary control.

motoneuron   Also called *motor neuron*. A neuron that transmits neural messages to muscles (or glands).

neuromuscular junction   The region where the motoneuron terminal and the adjoining muscle fiber meet. It is the point where the nerve transmits its message to the muscle fiber.

acetylcholine (ACh)   A neurotransmitter that is produced and released by the autonomic nervous system, by motoneurons, and by neurons throughout the brain.

The movement of a limb is determined by the degree and rate of contraction in some muscles and relaxation in others, or we can lock a limb in position by contracting opposing muscles simultaneously.

The muscles that we use for movement of the skeleton are called **skeletal muscles**. Because they have a striped appearance on microscopic examination, due to overlapping layers of contractile proteins called *myosin* and *actin*, skeletal muscles are said to be made of **striate muscle**. (*Smooth muscle*, which has a different appearance and is found in visceral organs and blood vessels, is not generally involved in voluntary behavior, so we will not concern ourselves with it here.) Contraction of the muscle increases the overlap of the actin and myosin filaments within *muscle fibers*, and as they slide past each other, the muscle fiber shortens. Most muscles consist of a specific mixture of two types of fibers: *slow-twitch fibers* that contract with relatively low intensity but fatigue slowly, and *fast-twitch fibers* that contract strongly but fatigue quickly. Through training, endurance athletes enhance the slow-twitch properties of their muscles (Putman et al., 2004).

Muscles contract because **motoneurons** of the spinal cord and brainstem (see Figures 2.7 and 2.8) send action potentials along their axons and axon collaterals to terminate at specialized synapses, called **neuromuscular junctions**, that are found on muscle fibers (**FIGURE 5.18**). The production of an action potential by a motoneuron triggers a release of the neurotransmitter **acetylcholine (ACh)** at all of the motoneuron's axon terminals. The motoneuron, together with all of the muscle fibers it innervates, is known as a *motor unit;* the fibers respond to the release of ACh by triggering the molecular events that cause actin and myosin to produce contraction (see Figure 5.18).

Some large motor units—where motoneurons innervate thigh muscle, for example—may involve hundreds or thousands of muscle fibers. But muscles that require more precise control—muscles of the face, for example—tend to have much smaller motor units, with each motoneuron controlling only a few muscle fibers. Many people experience "jumping nerves" in the eyelids when they're fatigued (from studying biological psychology, maybe). This tiny but incredibly annoying twitch, called a *fascic-ulation*, is actually a misfiring facial motor unit. A fasciculation in the thigh, in contrast, produces a much larger twitch.

Within the spinal cord, motoneurons tend to have large cell bodies and very wide-spread dendritic fields because they receive and integrate inputs from so many differ-ent sources—incoming sensory inputs, as well as descending signals from the brain—that form thousands of synapses onto the motoneurons. Virtually all motoneuron axons are myelinated (Kaar and Fraher, 1985), so the action potentials that result from the integration process quickly reach their target muscles. In a somewhat dramatic turn of phrase, neuroscientists refer to motoneurons as the **final common pathway**: the sole route through which the spinal cord and brain can control our many muscles.

Sensory feedback from muscles, tendons, and joints governs movement

To produce rapid coordinated movements of the body, the brain and spinal cord con-tinually monitor the state of the muscles, the positions of the limbs, and the instructions being issued by the motor centers. This collection of information about body movements and positions is called **proprioception** (from the Latin *proprius*, "own," and *recipere*, "to receive"). Ian, whom we met at the start of this chapter, was attacked by a virus that selectively killed proprioceptive axons; his predicament illustrates how important this "sixth sense" is for movement. Let's consider the two special proprioceptors—muscle spindles and Golgi tendon organs—that monitor muscle length and muscle tension.

The **muscle spindle** is basically a capsule, buried within the other fibers of the muscle, that contains a special kind of muscle fiber called an **intrafusal fiber** (from the Latin *intra*, "within," and *fusus*, "spindle") (**FIGURE 5.19A**). When a muscle is stretched—imagine someone handing you a heavy book, causing your arm to bend

final common pathway   The moto-neurons of the brain and spinal cord, so called because they receive and integrate all motor signals from the brain to direct movement.

proprioception   Body sense; informa-tion about the position and movement of the body.

muscle spindle   A muscle receptor that lies parallel to a muscle and sends impulses to the central nervous system when the muscle is stretched.

intrafusal fiber   Any of the small muscle fibers that lie within each muscle spindle.

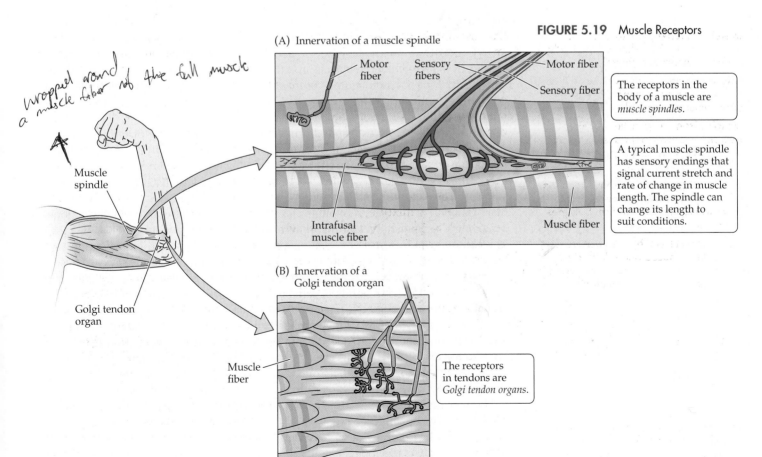

FIGURE 5.19 Muscle Receptors

(A) Innervation of a muscle spindle

Motor fiber — Sensory fibers — Motor fiber — Sensory fiber

Intrafusal muscle fiber — Muscle fiber

The receptors in the body of a muscle are *muscle spindles*.

A typical muscle spindle has sensory endings that signal current stretch and rate of change in muscle length. The spindle can change its length to suit conditions.

Muscle spindle

Golgi tendon organ

(B) Innervation of a Golgi tendon organ

Muscle fiber

The receptors in tendons are *Golgi tendon organs*.

wrapped around a muscle fiber not the full muscle

FIGURE 5.20 Activation of Muscle Receptors

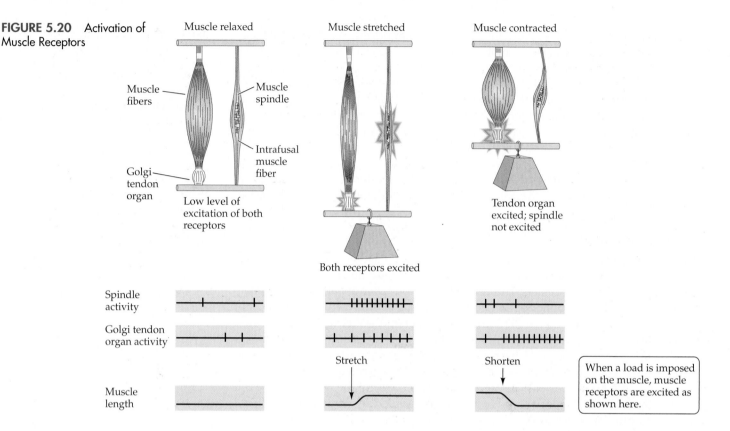

downward and stretching the biceps muscle—sensory endings within the spindle fiber become excited and trigger action potentials in afferent sensory nerves. This proprioceptive signal informs the spinal cord and brain about the extent and rate of change in the stretch of the muscle, and therefore about the load being imposed. Interestingly, a special motoneuron controls the length of the intrafusal fiber, adjusting it according to the movements being planned by the brain—in a sense, calibrating the muscle spindle to the *expected* limb position.

While muscle spindles respond primarily to *stretch*, the other proprioceptive receptors for muscle—**Golgi tendon organs**—are especially sensitive to muscle *tension*. Loads that are strong enough to stretch the tough tendon are sensed by the nerve endings of the Golgi tendon organ that weave through the tendon (**FIGURE 5.19B**). It takes a pretty strong load to stretch a tendon to this degree, so it makes sense that the primary function of Golgi tendon organs is to monitor the force of muscle contractions, providing a second source of sensory information about the muscles (**FIGURE 5.20**). This arrangement makes the Golgi tendon organs useful in another important way: they detect overloads that threaten to tear muscles and tendons, and they can cause a reflexive relaxation of the affected muscles, protecting the muscles (and causing you to drop that book). Another familiar example of a stretch reflex is the knee-jerk (or patellar) reflex that we discussed in Chapter 3 (see Figure 3.14).

Classic studies in physiology emphasized the importance of information from muscle spindles and Golgi tendon organs for controlling movement. Mott (1895) and Sherrington (1898) showed that severing the sensory fibers from a monkey's arm muscles causes the monkey to stop using the affected limb, even if the connections from motoneurons to the muscles are preserved. The arm dangles, apparently useless. But if the good arm is restrained, the animal soon learns to use the affected arm, and indeed it can become quite dexterous (Taub, 1976). Monkeys manage to do this the same way Ian does, by guiding their movements with visual feedback about how the arm is moving. In fact, we all supplement our proprioceptive information with feedback from other sensory channels, like vision.

Golgi tendon organ A type of receptor found within tendons that sends impulses to the central nervous system when a muscle contracts.

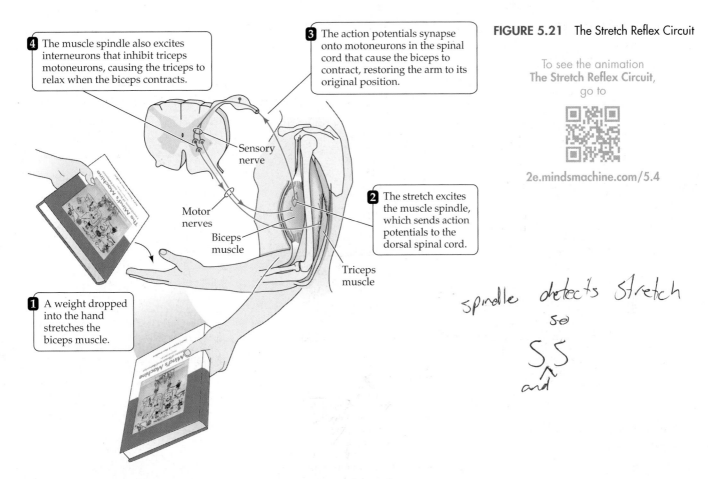

FIGURE 5.21 The Stretch Reflex Circuit

4 The muscle spindle also excites interneurons that inhibit triceps motoneurons, causing the triceps to relax when the biceps contracts.

3 The action potentials synapse onto motoneurons in the spinal cord that cause the biceps to contract, restoring the arm to its original position.

Sensory nerve

Motor nerves

Biceps muscle

Triceps muscle

2 The stretch excites the muscle spindle, which sends action potentials to the dorsal spinal cord.

1 A weight dropped into the hand stretches the biceps muscle.

To see the animation
The Stretch Reflex Circuit,
go to

2e.mindsmachine.com/5.4

spindle detects stretch so
S,S
and

The spinal cord mediates "automatic" responses and receives inputs from the brain

To really understand the physiology of movement, we need to understand how the "final common pathway" is controlled by the CNS. The lowest level of this hierarchy is the spinal cord, where relatively simple circuits produce reflexive behavioral responses to sensory stimuli. A straightforward example is the **stretch reflex**, illustrated in **FIGURE 5.21**, that can be elicited by stretching any muscle. In this case, dropping a load into the outstretched hand causes a sudden stretch of the biceps muscle, which is detected by muscle spindles. In the spinal cord, the incoming sensory information from the spindles has two immediate effects: it stimulates motoneurons of the biceps, causing a contraction, and it simultaneously inhibits the antagonistic motoneurons that connect to the triceps muscle on the back of the arm. The reflex thus generates a compensatory movement to bring the hand and arm back to their intended position. Not all spinal circuits are quite this simple; for example, the rhythmic movements of walking are governed by spinal circuits that may involve many neurons across multiple spinal segments.

Although muscles of the head are controlled *directly* by the brain, via the cranial nerves (see Figure 2.7), the muscles of the rest of the body are ultimately controlled by commands from the brain via the spinal cord. The brain sends these commands to the spinal cord through two major pathways: the pyramidal system and the extrapyramidal system. The **pyramidal** (or *corticospinal*) **system** consists of neuronal cell bodies within the frontal cortex and their axons, which pass through the brainstem, forming the pyramidal tract to the spinal cord (**FIGURE 5.22A**). In a cross section of the medulla, the tract is a wedge-shaped anterior protuberance (pyramid) on each side of the midline. Because the left and right pyramidal tracts each cross over to the other side, the right cortex controls the left side of the body while the left cortex controls the right. Lesions anywhere in the pyramidal tract will cause paralysis in the muscle targets of the damaged axons.

stretch reflex The contraction of a muscle in response to stretch of that muscle.

pyramidal system Also called *corticospinal system*. The motor system that includes neurons within the cerebral cortex and their axons, which form the pyramidal tract.

PMC → White matter → medulla → Pyramid → Cross → Cortical spinal track

cortex/spine

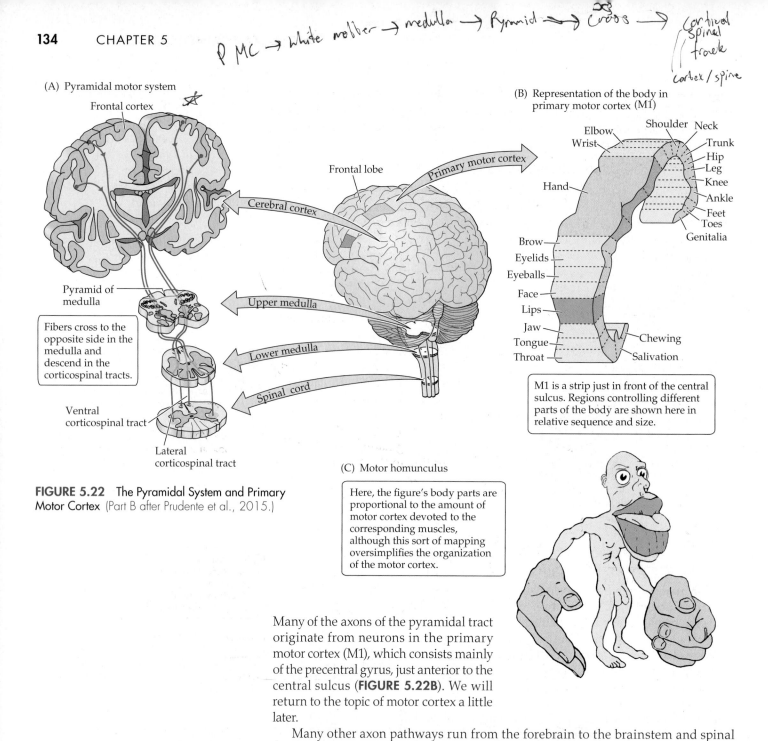

(A) Pyramidal motor system

Frontal cortex

Frontal lobe

Primary motor cortex

Cerebral cortex

Pyramid of medulla

Upper medulla

Fibers cross to the opposite side in the medulla and descend in the corticospinal tracts.

Lower medulla

Ventral corticospinal tract

Spinal cord

Lateral corticospinal tract

(B) Representation of the body in primary motor cortex (M1)

Elbow — Shoulder — Neck
Wrist — Trunk
— Hip
— Leg
— Knee
Hand — Ankle
— Feet
— Toes
— Genitalia
Brow
Eyelids
Eyeballs
Face
Lips
Jaw — Chewing
Tongue
Throat — Salivation

M1 is a strip just in front of the central sulcus. Regions controlling different parts of the body are shown here in relative sequence and size.

FIGURE 5.22 The Pyramidal System and Primary Motor Cortex (Part B after Prudente et al., 2015.)

(C) Motor homunculus

Here, the figure's body parts are proportional to the amount of motor cortex devoted to the corresponding muscles, although this sort of mapping oversimplifies the organization of the motor cortex.

Many of the axons of the pyramidal tract originate from neurons in the primary motor cortex (M1), which consists mainly of the precentral gyrus, just anterior to the central sulcus (**FIGURE 5.22B**). We will return to the topic of motor cortex a little later.

Many other axon pathways run from the forebrain to the brainstem and spinal cord. Because these tracts are outside the pyramids of the medulla, they and their connections are lumped together as the **extrapyramidal system**. In general, lesions of the extrapyramidal system do not prevent the movement of individual joints and limbs, but they do interfere with spinal reflexes, usually exaggerating them, and with systems that regulate and fine-tune motor behavior. Many of these extrapyramidal projections pass to the spinal cord via specialized motor regions (the reticular formation and red nucleus) of the midbrain and brainstem; as we'll see shortly, the basal ganglia are an important point of origin for extrapyramidal projections.

Spinal injuries due to vehicular accidents, violence, falls, and sports injuries are all too common, and they often cause heartbreaking disabilities. Because the spinal cord carries all of the instructions from the brain to the muscles, an injury that completely severs the cord results in immediate and permanent paralysis below the level of injury. Depending on the extent of destruction of the spinal cord below the injury site, spinal reflexes may or may not be lost as well (in fact, reflexes may become *stronger* because of the loss of descending inhibition from the brain). An estimated 250,000–400,000

extrapyramidal system A motor system that includes the basal ganglia and some closely related brainstem structures. Axons of this system pass into the spinal cord outside the pyramids of the medulla.

individuals in the United States have spinal cord injuries, and thousands more occur each year, mostly in young people. Although much remains to be discovered, the hope of reconnecting the injured spinal cord no longer seems far-fetched, as discussed in **A STEP FURTHER 5.3**, on the website.

Motor cortex plans and executes movements—and more

The **primary motor cortex** of humans—**M1**—is a major source of axons forming the pyramidal tract. Like S1, the primary somatosensory cortex that we discussed earlier in the chapter, M1 occupies a single large cortical gyrus: the **precentral gyrus**, located immediately in front of the central sulcus (M1 is thus a part of the frontal lobe) (see Figure 5.22B). And like S1, M1 is organized as a map of the contralateral side of the body. So, electrical stimulation of a discrete region of the left M1 will cause movement in the corresponding region of the right side of the body. Once again, the map is distorted, in the sense that the parts of the body that we control most precisely—hands, lips, tongue— are overrepresented in M1. **FIGURE 5.22C** shows the *motor homunculus*, a figure drawn using the body proportions represented in M1. But although the M1 map helps us understand the basic organization of motor cortex, recent research indicates that the map is really an oversimplification. The mapping of individual body regions in M1 isn't nearly as clear-cut and discrete as traditional M1 maps suggest. In fact, there is a fair bit of intermingling of body regions in the map, because many body parts coordinate with one another across regions of M1 (Schieber and Hibbard, 1993; Rathelot and Strick, 2006).

By recording from M1 neurons in monkeys making arm movements (Georgopoulos et al., 1993), we can eavesdrop on the commands originating there (**FIGURE 5.23**). Many M1 cells change their firing rates according to the direction of the

primary motor cortex (M1) The apparent executive region for the initiation of movement. It is primarily the precentral gyrus.

precentral gyrus The strip of frontal cortex, just in front of the central sulcus, that is crucial for motor control.

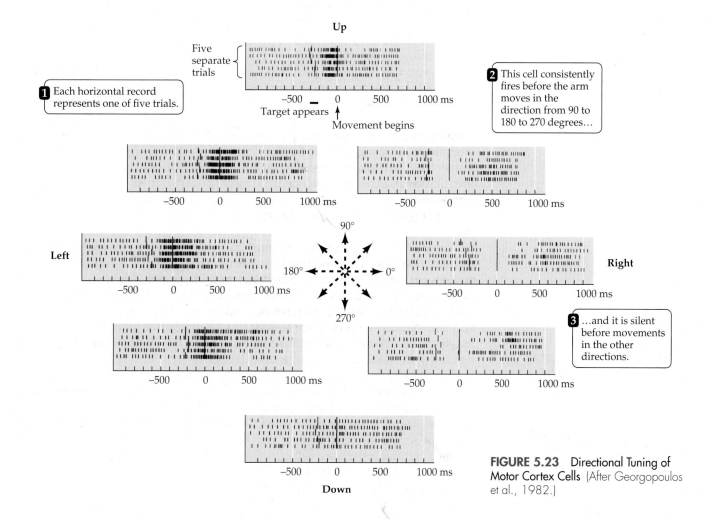

FIGURE 5.23 Directional Tuning of Motor Cortex Cells (After Georgopoulos et al., 1982.)

(A) Before training

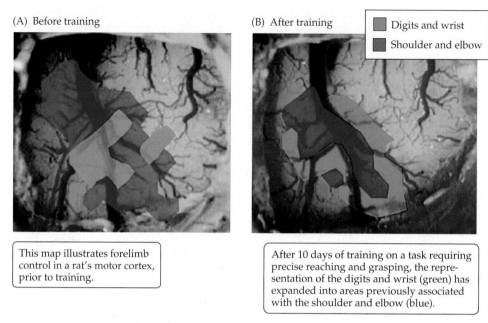

(B) After training

Digits and wrist
Shoulder and elbow

This map illustrates forelimb control in a rat's motor cortex, prior to training.

After 10 days of training on a task requiring precise reaching and grasping, the representation of the digits and wrist (green) has expanded into areas previously associated with the shoulder and elbow (blue).

FIGURE 5.24 Motor Learning Causes Remapping of Motor Cortex (Courtesy of J. Kleim.)

movement, but for any one cell, discharge rates are highest in one particular direction. Only by averaging the activity of hundreds of M1 neurons at once can we predict the direction of arm movements with reasonable accuracy. But of course, *millions* of M1 neurons are available, so in principle a larger sampling would provide a more accurate prediction.

Motor representations in M1 are not static; they change as a result of training. For example, M1 is wider in piano players, especially in the hand area, than in nonmusicians. The younger the musician was at the start of musical training, the larger the gyrus is in adulthood (Amunts et al., 1997), so this expansion of M1 seems to be in response to the experience of musical training. Studies using transcranial magnetic stimulation (TMS) (see Chapter 2) to noninvasively stimulate cortical neurons have shown that the movements produced by a patch of M1 may change with repeated use (Classen et al., 1998) or as a result of motor learning. In rats, this cortical plasticity associated with motor learning has been directly observed by means of sophisticated mapping of the motor cortex before and after extended training of a new skill (Monfils et al., 2005) (**FIGURE 5.24**).

Just anterior to M1 are cortical regions, collectively known as **nonprimary motor cortex**, that make additional crucial contributions to motor control. Nonprimary motor systems can contribute to behavior directly, through communication with lower levels of the motor hierarchy in the brainstem and spinal cord systems, as well as indirectly, through M1. The traditional account of nonprimary motor cortex emphasizes two main regions: the **supplementary motor area (SMA)**, which lies mainly on the medial aspect of the hemisphere, and the **premotor cortex**, which is anterior to the primary motor cortex (**FIGURE 5.25**).

The SMA seems important for the *initiation* of movement sequences, especially when they're being executed according to an internal preprogrammed plan (Tanji, 2001). In contrast, the premotor cortex seems to be activated when motor sequences are guided by *external* events (Halsband et al., 1994; Larsson et al., 1996). However, evidence is mounting that premotor cortex is not a single system, but really a mosaic of different units, controlling groups of motor behaviors that cluster together

nonprimary motor cortex Frontal lobe regions adjacent to the primary motor cortex that contribute to motor control and modulate the activity of the primary motor cortex.

supplementary motor area (SMA) A region of nonprimary motor cortex that receives input from the basal ganglia and modulates the activity of the primary motor cortex.

premotor cortex A region of nonprimary motor cortex just anterior to the primary motor cortex.

(A) Lateral view

(B) Medial view

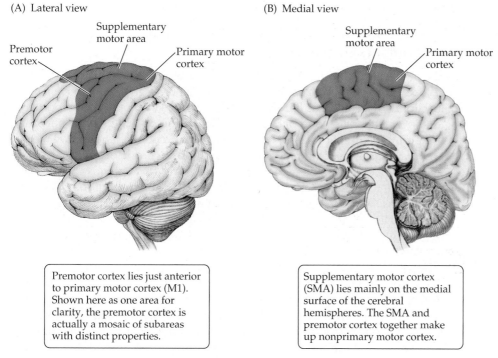

Premotor cortex lies just anterior to primary motor cortex (M1). Shown here as one area for clarity, the premotor cortex is actually a mosaic of subareas with distinct properties.

Supplementary motor cortex (SMA) lies mainly on the medial surface of the cerebral hemispheres. The SMA and premotor cortex together make up nonprimary motor cortex.

FIGURE 5.25 Human Motor Cortical Areas

into major categories: defensive movements, feeding behavior, and so on (Graziano, 2006; Graziano and Aflalo, 2007). This organization suggests that motor and premotor areas mostly map *behaviors*, rather than mapping specific *movements*, as in M1.

Strokes or other injuries in motor areas of the cortex tend to result in paralysis (**plegia**) or weakness (**paresis**) of voluntary movements, usually on the contralateral side of the body (*hemiplegia* or *hemiparesis*). Damage to nonmotor zones of the cerebral cortex, such as some regions of parietal or frontal association cortex, produces more-complicated changes in motor control, such as **apraxia** (from the Greek *a*, "not," and *praxis*, "action"), the inability to carry out complex movements even though paralysis or weakness is not evident and language comprehension and motivation are intact. There are several subtypes of apraxia, but in general it's as though the patient is unable to work out the sequence of movements required to perform a desired behavior—a high-level motor-programming problem.

plegia Paralysis, the loss of the ability to move.

paresis Muscular weakness, often the result of damage to motor cortex.

apraxia An impairment in the ability to carry out complex movements, even though there is no muscle paralysis.

how's it going ❓

1. Describe the arrangement of muscles and joints that allows movement.
2. Briefly describe the main components of a motor unit.
3. Define *proprioception*. Explain how two specialized sensors in muscle provide feedback about the muscle's current state.
4. Provide a summary of the path taken by motor fibers innervating the skeletal musculature—from the level of the brain, through the spinal cord, to the muscle targets.
5. Where is primary motor cortex located, and how is it organized?
6. Distinguish between the pyramidal and extrapyramidal systems.
7. What are some of the contributions of nonprimary motor cortex?

Mirror neurons in premotor cortex track movements in others

Recent research has discovered that a subregion of premotor cortex (called *F5*) may contain a population of remarkable neurons that seem to fulfill two functions. These neurons fire shortly before a monkey makes a very particular movement of the hand and arm to reach for an object; different neurons fire during different reaching movements. The data thus suggest that these neurons trigger specific movements. But these neurons also seem to fire whenever the monkey sees *another* monkey (or a human) make that same movement (**FIGURE 5.26**). These cells are called **mirror neurons** because they fire as though the monkey were imagining doing the same thing as the other individual. Mirror neurons are also found in adult humans (Buccino et al., 2004) and children (Lepage and Theoret, 2006), both in premotor and in other cortical locations.

Because the activity of these neurons suggests that they are important in the understanding of other individuals' actions (Rizzolatti and Craighero, 2004), an intriguing notion is that mirror neurons could be part of a neural system for empathy. Thus, there has been a great deal of speculation about the function of mirror neurons in the imitating behavior of human infants, the evolution of language, and other behavior (Gallese and Sinigaglia, 2011). Some have speculated that people with autism spectrum disorder, which is characterized by a failure to anticipate other people's thinking and actions, may have a deficit in mirror neuron activity (J. H. Williams et al., 2006). Note, however, that the specific functions ascribed to mirror neurons remain somewhat controversial (Caramazza et al., 2014).

FIGURE 5.26   Mirror Neurons

■ **Question**

The researchers hypothesized that neurons of the premotor cortex, in a ventral subregion called F5, encode specific and detailed movements rather than muscle contractions.

■ **Experiment**

The activity of single F5 neurons was recorded while the monkey made reaching movements.

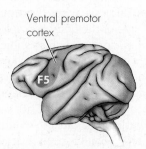

Ventral premotor cortex

F5

Extrapyramidal systems regulate and fine-tune motor commands

Earlier we noted that extrapyramidal projections—the motor fibers outside the pyramidal tracts—are especially important in modulation and ongoing control of movement. Two of the most important sources of extrapyramidal fibers are the basal ganglia and the cerebellum.

As we saw in Chapter 2, the **basal ganglia** are a group of several interconnected forebrain nuclei (especially the caudate nucleus, putamen, and globus pallidus), with strong inputs from the substantia nigra and the subthalamic nucleus. The basal ganglia receive inputs from wide expanses of the cortex via the thalamus, forming a loop from the cortex through the basal ganglia and thalamus and back to the cortex. The

mirror neuron   A neuron that is active both when an individual makes a particular movement and when that individual sees another individual make the same movement.

basal ganglia   A group of forebrain nuclei, including caudate nucleus, globus pallidus, and putamen, found deep within the cerebral hemispheres.

■ **Result**

The neurons fired shortly before the monkey made a specific movement, in accordance with the initial hypothesis. But to the experimenters' surprise, the neurons also became active when the monkey simply watched an experimenter perform the same movement, as if the monkey was *imagining* making the movement.

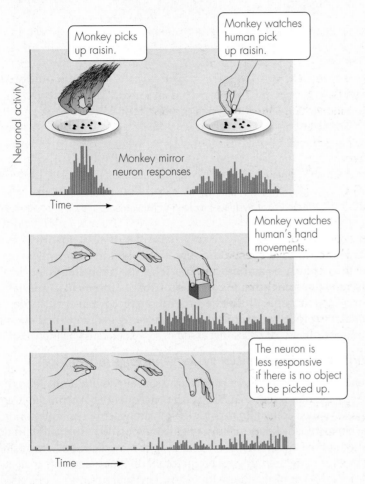

■ **Conclusion**

These "mirror neurons" may be part of a system for analyzing the behavior of others (Umilta et al., 2001).

basal ganglia help control the amplitude and direction of movement, and changes in activity in regions of the basal ganglia appear to be important for the initiation of movement (**FIGURE 5.27**). Much of the motor function of the basal ganglia appears to be the modulation of activity started by other brain circuits, such as the motor pathways of the cortex (see Figure 5.22). The basal ganglia are especially important for movements performed by memory, in contrast to those guided by sensory control (Graybiel et al., 1994).

Inputs to the **cerebellum** come both from sensory sources and from other brain motor systems. Sensory inputs include the muscle and joint receptors and the vestibular, somatosensory, visual, and auditory systems. Both pyramidal and nonpyramidal

cerebellum A structure located at the back of the brain, dorsal to the pons, that is involved in the central regulation of movement, and in some forms of learning.

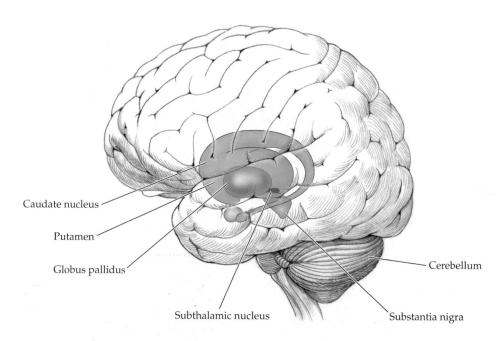

Caudate nucleus

Putamen

Globus pallidus

Cerebellum

Subthalamic nucleus

Substantia nigra

FIGURE 5.27 Subcortical Systems Involved in Movement

pathways contribute inputs to the cerebellum and in turn receive outputs—all of which are inhibitory—from the deep nuclei of the cerebellum. The cerebellum helps establish and fine-tune neural programs for *skilled* movements, especially the kinds of rapid, repeated movements that become automatic (Y. Liu et al., 1999). Also, we now know that the cerebellum is crucial for a wide variety of motor and nonmotor learning (Katz and Steinmetz, 2002), as discussed in more detail in Chapter 13.

Damage to extrapyramidal systems impairs movement

Different constellations of symptoms are associated with damage to the various extrapyramidal motor structures. The exact consequences of cerebellar damage depend on the part of the cerebellum that has been damaged, but common motor symptoms include characteristic abnormalities of gait and posture, especially **ataxia** (loss of coordination) of the legs. Other cerebellar lesions may cause **decomposition of movement** (in which gestures are broken up into individual segments instead of being executed smoothly) or difficulties with gaze and visual tracking of objects. The anatomy of the cerebellum and the symptomatology of cerebellar disease are discussed in more detail in **A STEP FURTHER 5.4**, on the website.

Two diseases that target the basal ganglia reveal important aspects of extrapyramidal contributions to motor control. Patients with **Parkinson's disease** show progressive degeneration of dopamine-containing cells in the **substantia nigra**. Loss of these neurons, which project to the caudate nucleus and putamen, is associated with a cluster of symptoms that are all too familiar: slow movement, tremors of the hands and face while at rest, a rigid bearing, and diminished facial expressions. Patients who suffer from Parkinson's show few spontaneous actions and have great difficulty in all motor efforts, no matter how routine.

Whereas damage to the basal ganglia in Parkinson's disease *slows* movement, other kinds of basal ganglia disorders cause *excessive* movement. The first symptoms of **Huntington's disease** are subtle behavioral changes: clumsiness, and twitches in the fingers and face. Subtlety is rapidly lost as the illness progresses; a continuing stream of involuntary jerks engulfs the entire body. Aimless movements of the eyes, jerky leg movements, and writhing of the body make even routine activity a major

ataxia A loss of movement coordination, often caused by disease of the cerebellum.

decomposition of movement Difficulty of movement in which gestures are broken up into individual segments instead of being executed smoothly; it is a symptom of cerebellar lesions.

Parkinson's disease A degenerative neurological disorder, characterized by tremors at rest, muscular rigidity, and reduction in voluntary movement, caused by loss of the dopaminergic neurons of the substantia nigra.

substantia nigra A brainstem structure that innervates the basal ganglia and is a major source of dopaminergic projections.

Huntington's disease A genetic disorder, with onset in middle age, in which the destruction of basal ganglia results in a syndrome of abrupt, involuntary writhing movements and changes in mental functioning.

challenge, exacerbated in later stages of the disease by intellectual deterioration. The neuroanatomical basis of this disorder is the progressive destruction of the basal ganglia, especially the caudate nucleus and the putamen, as well as impairment of the cerebral cortex.

The sensorimotor system is a very large and complex network that is vulnerable to a wide variety of injuries and diseases. Disease processes that target various parts of the network—ranging from viral infections of motoneurons, to autoimmune, genetic, and other pathological processes that attack the brain and spinal cord motor system— are covered in detail in **A STEP FURTHER 5.5**, on the website.

Although much remains to be discovered, there is more reason than ever to look forward to the introduction of effective treatments for motor disorders. Scientists are learning more and more about what goes wrong in Parkinson's and Huntington's diseases, and their continuing research efforts may pave the way to new therapies. Furthermore, it now seems likely that people with paralyzing spinal cord injuries will one day be able to manipulate robotic arms and legs with their minds.

SIGNS & SYMPTOMS

Hazards of Painlessness

Gabby was a happy baby; perhaps too happy. Minor accidents didn't seem to bother her. In fact, she not only didn't cry when she poked herself in the eye, she kept sticking her fingers in her eyes until eventually one had to be surgically removed (**FIGURE 5.28**). Her folks began restraining her hands, which turned out to be a good thing because when Gabby's teeth came in, she would bite her fingers to the bone otherwise. The parents had those baby teeth removed too, figuring that when Gabby's adult teeth came in, they could communicate with her to be careful. Gabby's problem is congenital insensitivity to pain (CIP). Like the Pakistani boy we mentioned earlier, Gabby carries a gene that makes a dysfunctional version of the sodium channel (*SCN9A*) used by nociceptors to produce action potentials (Oppenheim, 2006).

FIGURE 5.28   Extra Protection
Gabby, shown here at age 3, cannot feel pain, so her parents had her wear goggles to keep her from poking out her eyes.

Thanks in large part to a superattentive family, Gabby has survived to become a cheerful, outgoing teenager. She has also learned to take care of herself. If Gabby impulsively plunges her hand into boiling water, she must remember to pull it out right away, but not as fast as most people would. Gabby will always have to be vigilant about avoiding injury and about looking out for any injuries that might have happened without her being aware of them. So the next time you burn your finger or stub your toe, feel free to curse if you want, but afterward comfort yourself with the knowledge that these little slings and arrows are great blessings in disguise.

how's it going ?

1. What are mirror neurons, and what is their significance?
2. What are the symptoms of Parkinson's disease, and what brain changes cause it?
3. What are the symptoms of Huntington's disease, and what brain changes cause it?
4. Children of people with Huntington's disease have a fifty-fifty chance of inheriting the gene causing it. If you had a parent with Huntington's, would you want to take the test to see if you carry the disease?

Recommended Reading

Ballantyne, J. C., and Fishman, S. M. (Eds.). (2010). *Bonica's Management of Pain* (4th ed.). Philadelphia, PA: Lippincott.

Basbaum, A. I., and Bushnell, M. C. (2008). *Science of Pain.* New York, NY: Academic Press.

Cytowic, R. E., and Eagleman, D. M. (2009). *Wednesday Is Indigo Blue.* Cambridge, MA: MIT Press.

Graziano, M. (2006)."The Organization of Behavioral Repertoire in Motor Cortex." *Annual Review of Neuroscience, 29,* pp. 105–134.

Lumpkin, E. A., and Caterina, M. J. (2007)."Mechanisms of Sensory Transduction in the Skin." *Nature, 445,* 858–865.

McMahon, C., Koltzenberg, M., Tracey, I., and Turk, D. C. (2013). *Wall and Melzack's Textbook of Pain* (6th ed.). Philadelphia, PA: Saunders.

Merchant, H., and Georgopoulos, A. P. (2006)."Neurophysiology of Perceptual and Motor Aspects of Interception."*Journal of Neurophysiology, 95,* 1–13.

Purves, D., Augustine, G. J., Fitzpatrick, D., Hall, W., et al. (Eds.). (2012). *Neuroscience* (5th ed.). Sunderland, MA: Sinauer. (See Unit III:"Movement and Its Central Control," Chapters 16–21.)

Wolfe, J. M., Kluender, J. R., Levi, D. M., Bartoshuk, L. M., et al. (2015). *Sensation & Perception* (4th ed.). Sunderland, MA: Sinauer.

VISUAL SUMMARY

5

You should be able to relate each summary to the adjacent illustration, including structures and processes. Go to the online version of this summary (scan the QR code above) for links to figures, animations, and activities that will help you consolidate the material.

1 **Receptor cells** in sensory organs furnish only selected information to the brain. For example, our skin contains four kinds of touch receptors, each specialized for detecting a specific **stimulus**. Species tend to sense only those environmental stimuli that are important to them. Review **Figures 5.1** and **5.3**, Animation 5.2, Activity 5.1

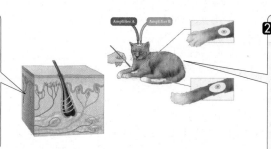

2 **Sensory transduction** is the conversion of an environmental stimulus into a **generator potential** in sensory receptors, leading to the activation of sensory neurons. The **receptive field** of a neuron is the region in space where a stimulus will change the firing of that cell. The receptive fields of neurons may be very different at successive levels of the sensory pathway. Review **Figures 5.4** and **5.5**, Animation 5.3

3 The succession of levels in a sensory pathway allows for increasingly elaborate kinds of processing. Touch information from the skin courses through a distinct spinal pathway, the **dorsal column system**. Review **Figure 5.7**

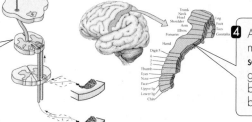

4 At the level of the cerebral cortex are multiple maps of the body surface. **Primary somatosensory cortex**, or S1, located in the postcentral gyrus, contains a map of the contralateral body, which overrepresents highly sensitive body regions. Review **Figures 5.8** and **5.9**

5 **Free nerve endings** detect mechanical damage or temperature changes because they have specialized receptor proteins that detect these conditions. **Pain**, **nociceptors**, temperature, and itch information enters the spinal cord, crosses the midline, and ascends through the **anterolateral system** (or spinothalamic system) to the brain. Review **Figures 5.11** and **5.12**, Activity 5.2

6 Pain sensation, detected by nociceptors, is subject to many modulating influences, including regions of the brain and spinal cord that employ **endogenous opioids**. Opioid systems are also active in the **placebo effect** and in other techniques, such as **acupuncture**. Review **Figure 5.14**

7 Motor control systems are organized into a hierarchy that consists of the skeletal system and associated muscles, the spinal cord, the brainstem, and various parts of the brain, including the **primary (M1)** and nonprimary motor cortices, the **cerebellum**, and the **basal ganglia**. Review **Figure 5.16**

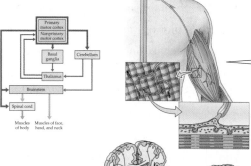

8 Muscles around a joint work in pairs. **Antagonists** work in opposition; **synergists** work together. Action potentials travel over motor nerve fibers (axons from **motoneurons**) and reach muscle fibers at the **neuromuscular junction**, releasing **acetylcholine (Ach)** to trigger muscle contraction. Review **Figures 5.18** and **5.19**

9 **Muscle spindles** and **Golgi tendon organs**—sensory receptors in the muscles and tendons, respectively—transmit crucial information about muscle activities to the central nervous system. The sensitivity of the muscle spindle can be adjusted by efferent impulses that control the length of the spindle. Review **Figures 5.20** and **5.21**, Animation 5.4

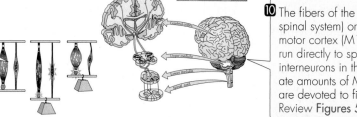

10 The fibers of the **pyramidal system** (or corticospinal system) originate mainly in the primary motor cortex (M1) and adjacent regions and run directly to spinal motoneurons or to interneurons in the spinal cord. Disproportionate amounts of M1, in the **precentral gyrus**, are devoted to finely controlled muscles. Review **Figures 5.22** and **5.25**

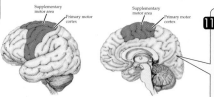

11 The **extrapyramidal system**, consisting of brain regions that modulate movement, includes the basal ganglia (caudate nucleus, putamen, globus pallidus, and **substantia nigra**), some major brainstem nuclei (thalamic nuclei, reticular formation, and red nucleus), and the cerebellum. Review **Figure 5.25**, Activity 5.3

chapter

6

Hearing, Balance, Taste, and Smell

Hold the Phone

It's like a classic horror movie scene: a scientist using amazing technology in an attempt to reanimate parts of dead bodies, seeking out nature's secrets. But when the young Hungarian engineer Georg von Békésy started experimenting with cadavers in the 1920s, he was not trying to create life. He was seeking an answer to a practical question: Why are human ears so much more sensitive than most microphones? Békésy thought that learning how the human ear detects sounds might allow him to make a better microphone for his employer, the Hungarian phone company. He gathered cadavers from local hospitals and came up with a clever dissection that would reveal the inner ear without destroying it. (His work was not always ap-

preciated by his fellow engineers; they didn't like finding their drill press full of human bone dust in the morning.)

Bringing his background in physics to bear on the question, Békésy devised exquisitely precise physical models and biophysical experiments that let him measure extremely brief, minuscule movements in the inner ear. His subsequent discoveries provided us with the key to understanding how we translate a stream of auditory data—sounds—into neural activity that the brain can understand.

In the end, Békésy did not come up with a better microphone, but his discoveries have helped restore hearing to thousands of people who once were deaf, as we'll see in this chapter.

To see the video
Inside the Ear,
go to

2e.mindsmachine.com/6.1

Your existence is directly attributable to the keen senses possessed by your distant ancestors—senses that enabled them to find food and mates, and to avoid predators and other dangers. In this chapter we consider some of the incredible sensors that let us monitor important signals from distant sources, especially sounds (by audition) and smells (by olfaction). We also discuss related systems for detecting position and movement of the body (the vestibular system, related to the auditory system) and tastes of foods (the gustatory or taste sense, which, like olfaction is a chemical sense). We begin with hearing, because audition evolved from special mechanical receptors related to the touch system that we discussed in Chapter 5.

To view the
Brain Explorer,
go to

2e.mindsmachine.com/6.2

decibel (dB) A measure of sound intensity, perceived as loudness.

hertz (Hz) Cycles per second, as of an auditory stimulus. Hertz is a measure of frequency.

pure tone A tone with a single frequency of vibration.

amplitude Also called *intensity*. The force that sound exerts per unit area, usually measured as dynes per square centimeter. In practical terms, amplitude corresponds to the volume of a sound.

PART I
Hearing and Balance

Hearing is vital for the survival of many animals. Humans can produce an impressive variety of vocalizations—from barely audible murmurs to soaring flights of song—but we especially rely on speech sounds to provide a basis for our social relations and for the transmission of knowledge between individuals. Across the animal kingdom, species produce and perceive sounds in wildly different ways, shaped by their unique adaptive needs. Birds sing and crickets chirp in order to attract mates; monkeys grunt and screech and burble to signal comfort, danger, and pleasure; owls and bats exploit the directional property of sound to locate prey and avoid obstacles in the dark; and whales produce sounds that can travel hundreds of miles in the ocean. In every case, we find that species have unique auditory capabilities that helped their ancestors solve important problems of survival.

Your auditory system detects changes in the vibration of air molecules that are caused by sound sources, sensing both the *intensity* of sounds (measured in **decibels [dB]** and perceived as loudness) and their *frequency* (measured in cycles per second, or **hertz [Hz]**, and perceived as pitch). Your ears are so good at detecting sounds in the range of normal speech that if they were any more sensitive, you would be distracted by the noise of air molecules bouncing against each other in your ear canals! **BOX 6.1** describes some basic properties of sound.

BOX 6.1 The Basics of Sound

We perceive a repetitive pattern of local increases and decreases in air pressure as sound. Usually this oscillation is caused by a vibrating object, such as a loudspeaker or a person's larynx during speaking. A single alternation of compression and expansion of air is called one *cycle*.

The figure illustrates the changes in pressure produced by a vibrating loudspeaker.

Because the sound produced by the loudspeaker here has only one frequency of vibration, it is called a **pure tone** and can be represented by a sine wave. A pure tone is described physically in terms of two measures:

1. **Amplitude**, or *intensity*, is usually measured as sound pressure, or force per unit area, in dynes per square centimeter (dyn/cm^2). Our perception of amplitude is termed *loudness*, expressed as decibels (dB). The decibel scale is logarithmic; one decibel is the threshold for human hearing, a whisper is about 20 dB, and a departing jet airliner a couple of hundred feet overhead—a sound a million times as intense—is about 120 dB.

2. **Frequency**, or the number of cycles per second, is measured in **hertz (Hz)**. For example, the musical note middle A has a frequency of 440 Hz; middle C is 361.6 Hz. Our perception of frequency is termed **pitch**.

Most sounds are more complicated than a pure tone. For example, a sound made by a musical instrument contains a fundamental frequency and harmonics. The **fundamental** is the basic frequency, and the **harmonics** are multiples of the fundamental. For example, if the fundamental is 440 Hz, the harmonics are 880, 1,320, 1,760, and so on. When different instruments play the same note, the notes differ in the relative intensities of the various harmonics; this difference is what gives each instrument its characteristic sound quality, or **timbre**.

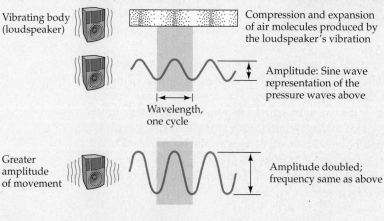

Amplitude and frequency of sound waves

Vibrating body (loudspeaker)

Compression and expansion of air molecules produced by the loudspeaker's vibration

Amplitude: Sine wave representation of the pressure waves above

Wavelength, one cycle

Greater amplitude of movement

Amplitude doubled; frequency same as above

Greater frequency of movement

Amplitude same as original; frequency doubled

Wavelength, one cycle

Each Part of the Ear Performs a Specific Function in Hearing

How do small vibrations in air become the speech, music, and other sounds we hear? The outer ears have been shaped through evolution to capture biologically important sound vibrations and direct them into the inner parts of the ear, where the mechanical force of sound is **transduced** into neural activity.

The external ear captures, focuses, and filters sound

The oddly shaped fleshy objects that most people call *ears* are properly known as **pinnae** (singular *pinna*). Aside from their occasional utility as handles and jewelry hangers, the pinnae funnel sound waves into the second part of the external ear: the **ear canal** (or *auditory canal*). The pinna is a distinctly mammalian characteristic, and mammals show a wide array of ear shapes and sizes. Furthermore, although only some humans can move their ears, and only enough to entertain children, many other mammals deftly shape and swivel their pinnae to help determine the source of a sound. Animals with exceptional auditory localization abilities, such as bats, may have especially mobile ears.

The "hills and valleys" of the pinna modify the character of sound that reaches the middle ear. Some frequencies of sound are enhanced; others are suppressed. For example, the shape of the human ear especially increases the reception of sounds

frequency   The number of cycles per second in a sound wave, measured in hertz.

pitch   A dimension of auditory experience in which sounds vary from low to high.

fundamental   The predominant frequency of an auditory tone. or a visual scene.

harmonic   A multiple of a particular frequency call the *fundamental*.

timbre   The characteristic sound quality of a musical instrument, as determined by the relative intensities of its various harmonics.

transduction   The conversion of one form of energy to another.

pinna   The external part of the ear.

ear canal   Also called *auditory canal*. The tube leading from the pinna to the tympanic membrane.

The Ears Have It The external ears, or pinnae, of mammals come in a variety of shapes, each adapted to a particular ecological niche. Many mammals can move their ears to direct them toward a particular sound. In such cases, the brain must account for the position of the ear to judge where a particular sound came from.

inner ear The cochlea and vestibular apparatus.

middle ear The cavity between the tympanic membrane and the cochlea.

tympanic membrane Also called *eardrum*. The partition between the external ear and the middle ear.

ossicles Three small bones (incus, malleus, and stapes) that transmit vibration across the middle ear, from the tympanic membrane to the oval window.

between 2,000 and 5,000 Hz—a frequency range that is important for speech perception. The shape of the external ear—and, in many species, the direction in which it is being pointed—provides additional cues about the direction and distance of the source of a sound, as we will discuss later in this chapter.

The middle ear concentrates sound energies

A collection of tiny structures made of membrane, muscle, and bone—essentially a tiny biological microphone—links the ear canal to the neural receptor cells of the **inner ear** (**FIGURE 6.1A**). This **middle ear** (**FIGURE 6.1B**) consists of the taut **tympanic membrane** (*eardrum*) sealing the end of the ear canal plus a chain of tiny bones, called **ossicles**, that mechanically couple the tympanic membrane to the inner ear at a specialized patch of membrane called the **oval window**. These ossicles, the smallest bones in the body, are called the **malleus** (Latin for "hammer"), the **incus** (Latin for "anvil"), and the **stapes** (Latin for "stirrup").

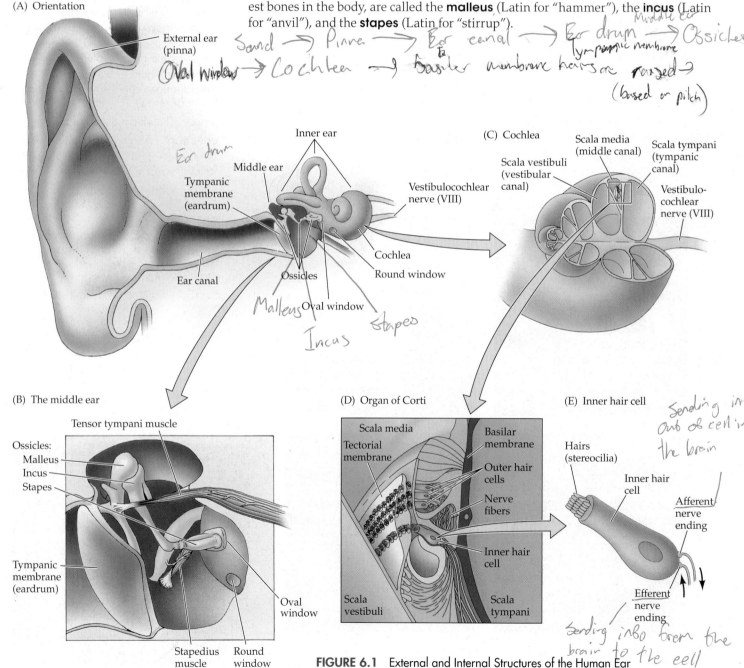

FIGURE 6.1 External and Internal Structures of the Human Ear

Sound waves in the air strike the tympanic membrane and cause it to vibrate with the same frequency as the sound; as a result, the ossicles start moving too. Because of how they are attached to the eardrum, the ossicles concentrate and amplify the vibrations, focusing the pressures collected from the relatively large tympanic membrane onto the small oval window. This amplification is crucial for converting vibrations in air into movements of fluid in the inner ear, as we'll see shortly.

The middle ear is equipped with the equivalent of a volume control, which helps protect against the damaging forces of extremely loud noises. Two tiny muscles—one called the *tensor tympani* and the other the *stapedius*, because of their proximity to the tympanic membrane and the stapes, respectively (see Figure 6.1B)—attach to the ends of the chain of ossicles. Within 200 milliseconds of the arrival of a loud sound, the brain signals the muscles to contract, which stiffens the chain of ossicles and reduces the effectiveness of the sounds. Interestingly, the middle-ear muscles activate just *before* we produce self-made sounds like speech or coughing, which is why we don't perceive our own sounds as distractingly loud.

The cochlea converts vibrational energy into neural activity

The part of the inner ear that ultimately converts vibrations from sound into neural activity—the coiled, fluid-filled **cochlea** (from the Greek *kochlos*, "snail")—is a marvel of miniaturization (**FIGURES 6.1C** and **D**). In an adult, the cochlea measures only about 4 millimeters in diameter—about the size of a pea. Fully unrolled, the cochlea would be about 35–40 millimeters long.

The cochlea is a coil of three parallel canals: (1) the **scala vestibuli** (also called the *vestibular canal*), (2) the **scala media** (*middle canal*), and (3) the **scala tympani** (*tympanic canal*). The oval window is adjacent to the *base* of the spiral; the distant end, like the tip of a snail's shell, is referred to as the *apex*. Because the canals are filled with noncompressible fluid, movement inside the cochlea in response to a push from the stapes onto the oval window requires a second membrane-covered window that can bulge outward a bit. That second membrane is the **round window**, which separates the tympanic canal from the middle ear (see Figure 6.1B).

The portion of the cochlea that converts sounds into neural activity is known as the **organ of Corti** (see Figure 6.1D). It consists of three main structures: (1) the sensory cells (**hair cells**) (**FIGURE 6.1E**), (2) an elaborate framework of supporting cells, and (3) the terminations of the auditory nerve fibers. The base of the organ of Corti is the **basilar membrane**. When the stapes moves in and out as a result of sound waves hitting the eardrum, it sets in motion waves or ripples in the fluid of the vestibular canal, which in turn cause the basilar membrane to ripple, like shaking out a rug. The basilar membrane is tapered—it's about 5 times wider at the apex of the cochlea than at the base. (It may seem counterintuitive for the basilar membrane to be narrowest at the *base* of the cochlea, where the canals are *widest*, but that's how it is. The different parts of the basilar membrane, as it gradually changes from narrow at the base of the cochlea to wide at the apex, show their strongest responses to different frequencies of sound. High frequencies have their greatest effect near the base, where the membrane is narrow and comparatively stiff; low-frequency sounds produce a larger response near the apex, where the membrane is wider and floppier (Ashmore, 1994).

oval window   The opening from the middle ear to the inner ear.

malleus   Latin for "hammer." A middle-ear bone that is connected to the tympanic membrane.

incus   Latin for "anvil." A middle-ear bone situated between the malleus and the stapes.

stapes   Latin for "stirrup." A middle-ear bone that is connected to the oval window.

cochlea   A snail-shaped structure in the inner ear that contains the primary receptor cells for hearing.

scala vestibuli   Also called *vestibular canal*. One of three principal canals running along the length of the cochlea.

scala media   Also called *middle canal*. The central of the three spiraling canals inside the cochlea, situated between the vestibular canal and the tympanic canal.

scala tympani   Also called *tympanic canal*. One of three principal canals running along the length of the cochlea.

round window   A membrane separating the tympanic canal from the middle ear.

organ of Corti   A structure in the inner ear that lies on the basilar membrane of the cochlea and contains the hair cells and terminations of the auditory nerve.

hair cell   One of the receptor cells for hearing in the cochlea, named for the stereocilia that protrude from the top of the cell and transduce vibrational energy in the cochlea into neural activity.

basilar membrane   A membrane in the cochlea that contains the principal structures involved in auditory transduction.

A Touching Friendship Helen Keller, who was both blind and deaf, said, "Blindness deprives you of contact with things; deafness deprives you of contact with people"—a poignant reminder of the importance of speech for our social lives. Here, Keller communicates with her young friend by feeling the boy's face as he speaks and makes facial expressions. Rather than living in sensory and social isolation, Keller honed her intact senses to such a degree that she was able to become a noted teacher and writer.

Phosphns di

researchers at work

Georg von Békésy and the cochlear wave

The discovery of the mechanics of the basilar membrane garnered a Nobel Prize for Georg von Békésy in 1961 (**FIGURE 6.2**).

To see the animation
Sound Transduction,
go to

2e.mindsmachine.com/6.3

FIGURE 6.2 Deformation of the Basilar Membrane Encodes Sound Frequencies

■ **Hypothesis**

Sound waves cause the basilar membrane to ripple.

■ **Experiment**

(1) Partially dissect a human cochlea, replacing fluid with a gel containing particles of coal and aluminum.

(2) Bounce intense flash of light off the basilar membrane during vibration at the oval window.
Record the displacement of aluminum particles to measure how much the basilar membrane has deformed.

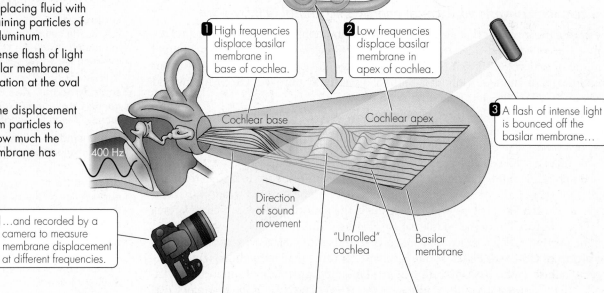

1 High frequencies displace basilar membrane in base of cochlea.

2 Low frequencies displace basilar membrane in apex of cochlea.

3 A flash of intense light is bounced off the basilar membrane...

4 ...and recorded by a camera to measure membrane displacement at different frequencies.

"Unrolling" of cochlea

Cochlear base   Cochlear apex

Direction of sound movement

"Unrolled" cochlea

Basilar membrane

400 Hz

Relative amplitude of movement (μm)

1600 Hz   400 Hz   25 Hz

Distance from stapes (mm)

■ **Result**

"Standing waves" are observed on the basilar membrane, their position corresponding to frequency of the vibration.

These experiments revealed an orderly map of frequencies along the tapering length of the basilar membrane, from low frequencies at the broad and floppy apex to high frequencies at the stiff and narrow base.

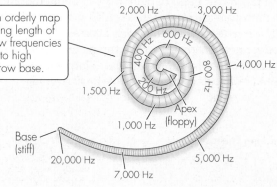

2,000 Hz   3,000 Hz
600 Hz
400 Hz   800 Hz   4,000 Hz
200 Hz
1,500 Hz
1,000 Hz   Apex (floppy)
Base (stiff)
20,000 Hz   5,000 Hz
7,000 Hz

■ **Conclusion**

The frequency of a sound is encoded by the specific location on the basilar membrane that shows the largest vibration in response to the sound. High frequencies displace the basilar membrane at the base of the cochlea; low frequencies cause displacement at the apex. This differential response depending on location along the membrane has become known as *place coding*.

FIGURE 6.3 How Auditory Stimulation Affects the Stereocilia on Cochlear Hair Cells (After Beurg et al., 2009, micrograph courtesy of A. J. Hudspeth.)

Tip links connect to large, nonselective ion channels on adjacent stereocilia.

Bending of the stereocilia tightens tip links and opens the channels, allowing K+ and Ca2+ to enter the stereocilia. The resulting depolarization opens Ca2+ channels in the cell's base, causing the release of neurotransmitter that excites nearby nerves, resulting in the transmission of action potentials to the brain.

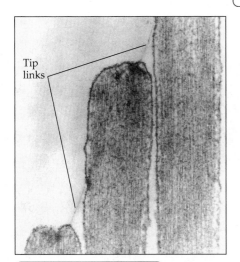

This micrograph of stereocilia shows the threadlike tip links.

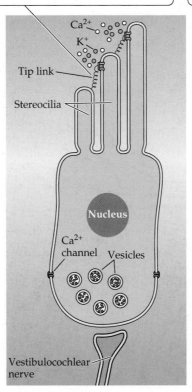

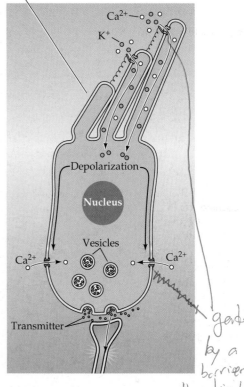

gated by a physical barrier and the tip link pulls it open when it stretches

The hair cells transduce movements of the basilar membrane into electrical signals

The rippling of the basilar membrane is converted into neural activity through the actions of hair cells. Each hair cell features a sloping brush of minuscule hairs (hence the name) called **stereocilia** (singular *stereocilium*) on its upper surface. In Figure 6.1D you'll notice that, although the bases of hair cells are implanted in the basilar membrane, the stereocilia nestle into hollows in the tectorial membrane that lies above. The hair cells—and especially the hairs themselves—thus form a mechanical bridge between the two membranes.

How do hair cells turn movement into neural activity? When sounds induce the basilar membrane to ripple, the movement of the hair cell bases relative to the tectorial membrane causes the stereocilia to bend. Even a tiny deflection of stereocilia is enough to produce a large depolarization of hair cells. This depolarization is the result of a remarkable structural feature: Each stereocilium contains one or two large, nonselective ion channels, resembling trap doors in the membrane. Fine, threadlike fibers called *tip links* connect the channels to the tips of neighboring stereocilia (**FIGURE 6.3**). As the stereocilia bend, even only very slightly, the tension on the tip links physically pops open the ion channels to which they are attached (Hudspeth, 1997; Hudspeth et al., 2000). The sudden opening of the channels allows potassium (K+) and calcium (Ca2+) ions to rush into the hair cell, causing a rapid depolarization. Just as it works in neurons (see Chapter 3), this depolarization leads to a rapid influx of Ca2+ at the base of the hair cell, which in turn causes synaptic vesicles there to fuse with the presynaptic membrane and release neurotransmitter, stimulating adjacent nerve fibers. The stereocilia channels snap shut again in a fraction of a millisecond as the hair cell sways back. This ability to rapidly switch on and off allows hair cells to accurately track the rapid oscillations of the basilar membrane with exquisite sensitivity.

stereocilium A tiny bristle that protrudes from a hair cell in the auditory or vestibular system.

FIGURE 6.4 Auditory Nerve Fibers and Synapses in the Organ of Corti

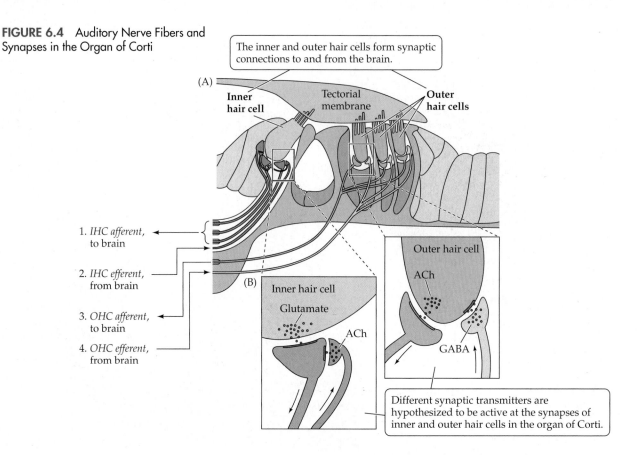

The inner and outer hair cells form synaptic connections to and from the brain.

(A)

Inner hair cell

Tectorial membrane

Outer hair cells

1. *IHC afferent,* to brain

2. *IHC efferent,* from brain

3. *OHC afferent,* to brain

4. *OHC efferent,* from brain

(B)

Inner hair cell

Glutamate

ACh

Outer hair cell

ACh

GABA

Different synaptic transmitters are hypothesized to be active at the synapses of inner and outer hair cells in the organ of Corti.

inner hair cell (IHC) One of the two types of receptor cells for hearing in the cochlea. Compared with outer hair cells, IHCs are positioned closer to the central axis of the coiled cochlea.

outer hair cell (OHC) One of the two types of receptor cells for hearing in the cochlea. Compared with inner hair cells, OHCs are positioned farther from the central axis of the coiled cochlea.

vestibulocochlear nerve Cranial nerve VIII, which runs from the cochlea to the brainstem auditory nuclei.

In the human cochlea, the hair cells are organized into a single row of about 3,500 **inner hair cells** (**IHCs;** called *inner* because they are closer to the central axis of the coiled cochlea) and about 12,000 **outer hair cells** (**OHCs**) in three rows (see Figure 6.1D). Fibers of the **vestibulocochlear nerve** (cranial nerve VIII) contact the bases of the hair cells (see Figures 6.1E and 6.3). Some of these fibers do indeed convey sound information to the brain, but the neural connections of the cochlea are a little more complicated than this. In fact, there are four kinds of neural connections with hair cells, each relying on a different neurotransmitter (Eybalin, 1993), as you can see in **FIGURE 6.4**.

The fibers are distinguished as follows:

1. *IHC afferents* convey to the brain the action potentials that provide the perception of sounds. IHC afferents make up about 95% of the fibers leading to the brain.

2. *IHC efferents* lead from the brain to the IHCs—through which the brain can control the responsiveness of IHCs.

3. *OHC afferents* are thought to convey information to the brain about the mechanical state of the basilar membrane, but *not* the perception of sounds themselves.

4. *OHC efferents* lead from the brain to the OHCs, allowing the brain to activate a remarkable property of OHCs. The OHCs don't detect sound. Instead, OHCs change their length almost instantaneously in response to commands from the brain (Zheng et al., 2000). Through this electromechanical action, OHCs continually modify the stiffness of regions of the basilar membrane, resulting in both sharpened tuning and pronounced amplification (Hudspeth, 2014).

If we record from IHC afferents sending auditory information to the brain, we find that each one has a maximum sensitivity to sound of a particular frequency, but will also respond to neighboring frequencies if the sound is loud enough. For example, the

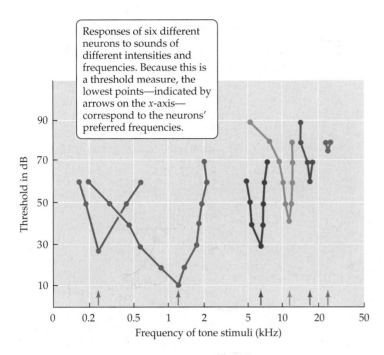

Responses of six different neurons to sounds of different intensities and frequencies. Because this is a threshold measure, the lowest points—indicated by arrows on the *x*-axis—correspond to the neurons' preferred frequencies.

FIGURE 6.5 Tuning Curves of Auditory Nerve Cells (After Kiang, 1965.)

auditory neuron whose responses are shown in red in **FIGURE 6.5** has its *best frequency* at 1,200 Hz (1.2 kHz)—that is, it is sensitive to even a very weak tone at 1,200 Hz—but for sounds that are 20 dB louder, the cell will respond to frequencies from 500 to 1,800 Hz. We call this the cell's *tuning curve*. If the brain received a signal from only one such fiber, it would not be able to tell whether the stimulus was a weak tone of 1,200 Hz or a stronger tone of 500 or 1,800 Hz, or any frequency in between. Instead, the brain analyzes the activity from thousands of such units simultaneously to calculate the intensity and frequency of each sound.

Auditory System Pathways Run from the Brainstem to the Cortex

Most of the auditory part of the vestibulocochlear nerve is made up of nerve fibers carrying information from the IHCs to the brainstem. Here the auditory fibers terminate in the (sensibly named) **cochlear nuclei**, where some initial processing occurs. Output from the cochlear nuclei projects to the **superior olivary nuclei**, each of which receives inputs from both right and left cochlear nuclei. This bilateral input makes the superior olivary nucleus the first brain site at which *binaural* (two-ear) processing occurs. As you might expect, this mechanism plays a key role in localizing sounds by comparing the two ears, as we'll discuss shortly.

The superior olivary nuclei pass information derived from both ears to the **inferior colliculi**, which are the primary auditory centers of the midbrain. Outputs of the inferior colliculi go to the **medial geniculate nuclei** of the thalamus. Pathways from the medial geniculate nuclei extend to several auditory cortical areas.

At every level of the auditory system, from cochlea to auditory cortex (as depicted in **FIGURE 6.6**), auditory pathways display **tonotopic organization**; that is, the pathways for the different *tones* are spatially arranged like a *topo*graphic map (*topos* is Greek for "place") from low frequency (sounds that we perceive as bass) to high frequency (perceived as treble). Furthermore, at the higher levels of the auditory system, auditory neurons are not only excited by certain frequencies, but also *inhibited* by neighboring frequencies, resulting in much sharper tuning of the frequency responses of these cells. This precision helps us discriminate tiny differences in the frequencies of sounds.

cochlear nucleus Either of two brainstem nuclei—left and right—that receive input from auditory hair cells and send output to the superior olivary nuclei.

superior olivary nucleus Either of two brainstem nuclei—left and right—that receive input from both right and left cochlear nuclei and provide the first binaural analysis of auditory information.

inferior colliculi Paired gray matter structures of the dorsal midbrain that process auditory information.

medial geniculate nucleus Either of two nuclei—left and right—in the thalamus that receive input from the inferior colliculi and send output to the auditory cortex.

tonotopic organization A major organizational feature in auditory systems, in which neurons are arranged as an orderly map of stimulus frequency, with cells responsive to high frequencies located at a distance from those responsive to low frequencies.

[Handwritten notes:]
Left hemisphere sounds are processed by the right
3 big steps
Into cochlea → Superior Olivary nucleus
→ Inferior Cochlea → Thalamus

Dorsal/Ventral Pathways
(auditory cortex)
Superior temporal cortex / Lobe

FIGURE 6.6 Auditory Pathways of the Human Brain

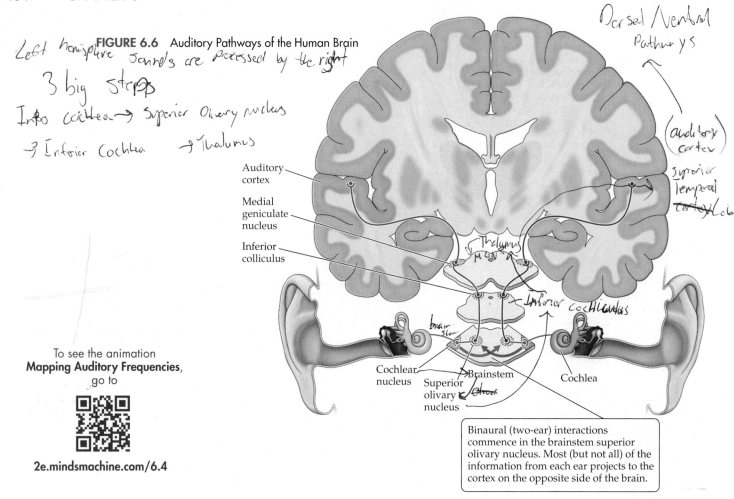

Auditory cortex

Medial geniculate nucleus

Inferior colliculus

Cochlear nucleus

Superior olivary nucleus

Brainstem

Cochlea

[Handwritten labels on figure:] Thalamus MGN Inferior Cochliculus brainstem

Binaural (two-ear) interactions commence in the brainstem superior olivary nucleus. Most (but not all) of the information from each ear projects to the cortex on the opposite side of the brain.

To see the animation
Mapping Auditory Frequencies,
go to

2e.mindsmachine.com/6.4

primary auditory cortex Also called *A1*. The cortical region, located on the superior surface of the temporal lobe, that processes complex sounds transmitted from lower auditory pathways.

Brain-imaging studies in humans have confirmed that many sounds (tones, noises, and so on) activate the **primary auditory cortex (A1)**, which is located on the upper surface of the temporal lobes (**FIGURE 6.7A**). Speech sounds produce similar activation, but they also activate other, more specialized auditory areas (**FIGURE 6.7B**). Interestingly, at least some of these regions are activated when hearing people try to lip-read—that is, to understand someone by watching that person's lips without auditory cues (Calvert et al., 1997; L. E. Bernstein et al., 2002). This suggests that the auditory cortex integrates other, nonauditory, information with sounds. (The organization of auditory cortical areas in other species is described in **A STEP FURTHER 6.1**, on the website.)

how's it going ?

1. Identify the major components of the external ear. What does the external ear do?

2. Identify the three ossicles, and explain their function. To what structures do the ossicles connect, and how is their action moderated?

3. Provide a brief description of the organ of Corti, naming the components that are most important for the perception of sound.

4. Explain how the movement of hair cells transduces sound waves into action potentials. Compare and contrast the functions of inner hair cells and outer hair cells.

5. Sketch the major anatomical components of the auditory projections in the brain. Where does binaural processing first occur? What is tonotopic organization? What kind of processing does auditory cortex perform?

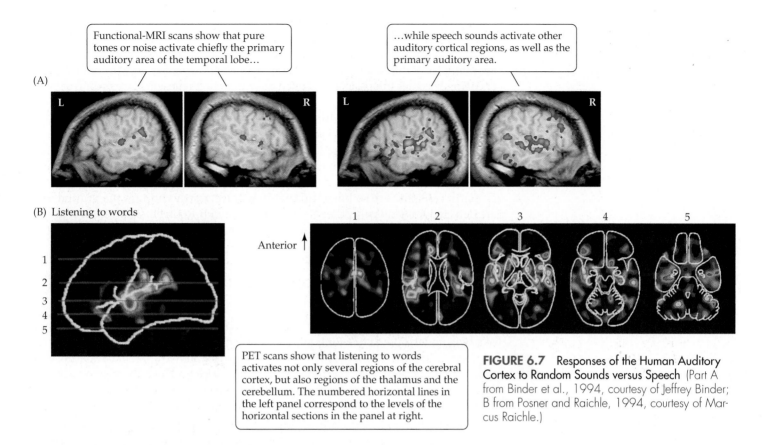

Functional-MRI scans show that pure tones or noise activate chiefly the primary auditory area of the temporal lobe…

…while speech sounds activate other auditory cortical regions, as well as the primary auditory area.

(A)

(B) Listening to words

Anterior ↑

PET scans show that listening to words activates not only several regions of the cerebral cortex, but also regions of the thalamus and the cerebellum. The numbered horizontal lines in the left panel correspond to the levels of the horizontal sections in the panel at right.

FIGURE 6.7 Responses of the Human Auditory Cortex to Random Sounds versus Speech (Part A from Binder et al., 1994, courtesy of Jeffrey Binder; B from Posner and Raichle, 1994, courtesy of Marcus Raichle.)

Our Sense of Pitch Relies on Two Signals from the Cochlea

At least when we're young, most of us can hear sounds ranging from 20 Hz to about 20,000 Hz, and within this range we can distinguish between sounds that differ by just a few hertz. Differences in frequency are important for our sense of pitch, but *pitch* and *frequency* are not synonymous. *Frequency* describes a *physical* property of sounds (see Box 6.1), but *pitch* relates solely to our subjective *perception* of those sounds. This is an important distinction because frequency is not the sole determinant of perceived pitch; at some frequencies, higher-intensity sounds may seem higher-pitched, and changes in pitch do not precisely parallel changes in frequency.

How do we distinguish pitches? Two signals from the cochlea appear to inform the brain about the pitch of sounds:

1. In **place coding**, the pitch of a sound is determined by the location of activated hair cells along the length of the basilar membrane, as we discussed in this chapter's "Researchers at Work" feature. So, activation of receptors near the base of the cochlea (which is narrow and stiff and responds to high frequencies) signals *treble*, and activation of receptors nearer the apex (which is wide and floppy and responds to low frequencies) signals *bass*.

2. A complementary process, known as **temporal coding**, encodes the frequency of auditory stimuli in the rate of firing of auditory neurons. For example, a 500 Hz sound might cause some auditory neurons to fire 500 action potentials per second. *Volleys* of action potentials being produced at this rate, by a number of neurons with similar tunings, provide the brain with a reliable additional source of pitch information.

Experimental evidence indicates that we rely on both of these processes to discriminate the pitch of sounds. Temporal coding is most evident at lower frequencies,

place coding Frequency discrimination in which the pitch of a sound is determined by the location of activated hair cells along the length of the basilar membrane.

temporal coding Frequency discrimination in which the pitch of a sound is determined by the rate of firing of auditory neurons.

up to about 4,000 Hz: Auditory neurons can fire a maximum of only about 1,000 action potentials per second, but to a limited extent they can encode sound frequencies that are multiples of the action potential frequency. Beyond about 4,000 Hz, however, this encoding becomes impossible, and pitch discrimination relies on place coding of pitch along the basilar membrane.

Some species are sensitive to sounds with very high frequencies (called *ultrasound*) or very low frequencies (*infrasound*) and make use of them in special ways. For example, many species of bats produce loud vocalizations in the range of 50,000–100,000 Hz, and they listen to the echoes reflected back from objects in order to navigate and hunt in the dark. At the other end of the spectrum, elephants emit ultra-low-frequency alarm calls that are so powerful that they travel partly through the ground and are detected *seismically* by other elephants (O'Connell-Rodwell, 2007) and yet are so nuanced that the elephants can distinguish human-related threats from bee-related threats (Soltis et al., 2014).

Brainstem Systems Compare the Ears to Localize Sounds

Being able to quickly identify where a sound is coming from—whether it is the crack of a twig under a predator's foot, or the sweet tones of a would-be mate—is a matter of great evolutionary significance. So it's no surprise that we are remarkably good at locating a sound source (our accuracy is about ±1 degree horizontally around the head, and many animals are even better). The auditory system accomplishes this feat by analyzing two kinds of binaural cues that signal the location of a sound source:

intensity difference A perceived difference in loudness between the two ears, which the nervous system can use to localize a sound source.

1. **Intensity differences** are differences in *loudness* at the two ears. Depending on the species—and the placement and characteristics of their pinnae—intensity differences occur because one ear is pointed more directly toward the sound source or because the head casts a sound shadow (**FIGURE 6.8A**), blocking sounds originating on one side (called *off-axis sounds*) from reaching both ears with equal loudness. The head shadow effect is most pronounced for higher-frequency sounds (**FIGURE 6.8B**).

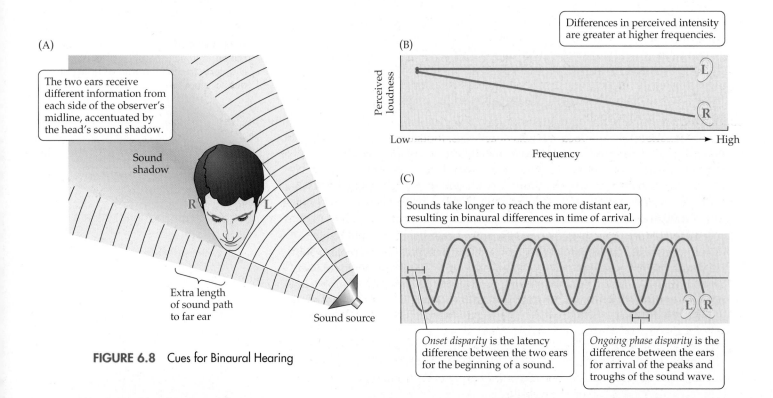

(A)

The two ears receive different information from each side of the observer's midline, accentuated by the head's sound shadow.

Sound shadow

R   L

Extra length of sound path to far ear

Sound source

(B)

Differences in perceived intensity are greater at higher frequencies.

Perceived loudness

Low ———————————————→ High

Frequency

L

R

(C)

Sounds take longer to reach the more distant ear, resulting in binaural differences in time of arrival.

L R

Onset disparity is the latency difference between the two ears for the beginning of a sound.

Ongoing phase disparity is the difference between the ears for arrival of the peaks and troughs of the sound wave.

FIGURE 6.8 Cues for Binaural Hearing

2. **Latency differences** are differences between the two ears in the *time of arrival* of sounds. They arise because one ear is always a little closer to an off-axis sound than the other ear is. Two kinds of latency differences are present in a sound: *onset disparity*, which is the difference between the two ears in hearing the beginning of the sound; and *ongoing phase disparity*, which is the continuing mismatch between the two ears in the arrival of all the peaks and troughs that make up the sound wave. These cues are illustrated in **FIGURE 6.8C**.

Both types of cues are employed in sound localization. At low frequencies, though, no matter where sounds are presented horizontally around the head, there are virtually no intensity differences between the ears. For these frequencies, differences in times of arrival are the principal cues for sound location (and at very low frequencies, neither cue is much help; this is why you can place the subwoofer of an audio system anywhere you want). At higher frequencies, however, the sound shadow cast by the head produces significant binaural intensity differences. Of course, we can't perceive which types of processing we're relying on for any given sound; in general, we are aware of the *results* of neural processing but not the processing itself. (You can learn about brain mechanisms of auditory localization in **A STEP FURTHER 6.2**, on the website.)

The structure of the external ear provides yet another localization cue. As we mentioned earlier, the hills and valleys of the external ear selectively reinforce some frequencies in a complex sound and diminish others. This process is known as **spectral filtering**, and the frequencies that are affected depend on the angle at which the sound arrives at those peaks and valleys. That angle varies, of course, depending on where the sound came from (Kulkarni and Colburn, 1998); these spectral cues provide critical information about the vertical localization (or elevation) of a sound source. Without them, you would have a hard time knowing whether a sound from straight in front of you came from the ground or from the treetops. Using recordings of the activity of single neurons in the auditory system, researchers have determined that the various binaural and spectral cues used for sound localization converge and are integrated in the inferior colliculus (Slee and Young, 2014).

The Auditory Cortex Specializes in Processing Complex Sound

In some sensory areas of the brain, lesions cause the loss of basic perceptions. For example, lesions of visual cortex result in blind spots, as we will discuss in Chapter 7. But the auditory cortex is different: researchers have long known that simple pure tones can be heard even after the entire auditory cortex has been surgically removed (Rosenzweig, 1946; Neff and Casseday, 1977). So if the auditory cortex is not involved in basic auditory perception, then what does it do? The auditory cortex seems to be specialized for the detection of more-complex "biologically relevant" sounds, of the sort we mentioned earlier—vocalizations of animals, footsteps, snaps, crackles, and pops—containing many frequencies and complex patterns (Masterton, 1997). In other words, the auditory cortex evolved to process the sounds of everyday life.

The unique capabilities of the auditory cortex result from a sensitivity that is fine-tuned by experience as we grow (Kandler et al., 2009). Human infants have diverse hearing capabilities at birth, but their hearing for complex speech sounds in particular becomes more precise and rapid through exposure to the speech of their families and other people. Newborns can distinguish all the different sounds that are made in any human language. But as they develop, they get better and better at distinguishing sounds in the language(s) they hear, and worse at distinguishing sounds that occur in other languages. Similarly, early experience with binaural hearing, compared with equivalent monaural (one-eared) hearing, has a significant effect on the ability of children to localize sound sources later in life (W. D. Beggs and Foreman, 1980). Studies with lab animals confirm that experience with sounds of a particular frequency can cause a rapid retuning of auditory neurons (**FIGURE 6.9**) (N. M. Weinberger, 1998; Fritz et al., 2003). You can learn more about the role of experience in auditory localization in owls in **A STEP FURTHER 6.3**, on the website.

latency difference A difference between the two ears in the time of arrival of a sound, which the nervous system can use to localize a sound source.

spectral filtering The process by which the hills and valleys of the external ear alter the amplitude of some, but not all, frequencies in a sound.

FIGURE 6.9 Long-Term Retention of a Trained Shift in the Tuning of an Auditory Receptive Field (From N. M. Weinberger, 1998.)

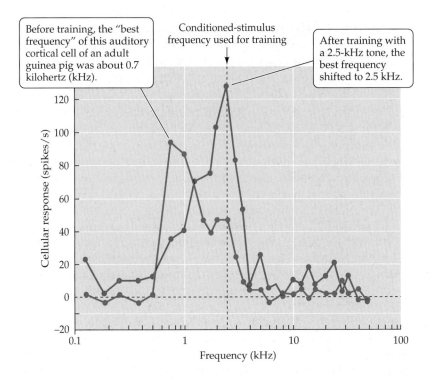

Before training, the "best frequency" of this auditory cortical cell of an adult guinea pig was about 0.7 kilohertz (kHz).

Conditioned-stimulus frequency used for training

After training with a 2.5-kHz tone, the best frequency shifted to 2.5 kHz.

Music also shapes the responses of auditory cortex. It might not surprise you to learn that the auditory cortex of trained musicians shows a bigger response to musical sounds than does the same cortex in nonmusicians. After all, when two people differ in any skill, their brains *must* be different in some way, and maybe people born with brains that are more responsive to complex sounds are also more likely to become musicians. The surprising part is that the extent to which a musician's brain is extra sensitive to musical notes is correlated with the age at which she began her serious training in music: the earlier the training began, the bigger the difference in auditory cortex in adulthood (Pantev et al., 1998). This finding indicates that intensive musical experience in development alters the functioning of auditory cortex later in life. By adulthood, the portion of primary auditory cortex where music is first processed, called *Heschl's gyrus*, is more than twice as large in professional musicians as

FIGURE 6.10 Brain Connections in Tone-Deaf People (From Loui et al., 2009, courtesy of Psyche Loui.)

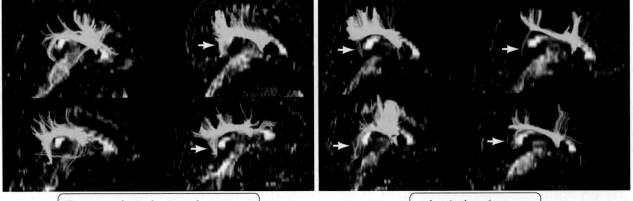

Diffusion tensor imaging (DTI) of axon projections reveals the arcuate fasciculus (yellow), a pathway connecting the frontal cortex, which is active during pitch discrimination, to the temporal lobe, where auditory processing begins (arrows).

The arcuate fasciculus is much more prominent in these four control cases...

...than in these four people who are tone-deaf.

in nonmusicians, and more than twice as strongly activated by music (P. Schneider et al., 2002). As we've discussed, at least part of this difference is attributable to early experience with music, but how much is innate?

Some people show a lifelong inability to discern tunes or sing (called **amusia**). Amusia is associated with subtly abnormal function in the right frontal lobe and impoverished connectivity between frontal and temporal cortex (**FIGURE 6.10**) (K. L. Hyde et al., 2006; Loui et al., 2009). The result is an inability to consciously access pitch information, even though cortical pitch-processing systems are intact (Zendel et al., 2015). Interestingly, studies of people with amusia indicate that when listening to music, we process pitch and rhythm quite separately (K. L. Hyde and Peretz, 2004).

Deafness Is a Widespread Problem

Whether it's a bit of trouble following quiet conversations, or the profound silence of total hearing loss, some degree of hearing difficulty affects as much as 15% of the population—some 37.5 million people in the United States alone (Blackwell et al., 2012). By now, you may have anticipated that there are three kinds of problems that can prevent sound waves in the air from being transformed into conscious auditory perceptions: a problem with sound waves reaching the cochlea, trouble converting sound waves into action potentials, and dysfunction of brain mechanisms of sound processing (**FIGURE 6.11**):

1. Before anything even happens in the nervous system, the ear may fail to convert the sound vibrations in air into waves of fluid within the cochlea. This form of hearing loss, called **conduction deafness** (**FIGURE 6.11A**), often comes about when the ossicles of the middle ear become fused together and vibrations of the eardrum can no longer be conveyed to the oval window of the cochlea.

2. Even if the vibration is successfully conducted to the cochlea, the sensory apparatus of the cochlea—the hair cells—may fail to respond to the ripples created in the basilar membrane and thus fail to create the action potentials to inform the brain about sounds. This form of hearing loss, termed **sensorineural deafness** (**FIGURE 6.11B**), is most often due to the permanent damage

amusia A disorder characterized by the inability to discern tunes accurately or to sing.

conduction deafness A hearing impairment in which the ears fail to convert sound vibrations in air into waves of fluid in the cochlea. It is associated with defects of the external ear or middle ear.

sensorineural deafness A hearing impairment most often caused by the permanent damage or destruction of hair cells or by interruption of the vestibulocochlear nerve that carries auditory information to the brain.

FIGURE 6.11 Types of Hearing Loss

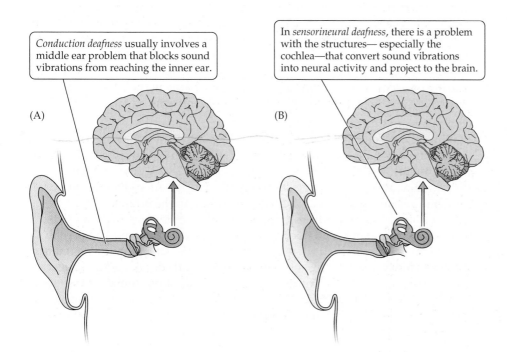

(A) *Conduction deafness* usually involves a middle ear problem that blocks sound vibrations from reaching the inner ear.

(B) In *sensorineural deafness*, there is a problem with the structures— especially the cochlea—that convert sound vibrations into neural activity and project to the brain.

(C) In *central deafness*, damage to auditory brain structures can affect hearing in various ways.

(A) Normal cochlea

In a normal cochlea, hair cells line the organ of Corti throughout its length, but exposure to excessively loud sounds can have rapid destructive effects. After exposure to excessive noise, a long section of the sound-damaged cochlea is completely missing its hair cells, resulting in deafness from the corresponding frequencies.

(B) Severe noise damage

Electron microscopy reveals that the orderly rows of stereocilia found in the organ of Corti in a normal cochlea are crushed and flattened by excessive noise exposure, like trees blown down in a windstorm.

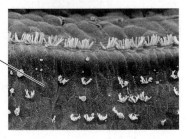

FIGURE 6.12   The Destructive Effects of Loud Noise (Micrographs by H. Engstrom and B. Engstrom, courtesy of Widex.)

tinnitus   A sensation of noises or ringing in the ears not caused by external sound.

central deafness   A hearing impairment in which the auditory areas of the brain fail to process and interpret action potentials from sound stimuli in meaningful ways, usually as a consequence of damage in auditory brain areas.

word deafness   A form of central deafness that is characterized by the specific inability to hear words although other sounds can be detected.

cortical deafness   A form of central deafness, caused by damage to both sides of the auditory cortex, that is characterized by difficulty in recognizing all complex sounds, whether verbal or nonverbal.

cochlear implant   An electromechanical device that detects sounds and selectively stimulates nerves in different regions of the cochlea via surgically implanted electrodes.

or destruction of hair cells by any of a variety of causes (**FIGURE 6.12**). Some people are born with genetic abnormalities that interfere with the function of hair cells (Cryns and Van Camp, 2004; Petit and Richardson, 2009). Many more people acquire sensorineural deafness during their lives as a result of being exposed to extremely loud sounds—overamplified music, nearby gunshots, and industrial noise are important examples—or because of medical problems such as infections and adverse drug effects (certain antibiotics, such as streptomycin, are particularly *ototoxic*). If you don't think it can happen to you, think again. Anyone listening to something for more than 5 hours per week at 89 dB or louder is already exceeding workplace limits for hearing safety (SCENIHR, 2008). Yet many personal music players can produce sound levels well in excess of 100 dB via headphones. Long-term exposure to loud sounds can cause hearing problems such as **tinnitus**—persistent ringing in the ears—and/or a profound loss of hearing for the frequencies being listened to at such high volumes.

3. For the action potentials sent from the cochlea to be of any use, the auditory areas of the brain must process and interpret them in meaningful ways. **Central deafness** (**FIGURE 6.11C**) occurs when auditory brain areas are damaged by, for example, strokes, tumors, or traumatic injuries. As you might expect from our earlier discussion of auditory processing in the brain, this type of deafness almost never involves a simple loss of auditory sensitivity. Afflicted individuals can often hear a normal range of pure tones but are impaired in the perception of complex, behaviorally relevant sounds. An example in humans is **word deafness**: selective trouble with speech sounds despite normal speech and normal hearing for nonverbal sounds. In **cortical deafness**—a rare syndrome involving bilateral lesions of auditory cortex—patients have more-complete impairment, struggling to recognize all complex sounds, whether verbal or nonverbal.

SIGNS **&** SYMPTOMS

Restoring Auditory Stimulation in Deafness

Although treatments for deafness have been greatly refined in the last few decades, they generally offer only partial restoration of the lost hearing. In cases of conduction deafness, it is sometimes possible to surgically free up the fused ossicles or replace them with Teflon prosthetics and thus restore the transmission of sound vibrations to the cochlea. But sensorineural deafness presents a much thornier problem, because neural elements have been destroyed (or were absent from birth). Can new hair cells be grown? Fishes and amphibians produce new hair cells throughout life, but mammals historically have been viewed as incapable of regenerating hair cells. This conclusion may have been too hasty, however (Brigande and Heller, 2009). Using several different strategies, researchers have succeeded in inducing the birth of new hair cells in cochlear tissues of lab animals (Izumikawa et al., 2005; Koehler et al., 2013), so there is reason to hope that an effective restorative therapy for deafness may be available some day.

For now, treatments for deafness focus on the use of prostheses. Traditional hearing aids detect and amplify sounds to provide extra stimulation to an impaired—but still functional—auditory system. More recently, implantable devices called **cochlear implants** have been used to directly stimulate the auditory nerve fibers of the cochlea, bypassing the ossicles and hair cells altogether and offering partial restoration of hearing even in cases of

complete deafness (**FIGURE 6.13**) (Loeb, 1990; J. M. Miller and Spelman, 1990). You may have had doubts about the value of Békésy's work with cadavers that we described at the start of this chapter. If so, consider this: the cochlear implants that have brought hearing to thousands of deaf people work by reproducing the phenomena Békésy discovered. In other words, the device sends information about low frequencies to electrodes stimulating nerves at the apex of the cochlea, and information about high frequencies to electrodes stimulating nerves at the base. As you might predict from our discussion of the importance of experience in shaping auditory responsiveness, the earlier in life that these devices are implanted, the better the person will be able to understand complex sounds later (Grieco-Calub et al., 2009). So in a sense, the success of these implants is due in part to the cleverness of the brain.

Within the deaf community, the use of cochlear implants is the subject of a lively controversy. While many hearing people might assume that a deaf person would automatically want to have his hearing restored, some deaf advocates are concerned that cochlear implants threaten a vibrant deaf culture. The solution may lie in ensuring through education that a child with cochlear implants retains a sense of belonging in both the hearing and deaf communities, rather than being forced to identify with one or the other.

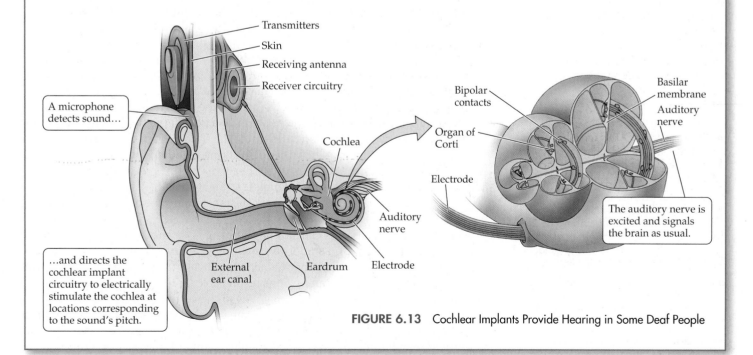

FIGURE 6.13   Cochlear Implants Provide Hearing in Some Deaf People

Handwritten margin annotations:

Linear movement

otoliths

Otolith membrane

cilia

Receptor cell

1. Compare and contrast the two important signals about pitch that the brain receives from the cochlea: place coding and temporal coding. How do they work together to give us our sense of pitch?
2. Discuss the sensory capabilities of different species as adaptations shaped by natural selection.
3. Provide an account of sound localization, identifying the several sources of information that we use to determine the source of a sound.
4. Discuss the types of processing that are performed by primary auditory cortex. Is experience with sound important for development of cortical auditory systems?
5. Name and describe the three major forms of deafness.

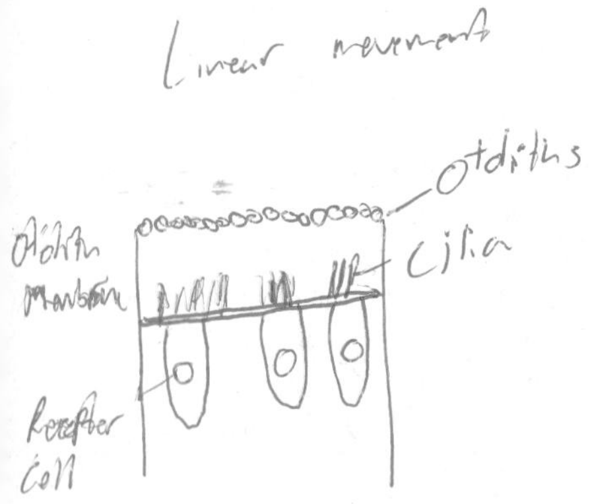

The Inner Ear Provides Our Sense of Balance

Without our sense of balance, it would be a challenge to simply stand on two feet. When you use an elevator, you clearly sense that your body is rising or falling, despite the sameness of your surroundings. When you turn your head, take a tight curve in your car, or bounce through the seas in a boat, your continual awareness of motion allows you to plan further movements and anticipate changes in perception due to movement of your head. And of course, too much of this sort of stimulation can make you feel distinctly queasy.

Like hearing, our sense of balance is the product of the inner ear, relying on several small structures that adjoin the cochlea and are known collectively as the **vestibular system** (from the Latin *vestibulum*, "entrance hall," reflecting the fact that the system lies in hollow spaces in the temporal bone). In fact, it is generally accepted that the auditory organ evolved from the vestibular system, although the ossicles probably evolved from parts of the jaw. The most obvious component of the vestibular system is the trio of fluid-filled **semicircular canals**, connected at their ends to two bulbs called the *saccule* and the *utricle* (**FIGURE 6.14A**). Notice that the three canals are oriented in the three different planes in which the head can rotate (**FIGURE 6.14B**)—nodding up and down (technically known as *pitch*, around the *x*-axis), shaking from side to side (*yaw*, around the *y*-axis), and tilting left or right (*roll*, around the *z*-axis).

The receptors of the vestibular system are hair cells—just like the ones in the cochlea—whose bending ultimately produces action potentials. The cilia of these hair cells are embedded in a gelatinous mass inside an enlarged chamber called the **ampulla** (plural *ampullae*) that lies at the base of the semicircular canals (see Figure 6.14B). Movement of the head in one axis sets up a flow of the fluid in the semicircular canal that lies in the same plane, deflecting the stereocilia in the ampulla and signaling the brain that the head has moved. Working together, the three semicircular canals accurately track the rotation of the head. Specialized receptors in the utricle and saccule provide the remaining signals—straight-line acceleration and deceleration—that the brain needs in order to track the precise position and movement of the body in three-dimensional space.

Vestibular information is crucial for planning body movements, maintaining balance against gravity, and smoothly directing sensory organs like the eyes and ears toward specific locations, and the nerve pathways from the vestibular system have strong connections to brain regions responsible for the planning and control of movement. On entering the brainstem, many of the vestibular fibers of the vestibulocochlear nerve (cranial nerve VIII) terminate in the **vestibular nuclei**, while some fibers project directly to the cerebellum to aid in motor programming there. Outputs from the vestibular nuclei project in a complex manner to motor areas throughout the brain, including motor nuclei of the eye muscles, the thalamus, and the cerebral cortex.

vestibular system   The sensory system that detects balance. It consists of several small inner-ear structures that adjoin the cochlea.

semicircular canal   Any one of the three fluid-filled tubes in the inner ear that are part of the vestibular system. Each of the tubes, which are at right angles to each other, detects angular acceleration in a particular direction.

ampulla   An enlarged region of each semicircular canal that contains the receptor cells (hair cells) of the vestibular system.

vestibular nucleus   A brainstem nucleus that receives information from the vestibular organs through cranial nerve VIII (the vestibulocochlear nerve).

(A) Vestibular system

External ear canal   Cochlea

Semicircular canals (handwritten)

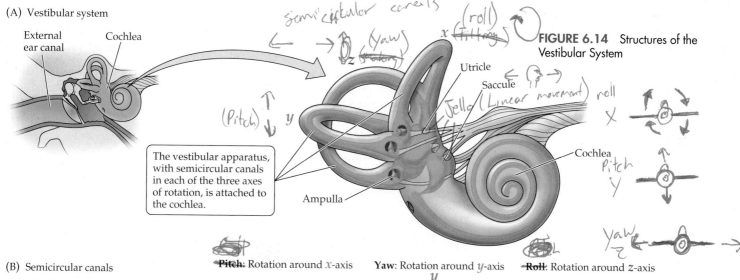

FIGURE 6.14 Structures of the Vestibular System

Utricle
Saccule
Ampulla
Cochlea

Jello (Linear movement) (handwritten) *roll* (handwritten)

The vestibular apparatus, with semicircular canals in each of the three axes of rotation, is attached to the cochlea.

(Pitch) y (handwritten)

roll X / Pitch Y / Yaw Z (handwritten)

Pitch: Rotation around x-axis   **Yaw:** Rotation around y-axis   **Roll:** Rotation around z-axis

(B) Semicircular canals

Head movement Movements of the head involve rotation around the three principal axes—x, y, and z.

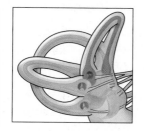

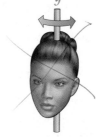

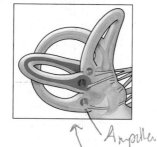

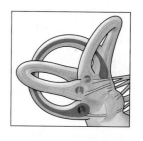

Vestibular response Each type of head movement is registered by a corresponding semicircular canal in the same plane of rotation. The rotation of the head is translated into movement of gel within the semicircular canals, which stimulates hair cells in ampullae at the base of the canals, exciting the vestibulocochlear nerve.

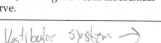
Vestibular system → (handwritten)

Ampulla (handwritten)

To see the animation
The Vestibular System,
go to

2e.mindsmachine.com/6.5

As fluid moves through the vest sys. it moves the hair cells in the Ampullas (handwritten)

those hair cells move like sea grass (handwritten)

Some Forms of Vestibular Excitation Produce Motion Sickness

There is one aspect of vestibular activation that many of us would gladly do without. Too much strong vestibular stimulation—think of boats and roller coasters—can produce the misery of **motion sickness**. Motion sickness is caused by movements of the body that we cannot control. For example, passengers in a car are more likely to suffer from motion sickness than is the driver.

Why do we experience motion sickness? According to the *sensory conflict theory,* we feel bad when we receive contradictory sensory messages, especially a discrepancy between vestibular and visual information. When an airplane bounces around in turbulence, for instance, the vestibular system signals that various changes in direction and accelerations are occurring, but as far as the visual system is concerned, nothing is happening; the plane's interior is a constant. One hypothesis is that the stimulation is activating a system that originally evolved to rid the body of swallowed poison (M. Treisman, 1977). According to this hypothesis, discrepancies in sensory information might normally signal a dangerous neurological problem, triggering dizziness and vomiting to get rid of potentially toxic food. While this action has obvious protective benefits in an ancestral environment, it sure can interfere with our enjoyment of more modern activities!

Sensory conflict theory = sickness (handwritten)

motion sickness The experience of nausea brought on by unnatural passive movement, as may occur in a car or boat.

[handwritten margin notes:]
5 tastes
Salty
Sour
Sweet
Bitter
Umami

1. Use a diagram to explain how the general layout of the vestibular system allows it to track movement in three axes. Where are the receptors for head movement located? Do they resemble other types of sensory receptors?
2. Where are the vestibular nuclei located? What nerve provides inputs to these nuclei?
3. How is vestibular information used in ongoing behavior?
4. Discuss the role of the vestibular system in motion sickness.

PART II
The Chemical Senses: Taste and Smell

Delicious foods, poisons, dangerous adversaries, and fertile mates—these are just a few of the sources of chemical signals in the environment. Being able to detect these signals is vital for survival and reproduction in organisms throughout the animal kingdom. The sense of taste provides an immediate assessment of foods (Lindemann, 1995). But it is the sense of smell that is critical for distinguishing individual foods, as well as discriminating the scent of friend versus foe. In this section we will look at how the senses of taste and smell work together to analyze the chemical signals all around us.

The Human Tongue Detects Five Basic Chemical Tastes

Most people derive great pleasure from eating delicious food, and because we recognize many substances by their distinct flavor, we tend to think that we can discriminate many tastes. In reality, though, humans detect only five basic **tastes**: salty, sour, sweet, bitter, and umami. (*Umami*, Japanese for "delicious taste," names the savory, meaty taste that is characteristic of gravy or soy sauce.) The sensations uniquely aroused by an apple, a steak, or an olive are **flavors** rather than simple tastes; they involve smell as well as taste. To understand the importance of smell to a flavor, block your nose while eating first some raw potato and then some apple, and you'll see that, without olfactory input, you can't tell the difference between them.

taste Any of the five basic sensations detected by the tongue: sweet, salty, sour, bitter, and umami.

flavor The sense of taste combined with the sense of smell.

papilla A small bump that projects from the surface of the tongue. Papillae contain most of the taste receptor cells.

taste bud A cluster of 50–150 cells that detects tastes. Taste buds are found in papillae.

Tastes excite specialized receptor cells on the tongue

Many people think that the myriad little bumps on their tongues are taste buds, but they aren't. They are actually **papillae** (singular *papilla*) (**FIGURE 6.15**), tiny lumps of tissue that increase the surface area of the tongue. There are three kinds of taste papillae—*circumvallate, foliate*, and *fungiform* papillae—occurring in different locations on the tongue (**FIGURE 6.16C**).

Taste buds, each consisting of a cluster of 50–150 taste receptor cells (**FIGURE 6.16A**), are found buried within the walls of the papillae (a single papilla may house several such taste buds; see Figure 6.15). Fine fibers, called *microvilli*, extend from the taste receptor cells into a tiny pore, where they can come into contact with substances that can be tasted, called *tastants* (**FIGURE 6.16B**). Each taste cell is sensitive to just one of the five basic tastes, and with a life span of only 10–14 days, taste cells are constantly being replaced. But as our varied experience with hot drinks, frozen flag poles, or spicy foods informs us, taste is not the only sensory capability of the tongue. It also possesses sensory cells for pain and touch.

Many books show a map of the tongue indicating that each taste is perceived mainly in one region (sweet at the tip of the

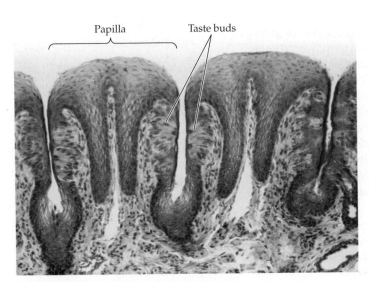

Papilla   Taste buds

FIGURE 6.15   A Cross Section of the Tongue

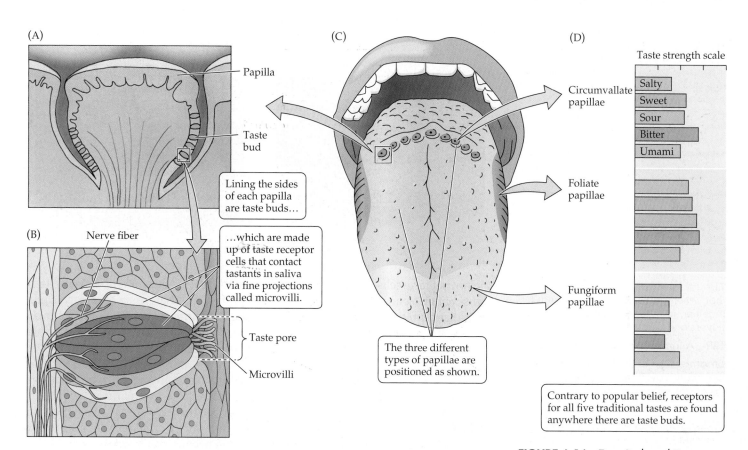

(A)

Papilla

Taste bud

Lining the sides of each papilla are taste buds…

…which are made up of taste receptor cells that contact tastants in saliva via fine projections called microvilli.

(B)

Nerve fiber

Taste pore

Microvilli

(C)

Circumvallate papillae

Foliate papillae

Fungiform papillae

The three different types of papillae are positioned as shown.

(D)

Taste strength scale

Salty
Sweet
Sour
Bitter
Umami

Contrary to popular belief, receptors for all five traditional tastes are found anywhere there are taste buds.

FIGURE 6.16 Taste Buds and Taste Receptor Cells (Part A after McLaughlin et al., 1994; C after Bartoshuk, 1993, and Chandrashekar et al., 2006.)

tongue, bitter at the back, and so on), but this map is a myth propagated by sloppy text-book writers (which we are not!). All five basic tastes can be perceived *anywhere* on the tongue where there are taste receptors (Chandrashekar et al., 2006). Those areas do not differ greatly in the strength of taste sensations that they mediate (**FIGURE 6.16D**).

The five basic tastes are signaled by specific sensors on taste cells

The tastes salty and sour are evoked when taste cells are stimulated by simple ions acting on ion channels in the membranes of the taste cells. Sweet and bitter tastes are perceived by specialized receptor molecules and communicated by second messengers. And at least two types of receptors may be involved in the perception of umami.

SALTY The transduction of the taste of salt (NaCl) is perhaps the easiest to understand because it relies mostly on ion channels of the sort we have seen in previous chapters. Sodium ions (Na^+) from salty food enter taste cells via sodium channels in the cell membrane, causing a depolarization of the cell and release of neurotransmitter. We know that this is a crucial mechanism for perceiving saltiness, because blocking the sodium channels with a drug greatly reduces—but does not eliminate—our ability to taste salt (Schiffman et al., 1986). The rest of our ability to taste salt comes from a second salt sensor, a variant of a receptor called *TRPV1* (*transient receptor potential vanilloid type 1*), that not only gives us a bit of extra sensitivity to Na^+, but also detects the positively charged ions (cations) of other salts in food, such as potassium (K^+) (Treesukosol et al., 2007). Depolarization of the salt-sensitive taste cells causes them to release neurotransmitters that stimulate afferent neurons that relay the information to the brain.

SOUR Acids in food taste sour—the more acidic the food, the sourer it tastes—but no one knows exactly how sour tastants are detected. Researchers think that the protons (also called hydrogen ions; H^+) that all acids release may interact with special acid-sensing ion channels (like the ionotropic receptors in Chapters 3 and 4) to change the

NaCl — relies mostly on ion channels

FIGURE 6.17 It's All a Matter of Taste **Buds** (From Bartoshuk and Beauchamp, 1994, courtesy of Linda Bartoshuk and the Bartoshuk Lab.)

Certain substances that taste unpleasantly bitter to most people can't be tasted at all by some people.

(A) Nontasters have far fewer papillae…

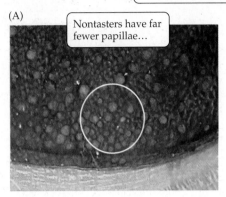

(B) …than supertasters, who are extra sensitive to bitter and sweet tastants.

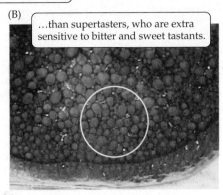

T1R A family of taste receptor proteins that, when particular members bind together, form taste receptors for sweet flavors and umami flavors.

T2R A family of bitter taste receptors.

umami One of the five basic tastes—the meaty, savory flavor. (The other four tastes are salty, sour, sweet, and bitter.)

polarity of taste cells and alter transmitter release. It seems that all sour-sensitive taste cells contain a particular type of ion channel protein and share an inward flow of protons that depolarizes the cell (Huang et al., 2006; Bushman et al., 2015). Interestingly, the same sensor appears to detect the sensation and taste of carbonation in drinks (Chandrashekar et al., 2009).

SWEET The receptors for sweet, bitter, and umami tastes are more like metabotropic receptors than ionotropic receptors (see Figure 4.2) because tastant molecules bind to a complex receptor protein on the taste cell's surface that activates a second messenger within the cell. These receptors are made up of simpler proteins belonging to two families—designated **T1R** and **T2R**—that are combined in various ways.

When two members of the T1R family—T1R2 and T1R3—combine (heterodimerize), they make a receptor that selectively detects sweet tastants (Nelson et al., 2001). Mice engineered to lack either T1R2 or T1R3 are insensitive to sweet tastes (Zhao et al., 2003). And if you've spent any time around cats, you may be aware that they couldn't care less about sweets. It turns out that all cats, from tabbies to tigers, share a mutation in the gene that encodes T1R2; their sweet receptors don't work (X. Li et al., 2009).

BITTER In nature, bitter tastes often signal the presence of toxins, so it's not surprising that a high sensitivity to different kinds of bitter tastes has evolved (McBurney et al., 1972; Lush, 1989), although individuals vary significantly in their taste sensitivity (**FIGURE 6.17**). Members of the T2R family of receptor proteins appear to function as bitter receptors (Adler et al., 2000; Chandrashekar et al., 2000). The T2R family has about 30 members, and this large number may reflect the wide variety of bitter substances encountered in the environment, as well as the adaptive importance of being able to detect and avoid them. Interestingly, each bitter-sensing taste cell produces most or all of the different types of T2R bitter receptors (Adler et al., 2000). So bitter-sensing taste cells exhibit broadly tuned sensitivity to *any* bitter-tasting substances (Brasser et al., 2005)—just what you'd expect in a sensory system that has evolved to act as a poison detector.

UMAMI The fifth basic taste, **umami**—the meaty, savory flavor—is detected by at least two kinds of receptors. One of these is a variant of the metabotropic glutamate receptor (Chaudhari et al., 2000; Maruyama et al., 2006) and most likely responds to the amino acid glutamate, which is found in high concentrations in meats, cheeses, kombu, and other savory foodstuffs (that's why MSG—monosodium *glutamate*—is used as a "flavor enhancer"). The second probable umami receptor, a combination of T1R1 and T1R3 proteins, responds to most of the dietary amino acids (Nelson et al., 2002). Given this receptor's similarity to the T1R2+T1R3 sweet receptor, there is reason to suppose that receptors for things that taste good may have shared evolutionary

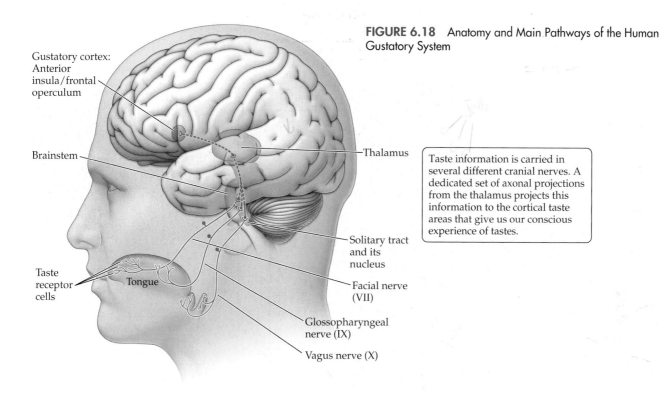

FIGURE 6.18 Anatomy and Main Pathways of the Human Gustatory System

Gustatory cortex: Anterior insula/frontal operculum

Brainstem

Thalamus

Taste information is carried in several different cranial nerves. A dedicated set of axonal projections from the thalamus projects this information to the cortical taste areas that give us our conscious experience of tastes.

Solitary tract and its nucleus

Taste receptor cells

Tongue

Facial nerve (VII)

Glossopharyngeal nerve (IX)

Vagus nerve (X)

origins. Consider the taste abilities of birds who, just like their house cat enemies, lack the T1R2 gene and thus ordinarily can't taste sweet. How then do hummingbirds sense the nectar they need for survival? It appears that evolution repurposed the hummingbird T1R1+T1R3 umami receptor into a new class of sweet receptor (Baldwin et al., 2014), allowing hummingbirds to thrive and spread.

Taste information is transmitted to several parts of the brain

Taste projections of the **gustatory system** (from the Latin *gustare*, "to taste") extend from the tongue to several brainstem nuclei, then to the thalamus, and ultimately to gustatory regions of the somatosensory cortex (**FIGURE 6.18**). Because there are only five basic tastes, and because each taste cell detects just one of the five, the encoding of taste perception could be quite straightforward, with the brain simply monitoring which specific axons are active in order to determine which tastes are present (Chandrashekar et al., 2006). In such a simple arrangement—sometimes called a *labeled-line system*—there is no need to analyze complex patterns of activity across multiple kinds of taste receptors (called *pattern coding*). Experimental evidence seemingly supports the conclusion that taste is a labeled-line system: selectively inactivating taste cells that express receptors for just one of the five tastes tends to completely eradicate sensitivity to that one taste while leaving the other four tastes unaffected (Huang et al., 2006). However, the same manipulation can also be viewed as knocking out one-fifth of any pattern of activity that would be normally present. From this perspective, the pattern-coding account cannot be ruled out. A precise understanding of the way in which the brain encodes taste information thus awaits future developments.

gustatory system The sensory system that detects taste.

how's it going ❓

1. What are the five basic tastes?
2. Generate a map of the human tongue, showing how sensitive each region is to the five basic tastes.
3. Compare and contrast taste buds and papillae.
4. Identify the cellular mechanisms underlying each of the five tastes. Discuss the evolution of taste sensitivity: How do these five tastes help us survive?

FIGURE 6.19 The Human Olfactory System

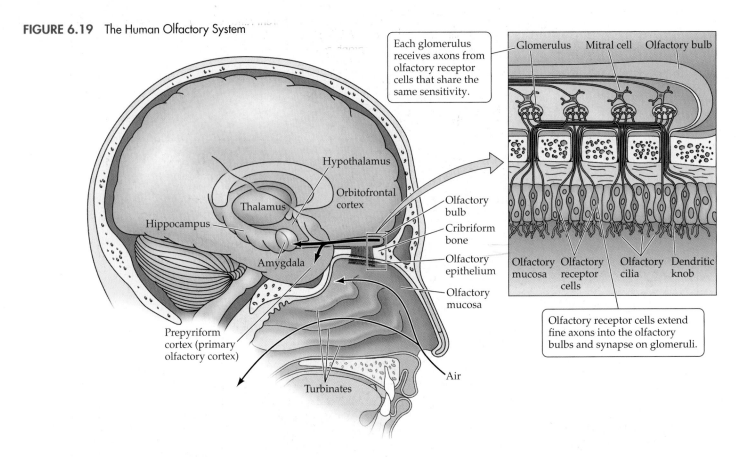

Each glomerulus receives axons from olfactory receptor cells that share the same sensitivity.

Hypothalamus

Orbitofrontal cortex

Thalamus

Hippocampus

Amygdala

Olfactory bulb

Cribriform bone

Olfactory epithelium

Olfactory mucosa

Prepyriform cortex (primary olfactory cortex)

Turbinates

Air

Glomerulus   Mitral cell   Olfactory bulb

Olfactory mucosa   Olfactory receptor cells   Olfactory cilia   Dendritic knob

Olfactory receptor cells extend fine axons into the olfactory bulbs and synapse on glomeruli.

odor The sensation of smell.

olfaction The sensory system that detects smell; the act of smelling.

anosmia The inability to detect odors.

olfactory epithelium A sheet of cells, including olfactory receptors, that lines the dorsal portion of the nasal cavities and adjacent regions.

To see the video
The Human Olfactory System,
go to

2e.mindsmachine.com/6.6

Chemicals in the Air Elicit Odor Sensations

Our ability to respond to many **odors**—it is estimated that humans can detect more than 10,000 different odors and can discriminate as many as 5,000 (Ressler et al., 1994)—is what produces the complex array of flavors that we normally think of as tastes. While those numbers may sound impressive, human **olfaction** (sense of smell) is far inferior to that of many other mammals, such as dogs and rabbits; in fact, partial **anosmia** (odor blindness) is quite prevalent among humans, with men exhibiting slightly worse olfaction than women (Gilbert and Wysocki, 1987). Most birds have only basic olfactory abilities, and dolphins don't have olfactory receptors at all (Freitag et al., 1998). These differences evolved along with differences between species in the importance of smell for survival and reproduction.

The sense of smell starts with receptor neurons in the nose

In humans, a sheet of cells called the **olfactory epithelium** (**FIGURE 6.19**) lines part of the nasal cavities. Within the 5–10 square centimeters of olfactory epithelium that we possess, three types of cells are found: supporting cells, basal cells, and about 10 million olfactory receptor cells. Each olfactory receptor cell is a complete neuron, with a long, slender apical dendrite that divides into branches (cilia) that extend into the moist mucosal surface. Substances that we can smell from the air that we inhale or sniff, called *odorants*, dissolve into the mucosal layer and interact with receptors studding the dendritic cilia of the olfactory neurons (Farbman, 1994). These olfactory receptor proteins are G protein–coupled receptors (GPCRs), much like the metabotropic neurotransmitter receptors found on neurons of the brain (see Figure 4.2), employing a second-messenger system to respond to the presence of odorants. However, despite these similarities, olfactory neurons differ from the neurons of the brain in several ways.

One way in which olfactory neurons are distinct from their cousins in the brain relates to the production of receptors: there is an incredible diversity of olfactory receptor protein subtypes. So, while there may be up to a dozen or so subtypes of receptors for

FIGURE 6.20   Different Kinds of Olfactory Receptor Molecules on the Olfactory Epithelium (After Vassar et al., 1993, photomicrograph courtesy of Robert Vassar.)

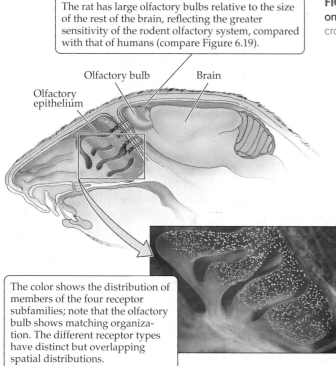

The rat has large olfactory bulbs relative to the size of the rest of the brain, reflecting the greater sensitivity of the rodent olfactory system, compared with that of humans (compare Figure 6.19).

Olfactory epithelium

Olfactory bulb

Brain

The color shows the distribution of members of the four receptor subfamilies; note that the olfactory bulb shows matching organization. The different receptor types have distinct but overlapping spatial distributions.

a given neurotransmitter in the brain, there are hundreds or even thousands of subtypes within the family of odorant receptors, depending on the species under study. The Nobel Prize–winning discovery of the genes encoding this odorant receptor superfamily (Buck and Axel, 1991) provided one of the most important advances in the history of olfaction research.

Mice have about 2 million olfactory receptor cells, each of which expresses only one of about 1,000 different receptor proteins. These receptor proteins can be divided into four different subfamilies of about 250 receptors each (Mori et al., 1999). Within each subfamily, members have similar structure and presumably recognize chemically similar odorants. Receptors of different subfamilies are expressed in separate bands of olfactory neurons within the olfactory epithelium (**FIGURE 6.20**) (Vassar et al., 1993). By comparison, humans make a total of about 350 different kinds of olfactory receptor proteins (Crasto et al., 2001; Glusman et al., 2001). That's still a large number, but in our case it looks like hundreds of additional olfactory receptor genes have become nonfunctional during the course of evolution, indicating that whatever they detected ceased to be important to our ancestors' survival and reproduction. To be able to discriminate 5,000 odors by using 350 kinds of functional olfactory receptors, we must recognize most odorants by their activation of a characteristic *combination* of different kinds of receptor molecules (Duchamp-Viret et al., 1999), an example of pattern coding.

Another big difference between olfactory neurons and brain neurons is that olfactory neurons die and are replaced in adulthood (Costanzo, 1991). This regenerative capacity is almost certainly an adaptation to the hazardous environment that olfactory neurons inhabit. If an olfactory neuron is killed—say, by the virus that gave you that darn head cold, or by a whiff of something toxic while you were cleaning out the shed, or by some other misadventure—an adjacent basal cell will soon differentiate into a neuron and begin extending a dendrite and an axon (Leung et al., 2007). Each olfactory neuron extends a fine, unmyelinated axon into the nearby **olfactory bulb** of the brain, where it terminates on a specific **glomerulus** (plural *glomeruli*) (from the Latin *glomus*, "ball")—a spherical structure that receives information from one specific class of odorant receptors (see Figure 6.19; we'll return to glomeruli shortly).

olfactory bulb   An anterior projection of the brain that terminates in the upper nasal passages and, through small openings in the skull, provides receptors for smell.

glomerulus   A complex arbor of dendrites from a group of olfactory cells.

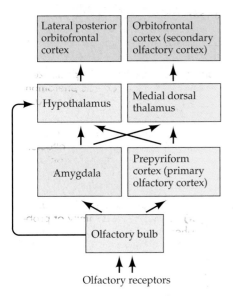

FIGURE 6.21 Components of the Brain's Olfactory System

No one knows exactly *how* the extending axon knows where to go to find its glomerulus, or how it knows where to form synapses within the glomerulus. One possibility is that olfactory receptor proteins that are found on the *axons* of these cells (as well as on the dendrites) guide the axons to their corresponding glomeruli (Barnea et al., 2004; Imai et al., 2009). But whatever may be the exact mechanisms of neuroplasticity in these cells, better understanding of the process of olfactory neuroregeneration may someday help us develop methods for restoring damaged regions of the brain and spinal cord.

Olfactory information projects from the olfactory bulbs to several brain regions

Each glomerulus within the olfactory bulb is a tight little sphere of neural circuitry that receives inputs exclusively from olfactory neurons that are expressing the same type of olfactory receptor. The glomerulus then actively tunes and sharpens the neural activity associated with the corresponding odorants (Aungst et al., 2003). The glomeruli are organized within the olfactory bulb according to an orderly, topographic map of smells, with neighboring glomeruli receiving inputs from receptors that are closely related. And, as Figure 6.19 shows, the spatial organization of glomeruli within the olfactory bulbs reflects the segregation of the four receptor protein subfamilies in the olfactory epithelium (Mori et al., 1999). In fact, this "olfactotopic" organization is maintained within the olfactory projections throughout the brain.

Olfactory information is conveyed to the brain via the axons of mitral cells (see Figure 6.19), which extend from the glomeruli in the olfactory bulbs to various regions of the forebrain (smell is the only sensory modality that synapses directly in the cortex rather than having to pass through the thalamus). Important targets for olfactory inputs include the hypothalamus, the amygdala, and the prepyriform cortex (**FIGURE 6.21**). These limbic structures are closely involved in memory and emotion, which may help explain the potency of odors in evoking nostalgic memories of childhood (M. Larsson and Willander, 2009).

Many vertebrates possess a vomeronasal system

Though many have tried, no human perfume inventor has concocted a fragrance with even a fraction of the attractive power that the natural scents governing mating and other behaviors in so many species have. The majority of terrestrial vertebrates—mammals, amphibians, and reptiles—possess a secondary chemical detection system that is specialized for detecting these **pheromones**. The system is called the **vomeronasal system** (**FIGURE 6.22**), and its receptors are found in a **vomeronasal organ** (**VNO**) of epithelial cells near the olfactory epithelium.

In rodents, the sensory neurons of the VNO make hundreds of different vomeronasal receptor proteins, belonging to two large families called V1R and V2R (Dulac and Torello, 2003). These receptors are extremely sensitive, able to detect very low levels of the pheromone signals—such as sex hormone metabolites and signals of genetic relatedness—that are released by other individuals (Leinders-Zufall et al., 2000; Loconto et al., 2003). From the VNO, information extends to the accessory olfactory bulb (adjacent to the main olfactory bulb), which projects to the medial amygdala and hypothalamus, structures that play crucial roles in governing emotional and sexual behaviors and in regulating hormone secretion. Hamsters and mice can distinguish relatives from nonrelatives just by smell, even if they were raised by foster parents (Mateo and Johnston, 2000; Isles et al., 2001), allowing these animals to select genetically compatible mates and avoid inbreeding.

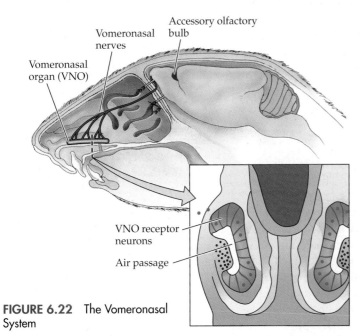

FIGURE 6.22 The Vomeronasal System

Do humans communicate via pheromones? Studies reporting pheromone-like phenomena in humans attract plenty of media attention because of the apparent link to our evolutionary past. Well-known examples include the finding that simple exposure to each other's bodily odors can cause women's menstrual cycles to synchronize (Stern and McClintock, 1998) and a report that exposure to female tears causes reductions in testosterone and sexual arousal in men (Gelstein et al., 2011). However, the VNO is either vestigial or absent in humans, and almost all of our V1R and V2R receptor genes have become nonfunctional "pseudogenes" over evolutionary time (Lübke and Pause, 2015). So, if humans do use olfactory communication, it is most likely accomplished using the main olfactory epithelium. In mice, receptors in the main olfactory epithelium called **TAARs** (for **trace amine–associated receptors**) reportedly respond to sex-specific pheromones instead of odorants (Liberles and Buck, 2006). Thus it seems the old notion that the olfactory epithelium detects odors while the VNO detects pheromones is an oversimplification, even in rodents. And because TAARs are also found in the human olfactory epithelium (Liberles, 2009), behavioral evidence indicating that humans respond to pheromones no longer presents a paradox. If rodents can detect pheromones through the olfactory epithelium, using TAARs or other yet-unknown mechanisms, then perhaps we can too.

pheromone A chemical signal that is released outside the body of an animal and affects other members of the same species.

vomeronasal system A specialized sensory system that detects pheromones and transmits information to the brain.

vomeronasal organ (VNO) A collection of specialized receptor cells, near to but separate from the olfactory epithelium, that detect pheromones and send electrical signals to the accessory olfactory bulb in the brain.

trace amine–associated receptor (TAAR) Any one of a family of probable pheromone receptors produced by neurons in the main olfactory epithelium.

---how's it going ❓

1. Discuss odor sensitivity in humans. How do we compare to other species?
2. Provide a brief sketch of the olfactory epithelium, showing the major cell types and their relationships to the brain.
3. Discuss the genetics of odor receptors, as well as their spatial organization in the nose and olfactory bulbs. What is a glomerulus?
4. Which regions of the brain receive strong olfactory inputs? What is the significance of this arrangement for the animal's behavior?
5. Discuss the structures and receptors associated with pheromone sensitivity, and speculate about the ecological importance of pheromone sensitivity in humans and other animals. Are humans sensitive to pheromones?

Recommended Reading

Bartoshuk, L. M., and Beauchamp, G. K. (2005). *Tasting and Smelling* (2nd ed.). New York, NY: Academic Press.

Doty, R. L. (2015). *Handbook of Olfaction and Gustation* (3rd ed.). New York, NY: Wiley-Blackwell.

Menini, A. (2009). *The Neurobiology of Olfaction*. Boca Raton, FL: CRC Press.

Musiek, F. E., and Baran, J. A. (2007). *The Auditory System: Anatomy, Physiology, and Clinical Correlates*. Boston, MA: Pearson.

Palmer, A., and Rees, A. (2010). *Oxford Handbook of Auditory Science*. Oxford, England: Oxford University Press.

Wilson, D. A., and Stevenson, R. J. (2006). *Learning to Smell: Olfactory Perception from Neurobiology to Behavior*. Baltimore, MD: Johns Hopkins University Press.

Wolfe, J. M., Kluender, K. R., Levi, D. M., Bartoshuk, L. M., and Herz, R. S. (2015). *Sensation & Perception* (4th ed.). Sunderland, MA: Sinauer.

Wyatt, T. D. (2014). *Pheromones and Animal Behavior: Chemical Signals and Signatures* (2nd ed.). Cambridge, England: Cambridge University Press.

Yost, W. A. (2006). *Fundamentals of Hearing* (5th ed.). San Diego, CA: Academic Press.

2e.mindsmachine.com/vs6

VISUAL SUMMARY

6

You should be able to relate each summary to the adjacent illustration, including structures and processes. Go to the online version of this summary (scan the QR code above) for links to figures, animations, and activities that will help you consolidate the material.

1 The **pinna** (external ear) captures, focuses, and filters sound. The sound arriving at the **tympanic membrane** (eardrum) is focused by the three **ossicles** of the **middle ear** onto the **oval window** to stimulate the fluid-filled **inner ear** (specifically, the **cochlea**). Review **Figure 6.1**, **Animations 6.2 and 6.3**

2 Sound arriving at the oval window causes traveling waves to sweep along the **basilar membrane** of the cochlea. For sounds of high **frequency**, the largest displacement of the basilar membrane is at the base of the cochlea, near the oval window; for low-frequency sounds, the largest **amplitude** is near the apex of the cochlea. Review **Figure 6.2**, **Box 6.1**

3 Movement of the **stereocilia** of the hair cells causes the opening and closing of ion channels, thereby **transducing** mechanical movement into changes in electrical potential. These changes in potential stimulate the nerve cell endings that contact the hair cells. Review **Figure 6.3**

4 The **organ of Corti** has both **inner hair cells** (IHCs, about 3,500 in humans) and **outer hair cells** (OHCs, about 12,000 in humans). The inner hair cells convey most of the information about sounds. The outer hair cells change their length under the control of the brain, amplifying the movements of the basilar membrane in response to sound and sharpening the frequency tuning of the cochlea. Review **Figure 6.4**, **Activity 6.1**

5 Afferents from the inner hair cells transmit auditory information to the **cochlear nuclei** of the brainstem. Cochlear neurons project bilaterally to the **superior olivary nucleus**, which in turn innervates the **inferior colliculus**. From there auditory information is relayed to the **medial geniculate nucleus** and then the primary auditory cortex in the temporal lobe. Review **Figure 6.6**, **Animation 6.4**

6 Auditory localization depends on differences in the sounds arriving at the two ears. For low-frequency sounds, differences in time of arrival at the two ears (**latency differences**) are especially important. For high-frequency sounds, **intensity differences** are especially important, and **spectral filtering** provides cues about elevation. Review **Figure 6.8**

7 Primary auditory cortex is specialized for processing complex, biologically important sounds, rather than **pure tones**. Experiences with sound early in life can influence later auditory localization and the responses of neurons in auditory pathways. Experiences later in life can also lead to changes in the responses of auditory neurons. Review **Figures 6.7 and 6.9**

8 Deafness can be caused by changes at any level of the auditory system. **Conduction deafness** consists of impairments in the transmission of sound through the external or middle ear to the cochlea. **Sensorineural deafness** arises in the cochlea, often because of the destruction of hair cells, or in the auditory nerve. **Central deafness** stems from brain damage. Review **Figure 6.11**

9 Some forms of deafness may be alleviated by direct electrical stimulation of the auditory nerve by a **cochlear implant**. Genetic manipulations can induce new hair cell growth in laboratory animals, raising hope of a gene therapy for sensorineural deafness. Review **Figure 6.13**

10 The receptors of the **vestibular system** that detect movement of the head lie within the inner ear next to the cochlea. In mammals the vestibular system consists of three **semicircular canals** plus the utricle and the saccule. The semicircular canals use hair cells to detect rotation of the body in three planes, and the utricle and saccule sense static positions and linear accelerations. Review **Figure 6.14**, **Animation 6.5**

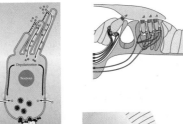
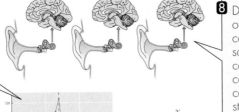
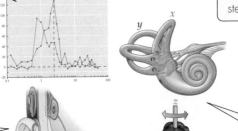

11 Humans detect only five main **tastes**—salty, sour, sweet, bitter, and **umami**—using taste receptor cells located in clusters called **taste buds**. Taste cells extend fine filaments into the taste pore of each bud, where tastants come into contact with them. The taste buds are situated on small projections from the surface of the tongue called **papillae**. The tastes of salty and sour are evoked primarily by the action of simple ions on ion channels in the membranes of taste cells. Sweet, umami, and bitter tastes are perceived by specialized receptor molecules belonging to the **T1R** and **T2R** families, which are coupled to G proteins. Review **Figure 6.16**, **Activity 6.2**

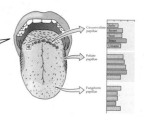

12 Each taste cell transmits information via cranial nerves to brainstem nuclei. This **gustatory system** extends from the taste receptor cells through brainstem nuclei to the thalamus and then to the cerebral cortex. Each taste axon responds most strongly to one category of tastes, providing a labeled line to the brain. Review **Figure 6.18**

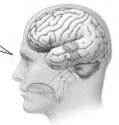

13 In contrast to being able to detect only a few tastes, humans can detect thousands of different **odors**. Olfactory receptor cells are small neurons whose dendrites in the **olfactory epithelium** express olfactory receptor proteins. The fine, unmyelinated axons of olfactory neurons run to the **olfactory bulbs** and synapse within **glomeruli**. If an olfactory receptor cell dies, an adjacent cell will replace it. Review **Figure 6.19**, **Animation 6.6**

14 There is a large family of odor receptor molecules, each of which utilizes G proteins and second messengers. Large subfamilies of receptors are synthesized in distinct bands of the olfactory epithelium. Review **Figure 6.20**

15 Outputs from the olfactory bulb extend to prepyriform cortex, amygdala, and hypothalamus, among other brain regions. Olfactory projections to the cortex maintain a stereotyped olfactory map of slightly overlapping projections from the glomeruli. Review **Figure 6.21**

16 The **vomeronasal organ** (VNO) contains receptors to detect pheromones released from other individuals of the species. These receptors transmit signals to the accessory olfactory bulb, which in turn communicates with the amygdala. **Pheromones** can also be detected by specialized receptors in the main olfactory epithelium. Review **Figure 6.22**

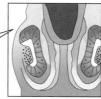

Go to **2e.mindsmachine.com** for study questions, quizzes, flashcards, and other resources.

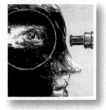
Vision
From Eye to Brain

When Seeing Isn't Seeing

It was cold in the bathroom, so the young woman turned on a small heater before she got in the shower. She didn't know that the heater was malfunctioning, filling the room with deadly, odorless carbon monoxide gas. Her husband found her unconscious on the floor and called for an ambulance to rush her to the emergency room. When she regained consciousness, "D.F." seemed to have gotten off lightly, avoiding what could have been a fatal accident. She could understand the doctors' questions and reply sensibly, move all her limbs, and perceive touch on her skin. But something was wrong with her sight.

D.F. had lost the ability to identify things that she viewed. Even the faces of family members had become unfamiliar. More than a decade later, D.F. still cannot recognize commonplace objects, yet she is not entirely blind. Show her a flashlight and she can tell you that it's made of shiny aluminum with some red plastic, but she doesn't recognize it ("Is it a kitchen utensil?"). Without telling her what it is, if you ask her to pick it up, her hand goes directly to the flashlight and holds it exactly as one normally holds a flashlight. Show D.F. a slot in a piece of plastic, and she cannot tell you whether the slot is oriented vertically, horizontally, or diagonally; but if you hand her a disk and ask her to put it through the hole, D.F. invariably turns the disk so that it goes smoothly through the slot (Goodale et al., 1991).

Can D.F. see or not?

To see the video
Object Recognition,
go to

2e.mindsmachine.com/7.1

Many species rely on vision for finding food, avoiding predators, finding mates, and locating shelter. And we humans are especially visual, using our sense of sight to enjoy nature and art, to read and write, to watch films and TV. So it's no surprise that we put a lot of effort into improving vision, preventing its deterioration, or even restoring sight to the blind.

However, the sheer volume of visual information poses a serious problem. Viewing the surrounding world has been compared to drinking from a waterfall. How does the visual system avoid being overwhelmed by the flood of information entering the eyes? One answer is that visual systems are especially sensitive to detecting *movement*, because moving objects are the most likely to represent the danger of predators or the good fortune of prey or mates. So the visual system often ignores information about stationary objects, which is why we have a harder time seeing them.

Another way to deal with information overload is for each species to evolve visual capabilities that are tailored to that species' particular lifestyle. No species can see everything. Most nocturnal species have better night vision than do animals that are

retina The receptive surface inside the eye that contains photoreceptors and other neurons.

transduction The conversion of one form of energy to another, as converting light into neuronal activity.

cornea The transparent outer layer of the eye, whose curvature is fixed. The cornea bends light rays and is primarily responsible for forming the image on the retina.

refraction The bending of light rays by a change in the density of a medium, such as the cornea and the lens of the eyes.

lens A structure in the eye that helps focus an image on the retina.

ciliary muscle One of the muscles that control the shape of the lens inside the eye, focusing an image on the retina.

accommodation The process by which the ciliary muscles adjust the lens to focus a sharp image on the retina.

To view the
Brain Explorer,
go to

2e.mindsmachine.com/7.2

active during the day, like us. Most rodent species, such as rats and mice, which live in tunnels and close quarters, have poor vision for distant objects, while daytime hunters like hawks have incredibly keen distance vision. Birds and bees can detect ultraviolet light, allowing them to see patterns in flowers that we cannot. But even within our limits of sight, we humans process a remarkable amount of visual information, which keeps about one-third of our cerebral cortex busy analyzing it.

This chapter is organized according to the stages of visual processing. We'll begin by tracing the path of visual information into the brain, and then we'll move on to consider perceptions of form, color, and motion of objects. Later we'll turn to some of the enduring mysteries of visual perception, including how D.F. can see *where* an object is placed in front of her and describe in fine detail what it looks like, yet be unable to say *what* that object is.

The Visual System Extends from the Eye to the Brain

To understand how the visual system constructs our vision, let's follow the path of information triggered by light entering the visual system, starting with the optical properties of the eye. Next, we'll consider the specialized cells in the back of the eye (a region called the *retina*) that sense light across a huge range of intensities, from the dim interior of a movie theater to a sunny beach, and encode that light as neural activity. Then we'll trace the information flowing from the eye to the cortex, giving rise to our conscious visual perception of the world.

The vertebrate eye acts in some ways like a camera

The eye is an elaborate structure with optical functions, capturing light and projecting detailed images of the external world onto the back of the eye. There a layer of neurons, called the **retina**, turns the light into neural signals, in a process called **transduction**. So, good vision requires an accurate optical image focused on the retina. In other words, light from a point on a target object must end up as a point of light—rather than a blur—on the retina.

To produce this sharply focused optical image, the eye has many of the features of a camera, starting with the transparent outer layer of the eye, called the **cornea** (**FIGURE 7.1**). Light travels in a straight line until it encounters a change in the density of the

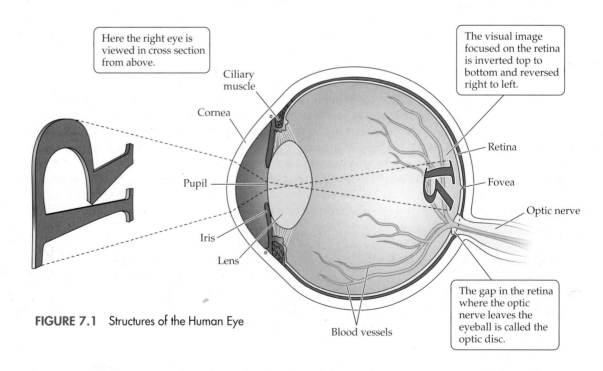

FIGURE 7.1 Structures of the Human Eye

Here the right eye is viewed in cross section from above.

The visual image focused on the retina is inverted top to bottom and reversed right to left.

Ciliary muscle

Cornea

Pupil

Iris

Lens

Retina

Fovea

Optic nerve

The gap in the retina where the optic nerve leaves the eyeball is called the optic disc.

Blood vessels

medium, which causes light rays to bend. This bending of light rays, called **refraction**, is the basis of such instruments as eyeglasses, telescopes, and microscopes. The curvature of the cornea, which does not change shape, refracts light rays and is primarily responsible for focusing on the retina. Light passing through the cornea is further refracted by the **lens**, which changes its shape to fine-tune that image on the retina.

The change in the shape of the lens is controlled by the **ciliary muscles** inside the eye. Changes in contraction of the ciliary muscles alter the focal distance of the eye, causing nearer or farther images to come into focus on the retina; this process is called **accommodation**. As mammals age, their lenses become less elastic and therefore less able to bring nearby objects into focus. Aging humans correct this problem either by holding books and menus farther away from their eyes, or by wearing reading glasses. In young people, the most common vision problem is **myopia** (nearsightedness), which is difficulty seeing distant objects. Myopia develops if the eyeball is too long, causing the cornea and lens to focus images in front of the retina rather than on it (**FIGURE 7.2**). Distance vision can be restored in such cases by glasses that correct refraction of the visual image onto the retina.

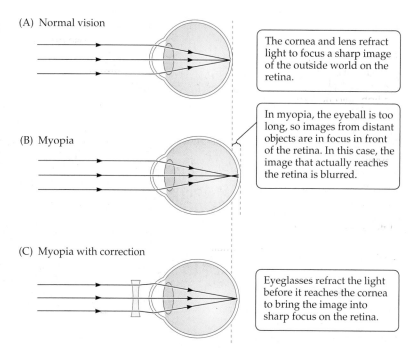

(A) Normal vision

The cornea and lens refract light to focus a sharp image of the outside world on the retina.

(B) Myopia

In myopia, the eyeball is too long, so images from distant objects are in focus in front of the retina. In this case, the image that actually reaches the retina is blurred.

(C) Myopia with correction

Eyeglasses refract the light before it reaches the cornea to bring the image into sharp focus on the retina.

FIGURE 7.2 Focusing Images on the Retina

If you've ever played around with a magnifying glass, you've probably noticed that if you hold the lens at arm's length, you can see a clearly focused image of a distant scene through the glass but that scene is upside down and reversed. Like a magnifying glass, the biconvex (bulging on both sides) shape of the lens of the eye causes the visual scene that is focused on the retina to be upside down and reversed compared with the real world (see Figure 7.1).

The movement of the eyes is controlled by the **extraocular muscles**, three pairs of muscles that extend from the outside of the eyeball to the bony socket of the eye. Fixating still or moving targets requires delicate control of these muscles to anchor the visual image on the retina. Let's talk about how that sharply focused visual image is processed in the retina.

Visual processing begins in the retina

The first stages of visual information processing occur in the retina, the receptive surface inside the back of the eye. The retina is only 200–300 micrometers thick—not much thicker than the edge of a razor blade—but it contains several types of cells in distinct layers (**FIGURE 7.3A**). Sensory neurons that detect light are called **photoreceptors**. There are two types of photoreceptors in the retina, called **rods** and **cones** in recognition of their respective shapes (**FIGURE 7.3B**). Cones come in several different varieties, which respond differently to light of varying wavelengths, providing us with color vision (as described later in the chapter). Rods respond to any visible light, regardless of wavelength.

Both rod and cone photoreceptors release neurotransmitter molecules into synapses on the **bipolar cells**, controlling their activity (see Figure 7.13). The bipolar cells, in turn, connect with **ganglion cells**. The axons of the ganglion cells form the **optic nerve**, which carries information to the brain. Two additional types of cells—**horizontal cells** and **amacrine cells**—are especially significant in interactions within the retina. The horizontal cells make contacts among the receptor cells and bipolar cells; the amacrine cells contact both the bipolar cells and the ganglion cells.

myopia Nearsightedness; the inability to focus the retinal image of objects that are far away.

extraocular muscle One of the muscles attached to the eyeball that control its position and movements.

photoreceptor A neural cell in the retina that responds to light.

rod A photoreceptor cell in the retina that is most active at low levels of light.

cone Any of several classes of photoreceptor cells in the retina that are responsible for color vision.

bipolar cell An interneuron in the retina that receives information from rods and cones and passes the information to retinal ganglion cells.

ganglion cell Any of a class of cells in the retina whose axons form the optic nerve.

optic nerve Cranial nerve II; the collection of ganglion cell axons that extend from the retina to the brain.

horizontal cell A specialized retinal cell that contacts both photoreceptors and bipolar cells.

amacrine cell A specialized retinal cell that contacts both bipolar cells and ganglion cells and is especially significant in inhibitory interactions within the retina.

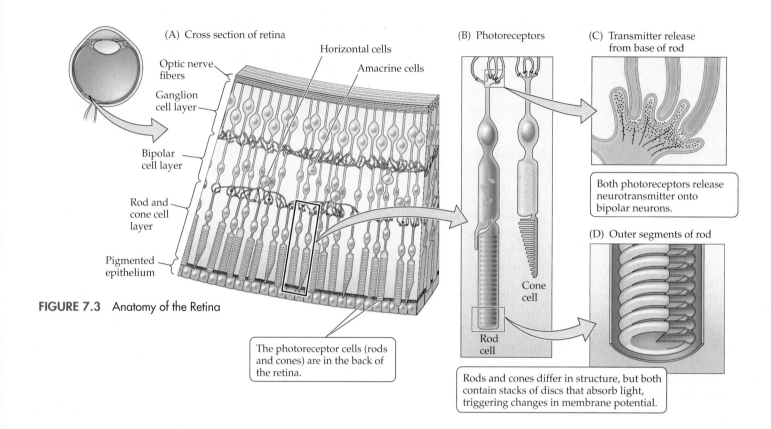

FIGURE 7.3 Anatomy of the Retina

(A) Cross section of retina

Optic nerve fibers
Ganglion cell layer
Bipolar cell layer
Rod and cone cell layer
Pigmented epithelium

Horizontal cells
Amacrine cells

The photoreceptor cells (rods and cones) are in the back of the retina.

(B) Photoreceptors

Cone cell
Rod cell

Rods and cones differ in structure, but both contain stacks of discs that absorb light, triggering changes in membrane potential.

(C) Transmitter release from base of rod

Both photoreceptors release neurotransmitter onto bipolar neurons.

(D) Outer segments of rod

Interestingly, the rods, cones, bipolar cells, and horizontal cells generate only graded, local potentials; they do not produce action potentials. Unlike most neurons, these cells affect each other through the *graded* release of neurotransmitters in response to *graded* changes in electrical potentials. The ganglion cells, on the other hand, conduct action potentials in the same way that most other neurons do. From the receptor cells to the ganglion cells, enormous amounts of data converge and are compressed; the human eye contains about 100 million rods and 4 million cones, but there are only 1 million ganglion cells to transmit that information to the brain. Thus, a great deal of information processing is done inside the eye, as information from over 100 million photoreceptors is compressed into the action potentials of 1 million ganglion cells.

The two different populations of photoreceptors (rods and cones) provide input to two different functional systems in the retina. A rod-based system, called the **scotopic system** (from the Greek *skotos*, "darkness," and *ops*, "eye"), is very sensitive and thus works especially well in low light—we use rod vision to detect objects in dim light—but it is insensitive to color. That's why in the darkness of night, when only our rods can detect light, we can't tell colors apart. There is a lot of **convergence** in the scotopic system, because the information from many rods *converges* onto each ganglion cell.

The other system uses cones, which are less sensitive than rods (that is, they have a higher threshold to respond) and therefore requires more light to function. This **photopic system** (which, like the term *photon*, gets its name from the Greek *phos*, "light") shows differential sensitivity to wavelengths, enabling our color vision. Compared with the scotopic system, the photopic system has less convergence, with some ganglion cells reporting information from only a single cone. At moderate levels of illumination, both the rods and the cones function, and some ganglion cells receive input from both types of receptors. **TABLE 7.1** summarizes the characteristics of the photopic and scotopic systems.

scotopic system A system in the retina that operates at low levels of light and involves the rods.

convergence The phenomenon of neural connections in which many cells send signals to a single cell.

photopic system A system in the retina that operates at high levels of light, shows sensitivity to color, and involves the cones.

TABLE 7.1 Properties of the Human Photopic and Scotopic Visual Systems

Property	Photopic system	Scotopic system
Receptors	Cones	Rods
Approximate number of receptors per eye	4 million	100 million
Photopigments	Three classes of cone opsins; the basis of color vision	Rhodopsin
Sensitivity	Low; needs relatively strong stimulation; used for day vision	High; can be stimulated by weak light intensity; used for night vision
Location in retina	Concentrated in and near fovea; present less densely throughout the retina	Outside fovea
Receptive-field size and visual acuity	Small in fovea, so acuity is high; larger outside fovea	Larger, so acuity is lower
Response time	Relatively rapid	Slow

Photoreceptors respond to light by releasing less neurotransmitter

Rods and cones owe their extraordinary sensitivity to their unusual structure and biochemistry (**FIGURE 7.3B–D**). A portion of their structure, when magnified, looks like a large stack of membranous pancakes. Because light is reflected in many directions by the various parts of the eye, only a fraction of the light that strikes the cornea actually reaches the retina. The stacking of the discs increases the probability that they will capture the light particles that make it to the retina.

The light particles, called *quanta* or *photons*, that strike the discs are captured by special photopigment receptor molecules. In the rods this photopigment is **rhodopsin** (from the Greek *rhodon*, "rose," and *opsis*, "vision"). Cones use similar photopigments, as we'll see later. Curiously enough, photoreceptors in the dark continually release neurotransmitter onto bipolar cells. When light hits photopigment in the photoreceptor, it triggers a cascade of chemical reactions that *hyperpolarize* the cell, causing the cell to release *less* neurotransmitter onto bipolar cells (**FIGURE 7.4**). You can learn the details of this process in **A STEP FURTHER 7.1**, on the website. It may seem puzzling that light causes photoreceptors to release less neurotransmitter, but remember that the visual system responds to *changes* in light. Either an increase or a decrease in the intensity of light can stimulate the visual system, and hyperpolarization is just as much a neural signal as depolarization is.

This change of potential in photoreceptors is the initial electrical signal in the visual pathway. Stimulation of rhodopsin by light hyperpolarizes the rods, just as stimulation of the cone pigments by light

rhodopsin   The photopigment in rods that responds to light.

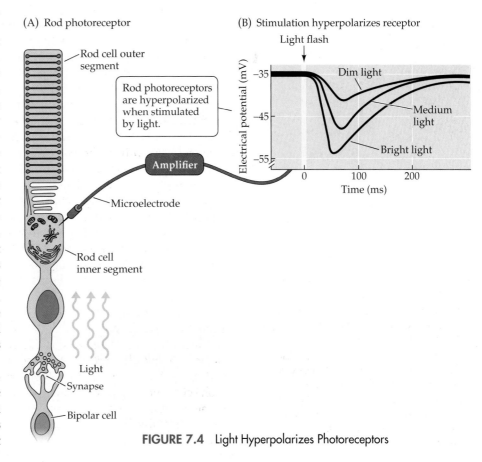

FIGURE 7.4   Light Hyperpolarizes Photoreceptors

FIGURE 7.5 The Wide Range of Sensitivity to Light Intensity

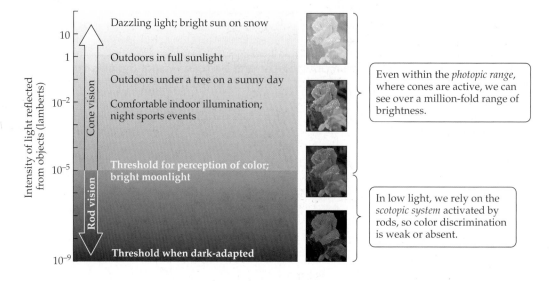

Intensity of light reflected from objects (lamberts)

Cone vision

Rod vision

Dazzling light; bright sun on snow
Outdoors in full sunlight
Outdoors under a tree on a sunny day
Comfortable indoor illumination; night sports events

Threshold for perception of color; bright moonlight

Threshold when dark-adapted

Even within the *photopic range*, where cones are active, we can see over a million-fold range of brightness.

In low light, we rely on the *scotopic system* activated by rods, so color discrimination is weak or absent.

pupil The opening, formed by the iris, that allows light to enter the eye.

iris The circular structure of the eye that provides an opening to form the pupil.

range fractionation The means by which sensory systems cover a wide range of intensity values, as each sensory receptor cell specializes in just one part of the overall range of intensities.

hyperpolarizes the cones. For both rods and cones, the size of the hyperpolarizing photoreceptor potential determines how much less transmitter will be released (see Figure 7.4). Another important feature of photoreceptors is that their sensitivity to light is constantly changing, depending on how much light is available. This feature, called *photoreceptor adaptation*, allows us to see over wide ranges of light, as discussed next.

Different mechanisms enable the eyes to work over a wide range of light intensities

Our visual system must be able to respond to stimuli of vastly different intensities: a very bright light is about 10 billion times as intense as the weakest lights we can see (**FIGURE 7.5**). One way the visual system deals with this large range of intensities is by adjusting the size of the **pupil**, which is an opening in the colorful disc called the **iris** (see Figure 7.1). In Chapter 2 we mentioned that dilation (opening) of the pupils is controlled by the sympathetic division of the autonomic system and that constriction is triggered by the parasympathetic division. Because usually both divisions are active, pupil size reflects a balance of influences. During an eye exam, the doctor may use a drug to block acetylcholine transmission in the parasympathetic synapses of your iris; this drug relaxes the muscle fibers and opens the pupil widely. One drug that has this effect—belladonna—got its name (Italian for "beautiful woman") because it was thought to make a woman more beautiful by giving her the wide-open pupils of an attentive person. Other drugs, such as morphine, constrict the pupils.

In bright light the pupil contracts quickly to admit only about one-sixteenth as much light as when illumination is dim (**FIGURE 7.6**). Although rapid, the 16-fold difference in light controlled by the pupil doesn't come close to accounting for the *billion*-fold range of visual sensitivity. Another mechanism for handling different light intensities is **range fractionation**, the handling of different intensities by different receptors—some with low thresholds (rods) and others with high thresholds (cones) (see Figure 7.5). But the

(A) Bright illumination

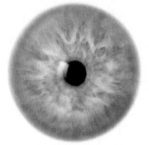

While the size of the pupil can change rapidly, it affects light entry by only about 16-fold. So it cannot possibly account for our ability to see over a billionfold range of illumination.

(B) Dark

FIGURE 7.6 The Iris Controls the Size of the Pupil Opening

main reason we can see over such a vast range of light is **photoreceptor adaptation**: each photoreceptor adjusts its sensitivity to match the average level of ambient illumination, over a tremendous range. Thus, the visual system is concerned with *differences*, or changes, in brightness—not with the absolute level of illumination.

At any given time, a photoreceptor operates over a range of intensities of about a hundred-fold; that is, it is completely depolarized by a stimulus about one-tenth the ambient level of illumination, and a light 10 times more intense than the ambient level will completely hyperpolarize it. The receptors constantly shift their whole range of response to work around the prevailing level of illumination. Further adaptation, controlled by neural circuits, occurs in the brain.

Acuity is best in foveal vision

Visual acuity, commonly known as the sharpness of vision, is a measure of how much detail we can see. Our visual acuity is especially fine in the center of the visual field and falls off rapidly toward the periphery. That's why when we want to look at something closely, we center our gaze on the object of interest.

The fine structure of the retina explains why our acuity is best in the center of the visual field in **FIGURE 7.7A**. Below that, notice how much more densely packed cones are in the fovea, where acuity is highest (**FIGURE 7.7B**), than in other parts of the retina. The central region of the retina, called the **fovea**, has a dense concentration of cones, absorbing so much light that the region looks dark in the photo. That is one

photoreceptor adaptation The tendency of rods and cones to adjust their light sensitivity to match current levels of illumination.

visual acuity Sharpness of vision.

fovea The central portion of the retina, which is packed with the highest density of photoreceptors and is the center of our gaze.

(A) Distributions of rods and cones across the retina

FIGURE 7.7 Densities of Retinal Receptors and Visual Acuity

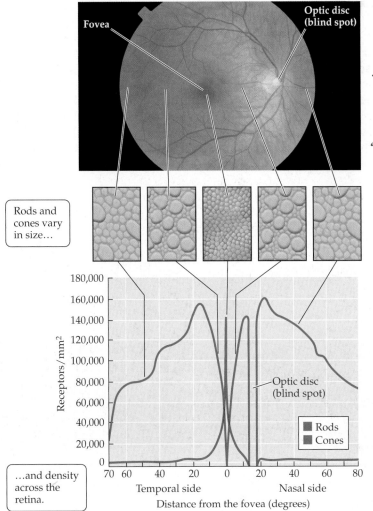

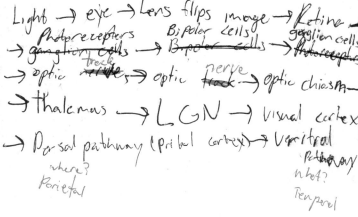

(B) Variation of visual acuity across the retina

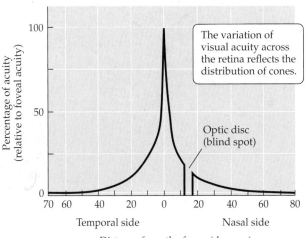

optic disc The region of the retina that is devoid of receptor cells because ganglion cell axons and blood vessels exit the eyeball there.

blind spot The portion of the visual field from which light falls on the optic disc.

reason visual acuity is so high in this region. People differ in their concentrations of cones (Curcio et al., 1987), and this variation may be related to individual differences in visual acuity. Species differences in visual acuity also reflect the density of cones in the fovea. For example, hawks, whose acuity is much greater than that of humans, have much narrower and more densely packed cones in the fovea than we do. In the human retina, both cones and rods are larger toward the periphery.

The rods show a different distribution from the cones: they are absent in the fovea but more numerous than cones in the periphery of the retina (see Figure 7.7A). This is why, if you want to see a dim star, you do best to search for it a little off to the side of your center of gaze. Not only are rods more sensitive than cones to dim light, but as we mentioned earlier, input from many rods converges on each ganglion cell in the scotopic system, further increasing the system's sensitivity to weak stimuli. But that greater convergence of rods comes at the cost of diminished acuity compared with the fovea. Rods provide high sensitivity with limited acuity; cones provide high acuity with limited sensitivity. Thus, really fine vision requires good lighting.

We noted earlier that rods cannot contribute to color vision. Yet the few cones scattered in the periphery of our gaze enable us to detect color there. That's why peripheral cones are large: so that they're likely to get some color information. We are not aware that most of our peripheral vision lacks color information, because the brain "fills in," assuming that the light striking a rod is of the same wavelength as light hitting a neighboring cone.

Another reason why acuity is greater in the fovea than elsewhere on the retina is that in this region light reaches the cones directly, without having to pass through other layers of cells and blood vessels (**FIGURE 7.8**). In the rest of the retina, many light particles hit those upper layers without reaching the photoreceptors. This is why the surface of the retina is depressed at the fovea (see Figure 7.1A), giving the structure its name (*fovea* means "pit" in Latin).

The **optic disc**, to the nasal side of the fovea, is where blood vessels and ganglion cell axons leave the eye (see Figure 7.7A). There are no photoreceptors at the optic disc, so there is a **blind spot** here that we normally do not notice. You can locate your blind spot, and experience firsthand some of its interesting features, with the help of **FIGURE 7.9**. The blind spot is much bigger than we usually appreciate; it is about 10 times larger than the image of a full moon, yet we typically don't even notice it!

FIGURE 7.8 An Unobstructed View

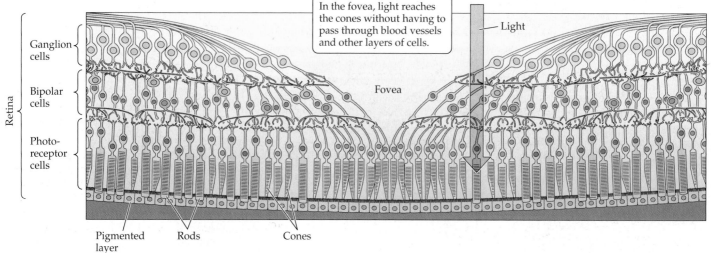

In the fovea, light reaches the cones without having to pass through blood vessels and other layers of cells.

Light

Fovea

Retina

Ganglion cells

Bipolar cells

Photo-receptor cells

Pigmented layer

Rods

Cones

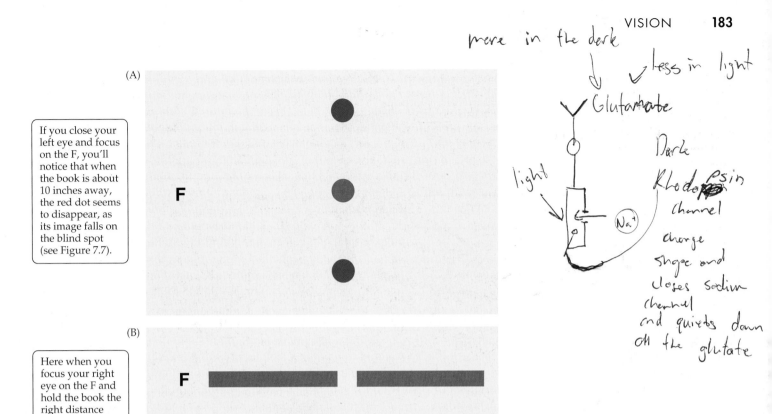

If you close your left eye and focus on the F, you'll notice that when the book is about 10 inches away, the red dot seems to disappear, as its image falls on the blind spot (see Figure 7.7).

Here when you focus your right eye on the F and hold the book the right distance away, the red line appears unbroken.

FIGURE 7.9 Experiencing the Blind Spot

[handwritten notes:] more in the dark ↓ ✓ less in light ✓ Glutamate light → Dark Rhodopsin channel change shape and closes sodium channel and quiets down off the glutate Na⁺

Again, brain systems "fill in" the missing information so that we perceive an uninterrupted visual scene.

Before we consider how information is processed at different levels of the visual system, we need to describe the pathway from the eye to the cortex, which we'll consider next.

how's it going ❓

1. Describe how structures of the eye refract light to focus an image on the retina.
2. How do the photopic and scotopic visual systems vary?
3. How are we able to discriminate differences in light over such a wide range of illumination?
4. Why is our vision so much more acute at the fovea than it is elsewhere?

To see the animation
Visual Pathways in the Human Brain,
go to

2e.mindsmachine.com/7.3

Neural signals travel from the retina to several brain regions

The ganglion cells in each eye produce action potentials that are conducted along their axons to send visual information to the brain. These axons make up the optic nerve (also known as cranial nerve II), which brings visual information into the brain on each side, eventually reaching the **occipital cortex** at the back of the brain.

In vertebrates, some or all of the axons of each optic nerve cross to the opposite cerebral hemisphere. The optic nerves cross the midline at the **optic chiasm** (named for the Greek letter X [chi] because of its crossover shape). In humans, axons from the half of the retina toward your nose (the *nasal hemiretina*) cross over to the opposite side of the brain. The half of the retina toward your temple (the *temporal hemiretina*) projects its axons to its own side of the brain (**FIGURE 7.10**).

occipital cortex Also called *visual cortex*. The cortex of the occipital lobe of the brain, corresponding to the primary visual area of the cortex.

optic chiasm The point at which parts of the two optic nerves cross the midline.

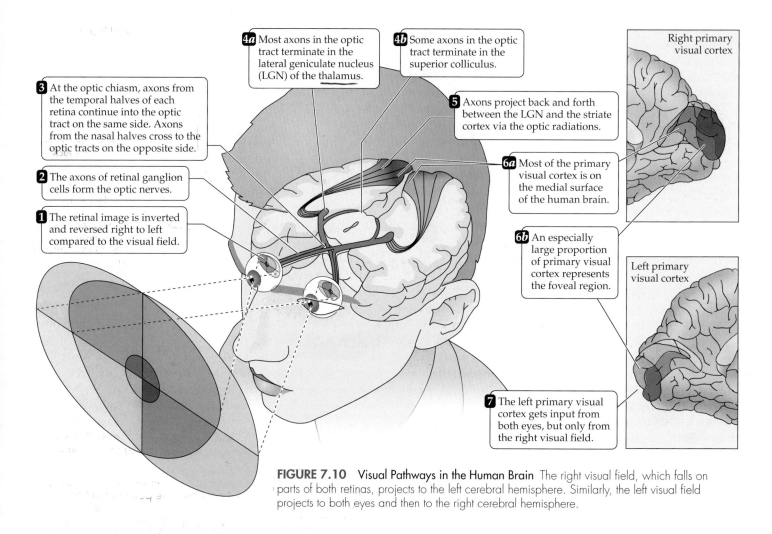

3 At the optic chiasm, axons from the temporal halves of each retina continue into the optic tract on the same side. Axons from the nasal halves cross to the optic tracts on the opposite side.

2 The axons of retinal ganglion cells form the optic nerves.

1 The retinal image is inverted and reversed right to left compared to the visual field.

4a Most axons in the optic tract terminate in the lateral geniculate nucleus (LGN) of the thalamus.

4b Some axons in the optic tract terminate in the superior colliculus.

5 Axons project back and forth between the LGN and the striate cortex via the optic radiations.

6a Most of the primary visual cortex is on the medial surface of the human brain.

6b An especially large proportion of primary visual cortex represents the foveal region.

7 The left primary visual cortex gets input from both eyes, but only from the right visual field.

Right primary visual cortex

Left primary visual cortex

FIGURE 7.10   Visual Pathways in the Human Brain The right visual field, which falls on parts of both retinas, projects to the left cerebral hemisphere. Similarly, the left visual field projects to both eyes and then to the right cerebral hemisphere.

optic tract   The axons of retinal ganglion cells after they have passed the optic chiasm. Most of these axons terminate in the lateral geniculate nucleus.

lateral geniculate nucleus (LGN) The part of the thalamus that receives information from the optic tract and sends it to visual areas in the occipital cortex.

optic radiation   Axons from the lateral geniculate nucleus that terminate in the primary visual areas of the occipital cortex.

primary visual cortex (V1)   Also called *striate cortex* or *area 17*. The region of the occipital cortex where most visual information first arrives.

binocular   Referring to two-eyed.

Proportionally more axons cross the midline in prey animals, such as rabbits, that have laterally placed eyes with little overlap in their fields of vision (**FIGURE 7.11**). This arrangement gives a prey animal an especially wide field of view (good for spotting threats) at the cost of poor depth perception (which predators gain by comparing the overlapping visual fields of their left and right eyes). After they pass the optic chiasm, the axons of the retinal ganglion cells are known collectively as the **optic tract**.

Most axons of the optic tract terminate on cells in the **lateral geniculate nucleus** (**LGN**) (see Figure 7.10, step 4a), which is the visual part of the thalamus. Axons of the LGN neurons form the **optic radiations** (step 5), which terminate in the **primary visual cortex (V1)** of the occipital cortex at the back of the brain (step 6). The primary visual cortex is sometimes called *striate cortex* because cross sections of brain tissue from this region feature a prominent stripe, or *striation*, corresponding to convergent **binocular** ("two-eyed") inputs. The binocular input to layer IV of the primary visual cortex is important for depth perception.

As Figure 7.10 shows, the visual cortex in the right cerebral hemisphere receives its input from the left half of the visual field, and the visual cortex in the left hemisphere receives its input from the right half of the visual field. Some retinal ganglion cells send their optic tract axons to the superior colliculus in the midbrain (step 4b). The superior colliculus helps coordinate rapid movements of the eyes toward a target, and it controls the pupil's response to light levels.

In addition to the primary visual cortex (V1) shown in Figure 7.10, numerous surrounding regions of the cortex are largely visual in function. These visual cortical areas outside the striate cortex are sometimes called **extrastriate cortex**. Working in parallel, these cortical regions process different aspects of visual perception, such as form, color, location, and movement, as we'll discuss later in this chapter. The striate cortex, as well as most extrastriate regions, contains a topographic projection of the retinas, which means there's a topographic projection of the visual field, discussed next.

The retina projects to the brain in a topographic fashion

The whole area that you can see without moving your head or eyes is called your **visual field**. The retina represents a two-dimensional map of the visual field. As this information courses through the brain, the point-to-point correspondence between neighboring parts of visual space is maintained, forming a maplike projection (see Figure 7.10). Much of this **topographic projection** of visual space is devoted to the foveal region (**FIGURE 7.12A**) (Tootell et al., 1982). Human V1 is located mainly on the medial surface of the cortex (**FIGURE 7.12B**; see also Figure 7.10). About half of the human V1 is devoted to the fovea and the retinal region just around the fovea, even though this represents a tiny fraction of the total retina. This disproportionate representation does not mean that our spatial perception is distorted. Rather, this representation makes possible the great acuity in the central part of the visual field. In other words, another reason why our vision is so much more acute in the foveal region is that we devote proportionally more brain regions to analyzing information from that region.

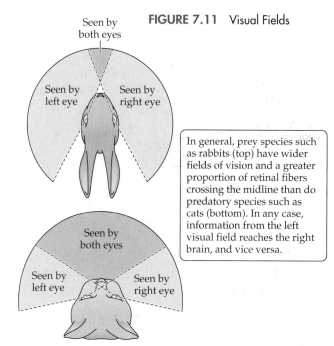

FIGURE 7.11 Visual Fields

In general, prey species such as rabbits (top) have wider fields of vision and a greater proportion of retinal fibers crossing the midline than do predatory species such as cats (bottom). In any case, information from the left visual field reaches the right brain, and vice versa.

extrastriate cortex Visual cortex outside of the primary visual (striate) cortex.

visual field The whole area that you can see without moving your head or eyes.

topographic projection A mapping that preserves the point-to-point correspondence between neighboring parts of space. For example, the retina extends a topographic projection onto the cortex.

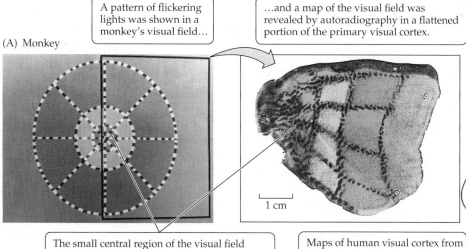

(A) Monkey

A pattern of flickering lights was shown in a monkey's visual field…

…and a map of the visual field was revealed by autoradiography in a flattened portion of the primary visual cortex.

1 cm

The small central region of the visual field projects to a large part of primary visual cortex.

Maps of human visual cortex from fMRI show primary visual cortex as the innermost yellow region in each of these medial views.

(B) Human

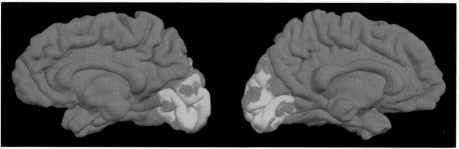

FIGURE 7.12 Location of the Primary Visual Cortex (Part A from Tootell et al., 1988; B from Tootell et al., 1998; both courtesy of Roger Tootell.)

scotoma A region of blindness within the visual fields, caused by injury to the visual pathway or brain.

blindsight The paradoxical phenomenon whereby, within a scotoma, a person cannot *consciously* perceive visual cues but may still be able to make some visual discrimination.

receptive field The stimulus region and features that affect the activity of a cell in a sensory system.

on-center bipolar cell A retinal bipolar cell that is excited by light in the center of its receptive field.

off-center bipolar cell A retinal bipolar cell that is inhibited by light in the center of its receptive field.

on-center ganglion cell A retinal ganglion cell that is activated when light is presented to the center, rather than the periphery, of the cell's receptive field.

off-center ganglion cell A retinal ganglion cell that is activated when light is presented to the periphery, rather than the center, of the cell's receptive field.

receptive field is going
a sensory cell area, that
is responsive to such as
light.

Because of the orderly mapping of the visual field (known as *retinotopic mapping*) at the various levels of the visual system, damage to parts of the visual system can be diagnosed from perceptual defects within the visual field. And if we know the site of injury in the visual pathway, we can predict the location of such a perceptual gap, or **scotoma**, in the visual field. Although the word *scotoma* comes from the Greek *skotos*, meaning "darkness," a scotoma is not perceived as a dark patch in the visual field; rather, it is a spot where nothing can be perceived, and usually rigorous testing is required to demonstrate its existence. As with the blind spots we all have, people may not be aware of scotomas that arise.

Within a scotoma, a person cannot *consciously* perceive visual cues, but some visual discrimination in this region may still be possible; this paradoxical phenomenon has been called **blindsight**. People with blindsight say they cannot see, but when asked to *guess* whether a stimulus is present, they're correct more often than could be expected by chance alone, or they may walk down a corridor strewn with objects without running into them (De Gelder et al., 2008).

─── how's it going ❓

1. Describe the path of information from the left visual field to the right side of the brain.
2. Name the structures that carry information from the eye to the brain.
3. Why is the proportion of primary visual cortex devoted to the fovea so large compared with other parts of the retina?

Neurons at Different Levels of the Visual System Have Very Different Receptive Fields

As we noted in Chapter 5, the **receptive field** of a sensory cell consists of the stimulus region and the features that excite or inhibit the cell. The nature of the receptive field of a cell gives us good clues about the cell's function(s) in perception. Cells in the retina or LGN can be activated by simple spots of light, but many cells in the visual cortex are more demanding and respond only to more-complicated stimuli. In this section we will see that neurons at lower levels in the visual system seem to account for some perceptual phenomena while neurons at higher levels account for others.

Photoreceptors excite some retinal neurons and inhibit others

At rest, both rod and cone photoreceptors steadily release the synaptic neurotransmitter glutamate. Light always hyperpolarizes the photoreceptors, causing them to release less glutamate. But the *response* of the bipolar cells that receive this glutamate differs, depending on the type of glutamate receptor they possess. Glutamate depolarizes one group of bipolar cells but hyperpolarizes another group.

One group of bipolar cells consists of **on-center bipolar cells**; turning on a light in the center of an on-center bipolar cell's receptive field *excites* the cell because it receives less glutamate, which *inhibits* this type of bipolar cell (**FIGURE 7.13** *left*). The second group consists of **off-center bipolar cells**; light hitting the center of an off-center bipolar cell's field *inhibits* it because glutamate *excites* this type of cell (see Figure 7.13 *right*). It's called an off-center bipolar cell because turning *off* a light in the center of its receptive field excites it.

Bipolar cells also *release* glutamate, which always depolarizes ganglion cells. Therefore, when light is turned on, on-center bipolar cells depolarize (excite) **on-center ganglion cells**; when light is turned off, off-center bipolar cells depolarize (excite) **off-center ganglion cells** (see Figure 7.13). The stimulated on-center and off-center ganglion cells then fire nerve impulses and report "light" or "dark" to higher visual centers.

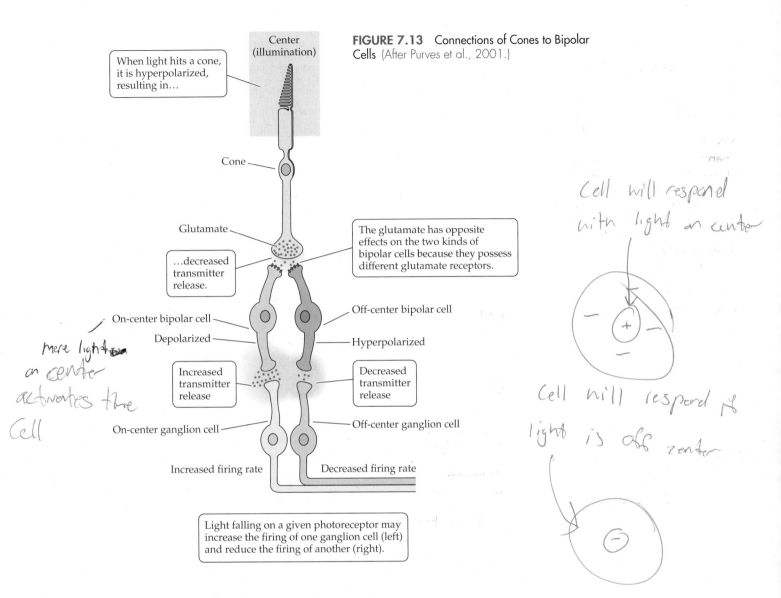

FIGURE 7.13 Connections of Cones to Bipolar Cells (After Purves et al., 2001.)

Handwritten annotations:

mere light on center activates the Cell

Cell will respond with light on center

Cell will respond if light is off center

Neurons in the retina and the LGN have concentric receptive fields

Scientists can record from single ganglion cells while moving a small spot of light across the visual field, keeping the animal's eye still. These studies show that the receptive fields of retinal ganglion cells are *concentric*, consisting of a roughly circular central area and a ring around it, which is usually called a *surround* because it surrounds the central patch. Through various retinal connections, the photoreceptors in the central area and those in the ring surrounding it tend to have opposite effects on the next cells in the circuit. Thus, both bipolar cells and ganglion cells have two basic types of retinal receptive fields: **on-center/off-surround** (**FIGURE 7.14A**) and **off-center/on-surround** (**FIGURE 7.14B**). These antagonistic effects of the center and its surround explain why uniform illumination of the visual field has little effect on ganglion cell activity, compared with a well-placed small spot of light on the cell's receptive field. Neurons in the LGN, which are stimulated by retinal ganglion cells, also have these concentric on-center/off-surround or off-center/on-surround receptive fields.

To understand why the effect of light falling on the surround of a firing ganglion cell is opposite to the effect of light falling in the center, we need to understand the concept of **lateral inhibition**, in which sensory receptor cells inhibit the reporting of

on-center/off-surround Referring to a concentric receptive field in which stimulation of the center excites the cell of interest while stimulation of the surround inhibits it.

off-center/on-surround Referring to a concentric receptive field in which stimulation of the center inhibits the cell of interest while stimulation of the surround excites it.

lateral inhibition The phenomenon by which interconnected neurons inhibit their neighbors, producing contrast at the edges of regions.

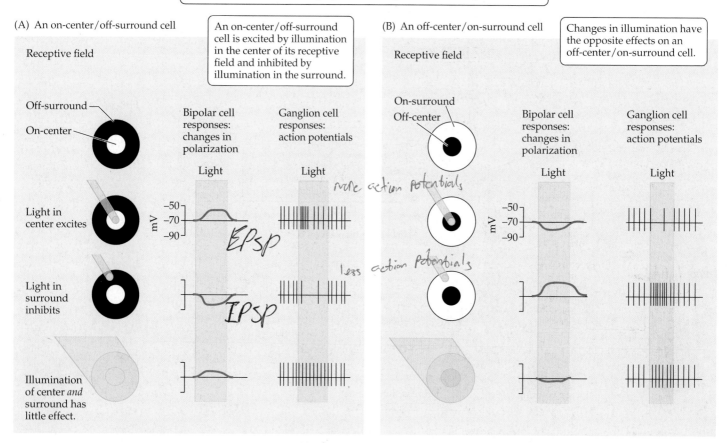

Each retinal bipolar cell and ganglion cell has a concentric receptive field, with antagonistic center and surround. Bipolar cells respond with changes in local membrane potentials, while ganglion cells respond with action potentials.

(A) An on-center/off-surround cell

An on-center/off-surround cell is excited by illumination in the center of its receptive field and inhibited by illumination in the surround.

(B) An off-center/on-surround cell

Changes in illumination have the opposite effects on an off-center/on-surround cell.

FIGURE 7.14 Receptive Fields of Retinal Cells

To see the animation
Receptive Fields in the Retina,
go to

2e.mindsmachine.com/7.4

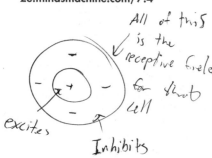

information from neighboring receptor cells. As illustrated in **FIGURE 7.15**, the bipolar cells that relay information from photoreceptors to ganglion cells also inhibit one another. Generally these connections are inhibitory synapses, so when one bipolar cell is active, it inhibits its neighbors.

Because of this lateral inhibition, the ganglion cells stimulated by the right-hand edge of each dark band in **FIGURE 7.16A** are inhibited by the neighboring photoreceptors stimulated by the lighter band next door. Thus, ganglion cells stimulated by the right edge of each bar report receiving less light than they actually do (that is, that edge looks darker to us). Conversely, the left edge of each bar looks lighter than the rest of the bar.

Again, in **FIGURE 7.16B** two indicated patches, which clearly differ in the brightness we perceive, *reflect the same amount of light.* If you use your finger to cover the edge where the two tiles meet, you'll see that the two patches are the same shade of gray. How are such puzzling effects produced? Although the contrast effect in Figure 7.16A is determined, at least in part, by lateral inhibition among adjacent retinal cells, the entire areas indicated in Figure 7.16B, not just the edges, appear different, so the effect must be produced higher in the visual system. One explanation is that we are accustomed to light sources coming from overhead (such as the sun, or a room light), so our brain assumes that the upper patch must actually be darker than the lower patch, because the upper one should be receiving more light than the lower one.

Whether or not that particular explanation is correct, the important point is that our *visual experience is not a simple reporting of the physical properties of light.* Rather,

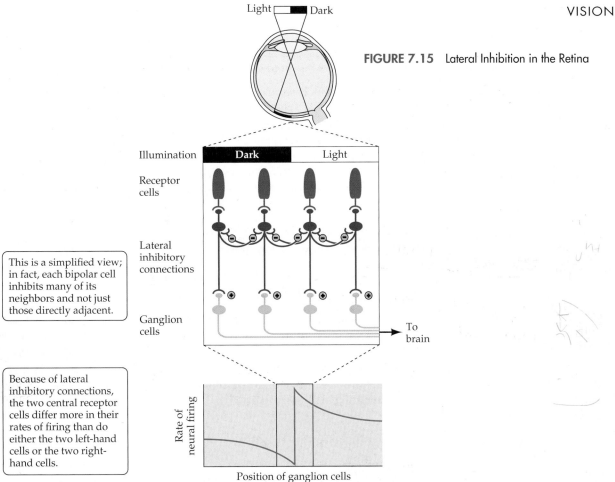

FIGURE 7.15 Lateral Inhibition in the Retina

Illumination

Receptor cells

Lateral inhibitory connections

This is a simplified view; in fact, each bipolar cell inhibits many of its neighbors and not just those directly adjacent.

Ganglion cells

To brain

Because of lateral inhibitory connections, the two central receptor cells differ more in their rates of firing than do either the two left-hand cells or the two right-hand cells.

Rate of neural firing

Position of ganglion cells

our perception of light versus dark is created by the brain in response to many factors, including surrounding stimuli. For example, if you read this book in bright sunlight, the black ink reflects far more light to your eyes than the blank parts of the page do indoors. Yet, whether you're in sunlight or indoors, you perceive the ink as black and the blank parts as white. Later in the chapter we'll find that our experience of color is also created by the visual system and that it is not a simple reporting of the wavelengths of light.

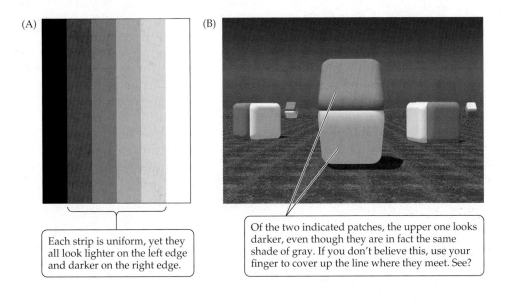

(A)

Each strip is uniform, yet they all look lighter on the left edge and darker on the right edge.

(B)

Of the two indicated patches, the upper one looks darker, even though they are in fact the same shade of gray. If you don't believe this, use your finger to cover up the line where they meet. See?

FIGURE 7.16 The Effect of Context on the Perception of Brightness (Part B from Purves and Lotto, 2011.)

how's it going ?

1. Given that *all* photoreceptors are hyperpolarized by light, how can the same photoreceptor excite some bipolar cells while inhibiting others?
2. What is a receptive field, and what two kinds of receptive fields are displayed by retinal ganglion cells?
3. Describe lateral inhibition in the retina and how it can sharpen our vision yet make us susceptible to the optical illusion we experience in Figure 7.16A.

researchers at work

Neurons in the visual cortex have varied receptive fields

Neurons from the LGN send their axons to cells in the primary visual cortex (V1), but the *spots* of light that are effective stimuli for LGN cells (**FIGURE 7.17A**; see also Figure 7.14A) are not very effective for cortical cells. In 1959, David Hubel and Torsten Wiesel reported that visual cortical cells require more-specific, elongated stimuli than those that activate LGN cells and ganglion cells.

FIGURE 7.17 Receptive Fields of Cells at Various Levels in the Cat Visual System

■ **Hypothesis**

Cells at higher levels of the visual system respond to progressively more complex stimuli.

■ **Test**

Compare receptive fields of neurons at each level, and see how they relate to one another.

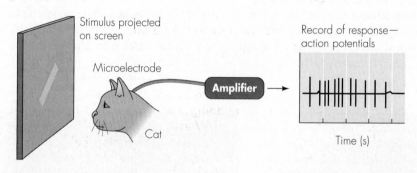

Stimulus projected on screen

Microelectrode

Amplifier →

Record of response— action potentials

Cat

Time (s)

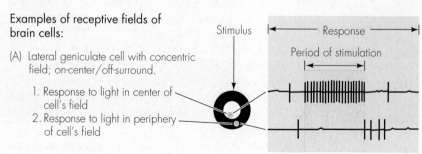

Examples of receptive fields of brain cells:

(A) Lateral geniculate cell with concentric field; on-center/off-surround.

1. Response to light in center of cell's field
2. Response to light in periphery of cell's field

Stimulus

Response

Period of stimulation

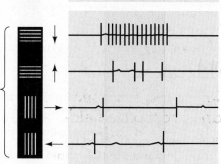

(B) A simple cortical cell is sensitive to orientation. This particular cell responds strongly only when the stimulus is a vertical stripe.

(C) A complex cortical cell is also sensitive to motion. This particular cell responds strongly only when the stimulus moves down. It responds weakly to upward motion and does not respond at all to sideways motion.

■ Result

(A) Visual cells in the LGN have concentric receptive fields.

(B) Visual cells in the cerebral cortex may show orientation specificity or respond only to motion, or...

(C) ...they may respond only to motion in a particular direction.

■ Conclusion

Neurons at each level of the visual system combine input from neurons at lower levels to make progressively more complex receptive fields. Thus, retinal and LGN neurons respond best to spots of light on the retina, while cortical cells respond best to lines of particular orientation, or lines that move in a particular direction.

Hubel and Wiesel categorized cortical cells according to the types of stimuli required to produce maximum responses. So-called **simple cortical cells** respond best to an edge or a bar that has a particular width and a particular orientation and location in the visual field (**FIGURE 7.17B**). These cells are therefore sometimes called *bar detectors* or *edge detectors*. Like the simple cells, **complex cortical cells** have elongated receptive fields, but they also require *movement* of the stimulus to make them respond actively. For some of these cells, any movement in their field is sufficient; others are more demanding, requiring motion in a specific direction (**FIGURE 7.17C**). Hubel and Wiesel's theoretical model can be described as hierarchical; that is, more-complex receptive fields are built up from inputs of simpler ones. For example, a simple cortical cell can be thought of as receiving input from a row of LGN cells (**FIGURE 7.18**), and a complex cortical cell can be thought of as receiving input from a row of simple cortical cells.

Other theorists extrapolated from this hierarchical model, suggesting that higher-order circuits of cells could detect any possible form. Thus it was suggested that, by integration of enough successive levels of analysis, a neuron could respond only to a person's grandmother, and such hypothetical "grandmother cells" were frequently mentioned in the literature. According to this view, whenever such a cell was excited, up would pop a mental picture of one's grandmother. This hypothesis was given as a possible explanation for facial recognition.

simple cortical cell Also called *bar detector* or *edge detector*. A cell in the visual cortex that responds best to an edge or a bar that has a particular width, as well as a particular orientation and location in the visual field.

complex cortical cell A cell in the visual cortex that responds best to a bar of a particular size and orientation anywhere within a particular area of the visual field and that needs movement to make it respond actively.

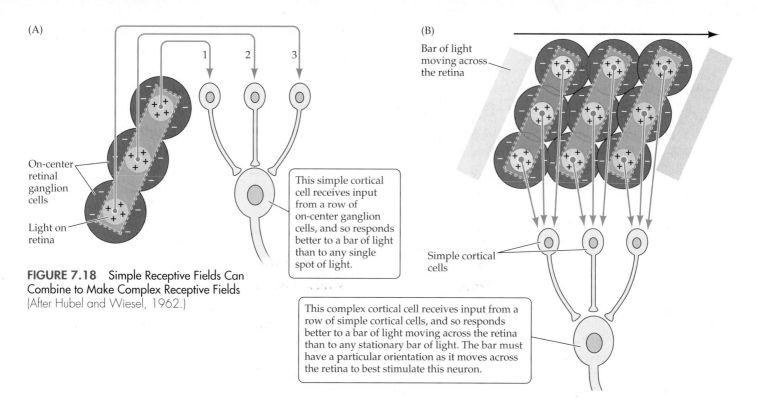

FIGURE 7.18 Simple Receptive Fields Can Combine to Make Complex Receptive Fields (After Hubel and Wiesel, 1962.)

On-center retinal ganglion cells

Light on retina

This simple cortical cell receives input from a row of on-center ganglion cells, and so responds better to a bar of light than to any single spot of light.

Bar of light moving across the retina

Simple cortical cells

This complex cortical cell receives input from a row of simple cortical cells, and so responds better to a bar of light moving across the retina than to any stationary bar of light. The bar must have a particular orientation as it moves across the retina to best stimulate this neuron.

spatial-frequency model A model of vision that emphasizes the analysis of different spatial frequencies, of various orientations and in various parts of the visual field, as the basis of visual perception of form.

The hierarchical model is supplanted by a more efficient analysis

Critics soon pointed out both theoretical and empirical problems with the hierarchical model. For one thing, a hierarchical system like this would require a *vast* number of cells—perhaps more neurons than the cortex possesses—in order to account for all the visual objects we might encounter. Although some neurons are indeed activated by the sight of very specific faces (e.g., "Halle Berry neurons" were found, which were activated by photos of that actress) in both humans (Pedreira et al., 2010) and monkeys (Freiwald et al., 2009), these neurons do not respond to specific features of the face, as we would expect if they were built up from feature detectors. Rather, they respond only when the *whole face* or most of the face is presented (Freiwald et al., 2009).

Faced with the inadequacies of the hierarchical model, scientists proposed an alternative account of vision, known as the **spatial-frequency model** (F. W. Campbell and Robson, 1968; R. L. De Valois and De Valois, 1988), that is more powerful but less intuitive. This model proposes that the visual system analyzes the number of cycles of light-dark (or color) patches in any stimulus. Some cycles are narrow, others broad. Some cycles of light-dark are oriented vertically, others horizontally, and others somewhere in between. If cortical neurons are indeed optimized to detect light-dark cycles, then they should respond to repeating bars of light, as in the examples shown in **FIGURE 7.19**, even better than to a single bar of light. And that's precisely what researchers found.

To see the animation **Spatial Frequencies**, go to

2e.mindsmachine.com/7.5

FIGURE 7.19 Examples of Spatial Frequencies in Vision The brain combines information about all these spatial frequencies to give us our perception of the scene.

Cortical cells respond even better to these repeating patterns of light than to single bars of light.

Each cortical cell fires best to such patterns in a particular orientation—vertical...

...horizontal...

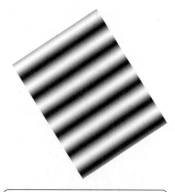

...or somewhere in between—with a particular frequency (narrow or wide bars), and in a particular part of the visual field.

The idea that the visual system processes spatial-frequency channels had a revolutionary impact because it led to entirely different conceptions of how the visual system might work. The idea suggests that, rather than specifically detecting such seminaturalistic features as bars and edges, the system is breaking down complex stimuli into their individual spatial-frequency components (R. L. De Valois and De Valois, 1988, p. 320). In such a system, we might require a view of the whole face, which includes the low-frequency components, for recognition. This could explain why "Halle Berry neurons" do not respond to small portions of a face, because such snippets contain only high-frequency components. The spatial-frequency approach has proven useful in the analysis of many aspects of human pattern vision (K. K. De Valois et al., 1979), and it provides the basis of high-definition television (HDTV). If you would like to know more about how the spatial-frequency model works, see **A STEP FURTHER 7.2**, on the website.

Neurons in the visual cortex beyond area V1 have complex receptive fields and help identify forms

Area V1 represents only a small fraction of the total amount of cortex that is devoted to vision. From area V1, axons extend to cortical areas involved in the perception of form: V2, V4, and the inferior temporal area (**FIGURE 7.20A–C**). The receptive fields of the cells in many of these extrastriate visual areas are even more complex than those of area V1.

The visual areas of the human brain (**FIGURE 7.20D**) have been less thoroughly mapped than those of the monkey brain, and mainly by neuroimaging (the spatial

FIGURE 7.20 Main Visual Areas in Monkey and Human Brains (Parts A–C after Van Essen and Drury, 1997; D from Tootell et al., 1998, courtesy of Roger Tootell.)

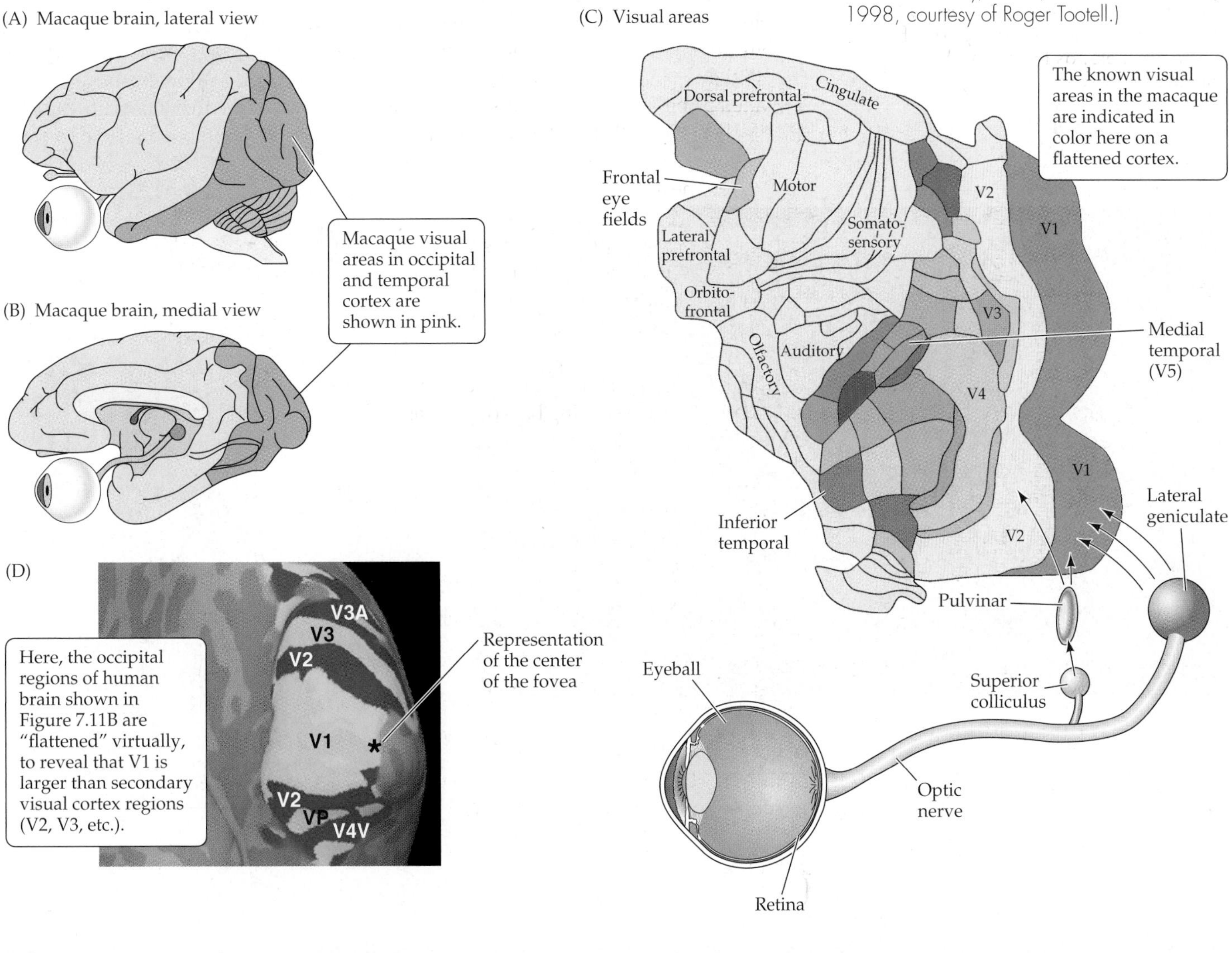

(A) Macaque brain, lateral view

(B) Macaque brain, medial view

Macaque visual areas in occipital and temporal cortex are shown in pink.

(D) Here, the occipital regions of human brain shown in Figure 7.11B are "flattened" virtually, to reveal that V1 is larger than secondary visual cortex regions (V2, V3, etc.).

V3A
V3
V2
V1 *
V2
VP
V4V

Representation of the center of the fovea

(C) Visual areas

The known visual areas in the macaque are indicated in color here on a flattened cortex.

Dorsal prefrontal
Cingulate
Frontal eye fields
Motor
Lateral prefrontal
Somato-sensory
Orbito-frontal
Olfactory
Auditory
Inferior temporal
V2
V1
V3
V4
V1
V2
Medial temporal (V5)
Pulvinar
Eyeball
Superior colliculus
Lateral geniculate
Optic nerve
Retina

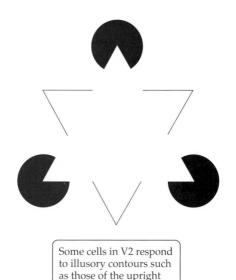

Some cells in V2 respond to illusory contours such as those of the upright triangle shown here.

FIGURE 7.21 A Geometric Figure with "Illusory" or "Subjective" Contours

resolution of which is not as fine as that of the electrophysiological recording used in the monkey brain), but the general layout appears similar in the two species, especially for V1 (Tootell et al., 2003).

An astonishing proportion of primate cortex analyzes visual information. The areas that are largely or entirely visual in function occupy about 55% of the surface of the macaque cortex (Van Essen and Drury, 1997) and about 30% of human cortex (Tootell et al., 2003). We will discuss only a few of the main visual cortical areas and their functions.

Area V2 is adjacent to V1, and many of its cells have receptive fields similar to those of V1 cells. Many V2 cells can respond to illusory contours, such as the boundaries of the upright triangle in **FIGURE 7.21** (Peterhans and von der Heydt, 1989). Clearly, such cells respond to complex relations among the parts of their receptive fields.

Area V4 cells generally give their strongest responses to the frequency gratings that we discussed earlier (see Figure 7.19). However, some V4 cells give even stronger responses to concentric and radial stimuli, such as those in **FIGURE 7.22A** (Gallant et al., 1993). Area V4 also has many cells that respond most strongly to color differences, as we will see later, when we discuss color vision.

The inferior temporal (IT) visual cortex has many cells that respond best to particular complex forms, sometimes combined with color or texture, such as those in **FIGURE 7.22B**. The complex receptive fields in IT cortex probably develop through experience and learning (Desimone et al., 1984). After a monkey was trained for a year to discriminate a set of 28 moderately complex shapes, 39% of the cells in its anterior IT cortex responded significantly to some of these shapes. In control monkeys, on the other hand, only 9% of the cells responded strongly to these forms (Kobatake and Tanaka, 1994).

The prefrontal cortex also contains a restricted region of neurons that are activated by faces but not by other visual stimuli (Scalaidhe et al., 1997; Ungerleider et al., 1998). These findings indicate that a pathway mediating visual recognition extends from V1 through temporal cortex to the prefrontal cortex. Later we'll see that this pathway was damaged in D.F., the woman described at the start of the chapter.

(A)

These concentric and radial stimuli evoke maximal responses (red) from some cells in visual cortical area V4.

(B)

Each of these stimuli evoked a maximal response from a different cell in the inferior temporal area.

FIGURE 7.22 Complex Stimuli Evoke Strong Responses in Visual Cortex (Part A from Gallant et al., 1993, courtesy of Jack Gallant; B from Tanaka, 1993, courtesy of Keiji Tanaka.)

Perception of visual motion is analyzed by a special system that includes cortical area V5

In area V5 (also called the *medial temporal area* or *MT*; see Figure 7.20C) of monkeys, all neurons respond to moving visual stimuli, indicating that they are specialized for the perception of motion and its direction. Imaging studies show that moving stimuli also evoke responses in human area V5.

Experimental lesions of area V5 in monkeys trained to report the direction of perceived motion impaired their performance, at least temporarily (Newsome et al., 1985). Conversely, electrically *stimulating* an area of V5 that normally responds to stimuli moving up caused monkeys to report that dots on the screen were moving up even when they were actually moving to the right. In other words, the electrical stimulation appeared to alter the monkey's *experience* of the motion.

One striking report described a woman who had lost the ability to perceive motion after a stroke damaged her area V5 (Zihl et al., 1983). The woman was unable to perceive continuous motion and saw only separate, successive still images. This impairment led to many problems in daily life. She had difficulty crossing streets because she could not follow the positions of automobiles in motion: "When I'm looking at the car at first, it seems far away. But then when I want to cross the road, suddenly the car is very near." She also complained of difficulties in following conversations because she could not see the movements of speakers' lips. Except for her inability to perceive motion, this woman's visual perception appeared normal.

---how's it going ?

1. How could information from LGN neurons with simple concentric receptive fields be combined in a cortical cell such that it would respond best to a line of light?

2. How could information from simple cortical cells be combined in another cortical cell so that it would respond best to a moving line of light?

3. Describe the spatial-frequency hypothesis of vision.

4. What are some examples of visual receptive fields outside of V1 that respond to very complex stimuli?

5. What kind of stimuli affect the firing of neurons in V5 (also called *area MT*)?

Color Vision Depends on Special Channels from the Retinal Cones through Cortical Area V4

For most people, color is a striking aspect of vision. We will discuss three stages of color perception. In the first stage the cones—the retinal receptor cells that are specialized to respond to certain wavelengths of light—receive visual information. In the second stage this information is processed by neurons in the local circuits of the retina, leading to retinal ganglion cells that are excited by light of some wavelengths and inhibited by light of other wavelengths. The ganglion cells send the wavelength information via their axons to the LGN. From there this information goes to area V1, which relays it to other visual cortical areas, where a third stage of color perception take place.

Color is created by the visual system

For most of us, the visible world has several distinguishable hues: blue, green, yellow, red, and their intermediates. These hues appear different because every light particle, or photon, vibrates as it travels across space, behaving like a sinusoidal wave. Photons vary in the frequency of vibration and therefore the **wavelength** (the distance between two adjacent peaks of the wave) of the light, and we can detect some of these differences, perceiving faster-vibrating (thus shorter-wavelength) photons as blue and green, and slower-vibrating (longer-wavelength) photons as more orange and red. The

wavelength The length between two peaks in a repeated stimulus such as a wave, light, or sound.

FIGURE 7.23 The Wavelengths of Light

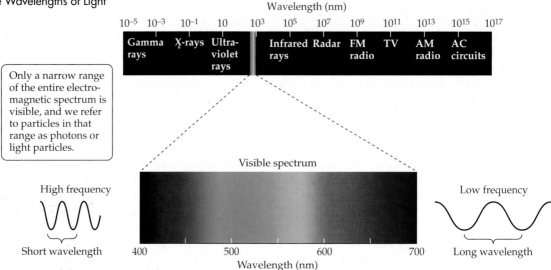

Only a narrow range of the entire electromagnetic spectrum is visible, and we refer to particles in that range as photons or light particles.

brightness One of three basic dimensions of light perception, varying from dark to light.

hue One of three basic dimensions of light perception, varying through the spectrum from blue to red.

saturation One of three basic dimensions of light perception, varying from rich to pale.

human visual system responds only to particles whose wavelengths lie within a very narrow section of the total electromagnetic range, from about 400 to 700 nanometers (nm), as **FIGURE 7.23** shows. If particles have shorter or longer wavelengths than this narrow range, we no longer call them photons but give them names like X-rays or radio waves. The color of an object depends on which wavelengths of light are absorbed versus which wavelengths are reflected. Our eyes detect the reflected wavelengths to distinguish different colors (**FIGURE 7.24**).

The three dimensions of color perception are:

1. **Brightness**, which varies from dark to light
2. **Hue**, which varies continuously through blue, green, yellow, orange, and red (and is what most people mean when they use the term *color*)
3. **Saturation**, which varies from rich, full colors to gray; for example, rich red through pink to gray as saturation decreases

It is important to understand that the perception of a particular hue is *not* a simple function of the wavelength of light. For example, a patch reflecting light of a particular wavelength is perceived as various different hues, depending on several factors, including the intensity of illumination, prior exposure to a different stimulus, and the surrounding field.

Sunlight consists of photons of all wavelengths.

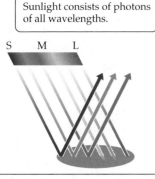

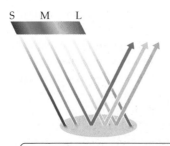

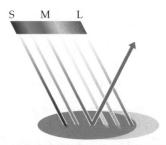

This patch looks "blue" because it absorbs/subtracts most of the long wavelengths and some of the medium wavelengths. The short- and medium-wavelength light that is reflected to the eye appears blue.

This patch looks "yellow" because it reflects best in the middle range of wavelengths and absorbs the other wavelengths.

Mix the two pigments together, and what you have left when each has absorbed its wavelengths are some remaining medium wavelengths that look "green."

FIGURE 7.24 Colored Objects Reflect Different Wavelengths of Light

As illumination fades, the blues in a painting or a rug appear more prominent and the reds appear duller, even though the wavelength distribution in the light reflecting off those objects has not changed. In addition, the hue perceived at a particular point is strongly affected by the pattern of wavelengths and intensities in other parts of the visual field, as **FIGURE 7.25** illustrates. To understand how the visual system creates our experience of color, we must understand how cone photoreceptors work.

Color perception requires receptor cells that differ in their sensitivities to different wavelengths

Artists have long known that all the hues can be obtained from a small number of primary colors. On the basis of observations of mixing pigments and lights, scientists at the start of the nineteenth century hypothesized that three separate kinds of receptors in the retina provide the basis for color vision. Endorsed in 1852 by the great physiologist-physicist-psychologist Hermann von Helmholtz, this **trichromatic hypothesis** (from the Greek *tri*, "three," and *chroma*, "color") became the dominant position.

Helmholtz predicted that blue-sensitive, green-sensitive, and red-sensitive receptors would be found, that each would be sharply tuned to its part of the spectrum, and that each type would have a separate path to the brain. The color of an object would be recognized, then, on the basis of which color receptor(s) were activated. This system would be like the mechanisms for discriminating touch and temperature on the basis of which skin receptors and labeled neural lines are activated (see Chapter 5).

Later in the nineteenth century, physiologist Ewald Hering proposed a different explanation. He argued, on the basis of visual experience, that there are *four* unique hues (blue, green, yellow, red) and three *opposed pairs of colors*—blue versus yellow, green versus red, and black versus white—and that three physiological processes with opposed positive and negative values must therefore be the basis of color vision. As we will see, both this **opponent-process hypothesis** and the trichromatic hypothesis are encompassed in current color vision theory, but neither of the old hypotheses is sufficient by itself.

Measurements of photopigments in cones have borne out the trichromatic hypothesis in part. Each cone of the human retina has one of three classes of pigments (each pigment has a name, but we'll just refer to them as *opsins*). The response of the cone depends on which wavelength of light its pigment absorbs to start the process depicted in Figure 7.4. These pigments do not, however, have the narrow spectral distributions that Helmholtz predicted.

Despite what you may have heard in other classes (or even read in other textbooks!), the human visual system does *not* have receptors that are sensitive to only narrow parts of the visible spectrum, such as "red" cones and "green" cones; instead, the receptor pigments exhibit broad sensitivities that substantially overlap. In fact, two of the three retinal cone pigments show some response to light of almost *any* wavelength. The pigments have different *peaks* of sensitivity, but even the peaks are not as far apart as Helmholtz predicted, and those peaks don't always correspond to a particular color. As **FIGURE 7.26** shows, the cone pigment peaks occur at about 420 nm (in the part of the spectrum where we usually see violet under daylight conditions), about 530 nm (where most of us see green), and about 560 nm (where most of us see yellow-green). Despite Helmholtz's prediction, *none* of the curves peak in the long-wavelength part of the spectrum, where most of us see red (about 630 nm).

Under ordinary conditions, almost any object, no matter what color it is, stimulates at least two kinds of cones, thus ensuring high visual acuity and good perception of form. It is the subsequent processing performed by the nervous system, comparing the *degree* of activation across receptors that differ in peak wavelength sensitivity, that extracts color information about the light falling on the retina. Thus, certain ganglion

The dotted squares in the center of each grid reflect the *same wavelength of light*, yet most of us perceive one as green and the other as yellow.

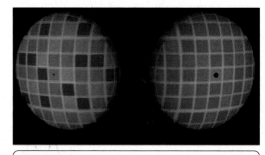

Color perception is affected by inferences about the color of light shining on an object. As the object on the right seems to be bathed in blue light, our visual system infers that the dotted square must actually be yellow, and so we perceive it as yellow.

FIGURE 7.25 Color Perception (From Purves and Lotto, 2011.)

trichromatic hypothesis A hypothesis of color perception stating that there are three different types of cones, each excited by a different region of the spectrum and each having a separate pathway to the brain.

opponent-process hypothesis A hypothesis of color perception stating that different systems produce opposite responses to light of different wavelengths.

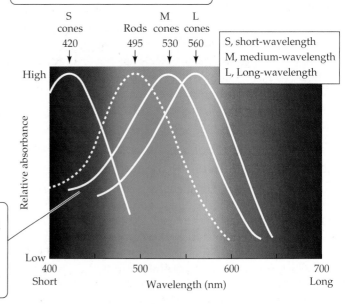

FIGURE 7.26 Spectral Sensitivities of Human Photopigments

Each pigment has a peak sensitivity but responds to a wide range of wavelengths.

S, short-wavelength
M, medium-wavelength
L, Long-wavelength

Knowing only that an M cone is active, you cannot tell whether it was stimulated by weak light at 530 nm ("green"), or by strong light anywhere from 450 nm ("blue") to 620 nm ("red"). Only by *comparing* responses of *different* cones can the brain extract color information.

cells and certain neurons at higher stations in the visual system are color-specific, even though the photoreceptors are not. In a similar manner, photoreceptors are not form-specific (they respond to single points of light), but form is detected later in the system, by comparison of the outputs of different receptors.

Because the cones are not color detectors, the most appropriate brief names for them can be taken from their peak areas of wavelength sensitivity: *short* (S) for the receptor with peak sensitivity at about 420 nm, *medium* (M) for 530 nm, and *long* (L) for 560 nm (see Figure 7.26). There are typically twice as many L as M receptors, but far fewer S receptors (Brainard et al., 2000; Carroll et al., 2000; Hagstrom et al., 1998); this difference explains why acuity is much lower with short-wavelength illumination (blue light) than in the other parts of the visible spectrum. In some insects, including bees, the short-wavelength receptors respond to ultraviolet wavelengths that we humans cannot see. This ability permits bees to see color patterns in flowers that are invisible to us (**FIGURE 7.27**). Most birds have not three but four different types of cones, and they can also detect ultraviolet light (Osorio and Vorobyev, 2008). Likewise, even those mammalian species with weak color vision can discriminate some colors, as we discuss in **BOX 7.1**.

When we humans look at flowers, we cannot see the reflected ultraviolet light…

…which may reveal patterns visible only to animals, like birds and bees, that detect light in that range.

FIGURE 7.27 How Flowers Look to the Birds and the Bees

BOX 7.1   Most Mammalian Species Have Some Color Vision

Animals exhibit different degrees of color vision. Many species of birds, fishes, and insects have excellent color vision. Humans and Old World monkeys also have an excellent ability to discriminate wavelengths. Many other mammals (e.g., cats) cannot discriminate wavelengths very well, but most mammals have at least some degree of color vision (G. H. Jacobs, 1993). Although only certain primates have good *trichromatic* color vision (vision based on *three* classes of cone photopigments), most mammalian species have at least *dichromatic* color vision (based on *two* classes of cone pigments). Most so-called color-blind people (actually color-*deficient*) have dichromatic vision and can distinguish short-wavelength stimuli (blue) from long-wavelength stimuli (not blue). When a gene carrying a third photopigment was introduced into photoreceptors of adult male squirrel monkeys with such dichromatic vision, they soon displayed excellent trichromatic vision (Mancuso et al., 2009). Likewise, introducing photopigment genes in mice enabled them to discriminate colors they normally cannot see (G. H. Jacobs et al., 2007), so it may be possible to correct dichromatic vision in humans.

There is a continuum of color vision capabilities, including at least four categories among mammalian species:

1. *Excellent trichromatic color vision* is found in diurnal primates such as humans and the rhesus monkey .

2. *Robust dichromatic color vision* is found in species that have two kinds of cone photopigments and a reasonably large population of cones, for example the dog and the pig. This type of vision is also found in the males of many South American monkeys, such as the squirrel monkey and the marmoset monkey.

3. *Feeble dichromatic color vision* occurs in species that have two kinds of cone pigments but very few cones. Examples include the domestic cat and the coati.

4. *Minimal color vision* is possessed by species that have only a single kind of cone pigment and that must rely on interactions between rods and cones to discriminate wavelength. Examples include the owl monkey and the raccoon. When you compare diurnal and nocturnal species of a given taxonomic family (e.g., the coati and the raccoon), the diurnal species (in this case the coati) usually has the better color vision.

Among those species of South American monkeys that are generally dichromats, some females are actually trichromatic. Why? Because the gene encoding one photopigment is on the X chromosome. Since females have two X chromosomes, if the two chromosomes carry *different* genes for the photopigment, then the female possesses a total of three different kinds of cones and therefore has trichromatic vision.

Unlike South American monkeys that carry a single photopigment gene on the X chromosome, many other primates, like humans, have two photopigment genes on the X chromosome: one for medium- and one for long-wavelength cones. (The short-wavelength photopigment gene is on an autosome rather than on the X.) Women are less likely than men to have color deficiency because if one of their X chromosomes has a defective gene for a photopigment, the copy of that gene on the other X is likely to be fine. Because men have only one X, if they have a defective gene for the medium- or long-wavelength pigment, there's no backup copy to compensate. They will see blue (light stimulating the short-wavelength cones) and not blue (light stimulating the other functioning cones), as the figure illustrates. A company selling special glasses that they claim correct this type of color deficiency explains the technology on their website (www.EnChroma.com).

(A)

(B)

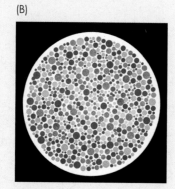

Simulating Color Blindness (A) The photograph on the right has been adjusted to simulate the experience of the most common form of color blindness in humans, which is the absence of cones sensitive to medium-wavelength light (M cones). For such individuals, the world's colors consist of blue (detected by short-wavelength photopigment encoded on the seventh chromosome) and not blue (detected by the long-wavelength photopigment encoded on the X chromosome). (B) In a typical test for color vision, dichromats may have a difficult time detecting the numerals displayed in figures like this. (A [right] was produced by software available from Vischeck [www.vischeck.com]).

The genes for wavelength-sensitive pigments in the retina have been analyzed, and the similarities in structure of the three genes suggest that they are all derived from a common ancestral gene (Nathans, 1987). In addition, the genes for the medium- and long-wavelength pigments occupy adjacent positions on the X chromosome and are much more similar to each other than either is to the gene for the short-wavelength pigment on chromosome 7. Probably our primate ancestors had only one photopigment gene on the X chromosome, which became duplicated. Then mutations caused the two genes to become more and more different, until their responses to various wavelengths of light were no longer the same. Thus, our ancestors went from having only two cone pigments (one on the X chromosome and the S pigment on chromosome 7) to three, with an associated improvement in color vision.

Certain evidence suggests that this evolution of a third photopigment may have happened recently (in evolutionary terms); for example, most South American monkeys have only a single longer-wavelength pigment. The fact that the genes for the M and L pigments are on the X chromosome also explains why defects of red-green color vision are much more frequent in human males (about 8%) than in human females (about 0.5%). Because males have only one X chromosome, mutations in the genes for the M and L pigments can impair color vision. But if a female has defective photopigment genes in one of her two X chromosomes, normal copies of the genes on the other X chromosome can compensate.

Interestingly, a woman may carry slightly different genes for the long-wavelength photopigment on her two X chromosomes (Neitz et al., 1998) and therefore have *four* different kinds of cones. Such "tetrachromats" tend to be very good at discriminating colors and very sensitive to clashing colors. It's interesting to speculate on whether such women have a different experience of, for example, green than those of us who are trichromats and dichromats. We will take up the question of our subjective experience of color again in Chapter 14.

Even though the term *color blindness* is commonly used, most people with impaired color vision are able to distinguish some hues. Complete color blindness can be caused by brain lesions or by the congenital absence of *any* cones, but in humans it is extremely rare.

Some retinal ganglion cells and LGN cells show spectral opponency

Recordings made from retinal ganglion cells in monkeys that can discriminate colors as humans do reveal the second stage in the processing of color vision. Most ganglion cells and LGN cells are excited and fire in response to some wavelengths and are inhibited by other wavelengths.

FIGURE 7.28A shows the response of an LGN cell as a large spot of light centered on its receptive field changes from one wavelength to another. Firing is stimulated by wavelengths above 600 nm, where the L cones are most sensitive; and then inhibited at shorter wavelengths, where the L cones are less sensitive than the M cones. A cell exhibiting this response

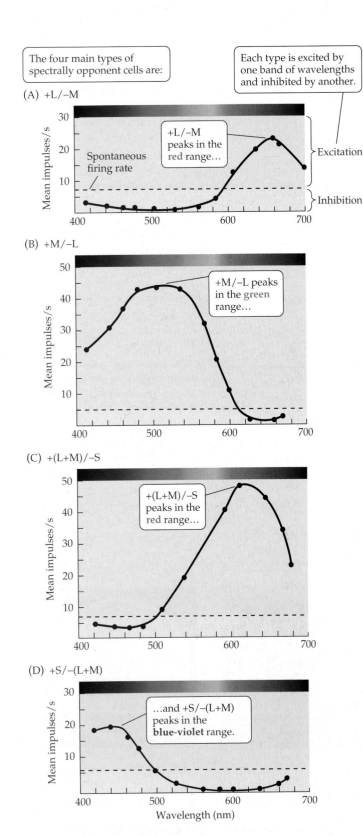

FIGURE 7.28 Responses by the Four Main Types of Spectrally Opponent Cells in Monkey LGN

pattern is therefore called a *plus L/minus M cell* (+L/–M). This is an example of a **spectrally opponent cell** (or *color-opponent cell*) because two regions of the spectrum have opposite effects on the cell's rate of firing. Figure 7.28 shows the responses of the four main kinds of spectrally opponent cells.

Each spectrally opponent ganglion cell receives input from two or three different kinds of cones through bipolar cells. The connections from at least one type of cone are excitatory, and those from at least one other type are inhibitory. The spectrally opponent ganglion cells thus record the *difference* in stimulation of different types of cones. For example, a +M/–L cell responds to the difference in the excitation of M and L cones.

The peaks of the sensitivity curves of the M and L cones are not very different (see Figure 7.26). However, whereas the M-minus-L *difference* curve (**FIGURE 7.28B**) shows a clear peak at about 500 nm (in the green part of the spectrum), the L-minus-M difference function (see Figure 7.28A) shows a peak at about 650 nm (in the red part of the spectrum). Thus, +M/–L and +L/–M cells yield distinctly different neural response curves. LGN cells that are excited by the L and M cells but inhibited by S cells—that is, +(L+M)/–S cells—peak in the red range (**FIGURE 7.28C**), while cells excited by S but inhibited by L and M—that is, +S/–(L+M) cells—peak in the blue-violet range (**FIGURE 7.28D**).

Spectrally opponent neurons are the second stage in the system for color perception, but they still cannot be called *color cells*, because (1) they send their outputs into many higher circuits—for detection of form, depth, and movement, as well as hue; and (2) their peak wavelength sensitivities do not correspond precisely to the wavelengths that we see as the principal hues. **FIGURE 7.29** diagrams the presumed inputs to not only the four kinds of spectrally opponent ganglion cells, but also the ganglion cells that detect brightness and darkness. The brightness detectors receive stimulation from both M and L cones (+M/+L); the darkness detectors are inhibited by those same cones (–M/–L).

In the monkey LGN, 70–80% of the cells are spectrally opponent; in the cat, very few spectrally opponent cells are found—only about 1%. This difference explains why monkeys so easily distinguish between colors and it's so difficult to train cats to discriminate even large differences in color.

Some visual cortical cells and regions appear to be specialized for color perception

In the cortex, spectral information appears to be used for various kinds of information processing. Forms are segregated from their background by differences in color or intensity (or both). The most important role that color plays in our perception is to denote which parts of a complex image belong to one object and which belong to another. Some animals use displays of brightly colored body parts to call attention to themselves, but color can also be used as camouflage.

Some spectrally opponent cortical cells contribute to the perception of color, providing the third stage of the color vision system. These cells are not just responding to the differences between two types of cones, as retinal ganglion cells and LGN cells do. Rather, they are responding to differences in colors that we *perceive*; in other words, they are perceptually opponent: red versus green, blue versus yellow, and black versus white (R. L.

spectrally opponent cell Also called *color-opponent cell*. A visual system neuron that has opposite firing responses to different regions of the spectrum.

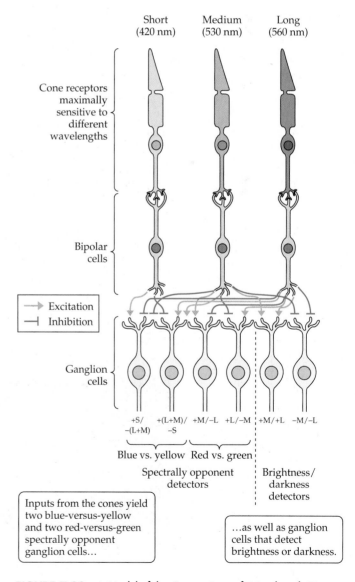

FIGURE 7.29 A Model of the Connections of Wavelength Discrimination Systems in the Primate Retina (After R. L. De Valois and De Valois, 1980.)

optic ataxia   Spatial disorientation in which the patient is unable to accurately reach for objects using visual guidance.

amblyopia   Reduced visual acuity that is not caused by optical or retinal impairments.

De Valois and De Valois, 1993). The spectral responses of these cells correspond to the wavelengths of the principal hues specified by human observers, and their characteristics also help explain other color phenomena.

Visual cortical region V4 is particularly rich in color-sensitive cells; each of these cells truly responds best to a particular hue, including the four that Hering postulated (blue, green, yellow, red). V4 cells respond best if the color outside the receptive field is different from the color preferred inside the receptive field (Schein and Desimone, 1990). These cells provide another stage of color perception that may be important for color constancy and for discrimination between a figure and background (Zeki et al., 1991). It would probably be wrong, however, to think of area V4 as devoted *exclusively* to color perception. Cells in V4 also show spatial sensitivity, responding to orientation and spatial frequency information (Tanigawa et al., 2010).

---

how's it going ❓

1. What two main hypotheses were developed to explain our ability to discriminate colors? Which aspects of the visual system appear to match each hypothesis?
2. Describe some examples in which our perception of color is not simply the detection of particular wavelengths of light.
3. Why do we label cones as S cones, M cones, and L cones rather than blue, green, and red cones?
4. Why are men more likely than women to have difficulty distinguishing some colors?

The Many Cortical Visual Areas Are Organized into Two Major Streams

Many investigators have wondered why primate visual systems contain so many distinct regions. Certain regions specialize in processing different attributes of visual experience (such as shape, location, color, motion, and orientation). But there are more than 30 cortical areas analyzing vision, which is a larger number than the basic attributes we can identify. Perhaps the reason so many separate visual regions have been found is simply that investigators, being visually oriented primates themselves, have lavished special attention on the visual system.

Mortimer Mishkin and Leslie Ungerleider (1982) proposed that primates have two main cortical processing streams, both originating in primary visual cortex: a ventral processing stream responsible for visually *identifying* objects, and a dorsal stream responsible for appreciating the spatial *location* of objects and for visually guiding our movement toward them (**FIGURE 7.30**). They called these processing streams, respectively, the *what* and *where* streams.

PET studies, as well as brain lesions in patients, indicate that the human brain possesses *what* and *where* visual processing streams similar to those that have been found in monkeys (Ungerleider et al., 1998). The two streams are not completely separate, because there are normally many cross connections between them. In the ventral stream, including regions of the occipitotemporal, inferior temporal, and inferior frontal areas, information about faces becomes more specific as one proceeds farther forward. PET studies show that, whereas general information about facial features and

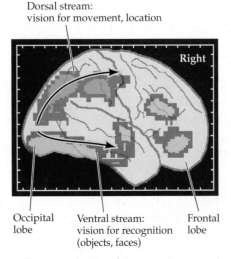

Dorsal stream: vision for movement, location

Right

The ventral (*what*) pathway, shown in yellow and red, and the dorsal (*where*) pathway, shown in green and blue, serve different functions.

Occipital lobe

Ventral stream: vision for recognition (objects, faces)

Frontal lobe

FIGURE 7.30   Parallel Processing Pathways in the Visual System (Courtesy of Leslie Ungerleider.)

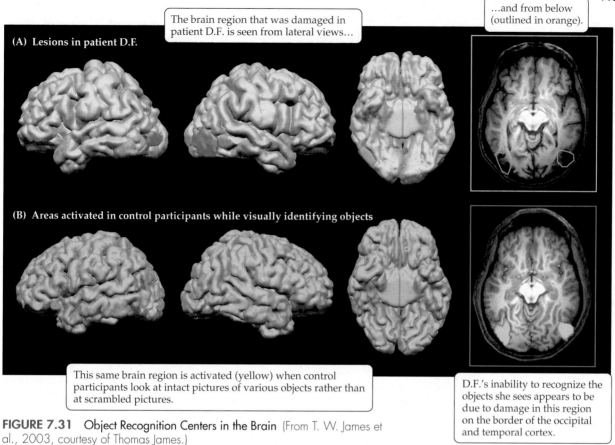

(A) Lesions in patient D.F.

The brain region that was damaged in patient D.F. is seen from lateral views…

…and from below (outlined in orange).

(B) Areas activated in control participants while visually identifying objects

This same brain region is activated (yellow) when control participants look at intact pictures of various objects rather than at scrambled pictures.

D.F.'s inability to recognize the objects she sees appears to be due to damage in this region on the border of the occipital and temporal cortex.

FIGURE 7.31 Object Recognition Centers in the Brain (From T. W. James et al., 2003, courtesy of Thomas James.)

gender is extracted more posteriorly, the more anterior parts of the stream provide representations of individual faces (Courtney et al., 1996).

Discovery of these separate visual cortical streams helps us understand the case of patient D.F., described at the start of this chapter. Recall that, after carbon monoxide poisoning, D.F. lost the ability to perceive faces and objects while retaining the ability to reach and grasp those same objects under visual control. The investigators who studied her (A. D. Milner et al., 1991) hypothesized that D.F.'s visual ventral (*what*) stream had been devastated but that her dorsal (*where*) stream was unimpaired. An opposite kind of dissociation had already been reported: damage to the dorsal parietal cortex often results in **optic ataxia**, in which patients have difficulty using vision to reach for and grasp objects, yet some of these patients can still identify objects correctly (Perenin and Vighetto, 1988).

Studies of D.F. support the idea that her ventral stream is impaired but her dorsal stream remains relatively normal (T. W. James et al., 2003). High-resolution MRI of her brain (**FIGURE 7.31A**) reveals diffuse damage concentrated in the ventrolateral occipital cortex. Throughout the brain there is evidence of atrophy, indicated by shrunken gyri and enlarged sulci. **FIGURE 7.31B** shows the area activated in fMRI recordings when healthy participants viewed pictures of objects; it corresponds to D.F.'s lateral occipital lesion. When D.F. reached for and grasped objects, her fMRI activation in the parietal lobe was similar to that of control participants, indicating that her dorsal stream is largely intact. D.F.'s intact dorsal pathway not only tells her where objects are but also guides her movements to use these objects properly.

It is still puzzling that one part of D.F. knows exactly how to grasp a pencil held in front of her, yet another part of her—the part that talks to you—has no idea whether the object she's holding is a pencil, a ruler, or a bouquet of flowers. This condition is reminiscent of the cortical damage that causes blindsight, mentioned earlier: people with such damage report being unable to see, but they show evidence that they can.

In Chapter 14 we'll learn about other people who can see only one thing at a time, or who can see faces but cannot recognize anyone. Imagining what such disjointed visual experience must be like helps us appreciate how effortlessly our brains usually bind together information with our marvelous sense of sight.

Visual Neuroscience Can Be Applied to Alleviate Some Visual Deficiencies

Vision is so important that many investigators have sought ways to prevent its impairment, to improve inadequate vision, and to restore sight to the blind. In the United States, half a million people are blind. Recent medical advances have reduced some causes of blindness but have increased blindness from other causes. For example, medical advances permit people with diabetes to live longer, but because we don't know how to prevent blindness associated with diabetes, there are more people alive today with diabetes-induced blindness. In the discussion that follows, we will first consider ways of avoiding the impairment of vision. Then we will take up ways of improving an impaired visual system.

Impairment of vision often can be prevented or reduced

Studies of the development of vision show that the incidence of myopia (nearsightedness) can be reduced. Myopia develops if the eyeball is too long, causing the eye to focus images in front of the retina rather than on the retina (see Figure 7.2). As a result, distant objects appear blurred. Considerable evidence suggests that the reason some children develop myopia is that certain environmental factors cause the eyeball to grow excessively. Previously it was thought that the modern habit of looking closely at nearby objects (books, computer screens, and so on) might be responsible for myopia (Marzani and Wallman, 1997), but mounting evidence suggests that indoor lighting may be to blame.

Before civilization, most people spent the bulk of their time outdoors, looking at objects illuminated by sunlight. But with the advent of indoor lighting, we've come to spend a lot of time looking at things with light that, while containing many wavelengths, does not exactly match the composition of sunlight. Several studies have found that children with myopia spend less time outdoors than do other children, but that correlation could be caused by genes that favor both myopia and indoor activities, like reading. Indeed, the advent of public schools in various nations is accompanied by increased rates of myopia. However, one of these studies focused on people of Chinese origin who lived in either Singapore, where crowded conditions mean that people spend little time outdoors, or Sydney, Australia. Even though these populations should be genetically very similar, 30% of the Chinese children living in Singapore, who average only 30 minutes a day outdoors, were myopic, versus only 3% of those living in Sydney, who average 2 hours a day outdoors (Rose et al., 2008). What's more, in these populations myopia correlates much more strongly with time spent indoors than with time spent reading.

Of course, too much sunlight can be a bad thing, especially for our skin. But the researchers note that almost all children in Australia wear hats to shield their faces when outdoors, yet they still benefit from being outdoors in terms of avoiding myopia. Likewise, there's no evidence that wearing glasses blocks the benefit of light from the sun. The next challenge will be determining what it is about indoor lighting, as opposed to sunlight, that encourages the eyeball to grow excessively in children, leading to myopia.

Increased exercise can restore function to a previously deprived or neglected eye

The misalignment of the two eyes (*lazy eye*) can lead to a condition called **amblyopia**, in which acuity is poor in one eye, even though the eye and retina are normal.

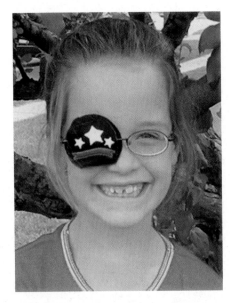

Hey There, You with the Stars over Your Eye As a treatment for amblyopia, this girl is wearing a patch over her "good" eye—the one she has been relying on while ignoring information from her other, "weak" eye. Increased visual experience through the weak eye will strengthen its influence on the cortex.

If the two eyes are not aligned properly during the first few years of life, the primary visual cortex of the child tends to suppress the information traveling to the cortex from one eye, and that eye becomes functionally blind. Studies of the development of vision in children and other animals show that most cases of amblyopia are avoidable.

The balance of the eye muscles can be surgically adjusted to bring the two eyes into better alignment. Alternatively, if the weak eye is given regular practice, with the good eye covered, vision can be preserved in both eyes. Attempts to alleviate amblyopia by training alone, however, have produced mixed results. The optimal treatment appears to be a combination of both surgical correction *and* eye patches and visual exercises (Pediatric Eye Disease Investigator Group, 2005).

SIGNS & SYMPTOMS

Robot Eyes?

Diseases such as macular degeneration damage the rods and cones. These diseases leave the ganglion cells and the higher neural pathways largely intact. Scientists are working on arrays of electrodes to be placed within the eye, in contact with the ganglion cells. These electrodes would be stimulated by a small camera and radio frequency transmitter lodged in the patient's eyeglasses (**FIGURE 7.32**). The Argus II retinal prosthesis system was approved for use in the United States in 2013 (Stronks and Dagnelie, 2014). It consists of 60 electrodes on the retina, providing only 60 pixels. But that is enough to allow patients to walk around objects, and efforts are underway to produce a model with 1,000 electrodes, which might make it possible to recognize faces. In theory, enough electrodes might restore normal vision for thousands of people.

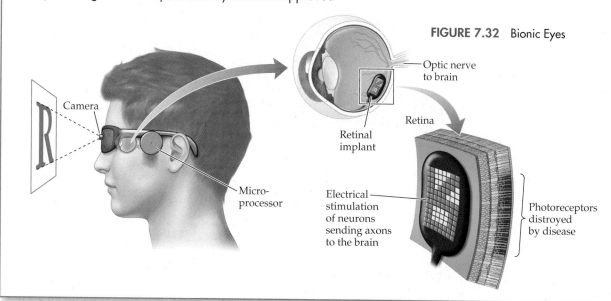

FIGURE 7.32 Bionic Eyes

Camera

Microprocessor

Optic nerve to brain

Retinal implant

Retina

Electrical stimulation of neurons sending axons to the brain

Photoreceptors distroyed by disease

how's it going ❓

1. What are the two main streams of visual processing in the cortex, and what aspects of vision does each stream support?

2. Describe D.F.'s symptoms, and relate them to the brain damage revealed by MRIs.

3. What is the evidence that indoor lighting may cause myopia in children?

Recommended Reading

De Valois, R. L., and De Valois, K. K. (1988). *Spatial Vision.* New York, NY: Oxford University Press.

Gregory, R. L. (2015). *Eye and Brain: The Psychology of Seeing* (5th ed.). Princeton, NJ: Princeton University Press.

Ings, S. (2008). *A Natural History of Seeing: The Art and Science of Vision.* New York, NY: Norton.

Purves, D., and Lotto, R. B. (2011). *Why We See What We Do Redux: A Wholly Empirical Theory of Vision.* Sunderland, MA: Sinauer.

Rodieck, R. W. (1998). *The First Steps in Seeing.* Sunderland, MA: Sinauer.

Wolfe, J. M., Kluender, K. R., Levi, D. M., Bartoshuk, L. M., et al. (2015). *Sensation & Perception* (4th ed.). Sunderland, MA: Sinauer.

2e.mindsmachine.com/vs7

VISUAL SUMMARY
7

You should be able to relate each summary to the adjacent illustration, including structures and processes. Go to the online version of this summary (scan the QR code above) for links to figures, animations, and activities that will help you consolidate the material.

1 Light is bent, or **refracted**, by the transparent outer layer of the eye, the **cornea**, focusing an image on the **retina** in the back of the eye. We vary the thickness of the **lens** to fine-tune the image. The refracted light forms an image on the retina that is upside down and reversed. Review **Figures 7.1 and 7.2, Animation 7.2, Activity 7.1**

2 The retina contains two different types of **photoreceptors** to detect light forming the focused image. **Rods** are very sensitive, working even in very low light, and they respond to light of any **wavelength**. Rods drive the **scotopic** system, which can work in dim light. Each of the three different types of **cones** responds better to some wavelengths of light than others, allowing us to detect colors. The cones provide information for the **photopic** system, which needs more light to function. Photoreceptors **adapt** to function across a wide range of light intensities. Review **Figures 7.3–7.6, Table 7.1**

3 The retina consists of layers of neurons, with the photoreceptors in the very back stimulating **bipolar cells**, which stimulate **ganglion cells**. The ganglion cells of the retina project their axons to the brain via the **optic nerve**. **Amacrine cells** and **horizontal cells** communicate across the retina, using processes such as **lateral inhibition** to analyze **brightness**. Review **Figures 7.3, 7.14 and 7.15**

4 The center of our **visual field** lands on the **fovea**, the portion of the retina with the greatest density of photoreceptors, an absence of overlying cell layers, and more direct synaptic connections to ganglion cells, providing us with our greatest **visual acuity** (sharpness of vision). Cones are concentrated in the fovea, and rods are concentrated in the peripheral retina, so our peripheral vision is best for seeing dim objects, but it provides no color information. Review **Figures 7.7–7.9**

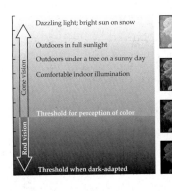

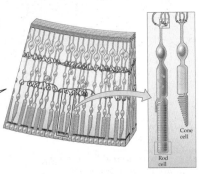

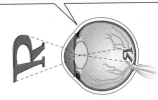

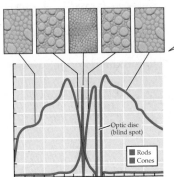

Dazzling light; bright sun on snow

Outdoors in full sunlight

Outdoors under a tree on a sunny day

Comfortable indoor illumination

Threshold for perception of color

Threshold when dark-adapted

Cone vision

Rod vision

Photopic range

Scotopic range

Cone cell

Rod cell

Optic disc (blind spot)

■ Rods
■ Cones

5 The left visual field falls on the nasal retina of the left eye and the temporal retina of the right eye. Only nasal retinal ganglion cells of each eye send their axons across the midline, forming the **optic chiasm**, so the left visual field projects to the right hemisphere and the right visual field projects to the left hemisphere. Review **Figures 7.10** and **7.12**, **Animation 7.3**

6 Most ganglion cells of the retina synapse on neurons in the **lateral geniculate nucleus (LGN)** of the thalamus. The LGN neurons send axons to synapse on neurons in layer IV of the **primary visual cortex (V1)** in the **occipital cortex**. V1 sends information to many different cortical areas, called **extrastriate cortex** (nearly one-third of the human cortex), to further analyze visual information. Review **Figures 7.10** and **7.20**

7 The **receptive fields** of bipolar cells and ganglion cells consist of a circular center and a surround that have opposing effects: either **on-center/off-surround** or **off-center/on-surround**. Review **Figures 7.13** and **7.14**, **Animation 7.4**

8 Receptive fields of cells at successively higher levels in the visual cortex change in two main ways: (1) they become larger (occupy larger parts of the visual field), and (2) they require increasingly specific stimuli to evoke responses. For example, they respond best to a bar of light at a particular angle, or to bars that move in a particular direction. Review **Figures 7.17–7.22**, **Animation 7.5**

9 Rods detect light using a pigment called **rhodopsin**. Our detection of **hue** (color) depends on the three different cone photopigments (opsins). Each cone responds to wide range of wavelengths, not just a single color. Our perception of hue results from the *relative activity* of each type of cone. One way of assessing this relative activity is by retinal connections that yield **spectrally opponent** neurons. Review **Figures 7.23–7.29**

10 Visual cortical areas are organized into two main streams: a ventral *what* stream that serves in the recognition of faces and objects, and a dorsal *where* stream that serves in location and visuomotor skills. Review **Figures 7.30** and **7.31**

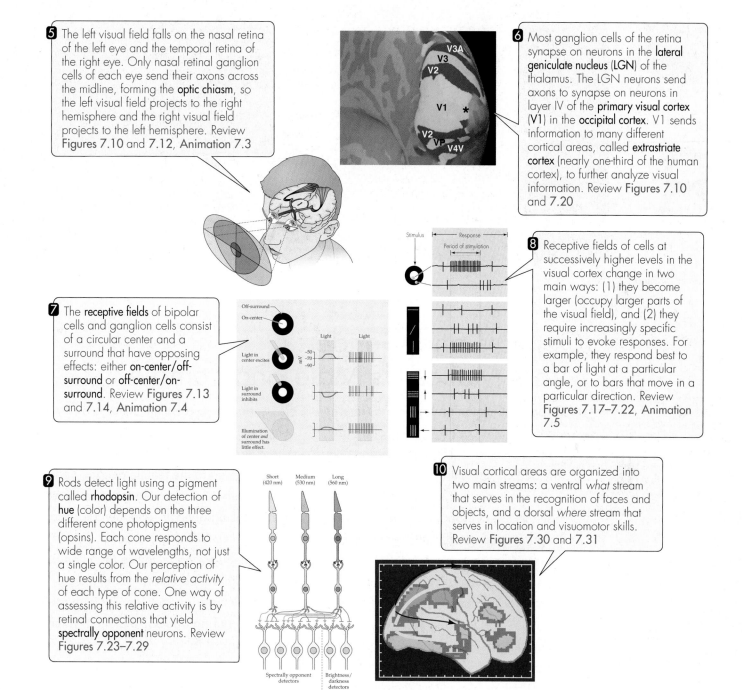

Go to **2e.mindsmachine.com** for study questions, quizzes, flashcards, and other resources.

chapter

8

Hormones and Sex

Genitals and Gender: What Makes Us Male and Female?

No aspects of human biology are as impressive and humbling as the making of a baby; it is a developmental ballet of staggering complexity and critical timing. Given the countless processes that must unfold perfectly and in precisely the right order, it is a marvel that, in the great majority of cases, development proceeds without a hitch. Inevitably, though, there are times when a crucial part of the program is derailed along the way and a baby is born with a heartbreaking deformity.

Such is the case with **cloacal exstrophy**, a developmental defect that affects about one in 400,000 babies. As a consequence of abnormal development of the pelvic region, a genetic male with this condition is typically born with normal testes but a very short, split penis or no penis at all. It isn't really possible to surgically fashion a normal penis in these cases, so the parents are faced with a terrible dilem-

ma. Is it better to raise the child as a boy without a penis, despite the emotional costs of the deformity? Or would it be better for the child to be assigned to the female gender, undergo early surgery to remove the testes and fashion female-looking genitals, and then be raised as a girl? Which would you choose?

Arguments for each course of action boil down to different opinions about the extent to which our gender identity is shaped through nurturing and socialization, rather than biological factors. In other words, we need to consider the larger question of why men and women behave differently. Is it because as boys and girls they were treated differently and trained to grow into their gender roles, or do the forces that provide a fetus with testes or ovaries also induce the developing brain to take on a masculine or feminine form?

To see the video
Gender,
go to

2e.mindsmachine.com/8.1

In this chapter we'll discuss research that informs us about this long-standing question: How do biological and social forces combine to direct development in male-typical and female-typical ways? In Part I we'll learn how hormones can affect the brain to influence behavior. Part II will explain how hormones act on different parts of the brain to influence sexual behavior and parental behavior in particular. Part III will take up the question of how the fetus normally develops into either a male or a female form, not just in terms of the body, but also in terms of the brain and behavior. In animals, prenatal hormones have a tremendous influence on the brain and sexual behaviors. We'll close by reviewing growing evidence that those same prenatal hormones also affect our development into boys or girls, men or women, as well as our sexual orientation.

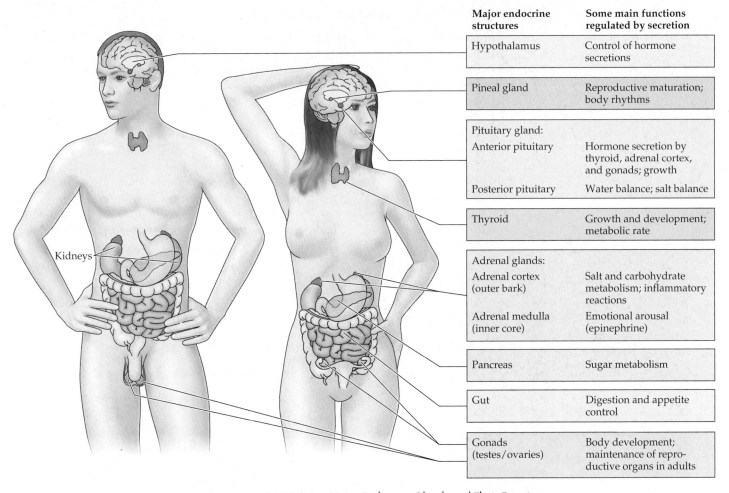

Major endocrine structures	Some main functions regulated by secretion
Hypothalamus	Control of hormone secretions
Pineal gland	Reproductive maturation; body rhythms
Pituitary gland:	
Anterior pituitary	Hormone secretion by thyroid, adrenal cortex, and gonads; growth
Posterior pituitary	Water balance; salt balance
Thyroid	Growth and development; metabolic rate
Adrenal glands:	
Adrenal cortex (outer bark)	Salt and carbohydrate metabolism; inflammatory reactions
Adrenal medulla (inner core)	Emotional arousal (epinephrine)
Pancreas	Sugar metabolism
Gut	Digestion and appetite control
Gonads (testes/ovaries)	Body development; maintenance of reproductive organs in adults

FIGURE 8.1   Major Endocrine Glands and Their Functions

cloacal exstrophy   A rare medical condition in which XY individuals are born completely lacking a penis.

hormone   A chemical, usually secreted by an endocrine gland, that is conveyed by the bloodstream and regulates target organs or tissues.

endocrine gland   A gland that secretes hormones into the bloodstream to act on distant targets.

castration   Removal of the gonads, usually the testes.

To view the
Brain Explorer,
go to

2e.mindsmachine.com/8.2

PART I
The Endocrine System

The body's cells use chemicals to communicate, including an extensive array of **hormones**, chemicals secreted by one group of cells and carried through the bloodstream to other parts of the body, where they act on specific target tissues to produce physiological effects. Changes in hormone levels can produce striking changes in brain function. Cognitive abilities, emotions, our appetite for food or drink or sex, our aggressiveness or submissiveness, our care for children—the scope of hormonal influences on behavior is vast. The ancient Greeks believed that the balance of four body "humors," or fluids, explained our temperament and emotions. Today we know there are a lot more than four hormones.

Hormones Act in a Great Variety of Ways throughout the Body

Most hormones are produced by **endocrine glands** (from the Greek *endon*, "within," and *krinein*, "to secrete"), so called because they release their hormones *within* the body (**FIGURE 8.1**). Endocrine glands are sometimes contrasted with *exocrine glands* (tear glands, salivary glands, sweat glands), which use ducts to secrete fluid *outside* the body.

Our current understanding of hormones developed in stages

Ancient civilizations understood the importance of hormones. In the fourth century BCE, Aristotle described the effects of **castration** (removal of the testes) in chickens and compared them with the effects in eunuchs (castrated men). The first major endocrine experiment, carried out in 1849 by German physician Arnold Berthold (1803–1861), followed up Aristotle's report that when roosters are castrated as juveniles, they fail to develop normal reproductive behavior and secondary sexual characteristics, such as the rooster's comb, in adulthood. Berthold observed, however, that returning one testis back into the body cavity of the young birds allowed them to develop normal male anatomy and behavior. In adulthood, these animals showed the usual male sexual behaviors of mounting hens, fighting, and crowing (**FIGURE 8.2**). Because no nerves

had reestablished contact with the transplanted testis, the organ could not be communicating to the brain through nerves. Berthold (1849) concluded that the testes release a chemical into the blood that affects both male behavior and male body structures. Today we know that the testes make and release the hormone testosterone, which exerts these effects.

Although Berthold didn't know it, experiments like this also illustrate another principle of hormone action. If he had waited until the castrated chicks were adults before returning their testes, Berthold would have seen little effect. The testosterone must be present *early* in life to have such dramatic effects on the body and behavior. We'll return to this point later in this chapter. For now, let's see how hormones fit into the grand scheme of chemical signaling in the body.

FIGURE 8.2 Berthold's (1849) Experiment Demonstrated the Importance of Hormones for Behavior

■ **Question**

Male chicks that are castrated grow up to have small wattles and combs, and they show little interest in mounting hens, fighting, or crowing. What causes these changes—the loss of a nerve connection between the testes and the body, or the loss of a chemical signal released from the testes?

■ **Experiment**

Berthold removed the testes from their normal position but then reimplanted them elsewhere in the abdomen, disconnected from normal innervation.

	Group 1	Group 2	Group 3
	Left undisturbed, young roosters grow up to have large red wattles and combs, to mount and mate with hens readily, and to fight one another and crow loudly.	Males whose testes were removed during development displayed neither the appearance nor the behavior of normal roosters as adults.	However, if one of the testes was reimplanted into the abdominal cavity immediately after its removal, the rooster developed normal wattles and normal behavior.
Comb and wattles:	Large	Small	Large
Mount hens?	Yes	No	Yes
Aggressive?	Yes	No	Yes
Crowing?	Normal	Weak	Normal

■ **Outcome**

The animals with the reimplanted testes grew up to look and act like normal males. Berthold reasoned that the testes must have secreted a signal, which today we would call a hormone, that has widespread effects on the body and brain. Today we know that the hormone is testosterone.

(A) Endocrine function

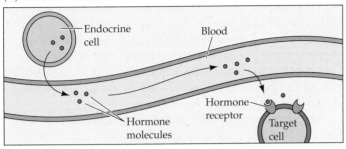

(B) Neural function (synaptic transmission)

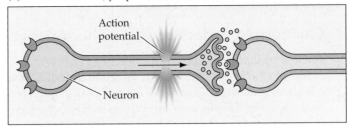

(C) Pheromone function

(D) Allomone function

FIGURE 8.3   Chemical Communication Systems

To see the animation
Chemical Communication Systems,
go to

2e.mindsmachine.com/8.3

endocrine   Referring to glands that release chemicals to the interior of the body. These glands secrete the principal hormones used by the body.

synapse   The cellular location at which information is transmitted from a neuron to another cell.

pheromone   A chemical signal that is released outside the body of an animal and affects other members of the same species.

allomone   A chemical signal that is released outside the body by one species and affects the behavior of other species.

peptide hormone   Also called *protein hormone.* A hormone that consists of a string of amino acids.

amine hormone   Also called *monoamine hormone.* A hormone composed of a single amino acid that has been modified into a related molecule, such as melatonin or epinephrine.

Hormones are one of several types of chemical communication

People had long suspected that special substances circulate to carry messages through the body, but as we discussed above, it wasn't until nineteenth-century scientists started experimenting with hormone-secreting glands that details of chemical communication began to emerge. Reviewing several categories of chemical signals used by the body, we can compare hormonal communication with other methods of communication:

- *Endocrine communication.* In **endocrine** communication, our topic for this chapter, the chemical signal is a hormone released into the bloodstream to selectively affect distant target organs (**FIGURE 8.3A**).

- *Synaptic communication.* Communication via **synapses** was described in Chapters 3 and 4. In typical synaptic transmission, the released chemical signal diffuses a tiny distance across the synaptic cleft and causes a change in the postsynaptic membrane (**FIGURE 8.3B**).

- *Pheromone communication.* Chemicals can be used for communication not only within an individual, but also *between* individuals. **Pheromones** are chemicals that are released outside the body to affect other individuals of the same species (**FIGURE 8.3C**). For example, ants produce pheromones that identify the route to a rich food source (to the annoyance of picnickers). Dogs and wolves urinate on landmarks to designate their territory. In Chapter 6 we discuss pheromones in more detail.

- *Allomone communication.* Some chemical signals are released by members of one species to affect the behavior of individuals of *another* species. These substances are called **allomones** (**FIGURE 8.3D**). Flowers exude scented allomones to attract insects and birds in order to distribute pollen. And the bolas spider—nature's femme fatale—releases a moth sex pheromone to lure male moths to their doom (Haynes et al., 2002).

Let's review the basic types of hormones and how they influence cells.

Hormones can be classified by chemical structure

Most hormones fall into one of three categories: peptide hormones, amine hormones, or steroid hormones. Peptides are simply small protein molecules, so, like any other protein, a molecule of **peptide hormone** is made up of a short string of amino acids (**FIGURE 8.4A**). Different peptide hormones consist of different combinations of amino acids. **Amine hormones** are smaller and simpler, consisting of a modified version of a single amino acid (hence their alias, *monoamine hormones*) (**FIGURE 8.4B**). The amine hormone melatonin is discussed in **A STEP FURTHER 8.1**, on the website.

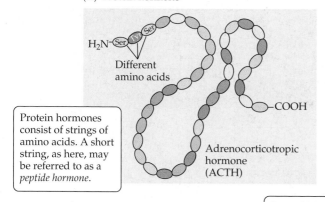

(A) Protein hormone

H_2N—Ser—Tyr—Ser

Different amino acids

Protein hormones consist of strings of amino acids. A short string, as here, may be referred to as a *peptide hormone*.

—COOH

Adrenocorticotropic hormone (ACTH)

FIGURE 8.4 *Chemical Structures of the Three Main Hormone Types*

(B)

Amine hormones such as thyroxine are modified single amino acids.

HO—⬡—O—⬡—$CH_2CHCOOH$

NH_2

Thyroxine (tetraiodothyronine)

(C)

Steroid hormones are derived from cholesterol and consist of four interconnected rings of carbon atoms, with various chemical attachments.

CH_3 OH

HO—

Estradiol

Steroid hormones are derivatives of cholesterol and thus share its structure of four rings of carbon atoms (**FIGURE 8.4C**). Different steroid hormones vary in the number and kinds of atoms attached to the rings. Because steroids dissolve readily in lipids, they can pass through membranes easily (recall from Chapter 2 that cell membranes consist of a lipid bilayer). **TABLE 8.1** gives examples of each class of hormones.

The distinction between peptide or amine hormones and steroid hormones is important because the different types of hormones interact with different types of receptors, as we discuss next.

steroid hormone Any of a class of hormones, each of which is composed of four interconnected rings of carbon atoms.

TABLE 8.1 Examples of Major Classes of Hormones

Class	Hormone
Peptide hormones	Adrenocorticotropic hormone (ACTH)
	Follicle-stimulating hormone (FSH)
	Luteinizing hormone (LH)
	Thyroid-stimulating hormone (TSH)
	Growth hormone (GH)
	Prolactin
	Insulin
	Glucagon
	Oxytocin
	Vasopressin (arginine vasopressin, AVP; antidiuretic hormone, ADH)
	Releasing hormones, such as:
	Corticotropin-releasing hormone (CRH)
	Gonadotropin-releasing hormone (GnRH)
Amine hormones	Epinephrine (adrenaline)
	Norepinephrine (NE)
	Thyroid hormones (e.g., thyroxine)
	Melatonin
Steroid hormones	Estrogens (e.g., estradiol)
	Progestins (e.g., progesterone)
	Androgens (e.g., testosterone, dihydrotestosterone)
	Glucocorticoids (e.g., cortisol)
	Mineralocorticoids (e.g., aldosterone)

FIGURE 8.5 Two Main Mechanisms of Hormone Action

To see the animation
Mechanisms of Hormone Action,
go to

2e.mindsmachine.com/8.4

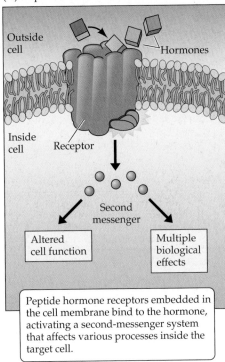

(A) Peptide hormone action

Outside cell

Hormones

Inside cell

Receptor

Second messenger

Altered cell function

Multiple biological effects

Peptide hormone receptors embedded in the cell membrane bind to the hormone, activating a second-messenger system that affects various processes inside the target cell.

(B) Steroid hormone action

Steroid hormone

Steroid receptor

DNA   Nucleus

New protein production and multiple biological effects

Steroid hormones diffuse passively in, binding to large receptor molecules inside target cells. The steroid-receptor complex then binds to DNA, altering the expression of certain genes—a so-called genomic effect.

second messenger   A slow-acting substance in a target cell that amplifies the effects of synaptic or hormonal activity and regulates activity within the target cell.

Hormones Act on a Wide Variety of Cellular Mechanisms

To prepare for later discussions of specific hormonal effects on behavior, let's look briefly at two aspects of hormone activity: first the mechanisms of hormone action, then the types of changes that hormones cause in target cells, including neurons.

Hormones initiate actions by binding to receptor molecules

The three classes of hormones exert their influences on target organs in two different ways. As we'll see next, peptide and amine hormones use one mode of action, while steroid hormones use another.

PEPTIDE AND AMINE HORMONES   Peptide and amine hormones bind to specific receptor proteins *on the surface* of the target cell and activate chemical signals inside the cell that are called **second messengers** (see Chapter 4) (**FIGURE 8.5A**). What determines whether a cell responds to a particular peptide hormone? Receptors are highly specific, so only those cells that produce the appropriate receptor proteins for a hormone can respond to that hormone. As we saw with neurotransmitter receptors in Chapters 3 and 4, the receptor protein spans the cellular membrane. The specific effect of the hormone depends in large part on the receptor it activates. Peptide hormones usually act relatively rapidly, within seconds to minutes. (Although rapid for a hormone, this action is much slower than neural activity.)

STEROID HORMONES   We mentioned earlier that steroid hormones easily pass through cell membranes, so their receptors are generally located *inside* the target cell. Different classes of steroids have their own specific receptors; for example, estrogens selectively interact with estrogen receptors, and androgens like testosterone bind to androgen receptors. When a steroid molecule and a receptor molecule combine, the steroid-receptor complex enters the nucleus of the cell and binds to the DNA, controlling the expression of specific genes (**FIGURE 8.5B**), increasing or decreasing the rate of protein production (see the Appendix). Because they involve multiple steps and the synthesis of large new molecules, steroid hormones are typically slower-acting than peptide or amine hormones. Steroid effects may take hours, days, or even years to fully unfold.

We can study where a steroid hormone is active by injecting radioactively tagged molecules of the steroid and observing where they accumulate. For example, tagged estrogens accumulate not only in the uterus (as you might expect), but also in the nuclei of some neurons throughout the hypothalamus. Because neurons that produce hormone receptors are found in only a limited number of brain regions, we can begin to learn how hormones affect behavior by finding those brain sites and asking what happens when the hormone arrives there. This strategy for learning about hormones and behavior is discussed in **BOX 8.1**.

knockout organism An individual in which a particular gene has been disabled by an experimenter.

BOX 8.1 Techniques of Modern Behavioral Endocrinology

To establish that a particular hormone affects behavior, investigators usually begin with the type of experiment that Arnold Berthold performed in the nineteenth century: observing the behavior of the intact animal, and then removing the endocrine gland and looking for a change in behavior (see Figure 8.2). Berthold was limited to this type of experiment, but modern scientists have many additional options available. Let's imagine that we're investigating a particular effect of hormones on behavior to see how we might proceed.

Which Hormones Affect Which Behavior?

First we must carefully observe the behavior of several individuals, seeking ways to classify and quantify the different types of behavior and to place them in the context of the behavior of other individuals. For example, most adult male rats will try to mount and copulate with receptive females placed in their cages. If the testes are removed from a male rat, he will eventually stop copulating with females. We know that one of the hormones produced by the testes is testosterone. Is it the loss of testosterone that causes the loss of male copulatory behavior?

To explore this question, we inject some testosterone into castrated males and observe whether the copulatory behavior

1 A rat is injected with molecules of testosterone (an androgen) that have been radioactively labeled.

2 The testosterone molecules enter the bloodstream and accumulate in those cells that have androgen receptors.

3 The brain is removed and frozen to keep the testosterone molecules inside the target cells.

Film

4 The brain is thinly sliced and film is placed on top in the dark. The radioactive molecules release particles that "expose" the film just as light would.

Exposure

5 When the film is developed, small black dots form on the film where the testosterone has accumulated in target cells.

(A) Steps in steroid autoradiography.

returns. (It does.) Another way to ask whether a steroid hormone is affecting a particular behavior is to examine the behavior of animals that lack the receptors

for that steroid. We can delete the gene for a given hormone receptor, making a **knockout organism** (so called because the gene for the receptor has been "knocked out"), and ask which behaviors are different in the knockouts versus normal animals.

Next we might examine individual male rats and ask whether the ones that copulate a lot have more testosterone circulating in their blood than those that copulate only a little. To investigate this question, we measure individual differences in the amount of copulatory behavior, take a sample of blood from each individual, and measure levels of testosterone. It turns out that individual differences in the sexual behavior of normal male rats (and normal male humans) do *not* correlate with differences in testosterone levels in the blood. In both rats and humans, a drastic loss of testosterone, as after castration, results in a gradual decline in sexual behavior. All normal males, however, appear to make more than enough testosterone to maintain sexual behavior, so something else must modulate this behavior. In other words, the hormone acts in a *permissive* manner: it permits the display of the behavior, but something else determines how much of the behavior each individual exhibits (see Figure 8.18).

continued

BOX 8.1 Techniques of Modern Behavioral Endocrinology *continued*

Where Are the Target Cells?

What does testosterone do to permit sexual behavior? One step toward answering this question is to ask another question: Which parts of the brain are normally affected by this hormone? We have several methods at our disposal for investigating this question.

First we might inject a castrated animal with radioactively labeled testosterone and wait for the hormone to accumulate in the brain regions that have receptors for the hormone. Then we could sacrifice the animal, remove the brain, freeze it, cut thin sections from it, and place the thin sections on photographic film. Radioactive emissions from the tissue would expose the film, revealing which brain regions accumulated the most labeled testosterone. This method is known as **autoradiography** because the tissue "takes its own picture" with radioactivity (**FIGURE A**).

When the labeled hormone is a steroid like testosterone, the radioactivity accumulates in the nuclei of neurons and leaves small black specks on the film (**FIGURE B**). When the radiolabeled hormone is a peptide hormone such as oxytocin, the radioactivity accumulates in the membranes of cells and appears in particular layers of the brain. Computers can generate color maps that highlight regions with high densities of receptors (**FIGURE C**).

Another method for detecting hormone receptors is **immunocytochemistry (ICC)** (described in more detail in Box 2.1). ICC enables us to map the distribution of hormone receptors in the brain. We allow specific antibodies to seek out and bind to receptors on slices of brain tissue, and then we use chemical methods to make the antibodies visible, leaving a dark color in the nuclei of target brain cells (**FIGURE D**). We can also use **in situ hybridization** (see Box 2.1) to look for the neurons that make the mRNA for the steroid receptor. Because these cells make the transcript for the receptor, they are likely to possess the receptor protein itself.

Once we've used autoradiography, immunocytochemistry, or *in situ* hybridization (or, better yet, all three) to identify brain regions that have receptors for the hormone, those regions become candidates for the places at which the hormone works to change behavior. Now we can take castrated males, implant tiny pellets of testosterone into one of those brain regions, and see whether the behavior is restored. If not, then in other animals we can implant pellets in a different region or try placing implants in a combination of brain sites.

It turns out that such implants can restore male sexual behavior in rats only if they are placed in the medial preoptic area (mPOA) of the hypothalamus. So far, we've found that testosterone does something to the mPOA to permit individual males to display sexual behavior. Now we can examine the mPOA in detail to learn what changes in the anatomy, physiology, or protein production of this region are caused by testosterone. And with that, we have more or less caught up to modern-day scientists who work on this very question, as described in this chapter. (Figure C courtesy of Bruce McEwen; D courtesy of Cynthia Jordan.)

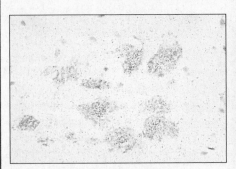

(B) An autoradiogram showing that spinal motoneurons (purple cell profiles) accumulate radioactive testosterone (small dots).

(C) An autoradiogram showing the concentration of oxytocin receptors (orange) in the ventromedial hypothalamus (oval outlines).

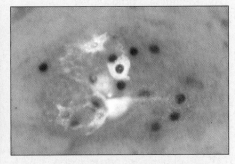

(D) Immunocytochemistry revealing cells with nuclei that contain androgen receptors (dark circles), to which testosterone can bind. The cell bodies of these neurons have been labeled with two different tracers, one white and the other red.

autoradiography A staining technique that shows the distribution of radioactive chemicals in tissues.

immunocytochemistry (ICC) A method for detecting a particular protein in tissues in which an antibody recognizes and binds to the protein and then chemical methods are used to leave a visible reaction product around each antibody.

Hormones can have different effects on different target organs

Virtually all hormones, whether peptide or steroid, act on more than one target organ. What's more, a given hormone may have one type of effect on one organ, and a quite different effect on another organ. This means hormones often act to coordinate different parts of the body, causing diverse changes to several different parts of the body, all of which prepare the animal for a particular activity. For example, the testes secrete testosterone, which acts in the testes themselves to drive sperm production but also acts to masculinize the body in diverse ways, favoring muscle development and, in humans, beard growth. In this and numerous other cases we'll discuss, the same hormone also acts on the brain—in the case of testosterone, to promote sexual behavior and aggression in many animal species.

How is the same hormone able to cause many different responses in different organs? First, often more than one receptor responds to a given hormone (**FIGURE 8.6**). For example, there are at least two different receptors for estrogens, and for other hormones there may be four or more. In addition, sometimes the same receptor, in response to the same hormone, will have a different effect because the target cell responds differently.

As we'll see later, the brain maintains strict control over all hormone secretion, which means, among other things, that the brain has receptors to detect almost all hormones, in order to monitor their release.

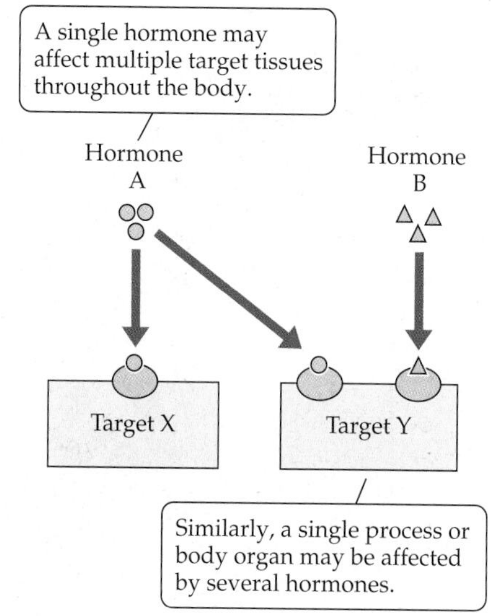

---

how's it going ❓

1. What are hormones, and how do they act?
2. Describe Berthold's experiment with young roosters, and explain how it indicated hormonal effects.
3. Compare and contrast the mechanisms by which peptide/amine hormones versus steroid hormones act on cells.
4. Why are hormones effective for coordinating different changes in different parts of the body?

Each Endocrine Gland Secretes Specific Hormones

Now we'll discuss the specific hormones that endocrine glands secrete. To cover all the hormones would require an entire book (e.g., Norman and Henry, 2014), so we will restrict our discussion to only some of the main endocrine glands. **A STEP FURTHER 8.2**, on the website, gives a fuller (though far from complete) listing of hormones and their functions. Hormones involved in thirst and hunger are discussed in Chapter 9.

We must begin with the pituitary because it regulates so many other endocrine glands. Resting in a socket in the base of the skull, the **pituitary gland** is about the size of a garden pea, weighing about 1 gram (see Figure 8.1). The hypothalamus sits just above the pituitary and is connected to the gland by a slender thread called the **pituitary stalk**. The term *pituitary* comes from the Latin *pituita*, "mucus," reflecting the outmoded belief that waste products dripped down from the brain into the pituitary, which then secreted them out through the nose. (The ancients may have thought you could literally sneeze your brains out!) Because the pituitary regulates most other endocrine glands, it is sometimes referred to as the *master gland*. But the pituitary is itself enslaved by the hypothalamus above it, as we'll see.

To understand how the pituitary works, we need to consider a special category of cells that are something of a blend of neuronal cells and endocrine cells, called **neuroendocrine cells**. On the one hand, neuroendocrine cells receive synaptic input from other neurons and, if they are excited past threshold, produce action potentials. But unlike regular neurons, which release a neurotransmitter into a synapse when they fire, neuroendocrine cells release a hormone into the bloodstream (**FIGURE 8.7**).

FIGURE 8.6 The Multiplicity of Hormone Action

---

in situ hybridization A method for detecting particular RNA transcripts in tissue sections by providing a nucleotide probe that is complementary to, and will therefore hybridize with, the transcript of interest.

pituitary gland A small, complex endocrine gland located in a socket at the base of the skull.

pituitary stalk A thin piece of tissue that connects the pituitary gland to the hypothalamus.

neuroendocrine cell A neuron that releases hormones into local or systemic circulation.

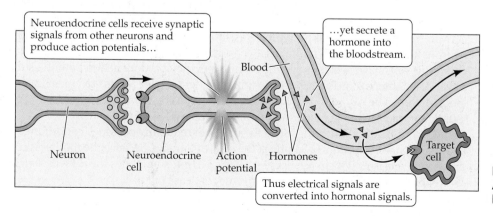

FIGURE 8.7 Neuroendocrine Cells Are the Interface between Neurons and Endocrine Glands

FIGURE 8.8 Hormone Production by the Posterior Pituitary

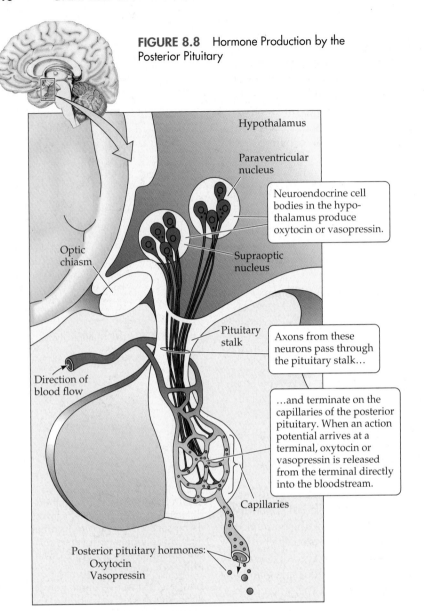

Hypothalamus

Paraventricular nucleus

Neuroendocrine cell bodies in the hypothalamus produce oxytocin or vasopressin.

Optic chiasm

Supraoptic nucleus

Pituitary stalk

Axons from these neurons pass through the pituitary stalk…

Direction of blood flow

…and terminate on the capillaries of the posterior pituitary. When an action potential arrives at a terminal, oxytocin or vasopressin is released from the terminal directly into the bloodstream.

Capillaries

Posterior pituitary hormones:
Oxytocin
Vasopressin

To understand the pivotal role of such neuroendocrine cells in hormone release from the pituitary, we need to consider separately the two parts of the pituitary: the *anterior pituitary* and the *posterior pituitary*. These two halves of the pituitary develop from different embryonic tissues and are completely separate in function. The mechanism of hormone release is simpler in the posterior pituitary, so let's discuss that first.

The posterior pituitary releases two hormones directly into the bloodstream

Although it is called a gland, the **posterior pituitary** does not itself produce hormones. Rather, the hormones are produced by neuroendocrine cells in two hypothalamic regions: the *supraoptic* and *paraventricular nuclei*. These neuroendocrine cells transport hormones down their axons, which extend through the pituitary stalk to terminate in the posterior pituitary. When the hypothalamic neuroendocrine cells are excited by synaptic input, they produce action potentials that travel down the axons and release hormone directly onto capillaries (small blood vessels) in the posterior pituitary, allowing the hormone to enter circulation immediately (**FIGURE 8.8**).

In this fashion, the neuroendocrine cells in the hypothalamus produce and release two peptide hormones from the posterior pituitary: **oxytocin** and **vasopressin**. Some of the signals that activate the nerve cells of the supraoptic and paraventricular nuclei are related to thirst and water regulation; vasopressin is involved in these interactions, which will be discussed in Chapter 9. Oxytocin is involved in many aspects of reproductive and parental behavior. One of its functions is to stimulate contractions of the uterus in childbirth. Injections of oxytocin (or a synthetic version) are frequently used in hospitals to induce or accelerate labor and delivery.

Oxytocin also triggers the **milk letdown reflex**, the contraction of mammary gland cells that ejects milk into the breast ducts. This reflex exemplifies the reciprocal relationship between behavior and hormone release. When an infant first begins to suckle, the arrival of milk at the nipple is delayed by 30–60 seconds. This delay is caused by the sequence of steps that precedes letdown. Stimulation of the nipple activates receptors in the skin, which transmit this information through a chain of neurons and synapses to hypothalamic cells that contain oxytocin. Once these neuroendocrine cells have been sufficiently stimulated, they produce action potentials that travel down their axons to the posterior pituitary, where they release oxytocin into the bloodstream. The oxytocin reaches muscle tissue in the mammary glands, and the muscle contracts to make milk available at the nipple (**FIGURE 8.9**).

For mothers, this reflex response to suckling frequently becomes conditioned to baby cries, so milk appears promptly at the start of nursing. Because the mother is conditioned to release oxytocin *before* the suckling begins, sometimes the cries of *someone else's* baby in public may trigger an inconvenient release of milk.

posterior pituitary The rear division of the pituitary gland.

oxytocin A hormone, released from the posterior pituitary, that triggers milk letdown in the nursing female, but is also associated with a variety of complex behaviors.

vasopressin Also called *arginine vasopressin* or *antidiuretic hormone*. A peptide hormone from the posterior pituitary that promotes water conservation and increases blood pressure.

milk letdown reflex The reflexive release of milk by the mammary glands of a nursing female in response to suckling or to stimuli associated with suckling.

pair-bond A durable and exclusive relationship between two individuals.

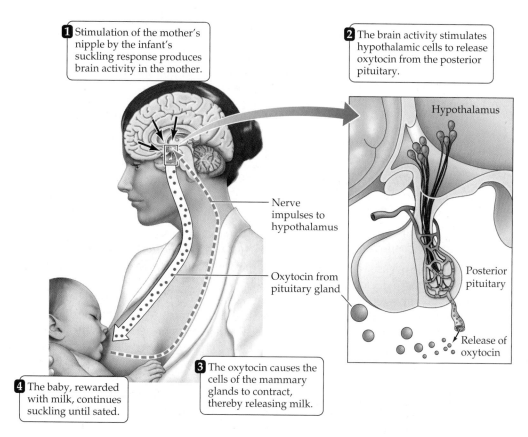

1 Stimulation of the mother's nipple by the infant's suckling response produces brain activity in the mother.

2 The brain activity stimulates hypothalamic cells to release oxytocin from the posterior pituitary.

Hypothalamus

Nerve impulses to hypothalamus

Oxytocin from pituitary gland

Posterior pituitary

Release of oxytocin

4 The baby, rewarded with milk, continues suckling until sated.

3 The oxytocin causes the cells of the mammary glands to contract, thereby releasing milk.

FIGURE 8.9   The Milk Letdown Reflex

Posterior pituitary hormones can affect social behavior

We've already seen the role of oxytocin in the interaction of nursing babies and their mothers (see Figure 8.9). It turns out that this hormone is involved in several other social behaviors too. For one thing, a pulse of oxytocin is released during orgasm in both men and women (Carmichael et al., 1994), adding to the pleasurable feelings accompanying sexual encounters.

In nonhuman animals, oxytocin and vasopressin facilitate many social processes (Lim and Young, 2006). Rodents given supplementary doses of oxytocin spend more time in physical contact with one another (Carter, 1992). Male mice that have the oxytocin gene knocked out and are therefore unable to produce the hormone display social amnesia: they seem unable to recognize the scents of female mice that they have met before (Ferguson et al., 2000). These oxytocin knockout males can be cured of their social amnesia with brain infusions of oxytocin (Winslow and Insel, 2002).

In prairie voles (*Microtus ochrogaster*), couples form stable **pair-bonds**, and oxytocin infusions in the brains of females help them bond to their mates. In *male* prairie voles, it is vasopressin rather than oxytocin that facilitates the formation of a preference for female partners. In fact, the distribution of vasopressin receptors in the brains of male prairie voles may be what makes them monogamous. Supporting this idea is the finding that the closely related meadow voles (*Microtus pennsylvanicus*), which do not form pair-bonds and instead have multiple mating partners, have far fewer vasopressin receptors in certain brain regions than prairie voles have (Lim et al., 2004). Thus, oxytocin and vasopressin regulate a range of social behaviors, and natural selection appears to alter the social behaviors of a species by changing the brain distribution of receptors for these peptides (Donaldson and Young, 2008) (**FIGURE 8.10**).

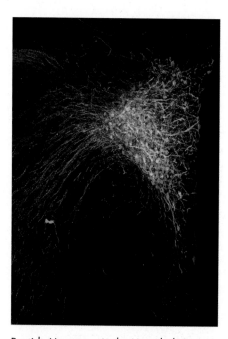

Peptide Hormones in the Hypothalamus
This section of the paraventricular nucleus reveals cells that make oxytocin (red) or vasopressin (green), hormones that are released from the posterior pituitary. (Courtesy of V. Tobin and M. Ludwig.)

(A)

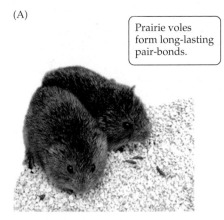

Prairie voles form long-lasting pair-bonds.

(B)

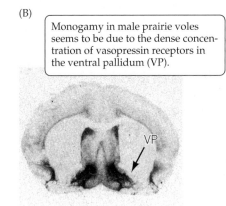

Monogamy in male prairie voles seems to be due to the dense concentration of vasopressin receptors in the ventral pallidum (VP).

VP

(C)

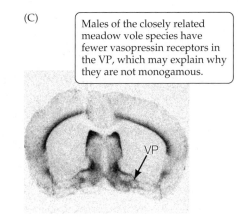

Males of the closely related meadow vole species have fewer vasopressin receptors in the VP, which may explain why they are not monogamous.

VP

FIGURE 8.10 Vasopressin and the Monogamous Brain (Photographs courtesy of Miranda Lim and Larry Young.)

negative feedback The property by which some of the output of a system feeds back to reduce the effect of input signals.

To see the animation **The Hypothalamus and Endocrine Function,** go to

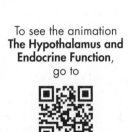

2e.mindsmachine.com/8.5

Feedback control mechanisms regulate the secretion of hormones

All hormone release is carefully controlled by the brain, which monitors internal and external cues to decide whether and how much hormone should be released. Brain regulation of posterior pituitary hormone release is readily understood, because the neuroendocrine cells secrete hormone only when they are excited synaptically. For example, sensory information from a child's suckling at the breast reaches the brain and excites hypothalamic neuroendocrine cells, which fire action potentials and release oxytocin. Once the baby is satisfied, the suckling stops, so the brain stops exciting the neuroendocrine cells and oxytocin release ceases.

This is an example of the basic mechanism that regulates all hormone secretion, called **negative feedback**: output of the hormone *feeds back* to inhibit the drive for more of that same hormone (**FIGURE 8.11A**). This negative feedback action of a hormonal system is like that of a thermostat, and just as the thermostat can be set to different temperatures at different times, the set points of a person's endocrine feedback systems can change to meet varying circumstances. We'll discuss negative feedback regulation of other processes in Chapter 9 (see Figure 9.1).

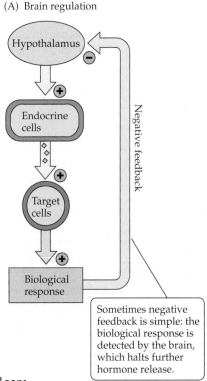

(A) Brain regulation

Hypothalamus
−
+
Endocrine cells
+
Target cells
+
Biological response

Negative feedback

Sometimes negative feedback is simple: the biological response is detected by the brain, which halts further hormone release.

(B) Brain and pituitary regulation

Hypothalamus
−
Releasing hormone
+
Anterior pituitary
−
Tropic hormone
+
Endocrine cells
+
Target cells
+
Biological response

Negative feedback

For the anterior pituitary, hormones from the endocrine gland have a negative feedback effect on both the hypothalamus and pituitary.

FIGURE 8.11 Endocrine Feedback Loops

Hormone secretion from the *anterior* pituitary is also regulated by negative feedback, but the mechanism is a bit more complicated, as we'll see next.

Hypothalamic releasing hormones govern the anterior pituitary

The **anterior pituitary** consists of many different endocrine cells, each secreting a different peptide hormone. So, unlike the posterior half of the pituitary, the anterior pituitary actually synthesizes the hormones it releases. The anterior pituitary hormones are called **tropic hormones**. (The *o* in *tropic* is pronounced "oh"; there is nothing "tropical" about these hormones.) The term *tropic* means "directed toward," and each tropic hormone acts on a different endocrine gland, such as the thyroid or ovaries, as if the tropic hormone were directed toward that gland. Actually the tropic hormone travels throughout the bloodstream, reaching *all* glands, but only the *target* glands have the appropriate receptors to respond to it. Once the tropic hormone reaches a target gland, it drives the gland to produce its own hormone. For example, one anterior pituitary tropic hormone acts on the thyroid gland to make it secrete thyroid hormones.

To regulate secretions of tropic hormones from the anterior pituitary, the hypothalamus uses another whole set of hormones, called **releasing hormones**. The cells that synthesize the different releasing hormones are neuroendocrine cells residing in various regions of the hypothalamus (**FIGURE 8.11B**). The axons of these neuroendocrine cells converge on the **median eminence**, just above the pituitary stalk. This region contains an elaborate profusion of blood vessels that form the **hypothalamic-pituitary portal system**. Here, in response to inputs from the rest of the brain, the axon terminals of the hypothalamic neuroendocrine cells secrete their releasing hormones into the *local* bloodstream (**FIGURE 8.12**). Blood carries the various releasing hormones only a very short distance, into the anterior pituitary. The rate at which

anterior pituitary   The front division of the pituitary gland. It secretes tropic hormones.

tropic hormone   Any of a class of anterior pituitary hormones that affect the secretion of hormones by other endocrine glands.

releasing hormone   Any of a class of hormones, produced in the hypothalamus, that traverse the hypothalamic-pituitary portal system to control the pituitary's release of tropic hormones.

median eminence   A midline feature on the base of the brain that marks the point at which the pituitary stalk exits the hypothalamus to connect to the pituitary. The median eminence contains elements of the hypothalamic-pituitary portal system.

hypothalamic-pituitary portal system   An elaborate bed of blood vessels leading from the hypothalamus to the anterior pituitary.

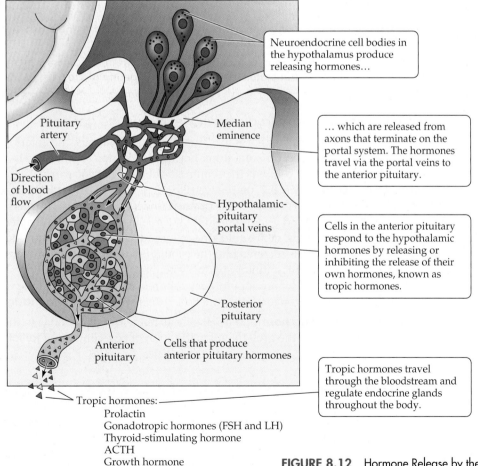

Neuroendocrine cell bodies in the hypothalamus produce releasing hormones…

Pituitary artery

Median eminence

Direction of blood flow

… which are released from axons that terminate on the portal system. The hormones travel via the portal veins to the anterior pituitary.

Hypothalamic-pituitary portal veins

Cells in the anterior pituitary respond to the hypothalamic hormones by releasing or inhibiting the release of their own hormones, known as tropic hormones.

Posterior pituitary

Anterior pituitary

Cells that produce anterior pituitary hormones

Tropic hormones travel through the bloodstream and regulate endocrine glands throughout the body.

Tropic hormones:
Prolactin
Gonadotropic hormones (FSH and LH)
Thyroid-stimulating hormone
ACTH
Growth hormone

FIGURE 8.12   Hormone Release by the Anterior Pituitary

releasing hormones arrive at the anterior pituitary controls the rate at which the anterior pituitary cells, in turn, release their tropic hormones into the *general* circulation. These tropic hormones then regulate the activity of major endocrine organs throughout the body. Thus, the brain's releasing hormones affect the anterior pituitary's tropic hormones, which affect the release of hormones from endocrine glands.

The hypothalamic neuroendocrine cells that synthesize the releasing hormones are themselves subject to two kinds of influences. First, they are directly affected by *circulating messages*, such as other hormones, especially hormones that were secreted in response to tropic hormones (see Figure 8.11B). This hormone sensitivity of the hypothalamic neurons is an important part of the negative feedback we mentioned earlier, because typically the hormones secreted from an endocrine gland feed back to inhibit the secretion of releasing hormones and tropic hormones. Negative feedback in this case goes from the hormone of the endocrine gland to both the hypothalamus and the anterior pituitary.

Second, the hypothalamic neuroendocrine cells that provide releasing hormones also receive *synaptic inputs* (either excitatory or inhibitory) from many other brain regions. As a result, the release of hormones by the anterior pituitary is coordinated with ongoing events, such as time of day, time of the year, safety of the individual, and so on. For example, if a child is living in stressful, abusive conditions, the brain monitors these conditions and reduces the production of releasing hormones that stimulate the anterior pituitary release of **growth hormone** (**GH**). We'll return to this topic at the end of the chapter.

Thus, the hypothalamic-releasing-hormone system exerts high-level control over endocrine organs throughout the body, translating brain activity into hormonal action. Cutting the pituitary stalk interrupts the portal blood vessels and the flow of releasing hormones, leading to profound atrophy of the pituitary, as well as major hormonal disruptions.

how's it going

1. How do hormones and behaviors interact in the course of the milk letdown reflex?
2. Describe the system regulating hormone release from the anterior pituitary, and explain how that system controls other endocrine glands.

Two anterior pituitary tropic hormones act on the gonads

Driven by various releasing hormones from the hypothalamus, the anterior pituitary gland secretes at least six different tropic hormones. We don't really need to go into all these hormones and the glands they control right now, but you can learn more about them in **A STEP FURTHER 8.3**, on the website. Here we will concentrate on the tropic hormones that affect male and female **gonads** (the *testes* and *ovaries*, respectively), because the hormones produced by gonads play a role in the rest of this chapter.

In the hypothalamus, neuroendocrine cells produce **gonadotropin-releasing hormone** (**GnRH**), which is secreted into the capillaries of the median eminence, traveling via the hypothalamic-pituitary portal system to arrive at the anterior pituitary. In response to this GnRH, anterior pituitary cells release one or both of the tropic hormones that act on the gonads, which are thus collectively known as **gonadotropins**:

1. **Follicle-stimulating hormone** (**FSH**) gets its name from its actions in the ovary, where it stimulates the growth and maturation of egg-containing **follicles** and the secretion of estrogens from the follicles. In males, FSH governs sperm production.

2. **Luteinizing hormone** (**LH**) stimulates the follicles of the ovary to rupture, release their eggs, and form into structures called **corpora lutea** (singular *corpus luteum*) that secrete the sex steroid hormone progesterone. In males, LH stimulates the testes to produce testosterone.

Since both of the gonadotropins drive the release of gonadal steroids, we'll turn our attention to those hormones next.

growth hormone (GH)   Also called *somatotropin* or *somatotropic hormone*. A tropic hormone, secreted by the anterior pituitary, that promotes the growth of cells and tissues.

gonad   Any of the sexual organs (ovaries in females, testes in males) that produce gametes for reproduction.

gonadotropin-releasing hormone (GnRH)   A hypothalamic hormone that controls the release of luteinizing hormone and follicle-stimulating hormone from the pituitary.

gonadotropin   An anterior pituitary hormone that selectively stimulates the cells of the gonads to produce sex steroids and gametes.

follicle-stimulating hormone (FSH)   A gonadotropin, named for its actions on ovarian follicles.

follicle   The structure of the ovary that contains an immature ovum (egg).

luteinizing hormone (LH)   A gonadotropin, named for its stimulatory effects on the ovarian corpora lutea.

corpus luteum   The structure that forms from the collapsed ovarian follicle after ovulation. The corpora lutea are a major source of progesterone.

testes   The male gonads, which produce sperm and androgenic steroid hormones.

testosterone   A hormone, produced by male gonads, that controls a variety of bodily changes that become visible at puberty. It is one of a class of hormones called *androgens*.

androgen   Any of a class of hormones that includes testosterone and other male hormones.

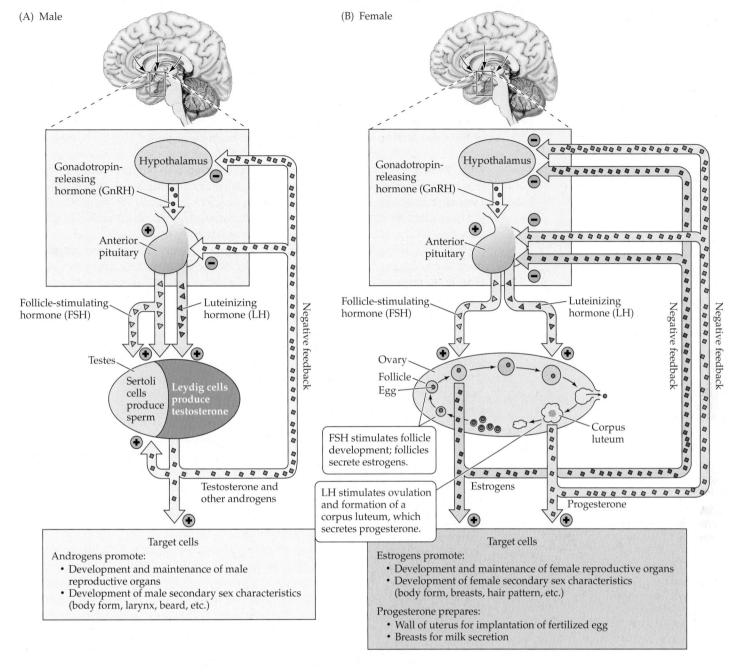

(A) Male

Gonadotropin-releasing hormone (GnRH)

Hypothalamus

Anterior pituitary

Follicle-stimulating hormone (FSH)

Luteinizing hormone (LH)

Testes

Sertoli cells produce sperm

Leydig cells produce testosterone

Testosterone and other androgens

Negative feedback

Target cells

Androgens promote:
- Development and maintenance of male reproductive organs
- Development of male secondary sex characteristics (body form, larynx, beard, etc.)

(B) Female

Gonadotropin-releasing hormone (GnRH)

Hypothalamus

Anterior pituitary

Follicle-stimulating hormone (FSH)

Luteinizing hormone (LH)

Ovary

Follicle

Egg

Corpus luteum

FSH stimulates follicle development; follicles secrete estrogens.

LH stimulates ovulation and formation of a corpus luteum, which secretes progesterone.

Estrogens

Progesterone

Negative feedback

Target cells

Estrogens promote:
- Development and maintenance of female reproductive organs
- Development of female secondary sex characteristics (body form, breasts, hair pattern, etc.)

Progesterone prepares:
- Wall of uterus for implantation of fertilized egg
- Breasts for milk secretion

FIGURE 8.13 Regulation of the Gonadal Steroid Hormones

The gonads produce steroid hormones, regulating reproduction

Almost all aspects of reproductive behavior, including mating and parental behaviors, depend on hormones, as we'll see later in this chapter. Each ovary or testis consists of two different subcompartments—one to produce hormones (the sex steroids we mentioned earlier) and another to produce *gametes* (eggs or sperm). The gonadal hormones are critical for triggering both reproductive behavior controlled by the brain and gamete production.

THE TESTES   Within the **testes** (singular *testis*) are Sertoli cells, which produce sperm, and Leydig cells, which produce and secrete the steroid **testosterone**. Testosterone and other male hormones are called **androgens** (from the Greek *andro*, "man," and *gennan*, "to produce").

Testosterone controls a wide range of bodily changes that become visible at puberty, including changes in voice, hair growth, and genital size. In species that breed only in certain seasons of the year, testosterone has especially marked effects on appearance and behavior—for example, the antlers and fighting between males that are displayed by many species of deer. **FIGURE 8.13A** summarizes the regulation of testosterone secretion. As men age, testosterone levels tend to decline. Although elderly men who happen to maintain high levels of circulating testosterone perform better on tests

ovaries The female gonads, which produce eggs (ova) for reproduction.

progestin Any of a major class of steroid hormones that are produced by the ovary, including progesterone.

estrogen Any of a class of steroid hormones, including estradiol, produced by female gonads.

estradiol The primary type of estrogen that is secreted by the ovary. Its formal name is *17-beta-estradiol*.

progesterone The primary type of progestin secreted by the ovary.

ovulatory cycle The periodic occurrence of ovulation in females.

oral contraceptive A birth control pill, typically consisting of steroid hormones to prevent ovulation.

of memory and attention than do those with low levels (Yaffe et al., 2002), there have been too few studies to tell whether taking supplemental testosterone actually helps aging men (Gold and Voskuhl, 2006; Nair et al., 2006). Furthermore, taking supplemental testosterone can sometimes increase aggressive or manic behaviors (Pope et al., 2000), as well as possibly increasing prostate cancer risk.

THE OVARIES The paired female gonads, the **ovaries**, also produce both the mature female gametes—called *ova* (singular *ovum*) or eggs—and sex steroid hormones. However, hormone secretion is more complicated in ovaries than in testes. Ovarian hormones are produced in cycles, the duration of which varies with the species. Human ovarian cycles last about 4 weeks; rat cycles last only 4 days.

Normally the ovary produces two major classes of steroid hormones: **progestins** (from the Latin *pro*, "favoring," and *gestare*, "to bear," because these hormones help to maintain pregnancy) and **estrogens** (from the Latin *oestrus*, "frenzy"—*estrus* is the scientific term for the periodic sexual receptivity of females in many species). The most important naturally occurring estrogen is **estradiol** (specifically, 17-beta-estradiol). The primary progestin is **progesterone**.

The **ovulatory cycle** begins when FSH stimulates ovarian follicles to grow and secrete estrogens (**FIGURE 8.13B**). The estrogens induce the hypothalamus and pituitary to release LH, which triggers the release of an egg from a follicle (ovulation) and causes the follicle to develop as a corpus luteum. The corpus luteum then secretes progesterone to maintain the uterus for pregnancy. If the female does not become pregnant, the cycle starts over again.

Oral contraceptives contain small doses of synthetic estrogens and/or progestins, which exert a negative feedback effect on the hypothalamus, inhibiting the release of GnRH. The lack of GnRH prevents the release of FSH and LH from the pituitary, and therefore the ovary fails to release an egg for fertilization.

Estrogens may improve aspects of cognitive functioning (Maki and Resnick, 2000), although this topic is still debated (Dohanich, 2003; Sherwin, 2009). Estrogens may also protect the brain from some of the effects of stress and stroke (S. Suzuki et al., 2009; Petrone et al., 2014). For these reasons and others, estrogen replacement therapy has been a popular postmenopausal treatment, but the possibility that these treatments increase the risk of serious diseases like cancer and heart disease (Turgeon et al., 2004; Prentice, 2014) makes the decision of whether to take the hormones difficult for postmenopausal women.

RELATIONS AMONG GONADAL HORMONES All steroid hormones—including androgens, estrogens, and progestins—are based on the chemical structure of cholesterol (see Figure 8.4C). Glands manufacture steroid hormones by using enzymes to modify cholesterol, step by step, into different steroids. For example, ovaries first convert cholesterol into progestins, and then they convert those progestins into androgens, which are then converted into estrogens.

Different organs—and the two sexes—differ in the *relative* amounts of gonadal steroids that they produce. For example, whereas the testes convert only a relatively small proportion of testosterone into estradiol, the ovaries convert most of the testosterone they make into estradiol. What's important to understand is that *no steroid is found exclusively in either males or females*.

Hormonal and neural systems interact to produce integrated responses

The endocrine system and the nervous system work together, each affecting the other, seamlessly integrating various body systems to produce adaptive responses to the environment. So, for example, if our sensory system tells us that a stimulus calls for action—perhaps that faint buzzing sound you're hearing turns out to be coming from a nest of angry wasps—hormones can be released to provide energy to fuel appropriate behaviors (sprinting away, yelling, cursing maybe).

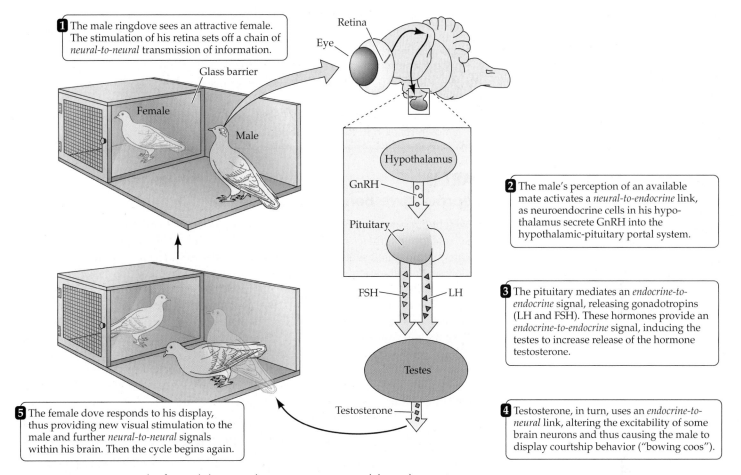

1 The male ringdove sees an attractive female. The stimulation of his retina sets off a chain of *neural-to-neural* transmission of information.

2 The male's perception of an available mate activates a *neural-to-endocrine* link, as neuroendocrine cells in his hypothalamus secrete GnRH into the hypothalamic-pituitary portal system.

3 The pituitary mediates an *endocrine-to-endocrine* signal, releasing gonadotropins (LH and FSH). These hormones provide an *endocrine-to-endocrine* signal, inducing the testes to increase release of the hormone testosterone.

4 Testosterone, in turn, uses an *endocrine-to-neural* link, altering the excitability of some brain neurons and thus causing the male to display courtship behavior ("bowing coos").

5 The female dove responds to his display, thus providing new visual stimulation to the male and further *neural-to-neural* signals within his brain. Then the cycle begins again.

FIGURE 8.14   Four Kinds of Signals between the Nervous System and the Endocrine System

Four kinds of signals are possible between neurons and endocrine cells: neural-to-neural, neural-to-endocrine, endocrine-to-endocrine, and endocrine-to-neural. All four types are illustrated in the courtship behavior of the ringdove. The visual processing that occurs when a male dove sees an attractive female involves neural-to-neural transmission (**FIGURE 8.14**, step 1). The details of the particular visual stimulus—namely, an opportunity to mate—activate a neural-to-endocrine link (step 2), which causes neuroendocrine cells in the male's hypothalamus to secrete GnRH. The GnRH provides an endocrine-to-endocrine signal (step 3), stimulating the pituitary to release gonadotropins, which induce the testes to release more testosterone. Testosterone in turn alters the excitability of neurons in the male's brain through an endocrine-to-neural link (step 4), causing the male to display courtship behavior. The female dove responds to this display (step 5), thus providing new visual stimulation to the male, which triggers another cycle of signaling in him.

The interactions between endocrine activity and behavior are cyclical, as depicted by the circle schema in **FIGURE 8.15**. The levels of circulating hormones can also be altered by experience, which in turn can affect future behavior and future experience. For example, men rooting for a sports team will produce more testosterone if their team wins (Bernhardt et al., 1998). Physical stresses, pain, and unpleasant emotional situations trigger the release of steroids from the adrenal gland (see Chapter 11).

Conversely, each of these hormonal events will affect the brain, shaping behavior, which will once more affect the person's future hormone production, and so on. It will be important to keep in mind these interactions between hormones and behavior as we consider reproductive behavior in the next section.

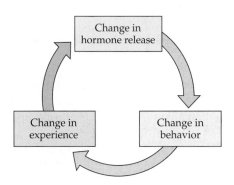

FIGURE 8.15   The Reciprocal Relations between Hormones and Behavior

1. Describe the role of the hypothalamus and the anterior pituitary in regulating gonadal hormones.
2. What hormones are found in birth control pills, and how do they prevent pregnancy?
3. Describe the interplay of neural and hormonal signaling in the human milk letdown reflex and the male ringdove's courtship behavior.

PART II
Reproductive Behavior

Sexual behaviors are almost as diverse as the species that employ them. But in every case, males and females must produce a specific set of behaviors, in a precise and intricately coordinated sequence, to reproduce successfully. In Part II of this chapter we review sexual behaviors, which include the sex act itself (copulation), as well as the parental behaviors that the newborns of many species require to survive. Part III of this chapter will cover sexual differentiation, the process by which an individual's body and brain develop in a male or female fashion.

We wish we could explain exactly why and how humans and other animals engage in the three Cs—courting, copulating, and cohabiting—but relatively little practical knowledge of such matters exists. Two barriers have blocked our understanding of sexual behavior: (1) cultural barriers to disseminating information about sexual behavior and (2) the remarkable variety of sexual behaviors.

Reproductive Behavior Can Be Divided into Four Stages

Taking a broad view of reproductive behavior, we recognize four distinct stages: (1) sexual attraction, (2) appetitive behavior, (3) copulation, and (4) postcopulatory behavior (**FIGURE 8.16**).

Sexual attraction is the first stage in bringing males and females together. In many species, sexual attraction is closely synchronized with physiological readiness to reproduce. Most male mammals are attracted by particular female odors, which tend to reflect estrogen levels. Because estrogen secretion is associated with the release of eggs, these mechanisms tend to synchronize female sexual attractiveness with peak fertility. Of course, the female may find a particular male unattractive and refuse to mate with him. Although apparent rape has been described in some nonhuman species (Maggioncalda and Sapolsky, 2002), for most species copulation is not possible without the female's active cooperation.

If the animals are mutually attracted, they may progress to the next stage: **appetitive behaviors**—species-specific behaviors that establish, maintain, or promote sexual interaction. A female displaying these behaviors is said to be **proceptive**: she may approach males, remain close to them, or show alternating approach and retreat behaviors. Proceptive female rats typically exhibit "ear wiggling" and a hopping and darting gait to induce a male to mount. Male appetitive behaviors usually consist of staying near the female. In many mammals the male may sniff around the female's face and vagina. Male birds may engage in elaborate songs or nest-building behaviors, as illustrated for the ringdove in Figure 8.14.

If both animals display appetitive behaviors, they may progress to the third stage of reproduction: **copulation**, also known as *coitus*. In many vertebrates, including all mammals, copulation involves one or more **intromissions**, in which the male inserts his penis into the female's **vagina**, followed by a variable amount of stimulation, usually through pelvic thrusting. When stimulation reaches a threshold level, the male **ejaculates** sperm-bearing **semen** into the female; the length of time and quantity of stimulation that are required vary greatly between species and between individuals.

sexual attraction   The first step in the mating behavior of many animals, in which animals emit stimuli that attract members of the opposite sex.

appetitive behavior   The second stage of mating behavior. It helps establish or maintain sexual interaction.

proceptive   Referring to a state in which a female advertises its readiness to mate through species-typical behaviors.

copulation   Also called *coitus*. The sexual act.

intromission   Insertion of the penis into the vagina during copulation.

vagina   The opening from the outside of the body to the cervix and uterus in females.

ejaculation   The forceful expulsion of semen from the penis.

semen   A mixture of fluid and sperm that is released during ejaculation.

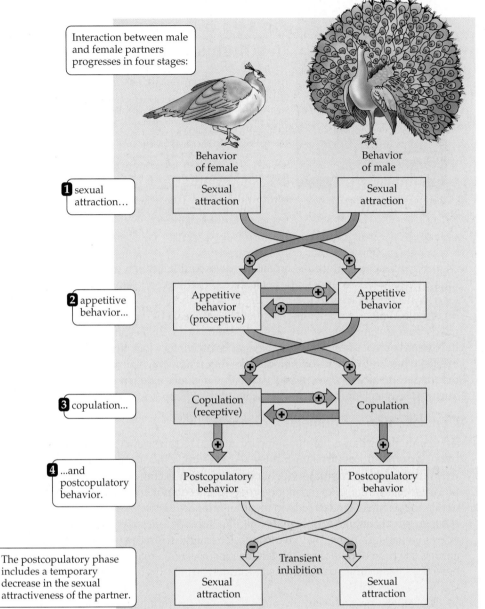

FIGURE 8.16 Stages of Reproductive Behavior (After Beach, 1977.)

Interaction between male and female partners progresses in four stages:

1 sexual attraction...

2 appetitive behavior...

3 copulation...

4 ...and postcopulatory behavior.

The postcopulatory phase includes a temporary decrease in the sexual attractiveness of the partner.

Behavior of female | Behavior of male

Sexual attraction | Sexual attraction

Appetitive behavior (proceptive) | Appetitive behavior

Copulation (receptive) | Copulation

Postcopulatory behavior | Postcopulatory behavior

Transient inhibition

Sexual attraction | Sexual attraction

After one bout of copulation, the animals will not mate again for a period of time, which is called the **refractory phase**. The refractory phase varies from minutes to months, depending on the species and circumstances. Many animals will resume mating sooner if they are provided with a new partner—a phenomenon known as the **Coolidge effect** (named after an old joke about U.S. President Calvin Coolidge [Google it]).

The female often appears to be the one to choose whether copulation will take place; when she is willing to copulate, she is said to be **sexually receptive**, in heat, or in **estrus**. In some species the female may show proceptive behaviors days before she will participate in copulation itself. In most (but not all) species, females are receptive only when mating is likely to produce offspring. Most species are seasonal breeders, with females that are receptive only during the breeding season; some—such as salmon, octopuses, and cicadas—reproduce only once, at the end of life.

Finally, the fourth stage of reproductive behavior consists of **postcopulatory behaviors**. These behaviors are especially varied, having been strongly shaped by diverse evolutionary pressures related to the different species' mating systems. For

refractory phase A period following copulation during which an individual does not recommence copulation.

Coolidge effect The propensity of an animal that appears sexually satisfied with a current partner to resume sexual activity when provided with a new partner.

sexually receptive Referring to the state in which an individual (in mammals, typically the female) is willing to copulate.

estrus The period during which female animals are sexually receptive.

postcopulatory behavior The final stage in mating behavior. Species-specific postcopulatory behaviors include rolling (in the cat) and grooming (in the rat).

The Coolidge Effect By reducing the refractory phase, when a sexually exhausted male encounters an unfamiliar female, the Coolidge effect permits him to take advantage of a new reproductive opportunity and sire more offspring. (Of course, encountering 24 lovelorn females at once is a situation few males—guinea pig or otherwise—could even dream of.)

Guinea pig Don Juan sires 43 offspring in 2 nights

PONTYPRIDD, WALES, 1 DECEMBER 2000

HAVING ESCAPED from captivity at Little Friend's Farm earlier this year, a male guinea pig named Sooty chose to re-enter captivity immediately—in the nearby cage housing 24 females. Two months later he is now the father of 43 offspring.

According to his owner, Carol Feehan, Sooty was missing for two whole days before the staff checked the females' pen. "We did a head count and found 25 guinea pigs," she told the press. "Sooty was fast asleep in the corner.

"He was absolutely shattered. We put him back in his cage and he slept for two days."

Sooty enjoyed two nights of passion among 24 females.

gamete A sex cell (sperm or ovum) that contains only unpaired chromosomes and therefore has only half of the usual number of chromosomes.

sperm The gamete produced by males for the fertilization of eggs (ova).

ovum An egg, the female gamete.

zygote The fertilized egg.

ovulation The production and release of an egg (ovum).

lordosis A female receptive posture in four-legged animals in which the hindquarters are raised and the tail is turned to one side, facilitating intromission by the male.

example, in some mammals, including dogs and southern grasshopper mice, the male's penis swells so much after ejaculation that he can't remove it from the female for a while. In species like these, where multiple males may copulate with an ovulating female in quick succession, this phenomenon, called *copulatory lock*, prevents other males from mating, at least for a while (Dewsbury, 1972). Despite wild stories you may have heard or read, humans *never* experience copulatory lock; that urban myth started in 1884 when a physician submitted a fake report as a practical joke (Nation, 1973). For mammals and birds, postcopulatory behavior includes extensive parental behaviors to nurture the offspring, as we describe later in this chapter.

Copulation brings gametes together

All mammals, birds, and reptiles employ internal fertilization: the fusion of their **gametes**—**sperm** and **ovum**—within the female's body to form a **zygote**. Most of what we know about the copulatory behavior of mammals comes from studies of lab animals, especially rats. Like most other rodents, rats do not engage in lengthy courtship, nor do the partners tend to remain together after copulation. Rats are attracted to each other largely through odors. Female rats, like humans, are spontaneous ovulators; that is, even when left alone, they **ovulate** (release eggs from the ovary). For a few hours around the time of ovulation, the female rat seeks out a male and displays proceptive behaviors, including vocalizations at frequencies too high for humans to detect but audible to other rats.

These behaviors prompt the male to mount the female from the rear, grasp her flanks with his forelegs, and rhythmically thrust his hips against her rump. If she is receptive, the female adopts a stereotyped posture called **lordosis** (**FIGURE 8.17**), elevating her rump and moving her tail to one side to allow intromission. Once intromission has been achieved, the male rat makes a single deep thrust and then springs back off the female. During the next 6–7 minutes the male and female orchestrate seven to nine such intromissions; then, instead of springing away, the male raises the front half of his body up for a second or two while he ejaculates. Finally, he falls backward off the female.

After copulation, the male and female separately engage in grooming their own genitalia, and the male pays little attention to the female for the next 5 minutes or so, until, often in response to the female's proceptive behaviors, the two engage in another bout of intromissions and ejaculation. The cycle may repeat five or six times in one mating session.

Hormones play an important role in rat mating behaviors. Testosterone drives the male's interest in copulation: if he is castrated (his testes are removed), he will stop ejaculating within a few weeks and will eventually stop mounting receptive females. Although testosterone disappears from the bloodstream within a few hours after castration, the hormone's effects on the nervous system take days or weeks to dissipate. Treating a castrated male with testosterone eventually restores mating behavior; if

The raised rump and deflected tail of the female (the lordosis posture) make intromission possible in rats.

FIGURE 8.17 Copulation in Rats

testosterone treatment is stopped, the mating behavior fades away again. This is an example of a hormone exerting an **activational effect**: the hormone transiently promotes certain behaviors. In normal development, the rise of androgen secretion at puberty activates masculine behavior in males.

activational effect   A temporary change in behavior resulting from the administration of a hormone to an adult animal.

researchers at work

Gonadal steroids activate sexual behavior

Although individual male rats and guinea pigs differ considerably in how eagerly they will mate, blood levels of testosterone clearly are *not* responsible for these differences. For one thing, animals displaying different levels of sexual vigor do not show reliable differences in blood levels of testosterone. Furthermore, when these males are castrated and subsequently all treated with exactly the same doses of testosterone, their precastration differences in sexual activity persist, as we'll see next (**FIGURE 8.18**).

FIGURE 8.18   Androgens Permit Male Copulatory Behavior (After Grunt and Young, 1953.)

■ **Hypothesis**

Individual differences in the vigor with which male guinea pigs mate are caused by differences in testosterone secretion.

■ **Test**

Classify individual males by mating vigor, then castrate and provide them all with the same dose of testosterone.

■ **Result**

A few weeks after castration, all males stopped mating. But when provided the same dose of testosterone, males returned to their previous levels of mating vigor. Giving a higher dose of testosterone did not eliminate these differences.

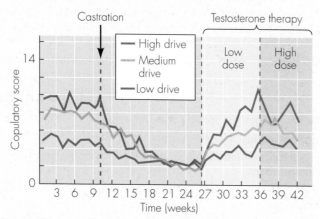

■ **Conclusion**

Although androgens—especially testosterone—are important for normal male sexual function, individual differences in sexual activity are not determined by differences in androgen levels.

Furthermore, it turns out that a very small amount of testosterone—one-tenth the amount normally produced by the animals—is enough to fully maintain the mating behavior of male rats (Damassa et al., 1977). Thus, since all male rats make more testosterone than is required to maintain their copulatory behavior, some other factor, which we can call *drive*, must differ across individual males.

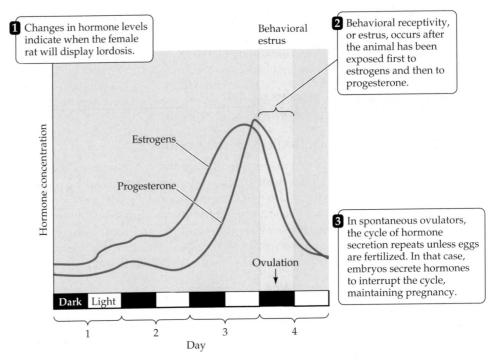

FIGURE 8.19 The Ovulatory Cycle of Rats

ventromedial hypothalamus (VMH)
A hypothalamic region involved in sexual behaviors, eating, and aggression.

Estrogens secreted at the beginning of the 4- to 5-day ovulatory cycle facilitate proceptive behavior in female rats, and the subsequent production of progesterone increases proceptive behavior and activates receptivity (**FIGURE 8.19**). An adult female whose ovaries have been removed will show neither proceptive nor receptive behaviors. However, 2 days of estrogen treatment followed by a single injection of progesterone will, about 6 hours later, make the female rat proceptive and receptive for a few hours. Only the correct combination of estrogens and progesterone will fully activate copulatory behaviors in female rats—another example of the activational effects of gonadal steroids.

Next we'll discuss how steroid hormones affect the brain to activate mating behavior.

how's it going ?

1. What are the four stages of reproductive behavior?
2. Describe a typical mating session in laboratory rats.
3. Describe the activational effects of gonadal steroids on mating behaviors in male and female rodents.
4. Are differences in testosterone secretion responsible for individual differences in male mating vigor?

The Neural Circuitry of the Brain Regulates Reproductive Behavior

Although most of what we know about the neural circuitry of sexual behavior comes from studies of rats, steroid receptors are found in the same specific brain regions across a wide variety of vertebrate species. Steroid-sensitive regions include the cortex, brainstem nuclei, medial amygdala, hippocampus, and many others. And, as we'll see, the hypothalamus plays a particularly important role in regulating copulatory behavior.

Estrogen and progesterone act on a lordosis circuit that spans from brain to muscle

Scientists have exploited the steroid sensitivity of the rat lordosis response to develop a map of the neural circuitry that controls this behavior. Using steroid autoradiography (see Box 8.1), investigators identified hypothalamic nuclei containing many estrogen- and progesterone-sensitive neurons. In particular, the **ventromedial hypothalamus (VMH)** is crucial for lordosis because lesions there abolish the response. Furthermore, tiny quantities of estradiol implanted directly into the brain can induce receptivity in females, but only when the hormone is placed in the VMH (Lisk, 1962).

One action of estrogen treatment is to cause dendrites of VMH neurons to grow and become more complex (Meisel and Luttrell, 1990). Another important action of estrogens is to stimulate the production of progesterone receptors so that the animal will become more responsive to that hormone. Activated progesterone receptors in turn help mediate the lordosis reflex (Mani et al., 2000).

The VMH sends axons to the **periaqueductal gray** region of the midbrain, where again, lesions greatly diminish lordosis. The periaqueductal gray neurons project to other brain regions and the spinal cord. In the spinal cord the sensory information provided by the mounting male will now evoke the motor response of lordosis when the female's estrogen and progesterone levels are right. Thus, the role of the VMH is to monitor steroid hormone concentrations and, at the right time in the ovulatory cycle, activate a neural circuit that allows a lordosis response to occur in response to a mounting male (Pfaff, 1997). **FIGURE 8.20B** schematically represents this neural pathway and its steroid-responsive components.

Androgens act on a neural system for male reproductive behavior

As with the lordosis circuit, mapping the sites of steroid action has provided important clues about the neural circuitry controlling male copulatory behavior (**FIGURE 8.20A**). The hypothalamic **medial preoptic area (mPOA)** is chock-full of steroid-sensitive neurons, and lesions of the mPOA abolish male copulatory behavior in a wide variety of vertebrate species (Meisel and Sachs, 1994). Note that lesions of the mPOA do not interfere with males' *motivation* for females; males will still press a bar to gain access to a receptive female (Everitt and Stacey, 1987), but they seem unable to commence mounting. Furthermore, mating can be reinstated in castrated males by small implants of testosterone in the mPOA, but not in other brain regions. Thus, the mPOA seems to provide "higher-order" control of male copulatory behaviors.

The mPOA coordinates copulatory behavior by sending axons to the ventral midbrain (which innervates several brain regions to coordinate mounting behaviors) and, via a multisynaptic pathway, to the spinal cord (Hamson and Watson, 2004), which mediates various genital reflexes, such as ejaculation. Brainstem projections of serotonergic fibers to the spinal cord normally hold the penile erection reflex in check (McKenna, 1999). Antidepressant drugs that boost serotonergic activity in the brain—for example, selective serotonin reuptake inhibitors like Prozac (see Chapter 12)—can produce side effects that include difficulty achieving erection, ejaculation, and/or orgasm, probably by enhancing serotonergic inhibition of the spinal cord.

Because the reflex circuits and ejaculation generator are in the lumbar spinal cord, men with damage at higher levels of the spinal cord often remain capable of copulation and ejaculation. (If you're wondering, the drug sildenafil, better known as Viagra, acts directly on tissue in the penis, not in the spinal cord or brain, to promote erection [Boolell et al., 1996].)

We can also learn about male copulatory mechanisms by tracing a sensory system that activates male arousal in rodents: the vomeronasal system. The **vomeronasal organ**, or **VNO** (see Chapter 9), consists of specialized receptor cells near to, but separate from, the olfactory epithelium. These sensory cells detect chemicals, called *pheromones* (see Figure 8.3C and Chapter 6) that are released by other individuals. Vomeronasal receptor neurons send their axons to the accessory olfactory bulb in the brain,

periaqueductal gray A midbrain region involved in pain perception.

medial preoptic area (mPOA) A region of the anterior hypothalamus implicated in the control of many behaviors, including sexual behavior, gonadotropin secretion, and thermoregulation.

vomeronasal organ (VNO) A collection of specialized receptor cells, near to but separate from the olfactory epithelium, that detect pheromones and send electrical signals to the accessory olfactory bulb in the brain.

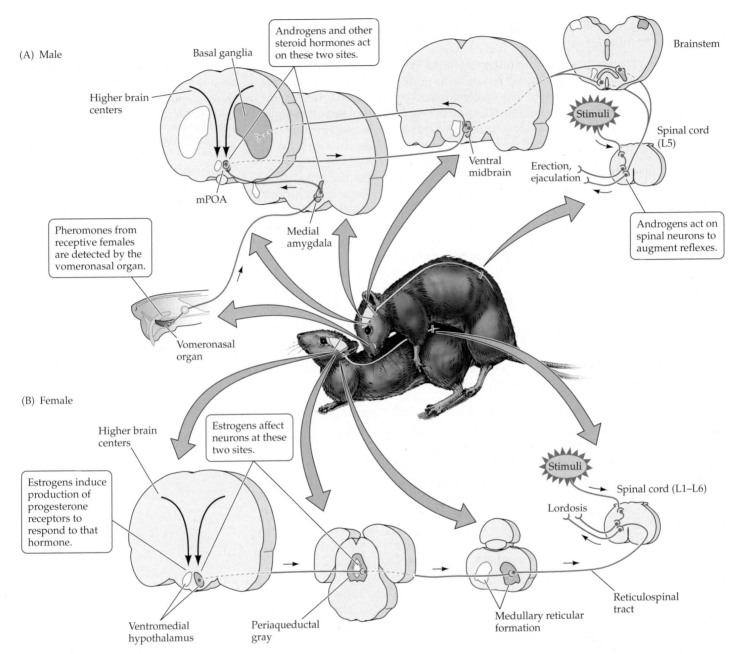

(A) Male

Higher brain centers

Basal ganglia

Androgens and other steroid hormones act on these two sites.

Brainstem

Ventral midbrain

Stimuli

Erection, ejaculation

Spinal cord (L5)

mPOA

Pheromones from receptive females are detected by the vomeronasal organ.

Medial amygdala

Androgens act on spinal neurons to augment reflexes.

Vomeronasal organ

(B) Female

Higher brain centers

Estrogens affect neurons at these two sites.

Estrogens induce production of progesterone receptors to respond to that hormone.

Stimuli

Spinal cord (L1–L6)

Lordosis

Ventromedial hypothalamus

Periaqueductal gray

Medullary reticular formation

Reticulospinal tract

FIGURE 8.20 Neural Circuits for Reproduction in Rodents (After Pfaff, 1980.)

medial amygdala A portion of the amygdala that receives olfactory and pheromonal information.

parental behavior Behavior of adult animals that has the goal of enhancing the well-being of their own offspring, often at some cost to the parents.

parabiotic Referring to a surgical preparation that joins two animals to share a single blood supply.

which then projects elsewhere in the brain to govern sexual behavior. For example, receptive female rats release pheromones that male rats find arousing, as evidenced by penile erections. (You can learn more about the role of pheromones in reproductive behavior in **A STEP FURTHER 8.4**, on the website.) Humans also respond to pheromones, but we detect them with our olfactory system, since the vomeronasal organ is absent in our species.

In rats, an important target for vomeronasal information is the **medial amygdala**, which depends on adult circulating levels of sex steroids to maintain a masculine form and function (Cooke et al., 2003). Lesions here will abolish the penile erections that normally occur around receptive females (Kondo et al., 1997). The medial amygdala, in turn, sends axons to the mPOA. So the mPOA appears to integrate hormonal and sensory information, such as pheromones, and to coordinate the motor patterns of copulation. Figure 8.20A summarizes the neural circuitry for male rat copulatory behavior.

We will see later that testosterone activates sexual arousal in humans too.

Parental behaviors are governed by several sex-related hormones

In many vertebrate species, copulation is not enough to ensure reproduction. Many young vertebrates, and all newborn mammals, need parental attention to survive. Earlier we discussed the milk letdown reflex, when the infant's suckling on the nipple triggers the secretion of oxytocin to promote the release of milk (see Figure 8.9). In rats, the pregnant female prepares for her pups by licking all of her nipples. Doing so probably helps clean the nipples before the pups arrive, but it also makes them more sensitive to touch. This self-grooming actually expands the amount of sensory cortex that responds to skin surrounding the nipples (Xerri et al., 1994), setting the stage for the letdown reflex. This is a wonderful example of an animal's behavior altering its own brain and therefore changing its future behavior.

Rat mothers (called *dams*) show four easily measured **parental behaviors**: nest building, crouching over pups, retrieving pups, and nursing (**FIGURE 8.21**). Neither virgin female rats nor male rats normally show these behaviors toward rat pups. In fact, a virgin female finds the smell of newborn pups aversive. Information about the odor from pups projects via the olfactory bulb to the medial amygdala and on to the VMH. Lesions anywhere along that path will cause a virgin rat to show maternal behavior right away (Numan and Numan, 1991), because she will no longer detect the smell.

However, if a virgin female is exposed to newborn pups a few hours a day for several days in a row, she (or almost any adult rat—male or female) will start building a nest, crouching over pups, and retrieving them. As the rat gradually habituates to the smell of the pups, it starts taking care of them. But the rat dam that gives birth to her first litter will *instantly* show these behaviors. It turns out that the rather complicated pattern of hormones during pregnancy shapes her brain to display maternal behaviors *before* she is exposed to the pups.

The effect of pregnancy hormones on a rat's maternal behaviors is demonstrated by a **parabiotic** preparation in which two female rats are surgically joined, sharing a single blood supply, such that each is exposed to any hormones secreted by the other (**FIGURE 8.22**). If one of those females is pregnant, then at the end of her pregnancy the other female, who was never pregnant but was exposed to the pregnant rat's hormones, will also immediately show maternal behavior (Terkel and Rosenblatt, 1972). Which hormone is responsible for promoting maternal behavior? No single hormone alone can do it; the combination of several hormones, including estrogens, progesterone, oxytocin, and prolactin, is required.

Rat dams clean their pups…

…crouch over them to allow them to nurse…

…and will retrieve them if they stray from the nest.

FIGURE 8.21 Parental Behavior in Rats

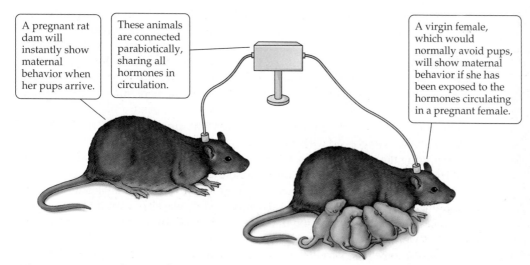

A pregnant rat dam will instantly show maternal behavior when her pups arrive.

These animals are connected parabiotically, sharing all hormones in circulation.

A virgin female, which would normally avoid pups, will show maternal behavior if she has been exposed to the hormones circulating in a pregnant female.

FIGURE 8.22 Parabiotic Exchange Facilitates Maternal Behavior (After Terkel and Rosenblatt, 1972.)

orgasm The climax of sexual behavior, marked by extremely pleasurable sensations.

phallus The clitoris or penis.

penis The male phallus.

clitoris The female phallus.

There is a growing realization that parenting behaviors shape the brains of male and female rodents (Kinsley and Lambert, 2006; Franssen et al., 2011). There is ample evidence that the hormones of pregnancy also prepare human mothers to nurture their newborns (Fleming et al., 2002). Compared with other women, mothers who have recently given birth are better at distinguishing odors from different newborns, and they can even recognize odors from their own newborn. This increased ability to discriminate odors correlates with hormone levels after delivery (Fleming et al., 1997). Pregnancy also has long-lasting effects on cognitive processing in women (Vanston and Watson, 2005; Henry and Sherwin, 2012).

The Hallmark of Human Sexual Behavior Is Diversity

How much of what we've described so far about sexual behavior in animals is relevant to human sexuality? Until the 1940s, when biology professor Alfred Kinsey began to ask friends and colleagues about their sexual histories, there was virtually no scientific study of human sexual behavior. Kinsey constructed a standardized set of questions and procedures to get information. Eventually, he and his collaborators published extensive surveys (based on tens of thousands of respondents) on the sexual behavior of American males (Kinsey et al., 1948) and females (Kinsey et al., 1953).

Controversial in their time, these surveys indicated that nearly all men masturbated, that college-educated people were more likely to engage in oral sex than were non-college-educated people, that many people had at one time or another engaged in homosexual behaviors, and that a stable proportion of the population preferred homosexual sex over heterosexual sex.

Another way to investigate human sexual behavior is to make behavioral and physiological observations of people engaged in sexual intercourse or masturbation, but the squeamishness of the general public impeded such research for many years. Finally, after Kinsey's surveys were published, physician William Masters and psychologist Virginia Johnson began a large, famous project of this kind (Masters and Johnson, 1970; Masters et al., 1994), documenting the impressively diverse sexuality of humans.

Among most mammalian species, including most nonhuman primates, the male mounts the female from the rear; but among humans, face-to-face postures are most common. A great variety of coital positions have been described, and many couples vary their positions from session to session or even within a session. It is this variety in reproductive behaviors, rather than differences in reproductive anatomy, that distinguishes human sexuality from that of most other species.

Another difference between species is that, unlike other animals, humans can report their subjective reactions to sexual behavior—specifically **orgasm**, the brief, extremely pleasurable sensations experienced by most men during ejaculation and by most women during copulation. In the original conceptual model of human sexuality, Masters and Johnson (1966) summarized the typical response patterns of both men and women as consisting of four phases: increasing excitement, plateau, orgasm, and resolution (**FIGURE 8.23**).

During the excitement phase, the **phallus** (the **penis** in men, the **clitoris** in women) becomes engorged with blood, making it erect. In women, parasympathetic activity during the excitement phase causes changes in vaginal blood vessels, producing lubricating fluids that facilitate intromission. Stimulation of the penis, clitoris, and vagina during rhythmic thrusting accompanying intromission may lead to orgasm. In both men and women, orgasm is accompanied by waves of contractions of genital muscles (mediating ejaculation in men and contractions of the uterus and vaginal opening in women).

In spite of some basic similarities, the sexual responses of men and women differ in important ways. For one thing, women show a much greater variety of commonly observed copulatory sequences. Whereas men have only one basic pattern, captured

Masters of Sex William H. Masters and Virginia E. Johnson.

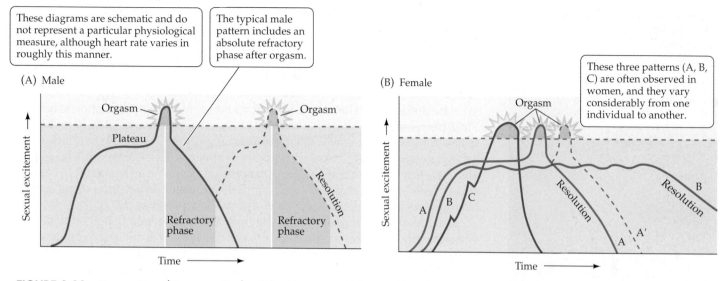

These diagrams are schematic and do not represent a particular physiological measure, although heart rate varies in roughly this manner.

The typical male pattern includes an absolute refractory phase after orgasm.

These three patterns (A, B, C) are often observed in women, and they vary considerably from one individual to another.

FIGURE 8.23 Human Sexual Response Cycles (After Masters and Johnson, 1966.)

by the linear model of Masters and Johnson (see Figure 8.23A), women have at least three typical patterns (see Figure 8.23B). Another important aspect of human sexuality is that most men, but not most women, have an absolute refractory phase following orgasm (also illustrated in Figure 8.23A). That is, most men cannot achieve full erection and another orgasm until some time has elapsed—the length of time varying from minutes to hours, depending on individual differences and other factors. Many women, on the other hand, can have multiple orgasms in rapid succession. Functional imaging of the brains of men and women during sexual activity suggests that, although the brain circuitry associated with orgasm itself is quite similar between the sexes, substantially different networks are active in men's and women's brains during sexual activity *prior* to orgasm (Georgiadis et al., 2009).

Taking a broader perspective on sexuality reveals additional distinctions between men and women (Peplau, 2003). Basic sex drive seems to be greater in men, reflected in more frequent masturbation, sexual fantasies, and pursuit of sexual contacts. Emotional components and cognitive factors play a stronger role in women's sex lives than in men's. In addition to having a somewhat flexible sexuality that adapts to new experiences and situations over time, women place more emphasis on sexual intimacy within the context of committed relationships. These observations and others have led sex researchers to adopt a more nuanced view of female sexuality. While Masters and Johnson simply adapted the linear model of male sexuality to women, the more modern perspective views women's sexuality as a cycle, governed in large measure by emotional factors (Basson, 2001, 2008). According to this model, emotional intimacy and desire (more than physiological arousal) are crucial in the initiation of sexual responses, and following a sexual encounter, a combination of both emotional and physical satisfaction affects the likelihood of subsequent sexual activity.

Although male and female sexuality may bear the imprint of our evolutionary history, on an individual basis it is also shaped by sociocultural pressures and experience. Sexual therapy, for example, usually consists of helping the person to relax, to recognize the sensations associated with coitus, and to learn the behaviors that produce the desired effects in both partners. Masturbation during adolescence, rather than being harmful as suggested in previous times, may help avoid sexual problems in adulthood. As with other behaviors, practice, practice, practice helps.

Sexual behavior may also aid overall health; epidemiological studies indicate that men who have frequent sex tend to live longer than men who do not (Davey-Smith et al., 1997), and sexual experiences induce adult neurogenesis in the hippocampus in

Testosterone Patches These patches can revive libido in men who have lost their testes through accident or disease.

lab animals, suggesting that sexual behaviors may be directly beneficial to the brain (Leuner et al., 2010). Of course, it is also important to one's health to take precautions (such as using condoms) to avoid contracting sexually transmitted infections.

Hormones play only a permissive role in human sexual behavior

We have already seen that a little bit of testosterone must be in circulation to activate male-typical mating behavior in rodents. The same relation seems to hold for human males. For example, boys who fail to produce testosterone at puberty show little interest in dating unless they receive androgen treatments. These males, as well as men who have lost their testes as a result of cancer or accident, made it possible to conduct experiments showing that testosterone indeed stimulates sexual interest and activity in men, as well as a sense of heightened energy (Davidson et al., 1979).

Recall that, in rats, *additional* testosterone has no effect on the vigor of mating. Consequently, there is no correlation between the amount of androgens produced by an individual male rat and his tendency to copulate. In humans, too, just a little testosterone is sufficient to restore behavior fully, and there is no correlation between systemic androgen levels and sexual activity among men who have at least *some* androgen.

Some women experience sexual dysfunction after menopause, reporting decreased sexual desire and difficulty achieving comfortable coitus. There are many possible reasons for such a change, including several hormonal changes. Providing postmenopausal women with low doses of both estrogens and androgens can have beneficial effects on the genital experience of sex and also on women's sexual interest (Sherwin, 1998, 2002; Basson, 2008).

There have been several attempts to determine whether women's interest or participation in sexual behavior varies with the menstrual cycle. Some researchers have found a slight increase in sexual behavior around the time of ovulation, but the effect is small, and several studies have failed to see any significant change in interest in sex across the menstrual cycle.

Now that we've discussed some of the unique aspects of male and female sexuality, let's look at some of the developmental processes that cause the growing individual to take on male-typical or female-typical forms and functions in the first place.

how's it going

1. What brain regions appear to be involved in male and female sexual behaviors?
2. Compare and contrast the maternal behavior of a female rat that has just been pregnant with that of a naive female. What factor is responsible for the differences?
3. Compare and contrast the patterns of sexual arousal in men and women.
4. What are the effects of gonadal steroids on human sexual behavior?

PART III
Sexual Differentiation and Orientation

For species such as our own, in which the only kind of reproduction is sexual reproduction (so far), each individual must become either a male or a female to reproduce. **Sexual differentiation** is the process by which individuals develop either male or female bodies and behaviors. In mammals this process begins before birth and continues into adulthood. As we'll see, some people may be very malelike (masculine) in some parts of the body and very femalelike (feminine) in others; such people can't be said to be either male or female but may be a blend of the two sexes.

sexual differentiation The process by which individuals develop either male-like or femalelike bodies and behavior.

sex determination The process that normally establishes whether a fetus will develop as a male or a female.

indifferent gonads The undifferentiated gonads of the early mammalian fetus, which will eventually develop into either testes or ovaries.

Genetic and Hormonal Mechanisms Guide the Development of Masculine and Feminine Structures

In mammals, every egg carries an X chromosome from the mother; fusion with an X- or Y-bearing sperm is the key event in **sex determination**, the developmentally early event that establishes the course of subsequent sexual differentiation of the body. With only occasional exceptions, mammals that receive an X chromosome from the father will become females with an XX sex chromosome complement; those that receive the father's Y chromosome will become XY males.

In vertebrates, the first major consequence of sexual determination is in the gonads. Very early in development, each individual has a pair of **indifferent gonads**, glands that vaguely resemble both testes and ovaries. During the first month of gestation in humans, differential genetic instructions determine whether the indifferent gonads begin changing into ovaries or testes.

Sex chromosomes direct sexual differentiation of the gonads

In mammals, the Y chromosome contains the **SRY gene** (for *s*ex-determining *r*egion on the *Y* chromosome), which is responsible for the development of testes. If an individual has a Y chromosome, the cells of the indifferent gonad begin making the Sry protein, which induces the indifferent gonad to develop into a testis.

In XX individuals (or XY individuals with a dysfunctional *SRY* gene), no Sry protein is produced, and the indifferent gonad becomes an ovary. This early event of forming either testes or ovaries has a domino effect, setting off a chain of actions that usually results in either a male or a female.

Gonadal hormones direct sexual differentiation of the body

For all mammals, including humans, the gonads secrete hormones to direct sexual differentiation of the body. Fetal ovaries produce very little hormone, but fetal testes produce several hormones. If other embryonic cells are exposed to the testicular hormones, they begin developing masculine characters; if the cells are not exposed to testicular hormones, they develop feminine characters.

We can chart masculine or feminine development by examining the structures that connect the gonads to the outside of the body: these are quite different in adult males and females, but at the embryonic stage all individuals have the precursor tissues of both systems. The early fetus has a **genital tubercle** (a "bump" between the legs) that can form either a clitoris or a penis, as well as two sets of ducts that connect the indifferent gonads to the outer body wall: the **wolffian ducts** and the **müllerian ducts** (**FIGURE 8.24A**). In females, the müllerian ducts develop into the oviducts (or *fallopian tubes*), uterus, and inner vagina (**FIGURES 8.24B,C**, right), and only a remnant of the wolffian ducts remains. In males, hormones secreted by the testes orchestrate the converse outcome: each wolffian duct develops into an epididymis, vas deferens, and seminal vesicle (see Figures 8.24B,C, left), while the müllerian ducts shrink to mere remnants.

The system is masculinized by two testicular secretions: testosterone, which promotes development of the wolffian system; and **anti-müllerian hormone** (**AMH**), which causes regression of the müllerian system. In the absence of testosterone and AMH, the genital tract develops in a feminine pattern, in which the wolffian ducts regress and the müllerian ducts develop into components of the female internal reproductive tract.

Testosterone also masculinizes other structures, acting on the fetal genitalia to form a scrotum and penis. These effects are aided by the local conversion of testosterone into a more potent androgen, **dihydrotestosterone** (**DHT**), promoted by an enzyme that is found in the genital skin, **5-alpha-reductase**. We'll see later that without the local production of DHT, testosterone alone is able to masculinize the genitalia only partially. If androgens are absent altogether, the genital tissues grow into the female labia and clitoris.

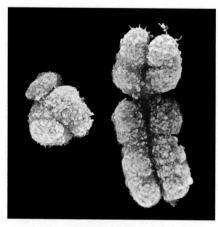

Mammalian Sex Chromosomes The Y chromosome (left) is much shorter than the X (right) because it carries far fewer genes. But one of those genes, *SRY*, is critical for masculine development.

SRY gene   A gene on the Y chromosome that directs the developing gonads to become testes. The name *SRY* stands for *s*ex-determining *r*egion on the *Y* chromosome.

genital tubercle   In the early fetus, a "bump" between the legs that can develop into either a clitoris or a penis.

wolffian duct   A duct system in the embryo that will develop into male reproductive structures (epididymis, vas deferens, and seminal vesicle) if androgens are present.

müllerian duct   A duct system in the embryo that will develop into female reproductive structures (oviducts, uterus, and upper vagina) if androgens are not present.

anti-müllerian hormone (AMH)   Also called *müllerian regression hormone.* A peptide hormone secreted by the fetal testes that inhibits müllerian duct development.

dihydrotestosterone (DHT)   The 5-alpha-reduced metabolite of testosterone. DHT is a potent androgen that is principally responsible for the masculinization of the external genitalia in mammals.

5-alpha-reductase   An enzyme that converts testosterone into dihydrotestosterone.

(A) 6 weeks of gestation (undifferentiated fetus)

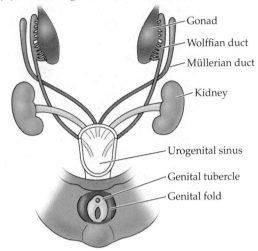

- Gonad
- Wolffian duct
- Müllerian duct
- Kidney
- Urogenital sinus
- Genital tubercle
- Genital fold

FIGURE 8.24 Sexual Differentiation in Humans

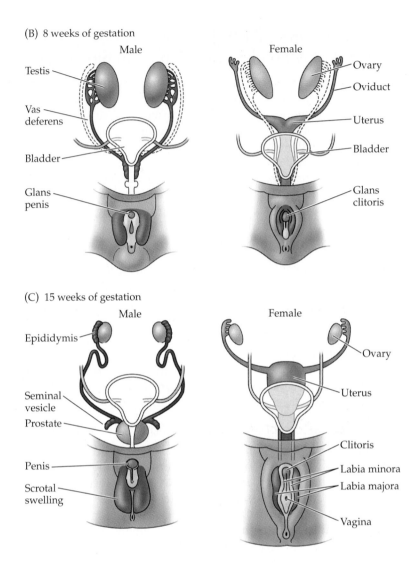

(B) 8 weeks of gestation

Male
- Testis
- Vas deferens
- Bladder
- Glans penis

Female
- Ovary
- Oviduct
- Uterus
- Bladder
- Glans clitoris

(C) 15 weeks of gestation

Male
- Epididymis
- Seminal vesicle
- Prostate
- Penis
- Scrotal swelling

Female
- Ovary
- Uterus
- Clitoris
- Labia minora
- Labia majora
- Vagina

Changes in sexual differentiation processes result in predictable changes in development

Some people have only one sex chromosome: a single X (embryos containing only a single Y chromosome do not survive). This genetic makeup results in **Turner's syndrome**, in which an apparent female has underdeveloped but recognizable ovaries, as you might expect because no *SRY* gene is available. In general, unless the indifferent gonad becomes a testis and begins secreting hormones, immature mammals develop as females in most respects, including women with Turner's syndrome. So the sex chromosomes determine the sex of the gonad, and gonadal hormones then drive sexual differentiation of the rest of the body (**FIGURE 8.25**).

Congenital adrenal hyperplasia (**CAH**) causes developing girls to be exposed to excess androgens before birth. In CAH, the adrenal glands produce considerable amounts of androgens, somewhere between those of normal females and males, so the newborn often has an **intersex** appearance: a phallus that is intermediate in size between a normal clitoris and a normal penis, and skin folds that resemble both labia and scrotum (**FIGURE 8.26**). Such individuals are usually recognizable at birth because, even in severe cases in which penis and scrotum appear well formed, no testes are present inside the "scrotum"; instead, these individuals have normal abdominal ovaries, as you would expect. Once born, CAH children are given medicine

Turner's syndrome A condition, seen in individuals carrying a single X chromosome but no other sex chromosome, in which an apparent female has underdeveloped but recognizable ovaries.

congenital adrenal hyperplasia (CAH) Any of several genetic mutations that can cause a female fetus to be exposed to adrenal androgens, resulting in partial masculinization at birth.

intersex Referring to an individual with atypical genital development and sexual differentiation that generally resembles a form intermediate between typical male and typical female genitalia.

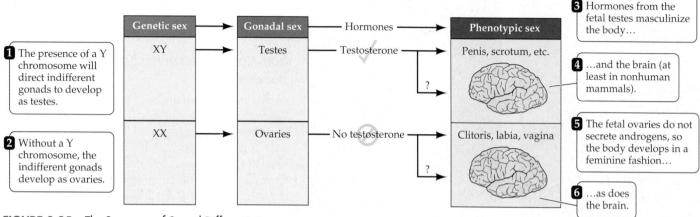

FIGURE 8.25 The Sequence of Sexual Differentiation

to prevent further androgen production. There is controversy, however, over whether the best course of action for the parents of CAH girls is to opt for immediate surgical modification of the genitalia or to wait until adulthood, when the CAH-affected individuals can decide for themselves whether to have surgery that is purely cosmetic.

In spotted hyenas, females are always exposed to prenatal androgens, resulting in highly masculinized genitalia. You can read about these fascinating animals in **A STEP FURTHER 8.5**, on the website.

Dysfunctional androgen receptors can block the masculinization of males

The importance of androgens for masculine sexual differentiation is illustrated by the condition known as **androgen insensitivity syndrome** (**AIS**). AIS results when an XY zygote inherits a dysfunctional gene for the androgen receptor, so the embryo's tissues cannot respond to androgenic hormones like testosterone. The gonads of people with AIS develop as normal testes (as directed by Sry), and the testes produce AMH (which inhibits müllerian duct structures) and plenty of testosterone.

In the absence of working androgen receptors, however, the wolffian ducts fail to develop and the external genital tissue forms labia and a clitoris. At puberty, AIS women develop breasts but fail to start menstruating, because neither ovaries nor

androgen insensitivity syndrome (AIS) A syndrome caused by a mutation of the androgen receptor gene that renders tissues insensitive to androgenic hormones like testosterone. Affected XY individuals are phenotypic females, but they have internal testes and regressed internal genital structures.

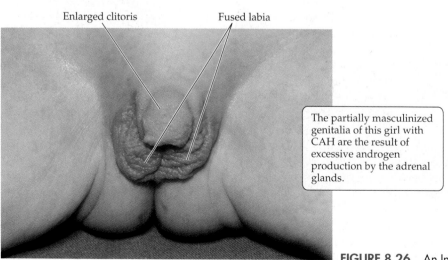

Enlarged clitoris   Fused labia

The partially masculinized genitalia of this girl with CAH are the result of excessive androgen production by the adrenal glands.

FIGURE 8.26 An Intersex Phenotype

Although these women have Y chromosomes and were born with testes, because they are insensitive to androgen their bodies developed in a feminine fashion. Their behavior is feminine too.

FIGURE 8.27 **Androgen-Insensitive Women** (Photos courtesy of Jane Goto and Cindy Stone.)

guevedoces Literally "eggs at 12" (in Spanish). A nickname for individuals who are raised as girls but at puberty change appearance and begin behaving as boys.

uterus are present. Women with AIS are infertile, but otherwise they look like other women (**FIGURE 8.27**) and behave like other women.

Some people seem to change sex at puberty

Babies are occasionally born with a rare genetic mutation that disables the enzyme (5-alpha-reductase) that converts testosterone to DHT. An XY individual with this condition will develop testes and normal male internal reproductive structures (because testosterone and AMH function normally), but the external genitalia will fail to masculinize fully (**FIGURE 8.28**). The reason for this failure is that the genital epithelium, which normally possesses 5-alpha-reductase, is unable to amplify the androgenic signal by converting testosterone to the more active DHT. Consequently, the phallus is only slightly masculinized and resembles a large clitoris, and the genital folds resemble labia, although they contain the testes. Usually there is no vaginal opening.

A particular village in the Dominican Republic is home to several families that carry the mutation causing 5-alpha-reductase deficiency. Children born with this appearance seem to be regarded as girls in the way they are dressed and raised (Imperato-McGinley et al., 1974). At puberty, however, the testes increase androgen production, and the external genitalia become more fully masculinized. The phallus grows into a small but recognizable penis; the body develops narrow hips and a muscular build, without breasts; and the individuals begin acting like young men. The villagers have nicknamed such individuals **guevedoces**, meaning "eggs (testes) at 12 (years)." These men never develop beards, but they usually have girlfriends, indicating that they are sexually interested in women. We will discuss the sexual behavior of *guevedoces*, as well as women with CAH or AIS, later.

(A) Newborn

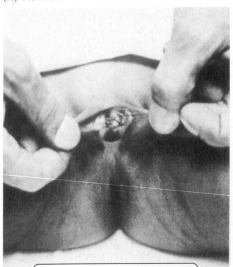

In the Dominican Republic, some individuals, called *guevedoces*, are born with ambiguous genitalia and are raised as girls.

(B) Adolescent

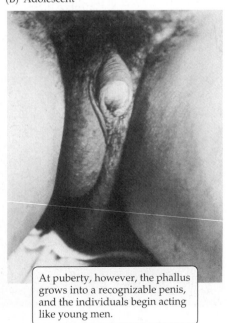

At puberty, however, the phallus grows into a recognizable penis, and the individuals begin acting like young men.

FIGURE 8.28 **Guevedoces** (Courtesy of Julianne Imperato-McGinley.)

How should we define gender—by genes, gonads, genitals?

Most humans are either male or female, and whether we examine their chromosomes, gonads, external genitalia, or internal structures, we see a consistent pattern: each one is either feminine or masculine in character. But as the various syndromes we've been discussing demonstrate, different physical features in a single person can be either masculine or feminine, so from a scientific perspective, legal efforts to categorize all people as either male or female are doomed to failure. Androgen-insensitive women have male XY sex chromosomes and internal testes, and like most males they do not have oviducts or a uterus. But they do have a vagina and breasts, and in most respects their behavior is typical of females: they dress like females, they are attracted to and marry males, and perhaps most important, even after they learn the details of their condition, they strongly identify themselves as women (Money and Ehrhardt, 1972). They are males in some respects, but females in others. If laws define marriage as only between a man and a woman, whom should these individuals, carrying a Y chromosome and born with testes, be allowed to marry? How about XY individuals with 5-alpha-reductase deficiency, born and raised as girls—with a birth certificate to prove it? Should they be restricted to marrying men, even if, at 12 years of age, they sprout a penis?

This recognition that a single individual may be masculine in some regards and feminine in others is especially important as we consider hormonal effects on the brain in the next section.

organizational effect   A permanent alteration of the nervous system, and thus permanent change in behavior, resulting from the action of a steroid hormone on an animal early in its development.

sensitive period   The period during development in which an organism can be permanently altered by a particular experience or treatment.

neonatal   Referring to newborns.

how's it going ❓

1. Describe the process of fetal sexual differentiation in males and females, especially the role of hormones.

2. Does it make sense to you that a single signal, such as androgens, would masculinize the entire body?

3. What are some of the syndromes that can affect sexual differentiation, and what do they tell us about the problems of defining a person's gender?

To see the animation
Organizational Effects of Testosterone,
go to

2e.mindsmachine.com/8.6

researchers at work

Gonadal hormones direct sexual differentiation of behavior and the brain

As scientists began discovering that testicular hormones direct masculine development of the fetal body, behavioral researchers found evidence for a similar influence on the fetal brain. A female guinea pig, like most other rodents, normally displays the lordosis posture in response to male mounting (see Figure 8.17) for only a short period around the time of ovulation, when her fertility is highest. If a male mounts her at other times, she does not show lordosis. Experimenters can induce female rodents to display lordosis by injecting ovarian steroids in the sequence they normally follow during ovulation—giving the animals estrogens for a few days and then progesterone (see Figure 8.19). A few hours after the progesterone injection, the female will display lordosis in response to male mounting.

Phoenix et al. (1959) exposed female guinea pigs to testosterone in utero. As adults, these females did *not* show lordosis. Even if their ovaries were removed and they were given the steroid regimen that reliably activated lordosis in normal females, these fetally androgenized females did not show lordosis. From these data the researchers inferred that the same testicular steroids that masculinize the genitalia during early development also masculinize the developing brain. In other words, they proposed that the brain was just one more target tissue that is masculinized by androgens acting early in life (see Figure 8.25). This type of lasting change due to steroid exposure is known as an **organizational effect**.

A steroid has an organizational effect only when present during a specific **sensitive period**, generally in early development. Unlike the transient nature that characterizes the activational effects of hormones, which we discussed earlier, the organizational effects of hormones tend to be permanent (**FIGURE 8.29A–C**). The exact boundaries of the sensitive period of development depend on which behavior and which species are being studied. For rats, androgens given just after birth (the **neonatal** period) can affect later behavior. Guinea pigs, however, must be exposed to androgens *before* birth

continued

for adult lordosis behavior to be affected. In mammals, puberty can be viewed as a second sensitive period; for example, steroid exposure during puberty causes the addition of new cells (an organizational effect) to sex-related brain regions of rats (Ahmed et al., 2008).

Early testicular secretions result in masculine behavior in adulthood

What has come to be called the *organizational hypothesis* provides a unitary explanation for sexual differentiation: that a single steroid signal (androgen) diffuses through all tissues, masculinizing the body, the brain, and behavior (see Figure 8.25). From this point of view the nervous system is just another type of tissue listening for the androgenic signal that will instruct it to organize itself in a masculine fashion. If the nervous system does not detect androgens, it will organize itself in a mostly feminine fashion.

What was demonstrated originally for the lordosis behavior of guinea pigs has been observed in a variety of vertebrate species and for many behaviors. Exposing female rat pups to testosterone either just before birth or during the first 10 days after birth greatly reduces their lordosis responsiveness as adults. This explains the observation that adult male rats show very little lordosis, even when given estrogens and progesterone. However, male

FIGURE 8.29 Organizational Effects of Testosterone on Rodent Behavior (After Phoenix et al., 1959.)

■ **Hypothesis**

Early in life, androgens organize the brain, and therefore adult behavior, in a masculine fashion.

■ **Test**

Manipulate androgen exposure in genetic male and female rodents early in life, then ask whether hormones in adulthood can elicit male and/or female behaviors.

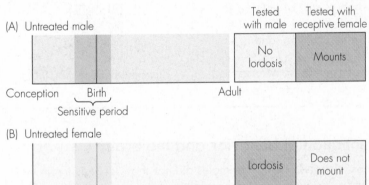

(A) Untreated male

Tested with male / Tested with receptive female

No lordosis | Mounts

Conception Birth Adult
Sensitive period

(B) Untreated female

Lordosis | Does not mount

(C) Treated female

Treated with testosterone during the sensitive period... ...and treated with steroids again in adulthood.

Testosterone

No lordosis | Mounts

(D) Castrated male

Castrated (i.e., deprived of testosterone) at birth.

Estradiol

Lordosis | Does not mount

Conception Birth Adult
Sensitive period

■ **Results**

• Normally male rodents eagerly mount receptive females but do not show lordosis behavior in response to another male, even when given ovarian steroids.

• Conversely, female rodents show little masculine sexual behavior, but they display lordosis when treated with an estrogen and progesterone in adulthood.

• However, female rodents exposed to testosterone early in life show little or no lordosis, even when given steroids that activate that behavior in normal females. When given testosterone, these androgenized females eagerly mount other females.

• Conversely, males deprived of androgens early in life are demasculinized as adults, showing little copulatory behavior, and are also feminized, displaying lordosis when given the proper steroids to activate that behavior.

■ **Conclusion**

When the developing brain is exposed to androgens, the animal's brain is organized in a masculine fashion, so that, as an adult, it is more likely to show malelike behaviors, and less likely to show femalelike behaviors.

rats that are castrated during the first week of life read-ily display lordosis responses in adulthood if injected with estrogens and progesterone (**FIGURE 8.29D**). In rats, many behaviors conform to the organizational hypoth-esis: animals exposed to androgens early in life behave like males, whereas animals not exposed to androgens early in life behave like females.

In most cases, full masculine behavior requires andro-gens both during development (to organize the nervous system to enable the later behavior) and in adulthood (to activate that behavior). Only animals exposed to androgen both in development and in adulthood show fully masculine behavior.

Several regions of the nervous system display prominent sexual dimorphism

The fact that male and female rats behave differently means that their brains must be different in some way, and according to the organizational hypothesis this differ-ence results primarily from androgenic masculinization of the developing brain. The exact form of these neural sex differences can be very subtle; the same basic circuit of neurons will produce very different behavior if the pattern of synapses varies. Sex differences in the number of synapses were identified in the preoptic area of the hy-pothalamus (Raisman and Field, 1971). But scientists soon found that there are much more obvious sex differences in the brain, including differences in the number, size, and shape of neurons. Darwin coined a term, **sexual dimorphism**, to describe the condition in which males and females show pronounced sex differences in struc-ture. In all species studied so far, androgens are responsible for the sexual dimor-phism seen in the brain: androgens masculinize the brain region, and the absence of androgens leads to a female-typical brain anatomy. We'll discuss two well-studied models.

THE PREOPTIC AREA OF RATS Roger Gorski et al. (1978) examined the preoptic area (POA) of the hypothalamus in rats because of the earlier reports that the number of synapses in this region was different in males and females and because lesions of the POA disrupt ovulatory cycles in female rats and, as we mentioned earlier, reduce copulatory behavior in males. Sure enough, the investigators found a nucleus within the POA that has a much larger volume in males than in females.

This nucleus, dubbed the **sexually dimorphic nucleus of the POA (SDN-POA)**, is much more evident in male rats than in females (**FIGURE 8.30**). The SDN-POA

sexual dimorphism The condition in which males and females of the same species show pronounced sex differ-ences in appearance.

sexually dimorphic nucleus of the preoptic area (SDN-POA) A region of the preoptic area that is 5 to 6 times larger in volume in male than in female rats.

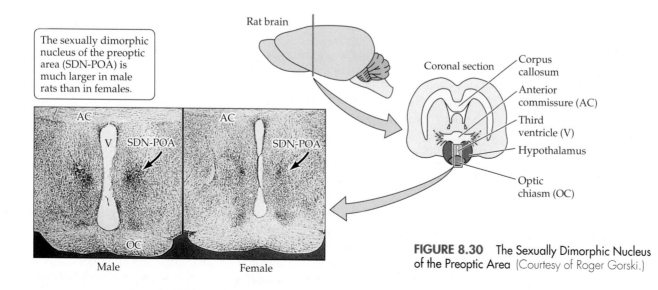

The sexually dimorphic nucleus of the preoptic area (SDN-POA) is much larger in male rats than in females.

Rat brain

Coronal section

Corpus callosum

Anterior commissure (AC)

Third ventricle (V)

Hypothalamus

Optic chiasm (OC)

AC

V SDN-POA

OC

Male

AC

SDN-POA

Female

FIGURE 8.30 The Sexually Dimorphic Nucleus of the Preoptic Area (Courtesy of Roger Gorski.)

FIGURE 8.31 Organization of the SDN-POA

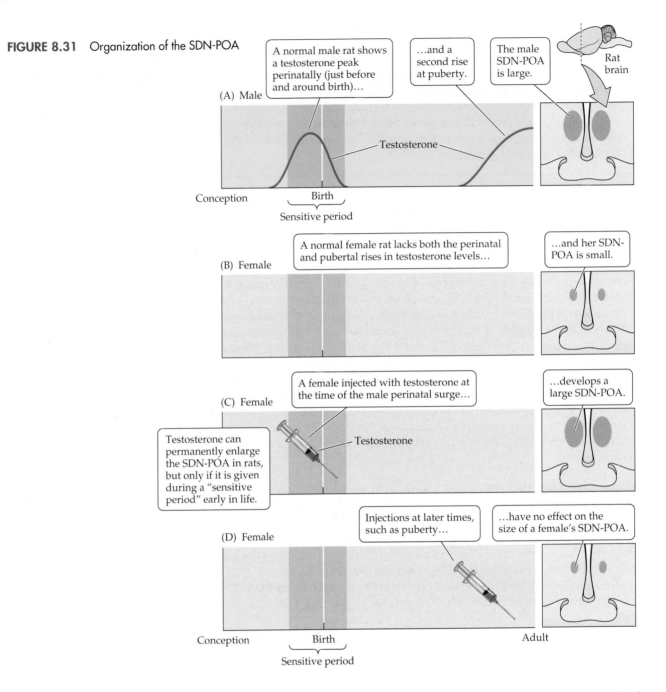

(A) Male

A normal male rat shows a testosterone peak perinatally (just before and around birth)…

…and a second rise at puberty.

The male SDN-POA is large.

Rat brain

Testosterone

Conception — Birth

Sensitive period

(B) Female

A normal female rat lacks both the perinatal and pubertal rises in testosterone levels…

…and her SDN-POA is small.

(C) Female

A female injected with testosterone at the time of the male perinatal surge…

…develops a large SDN-POA.

Testosterone can permanently enlarge the SDN-POA in rats, but only if it is given during a "sensitive period" early in life.

Testosterone

(D) Female

Injections at later times, such as puberty…

…have no effect on the size of a female's SDN-POA.

Conception — Birth — Adult

Sensitive period

conformed beautifully to the organizational hypothesis: males castrated at birth had much smaller SDN-POAs in adulthood, while females androgenized at birth had large, malelike SDN-POAs as adults. Castrating male rats in adulthood, however, did not alter the size of the SDN-POA. Thus, testicular androgens somehow alter the development of the SDN-POA, resulting in a nucleus permanently larger in males than in females (**FIGURE 8.31**).

One quirk of sexual dimorphism in the brains of some lab species, including rodents, is that testosterone reaching the brain is converted into estrogens that act on estrogen receptors, not androgen receptors, to masculinize the SDN-POA and some other brain regions. For example, XY rats that are androgen-insensitive (like the people with AIS discussed earlier) have testes but a feminine exterior. These rats have a masculine SDN-POA because their estrogen receptors are normal. Androgen-insensitive rats also do not display lordosis in response to estrogens and progesterone, because the testosterone that they secreted early in life was converted to an *estrogen* in the brain and masculinized their behavior (Olsen, 1979). Instead, the androgen-insensitive rats show normal male

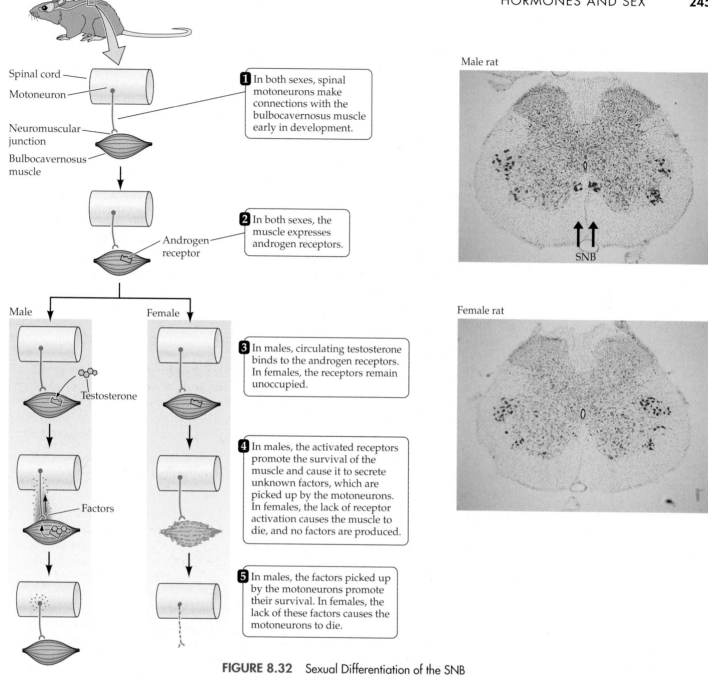

Spinal cord
Motoneuron

1 In both sexes, spinal motoneurons make connections with the bulbocavernosus muscle early in development.

Neuromuscular junction
Bulbocavernosus muscle

2 In both sexes, the muscle expresses androgen receptors.

Androgen receptor

Male

Female

3 In males, circulating testosterone binds to the androgen receptors. In females, the receptors remain unoccupied.

Testosterone

4 In males, the activated receptors promote the survival of the muscle and cause it to secrete unknown factors, which are picked up by the motoneurons. In females, the lack of receptor activation causes the muscle to die, and no factors are produced.

Factors

5 In males, the factors picked up by the motoneurons promote their survival. In females, the lack of these factors causes the motoneurons to die.

Male rat

SNB

Female rat

FIGURE 8.32 Sexual Differentiation of the SNB

attraction to receptive females, with whom they may attempt to mate, despite the lack of a penis (Hamson et al., 2009). Estrogenic metabolites of testosterone do not seem to play a role in masculinizing the primate brain (Grumbach and Auchus, 1999), so we won't deal with that mechanism any further, but you can learn more about it in **A STEP FURTHER 8.6**, on the website.

THE SPINAL CORD IN MAMMALS   In rats, the bulbocavernosus (BC) muscles that surround the base of the penis are innervated by motoneurons in the **spinal nucleus of the bulbocavernosus (SNB)**. Male rats have about 200 SNB cells, but females have far fewer motoneurons in this region of the spinal cord.

On the day before birth, female rats have BC muscles attached to the base of the clitoris that are nearly as large as the BC muscles of males and that are innervated by motoneurons in the SNB region (Rand and Breedlove, 1987). In the days just before and after birth, however, many SNB cells die, especially in females (Nordeen et al., 1985), and the BC muscles of females die (**FIGURE 8.32**).

spinal nucleus of the bulbocavernosus (SNB)   A group of motoneurons in the spinal cord of rats that innervate muscles controlling the penis.

Onuf's nucleus The human homolog of the spinal nucleus of the bulbocavernosus (SNB) in rats.

A single injection of androgens delivered to a newborn female rat permanently spares some SNB motoneurons and their muscles. Castration of newborn males, accompanied by prenatal blockade of androgen receptors, causes the BC muscles and SNB motoneurons to die as in females. The system also dies in newborn androgen-insensitive rats, so estrogens seem to be unimportant for masculine development of the SNB.

Androgens act on the BC muscles to prevent their demise, and this sparing of the muscles causes the innervating SNB motoneurons to survive (Fishman et al., 1990; C. L. Jordan et al., 1991). Thus the developmental rescue of SNB motoneurons is accomplished indirectly as a consequence of actions on muscle. Adult SNB neurons contain androgen receptors and retain androgen sensitivity throughout life. In adulthood, androgen acts directly on the neurons to cause them to grow (Watson et al., 2001) and to start expressing genes that help new connections to form (Monks and Watson, 2001).

In nonrodents, the BC motoneurons are found in a slightly different spinal location and are known as **Onuf's nucleus**. Surprisingly, most female mammals retain a BC muscle into adulthood; in women, for example, the bulbocavernosus (or *constrictor vestibule*) helps constrict the vaginal opening. But, as in rodents, the system is profoundly sexually dimorphic. Men have larger BC muscles and more Onuf's motoneurons than do women, probably because of androgen exposure during fetal development (Forger and Breedlove, 1986, 1987).

These systems and several others all demonstrate the power of androgens to affect the gender of the brain. By controlling the amount and timing of testosterone exposure, researchers can make the various brain structures as masculine or feminine as they like. However, despite the crucial role of androgens in masculinizing sexually dimorphic nuclei in the nervous system, there's evidence that experience affects them too.

Social influences also affect sexual differentiation of the nervous system

Environmental factors of many sorts, including social experience, can modulate the masculinization produced by steroids. The development of the SNB offers a clear example. Celia Moore et al. (1992) noticed that rat dams spend more time licking the anogenital regions of male pups than of females. If the dam is anosmic (unable to smell), she licks all the pups less and does not distinguish between males and females. Males raised by anosmic mothers thus receive less anogenital licking, and remarkably, fewer of their SNB cells survive the period around birth. The dam's stimulation of a male's anogenital region helps to masculinize his spinal cord.

On the one hand, this masculinization is still an effect of androgens, because the dam identifies male pups by detecting androgen metabolites in their urine. On the other hand, this effect is clearly the result of a social influence: the dam treats a pup differently because he's a male, and this differential treatment masculinizes his developing nervous system. Perhaps this example illustrates the futility of trying to distinguish "biological" from "social" influences.

Attention from the dam has a different organizing effect on female rat pups. In adulthood, females that were licked frequently as pups show enhanced estrogen and oxytocin sensitivity in brain regions associated with maternal behavior, and they tend to be attentive mothers themselves. Females that were licked less as pups are less attentive mothers later (Champagne et al., 2001).

What about humans? (No, no, not the licking part—the social influence part.) Humans are at least as sensitive to social influences as rats are. In every culture, most people treat boys and girls differently, even when they are infants. Such differential treatment undoubtedly has some effect on the developing human brain and contributes to later sex differences in behavior. Of course, this is a social influence, but testosterone instigated the influence when it induced the formation of a penis.

If prenatal androgens have even a very subtle effect on the fetal brain, then adults interacting with a baby might detect such differences and treat the baby differently.

Thus, originally subtle sex differences might be magnified by social experience, especially early in life. Such interactions of steroidal and social influences are probably the norm in the sexual differentiation of human behavior. So does fetal testosterone play a role in masculinizing human behavior? Let's examine that question by zeroing in on human sexual orientation, the final topic of this chapter.

how's it going

1. How does exposure to androgens early in life masculinize the brain and spinal cord in rats?
2. Why is it difficult to distinguish between social and biological influences on sexual differentiation?

Do Fetal Hormones Masculinize Human Behaviors in Adulthood?

As with rats and other animals, the fact that men and women behave differently implies that something about them, probably something about their brains, must also be different. Indeed, many parts of the brain are different between men and women (**FIGURE 8.33**). But are these sexual dimorphisms in the human brain caused by prenatal exposure to hormones, as in other animals, or by social influences? In other words, does prenatal exposure to steroids affect the adult behavior of humans? This is a tricky problem because although prenatal androgens may act on the human brain, they certainly act on the rest of the body too.

For example, recall that people with androgen insensitivity syndrome (AIS) are usually raised as girls, because their XY genotype is not discovered until puberty. We said earlier that androgen-insensitive women tend to be very feminine, including being sexually attracted to men and often seeking a family through adoption. Are they feminine because they received the social tutoring to be females, or because their brains, without androgen receptors, could not respond to testosterone? Their behavior is consistent with either hypothesis.

What about females with CAH, who are exposed to androgens before birth? In fact, girls with CAH are much more likely to be described by their parents (and

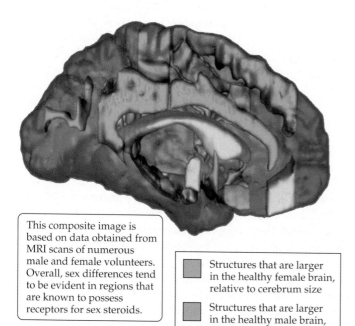

This composite image is based on data obtained from MRI scans of numerous male and female volunteers. Overall, sex differences tend to be evident in regions that are known to possess receptors for sex steroids.

Structures that are larger in the healthy female brain, relative to cerebrum size

Structures that are larger in the healthy male brain, relative to cerebrum size

FIGURE 8.33 Sexual Dimorphism in the Human Brain (After Goldstein et al., 2001, courtesy of Jill Goldstein.)

Training Girls to Be Verbal Adults tend to spend more time talking to a baby if they believe the baby is a girl (whether it actually is a girl or not) (Seavey et al., 1975).

themselves) as tomboys than are other girls, and they exhibit enhanced spatial abilities on cognitive tests that usually favor males (Berenbaum, 2001). In adulthood, most CAH females describe themselves as heterosexual, but they are more likely to report a homosexual orientation than are other women. Interestingly, as females with CAH grow older, the proportion who report homosexual attractions increases to as much as 40% (Dittmann et al., 1992), suggesting that they start off trying to follow the socially approved role of heterosexual female but then become more comfortable with a homosexual orientation later in life. Do women with CAH exhibit those behaviors because early androgens partially masculinized their brains? Or did their ambiguous genitalia cause parents and others to treat them differently from infancy?

The *guevedoces* of the Dominican Republic, who are raised as girls but grow a penis at puberty, behave like males as adults, dressing like men and seeking girlfriends. There are two competing explanations for why these people raised as girls later behave as men. First, prenatal testosterone may masculinize their brains; thus, despite being raised as girls, when they reach puberty, their brains lead them to seek out females for mates. This explanation suggests that the social influences of growing up—assigning oneself to a gender and mimicking role models of that gender, as well as gender-specific playing and dressing—are unimportant for later behavior and sexual orientation.

An alternative explanation is that early hormones have no effect—that the local culture simply recognizes and teaches children that some people can start out as girls and change to boys later. If so, then the social influences on gender role development might be completely different in this society from those in ours. Of course, a third option is that both mechanisms contribute to the final outcome; for example, early androgens may affect the brain to masculinize the child's behavior and predispose later sexual orientation toward females, and these masculine qualities of the child could alter the behavior of parents and others in ways that promote the emergence of male gender identity as the child develops.

At the opening of the chapter we discussed the dilemma of cloacal exstrophy, in which genetic boys are born with functional testes but without penises. Historically in these cases, neonatal sex reassignment has been recommended on the assumption that unambiguously raising these children as girls, and surgically providing them with the appropriate external genitalia, could produce a more satisfactory outcome. In a long-term follow-up of 14 such cases, however, Reiner and Gearhart (2004) found that 8 of these "girls" eventually declared themselves to be boys, even though several were unaware that they had ever been operated on. Although this finding indicates that prenatal exposure to androgens strongly predisposes subsequent male gender identity, 5 of the remaining 6 cases were apparently content with their female identities, suggesting that socialization can also play a strong role. In most people, gender identity is presumably established by nature and nurture working in conjunction.

Although far from conclusive, the studies of people with these various conditions are part of a growing body of evidence suggesting that prenatal testosterone does in fact influence sexual orientation in humans, as we'll see next.

What determines a person's sexual orientation?

There are two kinds of developmental influences that could shape human sexual orientation. Sociocultural influences may instruct developing children about how they should behave when they grow up (think of all those charming princes wooing princesses in Disney movies). But, as we discussed in the previous section, differences in fetal exposure to testosterone could also organize developing brains to be attracted to females or males in adulthood. For that great majority of people who are heterosexual, there's no way to distinguish between these two influences, because they both favor the same outcome. People who are homosexual provide a test, because homosexuality (and other sexual minorities) remains stigmatized by various social groups and cultural institutions (Herek and McLemore, 2013). Is there evidence that early hormones are responsible for causing some people to ignore society's prescription and become gay? If so, then maybe hormones play a role in heterosexual development too.

(A)

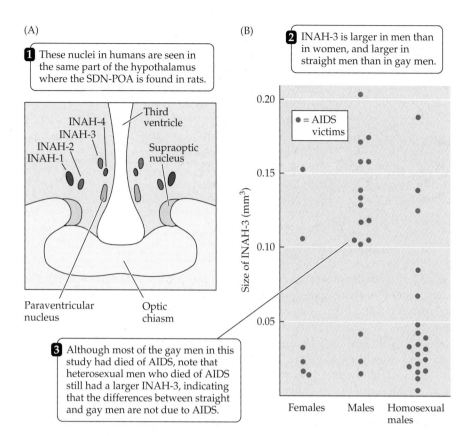

1 These nuclei in humans are seen in the same part of the hypothalamus where the SDN-POA is found in rats.

INAH-4
INAH-3
INAH-2
INAH-1

Third ventricle

Supraoptic nucleus

Paraventricular nucleus

Optic chiasm

3 Although most of the gay men in this study had died of AIDS, note that heterosexual men who died of AIDS still had a larger INAH-3, indicating that the differences between straight and gay men are not due to AIDS.

(B)

2 INAH-3 is larger in men than in women, and larger in straight men than in gay men.

● = AIDS victims

Size of INAH-3 (mm³)

0.20
0.15
0.10
0.05

Females   Males   Homosexual males

FIGURE 8.34 Interstitial Nuclei of the Anterior Hypothalamus

Homosexual behavior is certainly seen in other species—mountain sheep, swans, gulls, and dolphins, to name a few (Bagemihl, 1999). Interestingly, homosexual behavior is more common among apes and monkeys than in prosimian primates like lemurs and lorises (Vasey, 1995), so perhaps greater complexity of the brain makes homosexual behavior more likely. In the most-studied animal model—sheep—some rams consistently refuse to mount females but prefer to mount other rams. There is solid evidence of differences in the POA of "gay" versus "straight" rams (Roselli et al., 2004), apparently organized by testosterone acting on the brain during fetal development (Roselli and Stormshak, 2009).

Simon LeVay (1991) performed postmortem examinations of the POA in humans and found a nucleus (the third interstitial nucleus of the anterior hypothalamus, or INAH-3) (**FIGURE 8.34A**) that is larger in men than in women, and larger in heterosexual men than in homosexual men (**FIGURE 8.34B**). All but one of the gay men in the study had died of AIDS, but the brain differences could not be due to AIDS pathology, because straight men with AIDS still had a significantly larger INAH-3 than did the gay men (Byne et al., 2001). To the press and the public, this finding sounded like strong evidence that sexual orientation is "built in." It's still possible, however, that early social experience affects the development of INAH-3 to determine later sexual orientation. Furthermore, sexual experiences as an adult could affect INAH-3 structure, so the smaller nucleus in some homosexual men may be the *result* of their homosexuality, rather than the *cause*, as LeVay himself was careful to point out.

In women, purported markers of exposure to androgen as a fetus—sounds emitted from the ears (McFadden and Pasanen, 1998), finger length patterns (T. J. Williams et al., 2000; Grimbos et al., 2010), patterns of eye blinks (Rahman, 2005), and skeletal features (J. T. Martin and Nguyen, 2004)—all indicate that lesbians, on average, were exposed to slightly more fetal androgen than were heterosexual women. These findings suggest that fetal exposure to androgen increases the likelihood that a girl will grow up to be gay. There is always considerable overlap between the two groups, so you cannot use these features to predict whether a particular woman is gay, and clearly fetal androgens cannot account for all lesbians.

FIGURE 8.35 Older Brothers Increase the Chance That a Boy Will Grow Up Gay

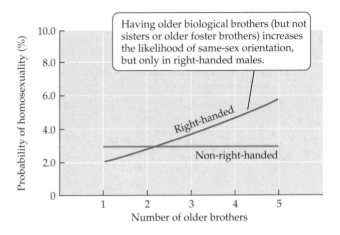

Having older biological brothers (but not sisters or older foster brothers) increases the likelihood of same-sex orientation, but only in right-handed males.

fraternal birth order effect A phenomenon in human populations, such that the more older biological brothers a boy has, the more likely he is to develop a homosexual orientation.

Those same markers of fetal androgen do not provide for a consensus about gay versus straight men; some markers suggest that gay men were exposed to less prenatal testosterone, and others suggest that they were exposed to *more* prenatal testosterone than were straight men. However, another nonsocial factor influences the probability of homosexuality in men: the more older brothers a boy has, the more likely he is to grow up to be gay (Blanchard et al., 2006). Your first guess might be that this is a social influence of older brothers, but it turns out that older stepbrothers that are raised with the boy have no effect, while biological brothers (sharing the same mother) increase the probability of the boy's being gay *even if they are raised apart* (Bogaert, 2006). Furthermore, this **fraternal birth order effect** is seen in boys who are right-handed, but not in left-handed boys (Blanchard et al., 2006; Bogaert, 2007), providing another indication of differences in early development between the two sexual-orientation groups (**FIGURE 8.35**). Statistically, the birth order effect is strong enough that about one in every seven homosexual men in North America—about a million people—is gay because his mother had sons before him (Cantor et al., 2002). One theory is that the immune system of a mother carrying a son is exposed for the first time to proteins from the Y chromosome, and it may produce antibodies that affect development of subsequent sons (Blanchard, 2012).

There is good evidence that human sexual orientation is at least partly heritable, reinforcing the notion that both biological and social factors have a say. About 50% of variability in human sexual orientation is accounted for by genetic factors, leaving ample room for early social influences. Monozygotic twins, who have exactly the same genes, do not always have the same sexual orientation (J. M. Bailey et al., 1993). In the unusual case of two nontwin brothers who are both homosexual, genetic evidence suggests that they are much more likely than chance would dictate to have both inherited the same X chromosome region (the Xq28 region) from their mother (Hamer et al., 1993); but again the genetic explanation accounts for only some, not all, of the cases. It seems clear that there are several different pathways to homosexuality.

From a political viewpoint, the controversy—whether sexual orientation is determined before birth or determined by early social influences—is irrelevant. Laws and prejudices against homosexuality are based primarily on religious views that homosexuality is a sin that some people "choose." But almost all homosexual and heterosexual men report that, from the beginning, their interests and romantic attachments matched their adult orientation. So any social influence would have to be acting very early in life and without any conscious awareness (do you remember "choosing" whom to find attractive?). Furthermore, despite extensive efforts, no one has come up with a reliable way to change sexual orientation (Spitzer, 2012). These findings, added to evidence that older brothers and prenatal androgens affect the probability of being gay, have convinced most scientists that we do not choose our sexual orientation.

SIGNS **&** SYMPTOMS

Psychosocial Dwarfism

Genie had a horrifically deprived childhood. For over 10 years, starting from the age of 20 months, she was isolated in a small, closed room, and much of the time she was tied to a potty chair. Her mentally ill parents provided food, but nobody held Genie or spoke to her. When she was released from her confinement and observed by researchers at the age of 13, her size made her appear only 6 or 7 years old (Rymer, 1993).

Other, less horrendous forms of family deprivation also result in failure of growth. This syndrome is referred to as **psychosocial dwarfism** to emphasize that the growth failure arises from psychological and social factors mediated through the central nervous system and its control over endocrine functions (W. H. Green et al., 1984). Often when children suffering from psychosocial dwarfism are removed from stressful circumstances, they begin to grow rapidly. The growth rates of five such children, before and after periods of emotional deprivation, are shown in **FIGURE 8.36** (Sirotnak et al., 2004).

How do stress and emotional deprivation impair growth? Growth impairments appear to be mediated by changed outputs of several hormones, including growth hormone (GH), cortisol, and other hormones, known as *somatomedins* (which are ordinarily released by the liver in response to GH). GH and the somatomedins normally stimulate cell growth.

Some children with psychosocial dwarfism show almost a complete lack of GH release (Albanese et al., 1994). Disturbed sleep has also been suggested as a cause of this failure, because GH is typically released during certain stages of sleep, and children under stress show disturbed sleep patterns (L. I. Gardner, 1972). Other children who exhibit psychosocial dwarfism show normal levels of GH but low levels of somatomedins, and these hormones, along with GH, appear to be necessary for normal growth. Still other children with this condition show elevated levels of cortisol, probably as a result of stress, that inhibit growth. Finally, some affected children show none of these hormonal disturbances, so there must also be other routes through which emotional experiences affect growth. For Genie, relief came in time to restore much of her *body* growth, but her mental development remains severely limited; she has never learned to say more than a few words, and now in her fifties, she lives in an institution.

psychosocial dwarfism A syndrome of stunted growth in children subjected to social stress, such as abusive caregivers.

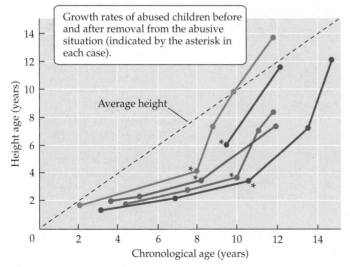

Growth rates of abused children before and after removal from the abusive situation (indicated by the asterisk in each case).

Average height

FIGURE 8.36 Psychosocial Dwarfism Growth rates of abused children before and after removal from the abusive situation. Removal from the abusive situation is indicated by the asterisk in each case.

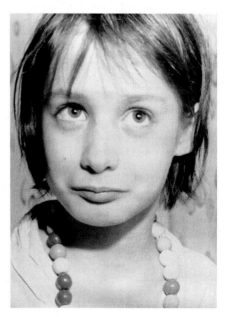

Genie

how's it going ❓

1. What does the sexual orientation of people with various syndromes of sexual differentiation suggest about hormonal influences on human sexual orientation?

2. If human sexual orientation were shown definitively to be influenced by prenatal factors such as hormones and the fraternal birth order effect, would you be more or less inclined to accept homosexuality?

Recommended Reading

Becker, J. B., Breedlove, S. M., Crews, D., and McCarthy, M. M. (Eds.). (2002). *Behavioral Endocrinology* (2nd ed.). Cambridge, MA: MIT Press.

Colapinto, J. (2000). *As Nature Made Him.* New York, NY: HarperCollins.

Eugenides, J. (2002). *Middlesex: A Novel.* New York, NY: Farrar, Straus, and Giroux.

LeVay, S., Baldwin, J., and Baldwin, J. (2015). *Discovering Human Sexuality* (3rd ed.). Sunderland, MA: Sinauer.

Nelson, R. J. (2011). *An Introduction to Behavioral Endocrinology* (4th ed.). Sunderland, MA: Sinauer.

Norman, A. W., and Henry, H. L. (2014). *Hormones* (3rd ed.). San Diego, CA: Academic Press.

2e.mindsmachine.com/vs8

VISUAL SUMMARY

8

You should be able to relate each summary to the adjacent illustration, including structures and processes. Go to the online version of this summary (scan the QR code above) for links to figures, animations, and activities that will help you consolidate the material.

1 Hormones are chemicals that are secreted by **endocrine glands** into the bloodstream and are taken up by receptor molecules in target cells. Unlike neurotransmitters in neuronal signaling, hormones spread more slowly and act throughout the body. Review **Figures 8.1–8.3, Table 8.1, Activity 8.1, Animations 8.2 and 8.3**

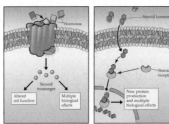

2 Most hormones act on receptors in a wide variety of cells, coordinating influences throughout the body. **Peptide hormones** and **amine hormones** bind to receptor molecules at the surface of the target cell membrane and activate **second-messenger** molecules inside the cell. **Steroid hormones** pass through the membrane and bind to receptor molecules inside the cell, ultimately regulating gene expression. Review **Figures 8.4–8.7, Animation 8.4**

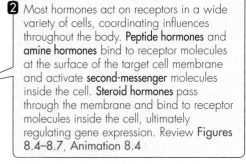

3 **Negative feedback** systems control hormone secretion. **Neuroendocrine cells** in the hypothalamus send axons down the **pituitary stalk** to the **posterior pituitary**, releasing two hormones—**oxytocin** and **vasopressin**—into the bloodstream. Release of the hormones is regulated by **synaptic** influences on those neuroendocrine cells. Review **Figures 8.8–8.10**

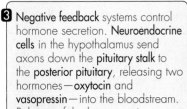

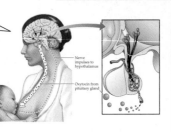

4 Other hormones are controlled by a **releasing hormone** from the hypothalamus that stimulates the anterior pituitary to release **tropic hormones**, which in turn control the secretion of hormones by endocrine glands. The endocrine gland hormone then provides negative feedback to the hypothalamus and pituitary. Review **Figures 8.11–8.13, Animation 8.5**

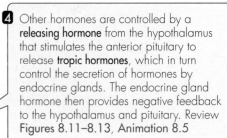

5 Many behaviors require the coordination of neural and hormonal components. Messages may be transmitted in the body via neural-to-neural, neural-to-endocrine, endocrine-to-endocrine, or endocrine-to-neural links. There are continual, reciprocal influences between the endocrine system and the nervous system: experience affects hormone secretion, and hormones affect behavior and therefore future experiences. Review Figures 8.14–8.16

6 In animals, reproductive behaviors are regulated by hormones. In female rats, a steroid-sensitive **lordosis** circuit extends from the **ventromedial hypothalamus (VMH)** to the spinal cord. In male rats, **medial preoptic area (mPOA)** neurons integrate inputs from the **medial amygdala** and **vomeronasal organ (VNO)**, and they project axons widely to regulate copulatory behavior. Review Figures 8.17–8.20

7 In humans, very low levels of **testosterone** are required for either men or women to display a full interest in sex, but additional testosterone has no additional effect. In animals, hormones significantly influence **parental behavior** by acting on the same brain regions that are important for sexual behavior (mPOA, VMH). Review Figures 8.18 and 8.21–8.23

8 Genes (XX or XY) **determine sex** by controlling whether the **indifferent gonads** of an individual develop as **testes** or **ovaries**. Then, hormonal secretions from the testes masculinize the rest of the body in males. This means that some people, for example, women who carry a Y chromosome but have **androgen insensitivity syndrome (AIS)**, may be masculine in some parts of the body but feminine in others. Review Figures 8.24–8.28

9 In animals, **androgens** also organize the developing brain, masculinizing regions such as the **sexually dimorphic nucleus of the preoptic area (SDN-POA)** and the **spinal nucleus of the bulbocavernosus (SNB)**. There is evidence that prenatal androgens also masculinize the human brain. Review Figures 8.29–8.33, Animation 8.6

10 Several regions of the human brain are **sexually dimorphic**, but we do not know whether these dimorphisms are organized by fetal steroids or by sex differences in the social environment. Research demonstrates that human sexual orientation is affected by prenatal influences and is not simply a matter of individual choice. Review Figures 8.33–8.35

Go to **2e.mindsmachine.com** for study questions, quizzes, flashcards, and other resources.

chapter
9

Homeostasis
Active Regulation of the Internal Environment

A Love-Hate Relationship with Food

Six-hundred ninety calories—that's what this milkshake represented to me.
But to Kitty it was the object of her deepest fear and loathing. "You're trying to make me fat,"
she said in a high-pitched, distorted voice that made the hairs on the back of my neck stand
up. She rocked, clutching her stomach, chanting over and over: "I'm a fat pig. I'm so fat."

Harriet Brown, "One Spoonful at a Time,"
New York Times Magazine, November 6, 2006

To see the video
Anorexia,
go to

2e.mindsmachine.com/9.1

How does food become an object of "fear and loathing" for some people? Harriet Brown paints a harrowing portrait of the "demon" of anorexia nervosa that seemed to possess her daughter, Kitty. This disorder completely deranges the sufferer's relationship with food. Obsession with food, lost appetite, starvation, fear, and distorted self-perception are hallmarks of anorexia nervosa. In Kitty's case, these symptoms developed like a gathering storm, from the first hints of unusual fascination with food, poring over *Gourmet* magazine, to

the loss of appetite and relentless exercising, until the day when 14-year-old Kitty lay in bed a mere 71 pounds, every bone sharply evident beneath her skin, her hair falling out in clumps, her breath laden with the pearlike scent of ketones as her body metabolized itself for fuel. Still, Kitty considered herself "a fat pig," sobbing and terrified at the prospect of drinking a milkshake or eating a piece of birthday cake. Kitty's parents faced uncertainty and difficult decisions. What was the best way to strike back at the demon?

BRuNo MaLLART

M illions of years of evolution have endowed our bodies with complex physiological mechanisms, and multiple backup systems, devoted to producing a stable internal environment. Food energy, body temperature, fluid balance, fat storage, nutrients—all of the conditions required for optimal cellular functioning—are carefully regulated by the brain at every stage. But in the context of modern society, some of these ancient systems are making trouble for us; obesity, for example, is reaching epidemic proportions and placing a severe burden on health care resources. The physiological and behavioral processes governing the internal environment, and their role when things go wrong, are our topic in this chapter.

To view the
Brain Explorer,
go to

2e.mindsmachine.com/9.2

homeostasis The active process of keeping a particular physiological parameter relatively constant.

motivation The psychological process that induces or sustains a particular behavior.

thermoregulation The active process of maintaining a constant internal temperature through behavioral and physiological adjustments.

endotherm An animal whose body temperature is regulated chiefly by internal metabolic processes.

ectotherm An animal whose body temperature is regulated by, and whose heat comes mainly from, the environment.

negative feedback The property by which some of the output of a system feeds back to reduce the effect of input signals.

set point The point of reference in a feedback system.

set zone The optimal range of a variable that a feedback system tries to maintain.

PART I
Principles of Homeostasis

Scientists in the nineteenth century realized that the body is a self-contained environment, providing optimal conditions for cells to live and grow. This perspective made it easier to understand many physiological and behavioral systems, whose job it is to prevent internal conditions from varying to any great extent. Variables such as acidity, saltiness, water level, oxygenation, temperature, and energy availability are closely monitored and maintained by elaborate physiological systems. Collectively, these systems are responsible for **homeostasis**, the active process of maintaining a relatively stable, balanced internal environment.

Alterations in the internal environment can have an effect on **motivation**, the psychological process that induces or sustains a particular behavior. According to this view, the mismatch between the actual internal state (e.g., becoming dehydrated) and the regulated, intended state produces a *drive* to restore balance (in this case, by having a drink of water). And as we all know, drive can rapidly escalate as the mismatch worsens, from a minor distraction (like the desire for a couple of sips if a glass of water happens to be present) to an overwhelming, all-consuming desire (like the raging thirst of someone lost in the desert). Our ability to regulate our internal resources is complicated by the fact that simply staying alive requires us to use up some of them.

Because it is a relatively simple system, we'll start by using **thermoregulation** (the regulation of body temperature) to look at some important general concepts of homeostasis: negative feedback, redundancy, behavioral homeostasis, and the concept of allostatic load. These topics will arise again when we talk about fluid balance, appetite, and body weight in the remainder of the chapter.

Homeostatic Systems Share Several Key Features

We mammals are **endotherms**, meaning that we make our own heat *inside* our bodies, using metabolism and muscular activity (and if our muscles aren't making enough heat, we can shiver them to make more). Endothermy gives us clear advantages over **ectotherms** (animals that get their heat mostly from *outside* the body—from the environment) by allowing us to roam more widely. Ectotherms, such as lizards and snakes, need to stay nearer sources of warmth. Furthermore, endothermy allows the muscles of mammals to work hard for longer periods of time: endothermic hares will always outrun ectothermic tortoises. So it's no surprise that our body temperature is carefully regulated. The systems that govern body temperature operate according to several general principles common to almost all homeostatic systems. (For more on the pros and cons of endothermy and ectothermy, see **A STEP FURTHER 9.1**, on the website.)

Internal states are governed through negative feedback

The homeostatic mechanisms that regulate temperature, body fluids, and metabolism are primarily **negative feedback** systems, where deviation from a desired value, called the **set point**, triggers a compensatory action of the system. Restoring the desired value turns *off* the response (this is why it is called *negative* feedback). A simple analogy for this mechanism is a household thermostat (**FIGURE 9.1**): a temperature drop below the set point activates

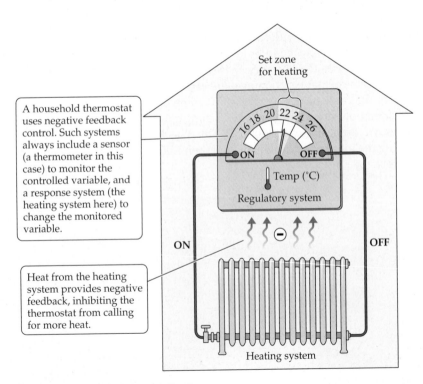

A household thermostat uses negative feedback control. Such systems always include a sensor (a thermometer in this case) to monitor the controlled variable, and a response system (the heating system here) to change the monitored variable.

Heat from the heating system provides negative feedback, inhibiting the thermostat from calling for more heat.

Set zone for heating

16 18 20 22 24 26

ON　OFF

Temp (°C)

Regulatory system

ON　OFF

Heating system

FIGURE 9.1 Negative Feedback

the thermostat, which turns on the heating system. The heat that is produced has a negative feedback effect on the thermostat, so it stops calling for heat. Most heating systems have at least a little bit of tolerance built in—otherwise the system would be going on and off too frequently—so there is generally a **set zone** rather than a rigid set *point*.

Just like a thermostat, your set zone for body temperature can be changed under certain circumstances. For example, your body temperature drops at night for much the same reason that people turn down their home thermostats at night: to conserve energy. Or your set zone may be temporarily elevated, producing a fever to help your body fight off an infection. But in either case there are narrow limits. Too hot, and proteins begin to lose their correct shape, link together, and malfunction (this modification of proteins is called *denaturing* or, if it is really hot, *cooking*), with lethal results if critical brain regions are compromised. If we are too cool, chemical reactions of the body occur too slowly; at very low body temperatures, ice crystals may disrupt cellular membranes, killing the cells.

Redundancy is a feature of many homeostatic systems

Just as engineers equip critical equipment with several backup systems, our bodies tend to have multiple mechanisms for monitoring our stores, conserving remaining supplies, obtaining new resources, and shedding excesses. Loss of function in one part of the system usually can be compensated for by the remaining parts. This redundancy attests to the importance of maintaining our inner environment, but it also complicates the lives of scientists who are trying to figure out exactly how the body normally regulates temperature, water balance, and food intake.

It has long been known that the hypothalamus senses and controls body temperature, but lesion experiments eventually showed that different hypothalamic sites control two separate thermoregulatory systems. Lesions in the preoptic area (POA) of rats impaired the physiological responses to cold, such as shivering and constriction of the blood vessels, but did not interfere with such behaviors as pressing levers to control heating lamps or cooling fans (Satinoff and Rutstein, 1970; Van Zoeren and Stricker, 1977). Lesions in the lateral hypothalamus of rats abolished behavioral regulation of temperature but did not affect the physiological responses (Satinoff and Shan, 1971; Van Zoeren and Stricker, 1977). This is a clear example of homeostatic redundancy: two different systems for regulating the same variable. Furthermore, there seems to be a hierarchy of thermoregulatory circuits—some located at the spinal level, some centered in the brainstem, and others in the hypothalamus (**FIGURE 9.2**).

To see the animation **Negative Feedback**, go to

2e.mindsmachine.com/9.3

Redundancy Engineers equip critical systems with multiple "failsafe" redundancies—for example, skydivers generally carry a secondary parachute—so that a backup always protects the critical system (the skydiver, in this example). Multiple redundancy is a feature of many of the body's homeostatic systems, protecting the constant internal environment that is crucial for survival.

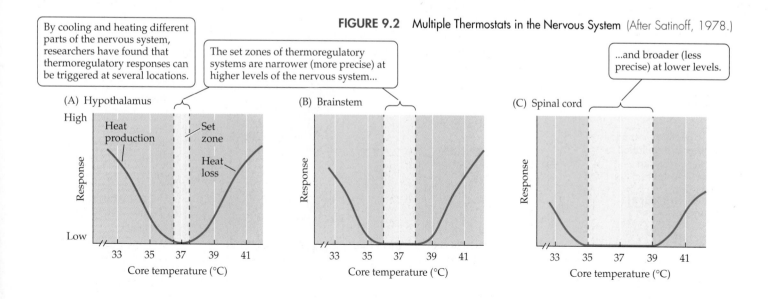

FIGURE 9.2 Multiple Thermostats in the Nervous System (After Satinoff, 1978.)

By cooling and heating different parts of the nervous system, researchers have found that thermoregulatory responses can be triggered at several locations.

The set zones of thermoregulatory systems are narrower (more precise) at higher levels of the nervous system...

...and broader (less precise) at lower levels.

(A) Hypothalamus — Response / Core temperature (°C) — High, Heat production, Set zone, Heat loss, Low — 33 35 37 39 41

(B) Brainstem — Response / Core temperature (°C) — 33 35 37 39 41

(C) Spinal cord — Response / Core temperature (°C) — 33 35 37 39 41

(A)

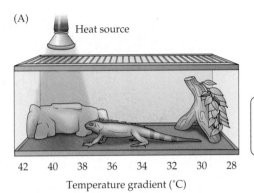

Heat source

The lizard controls its body temperature by moving around the cage.

42 40 38 36 34 32 30 28

Temperature gradient (°C)

(B)

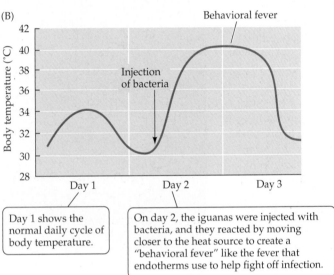

Behavioral fever

Injection of bacteria

Day 1 shows the normal daily cycle of body temperature.

On day 2, the iguanas were injected with bacteria, and they reacted by moving closer to the heat source to create a "behavioral fever" like the fever that endotherms use to help fight off infection.

FIGURE 9.3 Behavioral Thermoregulation in Bacteria-Challenged Lizards (After Kluger, 1978.)

allostasis The combined set of behavioral and physiological adjustments that an individual makes in response to current and predicted behavioral and environmental stressors.

Behavioral mechanisms are crucial for homeostasis

Organisms also use behavioral measures to help them regulate and acquire more heat, water, or food. In general, both ectotherms and endotherms deploy three kinds of temperature-regulating behavior: (1) behaviors that change *exposure* of the body surface—for example, by huddling or extending limbs; (2) behaviors that change external *insulation*, such as by using clothing or nests; and (3) behaviors that change *surroundings*, by moving into the sun, into the shade, or into a burrow (see **A STEP FURTHER 9.2**, on the website).

Because ectotherms generate little heat through metabolism, behavioral methods of thermoregulation are especially important to them. In the laboratory, iguanas carefully regulate their temperature by moving toward or away from a heat lamp, and when infected by bacteria they even produce a fever through such behavioral means (**FIGURE 9.3**), which helps them fight off an infection. Endotherms use internal processes to generate a fever when fighting infections, which boosts our immune system response. Unfortunately, sometimes the body goes too far, as a fever above 40°C (104°F) does more harm than good. **FIGURE 9.4** summarizes the basic mammalian thermoregulatory system: receptors in the skin, body core, and hypothalamus detect temperature and transmit that information to three neural regions (spinal cord, brainstem, and hypothalamus). If the body temperature moves outside the set zone, each of these neural regions can initiate physiological and behavioral responses to return it to the set zone.

A wide array of sensors continuously monitors the many internal and external threats to our physiological stability. At any given moment, depending on what's happening in the environment, simultaneous perturbations in multiple regulated systems cause a degree of physiological stress that ranges from mild and healthy to severe and overwhelming. Prior experience with challenges allows us to make predictions and activate a coordinated set of behavioral and physiological changes in order to ward off serious homeostatic deviations: this dynamic process is termed **allostasis** (McEwen and Wingfield, 2010). The associated cost of these responses—the wear and tear of daily life—is called *allostatic load*. Allostatic adjustments are a normal part of life, but the heavy allostatic load of chronically stressed individuals puts them at risk of pathology.

The regulation of internal resources is complicated by the fact that staying alive requires us to use up some of those resources. Our homeostatic mechanisms are thus

To see the animation **Thermoregulation in Humans,** go to

2e.mindsmachine.com/9.4

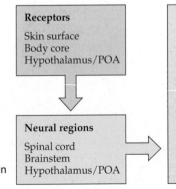

FIGURE 9.4 Basic Elements of Mammalian Thermoregulatory Systems

continually challenged by these unavoidable losses (sometimes called *obligatory losses*), which require us to regain the lost resources from the external environment. Restoring expended water and food takes up much of an animal's waking life, so we turn to those activities next.

how's it going ❓

1. Define *homeostasis*, and discuss its relation to the psychological concept of motivation. Why do homeostatic systems tend to have a set *zone* instead of a set *point*?

2. Distinguish between endotherms and ectotherms, and give a few examples of each.

3. Briefly describe each of the three key features of homeostatic systems that we have discussed, illustrated by examples. What is the significance of each?

4. Define *allostasis*. How does allostasis relate to homeostasis?

PART II
Fluid Regulation

The water you drink on a hot day is carefully measured and partitioned by the brain. A precise balance of fluids and dissolved salts bathes the cells of the body and enables them to function. The composition of this fluid provides an echo of our evolutionary past. The first living organisms on Earth were single-celled inhabitants of the ancient oceans, and it was in these simple organisms that the fundamental processes of cellular life were established. When multicellular organisms evolved much later and began to exploit opportunities on land and in the air, they had no choice but to bring along with them the watery environment that their cells needed to survive. For this reason, most organisms evolved homeostatic systems that ensure that the composition of their body fluids closely resembles dilute seawater (Bourque, 2008; **FIGURE 9.5**). Even relatively minor deviation from optimal water and salt balance can be lethal.

Behavioral Control of Body Temperature (A) A Galápagos marine iguana, upon emerging from the cold sea, raises its body temperature by hugging a warm rock and lying broadside to the sun. (B) Once its temperature is sufficiently high, the iguana reduces its surface contact with the rock and faces the sun to minimize its exposure. These behaviors control body temperature. (Photographs by Mark R. Rosenzweig.)

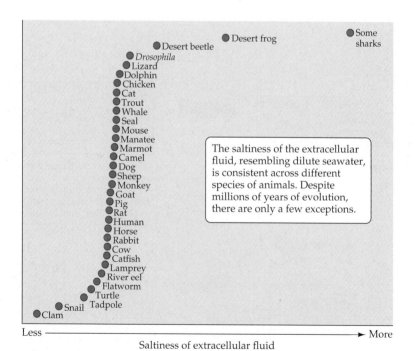

> The saltiness of the extracellular fluid, resembling dilute seawater, is consistent across different species of animals. Despite millions of years of evolution, there are only a few exceptions.

Less ⟶ More
Saltiness of extracellular fluid

FIGURE 9.5 Each Animal Contains a Tiny Sea (After Bourque, 2008.)

intracellular compartment The fluid space of the body that is contained within cells.

extracellular compartment The fluid space of the body that exists outside the cells.

diffusion The spontaneous spread of solute molecules through a solvent until a uniform solute concentration is achieved.

osmosis The passive movement of a solvent, usually water, through a semipermeable membrane until a uniform concentration of solute (often salt) is achieved on both sides of the membrane.

Because we cannot fully seal our bodies from the outside world, we experience constant obligatory losses of water and salts. Many body functions use up some water (and some salt molecules), as, for example, when we produce urine to get rid of waste molecules. In this section we discuss how the nervous system monitors and controls the precise composition of body fluids that cells require in order to function.

Water Moves between Two Major Body Compartments

Because we consist of trillions of cells living in a seawater bath, we can describe water balance by contrasting the inside with the outside of our cells. Most of the water in the body is contained within our cells; this water is collectively referred to as the **intracellular compartment**. The fluid that is outside of our cells, called the **extracellular compartment**, is divided between the *interstitial fluid* (the fluid between cells) and *blood plasma* (the protein-rich fluid that carries red and white blood cells). Water is continually moving back and forth between these compartments, in and out of cells.

To understand the forces driving the movement of water, we must understand diffusion and osmosis. In **diffusion**, molecules of a substance, like salt (a *solute*), that are dissolved in a quantity of another substance, such as water (a *solvent*), will passively spread through the solvent because of the random jiggling and movement of the molecules until they are more or less uniformly distributed throughout it. If we divide a container of water with a membrane that is impermeable to water and salt and we put salt in the water on one side, the salt molecules will diffuse only within the water on that side. If instead the membrane impedes salt molecules only a little, then the salt will distribute itself evenly within the water on the initial side but will also—more slowly—invade and distribute itself across the other side. A membrane that is permeable to some molecules but not others is referred to as *selectively permeable* or *semipermeable*. As we saw in Chapter 3, selective permeability of cell membranes is what lets neurons create and transmit electrical potentials.

Osmosis, illustrated in **FIGURE 9.6**, is the movement of water molecules that occurs when a semipermeable membrane separates solutions containing different

FIGURE 9.6 Osmosis

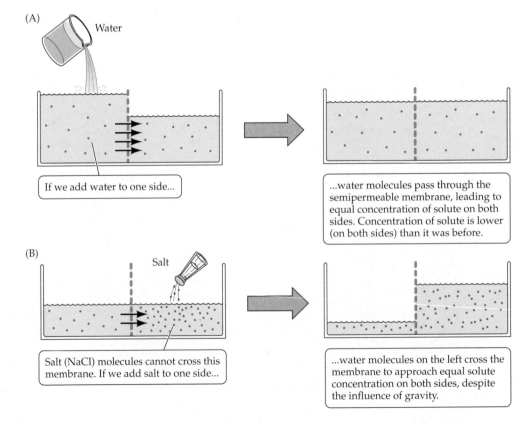

Salt water | Semipermeable membrane

Equal concentration of solute on both sides.

(A) Water

If we add water to one side...

...water molecules pass through the semipermeable membrane, leading to equal concentration of solute on both sides. Concentration of solute is lower (on both sides) than it was before.

(B) Salt

Salt (NaCl) molecules cannot cross this membrane. If we add salt to one side...

...water molecules on the left cross the membrane to approach equal solute concentration on both sides, despite the influence of gravity.

concentrations of solute and the solute cannot cross the membrane to spread itself evenly across both sides. This is the case we examine in Figure 9.6B, in which the semipermeable membrane blocks the passage of salt molecules. Here the water molecules are moving into the compartment where they are less concentrated (because the salt molecules are there), eventually resulting in equal concentrations of solution on both sides of the membrane. The physical force that pushes or pulls water across the membrane is called **osmotic pressure**.

Normally the concentration of salt (sodium chloride, or NaCl) in the extracellular fluid of mammals is about 0.9% (which means there's about 0.9 gram of NaCl for every 100 milliliters of water). A solution with this concentration of salt is called *physiological saline* or described as *isotonic*. Because water moves to produce uniform saltiness (see Figure 9.6B), cells will lose water if placed in a saltier solution and will gain water in a less salty solution. If excessive, this movement of water will damage or kill the cell. The extracellular fluid serves as a *buffer*, a reservoir of isotonic fluid that provides and accepts water molecules, so cells can maintain proper internal conditions and prevent such damage. The nervous system uses two cues to ensure that the extracellular compartment has about the right amount of water and solute to allow cells to absorb or shed water molecules readily, as we'll see next.

Two Internal Cues Trigger Thirst

The nervous system carefully monitors the quantity and concentration of the fluid in our bodies to determine whether we should seek additional water. Two different states can signal that more water is needed: (1) low extracellular volume (**hypovolemic thirst**), resulting from the loss of body fluids; or (2) high extracellular solute concentration (**osmotic thirst**) (**FIGURE 9.7**)—a consequence of our body fluids becoming too salty. We'll consider each in turn.

osmotic pressure   The tendency of a solvent to move across a membrane in order to equalize the concentration of solute on both sides of the membrane.

hypovolemic thirst   A desire to ingest fluids that is stimulated by a reduction in volume of the extracellular fluid.

osmotic thirst   A desire to ingest fluids that is stimulated by high concentration of solute (like salt) in the extracellular compartment.

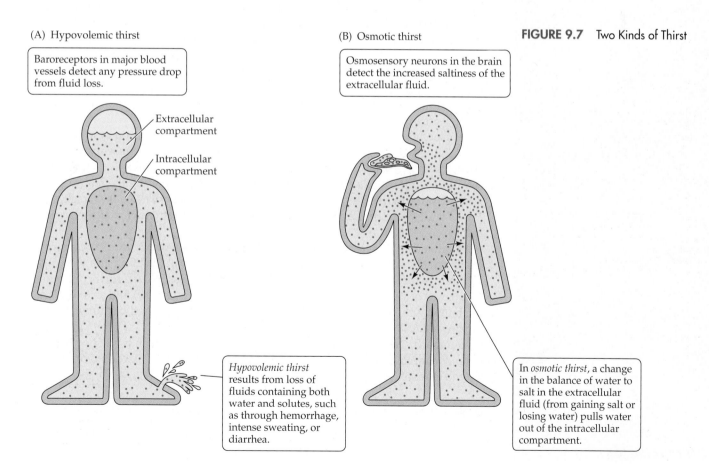

(A) Hypovolemic thirst

Baroreceptors in major blood vessels detect any pressure drop from fluid loss.

Extracellular compartment

Intracellular compartment

Hypovolemic thirst results from loss of fluids containing both water and solutes, such as through hemorrhage, intense sweating, or diarrhea.

(B) Osmotic thirst

Osmosensory neurons in the brain detect the increased saltiness of the extracellular fluid.

In *osmotic thirst*, a change in the balance of water to salt in the extracellular fluid (from gaining salt or losing water) pulls water out of the intracellular compartment.

FIGURE 9.7   Two Kinds of Thirst

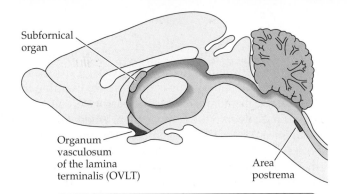

FIGURE 9.8 Circumventricular Organs

Subfornical organ

Organum vasculosum of the lamina terminalis (OVLT)

Area postrema

True to their name, the circumventricular organs lie in the walls of the ventricular system (blue). Thanks to a diminished blood-brain barrier, neurons of the circumventricular organs can monitor the concentration and composition of body fluids.

baroreceptor A pressure receptor in the heart or a major artery that detects a change in blood pressure.

atrial natriuretic peptide (ANP) A hormone, secreted by the heart, that normally reduces blood pressure, inhibits drinking, and promotes the excretion of water and salt at the kidneys.

vasopressin Also called *arginine vasopressin* (*AVP*) or *antidiuretic hormone* (*ADH*). A peptide hormone from the posterior pituitary that promotes water conservation and increases blood pressure.

angiotensin II A hormone that is produced in the blood by the action of renin and that may play a role in the control of thirst.

circumventricular organ Any of multiple distinct sites that lie in the wall of a cerebral ventricle and monitor the composition of the cerebrospinal fluid.

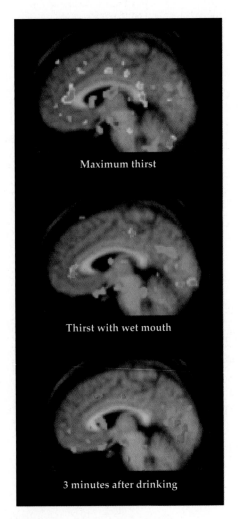

Maximum thirst

Thirst with wet mouth

3 minutes after drinking

Hypovolemic thirst is triggered by a loss of water volume

The example of hypovolemic thirst that is most easily understood is one we hope you never experience: serious blood loss (*hemorrhage*). Any animal that loses a lot of blood has a lowered total blood volume (*hypovolemic* means literally "low volume"). In this condition, blood vessels that would normally be full and slightly stretched no longer contain their full capacity. Blood pressure drops, and the individual becomes thirsty; in fact, powerful thirst is one of the most noticeable symptoms of serious blood loss. Note that losing fluids from blood loss (or from diarrhea or vomiting) does not change the *concentration* of the extracellular fluid, because salts and other ions are lost along with the water (**FIGURE 9.7A**).

The initial drop in extracellular volume is detected by pressure receptors, called **baroreceptors**, which are located in major blood vessels and in the heart. Reacting to the signal from the baroreceptors, the brain activates a variety of responses, such as thirst (to replace the lost water) and salt hunger (to replace the solutes that have been lost along with the water). Replacing the water without also replacing the salts would result in *hypotonic* (less salty than normal) extracellular fluid. The sympathetic nervous system also stimulates muscles in the artery walls to constrict, reducing the size of the vessels and partly compensating for the reduced volume.

Finally, several different organs respond by altering their hormonal signals. The heart decreases its secretion of **atrial natriuretic peptide** (**ANP**), which normally reduces blood pressure, inhibits drinking, and promotes the excretion of water and salt at the kidneys. The brain's posterior pituitary gland steps up its release of the hormone **vasopressin** (also called *arginine vasopressin* [*AVP*] or *antidiuretic hormone* [*ADH*]), which acts on the kidneys to slow the production of urine by increasing the reabsorption of water. And the kidneys trigger the production of **angiotensin II** (from the Greek *angeion*, "blood vessel," and the Latin *tensio*, "tension" or "pressure") from a precursor circulating in the bloodstream.

Angiotensin II (AII) has several water-conserving actions. By directly constricting blood vessels, AII increases blood pressure, ensuring that the brain and vital organs continue to receive essential materials for as long as possible. AII further stimulates the release of vasopressin and of aldosterone (discussed below) and acts directly on the brain, at the preoptic area (POA) and at distinct sites called **circumventricular organs** (**FIGURE 9.8**), to stimulate thirst and drinking behavior (Epstein et al., 1970; Fitzsimmons, 1998). Evidence suggests that the AII mechanism is just one of several redundant systems for provoking thirst and is not active under all conditions (McKinley and Johnson, 2004).

Ahhhhhhh! The experience of strong thirst, induced by the injection of hypertonic saline, is associated with activity in several brain regions, especially the cingulate cortex and cerebellum (*top*). Wetting the mouth reduces this activation only slightly (*middle*), but drinking a glass of water (*bottom*) reduces activation in these brain regions dramatically. (From Denton et al., 1999.)

Osmotic thirst occurs when the extracellular fluid becomes too salty

For most of us, severe hypovolemic thirst is a relatively uncommon event. Thirst is more commonly triggered by obligatory water losses—through respiration, urination, and so on—in which more water is lost than salt. In this case, not only is the *volume* of the extracellular fluid *decreased*, triggering the responses described in the previous section, but also the solute *concentration* of the extracellular fluid is *increased*. As a result of the increase in extracellular saltiness, water is pulled out of cells through osmosis.

Another thing that can make the extracellular fluid more concentrated is eating a lot of salty food. Once again, water will be drawn out of cells through osmosis. This loss of intracellular water triggers osmotic thirst (**FIGURE 9.7B**): we want to drink water in order to return the extracellular fluid to a comfortable isotonic state.

Specialized **osmosensory neurons**—neurons that specifically monitor the concentration of the extracellular fluid—are found in numerous regions of the hypothalamus, including the preoptic area, the anterior hypothalamus, and the supraoptic nucleus. Osmosensory neurons are also found in the organum vasculosum of the lamina terminalis (OVLT), a circumventricular organ (see Figure 9.8).

Thirst is a homeostatic signal that intrudes forcefully into consciousness, with associated strong activation of certain brain regions, particularly in the limbic system (Denton et al., 1999). The two types of thirst (hypovolemic and osmotic), the two fluid compartments (extracellular and intracellular), and the multiple redundant methods to conserve water make for a fairly complicated system that is not yet fully understood. The current conceptualization of this system is depicted in **FIGURE 9.9**.

We don't stop drinking just because the throat and mouth are wet

Although plausible, the most obvious explanation of why we stop drinking—that we have dampened our previously dry throat and mouth—is all wet (sorry). In a classic test of this idea, thirsty animals were allowed to drink water, but the water they consumed was diverted out of the esophagus through a small tube. They remained thirsty and continued drinking.

Furthermore, we stop drinking before water has left the gastrointestinal tract and entered the extracellular compartment. Somehow we monitor how much water we have ingested and stop in *anticipation* of correcting the extracellular volume and/or the concentration of solutes. Experience may teach us and other animals how to gauge accurately whether we've ingested enough to counteract our thirst. Normally, all the signals—blood volume, solute concentration, moisture in the mouth, estimates of the amount of water we've ingested that's "on the way"—agree, but the cessation of one signal alone will not stop thirst; in this way, animals insure against dehydration.

Water Balance Depends on the Regulation of Salt in the Body

Salt (NaCl) is crucial for fluid balance. We cannot maintain water in the extracellular compartment without solutes; if the extracellular compartment contained pure water, osmotic pressure would drive it into the cells until they ruptured and died. The amount of water that we can retain is determined primarily by the number of Na^+ ions we possess. That's why thirst is quenched more effectively by very slightly salty drinks (like sports drinks) than by pure water. But saltier water, like seawater, has the reverse

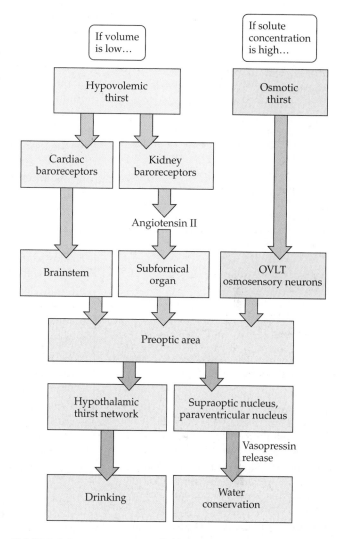

FIGURE 9.9 An Overview of Fluid Regulation

osmosensory neuron A specialized neuron that monitors the concentration of the extracellular fluid by measuring the movement of water into and out of the intracellular compartment.

effect. Seawater is hypertonic (saltier than our body fluids), so just like eating salty food, drinking seawater causes ever-worsening osmotic thirst. Lacking the specialized salt-excreting organs that marine animals have evolved, we simply can't get rid of excess salt fast enough to survive on seawater.

Some Na^+ loss is inevitable, as during urination or sweating. But when water is at a premium, the body tries to conserve Na^+ in order to retain water. In addition to its effects on thirst and vasopressin secretion, angiotensin II stimulates the release of the steroid hormone **aldosterone** from the adrenal glands. Aldosterone directly stimulates the kidneys to conserve Na^+. Nonetheless, animals must find additional salt in their environments in order to retain sufficient water to survive, and they will travel long distances to get it.

Our need to compensate for obligatory losses is also crucial to understanding energy regulation, as we'll see in the next section.

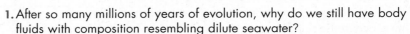
how's it going ❓

1. After so many millions of years of evolution, why do we still have body fluids with composition resembling dilute seawater?

2. Briefly describe diffusion and osmosis, and define *osmotic pressure*. How do these relate to the composition of fluids in the intracellular versus extracellular compartments?

3. What is the normal concentration of salt in extracellular fluid? What happens to cells if the saltiness of the extracellular fluid increases or decreases?

4. Distinguish between hypovolemic and osmotic thirst, and identify the physiological sensors that detect each condition.

5. Why do we sometimes get hungry for salt?

PART III
Food and Energy Regulation

Feast or famine—these are poles of human experience. Hunger for the food that we need to build, maintain, and fuel our bodies is a compelling drive, and flavors are powerful reinforcers. The behaviors involved in obtaining and consuming food shape our daily schedules, and mass media feed us a steady diet of information about food: crop reports, stories about famines and droughts, cooking shows, and restaurant ads.

Our reliance on food for energy and nutrition is shared with all other animals. In the remainder of this chapter we look at the regulation of feeding and energy expenditure, as well as some aspects of food-related behavior.

Nutrient Regulation Helps Prepare for Future Needs

The regulation of eating and of body energy, compared with regulation of drinking, involves more redundant mechanisms and a more complex set of homeostatic mechanisms. The added complexity is due to the fact that the brain must monitor and regulate a wide range of **nutrients** (chemicals required for the effective functioning, growth, and maintenance of the body), including more than 20 amino acids (of which 9 *essential amino acids* cannot be synthesized by the body) and a variety of vitamins and minerals, as well as carbohydrates (sugars and starches) for energy. No animal can afford to run out of energy or nutrients, so we need systems to anticipate future needs and keep a reserve on hand (but not *too* much!). (For more information about the complexity and variation in the nutrient needs of different species, see **A STEP FURTHER 9.3**, on the website.)

The principal fuel for the cells of the body is **glucose**, a simple sugar that is obtained through the breakdown of more-complex molecules. Because we need a steady supply of glucose between meals and may also experience elevated demand for fuel

aldosterone A mineralocorticoid hormone, secreted by the adrenal cortex, that promotes the conservation of sodium by the kidneys.

nutrient A chemical that is needed for growth, maintenance, and repair of the body but is not used as a source of energy.

glucose An important sugar molecule used by the body and brain for energy.

glycogen A complex carbohydrate made by the combining of glucose molecules for a short-term store of energy.

insulin A pancreatic hormone that lowers blood glucose, promotes energy storage, and facilitates glucose utilization by cells.

glucagon A pancreatic hormone that converts glycogen to glucose and thus increases blood glucose.

lipid A large molecule (frequently a *fat*) that consists of fatty acids and glycerol. Lipids are insoluble in water.

adipose tissue Commonly called *fat tissue*. Tissue made up of fat cells.

ketone A compound, liberated by the breakdown of body fats and proteins, that is a metabolic fuel source.

basal metabolism The consumption of energy to fuel processes such as heat production, maintenance of membrane potentials, and all the other basic life-sustaining functions of the body.

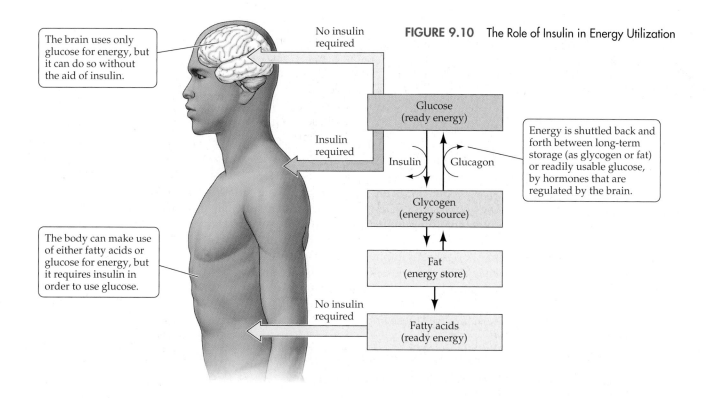

The brain uses only glucose for energy, but it can do so without the aid of insulin.

No insulin required

FIGURE 9.10   The Role of Insulin in Energy Utilization

Glucose (ready energy)

Insulin required

Insulin   Glucagon

Energy is shuttled back and forth between long-term storage (as glycogen or fat) or readily usable glucose, by hormones that are regulated by the brain.

The body can make use of either fatty acids or glucose for energy, but it requires insulin in order to use glucose.

Glycogen (energy source)

Fat (energy store)

No insulin required

Fatty acids (ready energy)

at other times—for example, during intense physical activity—several mechanisms have evolved for short- and long-term storage of excess fuel. For shorter-term storage, glucose can be converted into a more complex molecule called **glycogen** and stored as reserve fuel in several locations, most notably the liver and skeletal muscles. This process is promoted by the hormone **insulin**, which is synthesized and released by the pancreas. When blood glucose levels drop too low, a second pancreatic hormone, **glucagon**, converts glycogen back into glucose (**FIGURE 9.10**). For longer-term storage, molecules of **lipid** (fat) from dietary sources or created from surplus sugars and other nutrients are stored in **adipose tissue** (commonly called *fat tissue*). Under conditions of prolonged food deprivation, body fat can be converted into glucose and a secondary form of fuel, called **ketones**, which can similarly be utilized by the body and brain.

Only about 10–20% of the energy in food is used for active behavioral processes. The majority of food energy is spent on **basal metabolism**: the basic physiological processes of life, like heat production (the price we pay for endothermy), cellular activity, and maintenance of membrane potentials. Metabolism is under homeostatic control and can be adjusted to a surprising extent. An all-too-familiar consequence of this metabolic flexibility is that people and animals tend to resist either losing or gaining weight following dietary changes (**FIGURE 9.11**). To the frustration of dieters everywhere, many studies show that a calorie-reduced diet prompts the body to reduce its basal metabolic rate in order to prevent the loss of weight, regardless of whether the dieter

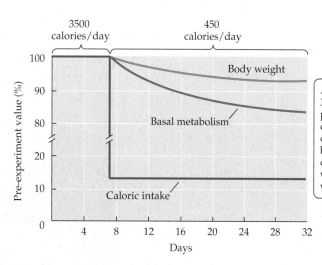

After 7 days on a rich 3,500-calorie diet, six obese participants reduced their daily calories by 87%, to 450 calories per day. However, basal metabolism also declined by 15%; so after 3 weeks on the new diet, body weight had declined by only

FIGURE 9.11   Why Losing Weight Is So Difficult (After Bray, 1969.)

diabetes mellitus   A condition, characterized by excessive glucose in the blood and urine and by reduced glucose utilization by body cells, that is caused by the failure of insulin to induce glucose absorption.

is initially of normal weight or obese (Bray, 1969; C. K. Martin et al., 2007). Mice whose basal metabolic rate has been increased (by an induced increase in the energy used by mitochondria) eat more yet weigh less than normal mice, without increased locomotor activity (Clapham et al., 2000). Perhaps someday a drug will be developed to exert this effect on human mitochondria and produce such wonderful results in humans as well.

In the meantime, the debate about the most effective ways to decrease fat deposition through dieting continues to be immensely popular in the mass media. Although it is counterintuitive, some evidence suggests that diets high in fats and proteins, and correspondingly low in carbohydrates, can help people lose weight and also may increase serum levels of "good" cholesterol while decreasing serum fats (G. D. Foster et al., 2003; Samaha et al., 2003). However, long-term studies will be required to establish the overall safety of low-carbohydrate diets; after all, plenty of evidence already indicates that people with diets high in fat have more heart disease. The only certain way to lose weight is to decrease the number of calories eaten and/or increase the number of calories spent in physical activity, and for the weight loss to be permanent, these changes in diet and activity must be permanent too. But as an added bonus, research with monkeys suggests that long-term restriction of caloric intake can slow the aging process and reduce the prevalence of disease (Colman et al., 2014). It remains to be seen how well these findings translate to humans.

Insulin Is Essential for Obtaining, Storing, and Using Food Energy

Each time you eat a meal, the foods are broken down and glucose is released into the bloodstream. We have already mentioned that insulin converts surplus glucose into glycogen, which can then be stored for later use. But insulin has another, more immediately critical function: your body needs insulin in order to make any use of the circulating glucose. That's because *glucose transporters*—the membrane-spanning proteins that most cells use to import glucose from the blood—need insulin in order to function properly (brain cells are an important exception; they can use glucose without the aid of insulin). The disease **diabetes mellitus** results from a lack of insulin production (in the *type 1*, or *juvenile-onset*, variety of the disease) or from greatly reduced tissue sensitivity to insulin (*type 2*, or *adult-onset*, diabetes, which is often associated with obesity). Although the brain can still make use of glucose from the diet, the rest of the body cannot and is forced to use energy from fatty acids. An untreated person with diabetes may eat a great deal because the body cannot make efficient use of the ingested food, and the long-term reliance on fatty acids for energy can result in severe damage to various tissues.

Insulin release around mealtimes is so important that it is triggered by several different mechanisms at different points in time. First, simply seeing, smelling,

(A)   (B)

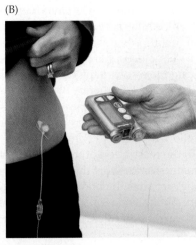

Not Too Sweet People with type 1 diabetes mellitus must receive insulin in order to utilize glucose. Taking insulin requires careful monitoring of blood glucose and, as shown here, self-administration via (A) daily injections or (B) drug pumps to infuse insulin throughout the day. Until the 1920s, when insulin was discovered as a result of experiments on dogs, this form of diabetes was a dreaded killer of children. In an especially dramatic moment in science, three of the discoverers of insulin—Frederick Banting, James Collip, and Charles Best—went through a hospital ward full of dying diabetic children, injecting them with the newly purified hormone. By the time the last child had been injected, the first was already waking up from a diabetic coma. The discovery saved millions of lives and garnered a Nobel Prize in 1923.

or tasting stimuli associated with food evokes a conditioned release of insulin in anticipation of glucose arrival in the blood. This release, because it is mediated by the brain, is called the *cephalic phase* of insulin release (recall that *cephalic* means "of the head"). Then, during the *digestive phase* (if you were lucky enough to get something to eat), food entering the digestive tract prompts an additional release of insulin. We now know that the digestive system contains the same sort of sweet taste receptors as are found on the tongue, and it uses them to help regulate insulin release (Kokrashvili et al., 2009). Finally, during the *absorptive phase*, as digested food is absorbed into the bloodstream, specialized liver cells called **glucodetectors** detect the increase in circulating glucose and signal the pancreas to release still more insulin. Information from the liver's glucodetectors is also conveyed directly to the hypothalamus via the vagus nerve (Powley, 2000), and the brain uses this information about circulating glucose to help control the pancreas and to stimulate feelings of hunger.

Given the crucial role of insulin in mobilizing and distributing food energy, you might think that the primary way the brain decides when it's time to eat (and when to stop) is by monitoring circulating insulin levels; high levels of insulin, signaling the presence of food in the gut, could provoke satiety (feeling "full"), and low levels of insulin between meals could make us feel hunger. Indeed, lowering an animal's blood insulin level causes it to become hungry and eat a large meal, and injecting some insulin causes the animal to eat much less. But injecting a larger dose of insulin doesn't produce fully satiated animals; instead, they become hungry again and eat a large meal! The reason for this surprising result is that the high insulin levels direct much of the blood glucose into storage, which means that there is *less* glucose in circulation. The brain learns of this glucose deficit, called *hypoglycemia*, directly via glucodetectors, leading to a hunger response.

Do these observations mean that the hunger signal might be circulating glucose rather than insulin? While it is certainly important, glucose can't be the sole appetite signal either, because people with untreated diabetes have very high levels of circulating glucose yet are constantly hungry. Somehow the brain integrates insulin and glucose levels with other sources of information to decide whether to initiate eating. As we'll see next, this has become a central theme in research on appetite control—that the brain integrates many different signals rather than relying exclusively on any single signal to trigger hunger.

glucodetector   A specialized type of liver cell that detects and informs the nervous system about levels of circulating glucose.

——————————how's it going

1. Provide a review of how glucose is used, stored, and retrieved from storage. Be sure to identify the roles of pancreatic hormones in each step.

2. What is basal metabolism? How is it affected by homeostatic processes, and how does the homeostatic regulation of metabolism frustrate efforts to lose weight?

3. Define the types of diabetes mellitus and discuss their causes. What are some of the ways that insulin release is normally controlled?

4. Discuss the evidence about whether blood levels of insulin and glucose directly control hunger.

The Hypothalamus Coordinates Multiple Systems That Control Hunger

Although no single brain region has exclusive control of appetite, decades of research has confirmed that the hypothalamus is critically important for regulating metabolic rate, food intake, and body weight. In classic research, scientists found that lesions in the hypothalamus could induce either chronic hunger and massive weight gain, or chronic satiety and severe weight loss, depending on the location of the lesion. Next we'll review the experiments that led to these conclusions.

Lesion studies showed that the hypothalamus is crucial for appetite

Early researchers made discrete bilateral lesions in the hypothalamus—in either the **ventromedial hypothalamus (VMH)** or the **lateral hypothalamus (LH)** (**FIGURE 9.12A**). After recovery, VMH-lesioned rats ate to excess and became obese (Hetherington and Ranson, 1940), leading researchers to suggest that the VMH is the *satiety center* of the brain (because the rats ceased to experience satiety once the VMH was gone). Rats with LH lesions, conversely, ceased eating and rapidly lost weight, suggesting that the LH acts as a *hunger center* (Anand and Brobeck, 1951). So, an early model of feeding behavior featured the VMH and LH acting in opposition to control appetite.

Subsequent research showed that the initial model of appetite was too simple. For one thing, although the VMH was identified as a satiety center, its destruction did not create out-of-control feeding machines.

Instead, VMH-lesioned animals exhibited a period of rapid weight gain but then stabilized at a new, higher body weight. When obese VMH-lesioned animals were forced to either gain or lose weight through dietary manipulation, they returned to their new "normal" weight as soon as they were allowed to eat freely again (**FIGURE 9.12B**). So, because VMH-lesioned rats experienced satiety, the VMH cannot be the sole satiety controller.

Similarly, the LH can't be the sole hunger center. Although they initially stopped eating, LH-lesioned rats that were kept alive with a feeding tube soon resumed eating and drinking and eventually stabilized their body weight at a new, lower level. As with the VMH-lesioned animals, LH-lesioned animals that were forced to gain weight would swiftly return to their new, lower set point for body weight after they returned to eating at will (see Figure 9.12B) (Keesey, 1980).

FIGURE 9.12 Lesion Studies Revealed That the Hypothalamus Is Involved in Appetite

■ **Hypothesis**

The hypothalamus contains discrete systems for controlling hunger and satiety.

■ **Test**

Place small lesions in target areas within the hypothalamus.

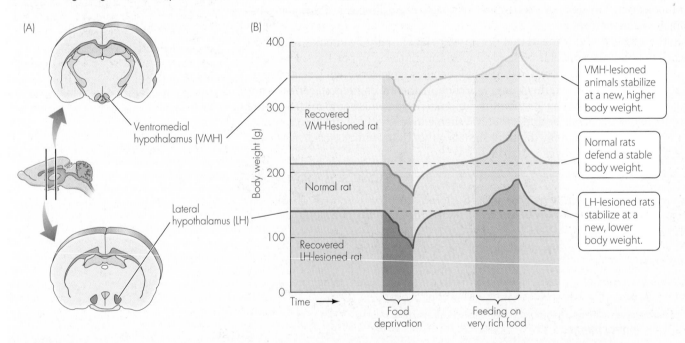

■ **Result**

Animals with lesions of the lateral hypothalamus (LH) decrease their food intake and rapidly lose weight, but they eventually stabilize at a new, lower weight. Following recovery, LH-lesioned animals forced to gain or lose weight return to the new lower weight when allowed to feed freely. Animals with lesions of the ventromedial hypothalamus (VMH) increase their food intake and rapidly gain weight, but they eventually stabilize at a new, higher weight. Following recovery, VMH-lesioned animals forced to gain or lose weight return to their new higher weight when allowed to feed freely.

■ **Conclusion**

The LH and VMH appear to play a role in appetite and body weight control, but because LH- and VMH-lesioned animals eventually show hunger and satiety, these two hypothalamic centers alone cannot constitute the entire appetite control system.

By demonstrating that the hypothalamus contains distinct components of an appetite control network, the early research on hunger and satiety provided a framework for subsequent work. For example, fMRI studies show that elevations in circulating glucose after a period of fasting produce large changes in the activity of the human hypothalamus (**FIGURE 9.13**) (Y. Liu et al., 2000), probably acting via hypothalamic glucodetector neurons that directly monitor blood levels of glucose (Parton et al., 2007).

Today it is clear that hypothalamic control of feeding is quite complicated and, like other homeostatic systems, exhibits redundancy as a safety measure. However, researchers have uncovered many of the details of the hypothalamic appetite control network and its integration of multiple signals, as we'll see next.

Hormones from the body drive a hypothalamic appetite controller

A spate of discoveries has sharpened our understanding of the hypothalamic control of appetite. This evidence indicates that a circuit within the **arcuate nucleus** of the hypothalamus is the key element in a highly specialized appetite network integrating peptide hormone signals from several sites in the body. One important source of information about energy stores is the pancreas; we have already discussed how the pancreatic hormone insulin signals the state of glucose circulating in the blood. Other information about energy balance—especially short-term and long-term reserves—comes in the form of hormonal secretions from elsewhere in the body, particularly the digestive organs and fat tissue.

You may be surprised to learn that the fat cells that make up adipose tissue are endocrine; in fact, they release a hormone called **leptin** (from the Greek *leptos*, "thin") into the bloodstream (Y. Zhang et al., 1994). Genetically modified mice carrying two copies of the aptly named *obese* gene fail to produce leptin—and they do indeed become obese (**FIGURE 9.14**). The fat mice will stay that way even if they are given only unpalatable food, or if they are required to work hard to obtain food (Cruce et al., 1974). And mice that make normal amounts of leptin but have nonfunctional leptin receptors likewise become obese (L. M. Zucker and Zucker, 1961; al-Barazanji et al., 1997). It appears that the brain senses circulating leptin levels to monitor the body's longer-term energy reserves in the form of fat. Defective leptin production or impaired

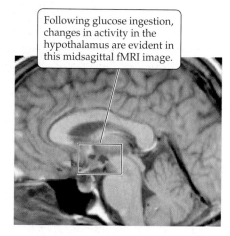

Following glucose ingestion, changes in activity in the hypothalamus are evident in this midsagittal fMRI image.

FIGURE 9.13 Sweet Spot (From Y. Liu et al., 2000.)

ventromedial hypothalamus (VMH) A hypothalamic region involved in eating and sexual behaviors.

lateral hypothalamus (LH) A hypothalamic region involved in the control of appetite and other functions.

arcuate nucleus An arc-shaped hypothalamic nucleus implicated in appetite control.

leptin A peptide hormone released by fat cells.

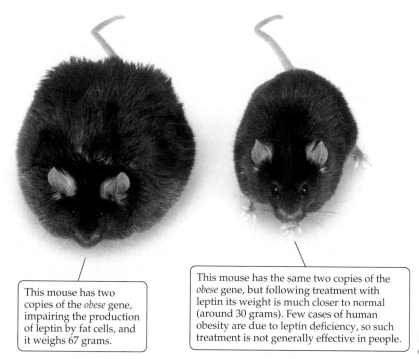

This mouse has two copies of the *obese* gene, impairing the production of leptin by fat cells, and it weighs 67 grams.

This mouse has the same two copies of the *obese* gene, but following treatment with leptin its weight is much closer to normal (around 30 grams). Few cases of human obesity are due to leptin deficiency, so such treatment is not generally effective in people.

FIGURE 9.14 Inherited Obesity Can Be Overcome

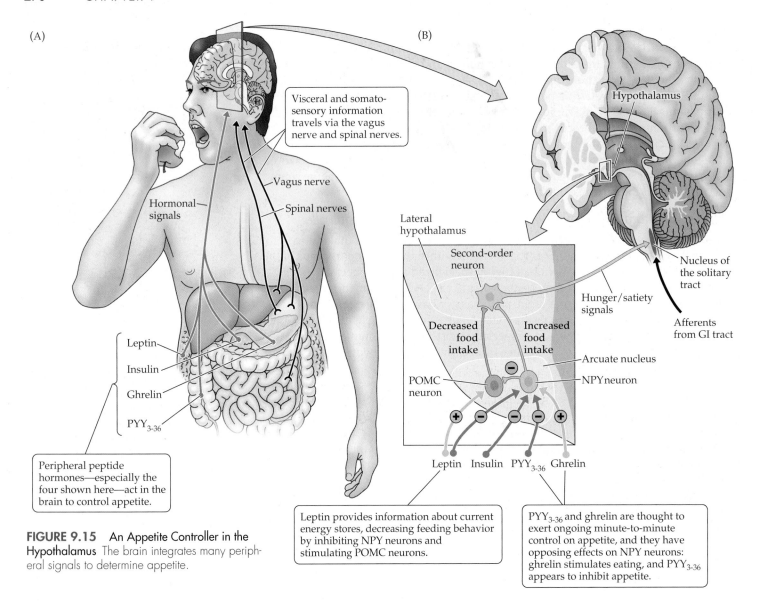

(A)

Visceral and somato-sensory information travels via the vagus nerve and spinal nerves.

(B)

Hypothalamus

Vagus nerve

Spinal nerves

Hormonal signals

Lateral hypothalamus

Second-order neuron

Decreased food intake

Increased food intake

Hunger/satiety signals

Nucleus of the solitary tract

Afferents from GI tract

Arcuate nucleus

POMC neuron

NPY neuron

Leptin

Insulin

Ghrelin

PYY_{3-36}

Leptin   Insulin PYY_{3-36} Ghrelin

Peripheral peptide hormones—especially the four shown here—act in the brain to control appetite.

FIGURE 9.15   An Appetite Controller in the Hypothalamus The brain integrates many peripheral signals to determine appetite.

Leptin provides information about current energy stores, decreasing feeding behavior by inhibiting NPY neurons and stimulating POMC neurons.

PYY_{3-36} and ghrelin are thought to exert ongoing minute-to-minute control on appetite, and they have opposing effects on NPY neurons: ghrelin stimulates eating, and PYY_{3-36} appears to inhibit appetite.

ghrelin   A peptide hormone produced and released by the gut.

PYY_{3-36}   A peptide hormone, secreted by the intestines, that probably acts on hypothalamic appetite control mechanisms to suppress appetite.

leptin sensitivity causes a false underreporting of body fat, leading the animals to overeat, especially high-fat or sugary foods.

Shorter-term energy balance—the presence or absence of food in the gut—is reported by hormones from the digestive organs. Two such hormones that seem to be especially important for appetite control are **ghrelin**, synthesized and released into the bloodstream by endocrine cells of the stomach, and the awkwardly named **PYY_{3-36}**, a small peptide released by intestinal cells. Ghrelin reaches high concentrations during fasting and powerfully stimulates appetite (Wren et al., 2000, 2001), dropping sharply after a meal is eaten (Nakazato et al., 2001). PYY_{3-36} shows the converse pattern, spiking to higher levels on ingestion of a meal and providing a potent appetite-*suppressing* satiety signal (see Karra et al., 2009, for a review). Injections of ghrelin cause increased appetite and feeding in rats or humans, and injections of PYY_{3-36} into the bloodstream or directly into the arcuate nucleus curb appetite (Batterham and Bloom, 2003; Chelikani et al., 2005; Baynes et al., 2006). Furthermore, ghrelin is chronically slightly elevated in obese people, and PYY_{3-36} is chronically lowered—possibly causing continual hunger (English et al., 2002). All of these hormone signals appear to converge on an appetite controller in the arcuate nucleus, so next we'll have a look at how that system seems to work.

A simplified view of the organization of appetite control circuitry in the arcuate nucleus is illustrated in **FIGURE 9.15**. The system relies on two types of neurons with

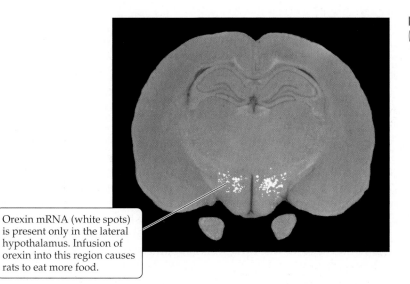

FIGURE 9.16 Neuropeptides That Induce Hunger? (Courtesy of Masashi Yanagisawa.)

Orexin mRNA (white spots) is present only in the lateral hypothalamus. Infusion of orexin into this region causes rats to eat more food.

opposite effects, named after the types of signaling substances they produce. Neurons that produce the peptides pro-opiomelanocortin (POMC) and cocaine- and amphetamine-regulated transcript—**POMC neurons**—act as satiety neurons when activated, inhibiting appetite and increasing metabolism. Neurons that produce the peptides neuropeptide Y and agouti-related peptide—**NPY neurons**—act as hunger neurons when activated, stimulating appetite directly, inhibiting POMC neurons (thereby blocking satiety signals) and reducing metabolism. Projections to regions outside the arcuate nucleus ultimately control food intake (see Figure 9.15B), as we'll see shortly. But first, let's consider what is known about how the peripheral hormones interact with this appetite controller.

As we've discussed, leptin (and to a lesser extent, insulin) conveys information about the body's longer-term energy reserves, stored away within fat cells. Leptin affects both types of arcuate appetite neurons, but in opposite ways. High circulating levels of leptin activate the POMC satiety neurons but simultaneously inhibit the NPY hunger neurons—so in both systems leptin is working to suppress hunger. In contrast to leptin, ghrelin and PYY_{3-36} provide shorter-term hour-to-hour hunger signals from the gut. Both of these peptides act primarily on the appetite-stimulating NPY neurons of the arcuate nucleus. Ghrelin stimulates these cells, leading to a corresponding increase in appetite. PYY_{3-36} works in opposition, inhibiting the same cells to *reduce* appetite. Thus, short-term control of appetite seems to reflect a balance between ghrelin and PYY_{3-36} concentrations in circulation.

The net result of all this is a constant balancing act between the appetite-stimulating effects of the NPY system and the appetite-suppressing effects of the POMC system, spread across the hypothalamus. The peptide **orexin**, produced by neurons in the lateral hypothalamus, appears to participate in the subsequent control of feeding behavior (**FIGURE 9.16**) (Sakurai et al., 1998).

Other systems also play a role in hunger and satiety

Appetite signals from the hypothalamus converge on the **nucleus of the solitary tract** (**NST**) in the brainstem (see Figure 9.15B). The NST can be viewed as part of a common pathway for feeding behavior, receiving appetite signals from a variety of sources in addition to the hypothalamus. For example, the sensation of hunger is affected by a wide variety of peripheral sensory inputs, such as oral stimulation and the feeling of stomach distension, transmitted via spinal and cranial nerves. Information about nutrient levels is conveyed directly from the body to the NST via the vagus nerve (Tordoff et al., 1991). For example, the gut peptide **cholecystokinin** (**CCK**), released by the gut after feeding, provides yet another appetite-suppressant signal to the brain directly via the vagus (H. Fink et al., 1998).

POMC neuron A neuron, involved in the hypothalamic appetite control system, that produces both pro-opiomelanocortin and cocaine- and amphetamine-related transcript.

NPY neuron A neuron, involved in the hypothalamic appetite control system, that produces both neuropeptide Y and agouti-related peptide.

orexin Also called *hypocretin*. A neuropeptide produced in the hypothalamus that is involved in switching between sleep states, in narcolepsy, and in the control of appetite.

nucleus of the solitary tract (NST) A complicated brainstem nucleus that receives visceral and taste information via several cranial nerves.

cholecystokinin (CCK) A peptide hormone that is released by the gut after ingestion of food that is high in protein and/or fat.

endocannabinoid An endogenous ligand of cannabinoid receptors, thus a marijuana analog that is produced by the brain.

epigenetic transmission The passage from one individual to another of changes in the expression of targeted genes, without modifications to the genes themselves.

In keeping with the concept of multiple redundancy that we discussed earlier in the chapter, a variety of additional signals and brain locations also participate in feeding behavior, either directly or through indirect effects on other processes. For example, as you might expect, the brain's reward system appears to be intimately involved with feeding. Activity of a circuit including the amygdala and the dopamine-mediated reward system in the nucleus accumbens (see Chapter 4) is hypothesized to mediate pleasurable aspects of feeding (Ahn and Phillips, 2002; Volkow and Wise, 2005).

The **endocannabinoid** system (see Chapter 4) likewise has a potent effect in appetite and feeding. Endocannabinoids, such as anandamide, are endogenous substances that act much like the active ingredient in marijuana (*Cannabis sativa*) and, like marijuana, can potently stimulate hunger. Acting both in the brain and in the periphery, endocannabinoids might stimulate feeding by affecting the mesolimbic dopamine reward system. However, injection of anandamide into the hypothalamus stimulates eating (C. D. Chapman et al., 2012), confirming that endocannabinoids act directly on hypothalamic appetite mechanisms, while also inhibiting satiety signals from the gut (Di Marzo and Matias, 2005).

Hypothalamic feeding control must be strongly influenced by inputs from higher brain centers, but little is known about these mechanisms. During development, for example, our feeding patterns are increasingly influenced by social factors such as parental and peer group pressures (Birch et al., 2003). Understanding the nature of cortical influences on feeding mechanisms is a major challenge for the future. The list of participants in appetite regulation is long and growing longer, revealing overlapping and complex controls with a high degree of redundancy, as befits a behavioral function of such critical importance to health and survival. With each new discovery, we draw nearer to finally developing safe and effective treatments for eating disorders, as we discuss next.

how's it going ❓

1. Review the early work implicating the LH and VMH in appetite control. Why did researchers abandon the view that the LH and VMH were the sole controllers of appetite and satiety?

2. Sketch or briefly describe the major components of the hypothalamic appetite controller, as currently understood.

3. Describe how the hypothalamic appetite controller functions during hunger, in contrast to just after a meal.

4. What is leptin? Discuss its origins in the body and its effects on feeding behavior.

5. What are some of the gut hormones that may be involved in appetite control, and what are they thought to do?

6. Briefly describe some of the additional (redundant) mechanisms of appetite control that supplement the hypothalamic appetite system. What is the basis of the increased appetite frequently experienced after use of marijuana?

Obesity Is Difficult to Treat

Unfortunately, effective treatments to aid weight loss have been elusive. In our modern world, with its plentiful calories and sedentary lifestyles, the multiple redundant systems for appetite and energy management that evolved in our distant ancestors work all too well in preventing weight loss. Obesity has certainly reached epidemic proportions: a majority of the adults in the United States are overweight—about two-thirds—and about one in three qualify as obese (**FIGURE 9.17**; Flegal et al., 2002), as defined by body mass index (BMI; **TABLE 9.1**). Future health care systems will be burdened by obesity-related disease—cardiovascular disease, diabetes, etc.—and parental obesity may program metabolic disadvantages in offspring via **epigenetic transmission** (Ng et al., 2010).

In Lewis Carroll's *Alice's Adventures in Wonderland*, Alice quaffs the contents of a small bottle in order to shrink. The quest for a real-life shrinking potion—but one

TABLE 9.1 Body Mass Index (BMI)

Value	Body weight category
<15	Starvation
15–18.5	Underweight
18.5–25	Ideal weight
25–30	Overweight
30–40	Obese
>40	Morbidly obese

Note:

$$BMI = \frac{weight\,(kg)}{height \times height\,(m \times m)}$$

or

$$BMI = 703\frac{weight\,(lb)}{height \times height\,(in. \times in.)}$$

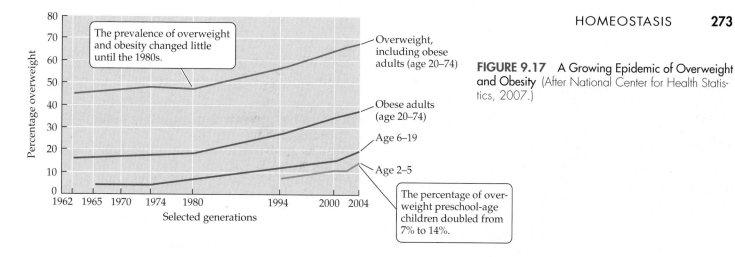

The prevalence of overweight and obesity changed little until the 1980s.

The percentage of overweight preschool-age children doubled from 7% to 14%.

FIGURE 9.17 A Growing Epidemic of Overweight and Obesity (After National Center for Health Statistics, 2007.)

that makes you thin rather than short—is the subject of intense scientific activity, and several major strategies or targets are emerging:

1. *Appetite control* Hopes are high that drugs designed to reset the hypothalamic appetite controller will be safe and potent obesity treatments. Alteration of leptin levels has not proven to be very effective (Montague et al., 1997). Drugs that directly interfere with endocannabinoid activity, which is normally regulated by leptin in the hypothalamus, effectively produce "anti-munchies"—the reverse of the hunger experienced by marijuana users (Van Gaal et al., 2005; Thornton-Jones et al., 2006). Significant mood problems (an "anti-high"?) also occur, however, so the search continues for drugs that can selectively modify the signaling systems in the arcuate appetite controller. Treatments that mimic other signaling hormones are promising, especially those that exploit the PYY_{3-36} satiety signal (Chelikani et al., 2005; Neary and Batterham, 2009). Simply spraying a PYY_{3-36} solution into the mouths of lab mice is apparently not aversive, yet it powerfully suppresses their appetite (Hurtado et al., 2013).

2. *Increased metabolism* An alternative approach to treating obesity involves treatments that cause the body's metabolic rate to increase and thus expend extra calories in the form of heat. For example, metabolic rate is controlled by the thyroid hormones, so scientists are trying to design drugs that will mimic some of the actions of thyroid hormones in order to increase the rate at which energy is used by cells, without producing harmful side effects (Grover et al., 2003). Another promising approach involves inducing fat tissue to start burning stored energy faster than normal (Boström et al., 2012; Kajimura and Saito, 2014).

3. *Inhibition of fat tissue* A third way to treat obesity involves blocking the formation of new fat tissue. For example, in order for fat tissue to grow, it must be able to recruit and develop new blood vessels. Drugs that block this process inhibit weight gain in mice and may have similar anti-obesity benefits in humans (Rupnick et al., 2002; Tam et al., 2009).

4. *Reduced absorption* One of the few currently approved obesity medications—orlistat (trade name Xenical)—works by interfering with the digestion of fat. However, this approach has generally produced only modest weight loss, and it often causes intestinal discomfort.

5. *Reduced reward* A different perspective on treating obesity focuses on the rewarding properties of food. Not only is food delicious, but "comfort foods" also directly reduce circulating stress hormones, thereby providing another reward. Drugs that affect the brain's reward circuitry (see Chapter 4), reducing the rewarding properties of food, may promote weight loss (Volkow and Wise, 2005).

6. *Anti-obesity surgery* The surgical removal of fat tissue, particularly through **liposuction**, has been a popular approach to controlling weight, but it is generally only moderately successful, and it is temporary. Surgical reduction of the stomach and intestine has thus become a popular alternative, as we discuss at the end of the chapter.

liposuction The surgical removal of fat tissue.

Changing Ideals of Female Beauty Actress Keira Knightley (*left*) exemplifies modern society's emphasis on thinness as an aspect of beauty. In contrast, Flemish painter Peter Paul Rubens' painting of his wife in *Helena Fourment as Aphrodite* (circa 1630; *right*) illustrates the very different ideal for the feminine form during her era. Are our modern weight-conscious notions of female beauty responsible for some cases of anorexia nervosa and bulimia?

anorexia nervosa A syndrome in which individuals severely deprive themselves of food.

bulimia Also called *bulimia nervosa*. A syndrome in which individuals periodically gorge themselves, usually with "junk food," and then either vomit or take laxatives to avoid weight gain.

binge eating The rapid intake of large quantities of food, often poor in nutritional value and high in calories.

Eating Disorders Can Be Life-Threatening

Sometimes people shun food, despite having no apparent aversion to it. Like Kitty, whom we met at the opening of the chapter, these people are usually young, become obsessed with their body weight, and become extremely thin—generally by eating very little and sometimes also by regurgitating food, taking laxatives, overexercising, or drinking large amounts of water to suppress appetite. This condition, which is more common in adolescent and adult women than in men, is called **anorexia nervosa**. The name of the disorder indicates (1) that the patients have no appetite (*anorexia*) and (2) that the disorder originates in the nervous system (*nervosa*).

People who suffer from anorexia nervosa tend to think about food a good deal, and physiological evidence suggests that they respond even *more* than normal people to the presentation of food (Broberg and Bernstein, 1989); for example, food stimuli provoke a large release of insulin, despite cognitive denial of any feelings of hunger. So, in a physiological sense their hunger may be normal or even exaggerated, but this hunger is somehow absent from their conscious perceptions and they refuse to eat. The idea that anorexia nervosa is primarily a nervous system disorder stems from this mismatch between physiology and cognition, as well as from the distorted body image of the patients (they may consider themselves fat when others see them as emaciated). There may also be abnormalities in the functioning of the dopamine-based reward system that signals pleasurable aspects of eating, persisting even after recovery (Kaye et al., 2009).

Anorexia nervosa is notoriously difficult to treat, because it appears to involve an unfortunate combination of genetic, endocrine, personality, cognitive, and environmental variables. One useful approach, called *family-based treatment* (FBT), is a therapy that deemphasizes the identification of causal factors and instead focuses on intensive, parent-led "refeeding" of the anorexic person (Le Grange, 2005). This approach (sometimes also called *Maudsley therapy* after the hospital where it was introduced) was effective in returning Kitty to her normal weight, and it remains one of the best established treatments for anorexia (Kass et al., 2013).

Bulimia (or *bulimia nervosa*, from the Greek *boulimia*, "great hunger") is a related disorder. Like those who suffer from anorexia nervosa, people with bulimia may believe themselves fatter than they are, but they periodically gorge themselves, usually with "junk food," and then either vomit the food or take laxatives to avoid weight gain. Also like sufferers of anorexia nervosa, people with bulimia may be obsessed with food and body weight, but not all of them become emaciated. Both anorexia nervosa and bulimia can be fatal because the patient's lack of nutrient reserves damages various organ systems and/or leaves the body unable to battle otherwise mild diseases.

In **binge eating**, people spontaneously gorge themselves with far more food than is required to satisfy hunger, often to the point of illness. Such people are often obese, and the causes of the bingeing are not fully understood. In susceptible people, the strong pleasure associated with food activates opiate and dopaminergic reward mechanisms to such an extent that bingeing resembles drug addiction.

Despite the epidemic of obesity in our society, or perhaps because of it, our present culture emphasizes that women, especially young women, must be thin to be attractive. This cultural pressure is widely perceived as one of the causes of eating disorders. In earlier times, however, when plump women were considered the most beautiful, some women still fasted severely and may have suffered from anorexia nervosa. The origins of these disorders remain elusive, and to date, the available therapies help only a minority of patients like Kitty.

SIGNS & SYMPTOMS

Fat-Busting Surgery

Because fat tissue tends to regrow after liposuction, some people are turning to a different strategy: **bariatric** surgeries that bypass part of the intestinal tract or stomach in order to reduce the absorptive capacity of the digestive system. Although gastric bypass surgery (**FIGURE 9.18A**) doesn't directly target appetite control mechanisms, alterations in appetite hormones such as ghrelin reportedly accompany the surgery (Baynes et al., 2006; D. E. Cummings, 2006). By producing significant and lasting weight loss, gastric bypass surgery can offer hope of substantial weight loss and reversal of comorbid conditions like type 2 diabetes and hypertension, but it is accompanied by risk of significant complications.

Less-invasive surgical procedures are under study, such as the use of liners that prevent the intestine from absorbing nutrients (**FIGURE 9.18B**), or gastric stimulators that activate the gut's satiety signals to reduce appetite (Miras et al., 2015). Curiously, simply implanting inert weights into the abdominal cavities of mice causes them to lose a proportionate amount of body weight, apparently by fooling the body into thinking it is fatter than it actually is (Adams et al., 2001). Perhaps some of us, someday, will be able to lose weight simply by taking on extra ballast!

bariatrics The branch of medicine that deals with the causes, prevention, and treatment of obesity.

FIGURE 9.18 Surgical Options for Obesity

(A) Gastric bypass

In gastric bypass (also called *Roux-en-Y bypass*), the stomach is surgically reduced to a small pouch and connected to the small intestine at some distance, thereby bypassing the initial stretch of small intestine. This reduces the ability of the digestive system to absorb nutrients from food.

(B) Implantation

A less invasive option is the implantation of a plastic liner into the small intestine. It acts as a barrier to prevent the absorption of food, fewer calories are absorbed from the diet. Both bypass and implantation result in weight loss and improvements in secondary problems like diabetes.

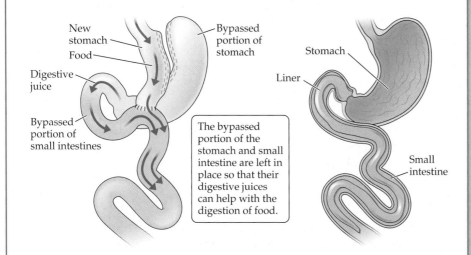

New stomach
Food
Bypassed portion of stomach
Digestive juice
Bypassed portion of small intestines

The bypassed portion of the stomach and small intestine are left in place so that their digestive juices can help with the digestion of food.

Stomach
Liner
Small intestine

how's it going ?

1. What proportion of adults in the United States are overweight or obese? How might homeostasis be part of the problem?

2. If you were designing drugs to combat obesity, what specific parts of the hypothalamic appetite controller might you target? Why?

3. Compare and contrast anorexia nervosa and bulimia. Are people with anorexia interested in food at all? Briefly discuss possible methods of treating anorexia.

4. Discuss the likely contributions of cultural versus biological factors in various eating disorders.

5. Discuss nonpharmacological methods of weight reduction. Is surgery a good option?

Recommended Reading

Agras, W. S. (Ed.). (2010). *Oxford Handbook of Eating Disorders*. New York, NY: Oxford University Press.

Blumberg, M. S. (2009). *Body Heat*. Cambridge, MA: Harvard University Press.

Brown, H. N. (2010). *Brave Girl Eating*. New York, NY: Morrow.

Kirkham, T., and Cooper, S. J. (Eds.). (2006). *Appetite and Body Weight: Integrative Systems and the Development of Anti-Obesity Drugs*. Burlington, MA: Academic Press.

McNab, B. K. (2012). *Extreme Measures: The Ecological Energetics of Birds and Mammals*. Chicago, IL: University of Chicago Press.

Power, M. L., and Schulkin, J. (2009). *The Evolution of Obesity*. Baltimore, MD: Johns Hopkins University Press.

Schulkin, J. (Ed.). (2012). *Allostasis, Homeostasis, and the Costs of Physiological Adaptation*. Cambridge, England: Cambridge University Press.

VISUAL SUMMARY

9

2e.mindsmachine.com/vs9

You should be able to relate each summary to the adjacent illustration, including structures and processes. Go to the online version of this summary (scan the QR code above) for links to figures, animations, and activities that will help you consolidate the material.

1 Homeostatic systems, such as the processes regulating body temperature (**thermoregulation**), work to maintain a constant internal environment. Like other homeostatic systems, thermoregulation employs **negative feedback** control: the resulting heat inhibits the system from calling for more. Review **Figure 9.1**, Animations **9.2** and **9.3**

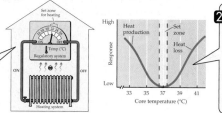

2 Critical homeostatic mechanisms tend to exhibit multiple redundancy. For example, redundant thermoregulatory controls with slightly different features can be found in the spinal cord, the brainstem, and the hypothalamus. Review **Figure 9.2**, Animation **9.4**

3 Homeostatic systems rely on specialized behaviors to help regulate physiological parameters. For example, most species have specialized behaviors to help warm or cool the body. Review **Figure 9.3**

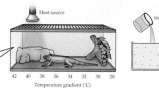

4 Our cells function properly only when the concentration of salt in the **intracellular compartment** of the body is within a critical range. The **extracellular compartment** is a source of replacement water for osmosis and a buffer between the intracellular compartment and the outside world. Review **Figure 9.6**

5 Thirst is a powerful motivator, triggered either by decreased volume of the extracellular fluid (**hypovolemic thirst**) or by increased extracellular saltiness (**osmotic thirst**). Because of the importance of solute concentration, we must regulate salt intake in order to regulate water balance effectively. Review **Figure 9.7**

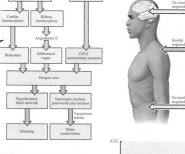

6 Specialized **osmosensory neurons** detect the concentration of extracellular fluid. **Baroreceptors** in the major blood vessels monitor blood pressure and volume. Review **Figure 9.8**

7 Hypovolemic and osmotic thirsts are triggered by different mechanisms and differ in their immediate effects, but both forms of thirst ultimately trigger a complex shared thirst network. Review **Figure 9.9**

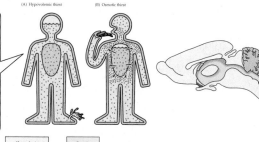

8 Our digestive system breaks down food and uses most of it for energy. **Insulin** helps most body cells to use **glucose** for fuel (the brain can use glucose directly) and promotes the storage of excess food energy. Another pancreatic hormone, **glucagon**, helps liberate glucose from storage. Review **Figure 9.10**

9 To the dismay of dieters, the body responds to caloric restriction by reducing metabolism, thus limiting weight loss. Review **Figure 9.11**

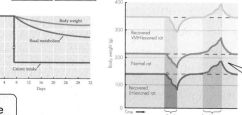

10 In **endotherms**, most food energy is used for **basal metabolism**. Metabolism readily shifts to compensate for changes in the availability of food. Review **Figure 9.12**

11 An appetite controller located in the **arcuate nucleus** of the hypothalamus responds to levels of several peptide gut hormones. When activated, arcuate **POMC neurons** act to decrease appetite, and arcuate **NPY neurons** act to stimulate appetite. **Leptin** and **insulin** provide important hormonal signals about longer-term energy storage. **Ghrelin** and **PYY**$_{3-36}$ provide more-acute signals from the gut. Ghrelin stimulates and PYY$_{3-36}$ inhibits the arcuate appetite control system. Review **Figure 9.15**, Activity **9.1**

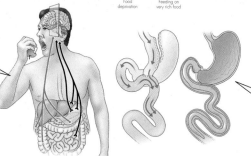

12 Obesity is a pervasive problem that is difficult to treat through diet, drugs, or surgery. The only long-lasting medical intervention for obesity is **bariatric** surgery, but several drug strategies based on a new understanding of appetite control offer promise. Review **Figure 9.18**

Go to **2e.mindsmachine.com** for study questions, quizzes, flashcards, and other resources.

chapter
10

Biological Rhythms and Sleep

When Sleep Gets Out of Control

Starting college always brings its share of new experiences and adjustments, but "Barry" knew something was wrong freshman year when he seemed to be sleepy all the time (S. Smith, 1997). Barry napped so often that his friends called him the hibernating bear. Of course, college can be exhausting, and many students seek refuge in long snooze sessions. But one day while Barry was camping with his pals, an even odder thing happened: "I laughed really hard, and I kind of fell on my knees … After that, about every week I'd have two or three episodes where if I'd laugh … my arm would fall down or my muscles in my face would get weak. Or if I was running around playing catch and someone said something, I would get weak in the knees. And there was a time there that my

friends kinda used it as a joke. If they're going to throw me the ball and they didn't want me to catch it, they'd tell me a joke and I'd fall down and miss it."

It was as if any big surge in emotion in Barry might trigger a sudden paralysis lasting anywhere from a few seconds to a few minutes, affecting either a body part or his whole body. Sex became something of a challenge because sometimes during foreplay, Barry's body would just collapse. "Luckily, you're probably laying down, so it's not that big a deal. But it just puts a damper on the whole thing."

What was happening to Barry? By the end of this chapter, we'll know a lot more about sleep and what went wrong in Barry's brain to cause these problems.

To see the video **Narcolepsy**, go to

2e.mindsmachine.com/10.1

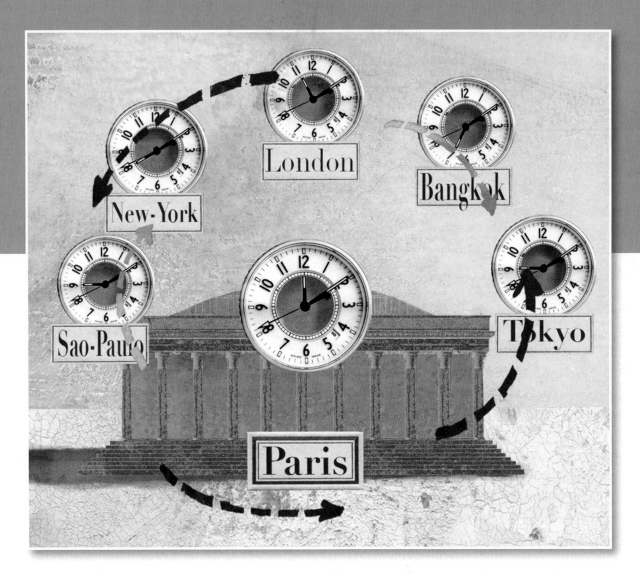

All living systems show repeating, predictable changes over time. Some rhythms, like brain potentials, are rapid; other rhythms, like annual hibernations, are slow. Daily rhythms, the topic of Part I of this chapter, have an intriguing clocklike regularity and are seen in virtually every physiological measure, including body temperature and hormone secretion. Part II of this chapter concerns that familiar daily rhythm known as the sleep-waking cycle. By age 60, most humans have spent 20 years asleep (some, alas, on one side or the other of the classroom podium). We'll find that sleep is not a passive state of "nonwaking," but rather the interlocking of several different states with distinct patterns of brain activity.

biological rhythm   A regular fluctuation in any living process.

circadian rhythm   A pattern of behavioral, biochemical, or physiological fluctuation that has a 24-hour period.

ultradian   Referring to a rhythmic biological event with a period shorter than a day, usually from several minutes to several hours long.

infradian   Referring to a rhythmic biological event with a period longer than a day.

PART I
Biological Rhythms

Biological rhythms are regular fluctuations in any living process. Almost all physiological processes—hormone levels, body temperature, drug sensitivity—change over the course of the day. Because such rhythms last about a day, they are called **circadian rhythms** (from the Latin *circa*, "about," and *dies*, "day"). Circadian rhythms are by far the most studied of the biological rhythms, and they will be our major concern in this chapter.

While circadian rhythms are the most familiar, you should know that some biological rhythms are shorter than a day. Such rhythms are referred to as **ultradian** (because they repeat more than once per day; the Latin *ultra* means "beyond"), and they vary from several minutes to hours long. Ultradian rhythms are seen in such behaviors as bouts of activity, feeding, and hormone release.

Biological rhythms that take *more* than a day are called **infradian** rhythms because they repeat less than once per day (the Latin *infra* means "below"). A familiar infradian rhythm is the 28-day human menstrual cycle. Many animal behaviors vary across the year; for example, most animals breed only during a particular season. There's also growing evidence of annual rhythms in the onset of human behavioral disorders, such as depression (see Chapter 12). (By the way, despite the urban myth that cases of depression peak around the holiday season, in fact they peak in the spring.) You might think that breeding seasons in animals would be triggered by changes in temperatures or food availability, but experiments suggest that the duration of light each day is the real trigger: in the laboratory, animals exposed to short days and long nights (mimicking wintertime conditions) reliably change to the nonbreeding condition.

Next we'll concentrate on the best-studied biological rhythms—circadian rhythms—in which light also plays a crucial role.

Many Animals Show Daily Rhythms in Activity

Humans and many other primates are *diurnal*—active during the day. But most other mammals, including most rodents, are *nocturnal*—active during dark periods. These circadian activities are extraordinarily precise: the beginning of activity may vary only a few minutes from one day to another. For humans equipped with watches and clocks, this regularity may seem uninteresting, but other animals achieve such remarkable regularity using only a built-in *biological clock*.

Circadian rhythms are generated by an endogenous clock

A favorite way to study circadian rhythms exploits rodents' love of running wheels. A switch attached to the wheel connects to a computer that registers each turn, revealing an activity rhythm as in **FIGURE 10.1A**. A hamster placed in a dimly lit room continues to show a daily rhythm in wheel running despite the absence of day versus night, suggesting that the animal has an internal clock. But even when the light is constantly dim, it is always possible that the animal detects other external cues (e.g., outside noises, temperature, barometric pressure—who knows?) signaling the time of day. Arguing for a biological clock, however, is the fact that in constant light or dark the circadian cycle is not *exactly* 24 hours: activity starts a few minutes later each day, so eventually

A Hamster for All Seasons Exposing Siberian hamsters to short day lengths mimicking autumn induces them to produce a silvery coat suitable for camouflage in snow. (Photo by Carol D. Hegstrom.)

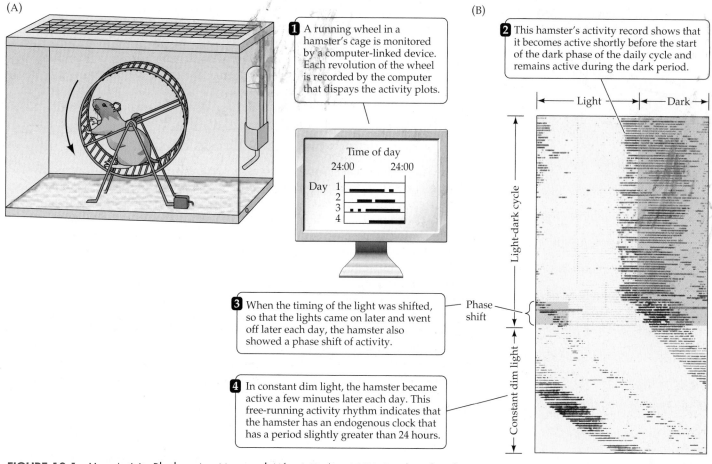

(A)

1 A running wheel in a hamster's cage is monitored by a computer-linked device. Each revolution of the wheel is recorded by the computer that dispays the activity plots.

Time of day

24:00   24:00

Day 1
     2
     3
     4

3 When the timing of the light was shifted, so that the lights came on later and went off later each day, the hamster also showed a phase shift of activity.

4 In constant dim light, the hamster became active a few minutes later each day. This free-running activity rhythm indicates that the hamster has an endogenous clock that has a period slightly greater than 24 hours.

(B)

2 This hamster's activity record shows that it becomes active shortly before the start of the dark phase of the daily cycle and remains active during the dark period.

|←— Light —→|←— Dark —→|

Light–dark cycle

Phase shift

Constant dim light

FIGURE 10.1 How Activity Rhythms Are Measured (After I. Zucker, 1976; Rusak and Zucker, 1979.)

the normally nocturnal hamster is active while it is daytime outside (**FIGURE 10.1B**, bottom). The animal is said to be **free-running**, maintaining its own personal cycle, which, in the absence of external cues, is a bit more than 24 hours long.

The free-running **period**, the time between two similar points of successive cycles (such as sunset to sunset), differs from one hamster to another. If two hamsters are placed in constant dim light next door to each other, eventually one may be active when the other is asleep—further evidence that they are not detecting some mysterious external cue. Rather, every animal has its own endogenous clock; periods vary from one animal to another within a species, and they vary across species.

Normally this internal clock is reset by light. If we expose a free-running nocturnal animal to periods of light and dark, the animal soon synchronizes its wheel running to the beginning of the dark period. The shift of activity produced by a synchronizing stimulus is referred to as a **phase shift** (see Figure 10.1B, middle), and the process of shifting the rhythm is called **entrainment**. Any cue that an animal uses to synchronize its activity with the environment is called a **zeitgeber** (German for "time giver"). Light acts as a powerful zeitgeber, and we can easily manipulate it in the lab. Because light stimuli can entrain circadian rhythms, the endogenous clock must receive input from the visual system, as we'll confirm shortly.

We humans experience a mismatch of internal and external time when we fly from one time zone to another. Flying three time zones east (say, from California to New York) means that sunlight arrives 3 hours sooner than our brain expects. The next morning we'll probably have a hard time waking up at 7:00 AM New York time,

To see the animation
Biological Rhythms,
go to

2e.mindsmachine.com/10.3

free-running Referring to a rhythm of behavior shown by an animal deprived of external cues about time of day.

period The interval of time between two similar points of successive cycles, such as sunset to sunset.

phase shift A shift in the activity of a biological rhythm, typically provided by a synchronizing environmental stimulus.

entrainment The process of synchronizing a biological rhythm to an environmental stimulus.

zeitgeber Literally "time-giver" (in German). The stimulus (usually the light-dark cycle) that entrains circadian rhythms.

suprachiasmatic nucleus (SCN)
A small region of the hypothalamus above the optic chiasm that is the location of a circadian clock.

because it's 4:00 AM California time. We need about one day per time zone to re-entrain after such travel, and in the meantime we experience jet lag, with symptoms such as insomnia and daytime fatigue.

In the absence of jet travel, the major value of circadian rhythms is obvious: they enable us to *anticipate* an event, such as sunrise or sunset, and to begin physiological and behavioral preparations *before* that event. Let's talk about how this circadian clock works.

The Hypothalamus Houses a Circadian Clock

Where is the biological clock that drives circadian rhythms, and how does it work? Early research showed that while removing various endocrine glands has little effect on the free-running rhythm of rats, large lesions of the hypothalamus interfere with circadian rhythms (Richter, 1967). It was subsequently discovered that a tiny subregion of the hypothalamus—the **suprachiasmatic nucleus** (**SCN**), named for its location above the optic chiasm—serves as a biological clock. Lesions confined to the SCN portion of the hypothalamus eliminate circadian rhythms of drinking and locomotor behavior (**FIGURE 10.2**) (Stephan and Zucker, 1972) and of hormone secretion (R. Y. Moore and Eichler, 1972).

The clocklike nature of the SCN is also evident in its metabolic activity (**FIGURE 10.3**). If we take SCN cells out of the brain and put them in a dish (Earnest et al., 1999; Yamazaki et al., 2000), their electrical activity continues to show a circadian rhythm for days or weeks. This striking evidence supports the idea that the SCN contains an endogenous clock. But even stronger proof that the SCN generates a circadian rhythm comes from transplanting the SCN from one animal to another, as we'll see next.

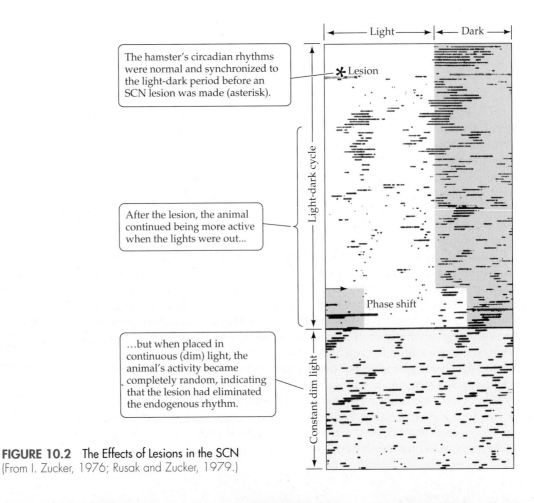

The hamster's circadian rhythms were normal and synchronized to the light-dark period before an SCN lesion was made (asterisk).

After the lesion, the animal continued being more active when the lights were out...

Phase shift

...but when placed in continuous (dim) light, the animal's activity became completely random, indicating that the lesion had eliminated the endogenous rhythm.

FIGURE 10.2 The Effects of Lesions in the SCN
(From I. Zucker, 1976; Rusak and Zucker, 1979.)

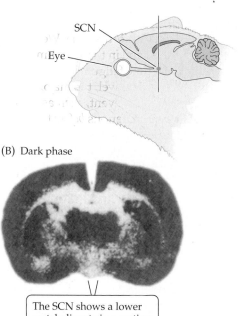

(A) Light phase

(B) Dark phase

In a section taken from an animal during a light phase, greater metabolic activity in the SCN is represented by the dark circles at the base of the brain.

The SCN shows a lower metabolic rate in a section taken from an animal during a dark phase.

FIGURE 10.3 The Circadian Rhythm of the SCN's Metabolic Activity (From Schwartz et al., 1979.)

researchers at work

Transplants prove that the SCN produces a circadian rhythm

Ralph and Menaker (1988) found a male hamster that exhibited an unusually short free-running activity rhythm in constant conditions. Normally, hamsters free-run at a period slightly longer than 24 hours, but this male showed a free-running period of 22 hours. Half of his offspring also had a shorter circadian rhythm, indicating that he had a genetic mutation affecting the endogenous clock. Grandchildren inheriting two copies of this mutation had an even shorter period: 20 hours. The mutation was named *tau*, after the Greek symbol used by scientists to represent the period (duration) of cyclical processes. These hamsters entrained to a normal 24-hour light-dark cycle just fine; their abnormal endogenous circadian rhythm was revealed only in constant conditions.

Dramatic evidence that the SCN is a master clock was provided by transplant experiments. When nonmutant hamsters with lesions of the SCN were placed in constant conditions, their circadian activity rhythms were abolished, as expected. The hamsters then received a transplant into the hypothalamus of an SCN taken from a fetal hamster with two copies of the mutant *tau* gene. About a week later the hamsters that had received the transplants began showing a free-running activity rhythm again, but the new rhythm matched that of the *donor* SCN: it was about 20 hours rather than the original 24.05 (**FIGURE 10.4**) (Ralph et al., 1990).

Reciprocal transplants gave comparable results: the endogenous rhythm following the transplant was always that of the *donor* SCN, not the recipient, so the SCN must be driving the circadian rhythms. This remains the only known case of transplanting brain tissue from one individual to another in which the recipient subsequently displayed the donor's behavior!

continued

researchers at work *continued*

FIGURE 10.4 Brain Transplants Prove That the SCN Contains a Clock (From Ralph et al., 1990.)

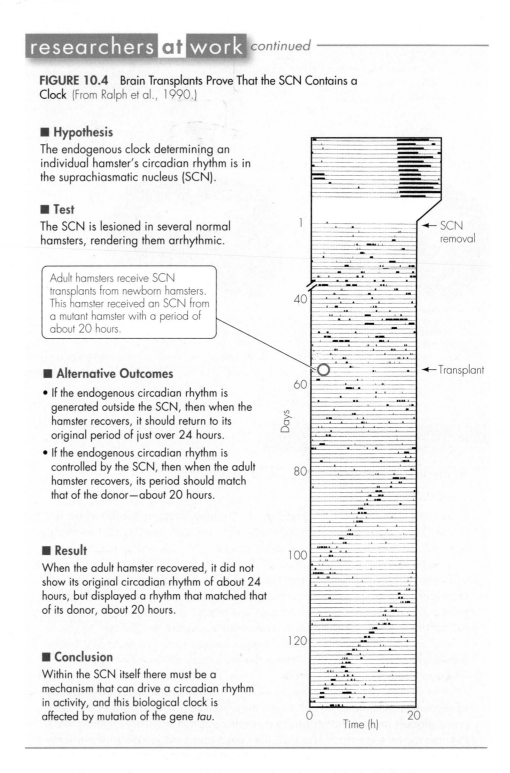

■ **Hypothesis**

The endogenous clock determining an individual hamster's circadian rhythm is in the suprachiasmatic nucleus (SCN).

■ **Test**

The SCN is lesioned in several normal hamsters, rendering them arrhythmic.

Adult hamsters receive SCN transplants from newborn hamsters. This hamster received an SCN from a mutant hamster with a period of about 20 hours.

■ **Alternative Outcomes**

• If the endogenous circadian rhythm is generated outside the SCN, then when the hamster recovers, it should return to its original period of just over 24 hours.

• If the endogenous circadian rhythm is controlled by the SCN, then when the adult hamster recovers, its period should match that of the donor—about 20 hours.

■ **Result**

When the adult hamster recovered, it did not show its original circadian rhythm of about 24 hours, but displayed a rhythm that matched that of its donor, about 20 hours.

■ **Conclusion**

Within the SCN itself there must be a mechanism that can drive a circadian rhythm in activity, and this biological clock is affected by mutation of the gene *tau*.

In mammals, light information from the eyes reaches the SCN directly

Most vertebrates have photoreceptors *outside the eye* that are part of the mechanism of light entrainment (Rusak and Zucker, 1979). For example, the pineal gland of some birds and amphibians is itself sensitive to light (Jamieson and Roberts, 2000) and helps entrain circadian rhythms to light. Because the skull over the pineal is especially thin in some species, we can think of those species as having a primitive "third eye" in the back of the head. (Some elementary school teachers also seem to have an eye in the back of the head, but this has not been proven to be the pineal gland.) At night, the pineal gland secretes a hormone, **melatonin**, that informs the brain about day length. For more on the nocturnal secretion of melatonin, see **A STEP FURTHER 8.1**.

melatonin An amine hormone that is secreted by the pineal gland at night, thereby signaling day length to the brain.

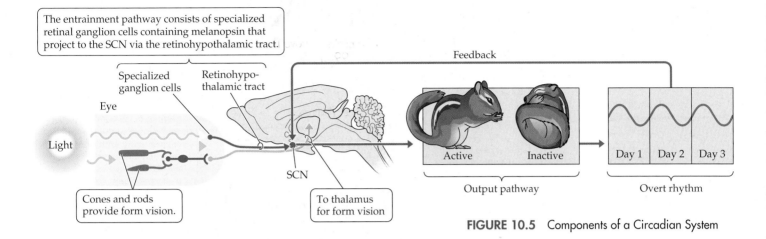

The entrainment pathway consists of specialized retinal ganglion cells containing melanopsin that project to the SCN via the retinohypothalamic tract.

Specialized ganglion cells

Retinohypo-thalamic tract

Eye

Light

Cones and rods provide form vision.

SCN

To thalamus for form vision

Feedback

Active   Inactive

Output pathway

Day 1 Day 2 Day 3

Overt rhythm

FIGURE 10.5 Components of a Circadian System

In mammals, however, cells in the eye tell the SCN when it is light out. Certain retinal ganglion cells send their axons along the **retinohypothalamic pathway**, veering out of the optic chiasm to synapse directly within the SCN. This short pathway carries information about light to the hypothalamus (R. Y. Moore, 1983) to entrain rhythms (**FIGURE 10.5**). Most of the retinal ganglion cells that extend their axons to the SCN do not rely on the traditional photoreceptors—rods and cones—to learn about light. Rather, these retinal ganglion cells themselves contain a special photopigment, called **melanopsin**, that makes them sensitive to light (Do et al., 2009). Transgenic mice that lack rods and cones, and so are blind in every other respect, will still entrain their behavior to light (Freedman et al., 1999) if the specialized melanopsin-containing ganglion cells are present.

Unfortunately, those melanopsin-containing retinal ganglion cells appear to be absent or dysfunctional in most totally blind humans, because people who are blind often show a free-running circadian rhythm, with difficulties getting to sleep at night and staying awake during the day (Sack, Lewy et al., 1992). Taking melatonin at bedtime, thus mimicking the normal nightly release of the hormone from the pineal gland, helps blind people to entrain to daylight (Sack et al., 2000). This result suggests that while humans rely primarily on light stimulation of the retinohypothalamic tract to the SCN in order to entrain to light, our brains have retained enough sensitivity to melatonin that we can use that cue in the absence of information about light.

Circadian rhythms have been genetically dissected in flies and mice

Like humans, the fruit fly *Drosophila melanogaster* displays a diurnal circadian rhythm in activity. Flies with a mutation that disabled the gene called *period* (*per*) were found to be arrhythmic when transferred to constant dim light, indicating that their internal clock wasn't running (Konopka and Benzer, 1971). Eventually, more genes were discovered that affect the circadian cycle in *Drosophila*, and mammals were found to have their own versions of each of those genes, paving the way for understanding the molecular basis of the circadian clock.

Cells in the mammalian SCN make two proteins, Clock and Cycle, that bind together to form a dimer (a pair of proteins attached to each other). The Clock/Cycle dimer then binds to the cell's DNA to promote the transcription of other genes, including *per*. The proteins made from these other genes go back to inhibit the action of Clock and Cycle, which started the whole process. Because those inhibitory proteins degrade with time, eventually the inhibition is lifted, starting the whole cycle over again (**FIGURE 10.6**). The entire cycle takes about 24 hours to complete, and it is this 24-hour molecular cycle that drives the 24-hour activity cycle of SCN cells. You can learn more about the molecular basis of the circadian clock in **A STEP FURTHER 10.1**, on the website.

retinohypothalamic pathway The route by which retinal ganglion cells send their axons to the suprachiasmatic nuclei.

melanopsin A photopigment found in those retinal ganglion cells that project to the suprachiasmatic nucleus.

FIGURE 10.6 A Molecular Clock in Flies and Mice This is a simplified view. The mammalian version of Cycle is Bmal1. (After Reppert and Weaver, 2002.)

To see the animation
A Molecular Clock,
go to

2e.mindsmachine.com/10.4

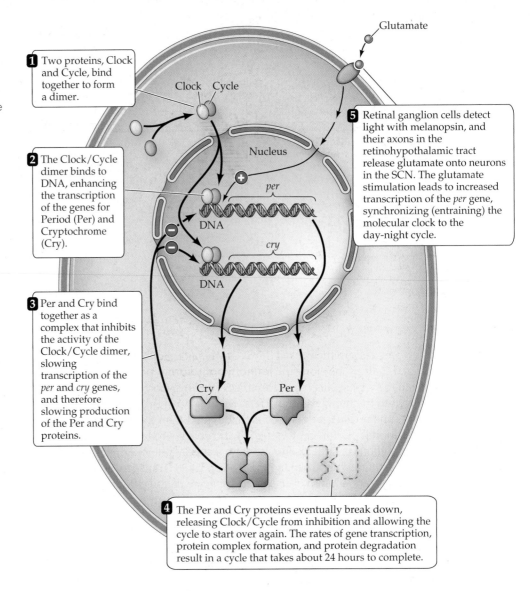

1 Two proteins, Clock and Cycle, bind together to form a dimer.

2 The Clock/Cycle dimer binds to DNA, enhancing the transcription of the genes for Period (Per) and Cryptochrome (Cry).

3 Per and Cry bind together as a complex that inhibits the activity of the Clock/Cycle dimer, slowing transcription of the *per* and *cry* genes, and therefore slowing production of the Per and Cry proteins.

4 The Per and Cry proteins eventually break down, releasing Clock/Cycle from inhibition and allowing the cycle to start over again. The rates of gene transcription, protein complex formation, and protein degradation result in a cycle that takes about 24 hours to complete.

5 Retinal ganglion cells detect light with melanopsin, and their axons in the retinohypothalamic tract release glutamate onto neurons in the SCN. The glutamate stimulation leads to increased transcription of the *per* gene, synchronizing (entraining) the molecular clock to the day-night cycle.

Glutamate

Clock   Cycle

Nucleus

per

DNA

cry

DNA

Cry   Per

One indication of the importance of the molecular clock in controlling circadian behavior is the effect of differences in the genes involved in the clock. We've already seen that hamsters with a mutation in *tau* have a free-running rhythm that is shorter than normal. Mice in which both copies of the *Clock* gene are disrupted show severe arrhythmicity under constant conditions (**FIGURE 10.7**). People who feel energetic in the morning ("larks") are likely to carry a different version of the *Clock* gene than "night owls" have (Katzenberg et al., 1998). Different versions of other genes in the molecular clock are also associated with being a lark versus a night owl (Carpen et al., 2005).

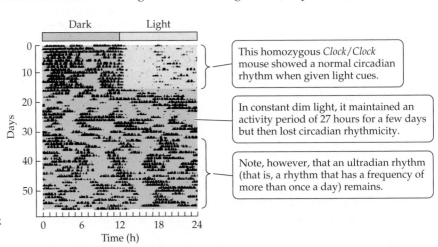

This homozygous *Clock/Clock* mouse showed a normal circadian rhythm when given light cues.

In constant dim light, it maintained an activity period of 27 hours for a few days but then lost circadian rhythmicity.

Note, however, that an ultradian rhythm (that is, a rhythm that has a frequency of more than once a day) remains.

FIGURE 10.7 When the Endogenous Clock Goes Kaput (From J. S. Takahashi, 1995.)

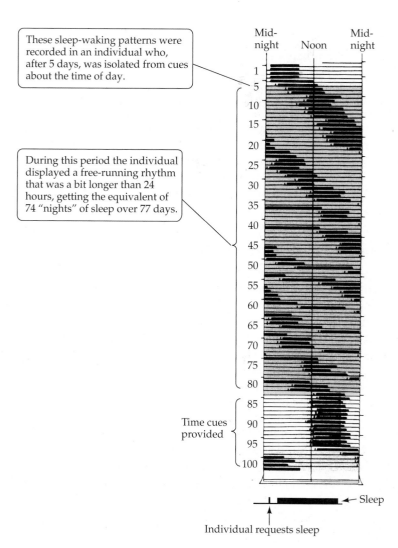

These sleep-waking patterns were recorded in an individual who, after 5 days, was isolated from cues about the time of day.

During this period the individual displayed a free-running rhythm that was a bit longer than 24 hours, getting the equivalent of 74 "nights" of sleep over 77 days.

Time cues provided

FIGURE 10.8 Humans Free-Run Too (From Weitzman et al., 1981.)

Sleep

Individual requests sleep

For humans, an important circadian rhythm is the sleep-waking cycle, the topic for the rest of this chapter. We noted earlier that totally blind humans often free-run. Sighted people also free-run, when they spend weeks in a dark cave with all cues to external time removed (Wever, 1979). They display a circadian rhythm of the sleep-waking cycle that slowly shifts from 24 to 25 hours (**FIGURE 10.8**), just as a hamster does (see Figure 10.1). Because the free-running period is greater than 24 hours, some people in these studies are surprised when they're told that the experiment has ended. They may have experienced only 74 sleep-waking cycles during a 77-day study.

Now that you understand some of the mechanisms that enforce our daily rhythm of sleep and waking, we'll spend the rest of the chapter exploring that mysterious state of consciousness called sleep.

how's it going ?

1. What are circadian rhythms, and how can they be studied and manipulated in lab animals?

2. Describe experiments that established which parts of the brain control circadian rhythms.

3. How does information about day and night reach the mammalian brain?

4. Give some examples of ultradian and infradian rhythms.

PART II
Sleeping and Waking

Perhaps the simplest explanation for why we sleep is that the brain gets tired from all that activity and just stops working for a while. But in fact our brains are quite active and go through many remarkable changes in the course of a night's sleep. We'll begin by describing the study of brain activity during sleep that revealed the different types of sleep. Then we'll talk about how sleep changes as we grow up (and grow old) before considering the effects of sleep deprivation and the perplexing question of why we sleep at all. That will lead to a discussion of the brain mechanisms underlying sleep, and we'll conclude by reviewing sleep disorders, including the one afflicting Barry, the sleepy college student we met at the start of the chapter.

Human Sleep Exhibits Different Stages

In the 1930s, experimenters found that brain potentials recorded from electrodes on the scalp by **electroencephalography** (**EEG**; see Figure 3.15A) provide a way to define, describe, and classify levels of arousal and states of sleep. In sleep studies, eye movements and muscle tension are monitored in addition to the EEG. Together, these measures led to the groundbreaking discovery that there are two distinct classes of sleep: **rapid-eye-movement (REM) sleep** (Aserinsky and Kleitman, 1953) and **non-REM sleep**. **TABLE 10.1** compares the properties of REM and non-REM sleep.

What are the electrophysiological distinctions that define different sleep states? Let's begin with the pattern of EEG activity in the brain of a fully awake, alert person. It is a mixture of low amplitude waves with many relatively fast frequencies (greater than 15–20 cycles per second, or hertz [Hz]). This pattern is sometimes referred to as *beta activity* or a **desynchronized EEG** (**FIGURE 10.9A**).

When you relax and close your eyes, a distinctive rhythm appears in the EEG, consisting of a regular oscillation at a frequency of 8–12 Hz, known as the **alpha rhythm**. As drowsiness sets in, the time spent in the alpha rhythm decreases, and the EEG shows waves of smaller amplitude and irregular frequency, as well as sharp waves called **vertex spikes**. This is the beginning of non-REM sleep, called **stage 1 sleep**

electroencephalography (EEG) The recording of gross electrical activity of the brain via large electrodes placed on the scalp.

rapid-eye-movement (REM) sleep Also called *paradoxical sleep*. A stage of sleep characterized by small-amplitude, fast EEG waves, no postural tension, and rapid eye movements. *REM* rhymes with "gem."

non-REM sleep Sleep, divided into stages 1–3, that is defined by the presence of distinctive EEG activity that differs from that seen in REM sleep.

desynchronized EEG Also called *beta activity*. A pattern of EEG activity comprising a mix of many different high frequencies with low amplitude.

alpha rhythm A brain potential of 8–12 hertz that occurs during relaxed wakefulness.

vertex spike A sharp-wave EEG pattern that is seen during stage 1 sleep.

stage 1 sleep The initial stage of non-REM sleep, which is characterized by small-amplitude EEG waves of irregular frequency, slow heart rate, and reduced muscle tension.

stage 2 sleep A stage of sleep that is defined by bursts of regular 14- to 18-hertz EEG waves called *sleep spindles*.

sleep spindle A characteristic 14- to 18-hertz wave in the EEG of a person said to be in stage 2 sleep.

K complex A sharp, negative EEG potential that is seen in stage 2 sleep.

stage 3 sleep Also called *slow wave sleep* (*SWS*). A stage of non-REM sleep that is defined by the presence of large-amplitude, slow delta waves.

delta wave The slowest type of EEG wave, characteristic of stage 3 sleep.

TABLE 10.1 Properties of REM Sleep and Non-REM Sleep

Property	REM sleep	Non-REM sleep
AUTONOMIC ACTIVITIES		
Heart rate	Variable, with high bursts	Slow decline
Respiration	Variable, with high bursts	Slow decline
Brain temperature	Increased	Decreased
Cerebral blood flow	High	Reduced
SKELETAL MUSCULAR SYSTEM		
Postural tension	Eliminated	Progressively reduced
Knee-jerk reflex	Suppressed	Normal
Twitches	Increased	Reduced
Eye movements	Rapid, coordinated	Infrequent, slow, uncoordinated
COGNITIVE STATE		
Dream state	Vivid dreams, well organized	Vague thoughts
HORMONE SECRETION		
Growth hormone secretion	Low	High in SWS
NEURAL FIRING RATES		
Cerebral cortex activity	Increased firing rates	Many cells reduced

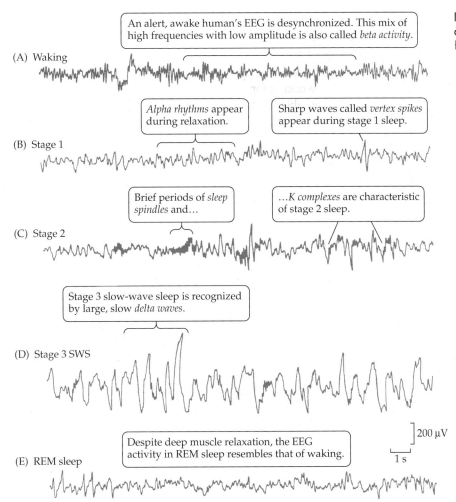

FIGURE 10.9 Electrophysiological Correlates of Sleep and Waking (After Rechtschaffen and Kales, 1968.)

(A) Waking

> An alert, awake human's EEG is desynchronized. This mix of high frequencies with low amplitude is also called *beta activity*.

(B) Stage 1

> *Alpha rhythms* appear during relaxation.

> Sharp waves called *vertex spikes* appear during stage 1 sleep.

(C) Stage 2

> Brief periods of *sleep spindles* and…

> …*K complexes* are characteristic of stage 2 sleep.

(D) Stage 3 SWS

> Stage 3 slow-wave sleep is recognized by large, slow *delta waves*.

(E) REM sleep

> Despite deep muscle relaxation, the EEG activity in REM sleep resembles that of waking.

200 μV

1 s

(**FIGURE 10.9B**), which is accompanied by slowing of the heart rate and relaxation of the muscles; in addition, under the closed eyelids the eyes may roll about slowly. Stage 1 sleep usually lasts several minutes and gives way to **stage 2 sleep** (**FIGURE 10.9C**), which is defined by waves of 12–14 Hz called **sleep spindles** that occur in periodic bursts, and by **K complexes**. If awakened during these first two stages of sleep, many people deny that they have been asleep, even though they failed to respond to signals while in those stages.

Stage 2 sleep leads to (can you guess?) **stage 3 sleep** (**FIGURE 10.9D**), which is defined by the appearance of large-amplitude, *very* slow waves (**delta waves**, about one per second). These waves give stage 3 sleep its other name—*slow wave sleep* (*SWS*). As the night progresses, the delta waves become even more prominent. (Previously, SWS with delta waves at least half the time was called *stage 4 sleep*, but that distinction is no longer made. Now all sleep with delta waves is called *stage 3* or *SWS*.) The slow waves of electrical potential that give SWS its name represent a widespread synchronization of cortical neuron activity (Poulet and Petersen, 2008) that has been likened to a room of people who are all chanting the same phrase over and over. From a distance you would be able to hear the rise and fall of the cadence of speech in a slow rhythm. Contrast this with a room full of people all saying something *different*. You would hear only a buzz—the rapid frequencies of many desynchronized speakers. This is like the desynchronized EEG of wakefulness, when many parts of the cortex are communicating different things and fulfilling different functions.

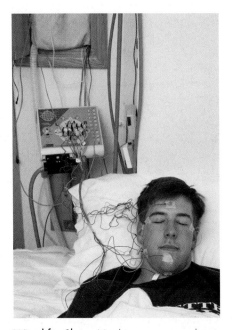

Wired for Sleep Machines measure electrical activity across the various electrodes to monitor EEG, eye movements, and muscle tension across sleep stages.

FIGURE 10.10 A Typical Night of Sleep in a Young Adult (After Kales and Kales, 1970.)

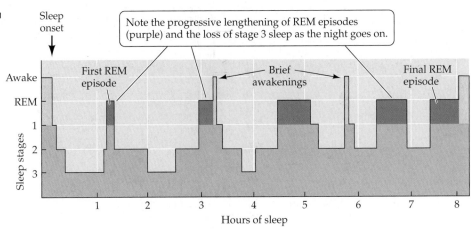

After about an hour—the typical time to progress through the SWS stage, with a brief return to stage 2—something totally different occurs: REM sleep. Quite abruptly, the EEG displays a pattern of small-amplitude, high-frequency activity similar in many ways to the pattern of an awake individual (**FIGURE 10.9E**), except the eyes are darting rapidly about under their lids (the *r*apid *e*ye *m*ovements that give REM sleep its name). Aside from those muscles moving the eyes, all other skeletal muscles are not just relaxed, but show a complete absence of muscle tone, called *atonia*. The active-looking EEG coupled with deeply relaxed muscles is typical of REM sleep. If you see a cat sleeping in the sitting, sphinx position, it cannot be in REM sleep; in REM, it will be sprawled limply on the floor. For the same reason, a student sleeping while sitting upright in class cannot be in REM sleep.

This flaccid muscle state appears, despite intense brain activity, because during this stage of sleep, brainstem regions are profoundly inhibiting motoneurons. This seeming contradiction—the brain waves look awake, but the muscles are flaccid and unresponsive—is what gives REM sleep its other name: *paradoxical sleep*. Unlike SWS, REM sleep is accompanied by irregular breathing and pulse rate, as in wakefulness. It is during REM sleep that we experience vivid dreams, as we'll discuss shortly.

The EEG portrait in Figure 10.9 shows that sleep consists of a complex series of brain states, not just an "inactive" period. The total sleep time of young adults usually ranges from 7 to 8 hours, about half of it in stage 2 sleep. REM sleep accounts for about 20% of total sleep. A typical night of adult human sleep shows repeating cycles approximately 90–110 minutes long, recurring four or five times in a night. These cycles change in a subtle but regular manner through the night. Stage 3 SWS is more prominent early in the night (**FIGURE 10.10**), and then it tapers off as the night progresses. In contrast, REM sleep is more prominent in the later cycles of sleep. The first REM period is the shortest, while the last REM period, just before waking, may last up to 40 minutes.

Brief arousals (yellow bars in Figure 10.10) occasionally occur immediately after a REM period, and the sleeper may shift posture at this time (Amici et al., 2014). The sleep cycle of 90–110 minutes suggests that there is a basic ultradian rest-activity cycle (Kaiser, 2013).

At puberty, most people shift their circadian rhythm of sleep so that they get up later in the day (**FIGURE 10.11**), but many school systems require students to come to school *earlier* in the day when they hit adolescence. One group of high schools shifted their morning start from 7:15 to 8:40 and noted improved student attendance and enrollment, with (big surprise!) less sleeping in class and, more important, a reduced incidence of depression (Wahlstrom, 2002). You can see that sleep is a remarkably complex, multifaceted set of behaviors. Now let's consider an even more fascinating aspect of sleep: dreaming.

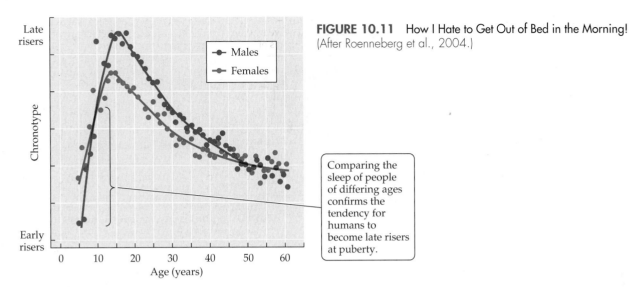

FIGURE 10.11 How I Hate to Get Out of Bed in the Morning! (After Roenneberg et al., 2004.)

Comparing the sleep of people of differing ages confirms the tendency for humans to become late risers at puberty.

We do our most vivid dreaming during REM sleep

We can record the EEGs of participants to monitor their sleep stages, awaken them at a particular stage (1, 2, 3, or REM), and question them about thoughts or perceptions they were having. Early studies of this sort suggested that dreams happen only during REM sleep, but we now know that dreams also occur in other sleep stages. What is distinctive about dreams during REM sleep is that they are characterized by visual imagery, whereas dreams during non-REM sleep are of a more "thinking" type. REM dreams are apt to include a story that involves odd perceptions and the sense that the dreamer "is there" experiencing sights, sounds, smells, and emotions. People awakened from non-REM sleep report thinking about problems rather than seeing themselves in a stage presentation. The dreams of these two states are so different that people can be trained to predict accurately whether a described dream occurred during REM sleep or SWS (Cartwright, 1979).

Almost everyone has terrifying dreams on occasion (Llewellyn and Hobson, 2015). **Nightmares** are defined as long, frightening dreams that awaken the sleeper from REM sleep. They are occasionally confused with **night terror**, which is a sudden arousal from stage 3 SWS marked by intense fear and autonomic activation. In night terror the sleeper does not recall a vivid dream but may remember a sense of a crushing feeling on the chest, as though being suffocated. Night terrors are common in children during the early part of an evening's sleep.

Many medications, including antidepressants and drugs that control blood pressure, make nightmares more frequent (Pagel and Helfter, 2003), but nightmares are quite prevalent even without such influences. At least 25% of college students report having one or more nightmares per month. Have you had the common one, which Sigmund Freud had, of suddenly remembering that you're supposed to be taking a final exam that is already in progress?

As fascinating as they are, we still do not know what function, if any, is fulfilled by dreams. The *activation-synthesis* theory suggests our experiences in REM sleep are the more or less random results of which neurons happen to get activated (Hobson and Friston, 2012). The brain strings together these disparate activated elements into a more or less coherent story, a narrative. Later we'll discuss evidence that at least some other animals experience dreaming, which suggests that dreaming either fulfills an important function or is an unavoidable consequence of some other function of REM sleep.

nightmare A long, frightening dream that awakens the sleeper from REM sleep.

night terror A sudden arousal from stage 3 sleep that is marked by intense fear and autonomic activation.

Night Terror This 1781 painting by Henry Fuseli is called *The Nightmare*. It also aptly illustrates night terror, or even sleep paralysis, discussed later in the chapter, as the demon crushes the breath from his victim.

FIGURE 10.12 Sleep in Marine Mammals
(After Mukhametov, 1984.)

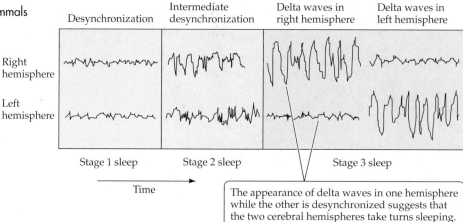

The appearance of delta waves in one hemisphere while the other is desynchronized suggests that the two cerebral hemispheres take turns sleeping.

Different species provide clues about the evolution of sleep

With the aid of behavioral and EEG techniques, sleep has been studied in a wide assortment of mammals and, to a lesser extent, in reptiles, birds, and amphibians (Lesku et al., 2009; Hartse, 2011). Nearly all mammalian species that have been investigated thus far, including our most distant relatives, such as the platypus (Siegel, Manger et al., 1999), display both REM sleep and SWS. Among the other vertebrates, only birds display clear signs of both SWS and REM sleep. These comparisons suggest that either REM sleep was present in an ancestor common to birds and mammals, or that REM sleep evolved independently in mammals and birds.

The absence of REM sleep in dolphins is probably a late adaptation that evolved when their land-dwelling ancestors took to the water, because they must come to the surface of the water to breathe. That requirement may be incompatible with the deep relaxation of muscles during REM sleep. Another dolphin adaptation to living in water is that only one side of the dolphin brain engages in SWS at a time (Mukhametov, 1984). It's as if one whole hemisphere is asleep while the other is awake (**FIGURE 10.12**). During these periods of "unilateral sleep," the animals continue to come up to the surface occasionally to breathe. Birds can also display unilateral sleep—one hemisphere sleeping while the other hemisphere watches for predators (Rattenborg, 2006). Unilateral sleep while gliding may also enable birds to fly long distances without stopping; for example, a bar-tailed godwit flew nonstop more than 10,000 miles, from Alaska to Australia, in a week (Gill et al., 2009). You can learn more about comparing patterns of sleep in different species in **A STEP FURTHER 10.2**, on the website.

how's it going ?

1. What are the different stages of sleep, and what measures define them?
2. What happens to our muscles during the sleep stage characterized by the most vivid dreams?
3. Describe one way that certain species manage to remain active around the clock and still sleep.

Our Sleep Patterns Change across the Life Span

How much sleep and what kind of sleep we get changes across our lifetime. As infants, we sleep a lot; as we grow, we sleep less and less until we hit adolescence, when once again sleep seems precious. After that, we sleep less and less as we age, sometimes to our disappointment. These changes as we grow up and grow old suggest that the function(s) of sleep are more important during some stages of life than others.

Sleeping on the Go A female bar-tailed godwit flew over 10,000 miles, nonstop, from Alaska to Australia in a week. Were both sides of its brain awake the entire time?

FIGURE 10.13 **The Trouble with Babies** This classic study may represent an extreme example of a baby slow to entrain to the day-night rhythm. (From Kleitman and Engelmann, 1953.)

Mammals sleep more during infancy than in adulthood

A clear cycle of sleeping and waking takes several weeks to become established in human infants (**FIGURE 10.13**). A 24-hour rhythm is generally evident by 16 weeks of age. Infant sleep is characterized by shorter sleep cycles than those of adults, probably reflecting the relative immaturity of the brain, since sleep cycles in prematurely born infants are even shorter than in full-term newborns.

Infant mammals also show a large percentage of REM sleep. In humans, for example, half of sleep in the first 2 weeks of life is REM sleep. The prominence of REM sleep is even greater in premature infants, accounting for up to 80% of total sleep. Unlike most adults, human infants can move directly from an awake state to REM sleep. The REM sleep of infants is quite active, accompanied by muscle twitching, smiles, grimaces, and vocalizations. The preponderance of REM sleep early in life (**FIGURE 10.14**) suggests that this state provides stimulation that is essential to maturation of the nervous system. By contrast, killer whales and bottlenose dolphins appear to spend little or no time in REM sleep (or any other sleep stage) for the first month of life (Lyamin et al., 2005), presumably because they have to surface often to breathe. So, either REM sleep does not fill a crucial need in all mammalian infants, or dolphin and whale infants have evolved a different way to fill that need.

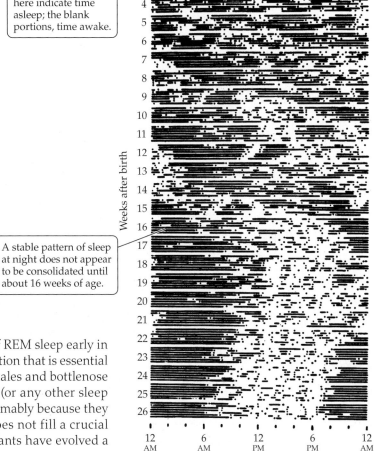

The dark portions here indicate time asleep; the blank portions, time awake.

Weeks after birth

A stable pattern of sleep at night does not appear to be consolidated until about 16 weeks of age.

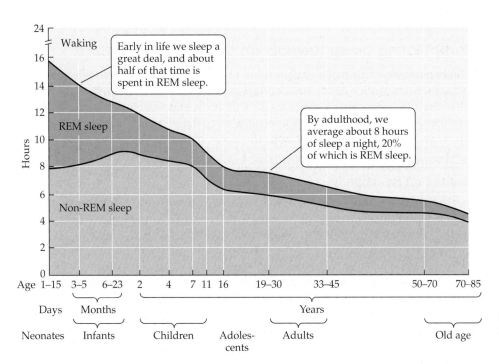

Early in life we sleep a great deal, and about half of that time is spent in REM sleep.

By adulthood, we average about 8 hours of sleep a night, 20% of which is REM sleep.

FIGURE 10.14 **Human Sleep Patterns Change with Age** (After Roffwarg et al., 1966.)

FIGURE 10.15 The Typical Pattern of Sleep in an Elderly Person Compare this recording with the young-adult sleep pattern shown in Figure 10.10. (After Kales and Kales, 1974.)

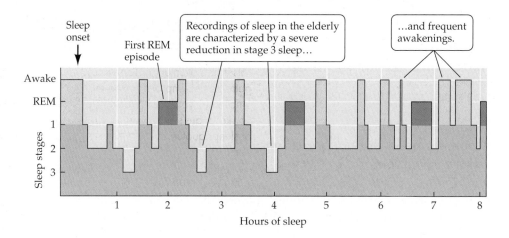

Most people sleep appreciably less as they age

The character of sleep changes in old age, though more slowly than in early development. **FIGURE 10.15** shows the sleep pattern typical of an elderly person. The total amount of sleep declines, while the number of awakenings increases (compare with Figure 10.10). Lack of sleep, or *insomnia* (which we discuss at the end of this chapter), is a common complaint of the elderly (Miles and Dement, 1980).

In humans and other mammals, the most dramatic decline is in stage 3 sleep; 60-year-old people spend only about half as much time in stage 3 as they did at age 20 (Bliwise, 1989). By 90 years of age, stage 3 sleep has disappeared. This decline in stage 3 sleep may be related to diminished cognitive functioning, since an especially marked reduction of stage 3 SWS characterizes the sleep of people who suffer from senile dementia. Growth hormone is secreted primarily during stage 3 SWS (see Table 10.1), so perhaps the loss of growth hormone due to disrupted sleep in the elderly leads to the cognitive deficits. Loss of SWS probably also impairs memory processes (discussed below) in older people and patients with dementia (Westerberg et al., 2012).

Most elderly people fall asleep easily enough, but then they may have a hard time staying asleep, which causes sleep "dissatisfaction." As in so many things, attitude may be important for how you experience sleep loss as you age. Objective measures of sleep suggest that elderly people who complain of poor sleep may actually sleep *more* than those who are satisfied with their sleep (McCrae et al., 2005). Perhaps if, as you grow older, you can regard waking up at 3:00 AM as a "bonus" (a little more time awake before you die), you will be more satisfied with the sleep you get.

Manipulating Sleep Reveals an Underlying Structure

Another persuasive clue that sleep is important is revealed when we go without it. First of all, as we'll see, our mental function is impaired. This is bad news for college students, who rarely get enough sleep, just when they're supposed to be learning how to make their way in the world. In addition, after sleep deprivation we tend to sleep more than we would have, as though catching up on something we need, as we'll see.

Sleep deprivation impairs cognitive functioning but does not cause insanity

Most of us at one time or another have been willing or not-so-willing participants in informal **sleep deprivation** experiments. Thus, most of us are aware of the primary effect of partial or total sleep deprivation: it makes us sleepy. It has other effects as well. Early reports from sleep deprivation studies emphasized a similarity between schizophrenia and "bizarre" behavior provoked by sleep deprivation. A frequent theme in this early work was the functional role of dreams as a "guardian of sanity." But examination of patients suffering from schizophrenia does not fit this view. For example, these patients can show sleep-waking cycles similar to those of typical adults, and sleep deprivation does not exacerbate their symptoms.

The behavioral effects of prolonged, total sleep deprivation vary appreciably and may depend on some general personality factors and on age. In studies employing

sleep deprivation The partial or total prevention of sleep.

BOX 10.1 Sleep Deprivation Can Be Fatal

Sleep that knits up the ravell'd sleave of care.
William Shakespeare,
Macbeth, Act II, Scene 2

Although some people seem to need very little sleep, most of us feel the need to sleep 7–8 hours a night. In fact, sustained sleep deprivation in rats causes them to increase their metabolic rate, lose weight, and, within an average of 19 days, die (Everson et al., 1989). Allowing them to sleep prevents their death.

After the fatal effect of sleep deprivation had been shown, researchers undertook studies in which they terminated the sleep deprivation before the fatal end point and looked for pathological changes in different organ systems (Rechtschaffen and Bergmann, 1995). No single organ system seems affected in chronically sleep-deprived animals, but early in the deprivation they develop sores on their bodies. These sores mark the beginning of the end; shortly thereafter, blood tests reveal infections from a host of bacteria, which probably enter through the sores (Everson, 1993).

These bacteria are not normally fatal, because the rat's immune system and body defenses keep the bacteria in check, but severely sleep-deprived rats fail to develop a fever in response to these infections. (Fever helps the body fight infection.) In fact, the sleep-deprived animals show a *drop* in body temperature, which probably speeds bacterial infections that in turn cause diffuse organ damage. The decline of these severely sleep-deprived rats is complicated, but it seems clear that getting sleep improves immune system function (Bryant et al., 2004). So perhaps Shakespeare's folk theory of the function of sleep, quoted at left, isn't so far from the truth.

Shortly, we'll learn of rare people who sleep only 1 or 2 hours a night. Why aren't their immune systems and inflammatory responses compromised? We don't know, but since the distinguishing trait of these people is that they don't need much sleep, perhaps their immune systems and inflammatory responses don't need much sleep either. Or perhaps the small amount of sleep they have almost every night is more efficient at doing whatever sleep does.

Some unfortunate humans inherit a defect in the gene for the prion protein, which can transmit mad cow disease, and although they sleep normally at the beginning of life, in midlife they simply stop sleeping—with fatal effect. People with this disease, called **fatal familial insomnia**, die 7–24 months after the insomnia begins (Medori et al., 1992; Mastrianni et al., 1999). Autopsy reveals degeneration in the cerebral cortex (shown in the figure) and thalamus, which may cause the insomnia (Almer et al., 1999). Like sleep-deprived rats, sleep-deprived humans with this disorder don't have obvious damage to any single organ system, but they suffer from diffuse bacterial infections. Apparently these patients die because they are chronically sleep-deprived, and these results, combined with research on rats, certainly support the idea that prolonged insomnia is fatal.

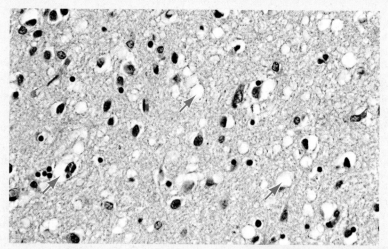

Fatal Sleeplessness Note the large holes (arrows) that have developed in this section of frontal cortex from a victim of fatal familial insomnia. (Micrograph courtesy of H. Budka.)

---

prolonged total deprivation—205 hours (8.5 days)—a few participants showed occasional episodes of hallucinations. But the most common behavior changes noted in these experiments were increases in irritability, difficulty in concentrating, and episodes of disorientation. The person's ability to perform tasks was summed up like this: "His performance is like a motor that after much use misfires, runs normally for a while, then falters again" (L. C. Johnson, 1969, p. 216).

You don't need to resort to total sleep deprivation to see effects. Moderate effects of sleep debt can accumulate with successive nights of little sleep. Voluntary experimental participants who got 6 or 4 hours of sleep per night for 2 weeks showed ever-mounting deficits in attention tasks and in speed of reaction, compared with those sleeping 8 hours per night (Van Dongen et al., 2003). Interestingly, the sleep-deprived participants often reported not feeling sleepy, yet they still exhibited behavioral deficits. By the end of the study, the people getting less than 8 hours of sleep per night had cognitive deficits equivalent to those of participants who had been totally sleep-deprived for 3 days!

Finally, it is clear that prolonged, total sleep deprivation in mammals compromises the immune system and leads to death (**BOX 10.1**). Even fruit flies will die without

fatal familial insomnia An inherited disorder in which humans sleep normally at the beginning of their life but in midlife stop sleeping and, 7–24 months later, die.

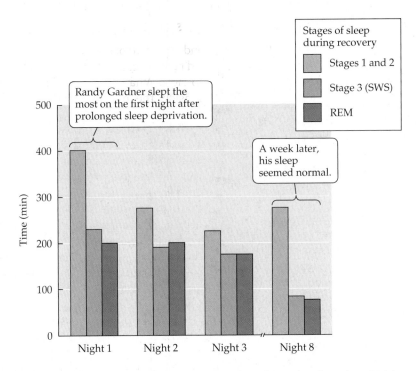

FIGURE 10.16 Sleep Recovery after 11 Days Awake (After Gulevich et al., 1966.)

sleep recovery The process of sleeping more than normally after a period of sleep deprivation, as though in compensation.

sleep, and a particular mutation of the *Cycle* gene (part of the circadian molecular clock; see Figure 10.6) causes flies to die after only 10 hours of sleep deprivation (P. J. Shaw et al., 2002).

Sleep recovery may take time

One of the most famous cases of sleep deprivation began as a high school student's science project. Researchers became involved only after Randy Gardner had started his deprivation schedule, which is why we have no data about his sleep before he decided to stay awake for, believe it or not, 11 days! As in other studies, Randy's performance on some tests was impaired, but he could still hold a conversation and was articulate and clear in a press conference at the end of his experiment. In other words, he showed no signs of insanity—he just acted really, really sleepy.

Randy's **sleep recovery** after 11 days of sleep deprivation, depicted in **FIGURE 10.16**, shows the same pattern of sleep recovery as in controlled studies with shorter periods of deprivation. In the first night of sleep recovery, stage 3 sleep shows the greatest relative difference from normal. This increase in stage 3 sleep is usually at the expense of stage 2 sleep. However, the added stage 3 sleep during recovery never completely makes up for the deficit accumulated over the deprivation period. In fact, Randy had no more additional stage 3 sleep than do people deprived of sleep for half as long. REM sleep in recovery nights is more "intense" than normal, with a greater number of rapid eye movements per period of time. So you never recover all the sleep time you lost, but you may make up for the loss by having more intense sleep for a while. The sooner you get to sleep, the sooner you recover.

The effects of sleep deprivation suggest that sleep plays an important function, or even several functions, that we'll consider next.

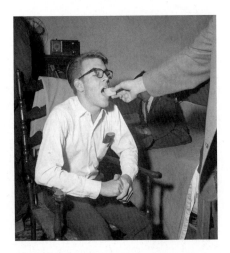

Skipping Sleep for Science As a young man, Randy Gardner decided to see how long he could stay awake as a science fair project. The answer? Just over 11 days. Is that a record for the most demanding science fair project?

how's it going ?

1. Describe how sleep changes as we grow up and grow old.
2. What happens when we are deprived of sleep?
3. Describe the outcome of Randy Gardner's famous self-study.

What Are the Biological Functions of Sleep?

Doesn't it seem like a big waste of time to spend one-third of our lifetime asleep? Most of us have fantasized about how great it would be if we could stay awake and chipper all the time, but we've seen that it's just not possible. What is so important about sleep that we can't seem to live without it? Let's consider the four functions that are most often ascribed to sleep:

1. Energy conservation
2. Niche adaptation
3. Body and brain restoration
4. Memory consolidation

Sleep conserves energy

We use up less energy when we sleep than when we're awake. For example, slow wave sleep is marked by reduced muscular tension, lowered heart rate, reduced blood pressure, reduced body temperature, and slower respiration. This diminished metabolic activity during sleep suggests that one role of sleep is to conserve energy. We can see the importance of this function by considering small animals. Small mammals and birds have very high metabolic rates (see Chapter 9), so activity for them is metabolically expensive. In general, the smaller the mammal, the higher its metabolic rate and the more time it spends asleep (Siegel, 2005). That correlation supports the idea that sleep helps conserve energy. But energy savings from sleep seem modest at best (Lesku et al., 2009).

Sleep enforces niche adaptation

Almost all animals are either nocturnal or diurnal. This specialization for either nighttime or daytime activity is part of each species' **ecological niche**, that unique assortment of environmental opportunities and challenges to which each organism is adapted. Thanks to these adaptations, each species is better at gathering food either at night or in the daytime, and it is also better at avoiding predators either during the day or at night. If you're a nocturnal mammal, like a mouse, you are adept at sneaking around in the dark, using your acute senses of hearing and smell to navigate and find food. The rest of the time, during daylight, you should spend holed up somewhere safe to stay away from keen-eyed predators. Sleep debt and the unpleasant feelings of sleepiness have the effect of enforcing the circadian rhythm characteristic of your species. So, one important function of sleep, or of the results of sleep deprivation, is to force the individual to conform to the particular ecological niche for which it is well adapted (Meddis, 1975), and natural selection must have played an important role in its evolution.

Sleep restores the body and brain

If someone asked you why you wanted to go to sleep, you might answer that you "feel worn out." Indeed, one of the proposed functions of sleep is simply the rebuilding or restoration of materials used during waking, such as proteins (Moruzzi, 1972; Pulak and Jensen, 2014). Maybe this is why most growth hormone release happens during slow wave sleep.

We've seen that prolonged and total sleep deprivation—either forced on rats or, in humans, as a result of inherited pathology—interferes with the immune system and leads to death (see Box 10.1). Even relatively mild deprivation, having sleep shortened or disrupted (e.g., by a nurse taking vital signs every hour), makes people more sensitive to pain the following day (Edwards et al., 2009). A study of over a million Americans found that those sleeping less than 6 hours per night were more likely to die over the next 6 years, although interestingly, people who slept *more than 8* hours per night were also at greater risk (Kripke et al., 2002). People who sleep less than 5 hours per night are more likely to develop diabetes (Gangwisch et al., 2007). Perhaps the most alarming link between sleep and health is the finding that people

Finding Your Niche in Life Species that can sleep in secure circumstances tend to sleep more than other species.

ecological niche   The unique assortment of environmental opportunities and challenges to which each organism is adapted.

who work at night and sleep in the daytime are more likely to develop cancer (Erren et al., 2009). So the widespread belief that sleep helps the body ward off illness is well supported by research (S. Cohen et al., 2009; Imeri and Opp, 2009).

There's also evidence that sleep may help "clean out" the brain. Glia control the flow of cerebrospinal fluid through a microscopic network of channels throughout the brain, collecting and disposing of toxins that build up. This flow is much faster during sleep than wakefulness (Xie et al., 2013), flushing out brain waste products as we snooze.

Sleep may aid memory consolidation

A peculiar property of dreams is that, unless we describe them to someone or write them down soon after waking, we tend to forget them as though the brain refuses to store anything that we experience during REM sleep. This seems like a good idea—why waste memory storage space on something that never happened? Similarly, and despite ads you might read in the backs of magazines, you cannot learn new material while you're sleeping (Druckman and Bjork, 1994). Putting a speaker under your pillow to recite material for a final exam will not help you, unless you stay awake to listen (J. M. Wood et al., 1992).

Sleep seems important for learning in another way, however. In 1924 an experiment suggested that sleep helps you learn or remember material or events experienced *before* you go to bed (Jenkins and Dallenbach, 1924). Some participants were trained in a verbal learning task at bedtime and tested 8 hours later on rising from sleep; other people were trained early in the day and tested 8 hours later (with no intervening sleep). The results showed better retention when a period of sleep intervened between a learning period and tests of recall. A recent surge of supporting evidence has shown that sleep helps with memory formation in many domains, not just verbal memory (U. Wagner et al., 2004; Ellenbogen et al., 2007; Korman et al., 2007).

When researchers deliberately deprive people of either REM sleep or SWS, results suggest that REM sleep improves performance on tests that involve learning perceptual skills. For example, humans learning to make difficult visual discriminations show little improvement in a single training session, but they show considerable improvement 8–10 hours after the session. If deprived of REM sleep after a training session, however, people fail to show the later improvement (Karni et al., 1994; C. Smith, 1995). In contrast, consolidation of declarative memory tasks (the kind of memories that you can declare to others; see Chapter 13), such as in the original 1924 study, and of complicated motor skills seems to benefit from SWS (Plihal and Born, 1999; Nishida and Walker, 2007). In fact, consolidation of a declarative memory task was even better if the person's cortical slow wave oscillations during SWS were boosted by electrically stimulating electrodes over the skull (Marshall et al., 2006). (Don't try this at home.)

However, one man who had brainstem injuries that seemed to eliminate REM sleep could still learn, and he earned a law school degree (P. Lavie, 1996). So even if REM sleep *aids* learning, clearly it is not absolutely *necessary* for learning.

Some humans sleep remarkably little, yet function normally

One challenge to all the theories about the function of sleep is the existence of a few people who seem perfectly healthy, yet sleep hardly at all. These cases are more than just folktales. A Stanford University professor slept only 3–4 hours a night for more than 50 years and lived to be 80 (Dement, 1974). Sleep researcher Ray Meddis (1977) found a cheerful 70-year-old retired nurse who said she had slept little since childhood. She was a busy person who easily filled up her 23 hours of daily wakefulness. During the night she sat in bed reading or writing, and at about 2:00 AM she fell asleep for an hour or so, after which she readily awakened.

For her first two nights in Meddis's laboratory, she did not sleep at all, because it was all so interesting to her. On the third night she slept a total of 99 minutes, and her sleep contained both SWS and REM sleep periods. In a later session her sleep

A Nonsleeper When Ray Meddis brought this 70-year-old nurse into the lab for sleep recording, he confirmed that she slept only about an hour per night. Yet she was a healthy and energetic person. Here she's touring a garden with Meddis's son. (Photo courtesy of Ray Meddis.)

was recorded for 5 days. She didn't sleep the first night, but on subsequent nights she slept an average of 67 minutes. She never complained about not sleeping more, and she did not feel drowsy during either the day or the night. Meddis described several other people who slept only an hour or two per night. Some of these people reported having parents who slept little, so there may be a genetic tendency for little sleep. Whatever the function of sleep is, such people have some way of fulfilling it with a brief nap. They show less stage 1 and stage 2 sleep, so perhaps they are more efficient sleepers. Importantly, though, no healthy person has ever been found who does not sleep at all.

how's it going ❓

1. Describe the four most prominent theories about the function of sleep and the evidence to support them.
2. What can we conclude about the function of sleep when we consider people who sleep very little?
3. What happens when we stop sleeping altogether?

At Least Four Interacting Neural Systems Underlie Sleep

At one time sleep was regarded as a passive state, as though most of the brain simply stopped working while we slept, leaving us unaware of events around us. We now know that sleep is an active state mediated by at least four interacting neural systems:

1. A *forebrain* system that generates SWS
2. A *brainstem* system that activates the sleeping forebrain into wakefulness
3. A *pontine* system that triggers REM sleep
4. A *hypothalamic* system that coordinates the other three brain regions to determine which state we're in

Let's examine each of these systems in some detail.

researchers at work

The forebrain generates slow wave sleep

Some of the earliest studies of sleep indicated that a system in the forebrain promotes SWS. These are experiments in which an animal's brain is transected—literally cut into two parts: an upper part and a lower part. The entire brain can be isolated from the body by an incision between the medulla and the spinal cord. This preparation was first studied by the Belgian physiologist Frédéric Bremer (1892–1982), who called it the **isolated brain** (Bremer, 1938).

The EEGs of such animals showed signs of waking alternating with sleep (**FIGURE 10.17A**). During EEG-defined wakeful periods, the pupils were dilated and the eyes followed moving objects. During EEG-defined sleep, the pupils were small, as in normal sleep. REM sleep can also be detected in the isolated brain. These results demonstrated that wakefulness, SWS, and REM sleep are all mediated by *networks within the brain*.

When Bremer made the transection higher along the brainstem—in the midbrain—a very different result was seen. Bremer referred to such a preparation as an **isolated forebrain**, and he found that the EEG from the brain in front of the cut displayed constant SWS (**FIGURE 10.17B**), with no indications of wakefulness or REM sleep. This result demonstrates that the forebrain alone can generate SWS, without contributions from the lower brain regions.

continued

isolated brain   An experimental preparation in which an animal's brainstem has been separated from the spinal cord by a cut below the medulla.

isolated forebrain   An experimental preparation in which an animal's nervous system has been cut in the upper midbrain, dividing the forebrain from the brainstem.

researchers at work *continued* ───────────────────────────

FIGURE 10.17   Transecting the Brain at Different Levels

■ **Hypothesis**

The forebrain contains a neural system promoting SWS.

■ **Test**

Bremer transected the brain at one of two levels: between the spinal cord and medulla, or between the midbrain and the forebrain. If neural systems in the forebrain generate SWS, then the isolated forebrain should show SWS.

■ **Result**

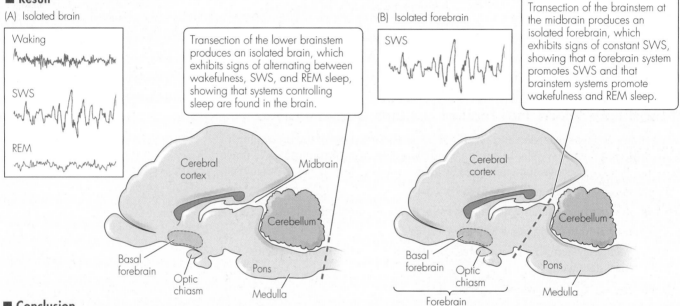

(A) Isolated brain

Waking

SWS

REM

Transection of the lower brainstem produces an isolated brain, which exhibits signs of alternating between wakefulness, SWS, and REM sleep, showing that systems controlling sleep are found in the brain.

(B) Isolated forebrain

SWS

Transection of the brainstem at the midbrain produces an isolated forebrain, which exhibits signs of constant SWS, showing that a forebrain system promotes SWS and that brainstem systems promote wakefulness and REM sleep.

Cerebral cortex   Midbrain

Cerebellum

Basal forebrain   Optic chiasm   Pons

Medulla

Cerebral cortex

Cerebellum

Basal forebrain   Optic chiasm   Pons

Medulla

Forebrain

■ **Conclusion**

The mechanism mediating SWS is in the forebrain and needs input from the brainstem to be roused from sleep.

The constant SWS seen in the cortex of the isolated forebrain appears to be generated by the **basal forebrain** in the ventral frontal lobe and anterior hypothalamus (**FIGURE 10.18**, step 1). Electrical stimulation of the basal forebrain can induce SWS activity (Clemente and Sterman, 1967), while lesions there suppress sleep (McGinty and Sterman, 1968). Neurons in this region become active at sleep onset and release gamma-aminobutyric acid (GABA) (Gallopin et al., 2000) to stimulate $GABA_A$ receptors in the nearby **tuberomammillary nucleus** in the posterior hypothalamus. These same $GABA_A$ receptors are stimulated by **general anesthetics**—drugs such as barbiturates and anesthetic gases that render people unconscious during surgery. Thus, general anesthetics produce slow waves in the EEG that resemble those seen in SWS (Franks, 2008).

So the basal forebrain promotes SWS by releasing GABA into the nearby tuberomammillary nucleus, and if left alone, this system would keep the cortex asleep forever. But as we'll see next, the brainstem contains a system that arouses the forebrain from slumber.

basal forebrain   A ventral region in the forebrain that has been implicated in sleep.

tuberomammillary nucleus   A region of the basal hypothalamus, near the pituitary stalk, that plays a role in generating slow wave sleep.

general anesthetic   A drug that renders an individual unconscious.

The reticular formation wakes up the forebrain

In the late 1940s, scientists found that they could wake sleeping animals by electrically stimulating an extensive region of the brainstem known as the **reticular formation** (**FIGURE 10.18**, step 2) (Moruzzi and Magoun, 1949). The reticular formation is a diffuse group of cells whose axons and dendrites course in many directions, extending from the medulla through the thalamus. Because electrical stimulation anywhere along this region activates the forebrain, the reticular formation is sometimes called the *reticular activating system* of the brainstem. Conversely, lesions of these regions produced persistent sleep in the animals. So the basal forebrain region actively imposes SWS on the brain, and the brainstem reticular formation seems to push the brain from SWS to wakefulness. What system imposes REM sleep?

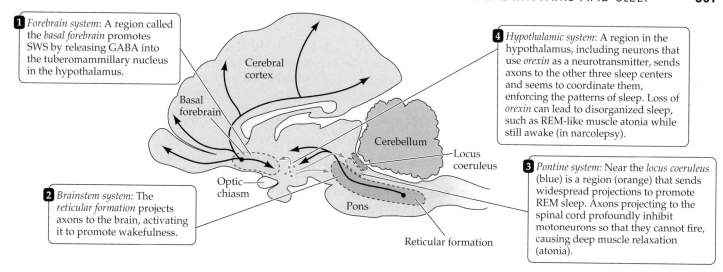

1 *Forebrain system:* A region called the *basal forebrain* promotes SWS by releasing GABA into the tuberomammillary nucleus in the hypothalamus.

4 *Hypothalamic system:* A region in the hypothalamus, including neurons that use *orexin* as a neurotransmitter, sends axons to the other three sleep centers and seems to coordinate them, enforcing the patterns of sleep. Loss of *orexin* can lead to disorganized sleep, such as REM-like muscle atonia while still awake (in narcolepsy).

2 *Brainstem system:* The *reticular formation* projects axons to the brain, activating it to promote wakefulness.

3 *Pontine system:* Near the *locus coeruleus* (blue) is a region (orange) that sends widespread projections to promote REM sleep. Axons projecting to the spinal cord profoundly inhibit motoneurons so that they cannot fire, causing deep muscle relaxation (atonia).

FIGURE 10.18 Brain Mechanisms Underlying Sleep

The pons triggers REM sleep

Several experiments indicated that a region of the pons is important for REM sleep. Lesions of the region just ventral to the **locus coeruleus** abolish REM sleep (**FIGURE 10.18**, step 3) (Friedman and Jones, 1984). Electrical stimulation of the same region, or pharmacological stimulation of this region with cholinergic agonists, can induce or prolong REM sleep. Finally, some neurons in this region seem to be active only during REM sleep (Siegel, 1994). So the pons has a REM sleep center near the locus coeruleus.

One important job of the pontine REM sleep center is to prevent motoneurons from firing. During REM sleep, the inhibitory transmitters GABA and glycine produce powerful inhibitory postsynaptic potentials (discussed in Chapter 3) in spinal motoneurons, preventing them from reaching threshold and producing an action potential (Kodama et al., 2003). Thus, the dreamer's muscles are not just relaxed, but flaccid. This loss of muscle tone during REM sleep can be abolished by small lesions that damage only a part of the REM center, suggesting that this subregion is what normally disables the motor system during REM.

Cats with such lesions seem to act out their dreams. They enter SWS as they normally would, but when they begin to display the EEG signs of REM sleep, instead of becoming completely limp as normal cats do, these cats stagger to their feet (A. R. Morrison, 1983; A. R. Morrison et al., 1995). Are they awake or in REM sleep? They move their heads as though visually tracking moving objects (that aren't there), bat with their forepaws at nothing, and ignore objects that are present (**FIGURE 10.19**). In addition, the cats' *inner eyelids*, the translucent nictitating membranes, partially cover

reticular formation Also called *reticular activating system.* An extensive region of the brainstem (extending from the medulla through the thalamus) that is involved in arousal.

locus coeruleus A small nucleus in the brainstem whose neurons produce norepinephrine and modulate large areas of the forebrain.

(A)

This cat received a small lesion near the locus coeruleus that blocked the completely limp muscles (atonia) that normally accompanies REM sleep.

Following a bout of SWS, the cat in REM rises up as though about to pounce, but its eyes are nearly closed and nothing is there.

(B)

The cat stands up wobbly, "looking" at something we cannot see. These behaviors indicate the cat is dreaming—seeing and interacting with objects that aren't really there.

To see the video **Animal Sleep Activity,** go to

2e.mindsmachine.com/10.5

FIGURE 10.19 Acting Out a Dream (From A. R. Morrison et al., 1995.)

narcolepsy A disorder that involves frequent, intense episodes of sleep, which last from 5 to 30 minutes and can occur anytime during the usual waking hours.

cataplexy Sudden loss of muscle tone, leading to collapse of the body without loss of consciousness. Cataplexy is sometimes a component of narcoleptic attacks.

the eyes. Thus, the cats appear to be in REM sleep, but motor activity is not being inhibited by the brain. These results strongly suggest that animals dream too. What do cats dream of? If their actions while sleeping are any indication, they dream of stalking prey, perhaps a mouse or a ball of yarn.

A hypothalamic sleep center was revealed by the study of narcolepsy

So far we've described three interacting brain systems controlling sleep: an SWS-promoting region in the forebrain, an arousing reticular formation in the brainstem, and a system in the pons that triggers REM sleep, including paralysis of the body during that state. There is a fourth important system, which seems to act as a "coordinating center" among these three centers, in the hypothalamus (**FIGURE 10.18**, step 4). To understand how we learned about this fourth system, we need to consider a rare but fascinating condition.

You might not consider getting lots of sleep an affliction, but some people are either drowsy all the time or suffer sudden attacks of sleep. At the extreme of such tendencies is **narcolepsy**, an unusual disorder in which the person is afflicted by frequent, intense attacks of sleep that last 5–30 minutes and can occur at any time during usual waking hours. These sleep attacks occur several times a day—usually about every 90 minutes (Dantz et al., 1994).

Most people display SWS for an hour or more before entering REM; individuals who suffer from narcolepsy, however, tend to enter REM in the first few minutes of sleep. People with this disorder exhibit an otherwise normal sleep pattern at night, but they suffer abrupt, overwhelming sleepiness during the day. Many people with narcolepsy also show **cataplexy**, a sudden loss of muscle tone, leading to collapse of the body *without loss of consciousness*. Cataplexy can be triggered by sudden, intense emotional stimuli, including both laughter and anger. Narcolepsy usually manifests itself between the ages of 15 and 25 years and continues throughout life. Remember Barry from the start of this chapter? His narcolepsy symptoms began in his freshman year of college, when he started showing the classic signs of excessive daytime sleepiness and cataplexy.

Several strains of dogs exhibit narcolepsy (Aldrich, 1993), complete with sudden collapse and very rapid sleep onset (**FIGURE 10.20**). Just like humans who suffer from

To see the video **Narcoleptic Dachshund**, go to

2e.mindsmachine.com/10.6

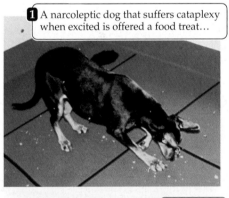

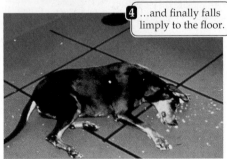

FIGURE 10.20 Canine Narcolepsy (Courtesy of Seiji Nishino.)

(A) Normal

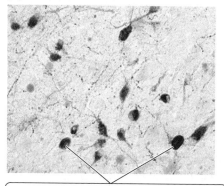

Immunocytochemistry reveals orexin containing neurons in the lateral hypothalamus of a person who did not have narcolepsy.

(B) Narcoleptic

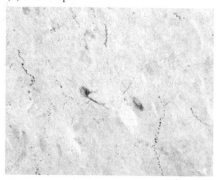

This same region of the brain from someone who suffered from narcolepsy has far fewer orexin neurons.

FIGURE 10.21 Neural Degeneration in Humans with Narcolepsy (Courtesy of Jerome Siegel.)

narcolepsy, these dogs often show REM signs immediately upon falling asleep. Abrupt collapse in these dogs is suppressed by the same drugs (discussed shortly) that are used to treat human cataplexy.

The mutant gene responsible for one of these narcoleptic strains of dogs was found to be the gene for receptors for the neuropeptide **orexin** (also known as *hypocretin*; see Chapter 9) (L. Lin et al., 1999). Mice with the *orexin* gene knocked out also display narcolepsy (Chemelli et al., 1999). Genetically normal rats can be made narcoleptic if injected with a toxin that destroys neurons possessing orexin receptors (Gerashchenko et al., 2001). The narcoleptic dogs show signs of neural degeneration in forebrain structures at about the age when symptoms of narcolepsy appear (Siegel, Nienhuis et al., 1999).

Similarly, humans with narcolepsy have lost about 90% of their orexin neurons (**FIGURE 10.21**) (Thannickal et al., 2000). This degeneration of orexin neurons seems to cause inappropriate activation of the cataplexy pathway that normally happens only during REM sleep. So, orexin normally keeps sleep at bay and prevents the transition from wakefulness directly into REM sleep.

The neurons that produce orexin are found almost exclusively in the hypothalamus. Where do these neurons send their axons to release the orexin? Not so coincidentally, the axons go to each of the three brain centers that we mentioned before: basal forebrain, reticular formation, and locus coeruleus (Sutcliffe and de Lecea, 2002). The orexin neurons also project axons to the hypothalamic tuberomammillary nucleus—the same structure that is inhibited by the basal forebrain to induce SWS. So, it looks as if the hypothalamus contains a orexin-based "switching station" (see Figure 10.18) that switches the brain between states, from wakefulness to non-REM sleep to REM sleep (Saper et al., 2010).

The traditional treatment for narcolepsy was the use of amphetamines in the daytime. The drug GHB (gamma-hydroxybutyrate, trade name Xyrem) helps some narcoleptics (although there are concerns about potential abuse of this drug [Tuller, 2002]). A newer drug, modafinil (Provigil), is sometimes effective for preventing narcoleptic attacks and has been proposed as an "alertness drug" for people with attention deficit hyperactivity disorder. There is also debate about whether modafinil should be available to anyone who feels sleepy or needs to stay awake (Pack, 2003), but at least one study found the drug no more effective than caffeine in this regard (Wesensten et al., 2002). Our friend Barry, from the chapter opener, eventually found a combination of mild stimulants that worked for him, and he ultimately earned an MD degree. Now that narcolepsy is known to be caused by a loss of orexin signaling, there is hope of developing synthetic drugs to stimulate orexin receptors, both for the relief of symptoms in narcolepsy and to combat sleepiness in people without narcolepsy.

orexin Also called *hypocretin*. A neuropeptide produced in the hypothalamus that is involved in switching between sleep states, in narcolepsy, and in the control of appetite.

sleep paralysis A state, during the transition to or from sleep, in which the ability to move or talk is temporarily lost.

Many people who do not suffer from narcolepsy nevertheless occasionally experience the cataplexy that accompanies narcolepsy (Fukuda et al., 1998). **Sleep paralysis** is the temporary inability to move or talk either just before dropping off to sleep or, more often, just after waking. In this state people may experience sudden sensory hallucinations (Cheyne, 2002), including the belief that something is crushing their chest. Sleep paralysis never lasts more than a few minutes, so it's best to relax and avoid panic. One hypothesis is that sleep paralysis results when the pontine center (see Figure 10.18, step 3) continues to impose paralysis for a short while after a person awakes from a REM episode.

how's it going ?

1. Describe the four brain systems that control different stages of the sleep-waking cycle, discussing the experiments that revealed each.
2. What happens when brainstem systems to inhibit movement during REM sleep are damaged?
3. What is narcolepsy, and what brain systems seem to be responsible for this disorder?

Sleep Disorders Can Be Serious, Even Life-Threatening

Narcolepsy is just one of several sleep disorders (**TABLE 10.2**) that have made sleep disorder clinics common in major medical centers. For some people, the peace and comfort of regular, uninterrupted sleep is routinely disturbed by the inability to fall asleep, by prolonged sleep, or by unusual awakenings.

To see the video
REM Behavior Disorder,
go to

2e.mindsmachine.com/10.7

TABLE 10.2 Classification of Sleep Disorders

DISORDERS OF INITIATING AND MAINTAINING SLEEP (INSOMNIA)
Ordinary, uncomplicated insomnia
Drug-related insomnia caused by:
Use of stimulants
Withdrawal of depressants
Chronic alcoholism
Insomnia associated with psychiatric disorders
Insomnia associated with sleep-induced respiratory impairment (sleep apnea)

DISORDERS OF EXCESSIVE DROWSINESS
Narcolepsy
Drowsiness associated with psychiatric problems
Drug-related drowsiness
Drowsiness associated with sleep-induced respiratory impairment (sleep apnea)

DISORDERS OF SLEEP-WAKING SCHEDULE
Temporary disruption caused by:
Time zone change by airplane flight (jet lag)
Shift work, especially night work
Persistent disruption (irregular rhythm)

DYSFUNCTIONS ASSOCIATED WITH SLEEP, SLEEP STAGES, OR PARTIAL AROUSALS
Sleepwalking (somnambulism)
Sleep enuresis (bed-wetting)
Night terror
Nightmares
Sleep-related seizures
Teeth grinding
REM behavior disorder (RBD)

Source: After Weitzman, 1981.

Some minor dysfunctions are associated with sleep

Some dysfunctions associated with sleep are much more common in children than in adults. Two sleep disorders in children—night terrors (described earlier) and **sleep enuresis** (bed-wetting)—are associated with SWS. Most people grow out of these conditions without intervention, but pharmacological approaches can be used to reduce the amount of stage 3 sleep (as well as REM time) while increasing stage 2 sleep. For sleep enuresis, some doctors prescribe a nasal spray of the hormone vasopressin (see Chapter 9) before bedtime, which decreases urine production.

Somnambulism (sleepwalking) consists of getting out of bed, walking around the room, and appearing awake. Although more common in childhood, it sometimes persists into adulthood. These episodes last a few seconds to minutes, and the person usually does not remember the experience. Because such episodes occur during stage 3 SWS, they are more common in the first half of the night when that stage predominates.

Insomniacs have trouble falling asleep or staying asleep

Almost all of us have trouble falling asleep on occasion, but many people persistently find it difficult to get as much sleep as they would like. Estimates of the prevalence of insomnia range from 15% to 30% of the adult population (Parkes, 1985). Insomnia is more commonly reported by older people, females, and users of drugs like tobacco, caffeine, and alcohol. It is not a trivial disorder; recall that adults who regularly sleep for short periods show a higher mortality rate than those who regularly sleep 7–8 hours each night (Kripke et al., 2002).

Insomnia seems to be the final common outcome for various conditions. Situational factors such as shift work, time zone changes, and changes in the daily routine (that hard motel bed) can lead to insomnia. Usually these conditions produce transient **sleep-onset insomnia**, a difficulty in falling asleep. Drugs, as well as neurological and psychiatric factors, seem to cause **sleep-maintenance insomnia**, a difficulty in remaining asleep. In this type of insomnia, sleep is punctuated by frequent nighttime arousals. This form of insomnia is especially evident in disorders of the respiratory system.

In some people, respiration becomes unreliable during sleep. Breathing may cease for a minute or so, or it may slow alarmingly; blood levels of oxygen drop markedly. This syndrome, called **sleep apnea**, arises either from the progressive relaxation of muscles of the chest, diaphragm, and throat cavity or from changes in the pacemaker respiratory neurons of the brainstem. In the former instance, relaxation of the throat obstructs the airway—a kind of self-choking. This mode of sleep apnea is common in very obese people, but it also occurs, often undiagnosed, in non-obese people. Sleep apnea is frequently accompanied by loud, interrupted snoring, so loud snorers should consult a physician about the possibility that they suffer from sleep apnea.

Investigators have speculated that **sudden infant death syndrome (SIDS**, or *crib death*) arises from sleep apnea as a result of immature systems that normally control respiration. Autopsies of SIDS victims reveal abnormalities in brainstem serotonin systems (Kinney, 2009); interfering with this system in mice renders them unable to regulate respiration effectively. What's more, some of these young mice spontaneously stop breathing and die (Audero et al., 2008). The incidence of SIDS has been cut almost in half by the Safe to Sleep campaign, which urges parents to place infants on their backs to sleep rather than on their stomachs. Placing the baby face down may lead to suffocation if the baby cannot regulate breathing or arouse properly. Exposure to cigarette smoke also increases the risk of crib death.

People with **sleep state misperception** (McCall and Edinger, 1992) *report* that they didn't sleep even when the EEG showed signs of sleep and they failed to respond to stimuli. They are sleeping without knowing it. Sometimes these people, upon learning that they really are sleeping, are more satisfied with the sleep they get.

Back to Sleep Placing infants on their backs for sleep reduces the risk of sudden infant death syndrome (SIDS) by half.

sleep enuresis Bed-wetting.

somnambulism Sleepwalking.

sleep-onset insomnia Difficulty in falling asleep.

sleep-maintenance insomnia Difficulty in staying asleep.

sleep apnea A sleep disorder in which respiration slows or stops periodically, waking the patient. Excessive daytime sleepiness results from the frequent nocturnal awakening.

sudden infant death syndrome (SIDS) Also called *crib death*. The sudden, unexpected death of an apparently healthy human infant who simply stops breathing, usually during sleep.

sleep state misperception Commonly, a person's perception that he has not been asleep when in fact he has. It typically occurs at the start of a sleep episode.

Although many drugs affect sleep, there is no perfect sleeping pill

Throughout recorded history, humans have reached for substances to enhance sleep. Ancient Greeks used opium from the juice of the poppy as well as products of the mandrake plant to aid sleep (Hartmann, 1978). The preparation of barbituric acid in the mid-nineteenth century started the development of many drugs—*barbiturates*—that were widely used to combat insomnia.

Most modern sleeping pills—including benzodiazepines (see Chapter 4) like triazolam (Halcion), and nonbenzodiazepine sedatives like zolpidem (Ambien) and eszopiclone (Lunesta)—bind to GABA receptors, inhibiting broad regions of the brain. But reliance on sleeping pills poses many problems (Rothschild, 1992). Viewed solely as a way to deal with sleep problems, current drugs fall far short of being a suitable remedy, for several reasons. First, none of them provide a completely normal night of sleep in terms of time spent in various sleep states, such as REM sleep. Second, continued use of sleeping pills causes them to lose effectiveness, and this declining ability to induce sleep often leads to increased self-prescribed dosages that can be dangerous. Another major drawback is that sleeping pills produce marked changes in the pattern of sleep, both while the drug is being used and for days afterward.

Use of sleeping pills may lead to a persistent "sleep drunkenness," coupled with drowsiness, that impairs waking activity, or to memory gaps about daily activity. Police have reported cases of "Ambien drivers," people who have taken a sleeping pill and then got up a few hours later to go for a spin, with sometimes disastrous results, while apparently asleep (Saul, 2006). In other cases, people taking such medicines eat snacks, shop over the Internet, or even have sex, with no memory of these events the next day (Dolder and Nelson, 2008).

Certainly, the treatment for insomnia that has the fewest side effects, and that is very effective for most people, is not to use any drug, but to develop a regular routine to exploit the body's circadian clock. The best advice for insomniacs is to use an alarm clock to wake up faithfully at the same time each day (weekends included) and then simply go to bed *once they feel sleepy* (Webb, 1992). They should also avoid daytime naps and having caffeine at night. Going through a bedtime routine in a quiet, dark environment can also help to condition sleep onset. Melanopsin, the retinal photopigment that tells the SCN about light and dark, is especially sensitive to bluish light (Gooley et al., 2010), such as the light from LCD screens. So avoiding the use of smart phones and laptops at bedtime (or at least dimming their light) can thus improve sleep (Bedrosian et al., 2013). These steps will let you get the sleep you need, and sleeping pills will not (no matter what the $600 million in annual pharmaceutical advertising might say).

Unfortunately, many college students adopt schedules that virtually guarantee they won't get enough sleep. Waking up early on Monday and Wednesday to attend one class, sleeping a bit later on Tuesday and Thursday, and then sleeping a *lot* later on weekends disrupts your circadian sleep cycle, making it hard to fall asleep when you should on the nights before an early class. It's unpopular advice, but if you want enough sleep, get up at the same time *every* day, not just the day you have that early class, and go to bed about the same time each night. True, you'll miss out on some late night activities with your friends, but you'll get to feel so self-righteous being awake and working while they are sleeping in. Plus you'll stay awake in lectures (we hope).

SIGNS & SYMPTOMS

REM Behavior Disorder

Most sleepwalkers are not acting out a dream (Parkes, 1985), but there is a disorder where people appear to be acting out a dream. **REM behavior disorder (RBD)** is characterized by organized behavior—such as fighting an imaginary foe, eating a meal, or acting like a wild animal—from a person who appears to be asleep (Schenck and Mahowald, 2002), like the gentleman in the low-light video on our website (**FIGURE 10.22**). Sometimes the person remembers a dream that fits well with his behavior (C. Brown, 2003). This disorder usually begins after the age of 50 and is more common in men than in women. Individuals with RBD are reminiscent of the cats with a lesion near the locus coeruleus, mentioned earlier, that were no longer paralyzed during REM and so acted out their dreams. Unfortunately, the onset of RBD is often followed by the early symptoms of Parkinson's disease and dementia (Abbott and Videnovic, 2014), suggesting that the disorder marks the beginning of widespread neurodegeneration. The breakdown appears to begin in the brainstem region that imposes muscle atonia (see Figure 10.18) (Peever et al., 2014). RBD may be controlled by antianxiety drugs (benzodiazepines like Valium) taken at bedtime.

People with RBD seem to be acting out a dream, often of running away from or fighting an unseen foe.

FIGURE 10.22   Battling in Your Dreams (From Mahowald and Schenck, 2005.)

how's it going ?

1. During what stage of sleep does sleepwalking tend to happen? During what time of night?
2. What are the different types of insomnia, and why are sleeping pills an imperfect long-term solution?
3. Describe REM behavior disorder.

REM behavior disorder (RBD)
A sleep disorder in which a person physically acts out a dream.

Recommended Reading

Aserinsky, E., and Kleitman, N. (1955). "Regularly Occurring Periods of Eye Motility, and Concomitant Phenomena, during Sleep." *Science, 118,* pp. 273–274.

Duff, K. (2014). *The Secret Life of Sleep.* New York, NY: Atria Books.

Dunlap, J. C., Loros, J. J., and DeCoursey, P. J. (2004). *Chronobiology: Biological Timekeeping.* Sunderland, MA: Sinauer.

Kryger, M. K., Roth, T., and Dement, W. C. (Eds.). (2011). *Principles and Practice of Sleep Medicine* (5th ed.). Philadelphia, PA: Saunders/Elsevier.

Max, D. T. (2007). *The Family That Couldn't Sleep: A Medical Mystery.* New York, NY: Random House.

VISUAL SUMMARY

10

You should be able to relate each summary to the adjacent illustration, including structures and processes. Go to the online version of this summary (scan the QR code above) for links to figures, animations, and activities that will help you consolidate the material.

1 Animals show **circadian rhythms** of activity that can be **entrained** by light. These rhythms synchronize behavior to changes in the environment. In constant dim light, animals **free-run**, displaying a period of about 24 hours. Review Figures 10.1 and 10.8, Animations 10.2 and 10.3

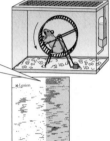
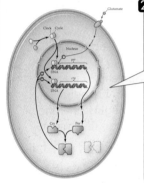

2 Lesions of the **suprachiasmatic nucleus (SCN)** abolish activity rhythms in constant conditions. Transplanting the SCN from one animal into another results in a free-running rhythm of the donor, demonstrating that the SCN contains a clock that can drive circadian activity. Several proteins (including Clock and Cycle) interact, increasing and decreasing in a cyclic fashion that takes about 24 hours. This molecular clock, pooled from many SCN neurons, drives circadian rhythms. Review **Figures 10.3–10.8, Animation 10.4**

3 Almost all mammals show two sleep states: **rapid-eye-movement (REM) sleep** and **non-REM sleep**. Human non-REM sleep has three distinct stages (**stages 1, 2, 3**) defined by **electroencephalography (EEG)** criteria, including **sleep spindles** and large, slow **delta waves** in stage 3. Review **Figure 10.9, Table 10.1, Activity 10.1**

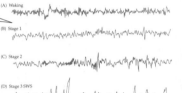
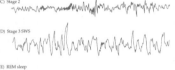
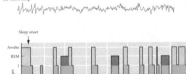

(A) Waking
(B) Stage 1
(C) Stage 2
(D) Stage 3 SWS
(E) REM sleep

4 REM sleep is characterized by rapid, low-amplitude EEG waves (almost like an EEG while awake), but also by profound muscle relaxation because motoneurons are inhibited. People awakened from REM frequently report vivid dreams, while people awakened from SWS report ideas or thinking. Review **Table 10.1**

5 Sleep stages cycle through the night, with stage 3 SWS prominent early, while REM predominates later. Infants sleep a lot, with lots of REM, but as we grow up we sleep less, with less REM. Elderly people sleep even less, and stage 3 sleep eventually disappears. Review **Figures 10.11, 10.12, and 10.15–10.17**

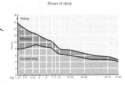

6 Four proposed functions of sleep are energy conservation, **ecological niche** adaptation, body and brain restoration, and memory consolidation. A few people sleep only an hour per night, suggesting that they accomplish the function(s) of sleep very efficiently, but all healthy people sleep. Review **Figures 10.17 and 10.18**

7 **Sleep deprivation** leads to impairments in vigilance and reaction times. It also incurs sleep debt, although the lost SWS and REM may be partially restored in subsequent nights. Prolonged sleep deprivation compromises the immune system and leads to death. Review **Box 10.1, Figure 10.16**

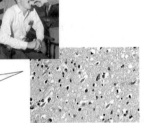
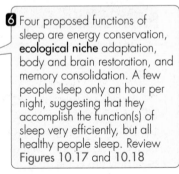

8 Four brain systems control sleep and waking. The **basal forebrain** promotes SWS, the brainstem **reticular formation** promotes arousal, a pontine system triggers REM sleep, and hypothalamic neurons releasing **orexin** regulate these three centers. Review **Figures 10.17 and 10.18, Activity 10.2, Video 10.5**

9 **Narcolepsy** is characterized by sudden, uncontrollable periods of sleep, which may be accompanied by **cataplexy**, paralysis while remaining conscious. Disruption of orexin signaling causes narcolepsy. Review **Figure 10.21, Video 10.6**

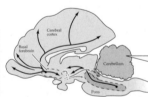

10 Sleep disorders fall into four categories: **sleep-onset** and **sleep-maintenance insomnia**; excessive drowsiness (e.g., narcolepsy); disruption of the sleep-waking schedule; and dysfunctions associated with sleep, sleep stages, or partial arousals (e.g., **somnambulism, RBD**). No pill guarantees a normal night's sleep. Review **Figure 10.20, Table 10.2, Video 10.7**

chapter

11

Emotions, Aggression, and Stress

Trouble in Paradise

You might think the life of a baboon on the Serengeti Plain of Africa is sheer bliss. After all, these highly social animals have so many vigilant eyes that predators almost never manage to sneak up close, and if one does, it may regret it when a whole troop of adults fights back. Food is relatively plentiful, so the baboons need to work only about 3 hours a day to get all the calories they need. Yet these baboons lead highly stressful lives. Why? Because, as Robert Sapolsky (2008) puts it, if you spend only 3 hours per day working, then "you've got nine hours of free time every day to devote to making somebody else just *miserable.*" Like modern humans, these individuals experience stress that comes almost exclusively from their own species.

When Sapolsky began studying a group of baboons he named the Forest Troop, the animals' life was dominated by the hyperaggressive interactions of the oldest, strongest, meanest males. These males were constantly picking fights with each other, resulting occasionally in injuries, even for the "winner." But it wasn't just the rowdy males who were

affected. When one of those males was in a bad mood—after losing a fight with another male, for example—he might attack females or even young baboons. These attacks, too, could be quite serious, taking off a strip of skin or part of an ear. The behavior of these snarly, aggressive males required everyone in the troop to be constantly vigilant, not for lions or food, but for other baboons.

Physiological examination of the baboons confirmed what behavioral observations had suggested: the whole troop was *stressed out.* Their blood had high levels of hormones that are released from the adrenal glands during times of stress. The blood samples also revealed that these baboons had fewer immune system cells to ward off illness (Sapolsky, 2001). That lowered immunity may have played a role in the disaster that befell the troop when it came across some tainted meat, which brought an agonizing death to some of the baboons (Angier, 2004).

As the survivors deal with the new social landscape, what will become of the Forest Troop?

To see the video
Stress,
go to

2e.mindsmachine.com/11.1

The sound of unexpected footsteps in the eerie quiet of the night brings fear to many of us. But a particular piece of music or a loved one's voice summons feelings of comfort or happiness. For some of us, feelings and emotions can become vastly exaggerated; fears may become paralyzing attacks of anxiety and panic. No story about human behavior is complete without considering these powerful feelings.

Our chapter begins with emotions, the bodily responses and brain mechanisms underlying happiness and joy, fear and loathing. We'll discuss both the development of emotions in individuals and the evolution of emotions. Part I's examination of brain mechanisms related to emotional states emphasizes fear and aggression because both are important for survival and they are readily studied in animals. That discussion will lead us to Part II where we'll see that fear and aggression are often associated with stress. Stress, in turn, is associated with a variety of health problems, because it involves not only the nervous and endocrine systems but also the immune system. We'll find that the nervous, endocrine, and immune systems interact extensively.

Baboons Experience Stress Too The baboons of the Forest Troop in Kenya went through a drastic social reorganization.

emotion A subjective mental state that is usually accompanied by distinctive behaviors as well as involuntary physiological changes.

sympathetic nervous system The part of the autonomic nervous system that acts as the "fight or flight" system, generally activating the body for action.

parasympathetic nervous system The part of the autonomic nervous system that generally prepares the body to relax and recuperate.

To view the
Brain Explorer,
go to

2e.mindsmachine.com/11.2

PART I
Emotional Processing

The topic of emotions is complicated by the fact that we apply the word *emotion* to several different things. Emotion is a private, subjective *feeling* that we may have without anyone else being aware of it. But the word emotional is also used to describe many *behaviors* that people show, such as fearful facial expressions, frantic arm movements, or angry shouting. Furthermore, during strong emotion we often experience *physiological* changes, such as a rapidly beating heart, shortness of breath, or excessive sweating. To encompass all three of these aspects, we will define **emotion** as a subjective mental state that is usually accompanied by distinctive behaviors as well as involuntary physiological changes.

Normally we display very distinctive facial expressions that differ for each of the emotions, and we'll see that people from very different cultures recognize these expressions. This near universal recognition of emotional faces suggests that emotions evolved and that the facial displays of emotional experience may help us get along with others. We'll conclude Part I by considering brain circuits for emotion, including a rather specialized brain circuit for fear. That topic will serve as a transition to Part II of the chapter, where we consider two consequences of fear: aggression and stress.

Broad Theories of Emotion Emphasize Bodily Responses

In many emotional states the heart races, the hands and face become warm, the palms sweat, and the stomach feels queasy. Common expressions capture this association: "my hair stood on end," "a sinking feeling in my stomach." These sensations are the result of activation of the autonomic nervous system—either the **sympathetic nervous system** (the "fight or flight" system that generally activates the body for action) or the **parasympathetic nervous system** (which generally prepares the body to relax and recuperate) (see Figure 2.9).

Several theories have tried to explain the close ties between the subjective feelings of emotions and the activity of the autonomic nervous system. Folk wisdom suggests that the autonomic reactions are *caused* by the emotion—"I was so angry, my stomach was churning"—as though the anger produces the churning sensation (**FIGURE 11.1A**). Yet research indicates that the relationship between emotion and physiological arousal is more subtle.

Do emotions cause bodily changes, or vice versa?

William James (1842–1910) and Carl Lange (1834–1900) turned the folk notion on its head, suggesting that the emotions we experience are caused by the bodily changes. From this perspective, we experience fear because we perceive the activity that dangerous conditions trigger in our body (**FIGURE 11.1B**). Different emotions thus feel different because they are generated by a different constellation of physiological responses.

The James-Lange theory inspired many attempts to link specific emotions to specific bodily responses. These attempts mostly failed because it turns out that there is no distinctive autonomic pattern for each emotion. Fear, surprise, and anger, for example, tend to be accompanied by sympathetic activation, while parasympathetic activation tends to accompany both joy and sadness.

In addition, the physiological reactions are rather slow, as physiologists Walter Cannon (1871–1945) and Philip Bard (1898–1977) pointed out (Cannon, 1929). In Cannon and Bard's view, it is the brain's job to decide which particular emotion is an

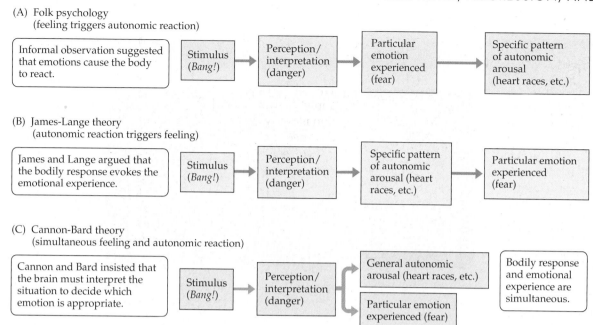

(A) Folk psychology
(feeling triggers autonomic reaction)

Informal observation suggested that emotions cause the body to react.

Stimulus (*Bang!*) → Perception/interpretation (danger) → Particular emotion experienced (fear) → Specific pattern of autonomic arousal (heart races, etc.)

(B) James-Lange theory
(autonomic reaction triggers feeling)

James and Lange argued that the bodily response evokes the emotional experience.

Stimulus (*Bang!*) → Perception/interpretation (danger) → Specific pattern of autonomic arousal (heart races, etc.) → Particular emotion experienced (fear)

(C) Cannon-Bard theory
(simultaneous feeling and autonomic reaction)

Cannon and Bard insisted that the brain must interpret the situation to decide which emotion is appropriate.

Stimulus (*Bang!*) → Perception/interpretation (danger) → General autonomic arousal (heart races, etc.) / Particular emotion experienced (fear) → Bodily response and emotional experience are simultaneous.

FIGURE 11.1 Different Views of the Chain of Events in Emotional Responses

appropriate response to the stimuli. According to this model, the cerebral cortex *simultaneously* decides on the appropriate emotional experience (fear, surprise, joy) and activates the autonomic nervous system to appropriately prepare the body, using either the parasympathetic system to help the body rest, or the sympathetic system to ready the body for action (**FIGURE 11.1C**). It is because the sympathetic system is activated by any threatening situation that so-called lie detectors are very poor at distinguishing liars from truthful people who are anxious (**BOX 11.1**).

polygraph Popularly known as a *lie detector*. A device that measures several bodily responses, such as heart rate and blood pressure.

BOX 11.1 Lie Detector?

One of the most controversial attempts to apply biomedical science is the so-called lie detector test. In this procedure, properly known as a **polygraph** test (from the Greek *polys*, "many," and *graphein*, "to write"), multiple physiological measures are recorded in an attempt to detect lying during a carefully structured interview. The test is based on the assumption that people have emotional responses when lying because they fear detection and/or feel guilty about lying. Emotions are usually accompanied by bodily responses that are difficult to control, such as changes in respiratory rate, heart rate, blood pressure, and skin conductance (a measure of sweating). In polygraph recordings like the one in **FIGURE A**, each wiggly line, or *trace*, provides a measurement of one of these physiological variables. Taken together, the measurements are assumed to track the physiological arousal, over time, of the person being tested. When a person

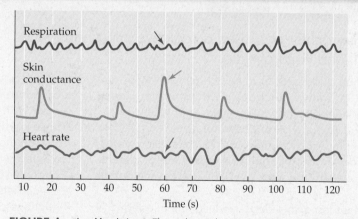

FIGURE A Are You Lying? The polygraph measures signs of arousal.

lies in response to a direct question (arrows), momentary changes in several of the measured variables may occur.

People who administer polygraph examinations for a living claim that polygraphs are accurate in 95% of tests, but the estimate from impartial research is an overall accuracy of about 65% (Nietzel, 2000; Eriksson and Lacerda, 2007).

Even if the higher figure were correct, the fact that these tests are widely used means that thousands of truthful people could be branded as liars and fired, disciplined, or not hired. On the other hand, many criminals and spies have been able to pass the tests without detection (Wollan, 2015). For example, longtime CIA

continued

BOX
11.1 Lie Detector? *continued*

agent Aldrich Ames, who was sentenced in 1995 to life in prison for espionage, successfully passed polygraph tests after becoming a spy. In the wake of the terrorist attacks of 2001, a federally appointed panel of scientists noted that even if polygraphs were correct 80% of the time (which is much higher than impartial research suggests), then giving the test to a group of 10,000 people that included 10 spies would condemn 1,600 innocent people, and let 2 spies go free (National Academy of Sciences, 2003)!

Some scientists believe that modern neuroscience may provide new methods of lie detection someday. Conscious lying may trigger unusual activation of executive control mechanisms of the prefrontal cortex (**FIGURE B**) (Abe et al., 2007). Fear results in activation of the amygdala (as we'll discuss later in this chapter) that, in the case of deception, also might be visible with fMRI (see Chapter 2).

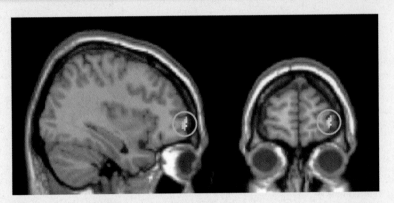

FIGURE B Your Cheatin' Brain PET images reveal selective activation (relative to a control scan) of prefrontal cortex in a study participant engaged in lying. (Courtesy of Nobuhito Abe.)

Functional MRI shows that the anterior cingulate cortex (another region associated with executive control) becomes more active when study participants are lying (Langleben et al., 2002). Although such initial results from brain-imaging studies of deception are intriguing, much more work will be required to establish that brain imaging can detect lies with enough reliability to be useful in making important decisions about individual people. And of course, even if they are validated, such lie detectors will be more costly and less widely available than polygraphs. (Figure B from Abe et al., 2007.)

researchers at work

Stanley Schachter proposed a cognitive interpretation of stimuli and visceral states

Like Cannon and Bard, Stanley Schachter (1975) emphasized cognitive mechanisms in emotion. Under Schachter's model, however, emotional labels (e.g., *anger, fear, joy*) are attributed to relatively nonspecific feelings of physiological arousal. Which emotion we experience depends on cognitive systems that assess the *context*—our current social, physical, and psychological situation.

In a famous test of this idea, students were injected with epinephrine (adrenaline) and told either that there would be no effect or that their hearts would race (Schachter and Singer, 1962).

Students who were warned of the reaction reported no emotional experience, but some participants who were not forewarned experienced emotions when their bodies responded to the drug.

However, *which* emotion was experienced could be affected by whether a confederate in the room acted angry or happy. The unsuspecting participants injected with epinephrine were much more likely to report feeling angry when in the presence of an "angry" confederate, and more likely to report feeling elated when with a "happy" confederate (**FIGURE 11.2A**). These findings contradict the James-Lange prediction that feelings of anger or elation should each be associated with a unique profile of autonomic reactions. Schachter and Singer concluded that the misinformed participants experienced their physiological arousal as whichever emotion seemed appropriate: "My heart's really pounding—I'm so angry!" or "My heart's really pounding—I'm so elated!" depending on the environment. Thus, they said, our emotional states are the results of interaction between physiological arousal and cognitive interpretation of that arousal. This cognitive theory of emotions emphasizes that our interpretation of the context, including social factors such as other people's emotions, determines which emotion we'll experience. The cognitive theory also suggests that our emotional experience at one time may affect how we interpret later events (**FIGURE 11.2B**).

FIGURE 11.2 The Classic Schacter and Singer Experiment

■ **Hypothesis**

We experience a particular emotion as a result of the activity of the autonomic nervous system (the James-Lange theory).

■ **Test**

Activate the sympathetic nervous system with an injection of epinephrine to see whether the participants, uninformed about the drug's effects, experience one particular emotion as they fill out some forms. To test the alternative hypothesis—that our emotional experience is determined by cognitive processes—expose the participants to a confederate who acts either angry or happy while filling out the forms.

(A)

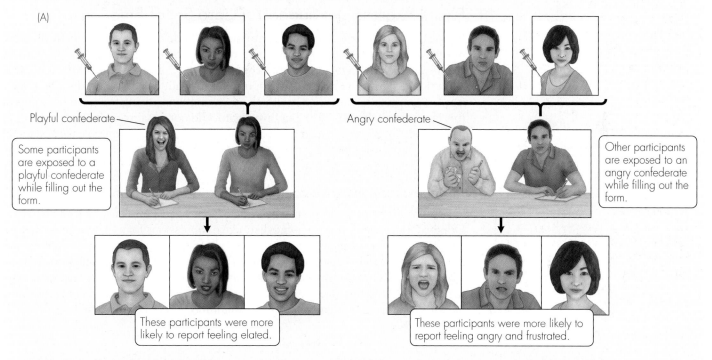

Playful confederate

Some participants are exposed to a playful confederate while filling out the form.

Angry confederate

Other participants are exposed to an angry confederate while filling out the form.

These participants were more likely to report feeling elated.

These participants were more likely to report feeling angry and frustrated.

■ **Outcome**

Participants who were warned that the injection might affect heart rate reported no emotional reaction. Participants who were not warned about the sympathetic arousal reported more intense emotional reactions than those who were given a control injection. However, among the participants who were not warned about the effects of the injection, *which* emotion they experienced (angry or happy) tended to match that of the confederate.

■ **Conclusion**

While autonomic responses can *intensify* our emotional experience, they cannot explain why we have different emotional experiences in different situations. Rather, our cognitive analysis of the environment affects which emotion we experience.

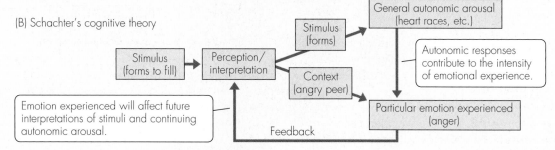

(B) Schachter's cognitive theory

General autonomic arousal (heart races, etc.)

Stimulus (forms)

Stimulus (forms to fill)

Perception/ interpretation

Context (angry peer)

Autonomic responses contribute to the intensity of emotional experience.

Emotion experienced will affect future interpretations of stimuli and continuing autonomic arousal.

Feedback

Particular emotion experienced (anger)

Another interesting outcome of the experiment is that the students receiving epinephrine reported experiencing *more intense* emotions than those given saline, as would be predicted by the James-Lange theory. The autonomic responses did not specify *which* emotion was being experienced, but our awareness of the body's autonomic responses intensifies our experience of emotion (G. W. Hohmann, 1966).

1. Compare and contrast the folk psychology view of bodily responses to emotions with the James-Lange theory.
2. What two findings cast doubt on the James-Lange theory of emotions?
3. Describe the results of Schachter and Singer's experiment. What do these findings suggest about how autonomic reactions, emotional experience, and cognitive processing are related?

Facial Expressions Suggest a Core Set of Emotions

Just as the colors of the spectrum combine into subtle hues, researchers think there may be a basic core set of emotions underlying the more varied and delicate nuances of our world of feelings. One theory (Plutchik, 1994) proposes there are eight basic emotions, grouped in four pairs of opposites—joy/sadness, affection/disgust, anger/ fear, and expectation/surprise—with all other emotions arising from combinations of this basic array (**FIGURE 11.3**). But investigators do not yet agree about the number of basic emotions (six, seven, eight?). While there is no way to determine once and for all the number of basic emotions, one clue comes from examining the number of different kinds of facial expressions that we produce and can recognize in others.

Facial expressions have complex functions in communication

How many different emotions can be detected in facial expressions? According to Paul Ekman and collaborators, there are distinctive expressions for anger, sadness, happiness, fear, disgust, surprise, contempt, and embarrassment (**FIGURE 11.4**) (Keltner and Ekman, 2000). Researchers suggested that Pixar Studios portray all eight emotions inside a little girl's head in the movie *Inside Out*, but the director felt the story could only handle 5–6 characters (Keltner and Ekman, 2015). Facial expressions of these emotions are interpreted similarly across many cultures without explicit training.

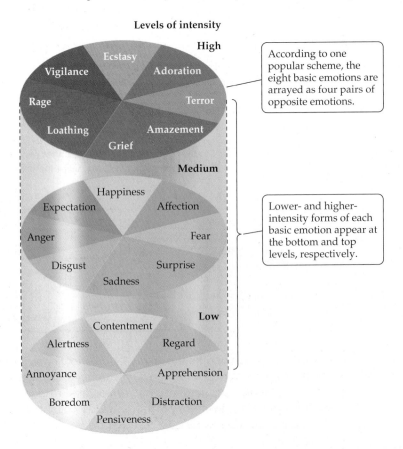

Levels of intensity

According to one popular scheme, the eight basic emotions are arrayed as four pairs of opposite emotions.

Lower- and higher-intensity forms of each basic emotion appear at the bottom and top levels, respectively.

FIGURE 11.3 One Classification of Basic Emotions (After Plutchik, 1994.)

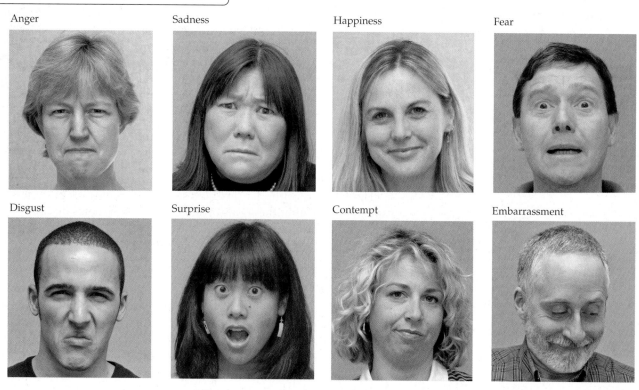

According to Paul Ekman and colleagues, the basic emotional facial expressions shown here are displayed in all cultures.

Anger   Sadness   Happiness   Fear

Disgust   Surprise   Contempt   Embarrassment

FIGURE 11.4   The Eight Universal Facial Expressions of Emotion

Cross-cultural similarity is also noted in the *production* of expressions specific to particular emotions. For example, people in a preliterate New Guinea society show emotional facial expressions like those of people in industrialized societies. However, facial expressions are not unfailingly universal. For example, although Russell (1994) found significant agreement across cultures in the recognition of most emotional states from facial expressions, isolated nonliterate groups did not agree with Westerners about recognizing expressions of surprise and disgust (**FIGURE 11.5**), suggesting that different cultures have adopted different ways to express those emotions.

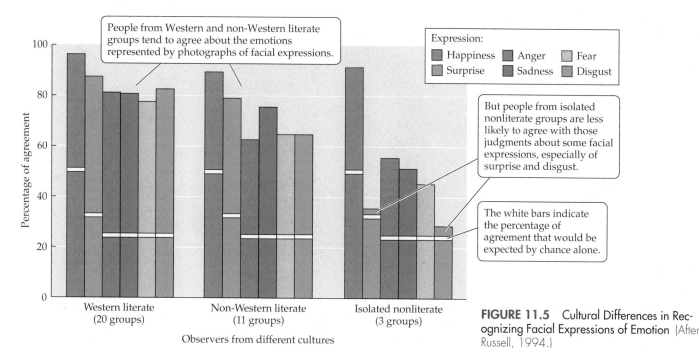

People from Western and non-Western literate groups tend to agree about the emotions represented by photographs of facial expressions.

Expression:
Happiness   Anger   Fear
Surprise   Sadness   Disgust

But people from isolated nonliterate groups are less likely to agree with those judgments about some facial expressions, especially of surprise and disgust.

The white bars indicate the percentage of agreement that would be expected by chance alone.

Percentage of agreement

Western literate
(20 groups)

Non-Western literate
(11 groups)

Isolated nonliterate
(3 groups)

Observers from different cultures

FIGURE 11.5   Cultural Differences in Recognizing Facial Expressions of Emotion (After Russell, 1994.)

FIGURE 11.6 A Model for Emotional Facial Expressions across Cultures

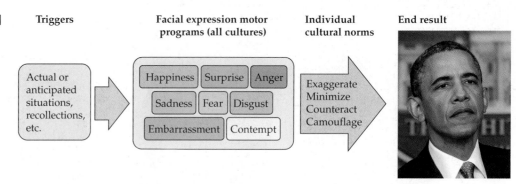

Triggers	Facial expression motor programs (all cultures)	Individual cultural norms	End result

Actual or anticipated situations, recollections, etc.

Happiness Surprise Anger
Sadness Fear Disgust
Embarrassment Contempt

Exaggerate
Minimize
Counteract
Camouflage

These subtle cultural differences suggest that cultures prescribe rules for facial expression, and that they control and enforce those rules by cultural conditioning. Everyone agrees that cultures affect the facial display of emotion; the remaining controversy is over the extent of that cultural influence (**FIGURE 11.6**).

Facial expressions are mediated by muscles, cranial nerves, and CNS pathways

The human face is a complicated object, a network of small muscles that are carefully controlled by the nervous system. We use subsets of those muscles to produce nuanced facial expressions, from grimace to grin, alongside less subtle facial behaviors, like eating and speaking. Facial muscles can be divided into two categories:

1. *Superficial facial muscles* mostly attach only between different points of facial skin (**FIGURE 11.7**), so when they contract, they change the shape of the mouth, eyes, or nose, or maybe create a dimple.

2. *Deep facial muscles* attach to bone and produce larger-scale movements, like chewing.

These facial muscles are innervated by two cranial nerves: (1) the facial nerve (VII), which innervates the superficial muscles of facial expression; and (2) the motor branch of the trigeminal nerve (V), which innervates muscles that move the jaw (see Figure 2.7). The activity of the cranial nerves is governed by the face area of motor cortex: this

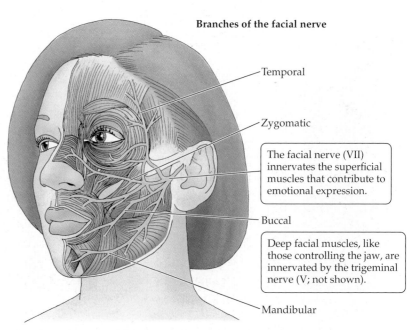

Branches of the facial nerve

Temporal

Zygomatic

The facial nerve (VII) innervates the superficial muscles that contribute to emotional expression.

Buccal

Deep facial muscles, like those controlling the jaw, are innervated by the trigeminal nerve (V; not shown).

Mandibular

FIGURE 11.7 Superficial Facial Muscles and Their Neural Control

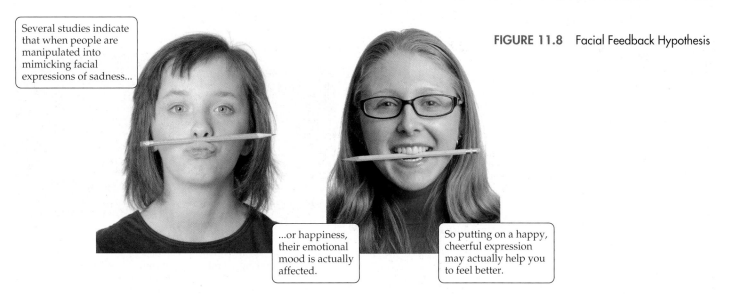

FIGURE 11.8 Facial Feedback Hypothesis

Several studies indicate that when people are manipulated into mimicking facial expressions of sadness...

...or happiness, their emotional mood is actually affected.

So putting on a happy, cheerful expression may actually help you to feel better.

is a disproportionately large brain region in humans (see Figure 5.9), probably reflecting the importance of emotional expression in our species.

Lending support to the James-Lange notion that sensations from our body inform us about our emotions, the **facial feedback hypothesis** suggests that sensory feedback from our facial expressions can affect our mood. In the best tests of this idea, people are given a task, such as holding a pencil either under their nose or between their teeth (**FIGURE 11.8**), that effectively has them take on a sad or happy face, respectively. The participants are then probed to see how happy or sad they feel, or how funny they find a cartoon. Typically, people who have been simulating a smile report more positive feelings than the folks who have been simulating a frown (Davis et al., 2009). So it is possible that forcing yourself to take on a cheerful expression may actually help you feel happier, and the old song that advises "just put on a happy face" may have its basis in fact. Conversely, people who receive Botox injections, paralyzing facial muscles to relieve wrinkles and look younger, experience emotions less intensely after the treatment than before (Davis et al., 2010).

How did emotion and emotional displays evolve?

In his book *The Expression of the Emotions in Man and Animals* (1872), Charles Darwin presented evidence that certain expressions of emotions are universal among people of all regions of the world, anticipating by more than a century the studies of facial expressions that we reviewed earlier (see Figure 11.5). Furthermore, Darwin asserted that some nonhuman animals show comparable expressions of some emotions, suggesting that aspects of emotional expression may have originated in a common ancestor.

Disputing earlier scholars, Darwin emphasized that nonhuman primates have the same facial muscles that humans have. A century later, Redican (1982) noted distinct facial expressions in nonhuman primates, including a *play face*, homologous to the human laugh (**FIGURE 11.9**). This connection may even extend beyond primates: for example, tickling and playing with rats can elicit ultrasonic vocalizations that may be analogous to laughter. Expression of positive emotions of this sort may facilitate social contact and learning in a variety of species (Panksepp, 2007; Burgdorf et al., 2008).

How do emotions and their expression help individuals survive and reproduce? Darwin (1872) offered several suggestions:

> The movements of expression in the face and body [are important] for our welfare. They serve as the first means of communication between the mother and her infant; she smiles approval, and thus encourages her child on the right path, or frowns disapproval. We readily perceive sympathy in others by their expression... and mutual good feeling is thus strengthened (p. 365).

facial feedback hypothesis The idea that sensory feedback from our facial expressions can affect our mood.

FIGURE 11.9 Facial Expression of Emotions in Nonhuman Primates (Photographs by Frans de Waal, from de Waal, 2003.)

(A)

An adult female chimpanzee screams at another female, who is pulling at her food. Screaming is used in submission and protest.

(B)

A juvenile chimpanzee shows a play face while being tickled. He also makes a guttural laughing sound.

(C)

A Tibetan macaque bares his teeth, grinning to signal submission to a dominant animal. In humans, teeth baring has gained a different, friendlier meaning.

Most of us have experienced the frightening nighttime perception of being stalked by a predator—real or imagined, human or nonhuman. Through natural selection, a program for dealing with this situation evolved: we call that program *fear*. The emotion of fear shifts our perception, attention, cognition, and action to focus on avoiding danger and seeking safety, while preparing us physiologically for fighting or for flight. Other activities, such as seeking food, sleep, or mates, are suppressed. In the face of an imminent threat to life, it is better to be afraid, calling on this recipe for action, developed and tested over the ages, than to ad-lib something.

Viewed in this way, emotions can be seen as evolved preprogramming that helps us deal quickly and effectively with a wide variety of situations. As another example, feelings of disgust for body fluids may help us avoid exposure to germs (Curtis et al., 2004), so it may be wise to recognize disgust in others. Our unfortunate tendency to make snap judgments about other people, based on their appearance and facial expressions, may be an overgeneralization of mechanisms that evolved to help us recognize signs of threat or danger from others (Todorov et al., 2008).

how's it going ?

1. List some examples of particular facial expressions that are associated with particular emotions.
2. What evidence suggests that facial expressions of emotional state are inherited rather than taught by culture?
3. Describe the facial feedback hypothesis of emotion, and give an example of an experiment supporting this hypothesis.
4. What is the evidence that emotions, and the facial expressions that accompany them, evolved by natural selection?

Do Distinct Brain Circuits Mediate Emotions?

The question of whether specific brain circuits have control over emotions has been explored in studies using either brain lesions or electrical stimulation. The evidence confirms not only that some brain regions do specialize in emotions, but also that the same regions may be involved in multiple emotions.

Electrical stimulation of the brain can produce emotional effects

One way to study the neuroanatomy of emotion is to electrically stimulate brain sites in conscious animals and then observe the effects on behavior. Classic work in the 1950s produced an intriguing finding: rats will enthusiastically press a lever in order to

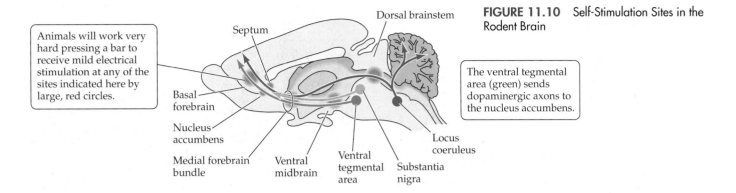

Animals will work very hard pressing a bar to receive mild electrical stimulation at any of the sites indicated here by large, red circles.

FIGURE 11.10   Self-Stimulation Sites in the Rodent Brain

The ventral tegmental area (green) sends dopaminergic axons to the nucleus accumbens.

Dorsal brainstem
Septum
Basal forebrain
Nucleus accumbens
Medial forebrain bundle
Ventral midbrain
Ventral tegmental area
Substantia nigra
Locus coeruleus

give themselves brief electrical stimulation in a brain region called the *septum* (**FIGURE 11.10**) (Olds and Milner, 1954). This phenomenon, called **brain self-stimulation**, can also happen in humans. Patients receiving electrical stimulation in the septum feel a sense of pleasure or warmth, or sometimes sexual excitement (Heath, 1972).

A rush of experimentation soon mapped brain sites that yield self-stimulation responses. Almost all of these sites are subcortical, especially in the hypothalamus and extending into the brainstem. We now know that a bundle of axons that ascends from the midbrain through the hypothalamus—the **medial forebrain bundle**—may be affected at many of the self-stimulation sites. One important target for the axons of the median forebrain bundle is the **nucleus accumbens**, a major component of the brain's reward circuitry (see Chapter 4). The release of dopamine into the nucleus accumbens appears to produce very pleasurable feelings.

One theory is that electrical stimulation taps into dopaminergic circuits that are normally activated by behaviors that produce pleasurable feelings, such as feeding or sexual activity (White and Milner, 1992). As we discussed in Chapter 4, researchers have proposed that drugs of abuse are addictive because they activate these same neural circuits with an artificial intensity (R. C. Pierce and Kumaresan, 2006).

Brain lesions also affect emotions

Early in the twentieth century, dogs in which the cortex had been removed were found to respond to routine handling with sudden intense **decorticate rage**—snarling, biting, and so on—sometimes referred to as *sham rage* because it lacks well-directed attack. Clearly, then, emotional behaviors of this type must be organized at a subcortical level, with the cerebral cortex normally *inhibiting* rage responses. On the basis of studies such as these, combined with observations from brain autopsies of people with emotional disorders, James Papez (1937) proposed a subcortical circuit of emotion. Noting associations between emotional changes and specific sites of brain damage, Papez (which rhymes with "capes") concluded that destruction of a set of interconnected pathways in the brain would impair emotional processes. These interconnected regions, known as the **limbic system** (MacLean, 1949), include the mammillary bodies of the hypothalamus, the anterior thalamus, the cingulate cortex, the hippocampus, the amygdala, and the fornix. The arrows in **FIGURE 11.11** schematically depict the interconnections of this circuit.

Early support for the limbic model of emotion came from studies of monkeys following removal of their temporal lobes (Klüver and Bucy, 1938). The animals' behavior changed dramatically after surgery; the highlight of this behavioral change was an extraordinary taming effect known as the **Klüver-Bucy syndrome**. Animals that had been wild and fearful of humans before surgery became tame and showed neither fear nor aggression afterward. In addition, they showed strong oral tendencies, eating a variety of objects, including rocks. Frequent and often inappropriate sexual behavior was also observed. Because this type of behavior is also seen in monkeys in which only the left and right amygdalas have been destroyed—without damaging any adjacent tissue

brain self-stimulation   The process in which animals will work to provide electrical stimulation to particular brain sites, presumably because the experience is very rewarding.

medial forebrain bundle   A collection of axons traveling in the midline region of the forebrain.

nucleus accumbens   A region of the forebrain that receives dopaminergic innervation from the ventral tegmental area, often associated with reward and pleasurable sensations.

decorticate rage   Also called *sham rage*. Sudden intense rage characterized by actions (such as snarling and biting in dogs) that lack clear direction.

limbic system   A loosely defined, widespread group of brain nuclei that innervate each other to form a network. These nuclei are implicated in emotions.

Klüver-Bucy syndrome   A condition, brought about by bilateral amygdala damage, that is characterized by dramatic emotional changes including reduction in fear and anxiety.

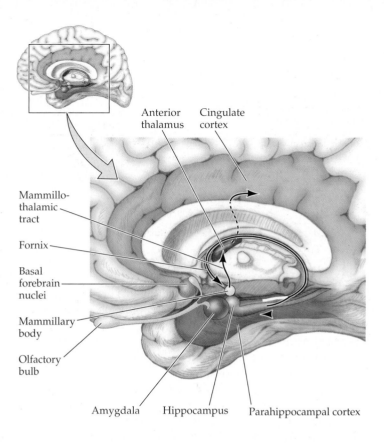

FIGURE 11.11   The Limbic System: Medial Brain Regions Involved in Emotions

Anterior thalamus

Cingulate cortex

Mammillo-thalamic tract

Fornix

Basal forebrain nuclei

Mammillary body

Olfactory bulb

Amygdala   Hippocampus   Parahippocampal cortex

fear conditioning   A form of classical conditioning in which a previously neutral stimulus is repeatedly paired with an unpleasant stimulus, like foot shock, until the previously neutral stimulus alone elicits the responses seen in fear.

amygdala   A group of nuclei in the medial anterior part of the temporal lobe.

(Emery et al., 2001)—it appears that the amygdala is a key structure in the behavioral changes in Klüver-Bucy syndrome, specifically the loss of fear. Let's take a closer look at the role of the amygdala in fear.

The amygdala governs a fear circuit

There is nothing subtle about fear. Conditions that provoke fear elicit similar behaviors from individuals of many different species. This lack of subtlety and the similarity of fear-related behavior across species may explain why we know much more about the neural circuitry of fear than of any other emotion (LeDoux, 1995). For example, it is very easy to reliably elicit fear by using classical conditioning, in which the person or animal is presented with a stimulus such as light or sound that is paired with a brief aversive stimulus such as mild electric shock (**FIGURE 11.12A**). After several such pairings, the response to the sound or light itself is the typical fear portrait, including freezing in position and autonomic signs such as more rapid heart rate and breathing.

Studies of such **fear conditioning** have provided a map of the neural circuitry, with the **amygdala** as a key structure in the mediation of fear (**FIGURE 11.12B**). Located at the anterior medial portion of each temporal lobe, the amygdala is composed of about a dozen different nuclei, each with a distinctive set of connections. Lesioning just the central nucleus of the amygdala in rats prevents blood pressure increases and freezing behavior in response to a conditioned fear stimulus.

On its way to the amygdala, via various sensory channels, information about fear-provoking stimuli reaches a fork in the road at the level of the thalamus (recall from Chapter 2 that the thalamus acts like a switchboard, directing sensory information to specific brain regions). A direct projection from the thalamus to the amygdala, nicknamed the "low road" for fear responses (LeDoux, 1996), bypasses conscious processing and allows for immediate reactions to fearful stimuli (Tamietto and de Gelder, 2010). An alternate "high road" pathway routes the incoming information through sensory cortex, allowing for processing that, while slower, is conscious, fine-grained, and integrated with higher-level cognitive processes, such as memory (**FIGURE 11.12C**). You can learn more details about the amygdala circuitry for fear and other emotions in **A STEP FURTHER 11.1**, on the website.

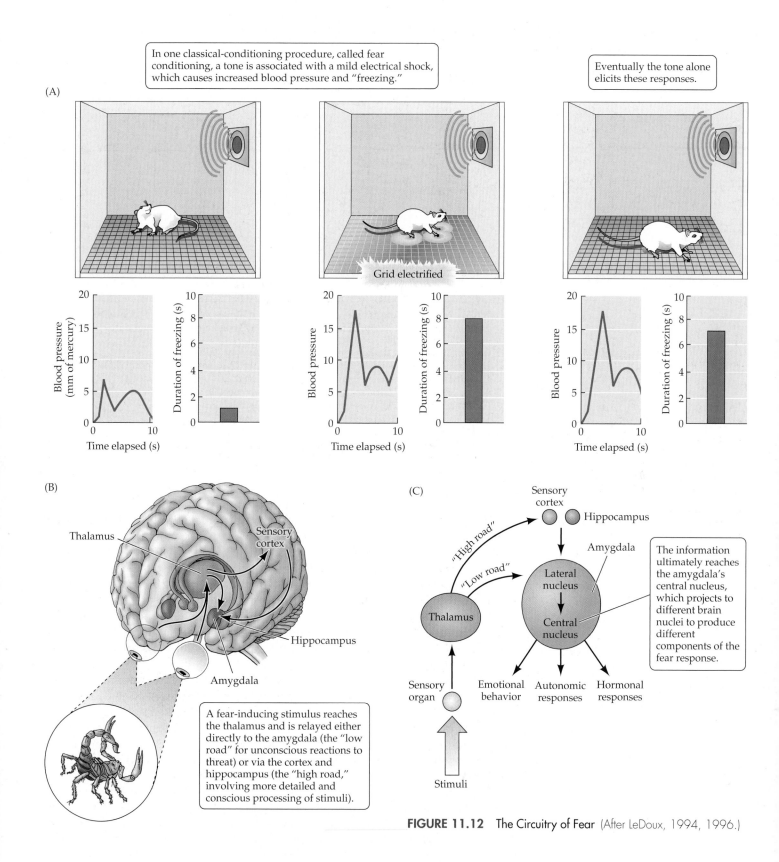

(A)

In one classical-conditioning procedure, called fear conditioning, a tone is associated with a mild electrical shock, which causes increased blood pressure and "freezing."

Eventually the tone alone elicits these responses.

Grid electrified

(B)

Thalamus

Sensory cortex

Hippocampus

Amygdala

A fear-inducing stimulus reaches the thalamus and is relayed either directly to the amygdala (the "low road" for unconscious reactions to threat) or via the cortex and hippocampus (the "high road," involving more detailed and conscious processing of stimuli).

(C)

Sensory cortex

Hippocampus

"High road"

"Low road"

Amygdala

Lateral nucleus

Central nucleus

Thalamus

Sensory organ

Stimuli

Emotional behavior

Autonomic responses

Hormonal responses

The information ultimately reaches the amygdala's central nucleus, which projects to different brain nuclei to produce different components of the fear response.

FIGURE 11.12 The Circuitry of Fear (After LeDoux, 1994, 1996.)

FIGURE 11.13 The Woman Who Was Never Afraid (After Feinstein et al., 2011.)

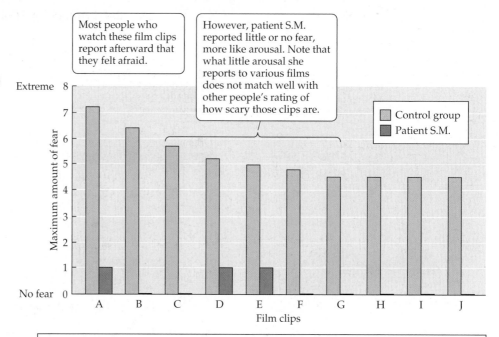

A – *The Ring* (2002) The ghost of a murdered child infiltrates the lives of her soon-to-be victims.
B – *Blair Witch Project* (1999) Campers are attacked by an unknown apparition during the middle of the night.
C – *CSI* (2009) A man struggles to survive after being buried alive.
D – *The English Patient* (1996) A man is tortured by the Germans during World War II.
E – *Se7en* (1995) A mutilated man awakes from the dead.
F – *Cry Freedom* (1987) Armed trespassers attack a woman who is home alone during the night.
G – *Arachnophobia* (1990) A large poisonous spider attacks a girl in the shower.
H – *Halloween* (1978) A woman is being chased by a murderer.
I – *The Shining* (1980) A young boy hears voices in the hallway of a haunted hotel.
J – *The Silence of the Lambs* (1991) A female FBI agent tries to capture a twisted serial killer who is hiding in a dark basement.

The data from rats and mice implicating the amygdala fit well with observations in humans. For example, when people are shown visual stimuli associated with pain or fear, blood flow to the amygdala increases (LaBar et al., 1998), even if the person is not consciously aware of the stimuli (Pegna et al., 2005), and studies of white matter projections confirm that like other mammals, humans have a direct connection between the thalamus and the amygdala that allows emotional visual stimuli to bypass conscious perception (Tamietto et al., 2012). People who suffer from temporal lobe seizures that include the amygdala commonly report that intense fear precedes the start of a seizure (Engel, 1992). Likewise, stimulation of temporal lobe sites may elicit feelings of fear in patients (Bancaud et al., 1994).

The case of a woman, referred to as S.M., who suffers a rare genetic disease that damaged both amygdalas, shows the value of being afraid (Feinstein et al., 2011). Not only is S.M. unafraid of snakes or spiders, but she once walked right up to a knife-wielding drug addict who was trying to rob her. He was so disquieted by her strange response that *he* was the one who ran away. Once she was nearly killed in an act of domestic violence. While her behavioral responses and self-report appear typical for other emotions, S.M. shows very little sympathetic response to fearful stimuli, and she produces almost no startle response to a sudden, loud noise. Shown movie clips that other people find frightening, S.M. reports being unmoved (**FIGURE 11.13**). S.M. is also very poor at recognizing the facial expressions of fear in other people, but she recognizes other emotional expressions—a pattern seen in other people with damaged amygdalas (Adolphs et al., 2005). Interestingly, when S.M. was asked to breathe air with a high concentration of carbon dioxide, she soon felt a panicky fear, flailing her hands about (Feinstein et al., 2013). This result suggests that some other brain region mediates the fear of suffocation.

Composite faces reveal differences between left and right in the level of intensity of emotional expression.

Photographs constructed from only the left side of the face are judged to be more emotional than either the original face or a composite based on just the right side of the face.

FIGURE 11.14 Emotions and Facial Asymmetry

(A) Left sides   (B) Original   (C) Right sides

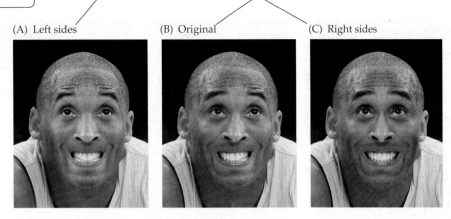

The two cerebral hemispheres process emotion differently

Experimental and clinical observations indicate that the two cerebral hemispheres play different roles in cognitive processes in humans (see Chapter 15). There appear to be similar hemispheric differences in emotion processing. Patients suffering strokes in the left cerebral hemisphere are more likely to have depressive symptoms, independent of any trouble they may be having with speech (Starkstein and Robinson, 1994). In contrast, patients with lesions of the right parietal or temporal cortex are described as unduly cheerful and indifferent to their loss. **A STEP FURTHER 11.2**, on the website, details some of the emotional changes resulting from brain damage. Injection of the anesthetic sodium amytal into just one side—either the left or the right carotid artery (the Wada test, described in Box 15.1)—causes one entire cerebral hemisphere to go to sleep for a short time. As with the people who have left-hemisphere damage, injection of anesthetic into the left hemisphere causes depressed mood, whereas right-sided injections elicit smiling and a feeling of euphoria (Terzian, 1964).

Differences in the contributions of the two hemispheres to emotion are also evident in the asymmetry of facial expressions. By cutting a photograph of a person who is displaying an emotion down the exact middle of the face, we can create two new composite photos: one that combines two left sides of the face (one of which is printed in mirror image), and another that combines two right sides. The results reveal that facial expressions are not symmetrical (**FIGURE 11.14**). Most observers judge the left-sides photos as more emotional than the right-sides photos. Because the left side of the face is controlled by the right hemisphere, this observation again suggests that the right hemisphere is especially important for emotional processing (Borod et al., 1997).

Different emotions activate different regions of the human brain

Several forebrain areas have been consistently implicated in various emotions. Bartels and Zeki (2000) recruited volunteers who professed to be "truly, deeply, and madly in love." Each participant furnished four color photographs: one photo of his or her boy- or girlfriend, and three photos of friends who were the same sex as the loved partner and were similar in age and length of friendship. Functional-MRI brain scans were made while each participant was shown counterbalanced sequences of the four photographs. Brain activity elicited by viewing the loved person was compared with that elicited by viewing friends. Love, compared with friendship, involved *increased* activity in the insula and anterior cingulate cortex, and *reduced* activity in the posterior

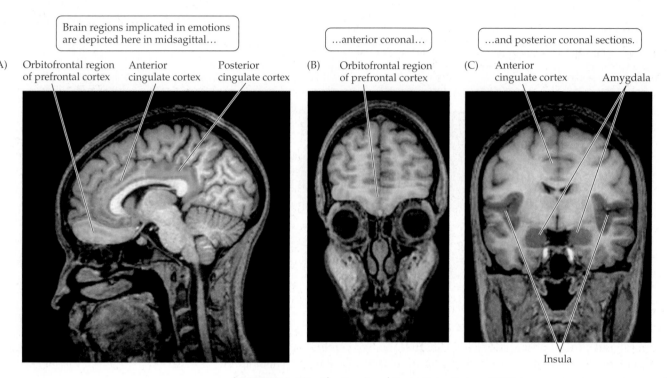

Brain regions implicated in emotions are depicted here in midsagittal...

...anterior coronal...

...and posterior coronal sections.

(A) Orbitofrontal region of prefrontal cortex | Anterior cingulate cortex | Posterior cingulate cortex

(B) Orbitofrontal region of prefrontal cortex

(C) Anterior cingulate cortex | Amygdala

Insula

FIGURE 11.15 The Emotional Brain (From Dolan, 2002.)

cingulate and prefrontal cortices (**FIGURE 11.15**). Given its role in fear, you won't be surprised to learn that the amygdala also showed *reduced* activity when people were contemplating their romantic partner.

Another study compared brain activation during four different kinds of emotion, and again the insula, cingulate cortex, and prefrontal cortex were among the regions implicated. These studies indicate that *there is no simple, one-to-one relation between a specific emotion and changed activity of a brain region*. There is no "happy center" or "sad center." Instead, each emotion involves differential patterns of activation across a network of brain regions associated with emotion, as you can see in **A STEP FURTHER 11.3**, on the website. For example, activity of the cingulate cortex is altered in sadness, happiness, and anger; and the left somatosensory cortex is deactivated in both anger and fear. Although different emotions are associated with different patterns of activation, there is a good deal of overlap among patterns for different emotions (A. R. Damasio et al., 2000).

In Part II of this chapter we'll focus on the darker side of human emotional experience: aggression, stress, and the toll these negative experiences take on our health.

how's it going ❓

1. Describe brain self-stimulation and what this phenomenon suggests about emotional experience.

2. What is the limbic system, and what happens when portions of this system are damaged, such as in Klüver-Bucy syndrome?

3. Describe fear conditioning and the evidence that the amygdala plays a role in this process.

4. What evidence suggests that the amygdala mediates fear in humans?

5. Describe the evidence that one hemisphere of the brain is more involved than the other in the display of emotions.

PART II
Aggression and Stress

Violence, assaults, and homicide exact a high price in modern society, and physical assault is not the only form of aggression. Verbal and symbolic aggression—name calling, horn honking, angry glares—also take their toll. We begin this section by examining the biological factors that influence aggression in animals and humans. Then we'll conclude the chapter by discussing a consequence of being the target of aggression: stress. As we noted at the start of the chapter, most of our stress today is caused by the behavior of other people. We'll learn that the body's hormonal responses to stress, an adaptation to the brief crises of our ancestral environment, can be disastrous in the face of the long-lasting stresses of modern-day life. In learning how chronic stress compromises the immune system, we'll learn that the nervous system, the endocrine system, and the immune system are constantly regulating each other to preserve our health.

aggression Behavior that is intended to cause pain or harm to others.

intermale aggression Aggression between males of the same species.

testosterone A hormone, produced by male gonads, that controls a variety of bodily changes that become visible at puberty; one of a class of hormones called *androgens*.

Neural Circuitry, Hormones, and Synaptic Transmitters Mediate Violence and Aggression

We can define **aggression** as behavior that is intended to cause pain or harm to others. In the discussion that follows, we focus primarily on physical aggression between individuals, excluding the aggression of predators toward their prey, which is perhaps better viewed as feeding behavior (Glickman, 1977). **Intermale aggression** (aggression between males of the same species) is observed in most vertebrates. The relevance to humans is reflected in the fact that males are 5 times as likely as females to be arrested on charges of murder in the United States. Further, aggressive behavior between boys, in contrast to that between girls, is evident early, in the form of vigorous and destructive play behavior (J. Archer, 2006). Whatever we may think about aggression, it seems clear that in many species aggressive behavior in males is adaptive for gaining access to food and mates. We'll find that the hormone that prepares males for reproduction—testosterone—also makes them more aggressive. We'll also discuss certain neurotransmitters that play a role in aggression, as well as the controversial idea of using surgery to control violence.

Androgens seem to increase aggression

At sexual maturity, as the testes begin secreting the steroid hormone **testosterone**, intermale aggression markedly increases in many species (McKinney and Desjardins, 1973). In seasonally breeding species as diverse as birds and primates, intermale aggression waxes and wanes in concert with seasonal changes in levels of testosterone (Wingfield et al., 1987). Conversely, castrating males to remove the source of testosterone usually reduces aggressive behavior profoundly. Treating castrated males with testosterone restores fighting behavior (**FIGURE 11.16**).

The relationship between testosterone and aggression in humans is less clear-cut. Treating adult volunteers with extra testosterone does not increase their aggression (O'Connor et al., 2004). Similarly, young men going through puberty experience a sudden large increase in circulating testosterone, yet they do not show a correlated increase in aggressive behavior (J. Archer, 2006). Nevertheless, some human studies report a positive correlation between testosterone levels and the magnitude of hostility, as measured by behavior rating scales (Dabbs and Morris, 1990). Nonaggressive tendencies in males are associated with satisfaction in family functioning and with lower levels of serum testosterone (Julian and McKenry, 1979). Among female convicts, testosterone concentrations are highest in women convicted of unprovoked violence and lowest among women convicted of defensive violent crimes (Dabbs et al., 1988; Dabbs and Hargrove, 1997).

Nature, Red in Tooth and Flipper Fighting male elephant seals draw blood. In most mammalian species, males must compete with one another, often in the form of physical aggression, for the chance to mate with females.

FIGURE 11.16 The Effects of Androgens on the Aggressive Behavior of Mice (After G. C. Wagner et al., 1980.)

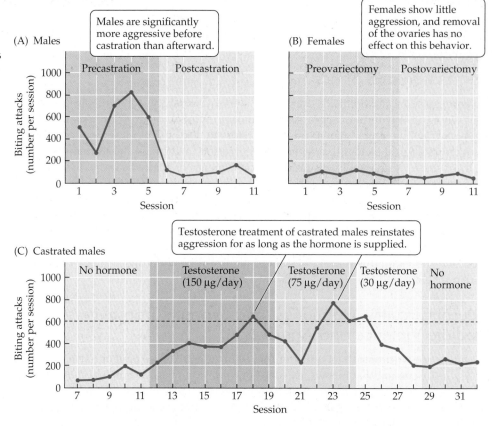

At least two variables confound the correlations between testosterone and aggression. First is the observation that experience can affect testosterone levels. In mice and monkeys, the loser in aggressive encounters shows reduced androgen levels (Lloyd, 1971; I. S. Bernstein and Gordon, 1974), so measured levels of testosterone sometimes may be a *result*, rather than a cause, of behavior. In men, testosterone levels rise in the winners and fall in the losers after competitions ranging from wrestling to chess (van Anders and Watson, 2006). Male sports fans even show a vicarious competition effect in response to simply watching "their" team win or lose a sporting event (Bernhardt, 1997), and during the 2008 U.S. presidential election, men who voted for John McCain experienced a sharp drop in circulating testosterone, compared with Obama backers (Stanton et al., 2009).

These observations suggest that a second confounding variable between testosterone and aggression is dominance (Mazur and Booth, 1998), since most chess players could hardly be said to be aggressive, at least not physically. According to this model, testosterone levels should be associated with behaviors that confer or protect the individual's social status (and thus reproductive fitness). These behaviors may *sometimes*, but not always, involve overt aggression. Despite the lack of a close relationship between aggression and androgens, people have tried to modify the behavior of male criminals by manipulating sex hormones, through surgical castration or "chemical" castration with drugs that block androgen receptors or testosterone production. While lowered testosterone may reduce violence in some sex offenders (Brain, 1994), the main effect is a reduction in sexual drive and interest more than a direct effect on aggression. Many ethical issues raised by this approach to the rehabilitation of sex offenders, not to mention the intricacies of such intervention, have yet to be worked out.

Brain circuits mediate aggression

Aggressive behavior in various animals, including humans, is modulated by activity in several neurotransmitter systems. In particular, studies suggest that the release of **serotonin** in the brain normally inhibits aggression. For example, Higley et al. (1992) observed aggressive behavior in 28 monkeys from a large, free-ranging colony, and

serotonin A synaptic transmitter that is produced in the raphe nuclei and is active in structures throughout the central nervous system.

they ranked the animals from least to most aggressive. When researchers gauged serotonin activity by measuring serotonin metabolites in the cerebrospinal fluid, they found evidence that the most aggressive monkeys had the lowest level of serotonin being released in the brain. Similarly, mice that lack one subtype of serotonin receptor, and thus are less sensitive to serotonin, are hyperaggressive (Bouwknecht et al., 2001)—just what we would expect if serotonin normally inhibits aggression. This inhibitory role of serotonin in aggression is probably evolutionarily ancient, since it is evident even in invertebrates like crayfish (Panksepp et al., 2003).

The **medial amygdala** analyzes olfactory and pheromonal information, allowing male rats and mice to distinguish between male rivals to be attacked and females to be courted. That information is relayed to the **ventromedial hypothalamus (VMH)**, where activation through **optogenetics**, the use of lasers to activate neurons in genetically modified mice, will cause males that have been mating with females to suddenly attack them (Lee et al., 2014). The VMH neurons triggering this aggression also contain estrogen receptors (Falkner and Lin, 2014) and project to the *periaqueductal gray* (PAG), a midbrain region where electrical stimulation also elicits aggression.

So far we've discussed aggression in males, but females are also aggressive at times, particularly when they are caring for their young. This **maternal aggression** is typically studied by introducing an intruder mouse, usually a male, into the cage of a mother nursing a litter. In such conditions, she may immediately attack the intruder. Maternal aggression is also controlled by neural circuits in the VMH, as well as other hypothalamic regions, including the preoptic area (POA) and the premammillary nucleus (Motta et al., 2013).

The biopsychology of human violence is a topic of controversy

Some forms of human violence are characterized by sudden, intense physical assaults. A long-standing controversy surrounds the idea that some forms of intense human violence are caused by temporal lobe disorders (Mark and Ervin, 1970). Aggression is sometimes a prominent symptom in patients with temporal lobe seizures, and a significant percentage of people arrested for violent crimes have abnormal EEGs or other indicators of temporal lobe dysfunction (D. Williams, 1969; D. O. Lewis et al., 1979; D. O. Lewis, 1990).

Psychopaths are often intelligent individuals with superficial charm who have poor self-control, a grandiose sense of self-worth, and little or no feelings of remorse (Hare et al., 1990) and who sometimes commit very violent acts. Compared with controls, psychopaths do not react as negatively to words about violence (Gray et al., 2003). PET studies suggest that psychopaths have reduced activity in the prefrontal cortex (Raine et al., 1998), and it is hypothesized that this lower activity may impair their ability to control impulsive behavior. An MRI follow-up indicated that the prefrontal cortex is smaller in psychopaths than in controls (Raine et al., 2000)—another finding that is consistent with this hypothesis.

Undoubtedly, human violence and aggression stem from many sources. Biological studies of aggression have been vigorously criticized by some politicians and social scientists. These critics argue that, as a result of emphasizing biological factors such as genetics or brain mechanisms, the *preventable* origins of human violence and aggression, such as poverty and child neglect, might be overlooked. But as we have seen throughout the book, the brain is a malleable organ that is shaped by experience—and violent behavior *must* have its origins in the brain—so in principle it could be possible to reshape or at least moderate brain mechanisms of violence. The quality of life of some violent persons might be significantly improved if biological problems could be identified and addressed. For example, treatments that enhance serotonin activity in the brain might be a useful adjunct to psychotherapeutic intervention (Hollander, 1999).

Next we'll consider one consequence of suffering from aggression: Stress.

medial amygdala   A portion of the amygdala that receives olfactory and pheromonal information.

ventromedial hypothalamus (VMH)   A hypothalamic region involved in sexual behaviors, eating, and aggression.

optogenetics   The use of lasers to excite or inhibit neurons expressing light-sensitive membrane channels, typically in transgenic mice.

maternal aggression   Aggression of a mother defending her nest or offspring.

psychopath   An individual incapable of experiencing remorse.

Psychopathic Impulsivity Serial killer Theodore Bundy displayed many characteristics of a psychopath. He was superficially charming and, as shown here acting out in the courtroom when the judge was away, impulsive in nature. This scene also hints that, like other psychopaths, Bundy felt little or no remorse for his actions.

Stress Activates Many Bodily Responses

We all experience stress, but what is it? Attempts to define this term have a certain vagueness. Hans Selye (1907–1982), whose work launched the modern field of stress research, broadly defined *stress* as "the rate of all the wear and tear caused by life" (Selye, 1956). Nowadays, researchers try to sharpen their focus by treating **stress** as a multidimensional concept that encompasses stressful stimuli, the stress-processing system (including cognitive assessment of the stimuli), and responses to stress.

The stress response progresses in stages

Selye called the initial response to stress the *alarm reaction*. As one part of the alarm reaction, the hypothalamus activates the sympathetic nervous system to ready the body for action; this is the "fight or flight" system we mentioned at the start of the chapter. The "fight or flight" response includes sympathetic stimulation of the core of the adrenal gland, which is called the **adrenal medulla**, to release the hormones **epinephrine** (also known as *adrenaline*) and **norepinephrine** (or *noradrenaline*). These hormones act on many parts of the body to boost heart rate, breathing, and other physiological processes that prepare the body for action. As another part of the alarm reaction, the hypothalamus stimulates the anterior pituitary to release a hormone that drives the outer layer of the adrenal gland, the **adrenal cortex**, to release **adrenal steroid hormones** such as **cortisol** (**FIGURE 11.17**). These hormones act more slowly than epinephrine, but they also ready the body for action, including releasing body stores of energy.

stress   Any circumstance that upsets homeostatic balance.

adrenal medulla   The inner core of the adrenal gland.

epinephrine   Also called *adrenaline*. A compound that acts both as a hormone (secreted by the adrenal medulla under the control of the sympathetic nervous system) and as a synaptic transmitter.

norepinephrine   Also called *noradrenaline*. A neurotransmitter produced and released by sympathetic postganglionic neurons to accelerate organ activity.

adrenal cortex   The steroid-secreting outer rind of the adrenal gland.

adrenal steroid hormone   A steroid hormone that is secreted by the adrenal cortex.

cortisol   A glucocorticoid stress hormone of the adrenal cortex.

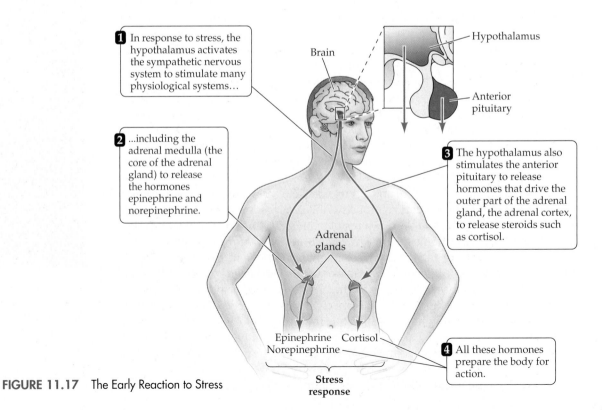

FIGURE 11.17   The Early Reaction to Stress

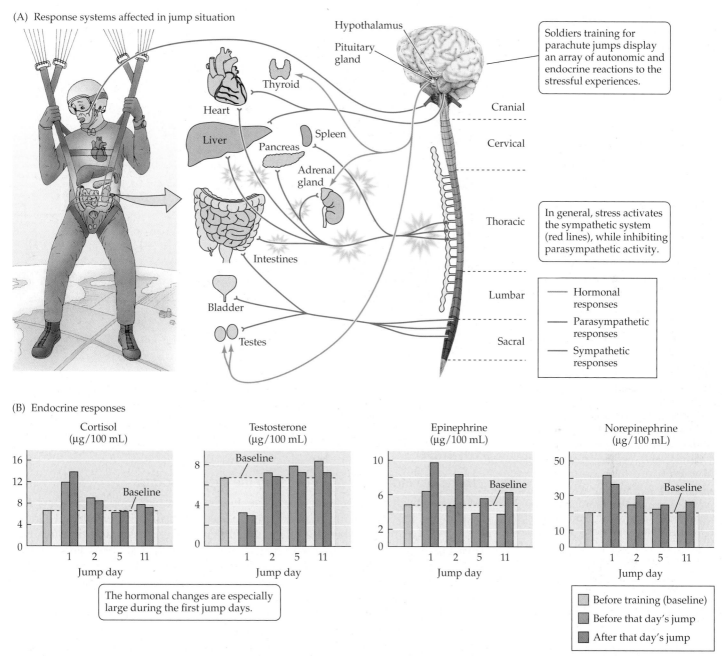

(A) Response systems affected in jump situation

Hypothalamus

Pituitary gland

Thyroid

Heart

Liver

Spleen

Pancreas

Adrenal gland

Intestines

Bladder

Testes

Cranial

Cervical

Thoracic

Lumbar

Sacral

Soldiers training for parachute jumps display an array of autonomic and endocrine reactions to the stressful experiences.

In general, stress activates the sympathetic system (red lines), while inhibiting parasympathetic activity.

— Hormonal responses
— Parasympathetic responses
— Sympathetic responses

(B) Endocrine responses

Cortisol (µg/100 mL)

Testosterone (µg/100 mL)

Epinephrine (µg/100 mL)

Norepinephrine (µg/100 mL)

Baseline

Jump day

The hormonal changes are especially large during the first jump days.

▢ Before training (baseline)
▨ Before that day's jump
▪ After that day's jump

FIGURE 11.18 Autonomic Activation during a Stress Situation (After Ursin et al., 1978.)

These stages of hormonal response to stress were studied in a group of young recruits in the Norwegian military both before and during scary parachute training (Ursin et al., 1978). On each jump day, the anterior pituitary released enhanced levels of hormones, and both the sympathetic and parasympathetic systems were activated (**FIGURE 11.18A**). Initially, cortisol levels were elevated in the blood before each jump, but with more and more successful jumps over successive days, the pituitary-adrenal response soon declined. Epinephrine and norepinephrine were also elevated before the first jumps, but eventually they returned to normal before jumps. Testosterone showed the reverse pattern, falling far below control levels on the first day of training, but returning to normal with subsequent jumps (**FIGURE 11.18B**). Once the soldiers mastered the jumps, they no longer showed increased hormonal responses, having adapted to the activity

Less-dramatic real-life situations also evoke clear endocrine responses (Frankenhaeuser, 1978). For example, riding in a commuter train provokes the release of epinephrine; the longer the ride and the more crowded the train, the greater the

FIGURE 11.19 Hormonal Changes in Humans in Response to Social Stresses (After Frankenhaeuser, 1978.)

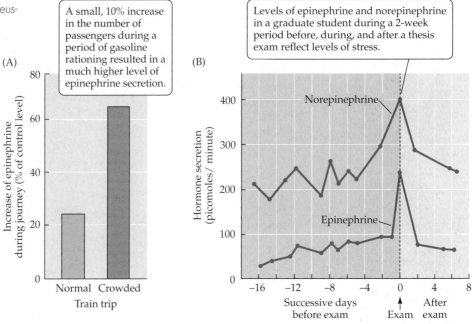

(A)

A small, 10% increase in the number of passengers during a period of gasoline rationing resulted in a much higher level of epinephrine secretion.

(B)

Levels of epinephrine and norepinephrine in a graduate student during a 2-week period before, during, and after a thesis exam reflect levels of stress.

stress immunization The concept that mild stress early in life makes an individual better able to handle stress later in life. The benefits seem to be due to effective comforting after stressful events, not the stressful events themselves.

epigenetic regulation Changes in gene expression that are due to environmental effects rather than to changes in the nucleotide sequence of the gene.

hormonal response (**FIGURE 11.19A**). Factory work likewise leads to the release of epinephrine; the shorter the work cycle—that is, the more frequently the person has to repeat the same operations—the higher the levels of epinephrine. The stress of a PhD oral exam leads to a dramatic increase in both epinephrine and norepinephrine (**FIGURE 11.19B**).

There are individual differences in the stress response

Why do individuals differ in their responses to stress? One hypothesis focuses on early experience. Rat pups clearly find it stressful to have a human pick them up and handle them. Yet rats that had been briefly handled as pups were less susceptible to adult stress than were rats that had been left alone as pups (Levine et al., 1967). For example, the previously handled rats secreted lower adrenal steroid amounts in response to a wide variety of adult stressors. This effect was termed **stress immunization** because a little stress early in life seemed to make the animals more resilient to later stress.

Follow-up research suggests that there is more to the story. The pups did not benefit because they were stressed; they benefited because their mothers *comforted* them *after* the stress. When pups are returned to their mother after a separation, she spends considerable time licking and grooming them. And she will lick the pups much longer if they were handled by humans during the separation. Michael Meaney and colleagues suggest that this gentle tactile stimulation from Mom is crucial for the stress immunization effect. They found that, even among undisturbed litters, the offspring of mother rats that exhibited more licking and grooming behavior were more resilient in their responses to adult stress than other rats were (D. Liu et al., 1997). So the "immunizing" benefit of early stressful experience happens only if the pups are promptly comforted after each stressful event.

If the pups are deprived of their mother for long periods, receiving very little of her attention, then as adults they exhibit a greater stress response, have difficulty learning mazes, and show reduced neurogenesis in the hippocampus (Mirescu et al., 2004). Maternal deprivation exerts this negative effect on adult stress responses by causing long-lasting changes in the expression of adrenal steroid receptors in the brain. This change is termed **epigenetic regulation** because it represents a change in the *expression* of the gene, rather than a change in the encoding region of the gene (see Figure 13.32).

Dramatic evidence for the same phenomenon has been seen in humans. Examination of the brains of suicide victims revealed the same epigenetic change in expression of the adrenal steroid receptor, but only in those victims who had a history of being abused or neglected as children (McGowan et al., 2009). The implication is that the early abuse epigenetically modified expression of the gene, making the person less able to handle stress and thus more likely to become depressed and commit suicide. Suicide victims who had no history of early neglect did not show the epigenetic change, so their depression may have been a response to other influences.

Prenatal stress also affects adult health. For example, people who were being carried by their mothers during the so-called Dutch famine in World War II had a greater risk of schizophrenia, depression, and type 2 diabetes as adults (Roseboom et al., 2011).

Stress and emotions influence the immune system

The field of **psychosomatic medicine** emphasizes that distinctive behaviors, psychological characteristics, and personality factors may affect either susceptibility or resistance to diverse illnesses. The related field called **health psychology** (or *behavioral medicine*) has developed to focus on identifying ways in which specific emotions and social contexts affect health outcomes and disease processes (Baum and Posluszny, 1999; Schwartzer and Gutiérrez-Doña, 2000).

The field of **psychoneuroimmunology** studies how the immune system—with its collection of cells that recognize and attack intruders—interacts with other organs, especially those of the hormonal systems and nervous system (Ader, 2001). Studies of both human and nonhuman subjects clearly show psychological and neurological influences on the immune system. For example, people with happy social lives are less likely to develop a cold when exposed to the virus (S. Cohen et al., 2006). Likewise, people who tend to feel positive emotions will also produce more antibodies in response to a flu vaccination (Rosenkranz et al., 2003), which should help them fight off sickness. These interactions go both ways: the brain influences responses of the immune system, and immune cells and their products affect brain activities, as **FIGURE 11.20** shows. You can learn details of how the immune system, endocrine system, and nervous system communicate with one another in **A STEP FURTHER 11.4**, on the website.

As an example, stressful exam periods usually suppress the immune system (Glaser et al., 1986). Importantly, the student's *perception* of the stress of the academic program is a predictor of immune system suppression: those who perceive a program as stressful show the most suppression. One experiment considered the effects of university examinations on wound healing in dental students (Marucha et al., 1998). Two small wounds were placed on the roof of the mouth of 11 dental students (sounds like revenge, doesn't it?). The first wound was timed during summer vacation; the second was inflicted 3 days before the first major examination of the term. Two independent daily measures showed that no student healed as rapidly during the exam period, when healing took 40% longer. One measure of immunological response declined 68% during the exam period. The experimenters concluded that even something as transient, predictable, and relatively benign (do students agree with this description?) as final exams can have significant consequences for wound healing.

Why does stress suppress the immune system?

Under stressful conditions, as noted earlier, the brain causes adrenal steroid hormones such as cortisol to be released from the adrenal cortex. These adrenal steroids directly suppress the immune system. But doesn't it seem like a bad idea to suppress immunity just when you are more likely to sustain an injury, and maybe an infection? Modern evolutionary theory offers some possible explanations for this seemingly maladaptive situation (for a very readable account, see Sapolsky, 2004).

psychosomatic medicine A field of study that emphasizes the role of psychological factors in disease.

health psychology Also called *behavioral medicine*. A field of study that focuses on psychological influences on health-related processes.

psychoneuroimmunology The study of the immune system and its interaction with the nervous system and behavior.

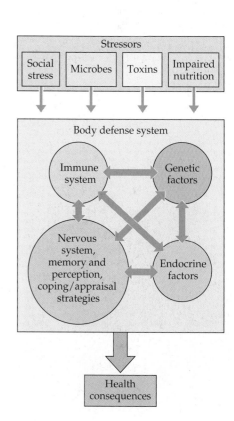

FIGURE 11.20 Factors That Interact during the Development and Progression of Disease

TABLE 11.1 The Stress Response and Consequences of Prolonged Stress

Principal components of the stress response	Common pathological consequences of prolonged stress
Mobilization of energy at the cost of energy storage	Fatigue, muscle wasting, diabetes
Increased cardiovascular and cardiopulmonary tone	Hypertension (high blood pressure)
Suppression of digestion	Ulcers
Suppression of growth	Psychogenic dwarfism, bone decalcification
Suppression of reproduction	Suppression of ovulation, loss of libido
Suppression of immunity and of inflammatory response	Impaired disease resistance
Analgesia (painkilling)	Apathy
Neural responses, including altered cognition and sensory thresholds	Accelerated neural degeneration during aging

Source: Sapolsky, 1992.

To the extent that stress might be a sudden emergency, the *temporary* suppression of immune responses makes some sense because the stress response demands a rapid mobilization of energy. Slow and long-lasting immune responses consume energy that otherwise could be used for dealing with the emergency at hand. A zebra wounded by a lion must first escape and hide, and only then does infection of the wound pose a threat. So the stress of the encounter first suppresses the immune system, conserving resources until a safe haven is found. Later the animal can afford to mobilize the immune system to heal the wound. The adrenal steroids also suppress the swelling (inflammation) of injuries, especially of joints, to help the animal remain mobile long enough to find refuge (Cox et al., 2014). It is precisely this action that makes adrenal steroids like prednisone such useful medicines.

In the wild, animals are under stress for only a short while; any animal stressed for a *prolonged* period dies. So natural selection has favored stress reactions as a drastic effort to deal with a short-term problem. What makes humans and baboons "special" is that, with our highly social lives and keen analytical minds, we are capable of experiencing stress for prolonged periods—months or even years. The bodily reactions to stress, which evolved to deal with short-term problems, become a handicap when extended too long (Sapolsky, 2004). **TABLE 11.1** lists a variety of stress responses that are beneficial in the short term but detrimental in the long term.

What we have described so far is a really depressing picture. If you are stressed for long periods of time, your health suffers, which brings another source of stress to your life. But don't give up hope. Remember the Forest Troop of baboons we met at the start of the chapter? Perhaps it was fortunate that only the meanest, most aggressive males of the troop got access to the tainted meat. It was mostly those males that died horrible, agonizing deaths, while about half the males and most of the females and youngsters survived. This loss of the most aggressive males had a remarkable effect on the social life of the Forest Troop. The surviving males, which had always been a little easier to get along with than most males, never ramped up their aggression to the levels that had been seen before. They might occasionally tussle with one another, but hostilities usually ended before anyone got hurt, and these baboons rarely took out their aggression on the innocent bystanders—the females and the young. Now, everyone spent much more time grooming each other and otherwise spending "quality baboon time" in peaceful pursuits. Twenty years later, members of the troop still showed fewer physiological signs of stress than before.

Perhaps the most important lesson these animals have for us is that most of the stress we experience is avoidable. If we all treat each other better, everyone can benefit. These animals also make it clear that aggression, fighting, and stress are not inevitable consequences of biology. If baboons can, in a short time, convert from a fractious, stressed-out society to a more peaceful, nurturing society, and can maintain that new social order for generations, shouldn't an even more intelligent species like humans be able to follow suit?

Baboons Can Peacefully Coexist

SIGNS & SYMPTOMS

Long-Term Consequences of Childhood Bullying

There is growing recognition that children who are bullied, subjected to verbal or physical assault by other children, are at greater risk for mental and physical disorders when they grow up. A British study of children born in 1958 first gathered reports of whether they were being bullied at ages 7 and 11, then followed their health until they were age 50. After adjusting for IQ and other factors, the researchers found that those bullied as children were at increased risk for anxiety disorders and depression, as well as suicide (**FIGURE 11.21**) (Takizawa et al., 2014). Another study of both British and American children confirmed that those who

were bullied were more likely to suffer these disorders. The authors were surprised to see that the effects of bullying were as strong as those of physical or sexual abuse (Lereya et al., 2015). Faced with such reports, schools are being encouraged to develop antibullying programs that teach children to recognize and report bullying and that train teachers to intervene rather than downplay bullying as a rite of passage or "normal" behavior. Several U.S. federal departments have collaborated to provide an online resource to increase understanding and combat bullying, including cyberbullying over Facebook and other media, at www.stopbullying.gov.

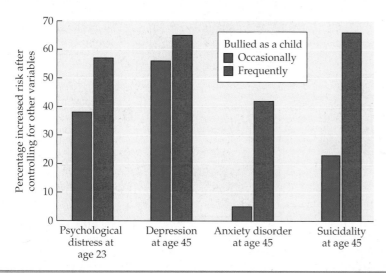

FIGURE 11.21 Being Bullied Is Bad For Your Health (After Takizawa et al., 2014)

how's it going ?

1. What are the hormonal responses to stressful events, and how do those change as individuals adapt to those events?
2. What is stress immunization, and how is it mediated by epigenetic events?
3. Why do we suppress the immune system in times of stress, and how does that suppression impair health in brainy, social animals like us?
4. In what ways is a childhood history of being bullied evident in the psychological health of adults?

Recommended Reading

Ekman, P. (2007). *Emotions Revealed: Recognizing Faces and Feelings to Improve Communication and Emotional Life* (2nd ed.). New York, NY: Owl Books.

Gross, J. J. (2015). *Handbook of Emotion Regulation* (2nd ed.). New York, NY: Guilford Press.

Keltner, D., Oatley, K., and Jenkins, J. M. (2013). *Understanding Emotions* (3rd ed.). New York, NY: Wiley.

Nelson, R. J. (2006). *The Biology of Aggression.* New York, NY: Oxford University Press.

Nettle, D. (2006). *Happiness: The Science behind Your Smile.* New York, NY: Oxford University Press.

Panksepp, J. (2004). *Affective Neuroscience: The Foundations of Human and Animal Emotions.* New York, NY: Oxford University Press.

Sapolsky, R. (2004). *Why Zebras Don't Get Ulcers* (3rd ed.). New York, NY: Holt.

VISUAL SUMMARY

2e.mindsmachine.com/vs11

11

You should be able to relate each summary to the adjacent illustration, including structures and processes. Go to the online version of this summary (scan the QR code above) for links to figures, animations, and activities that will help you consolidate the material.

1 Emotions are a constellation of feelings, behaviors, and physiological reactions. The James-Lange theory considered emotions to be the perceptions of stimulus-induced bodily changes. The Cannon-Bard theory emphasized simultaneous emotional experience and bodily response. In Schachter's cognitive theory, we attribute visceral arousal to specific emotions by analyzing the physical and social context. Review **Figures 11.1** and **11.2**, **Animation 11.2**

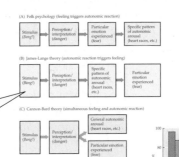

2 Distinct facial expressions represent anger, sadness, happiness, fear, disgust, surprise, contempt, and embarrassment, which are interpreted similarly across many cultures. **Polygraphs** actually measure activation of the **sympathetic nervous system**, and therefore reflect stress, not lying. Review **Figures 11.3–11.6, Box 11.1**

3 Facial expressions are controlled by distinct sets of facial muscles controlled by the facial and trigeminal nerves. Emotions evolved as adaptations, and our expression of emotions helps in social relations. The left side of the face, controlled by the right hemisphere of the brain, is more emotionally expressive than the right side of the face. Review **Figures 11.7–11.9**

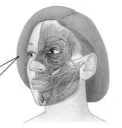

4 Lesions revealed an interconnected brain circuit, the **limbic system** (which includes the amygdala) that mediates and controls emotions. Electrical **self-stimulation** of some brain regions is rewarding. The right cerebral hemisphere is better than the left at interpreting emotional expressions and stimuli. Review **Figures 11.10** and **11.11**

5 Fear is mediated by circuitry involving the **amygdala**, which receives information both through a rapid direct route and via cortical sensory regions, allowing for both immediate responses and cognitive processing. Review **Figures 11.12** and **11.13, Activity 11.1**

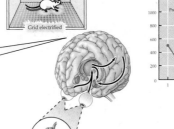

6 **Aggression** is increased by androgens such as **testosterone**. Stimulation of some limbic system regions elicits a species-typical pattern of aggression. The **ventromedial hypothalamus (VMH)** appears to play a central role in aggression in both sexes. Review **Figure 11.16**

7 Stress elevates levels of hormones from the **adrenal cortex (cortisol)** and the **adrenal medulla (epinephrine** and **norepinephrine)**, while suppressing other hormones (testosterone). While these responses to stress are adaptive in the short run, in socially complex species that can experience stress for long periods, these hormonal responses decrease immune system competence, damaging our health. Review **Figures 11.17–11.19, Table 11.1, Activity 11.2**

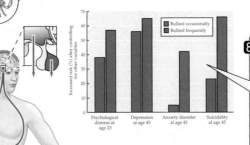

8 The nervous, endocrine, and immune systems interact reciprocally to monitor and maintain health. Childhood stress, including being subjected to bullying, seems to increase lifelong risk for schizophrenia, depression, and suicide. Review **Figures 11.20** and **11.21**

Go to **2e.mindsmachine.com** for study questions, quizzes, flashcards, and other resources.

chapter

12

Psychopathology
The Biology of
Behavioral Disorders

"My Lobotomy"

Howard Dully's biological mother died unexpectedly when he was 4, and 3 years later his father married a woman named Lucille. For whatever reasons, Howard and Lucille did not get along. Howard was rebellious in the way that virtually all kids are—sassing back, breaking curfew, skipping out on church. But Howard was never violent with his stepmother (or anyone else). He sometimes got in trouble at school, for not paying attention in class or smoking in the bathroom, but not for fighting or damaging school property. Howard's grades were erratic—an A on a test one day, an F on a test the next—but he was not flunking out.

Still, Lucille, frustrated with a headstrong boy in her house, took Howard to six different psychiatrists to find out "what was wrong with him." All concluded that his behavior was normal. But doctor number seven, the famous Walter Freeman, diagnosed the boy as schizophrenic. In 1960, Freeman gave 12-year-old Howard a lobotomy. First Freeman sedated the boy by giving him electroshocks—jolts of electricity across the skull that induce a seizure and render the patient

unconscious. Then he lifted the boy's upper eyelids and used a hammer on an ice pick–like device to punch holes in Howard's skull above each eye. He then inserted a device to cut off some of Howard's prefrontal cortex from the rest of his brain. Freeman was an old hand at the procedure, having lobotomized thousands of people, so the procedure itself took only 10 minutes. The total hospital charge was $200.

Family members report that Howard acted like a zombie for several days, and then he was so lethargic and disinterested in the events around him that, according to Freeman's notes, they called Howard "lazy, stupid, dummy, and so on." One aunt said he acted like he was permanently tranquilized. And yet Lucille *still* wanted Howard out of her house. Soon Howard was institutionalized, and he would spend decades in various mental wards. Not until he was 50 was Howard able to find out what had happened to him as a child, a journey he recounts movingly in his memoir, *My Lobotomy* (Dully and Fleming, 2007).

To see the video
Lobotomy,
go to

2e.mindsmachine.com/12.1

D ebilitating mental afflictions have plagued humankind throughout history, plunging their victims into an abyss of disordered thought and emotional chaos. We have made great progress in understanding the causes of mental health issues like schizophrenia, depression, and anxiety disorders and have developed a wide variety of treatments that are at least partly effective, but the emotional and economic costs of these illnesses remain great. And they are widespread: psychopathology affects hundreds of millions of people throughout the world.

In this chapter we survey the major categories of psychiatric disorders and explore their biological underpinnings. Although no single remedy has been found that cures all who suffer from any of these disorders, modern discoveries have helped millions of people maintain a healthy life.

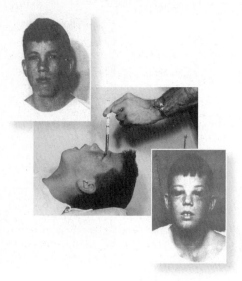

Changed for Life Twelve-year-old Howard Dully before, during, and after his trans-orbital lobotomy. The swelling around his eyes eventually went away, but Howard would spend the next four decades in various mental institutions.

To view the
Brain Explorer,
go to

2e.mindsmachine.com/12.2

The Toll of Psychiatric Disorders Is Huge

Because there are few objective and unambiguous physical markers of psychiatric disorders—they are diagnosed more on the basis of behaviors and reported feelings than on lab tests and physical exams—the classification of psychiatric disorders is an evolving science and subject to periodic revision. Currently in its fifth edition, the American Psychiatric Association's *Diagnostic and Statistical Manual of Mental Disorders,* generally called the *DSM-5,* provides a standardized system for diagnosing and classifying the major psychiatric illnesses according to current knowledge (APA, 2013). The psychiatric disorders that are described in the *DSM-5* are startlingly prevalent in modern society. Worldwide, between 15% and 50% of the population report psychiatric symptoms at some point in life, with North Americans positioned at the top end of this range (Kessler et al., 2007). About 19% of the adult population of the United States experiences psychiatric symptoms in the course of a year (Substance Abuse and Mental Health Services Administration, 2013), and of this number more than 4% (equating to almost 10 million people) are so ill that they are unable to carry out major life activities, like working or living independently. As shown in **FIGURE 12.1**, these rates are higher for females than for males, primarily because females are more likely to be depressed. (On the other hand, drug dependency and alcoholism, which are not reflected in Figure 12.1, are much more frequent in males.) Note also the high rates that are evident in 18- to 25-year-olds because certain psychiatric disorders—for example, schizophrenia—tend to appear in adolescence and young adulthood. Clearly, mental disorders exact an enormous toll on our lives.

The seeds for a biological perspective in psychiatry were sown at the start of the twentieth century. At that time, almost a quarter of the patients in mental hospitals suffered from so-called paralytic dementia, featuring sudden onset of **delusions** (false beliefs strongly held in spite of contrary evidence), grandiosity (boastful self-importance), euphoria, poor judgment, impulsive behavior, disordered thought, and physiological signs like abnormal pupillary constriction (Argyll-Robertson, 1869).

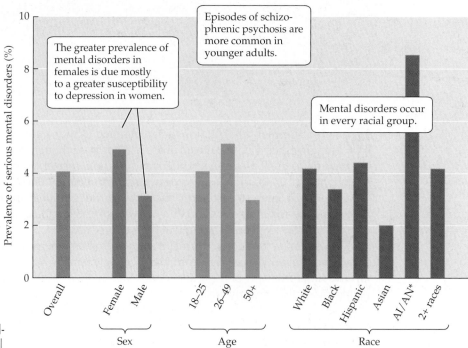

FIGURE 12.1 Prevalence of Serious Mental Illness among U.S. Adults in 2012 (After National Institute of Mental Health, 2012.)

The disorder was originally believed to be caused by "weak character," but analyses of the brains and behavior of the afflicted revealed that their illness had a physiological cause: syphilis. This finding opened the door to biological explanations for other mental illnesses.

Schizophrenia Is a Major Neurobiological Challenge in Psychiatry

Throughout the world and across the centuries, some people have been recognized as unusual because they hear voices that others don't, feel intensely frightened, sense persecution from unseen enemies, and generally act strangely (Bark, 2002; Heinrichs, 2003). For many, this disordered state—now known as **schizophrenia**—lasts a lifetime. For others, it appears and disappears unpredictably. Schizophrenia is also a public health problem because all too many of the people who suffer from it become homeless.

Schizophrenia is characterized by an unusual array of symptoms

The term *schizophrenia* (from the Greek *schizein*, "to split," and *phren*, "mind") was introduced early in the twentieth century to convey the idea that various functions of the mind—like memory, perception, and thinking—were split from each other (Bleuler, 1950, originally published in 1911). This poetic but vague description of schizophrenia was subsequently replaced with a more objective definition (K. Schneider, 1959) focusing on "first-rank symptoms," which include (1) auditory hallucinations, (2) highly personalized delusions, and (3) changes in affect (emotion). By the 1990s it became clear that many schizophrenia symptoms could be viewed as belonging to two general groups: positive and negative (Andreasen, 1991). **Positive symptoms** are abnormal behavioral states that have been *gained*; examples include hallucinations, delusions, and excited motor behavior. **Negative symptoms** are abnormalities resulting from the *loss* of normal functions—for example, slow and impoverished thought and speech, emotional and social withdrawal, or blunted affect.

Researchers now recognize that schizophrenia is a complex syndrome in which individuals exhibit varying degrees of four distinct but correlated categories of symptoms (Van Os et al., 2010):

1. Psychosis, such as hallucinations, delusions, and disorganization of speech and thought
2. Emotional ("affective") symptoms such as depression and reduced emotional expression
3. Motivational impairment
4. Cognitive impairments such as changes in memory, attention, and social perception

The contemporary view of the symptoms of schizophrenia (**TABLE 12.1**) retains the distinction between positive symptoms (psychosis) and negative symptoms (emotional and motivational impairments), but recognizes an additional dimension: cognitive impairment. Differences in symptoms between individuals with schizophrenia may therefore reflect individual differences across the four symptom categories. And the fact that the various categories of symptoms respond differently to drug treatments suggests that multiple neural mechanisms are involved in the disorder.

Schizophrenia has a heritable component

For many years, genetic studies of schizophrenia were controversial because some early researchers failed to understand that genes need not act in an all-or-none fashion. For any genotype there is often a large range of alternative outcomes determined by both developmental and environmental factors, as we'll see.

delusion A false belief that is strongly held in spite of contrary evidence.

schizophrenia A severe psychopathological disorder characterized by negative symptoms such as emotional withdrawal and flat affect, and by positive symptoms such as hallucinations and delusions.

positive symptom In psychiatry, an abnormal behavioral state. Examples include hallucinations, delusions, and excited motor behavior.

negative symptom In psychiatry, an abnormality that reflects insufficient functioning. Examples include emotional and social withdrawal, and blunted affect.

TABLE 12.1 Symptoms of Schizophrenia

Symptom dimension	Symptom category
POSITIVE SYMPTOMS Refers to symptoms that are present but should not be	**PSYCHOSIS** Hallucinations Delusions Disorganized thought and speech Bizarre behaviors
NEGATIVE SYMPTOMS Refers to characteristics of the individual that are absent but should be present	**EMOTIONAL DYSREGULATION** Lack of emotional expression Reduced facial expression (flat affect) Inability to experience pleasure in everyday activities (anhedonia) **IMPAIRED MOTIVATION** Reduced conversation (alogia) Diminished ability to begin or sustain activities Social withdrawal
COGNITIVE SYMPTOMS Refers to problems with processing and acting on external information	**NEUROCOGNITIVE IMPAIRMENT** Memory problems Poor attention span Difficulty making plans Reduced decision-making capacity Poor social cognition Abnormal movement patterns

FAMILY STUDIES   If schizophrenia is inherited, relatives of people with schizophrenia should show a higher incidence of the disorder than is found in the general population. In addition, the risk of schizophrenia among relatives should increase with the closeness of the relationship, because closer relatives share a greater number of genes. Indeed, parents and siblings of people with schizophrenia have a higher risk of becoming schizophrenic than do individuals in the general population (**FIGURE 12.2**) (Gottesman, 1991). However, the mode of inheritance of schizophrenia is not simple; that is, it does not involve a single recessive or dominant gene (Tamminga and Schulz, 1991). Rather, multiple genes play a role in the emergence of schizophrenia.

ADOPTION STUDIES   It is easy to find fault with family studies. They confuse hereditary and environmental factors because members of a family share both. But what about children who are not raised with their biological parents? In fact, studies of adopted people confirm a strong genetic factor in schizophrenia. The biological parents of adoptees who suffer from schizophrenia are far more likely to have suffered from this disorder than are the adopting parents (Kety et al., 1975, 1994).

TWIN STUDIES   In twins, nature provides researchers with an excellent opportunity for a genetic experiment. In identical (or *monozygotic*) twins, which derive from a single fertilized egg and thus share the same set of genes, if one of the twins develops schizophrenia, the other twin has a roughly fifty-fifty chance of also developing the disorder. But in fraternal (or *dizygotic*) twin pairs, which come from two fertilized eggs and thus share about 50% of their

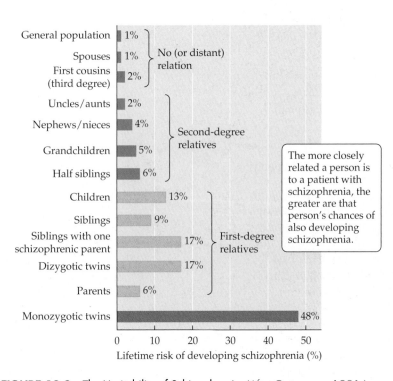

FIGURE 12.2   The Heritability of Schizophrenia (After Gottesman, 1991.)

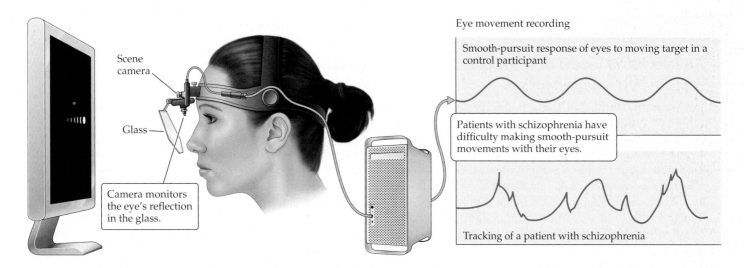

Eye movement recording

Smooth-pursuit response of eyes to moving target in a control participant

Patients with schizophrenia have difficulty making smooth-pursuit movements with their eyes.

Tracking of a patient with schizophrenia

Scene camera

Glass

Camera monitors the eye's reflection in the glass.

FIGURE 12.3 Eye Tracking in Patients with Schizophrenia versus Controls

genes, just like any pair of siblings, this **concordance** (sharing of a characteristic) drops to about 17% (see Figure 12.2) (Cardno and Gottesman, 2000). The higher concordance in the genetically identical twins is thus strong evidence of a genetic factor. Yet even with identical twins, the concordance rate for schizophrenia is only about 50% (see Figure 12.2), so genes alone cannot fully explain whether a person will develop schizophrenia. Presumably, other factors, especially environmental influences, account for the 50% of identical twin pairs that are discordant (only one twin develops the disorder). Often, the twin who goes on to develop schizophrenia has an abnormal developmental history, such as lower birth weight, more physiological distress in early life, and behavior that seems more submissive, tearful, and sensitive than that of the unaffected twin (Wahl, 1976; Torrey et al., 1994). Subtle neurological signs, such as impaired motor coordination and difficulty with smooth movements of the eyes to follow a moving target (**FIGURE 12.3**), are also common (Torrey et al., 1994; Avila et al., 2006). In short, the twin studies show that schizophrenia has both environmental and genetic origins.

concordance Sharing of a characteristic by both individuals of a pair of twins.

INDIVIDUAL GENES It has been difficult to identify any single gene that causes schizophrenia to develop or increases susceptibility (Mowry et al., 2004). In fact, genetic analyses suggest that genes influencing the development of schizophrenia are scattered across many different human chromosomes (Stefansson et al., 2009). Nonetheless, a few genes have been identified that appear to be abnormal in a small proportion of schizophrenia cases, including genes that are known to participate in synaptic plasticity (Kennedy et al., 2003; Mei and Xiong, 2008). In one large Scottish family, several members who had schizophrenia also carried a mutant, disabled version of a gene, which was therefore named *disrupted in schizophrenia 1* (*DISC1*). We'll discuss *DISC1* further a little later in the chapter.

An interesting *epigenetic* factor (see Chapter 13) in schizophrenia is paternal age: older men are more likely than younger men to have children with schizophrenia (Rosenfield et al., 2010). It is thought that, because they are the product of more cell divisions than the sperm of younger men, the sperm of older men have had more opportunity to accumulate mutations caused by errors in copying the chromosomes; these mutations may contribute to the development of schizophrenia in some cases.

Taken together, the studies make it clear that certain genes can indeed increase the risk of developing schizophrenia but that the environment also matters. As we'll see next, a big factor in whether a person will develop schizophrenia is stress.

An integrative model of schizophrenia emphasizes the interaction of multiple factors

We've established that there is genetic influence on schizophrenia but also that genes alone cannot account for the disorder. What environmental factors contribute to the probability of developing schizophrenia? Research suggests that a variety of stressful events significantly increase the risk. For example, schizophrenia usually appears during a time in life that many people find stressful—the transition from childhood to adulthood, when people deal with physical, emotional, and lifestyle changes (e.g., going away to college).

Another risk factor seen in multiple studies is the stress of city living. As **FIGURE 12.4** shows, people living in a medium-sized city are about 1½ times more likely to develop schizophrenia than are people living in the country. What's more, the earlier in life a person begins living in the city, the greater the risk. People living in a *big* city are even more likely to develop the disorder (Pedersen and Mortensen, 2001). Conversely, children who move from the city to the country have a *reduced* risk of developing schizophrenia (Van Os et al., 2010). We don't know what it is about living in a city that makes schizophrenia more likely. Pollutants, greater exposure to minor diseases, crowded conditions, tense social interactions—all of these could be considered stressful.

A different schizophrenia risk is present before a baby is born. Prenatal stress, such as infection during

FIGURE 12.4 City Living Increases the Risk of Schizophrenia

■ **Hypothesis**
Stress increases the risk of schizophrenia.

■ **Test**
Find a data set that lets you compare people from a relatively homogeneous culture and genetic background, such as Denmark. Assuming that city living is more stressful than living in the country, determine whether populations living in a big city, in other (smaller) cities, or in the country differ in the proportion of people diagnosed with schizophrenia.

■ **Result**

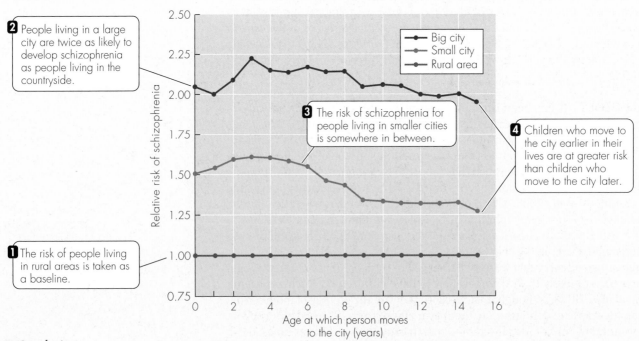

2 People living in a large city are twice as likely to develop schizophrenia as people living in the countryside.

3 The risk of schizophrenia for people living in smaller cities is somewhere in between.

4 Children who move to the city earlier in their lives are at greater risk than children who move to the city later.

1 The risk of people living in rural areas is taken as a baseline.

■ **Conclusion**
Some aspect of city living, perhaps generalized stress, increases the probability that a person will develop schizophrenia.

pregnancy, increases the likelihood that the baby will develop schizophrenia later in life. For example, the baby of a woman who contracted influenza in the first trimester of pregnancy is 7 times more likely to develop schizophrenia (P. H. Patterson, 2007). Several other maternal infections also increase the chances that the child will develop schizophrenia one day (A. S. Brown, 2011). This correlation may be why people born in late winter and early spring are more likely to develop schizophrenia (Messias et al., 2004): their mothers may be more likely to have gotten sick during the winter before, when the fetus was vulnerable. Likewise, if the mother and baby have incompatible blood types, or the mother becomes diabetic during pregnancy, or if there is a low birth weight for some reason, the baby is more likely to develop schizophrenia (King etal., 2010). Birth complications that deprive the baby of oxygen also increase the probability of schizophrenia (Clarke et al., 2011).

Because we know that some people are genetically more susceptible to schizophrenia than others, these findings suggest that relatively minor stress during development can make the difference in whether schizophrenia develops. It is fascinating, and frightening, to think that events in the womb or early childhood can affect the outcome 16 or 20 years later, when the schizophrenia appears.

Thus the evidence indicates that schizophrenia results from a complex interaction of genetic factors and stress. Each life stage has its own specific features that increase vulnerability to schizophrenia: infections before birth, complications at delivery, urban living in childhood and adulthood (Powell, 2010). From this perspective, the emergence of schizophrenia and related disorders depends on whether a genetically susceptible person is subjected to environmental stressors (**FIGURE 12.5**). The hope is that through the development of sensitive diagnostic approaches that combine brain imaging with genetic and behavioral measures, we will be able to accurately identify and understand the at-risk child early in life, when interventions to reduce stress might prevent schizophrenia later in life.

Once the interaction of genetic susceptibility and stress results in schizophrenia, the condition affects not only the person's behavior but also the physical state of the brain, as we'll see next.

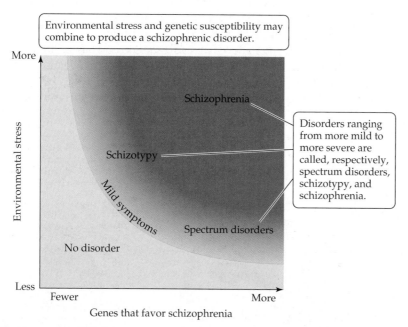

FIGURE 12.5   A Model of the Interaction between Stress and Genetic Influences in Schizophrenia (After Mirsky and Duncan, 1986.)

FIGURE 12.6 Identical Genes, Different Fates (After Torrey et al., 1994, MRIs courtesy of E. Fuller Torrey.)

Although the two members of each set of identical twins shown here have the same genes…

35-year-old female identical twins

28-year-old male identical twins

Well

Affected

Well

Affected

…only the twins with larger ventricles have schizophrenia.

The brains of some patients with schizophrenia show structural and functional changes

Because the symptoms of schizophrenia can be so marked and persistent, investigators hypothesized early on that the brains of people with this illness would show distinctive and measurable structural abnormalities. Later, CT and MRI scans confirmed this idea, revealing significant, consistent anatomical differences in the brains of many patients with schizophrenia (Trimble, 1991). Interestingly, these scans also confirm the idea that genes alone cannot account for whether a person will develop schizophrenia.

VENTRICULAR ABNORMALITIES   Most patients with schizophrenia have enlarged cerebral ventricles, especially the lateral ventricles (T. M. Hyde and Weinberger, 1990; Vita et al., 2000). An important distinction is that twins with schizophrenia have decidedly enlarged lateral ventricles compared with their well counterparts, whose ventricles are of normal size (**FIGURE 12.6**) (Torrey et al., 1994). Among people with schizophrenia, those with larger ventricles benefit less from antipsychotic drugs (Garver et al., 2000).

Recall that a disabled version of the gene *DISC1* is associated with schizophrenia in one large family. The DISC1 protein normally interacts with a bewildering array of other proteins during brain development (Brandon and Sawa, 2011). When researchers inserted the schizophrenia-associated mutant version of *DISC1* into mice, they found that the mice developed enlarged lateral ventricles (**FIGURE 12.7**) that were reminiscent of the enlarged ventricles in people with schizophrenia (Pletnikov et al., 2008).

Transgenic mice expressing the *DISC1* mutation associated with schizophrenia in humans develop enlarged lateral ventricles (green) reminiscent of those in people with schizophrenia.

Control

Mutant

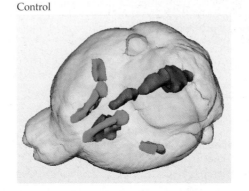
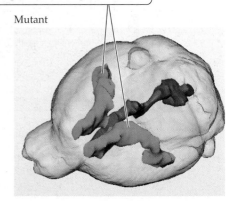

FIGURE 12.7 Enlarged Ventricles in a Mouse Model (From Pletnikov et al., 2008.)

What is the significance of enlarged ventricles? Because overall brain size does not seem to be affected in people with schizophrenia or in mice expressing mutant *DISC1*, the enlarged ventricles must come at the expense of brain tissue. Therefore, interest has centered on possible changes in brain structures that run alongside the lateral ventricles, as discussed in **A STEP FURTHER 12.1**, on the website.

CORTICAL ABNORMALITIES   People with schizophrenia differ from those without in the structure and functional activity of the corpus callosum (Rotarska-Jagiela et al., 2008). Furthermore, brain-imaging studies have revealed accelerated loss of gray matter at adolescence in schizophrenia patients compared with controls (P. M. Thompson et al., 2001; Karlsgodt et al., 2010) (**FIGURE 12.8**). Accelerated cortical thinning (and reduction in subcortical volume) emerges specifically as high-risk individuals progress to schizophrenia, and it occurs even in the absence of drug treatments, so it seems that the brain changes are central to the disease process (Cannon et al., 2015).

Although early researchers reported that the frontal lobes are especially different in people with schizophrenia, other studies have not found abnormalities there (Wible et al., 1995; Highley et al., 2001). So if the frontal lobes of people with schizophrenia are not very different in their *structure*, what about the *activity* of the frontal cortex?

DIFFERENCES IN BRAIN ACTIVITY   People with schizophrenia tend to be impaired on neuropsychological tests that are sensitive to frontal cortical lesions. These findings raised the possibility that frontal cortex activity is abnormal in schizophrenia. Early observations using PET found that, compared with nonschizophrenic controls, people with schizophrenia had reduced metabolic activity in the frontal lobes relative to other regions of the brain (Buchsbaum et al., 1984). This observation led to the **hypofrontality hypothesis** that the frontal lobes are underactive in people with schizophrenia. A review of many studies over the past 25 years seems to support this idea (Minzenberg et al., 2009).

In discordant identical twin pairs, where one twin is healthy and one is schizophrenic, reduced activity of the frontal cortex is evident only in the affected twin

hypofrontality hypothesis   The idea that schizophrenia may reflect underactivation of the frontal lobes.

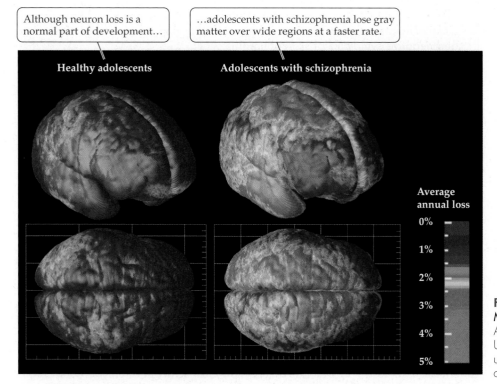

FIGURE 12.8 Accelerated Loss of Gray Matter in Adolescents with Schizophrenia Additional brain imagery is available at the USC Laboratory of NeuroImaging: www.loni.usc.edu. (From P. M. Thompson et al., 2001, courtesy of Paul Thompson.)

(A) At rest

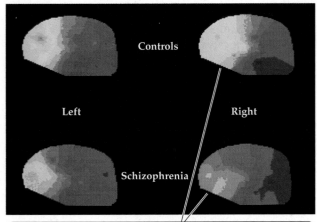

(B) During card-sorting task

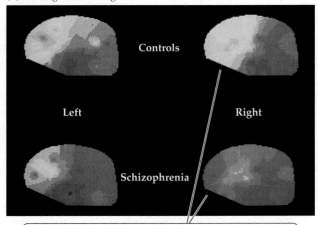

In these PET scans, cooler (blue-green) colors reflect less activation. In the twins with schizophrenia, the frontal cortex (left side of each brain profile) is less activated than in their unafffected twins, both at rest…

…and during the Wisconsin Card Sorting Task, which is very difficult for people with damage to the frontal lobes.

FIGURE 12.9 Hypofrontality in Schizophrenia (Courtesy of Karen Berman.)

lobotomy The surgical separation of a portion of the frontal lobes from the rest of the brain, once used as a treatment for schizophrenia and many other ailments.

(Morihisa and McAnulty, 1985; Andreasen et al., 1986). Behavioral evidence indicates that hypofrontality is especially problematic during difficult cognitive tasks that depend on the frontal lobes for accurate performance, such as the Wisconsin Card Sorting Task (**FIGURE 12.9**). Unlike control participants, people with schizophrenia show little increase in their prefrontal activation during the task (D. R. Weinberger et al., 1994). In many cases, drugs that alleviate symptoms of schizophrenia, discussed next, also increase the activation of frontal cortex (Honey et al., 1999).

how's it going ❓

1. What is the incidence of psychiatric illness? At what stage of life does schizophrenia usually appear?
2. What are the four major categories of symptoms in schizophrenia? Provide some examples of each symptom category. What do we mean by positive and negative symptoms of schizophrenia?
3. Review the evidence that heredity plays a role in schizophrenia.
4. What is the evidence that stress can affect whether a person will develop schizophrenia?
5. What are some of the ways in which the brains of people with schizophrenia differ from the brains of people without schizophrenia?

Antipsychotic medications revolutionized the treatment of schizophrenia

In the 1930s, there were no effective treatments for schizophrenia. Because patients were often unable to take care of themselves, they were placed in caregiving institutions. In many cases, the health and welfare of patients in these (poorly funded) institutions were badly neglected, leading to recurrent scandals. So perhaps it was in desperation that psychiatrists turned to **lobotomy**, the surgical separation of a portion of the frontal lobes from the rest of the brain, as a treatment for schizophrenia. Certainly there was little scientific evidence to think the surgery would be effective. But early practitioners reported nearly miraculous recoveries that, in retrospect, must be regarded as wishful thinking on the part of the physicians. The surgery may well have made the patients easier to handle, but they were rarely able to leave the mental institution. Used for almost any mental disorder, not just schizophrenia, lobotomies were performed on some 40,000 people in the United States alone (Kopell et al., 2005).

Another Victim of Lobotomy Although it's unclear what psychological problems she had, if any, Rosemary Kennedy was given a lobotomy at age 23. The younger sister of President John F. Kennedy, shown here at age 20 with her father Joseph P. Kennedy, Rosemary was permanently incapacitated and spent the rest of her long life in an institution.

BOX
12.1
Long-Term Effects of Antipsychotic Drugs

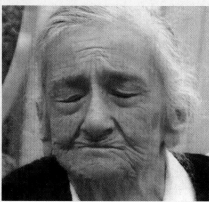

Tardive dyskinesia is a late onset of involuntary movements, often of the lower face. This woman with tardive dyskinesia suffers from involuntary facial movements such as her tongue popping out (*left*) and intense grimacing (*right*).

Few people would deny that antipsychotic drugs are "miracle drugs." With drug treatment, many people who might otherwise have been in mental hospitals their whole lives can take care of themselves in nonhospital settings.

Unfortunately, traditional antipsychotic drugs can have other, undesirable effects as well. Soon after beginning to take these drugs, some people develop maladaptive motor symptoms called **dyskinesia** (from the Greek *dys*, "bad," and *kinesis*, "motion"). Although many of these symptoms are transient and disappear when the dosage of drug is reduced, some drug-induced motor changes emerge only after prolonged drug treatment—after months, sometimes years—and are effectively permanent. This condition, called **tardive dyskinesia** (the Latin *tardus* means "slow"), is characterized by repetitive, involuntary movements, especially involving the face, mouth, lips, and tongue (see figure). Elaborate, uncontrollable movements of the tongue are particularly prominent, including incessant rolling movements, as well as sucking or smacking of the lips. Some patients show twisting and sudden jerking movements of the arms or legs (Casey, 1989). The atypical neuroleptics discussed in the text have fewer dyskinesia side effects than traditional

neuroleptics have, but unfortunately they are more likely to lead to weight gain.

The underlying mechanism for tardive dyskinesia continues to be a puzzle. It may arise from the chronic blocking of dopamine receptors, which results in what is called *receptor supersensitivity*. Tardive dyskinesia frequently takes a long time to develop and tends to be irreversible. Long-term treatment with traditional antipsychotic drugs can also have another undesirable effect. In some patients, discontinuation of the drugs or a lowering of the dosage results in a sudden, marked increase in positive

symptoms of schizophrenia, such as delusions or hallucinations. This **supersensitivity psychosis** can often be reversed by the administration of increased dosages of dopamine receptor–blocking agents.

To see the video
Tardive Dyskinesia,
go to

2e.mindsmachine.com/12.3

By the mid-twentieth century, more and more physicians were skeptical that lobotomy was effective for any disorder, and a drug discovered in the early 1950s—**chlorpromazine** (trade name Thorazine)—quickly replaced lobotomy as a treatment for schizophrenia. Although chlorpromazine was originally developed as an anesthetic, a lucky observation revealed that it could powerfully reduce the positive symptoms of schizophrenia. These symptoms—auditory hallucinations, delusions, and disordered thinking—were exactly the ones that kept people in mental institutions. So, the introduction of chlorpromazine truly revolutionized psychiatry, relieving symptoms for millions of sufferers and freeing them from long-term beds in psychiatric hospitals. Unfortunately, sometimes people taking antipsychotics develop undesirable side effects in movement (**BOX 12.1**).

Poor Howard Dully, whom we met at the start of the chapter, was very unlucky to run into a physician still performing lobotomies as late as the 1960s. Why didn't Dr. Freeman try giving Howard chlorpromazine? For one thing, the drug helps only positive symptoms, and Howard didn't have any of those. In fact, there's little reason to think the boy had *any* symptoms of schizophrenia (six psychiatrists had declared him "normal"). Unfortunately for Howard, his stepmother just happened upon the wrong physician at the wrong time.

THE DOPAMINE HYPOTHESIS   Chlorpromazine and other **antipsychotic** drugs (also known as *neuroleptics*) that came along a little later were eventually found to share a specific action: they block postsynaptic dopamine receptors, particularly dopamine D_2

dyskinesia Difficulty or distortion in voluntary movement.

tardive dyskinesia A disorder associated with typical antipsychotic use, and characterized by involuntary movements, especially of the face and mouth.

supersensitivity psychosis An exaggerated "rebound" psychosis that may emerge when doses of antipsychotic medication are reduced.

chlorpromazine An early antipsychotic drug that revolutionized the treatment of schizophrenia.

antipsychotic Also called *neuroleptic*. Any of a class of drugs that alleviate symptoms of schizophrenia, typically by blocking dopamine receptors.

FIGURE 12.10 Antipsychotic Drugs Block Dopamine D$_2$ Receptors (After Seeman and Tallerico, 1998.)

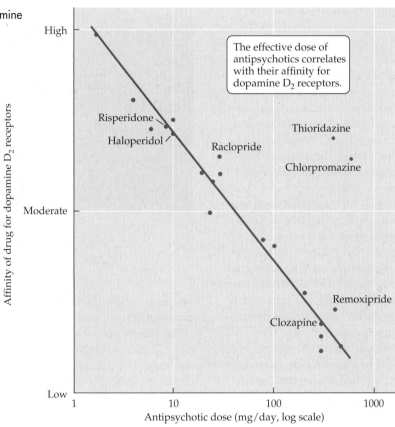

The effective dose of antipsychotics correlates with their affinity for dopamine D$_2$ receptors.

(Y-axis: Affinity of drug for dopamine D$_2$ receptors — High, Moderate, Low)

(Labels on plot: Risperidone, Haloperidol, Raclopride, Thioridazine, Chlorpromazine, Remoxipride, Clozapine)

(X-axis: Antipsychotic dose (mg/day, log scale) — 1, 10, 100, 1000)

dopamine hypothesis The idea that schizophrenia results from either excessive levels of synaptic dopamine or excessive postsynaptic sensitivity to dopamine.

typical antipsychotic Also called *typical neuroleptic*. An antischizophrenic drug that shows antagonist activity at dopamine D$_2$ receptors.

atypical antipsychotic Also called *atypical neuroleptic*. An antipsychotic drug that has primary actions other than or in addition to the dopamine D$_2$ receptor antagonism that characterizes the typical antipsychotics.

receptors. Because antipsychotic drugs all blocked dopamine D$_2$ receptors to some extent, researchers proposed the **dopamine hypothesis**: that people with schizophrenia suffer from an excess of either dopamine release or dopamine receptors. Interestingly, high doses of amphetamine cause an excess of dopamine to accumulate in synapses (see Chapter 4), resulting in a transient *amphetamine psychosis* that is strikingly similar to schizophrenia and is reversed by treatment with antischizophrenic medication. You might think that hallucinogenic drugs, like LSD, would similarly produce a schizophrenia-like state, but in fact there is little resemblance; for one thing, the effects of hallucinogens are primarily visual rather than auditory.

All of the various drugs that are now classified as **typical antipsychotics** (or *typical neuroleptics*) are D$_2$ receptor antagonists (**FIGURE 12.10**). In fact, the clinically effective dose of a typical antipsychotic can be predicted from its affinity for D$_2$ receptors, as the dopamine hypothesis would predict. For example, haloperidol, discovered a few years after chlorpromazine, has a greater affinity for D$_2$ receptors and quickly became the more widely used drug. Over the years, other clinical and experimental findings have bolstered the dopamine hypothesis; for example, treating patients who suffer from Parkinson's disease with L-dopa (the metabolic precursor of dopamine) may induce schizophrenia-like symptoms, presumably by boosting the synaptic availability of dopamine.

Although research has provided support for the dopamine hypothesis of schizophrenia, there are also several problems with the hypothesis. For example, there is no correspondence between the speed with which drugs block dopamine receptors (quite rapidly—within hours) and how long it takes for the symptoms to diminish (usually on the order of weeks). Thus, the relation of dopamine to schizophrenia is more complex than just hyperactive dopamine synapses. Furthermore, work with new types of antischizophrenic drugs, developed to reduce motor side effects (see Box 12.1), suggested that some symptoms of schizophrenia respond to modifications of other neurotransmitter systems. Called **atypical antipsychotics** (or *atypical neuroleptics*), these drugs generally have only moderate affinity for the D$_2$ dopamine receptors that are

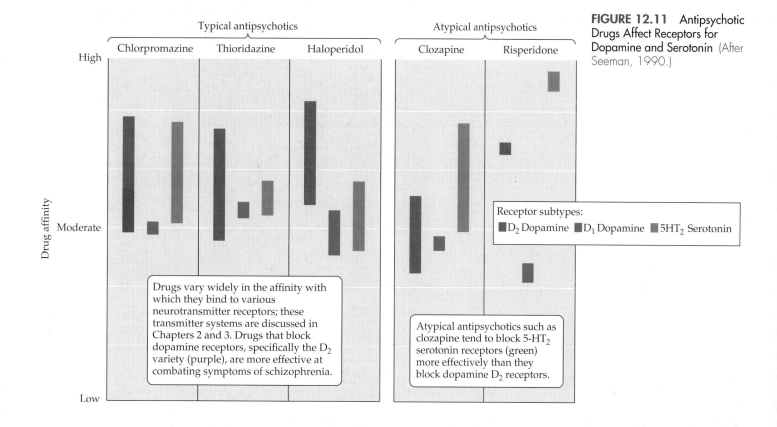

FIGURE 12.11 Antipsychotic Drugs Affect Receptors for Dopamine and Serotonin (After Seeman, 1990.)

Drugs vary widely in the affinity with which they bind to various neurotransmitter receptors; these transmitter systems are discussed in Chapters 2 and 3. Drugs that block dopamine receptors, specifically the D$_2$ variety (purple), are more effective at combating symptoms of schizophrenia.

Atypical antipsychotics such as clozapine tend to block 5-HT$_2$ serotonin receptors (green) more effectively than they block dopamine D$_2$ receptors.

Receptor subtypes:
D$_2$ Dopamine   D$_1$ Dopamine   5HT$_2$ Serotonin

the principal site of action of the typical antipsychotics. Instead, atypical antipsychotics have their highest affinity for other transmitter receptors: **clozapine**, for example, selectively blocks *serotonin* receptors (especially 5-HT$_{2A}$ receptors), as well as other receptor types (**FIGURE 12.11**).

Atypical antipsychotics are just as effective as the older generation of drugs for relieving the symptoms of schizophrenia. So, if the problem is as simple as an overstimulation of dopamine receptors, why are the atypical antipsychotics effective? For example, clozapine can *increase* dopamine release in frontal cortex (Hertel et al., 1999)—hardly what we would expect if excess dopaminergic activity lies at the root of schizophrenia. In fact, it seems that supplementing antipsychotic treatments with L-dopa (thereby increasing dopaminergic activity) actually helps reduce symptoms of schizophrenia (Jaskiw and Popli, 2004).

Until recently, almost all clinicians believed that atypical antipsychotics were more effective than typical antipsychotics for treating schizophrenia, especially for relieving negative symptoms. But a large British study comparing the outcome for patients who had been given the two types of drugs found no difference (Jones et al., 2006). Although the atypical antipsychotics are less likely than typical antipsychotics to cause side effects in motor function (see Box 12.1), they are more likely to cause weight gain (Sikich et al., 2008). So the overall outcome for quality of life appears equivalent for the two types of drugs (Heres et al., 2006).

THE GLUTAMATE HYPOTHESIS   Another drug that, like chlorpromazine, was initially developed as an anesthetic has a much different relationship to schizophrenia. **Phencyclidine (PCP)** was soon found to be a potent **psychotomimetic**; that is, PCP produces phenomena strongly resembling both the positive and negative symptoms of schizophrenia. Users of PCP often experience auditory hallucinations, strange depersonalization, and disorientation, and they may become violent as a consequence of their drug-induced delusions. Prolonged psychotic states can develop with chronic use of PCP.

clozapine   An atypical antipsychotic.

phencyclidine (PCP)   Also called *angel dust*. An anesthetic agent that is also a psychedelic drug. PCP makes many people feel dissociated from themselves and their environment.

psychotomimetic   A drug that induces a state resembling schizophrenia.

FIGURE 12.12   The Effects of PCP on the NMDA Receptor

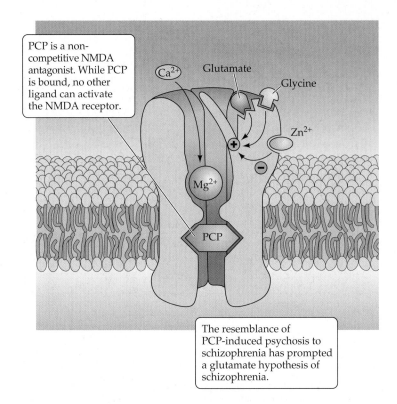

PCP is a non-competitive NMDA antagonist. While PCP is bound, no other ligand can activate the NMDA receptor.

The resemblance of PCP-induced psychosis to schizophrenia has prompted a glutamate hypothesis of schizophrenia.

ketamine   A dissociative anesthetic drug, similar to PCP, that acts as an NMDA receptor antagonist.

glutamate hypothesis   The idea that schizophrenia may be caused, in part, by understimulation of glutamate receptors.

depression   A psychiatric condition characterized by such symptoms as an unhappy mood; loss of interests, energy, and appetite; and difficulty concentrating.

As illustrated in **FIGURE 12.12**, PCP acts as an NMDA receptor antagonist. PCP blocks the NMDA receptor's central calcium channel, thereby preventing the endogenous ligand—glutamate—from having its usual effects. Treating monkeys with PCP for 2 weeks produces a schizophrenia-like syndrome, including poor performance on a test that is sensitive to prefrontal damage (Jentsch et al., 1997). Other antagonists of NMDA receptors, such as **ketamine**, have similar effects. These and other observations prompted researchers to propose a **glutamate hypothesis** of schizophrenia (Moghaddam and Adams, 1998), which suggests that schizophrenia results from an *under*activation of glutamate receptors (Coyle et al., 2003). If this hypothesis is true, you might ask whether compounds that increase glutamatergic activity would be effective antischizophrenic drugs. However, drugs that stimulate the ionotropic NMDA receptors tend to produce seizures, so NMDA receptor agonists are not an option. Instead, researchers are focusing on manipulations of the metabotropic glutamate receptors—mGluR's—of which there are at least eight different subtypes (Mueller et al., 2004). Although results from clinical trials with early candidate drugs have been disappointing (Stauffer et al., 2013), researchers hope that targeting the proper class of mGluR's with drugs that have novel modes of action may someday lead to a new generation of antipsychotics (Rook et al., 2015).

how's it going ❓

1. Evaluate the evidence supporting the dopamine hypothesis of schizophrenia, and contrast it with the evidence casting doubt on this hypothesis.
2. What distinguishes typical antipsychotics from atypical antipsychotics?
3. Which drugs can induce a psychosis resembling schizophrenia? What receptor(s) do these drugs act on?

Depression Is the Most Prevalent Disorder of Mood

Disturbances of mood are a fact of life for humans; most of us experience periods of unhappiness that we commonly describe as depression. But for some people, an unhappy mood state is more than a passing malaise. Clinically, **depression** is characterized by a combination of unhappy mood, loss of interests, reduced energy, changes in appetite and sleep patterns, and loss of pleasure in most things. Difficulty in concentration and restless agitation or torpor are common; the person may dwell on thoughts of death or even contemplate suicide. Pessimism seems to seep into every act (Solomon, 2001). Periods of such depression that alternates with normal emotional states, called *unipolar depression*, can occur with no readily apparent stress, and without treatment the depression often lasts several months (Cassens et al., 1990). Each year, about 7% of American adults experience at least one such major depressive episode (Substance Abuse and Mental Health Services Administration, 2013). This condition is more common in people over 40 years of age, especially women, but depression can afflict people of any age, race, or ethnicity (CDC, 2010).

Depression can be lethal, as it may lead to suicide: about 80% of all suicide victims are profoundly depressed. Whether or not the person is depressed, many suicides appear to be impulsive acts. For example, one classic study found that of the more than 500 people who were prevented from jumping off the Golden Gate Bridge in San Francisco, only 6% later went on to commit suicide (Seiden, 1978). Similarly, suicide rates went down by a third in Britain when that country switched from using coal gas, which contains lots of deadly carbon monoxide, to natural gas for heating. The suicide rate has remained at that reduced level in the 40+ years since. Apparently those thousands of Britons who would have found it easy to follow a suicidal impulse by turning on the kitchen oven did not kill themselves when more planning was required. Thus, it is important for society to erect barriers, either literally (e.g., on bridges) or metaphorically, to make it difficult for people to kill themselves. If suicide is averted the first time it is seriously attempted, the person is very unlikely to ever try it again.

Inheritance is an important determinant of depression

Genetic studies of depressive disorders reveal strong hereditary contributions. The concordance rate for identical twins (about 60%) is substantially higher than for fraternal twins (about 20%) (Kendler et al., 1999). The concordance rates for identical twins are similar whether the twins are reared apart or together. Although several early studies implicated specific chromosomes, subsequent research has failed to identify any particular gene (Risch et al., 2009). So, as is the case for schizophrenia, there probably is no single gene for depression. Rather, *many* genes contribute to making a person more or less susceptible, and environmental factors determine whether depression results.

The brain changes with depression

Most reports of differences in the brains of depressed people focus on functional changes as detected by PET or fMRI. Depressed patients show changes in activity in a number of brain regions, depending on whether the tasks being processed are principally cognitive or emotional in nature (S. M. Palmer et al., 2015). Compared with control individuals, increased activation in the amygdala is especially evident during emotional processing, and increased activity in the frontal lobes is evident during more cognitively demanding tasks (**FIGURE 12.13**). Decreased activity is evident in the parietal and posterior temporal cortex and in the anterior cingulate cortex—systems that have been implicated in attention (see Chapter 14) (Drevets, 1998). The increased activity in the amygdala—a structure involved in mediating fear (see Chapter 11)—persists even after the depression has lifted.

Descendants of people with severe depression also have a thinner cortex across large swaths of the right hemisphere than do control participants (B. S. Peterson et al., 2009), which might make them vulnerable to depression. There is also evidence that people who are depressed have difficulties regulating stress hormone release, as discussed in **A STEP FURTHER 12.2**, on the website. However, whether the change in stress hormone response is a *cause* or a *result* of depression is unknown.

Many studies report hippocampal volume is reduced in people with depression (Sexton et al., 2013), and there is reduced activation of the hippocampal region in depressed people during memory tasks (K. D. Young et al., 2012). But whether these changes in the hippocampus are present before the depression, and therefore may be a contributing cause of the disorder, or are a result of the depression remains unknown. In any case, there are effective treatments for depression, as we discuss next.

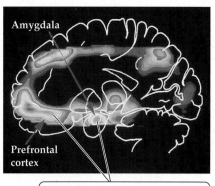

PET scans reveal increased activity in the prefrontal cortex and the amygdala of depressed patients.

FIGURE 12.13 Brain Activity Patterns in Depression (Courtesy of Wayne C. Drevets.)

A wide variety of treatments are available for depression

electroconvulsive shock therapy (ECT) A last-resort treatment for unmanageable depression, in which a strong electrical current is passed through the brain, causing a seizure.

repetitive transcranial magnetic stimulation (rTMS) A noninvasive treatment in which repeated pulses of focused magnetic energy are used to stimulate the cortex through the scalp.

monoamine oxidase (MAO) An enzyme that breaks down monoamine neurotransmitters, thereby inactivating monoamine transmitters.

selective serotonin reuptake inhibitor (SSRI) An antidepressant drug that blocks the reuptake of transmitter at serotonergic synapses.

Electroconvulsive shock therapy (ECT)—the intentional induction of a large-scale seizure (Weiner, 1994)—was originally a schizophrenia treatment, born of desperation during the 1930s. Although it proved to be of little help in schizophrenia, it soon became evident that ECT *could* rapidly reverse severe depression. The advent of antidepressant drugs has made ECT less common, but ECT remains an important tool for treating severe, drug-resistant depression (M. Fink and Taylor, 2007). A more modern technique for altering cortical electrical activity called **repetitive transcranial magnetic stimulation** (**rTMS**; see Chapter 2) is likewise being developed as a treatment for depression (D. R. Kim et al., 2009).

Today, the most common treatment for depression is the use of drugs that affect the monoamine transmitters: norepinephrine, dopamine, and serotonin. The first antidepressants were inhibitors of **monoamine oxidase** (**MAO**), the enzyme that normally inactivates monoamines (Schildkraut and Kety, 1967). This action of MAO inhibitors causes monoamine transmitters to accumulate to higher levels in synapses, so researchers proposed that depressed people do not get enough stimulation at monoamine synapses (this is sometimes called the *monoamine hypothesis of depression*). A second generation of antidepressants, called *tricyclics*, inhibits the reuptake of monoamines, which similarly boosts their synaptic activity. ECT may help depression by inducing the release of monoamines.

Among the monoamines, serotonin seems to play an especially important role in depression (Asberg et al., 1986; L. Du et al., 2000; Svenningsson et al., 2006). A major class of modern antidepressants, the **selective serotonin reuptake inhibitors** (**SSRIs**), such as Prozac (**TABLE 12.2**) (see Chapter 4), act to increase synaptic serotonin levels in the brain. In rats, SSRIs increase the birth of new neurons in the hippocampus (Sahay and Hen, 2007), which may mediate some of the mood effects of the drugs.

However, there are problems with the idea that reduced serotonin stimulation causes depression. We know that SSRI drugs increase synaptic serotonin within hours of administration. Yet it typically takes several weeks of SSRI treatment before people feel better. This paradox suggests that it is the brain's response to increased synaptic serotonin that relieves the symptoms, and that this response takes time. So even though boosting serotonin helps some people, their depression may originally have been caused by other factors in the brain. Furthermore, as we discuss in Signs & Symptoms, while there is no doubt that SSRIs do help many people who are depressed, evidence has accumulated that for a significant number of people, part of the benefit may actually be a placebo effect (Turner et al., 2008).

TABLE 12.2 Drugs Used to Treat Depression

Drug class	Mechanism of action	Examples[a]
Monoamine oxidase (MAO) inhibitors	Inhibit the enzyme monoamine oxidase, which breaks down serotonin, norepinephrine, and dopamine	Marplan, Nardil, Parnate
Tricyclics and heterocyclics	Inhibit the reuptake of norepinephrine, serotonin, and/or dopamine	Wellbutrin, Elavil, Aventyl, Ludiomil, Norpramin
Selective serotonin reuptake inhibitors (SSRIs)	Block the reuptake of serotonin, having little effect on norepinephrine or dopamine synapses	Prozac, Paxil, Zoloft
Atypical antidepressants and investigational drugs	Norepinephrine and dopamine reuptake inhibitors (NDRIs), serotonin and norepinephrine reuptake inhibitors (SNRIs), noradrenergic and specific serotonergic antidepressants (NaSSAs), serotonin antagonist and reuptake inhibitors (SARIs), opioid receptor modulators	Wellbutrin/Zyban (NDRI), Effexor (SNRI), Remeron (NaSSA), Oleptro (SARI), Buprenex (opioid receptor modulator)

[a]The names given are the more commonly used trade names rather than chemical names.

SIGNS & SYMPTOMS

Mixed Feelings about SSRIs

At their introduction, selective serotonin reuptake inhibitors (SSRIs) represented a major revolution in depression treatment. Heavily marketed to the public and to health professionals, SSRIs soon became one of the most widely prescribed medications, propelling an incredible 400% increase in antidepressant prescriptions by 2008. In fact, for 18- to 44-year-olds, antidepressants are prescribed more than any other drug; more than one in ten adult Americans is currently using antidepressant medication (Pratt et al., 2011). Needless to say, the development and sales of antidepressants have provided a huge windfall for the pharmaceuticals industry.

Now that SSRIs have been with us for more than 20 years, researchers have turned to retrospective analyses to reevaluate the efficacy of SSRIs. In part, these large-scale **meta-analyses** (analyses that combine the results of many previously published studies) have been prompted by the concern that for various reasons—public appetite, profit motives, the tendency of journals to publish only positive findings—studies that failed to find effects of SSRIs may historically have been underreported. The results of these meta-analyses have been mixed, but they at least give us cause to take a sober second look at SSRI usage.

As illustrated in **FIGURE 12.14**, one influential large-scale review concluded that only a minority of depressed patients, the 13% constituting the most severe cases, responded significantly better to SSRIs than to placebos (Fournier et al., 2010). Furthermore, only about half of the people getting the drug are completely "cured," and about 20% show no improvement at all. And while there are reservations about the efficacy of SSRIs for children or teenagers (Bower, 2006), millions of American children have been given prescriptions for SSRIs despite a reported increased risk of suicide in these age groups (Olfson et al., 2006).

However, other large meta-analytic studies, also based on multiple clinical trials, find evidence of clear beneficial effects of SSRIs relative to placebos for people of all ages and with all levels of severity of depression (Gibbons et al., 2012). These authors argue that the apparent relationship of severity and SSRI efficacy is a statistical artifact of the methodology employed in order to combine studies in a meta-analysis.

This controversy seems likely to rage on for a while. In the meantime, a prudent course of action is to deploy cognitive behavioral therapy as a first-rank treatment in moderate cases, supplemented with antidepressant medication in more severe or nonresponsive cases.

meta-analysis   A type of quantitative review of a field of research, in which the results of multiple previous studies are combined in order to identify overall patterns that are consistent across studies.

FIGURE 12.14   When Are Antidepressants More Effective Than Placebos? (After Fournier et al., 2010.)

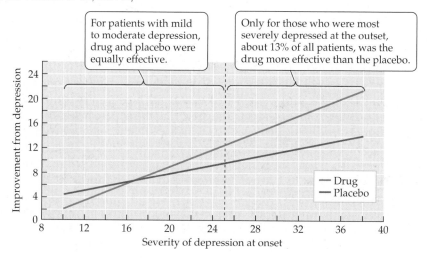

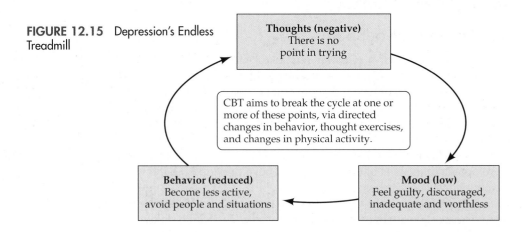

FIGURE 12.15 Depression's Endless Treadmill

Thoughts (negative)
There is no point in trying

CBT aims to break the cycle at one or more of these points, via directed changes in behavior, thought exercises, and changes in physical activity.

Behavior (reduced)
Become less active, avoid people and situations

Mood (low)
Feel guilty, discouraged, inadequate and worthless

cognitive behavioral therapy (CBT)
Psychotherapy aimed at correcting negative thinking and consciously changing behaviors as a way of changing feelings.

deep brain stimulation (DBS) Mild electrical stimulation through an electrode that is surgically implanted deep in the brain.

Despite the popularity of SSRIs for treating depression, treatment with **cognitive behavioral therapy** (**CBT**), a type of psychotherapy aimed at correcting negative thinking and improving interpersonal relationships, is about as effective as SSRI treatment (Butler et al., 2006). Furthermore, the rate of relapse is lower for CBT than for SSRI treatment (DeRubeis et al., 2008). Interestingly, CBT and SSRI treatment *together* are more effective in combating depression than either one is alone (March et al., 2004). Typically, CBT helps the client to recognize self-defeating modes of thinking and encourages breaking out of a cycle of self-fulfilling depression (**FIGURE 12.15**).

A number of additional treatments for depression are currently under study. For example, compounds being investigated as potential antidepressants include the glutamate receptor antagonist ketamine (see Chapter 4), which relieves depression almost instantly (Machado-Viera et al., 2009), and leptin (Lu, 2007), the hormone normally secreted by fat cells (see Chapter 9). An unusual treatment for depression involves a pacemaker that periodically applies mild electrical stimulation to the vagus nerve (cranial nerve X; see Chapter 2). This treatment is offered in cases where drugs or ECT have been ineffective, but it remains to be established whether vagal stimulation is a long-term solution (Boodman, 2006; Grimm and Bajbouj, 2010).

For extremely difficult cases of depression, researchers have turned again to psychosurgery, but nothing that resembles the ravages suffered by Howard Dully. In **deep brain stimulation** (**DBS**) surgery, delicate electrodes are surgically implanted in the cingulate cortex or other brain sites. Mild electrical stimulation applied through the electrodes reportedly relieves depression that resists other treatments (Kringelbach et al., 2007).

The effectiveness of DBS or vagal nerve stimulation for depression is difficult to evaluate, because trials have few participants and most studies have no placebo control (R. Robinson, 2009). The lack of such controls is unfortunate because we know depression is very susceptible to placebo effects.

Why do more females than males suffer from depression?

Studies all over the world show that more women than men suffer from major depression. In the United States, women are twice as likely as men to suffer major depression (CDC, 2010). Some researchers suggest that the apparent sex difference reflects patterns of help seeking by males and females—that women are willing to use health facilities, while men see that as a sign of weakness. But sex differences in the incidence of depression also are evident in door-to-door surveys (Robins and Regier, 1991), which would appear to rule out the simple explanation that women seek treatment more often than men do.

Some researchers have emphasized gender differences in endocrine physiology. The appearance of clinical depression often is related to events in the female reproductive cycle—for example, before menstruation, during use of contraceptive pills, following

Depression For reasons that are not understood, women are more likely than men to suffer from depression.

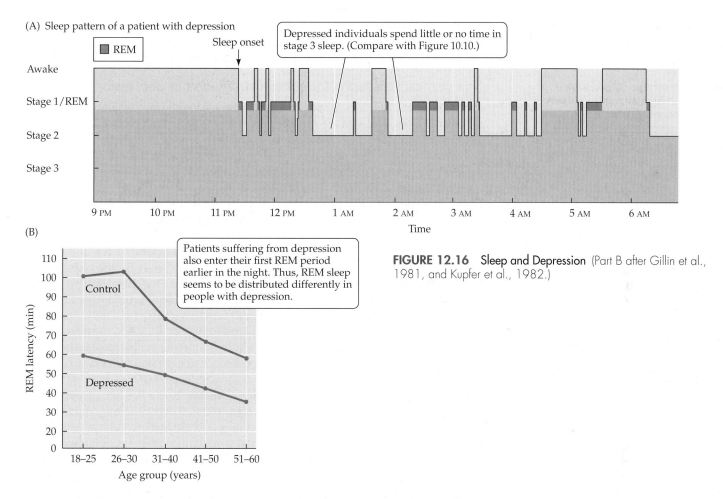

(A) Sleep pattern of a patient with depression

Awake | Stage 1/REM | Stage 2 | Stage 3

Sleep onset

Depressed individuals spend little or no time in stage 3 sleep. (Compare with Figure 10.10.)

9 PM 10 PM 11 PM 12 PM 1 AM 2 AM 3 AM 4 AM 5 AM 6 AM

Time

(B)

Patients suffering from depression also enter their first REM period earlier in the night. Thus, REM sleep seems to be distributed differently in people with depression.

Control

Depressed

REM latency (min)

Age group (years)

FIGURE 12.16 Sleep and Depression (Part B after Gillin et al., 1981, and Kupfer et al., 1982.)

childbirth, and during menopause. Although there is little relation between circulating levels of individual hormones and measures of depression, the phenomenon of **postpartum depression**, a bout of depression immediately preceding and/or following childbirth, suggests that some combination of hormones can precipitate depression. About one out of every seven pregnant women will show symptoms of depression (Dietz et al., 2007). Because postpartum depression may affect the mother's relationship with her child, resulting in long-lasting negative effects on the child's behavior (Tronick and Reck, 2009), there is growing concern about this problem. However, there is also evidence that SSRIs taken by pregnant women may affect the later behavior of their children (Oberlander et al., 2010), and it is uncertain whether exposure to SSRIs via breast milk will have a comparable long-term effect. Thus CBT offers the safest treatment for postpartum depression, and researchers continue to weigh the costs and benefits of supplementing that therapy with antidepressant medication (Bourke et al., 2014).

Sleep characteristics change in affective disorders

Difficulty falling asleep and inability to maintain sleep are common in depression. In addition, EEG sleep studies of depressed patients show certain abnormalities that go beyond difficulty falling asleep. The sleep of patients with major depressive disorders is marked by a striking reduction in stage 3 slow wave sleep (see Chapter 10) and a corresponding increase in stages 1 and 2 (**FIGURE 12.16A**). Similarly, depressed patients enter rapid-eye-movement sleep (REM sleep) much sooner after sleep onset (**FIGURE 12.16B**); the length of time before REM sleep begins correlates with the severity of depression. Furthermore, the distribution of REM sleep across the night is altered, with an increased amount of REM sleep occurring during the first half of sleep, as though REM sleep were displaced toward an earlier period in the night (Wehr et al., 1985).

postpartum depression A bout of depression that afflicts a woman either immediately before or after giving birth.

learned helplessness A learning paradigm in which individuals are subjected to inescapable, unpleasant conditions.

bipolar disorder A psychiatric disorder characterized by periods of depression that alternate with excessive, expansive moods.

As with these links between the daily rhythm of sleep and depression, seasonal rhythms have been implicated in a particular depressive condition known as *seasonal affective disorder* (*SAD*), which is described in **A STEP FURTHER 12.3**, on the website.

Scientists are still searching for animal models of depression

Everyone agrees that an animal model of depression could be invaluable, as it might reveal underlying mechanisms or offer a convenient way to screen potential treatments (Nestler and Hyman, 2010). But it's not clear that any animal model has lived up to this promise so far. If depression were caused by the mutation of a particular gene, it might be possible to create a model by introducing that mutated gene in mice. But in fact, it appears that human depression is influenced by many genes, each having a modest effect alone. Plus, some of the most powerful symptoms of depression are internal—apathy and a feeling of hopelessness—and thus are difficult to assess in species we can't talk to.

Still, many of the signs of depression—such as decreased social contact, problems with eating, and changes in activity—are observable behaviors. So researchers have used these behaviors to evaluate animal models of depression. In one type of stress model—**learned helplessness**—an animal is exposed to a repetitive stressful stimulus, such as an electrical shock, that it cannot escape. Like depression, learned helplessness has been linked to a decrease in serotonin function (Petty et al., 1994) and also to mechanisms that control the release of dopamine (B. Li et al., 2011), the main reward signal in the brain. Removing the olfactory bulb from rodents also creates a model of depression: the animals display irritability, preferences for alcohol, and elevated levels of corticosteroids—all of which are reversed by many antidepressants. A strain of rats created through selective breeding—the Flinders Sensitive Line—has been proposed as a model of depression because these animals show reduced overall activity, reduced body weight, increased REM sleep, learning difficulties, and exaggerated response to chronic stress (Overstreet, 1993). These varied animal models may be useful in finding the essential mechanisms that cause and maintain depression in humans.

In Bipolar Disorder, Mood Cycles between Extremes

Affecting about 2.6% of the U.S. population each year (Kessler et al., 2005), **bipolar disorder** is characterized by periods of depression alternating with periods of excessively expansive mood (or *mania*) that includes sustained overactivity, talkativeness, strange grandiosity, and increased energy (**FIGURE 12.17**). The rate at which

(A) Manic phase

(B) Depressive phase

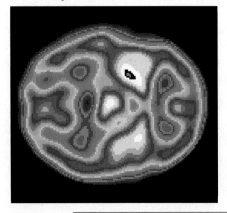 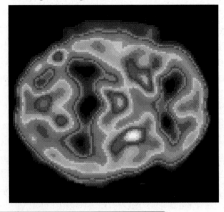

FIGURE 12.17 Functional Images of Bipolar Disorder (Courtesy of Dr. Robert G. Kohn, Brain-Spect.com.)

There are dramatic differences in brain activity between the manic (A) and depressive (B) phases of this person's bipolar disorder. White indicates the greatest brain activity, while blue indicates the lowest.

the alternation occurs varies between individuals: some patients exhibit *rapid-cycling* bipolar disorder, defined as consisting of four or more distinct cycles in one year (and some individuals have many more cycles than that; some may even show several cycles per *day*).

Men and women are equally affected by bipolar disorder, and the age of onset is usually much earlier than that of unipolar depression. Bipolar disorder is clearly heritable, with several different genes affecting the probability of the disorder (Smoller and Finn, 2003; Faraone et al., 2004).

The neural basis of bipolar disorder is not fully understood, but patients with bipolar disorder exhibit enlarged ventricles on brain scans (Arnone et al., 2009), as is seen in schizophrenia (see Figure 12.6). The more manic episodes the person has experienced, the greater the ventricular enlargement, suggesting a worsening of brain loss over time (Moorhead et al., 2007).

The observed pattern of changes in the brain and behavior of people with bipolar disorder has led to a recognition that bipolar disorder has more in common with schizophrenia than with unipolar depression, so the older term *manic depression* has been largely abandoned. For example, the self-aggrandizing ideas and extreme talkativeness of people in the manic phase of bipolar disorder (e.g., "The president called me this morning to thank me for my efforts") may resemble the frank delusions seen in schizophrenia. In addition, families in which some individuals have been diagnosed with bipolar disorder are more likely than other families to have individuals with a diagnosis of schizophrenia (Lichtenstein et al., 2009; Van Os and Kapur, 2009). And although typical antipsychotics do not seem to help people with bipolar disorder, the newer, atypical antipsychotics seem to help dampen the manic phase in people with bipolar disorder.

Most people suffering from bipolar disorder benefit from treatment with the element **lithium** (Kingsbury and Garver, 1998). The effect of lithium on bipolar disorder was discovered by accident when it was intended as an inert control for another drug, so the mechanism of action is not understood. Lithium has wide-ranging effects on the brain, including interacting with a protein that is part of the circadian molecular clock (see Chapter 10) (L. Yin et al., 2006). Because lithium has a narrow range of safe doses (see Chapter 4), care must be taken to avoid toxic side effects of an overdose. Nevertheless, well-managed lithium treatment produces marked relief for many patients and even has been reported to increase the volume of gray matter in their brains (G. J. Moore et al., 2000).

The fact that the manic phases blocked by lithium are so exhilarating may be the reason that some bipolar clients stop taking the medication. Unfortunately, doing so means that the depressive episodes return as well. As in unipolar depression, transcranial magnetic stimulation may provide a nonpharmacological treatment alternative in difficult cases of bipolar disorder (Michael and Erfurth, 2004). Furthermore, mounting evidence suggests that some forms of CBT for mild cases of bipolar disorder can be as effective as drug treatments (Hollon et al., 2002) and perhaps can be beneficially combined with other forms of treatment.

A Disturbance in the Force Actress Carrie Fisher, who played Princess Leia in the original *Star Wars* series, has written about her struggles with bipolar disorder.

lithium An element that, administered to patients, often relieves the symptoms of bipolar disorder.

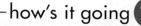

how's it going

1. What are the symptoms of depression, and how does depression differ from simple sadness?
2. What treatments for depression arose in the twentieth century, and which treatment is used most often today?
3. Summarize the evidence for and against the use of SSRIs in depression. Why is the use of SSRIs controversial?
4. What is bipolar disorder, and how does it compare with unipolar depression and with schizophrenia? How is it treated?

There Are Several Types of Anxiety Disorders

anxiety disorder   Any of a class of psychological disorders that includes recurrent panic states, generalized persistent anxiety disorders, and post-traumatic stress disorder.

benzodiazepine   Any of a class of antianxiety drugs that are agonists of GABA receptors in the central nervous system. One example is diazepam (Valium).

anxiolytic   A substance that is used to combat anxiety. Examples include alcohol, opiates, barbiturates, and the benzodiazepines.

All of us have at times felt apprehensive and fearful. But some people experience this state with an intensity that is overwhelming and includes irrational fears; a sense of terror; body sensations such as dizziness, difficulty breathing, trembling, and shaking; and a feeling of loss of control. Anxiety can be lethal: men with panic disorder are more likely than others to die from cardiovascular disease or suicide (Coryell et al., 1986).

The *DSM-5* distinguishes several major types of **anxiety disorders**: *Phobic disorders* are intense, irrational fears that become centered on a specific object, activity, or situation that the person feels compelled to avoid. Another type of anxiety disorder is *panic disorder*, characterized by recurrent transient attacks of intense fearfulness. In *generalized anxiety disorder*, persistent, excessive anxiety and worry are experienced for months. There is a strong genetic contribution to each of these disorders (Shih et al., 2004; Oler et al., 2010) and distinctive underlying neurobiological predispositions to the development of anxiety disorders (Shackman et al., 2013).

Some patients who suffer from recurrent panic attacks have temporal lobe abnormalities, especially in the left hemisphere (Vythilingam et al., 2000; Van Tol et al., 2010). Given the special role of the amygdala in mediating fear (see Chapter 11), changes may be particularly evident in the amygdala and associated circuitry within the temporal lobes (Rauch et al., 2003).

Drug treatments provide clues to the mechanisms of anxiety

Throughout history, people have consumed all sorts of substances in the hopes of controlling anxiety. The list includes alcohol, bromides, scopolamine, opiates, and barbiturates. In the 1950s the tranquilizing drug meprobamate (Miltown) was introduced, and it became an instant best seller, ushering in the modern age of anxiety pharmacotherapy. Soon researchers discovered a new class of drugs called **benzodiazepines**, which quickly replaced Miltown as the favored drugs for treating anxiety. One type of benzodiazepine—diazepam (trade name Valium)—is one of the most prescribed drugs in history. Other commonly prescribed benzodiazepines include Xanax, Halcion, and Ativan. Such drugs that combat anxiety are termed **anxiolytics** ("anxiety-dissolving"), although they also may have anticonvulsant and sleep-inducing properties. The anxiolytic drugs are also discussed in Chapter 4.

Anxiolytic benzodiazepines interact with binding sites that are part of GABA receptors, especially the GABA$_A$ receptors, where they act as noncompetitive agonists. Recall from Chapter 4 that GABA is the most common inhibitory transmitter in the brain. When GABA is released from a presynaptic terminal and activates postsynaptic receptors, it hyperpolarizes the target neuron and therefore inhibits it from firing. Benzodiazepines alone have little effect on the GABA$_A$ receptor, but when benzodiazepines are present, GABA produces a markedly enhanced hyperpolarization. In other words, benzodiazepines boost GABA-mediated postsynaptic inhibition, reducing the excitability of postsynaptic neurons.

Interestingly, the brain probably makes its own anxiety-relieving substances that interact with the benzodiazepine-binding site on the GABA receptor; the neurosteroid allopregnanolone is a prime candidate for this function. Drugs developed to act at this site are effective anxiolytics in both rats and humans (Rupprecht et al., 2009). As you can see in **FIGURE 12.18**, benzodiazepine/GABA$_A$ receptors are widely distributed throughout the brain, especially in the cerebral cortex and some subcortical areas, such as the hippocampus and the amygdala.

Although the benzodiazepines remain an important category of anxiolytics, especially for acute attacks, other anxiety-relieving drugs have been developed that lack the abuse potential of the

This PET scan of benzodiazepine receptors shows their wide distribution in the brain, especially the cortex. Highest concentrations are in red; lowest concentrations are in blue.

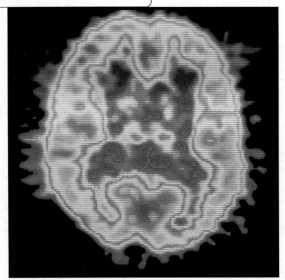

FIGURE 12.18   The Distribution of Benzodiazepine Receptors in the Human Brain (Courtesy of Goran Sedvall.)

benzodiazepines. A notable example is the drug buspirone (Buspar), an agonist at serotonin 5-HT$_{1A}$ receptors that can provide relief from anxiety. This effect is consistent with functional-imaging research that reveals an abnormal density of 5-HT$_{1A}$ receptors in the brains of people with anxiety disorders (Neumeister et al., 2004). SSRI antidepressants, such as paroxetine (Paxil) and fluoxetine (Prozac), which increase the stimulation of serotonin receptors, are also sometimes effective treatments for anxiety disorders.

In Posttraumatic Stress Disorder, Horrible Memories Won't Go Away

Some people experience especially awful moments in life that seem indelible, resulting in vivid impressions that persist the rest of their lives. The kind of event that seems particularly likely to produce subsequent stress disorders is intense and is usually associated with witnessing abusive violence and/or death. Examples include the sudden loss of a close friend, rape, torture, kidnapping, or profound social dislocation, such as in forced migration. In these cases, memories of horrible events intrude into consciousness and produce the same intense visceral arousal—the fear and trembling and general autonomic activation—that the original event caused. These traumatic memories are easily reawakened by stressful circumstances and even by harmless stimuli that somehow prompt recollection of the original event. An ever watchful and fearful stance becomes the portrait of individuals afflicted with what is called **posttraumatic stress disorder** (**PTSD**), formerly called *combat fatigue*, *war neurosis*, or *shell shock*. Although related in some ways to anxiety disorders, posttraumatic stress disorder is now recognized in the *DSM-5* as a separate entity.

Analysis of a random sample of Vietnam War veterans has indicated that 19% had PTSD at some point after service. This was the rate for *all* Vietnam veterans; when the researchers focused more specifically on veterans exposed to intense war zone stressors, more than 35% developed PTSD at some point, and most of them were still suffering from the disorder *decades* later (Dohrenwend et al., 2006). More recently, it has been found that Gulf War veterans likewise suffer from high rates of PTSD (Institute of Medicine, 2010).

Genetic factors affect vulnerability to PTSD, as indicated in twin studies of Vietnam War veterans who had seen combat, which showed that monozygotic twins were more similar than dizygotic twins. People who display combat-related PTSD show (1) memory changes such as amnesia for some war experiences, (2) flashbacks, and (3) deficits in short-term memory (Bremner et al., 1993). These memory disturbances suggest involvement of the hippocampus (see Chapter 13), and indeed the volume of the right hippocampus is smaller in combat veterans with PTSD than in those without it, with no differences in other brain regions (Bremner et al., 1995). It was once widely assumed that stressful episodes caused the hippocampus to shrink, but some veterans suffering PTSD left their monozygotic twins at home, and it turns out that the nonstressed twins without PTSD also tended to have a smaller hippocampus (Gilbertson et al., 2002). So some inherited characteristic that's associated with having a small hippocampus, and perhaps a reduced rate of adult neurogenesis (Snyder et al., 2011; Kheirbek et al., 2012), may increase susceptibility to developing PTSD if the person is exposed to stress. In Gulf War veterans with more severe PTSD, marked hippocampal size difference is associated with markers of inflammatory processes (O'Donovan et al., 2015), which can strongly contribute to neural degeneration and decreased neurogenesis.

A comprehensive psychobiological model of the development of PTSD draws connections from PTSD's memory disturbances to the neural mechanisms of fear conditioning, behavioral sensitization, and extinction (Charney et al., 1993). Work in animals has revealed that **fear conditioning**—memory for a stimulus that the animal has learned to associate with a negative event—is very persistent and involves the amygdala and brainstem pathways that are part of a circuit of startle response behavior (see

Delayed Reaction Many combat veterans suffer from PTSD for years afterward.

posttraumatic stress disorder (PTSD) A disorder in which memories of an unpleasant episode repeatedly plague the victim.

fear conditioning A form of classical conditioning in which fear comes to be associated with previously neutral stimuli.

FIGURE 12.19 A Neural Model of Post-Traumatic Stress Disorder

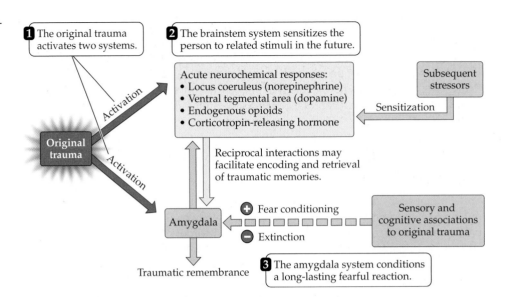

1 The original trauma activates two systems.

2 The brainstem system sensitizes the person to related stimuli in the future.

Acute neurochemical responses:
- Locus coeruleus (norepinephrine)
- Ventral tegmental area (dopamine)
- Endogenous opioids
- Corticotropin-releasing hormone

Subsequent stressors

Sensitization

Activation

Original trauma

Activation

Reciprocal interactions may facilitate encoding and retrieval of traumatic memories.

Amygdala

+ Fear conditioning

− Extinction

Sensory and cognitive associations to original trauma

Traumatic remembrance

3 The amygdala system conditions a long-lasting fearful reaction.

Chapter 11). The persistence of memory and fear in PTSD may depend on the failure of mechanisms to *forget*. There is also a hormonal link, because PTSD sufferers exhibit a paradoxical long-term *reduction* in cortisol (stress hormone) levels (Yehuda, 2002), perhaps due to persistent increases in *sensitivity* to cortisol. If, as a result, they feel the effect of stress hormones more strongly than other people do, that greater effect might repeatedly retrigger the fear response, making it harder for them to forget stressful events (**FIGURE 12.19**). In Chapter 13 we will discuss research-based methods that have been proposed to help people forget traumatic life events by having them recall the event while under the influence of a drug that dampens the stress response.

In Obsessive-Compulsive Disorder, Thoughts and Acts Keep Repeating

Most of us aspire to be neat and clean, especially when we discover a thick layer of dust under the furniture, or perhaps realize we've created yet another tottering pile of papers and bills. And of course, having certain small rituals in our lives—making coffee a certain way in the morning, wishing everyone good night before going to bed—can be a comfort amid the chaos of daily life. But when does orderliness and routine cross the line into pathology? People with **obsessive-compulsive disorder** (**OCD**) lead lives riddled with repetitive rituals and persistent thoughts that they feel powerless to control or stop, despite recognizing that the behaviors are abnormal. In people with OCD, routine acts that we all engage in, such as checking whether the door is locked when we leave our home, become *compulsions*, acts that are repeated over and over. Recurrent thoughts, or *obsessions*, such as fears of germs or other potential harms in the world, invade the consciousness. These symptoms progressively isolate a person from ordinary social engagement with the world. For many people with OCD, hours each day are consumed by compulsive acts such as repetitive hand washing. **TABLE 12.3** summarizes some of the symptoms of OCD.

Determining the number of people afflicted with OCD is difficult, especially because people with this disorder tend to hide their symptoms (Rapoport, 1989). It is estimated that nearly 1% of adults in the United States will suffer from "severe" OCD in any given year (Kessler et al., 2005). In many cases, the initial symptoms of this disorder appear in childhood; the peak age group for onset of OCD, however, is 25–44 years. People with OCD display increased metabolic rates in the orbitofrontal cortex, cingulate cortex, and caudate nuclei (Saxena and Rauch, 2000; Chamberlain et al., 2008).

Happily, OCD responds to treatment in most cases. OCD shows excellent response to cognitive behavioral therapy (Olatunji et al., 2013) and also to several drugs. What

obsessive-compulsive disorder (OCD) An anxiety disorder in which the affected individual experiences recurrent unwanted thoughts and engages in repetitive behaviors without reason or the ability to stop.

TABLE 12.3 Symptoms of Obsessive-Compulsive Disorder

Symptoms	Percentage of patients exhibiting symptom
OBSESSIONS (thoughts)	
Dirt, germs, or environmental toxins	40
Something terrible happening (e.g., fire, death or illness of self or loved one)	24
Symmetry, order, or exactness	17
Religious obsessions	13
Body wastes or secretions (urine, stool, saliva, etc.)	8
Lucky or unlucky numbers	8
Forbidden, aggressive, or perverse sexual thoughts, images, or impulses	4
Fear of harming self or others	4
Household items	3
Intrusive nonsense sounds, words, or music	1
COMPULSIONS (acts)	
Performing excessive or ritualized hand washing, showering, bathing, tooth brushing, or grooming	85
Repeating rituals (e.g., going in or out of a door, getting up from or sitting down on a chair)	51
Checking (doors, locks, stove, appliances, emergency brake on car, paper route, homework, etc.)	46
Engaging in miscellaneous rituals (such as writing, moving, speaking)	26
Decontaminating	23
Touching	20
Counting	18
Ordering or arranging	17
Preventing harm to self or others	16
Hoarding or collecting	11
Cleaning household or inanimate objects	6

Source: Swedo et al., 1989.

do effective OCD drugs—like fluoxetine (Prozac), fluvoxamine (Luvox), and clomipramine (Anafranil)—tend to have in common? They share the ability to inhibit the reuptake of serotonin at serotonergic synapses, thereby increasing the synaptic availability of serotonin. This observation suggests that the dysfunction of serotonergic neurotransmission plays a central role in OCD. Recall that we already discussed SSRIs like Prozac that inhibit the reuptake of serotonin when we discussed treatments for depression. How can the same drug help two disorders that seem so different? For one thing, depression often accompanies OCD, so the two disorders may be related. Furthermore, functional brain imaging suggests that the same SSRI drugs alter the activity of the orbitofrontal prefrontal cortex in people with OCD (Saxena et al., 2001) while affecting primarily ventrolateral prefrontal cortex in people with depression.

Many researchers also believe that OCD and another disease involving repetitive behaviors, Tourette's syndrome, are part of a spectrum of related disorders (Olson, 2004). OCD and Tourette's are often comorbid (occurring together), and both disorders involve abnormalities of the basal ganglia. However, drug therapy in Tourette's syndrome has typically focused on modifying the actions of dopamine rather than of serotonin (**BOX 12.2**).

There is a heritable genetic component to OCD; as with schizophrenia and depression, several genes may contribute to susceptibility to this disorder (Grados et al., 2003; Hall et al., 2003), including the gene encoding the serotonin 5-HT$_{2A}$ receptor (Enoch et al., 1998). There is also evidence that OCD can be triggered by infections. Upon observing that numerous children exhibiting OCD symptoms had recently been treated for strep throat, Dale et al. (2005) found that many children with OCD are producing antibodies to brain proteins. Perhaps, in mounting an immune response to

BOX 12.2 Tics, Twitches, and Snorts: The Unusual Character of Tourette's Syndrome

Their faces twitch in an insistent way, and every now and then, out of nowhere, they blurt out an odd sound. At times they fling their arms, kick their legs, or make violent shoulder movements. Sufferers of **Tourette's syndrome** are also supersensitive to tactile, auditory, and visual stimuli (A. J. Cohen and Leckman, 1992). Many patients report that an urge to emit verbal or phonic tics builds up and that giving in to the urge brings relief. Although popular media often portray people with Tourette's as shouting out insults and profanities (a symptom called *coprolalia*), verbal tics of that sort are rare.

Tourette's syndrome begins early in life; the mean age of diagnosis is 6–7 years (De Groot et al., 1995), and the syndrome is 3–4 times more common in males than in females. **FIGURE A** draws a portrait of the chronology of symptoms. Often people with Tourette's also exhibit attention deficit hyperactivity disorder (ADHD) or obsessive-compulsive disorder (OCD) (Park et al., 1993). Children with Tourette's display

a thinning of primary somatosensory and motor cortex representing facial, oral, and laryngeal structures (Sowell et al., 2008), suggesting that the tics mediated by these regions may be underinhibited by cortex.

Family studies indicate that genetics plays an important role in this disorder. Twin studies of the disorder reveal a concordance rate among monozygotic twins of 53–77%, contrasted with a concordance rate among dizygotic twins of 8–23% (T. M. Hyde et al., 1992). Among discordant monozygotic twin pairs, the twin with Tourette's has a greater density of dopamine D_2 receptors in the caudate nucleus of the basal ganglia than the unaffected twin has (Wolf et al., 1996). This observation suggests that differences in the dopaminergic system, especially in the basal ganglia, may be important (D_2 receptor binding in an affected twin is illustrated in **FIGURE B**). The contemporary view is that Tourette's syndrome is mediated in a complex manner by more than one gene, but the precise genes that

are involved remain unidentified (Abelson et al., 2005).

Treatment with haloperidol, a dopamine D_2 receptor antagonist that is better known as a typical antipsychotic drug, significantly reduces tic frequency and is a primary treatment for Tourette's syndrome. Unfortunately, this treatment can have unpleasant side effects (as noted in Box 12.1), but some people with Tourette's also respond well to the atypical antipsychotics, which may bring fewer side effects. Behavior modification techniques aimed at reducing the frequency of symptoms, especially tics, help some patients learn how to replace their obvious tics with behaviors that are more subtle and socially acceptable (Himle et al., 2006).

Deep brain stimulation (DBS), which we mentioned in the text as a treatment for depression, may also benefit people with Tourette's. In this case, battery-powered stimulating electrodes are aimed bilaterally at targets within the thalamus, in regions associated with the control of movement. Activation of the electrodes can bring dramatic and almost immediate relief from symptoms, and a postsurgical group of 15 people reported that their symptoms remained reduced 24 months later (Porta et al., 2009). (Figure B courtesy of Steven Wolf.)

(A) The chronology of Tourette's symptoms

Motor tics
- Eyes, face, head
- Shoulder, neck
- Arms, hands
- Trunk
- Legs

Vocal tics
- Low noises
- Loud noises
- Stuttering
- Repetition
- Obscenities
- Syllables
- Blocking
- Words out of context
- Repeating others

Compulsive actions
- Head banging
- Kissing
- Touching objects
- Kicking
- Tapping
- Touching self or others
- Biting self
- Touching sexual organs
- Mimicking others

Mean age at onset (years)

(B) D_2 receptor binding in Tourette's syndrome

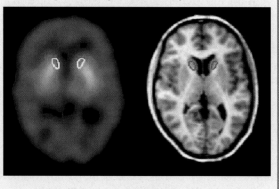

These MRIs show the brain of a patient who underwent a cingulotomy—the disruption of cingulate cortex connections—in an attempt to treat OCD.

(A) Horizontal view

(B) Sagittal view

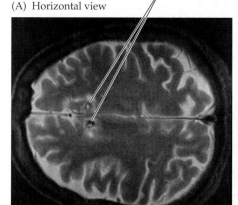
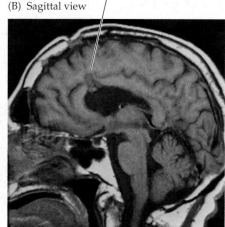

FIGURE 12.20 Neurosurgery to Treat Obsessive-Compulsive Disorder (From Martuza et al., 1990, courtesy of Robert L. Martuza.)

the streptococcal bacteria, these children also make antibodies that attack their own brains. The genetic link may be that some people are more likely than others to produce antibodies to the brain proteins.

For people with OCD that does not respond to any other treatments, psychosurgery may be a treatment of last resort. Unlike lobotomies, these surgeries target much smaller regions. In one study, about one-third of severely disabled OCD patients who underwent cingulotomy (making lesions that interrupt pathways in the cingulate cortex) (**FIGURE 12.20**) benefited substantially (S. Rasmussen et al., 2000; Shah et al., 2008). Other types of OCD psychosurgery place small lesions in the white matter of the anterior part of the brain, or place DBS electrodes in the region of the basal ganglia and nucleus accumbens, with good effect (Pepper et al., 2015). Frontal lobotomy, which causes much more extensive damage to the brain, is virtually never performed today. So while psychosurgery remains a viable option in severe psychiatric disturbance, we can be pretty confident that no one else will suffer the fate of Howard Dully, lobotomized for being a teenager.

Tourette's syndrome A disorder involving heightened sensitivity to sensory stimuli that may be accompanied by verbal or physical tics.

how's it going ❓

1. What are the main types of anxiety disorders?
2. What class of drugs is the most common anxiolytic, and what effect do these drugs have on transmitter systems?
3. Describe PTSD and the hypothesis that the disorder is a special case of fear conditioning.
4. What is OCD, and what treatments are available to combat it?

Recommended Reading

Charney, D. S., Buxbaum, J. D., Sklar, P., and Nestler, E. J. (Eds.). (2013). *Neurobiology of Mental Illness* (4th ed.). New York, NY: Oxford University Press.

Hersen, M., Beidel, D. C., and McCarthy, G. (2012). *Adult Psychopathology and Diagnosis* (6th ed.). New York, NY: Wiley.

Huettel, S. A., Song, A. W., and McCarthy, G. (2014). *Functional Magnetic Resonance Imaging* (3rd ed.). Sunderland, MA: Sinauer.

Martino, D., and Leckman, J. F. (Eds.). (2013). *Tourette Syndrome.* Oxford, England: Oxford University Press.

Meyer, J. S., and Quenzer, L. F. (2013). *Psychopharmacology: Drugs, the Brain, and Behavior* (2nd ed.). Sunderland, MA: Sinauer.

Solomon, A. (2001). *The Noonday Demon: An Atlas of Depression.* New York, NY: Scribner.

Steketee, G. (Ed.). (2011). *The Oxford Handbook of Obsessive Compulsive and Spectrum Disorders.* New York, NY: Oxford University Press.

VISUAL SUMMARY

12

You should be able to relate each summary to the adjacent illustration, including structures and processes. Go to the online version of this summary (scan the QR code above) for links to figures, animations, and activities that will help you consolidate the material.

1 Population studies find that psychiatric disorders are prevalent in modern society. Studies of families, twins, and adoptees demonstrate a strong role of genetic factors in schizophrenia. Rather than a single gene determining whether a person will develop **schizophrenia**, several genes contribute to the risk. Review **Figures 12.1** and **12.2**, **Table 12.1, Animation 12.2**

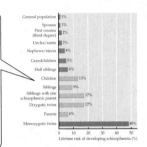

2 Structural changes in the brains of patients with schizophrenia—including enlarged ventricles—may arise from early developmental problems. The emergence of schizophrenia depends on the interaction of genes that make a person vulnerable to environmental stressors. Review **Figures 12.3–12.8**

3 The frontal lobes are less active in people with schizophrenia than in people without it. Biochemical theories of schizophrenia emphasize the importance of the dopamine, glutamate, and serotonin receptors. **Typical antipsychotics** block dopamine D_2 receptors, while atypical **antipsychotics** block serotonin 5-HT_{2A} receptors in addition to acting on dopamine receptors. Review **Figures 12.9–12.12, Box 12.1, Video 12.3**

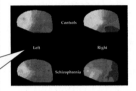

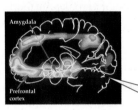

4 **Depression** also has a strong genetic factor. Serotonin has been implicated in this disorder. In general, females are more likely than males to suffer from depression. People suffering from depression show increased activity in the frontal cortex and the amygdala, as well as disrupted sleep patterns. Review **Figures 12.13–12.16**

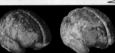

5 The most effective treatment for most cases of depression is a combination of **cognitive behavioral therapy (CBT)** and a **selective serotonin reuptake inhibitor (SSRI)**. Review **Table 12.2**

6 **Bipolar disorder** is characterized by extreme mood swings and subtle changes in the brain, and it has a complex genetic component. The disorder is commonly treated with **lithium**. Review **Figure 12.17**

7 Anxiety states are characterized by functional changes in the temporal lobes, particularly the amygdala. **Benzodiazepine** antianxiety drugs **(anxiolytics)** enhance the inhibitory effects of receptors for the transmitter GABA. Drugs affecting serotonergic synapses may also reduce anxiety. Review **Figure 12.18**

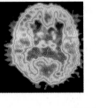

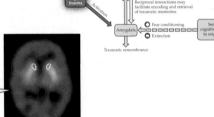

8 **Posttraumatic stress disorder (PTSD)** is characterized by an inability to forget horrible experiences. Temporal lobe atrophy in this disorder may be caused by chronic exposure to stress hormones. Review **Figure 12.19**

9 **Obsessive-compulsive disorder (OCD)** is characterized by changes in basal ganglia and frontal cortex and linked to serotonin. A restricted type of neurosurgery is sometimes used to treat the most severe cases of anxiety disorders. In **Tourette's syndrome**, overstimulation of dopamine receptors induces motor and verbal tics and compulsions. Review **Figure 12.20, Table 12.3, Box 12.2, Activity 12.1**

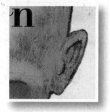

chapter
13

Memory, Learning, and Development

Trapped in the Eternal Now

Every day is alone in itself, whatever enjoyment I've had, and whatever sorrow I've had… Right now, I'm wondering, have I done or said anything amiss? You see, at this moment everything looks clear to me, but what happened just before? That's what worries me. It's like waking from a dream. I just don't remember.

(B. Milner, 1970, p. 37)

Henry Molaison, known as "patient H.M." in a classic series of research articles, was probably the most famous research participant in the history of brain science. Henry started to suffer seizures during adolescence, and by his late twenties, in 1953, his epilepsy was out of control. Because tests showed that Henry's seizures began in both temporal lobes, a neurosurgeon removed most of the anterior temporal lobe on both sides.

After the operation, Henry's seizures were milder, and they could be controlled by medication. But this relief came at a terrible, unforeseen price: He couldn't seem to form new memories (Scoville and Milner, 1957). For more than 50 years after the surgery, until his death in 2008, Henry could retain any new fact only briefly; as soon as he was distracted, the newly acquired information vanished. He didn't know his age or the current date, or that his parents

(with whom he lived well into adulthood) had died years previously. Henry knew that something was wrong with him, because he had no memories from the years since his surgery, or even memories from earlier the same day.

Henry's inability to form new memories meant that he couldn't have a lasting relationship with anybody new. No matter what experiences he might share with someone he met, Henry would have to start the acquaintance anew the following day, because he would have no recollection of ever meeting the person before. In some ways, this dreadful loss of memory ended Henry's journey as a human being—he could no longer grow in his experience of historical events, friendships, or even a sense of his own life story.

What happened to Henry, and what does his experience teach us about learning and memory?

To see the video
Memory,
go to

2e.mindsmachine.com/13.1

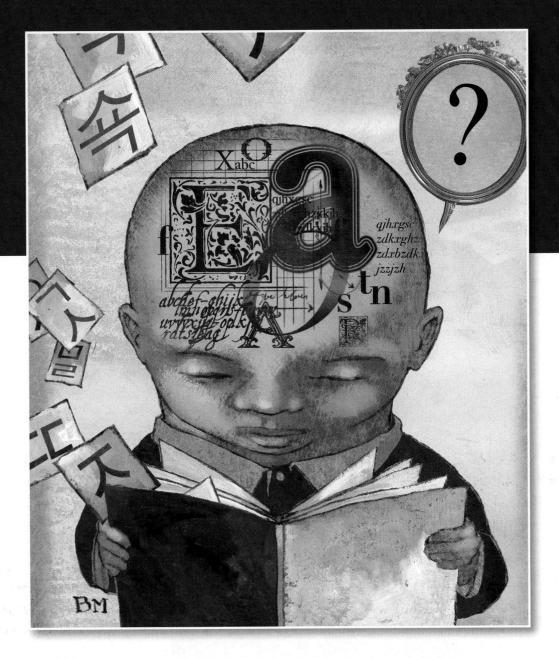

All the distinctively human aspects of our behavior are learned: the languages we speak, how we dress, the foods we eat and how we eat them, our skills, and the ways we reach our goals. So much of our own individuality depends on **learning**, the process of acquiring new information, and on **memory**, the ability to store and retrieve that information. We begin this chapter with a discussion of memory because research in the twentieth century revealed that there are fundamentally different types of memory. In Part II we delve into what we know about how learning alters the structure of the brain, which differs for different types of memory. Part III takes up the fascinating story of brain development, the stupendous rate at which our brains grow, and how we can learn and remember the masses of information that make us into unique human beings.

To view the
Brain Explorer,
go to

2e.mindsmachine.com/13.2

learning   The process of acquiring new and relatively enduring information, behavior patterns, or abilities, characterized by modifications of behavior as a result of practice, study, or experience.

PART I
Types of Learning and Memory

We can discover a great deal about **learning** and **memory** by examining how they fail. Clinical case studies teach us that memory can fail in very different ways, revealing different forms of learning and memory. The clinical research also guides studies with animal models and brain-imaging technology; together, these diverse approaches are providing a comprehensive picture of the brain's mechanisms of learning and memory.

There Are Several Kinds of Learning and Memory

The terms *learning*, the process of acquiring new information, and *memory*, the ability to store and retrieve that information, are so often paired that it sometimes seems as if one necessarily implies the other. We cannot be sure that learning has occurred unless a memory can be elicited later. Many kinds of brain damage, caused by disease or accident, impair learning and memory. We'll start by looking at some brain damage cases that revealed different classes of learning and memory.

For patient H.M., the present vanished into oblivion

Amnesia (Greek for "forgetfulness") is a severe impairment of memory, usually as a result of accident or disease. Loss of memories that formed prior to an event (such as surgery or trauma)—called **retrograde amnesia** (from the Latin *retro*, "backward," and *gradi*, "to go")—is not uncommon. After an accident that damages the brain, people often have retrograde amnesia with regard to events that happened a few hours or days before the accident, or even a year before, but despite dramatic depictions you may see on TV, it is unlikely that longer-term (or "complete") retrograde memory loss has ever occurred.

FIGURE 13.1   Brain Regions Crucial for Forming New Memories
(From Corkin et al., 1997, scans courtesy of Suzanne Corkin.)

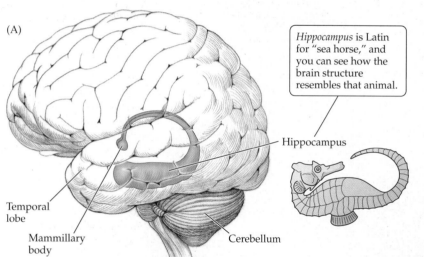

Hippocampus is Latin for "sea horse," and you can see how the brain structure resembles that animal.

As a young man, Henry Molaison (patient H.M.) had the hippocampus (H) from both hemispheres surgically removed, with disastrous results. The surgeons did not remove the cerebellum (Cer), but the loss of hippocampal inputs caused it to shrink as he aged.

Patient H.M.—Henry Molaison, whom we met at the start of the chapter—suffered from a far more unusual symptom. In Henry's case, most old memories remained intact, but he had difficulty recollecting any events that took place *after* his surgery. What's more, he was unable to retain any new material for more than a brief period. The inability to form new memories after an event is called **anterograde amnesia** (the Latin *antero* means "forward").

Over the very short term, Henry's memory was normal. If given a series of six or seven digits, he could immediately repeat the list without error. But if he was given a list of words to study and then tested on them after being distracted by another task, he could not repeat the list or even recall that there *was* a list. So Henry's case provided clear evidence that *short-term memory* differs from *long-term memory*—a distinction, long recognized by psychologists on behavioral grounds (W. James, 1890), that we will discuss in more depth later in this chapter.

Henry's surgery removed the amygdala, most of the hippocampus, and surrounding cortex from both temporal lobes (**FIGURE 13.1**). The memory deficit seemed to be caused by loss of the *medial temporal lobe*, including the **hippocampus**, because surgical patients who had only the lateral temporal cortex removed had no memory impairment. Despite his obvious memory problems, Henry showed noticeable improvement over days of practice on a mirror-tracing task (**FIGURE 13.2A**) (B. Milner, 1965). Each day, when asked if he remembered the test, Henry said no, yet his performance was better than at the start of the first day (**FIGURE 13.2B**). So, was Henry's memory loss limited to tasks that relied on verbal processing? Not quite. For example, patients with amnesia like Henry's can learn the skill of *reading* mirror-reversed text (**FIGURE 13.3**), which is a verbal task.

Henry Molaison Henry was a young man when he had the surgery that disrupted his memory, as described in the famous studies of "patient H.M." (Photo courtesy of Dr. Suzanne Corkin, MIT.)

memory 1. The ability to learn and neurally encode information, consolidate the information for longer term storage, and retrieve or reactivate the consolidated memory at a later time. 2. The specific information that is stored in the brain.

amnesia Severe impairment of memory.

retrograde amnesia Difficulty in retrieving memories formed before the onset of amnesia.

patient H.M. The late Henry Molaison, a patient who was unable to encode new declarative memories because of surgical removal of medial temporal lobe structures.

anterograde amnesia Difficulty in forming new memories beginning with the onset of a disorder.

hippocampus A medial temporal lobe structure that is important for learning and memory.

(A)

Henry was given this mirror-tracing task to test motor skill.

bi lateral medial temporal lobe surgery stopped epilepsy could not create new memories

consolidation - store memory

(B)

Although Henry never recognized the task, his performance progressively improved over successive days, demonstrating a type of long-term memory.

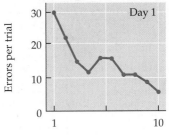
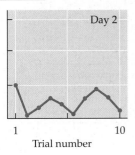
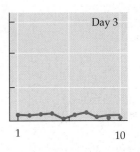

FIGURE 13.2 Henry's Performance on a Mirror-Tracing Task (After B. Milner, 1965.)

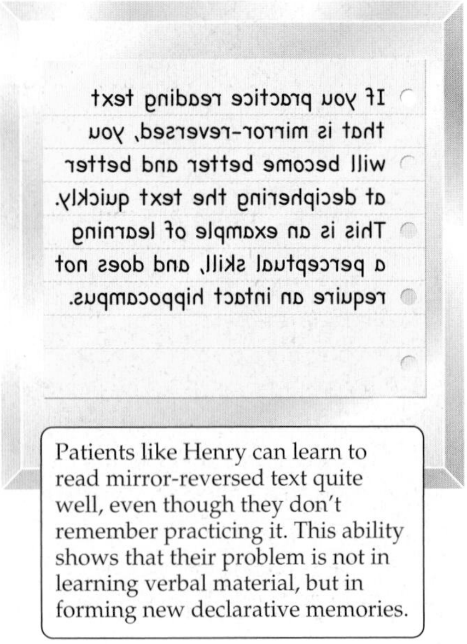

Long-term memory

Declarative: Things you know that you can tell others

Nondeclarative (procedural): Things you know that you can show by doing

Declarative memory can be tested readily in humans because they can talk. Henry Molaison was unable to form new declarative memories, indicating that the hippocampus is needed to form these memories, but that they must be stored elsewhere.

Nondeclarative memory can be tested readily in other animals, as well as in humans. Henry could form this type of memory, as when he learned the skill of mirror tracing.

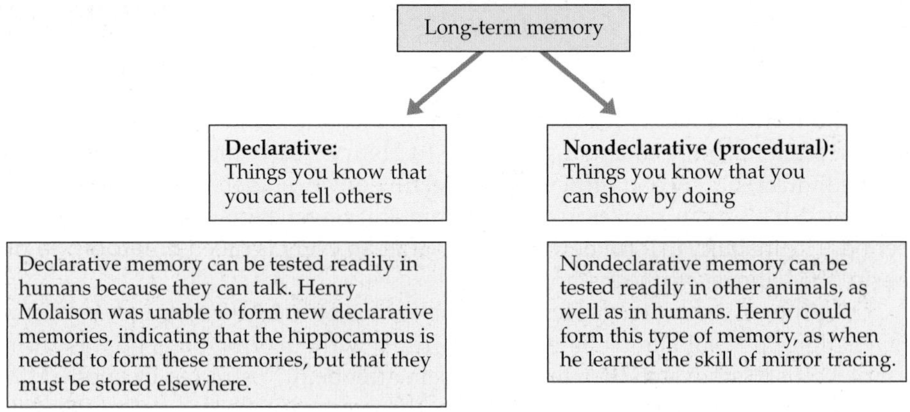

FIGURE 13.4   Two Main Kinds of Memory: Declarative and Nondeclarative

If you practice reading text that is mirror-reversed, you will become better and better at deciphering the text quickly. This is an example of learning a perceptual skill, and does not require an intact hippocampus.

Patients like Henry can learn to read mirror-reversed text quite well, even though they don't remember practicing it. This ability shows that their problem is not in learning verbal material, but in forming new declarative memories.

FIGURE 13.3   Reading Mirror-Reversed Text

D memory — stated or said

Nan D memory — shewn/skill performed

declarative memory   A memory that can be stated or described.

nondeclarative memory   Also called *procedural memory*. A memory that is shown by performance rather than by conscious recollection.

delayed non-matching-to-sample task   A test in which the subject must respond to the unfamiliar stimulus in a pair of stimuli.

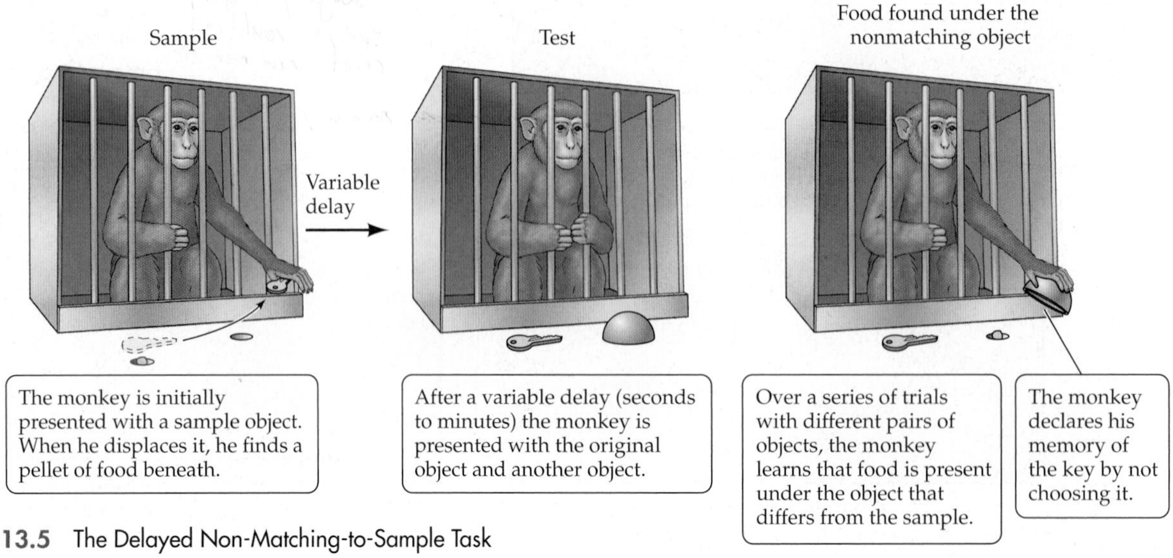

Subsequent research showed that the important distinction is not between motor and verbal performances, but rather between two general categories of memory:

1. **Declarative memory** is what we usually think of as memory: facts and information acquired through learning. It is memory we are aware of accessing, which we can *declare* to others. This is the type of memory that was so profoundly impaired by Henry's surgery. Tests of declarative memory take the form of requests for specific information that was learned previously, such as a story or word list. It is the type of memory we use to answer "what" questions—and thus is difficult to test in animals.

2. **Nondeclarative memory**, or *procedural memory*—that is, memory about perceptual or motor procedures—is shown by *performance* rather than by conscious recollection. Examples of procedural memory include learning the mirror-tracing task, at which Henry excelled, and the skill of mirror reading, as well as more familiar talents like riding a bike. It is the type of memory we use for "how" problems and is often (but not always) nonverbal. **FIGURE 13.4** illustrates this basic division in memory type.

Once researchers caught on to this distinction, they devised a clever way to measure declarative memory in monkeys. The **delayed non-matching-to-sample task** (**FIGURE 13.5**) is a test of *object recognition* memory that requires monkeys to declare what they remember. In this task the monkeys must identify which of two objects was *not* seen previously, with delays ranging from 8 seconds to 2 minutes (Spiegler and Mishkin, 1981). Monkeys that have extensive damage to the medial temporal lobe, and thus are similar to Henry, are severely impaired on this task, especially with the longer delays.

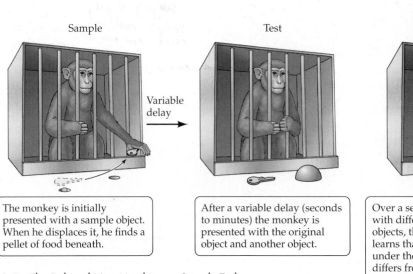

Sample

Variable delay

Test

Food found under the nonmatching object

The monkey is initially presented with a sample object. When he displaces it, he finds a pellet of food beneath.

After a variable delay (seconds to minutes) the monkey is presented with the original object and another object.

Over a series of trials with different pairs of objects, the monkey learns that food is present under the object that differs from the sample.

The monkey declares his memory of the key by not choosing it.

FIGURE 13.5   The Delayed Non-Matching-to-Sample Task

researchers at work

Which brain structures are important for declarative memory?

To determine which parts of the temporal lobe are crucial for declarative memory, researchers selectively removed specific parts of the medial temporal lobes of monkeys to confirm that the amygdala—one of the structures removed in Henry's surgery—is not crucial for performance on tests of declarative memory. However, removal of the adjacent hippocampus significantly impaired performance on these tests and, as shown in **FIGURE 13.6**, the deficit was even more pronounced when the hippocampal damage was paired with lesions of nearby cortical regions that communicate with the hippocampus: entorhinal, parahippocampal, and perirhinal cortices

(Zola-Morgan et al., 1994). Human patients similarly show larger impairments when both the hippocampus and medial temporal cortex are damaged (Zola-Morgan and Squire, 1986; Rempel-Clower et al., 1996). So Henry's symptoms were probably caused by loss of the medial temporal lobe on both sides of the brain.

The experiments with monkeys, together with Henry's case, indicate that we need at least one intact medial temporal lobe (including the hippocampus) in order to make new declarative memories. But the hippocampus isn't the only brain structure needed for new declarative memories, as we'll see next.

FIGURE 13.6 Memory Performance after Medial Temporal Lobe Lesions

■ **Hypothesis**
Particular portions of the medial temporal lobe are required for the formation of new declarative memories.

■ **Test**
Selectively remove different portions of the temporal lobe from both sides of the brain, and test for declarative memories using the delayed non-matching-to-sample task (see Figure 13.5).

■ **Result**

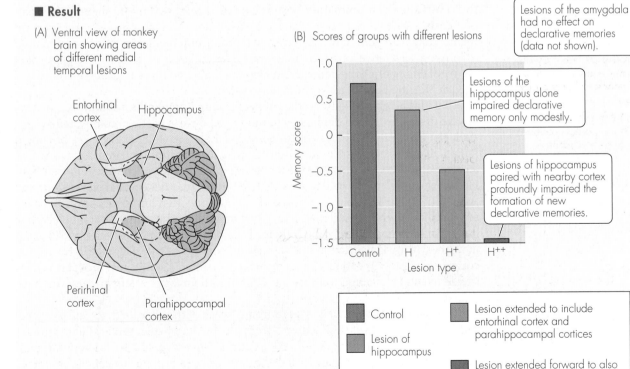

(A) Ventral view of monkey brain showing areas of different medial temporal lesions

Entorhinal cortex   Hippocampus

Perirhinal cortex   Parahippocampal cortex

(B) Scores of groups with different lesions

Lesions of the amygdala had no effect on declarative memories (data not shown).

Lesions of the hippocampus alone impaired declarative memory only modestly.

Lesions of hippocampus paired with nearby cortex profoundly impaired the formation of new declarative memories.

Memory score — Lesion type: Control, H, H⁺, H⁺⁺

■ Control
■ Lesion of hippocampus
■ Lesion extended to include entorhinal cortex and parahippocampal cortices
■ Lesion extended forward to also include the anterior entorhinal and perirhinal cortices

■ **Conclusion**
The severe disruption of new declarative memories in Henry Molaison and patients like him is due to damage to both the hippocampus itself and to nearby cortex.

FIGURE 13.7 The Brain Damage in Patient N.A. (From Squire et al., 1989, MRI scans courtesy of Larry Squire.)

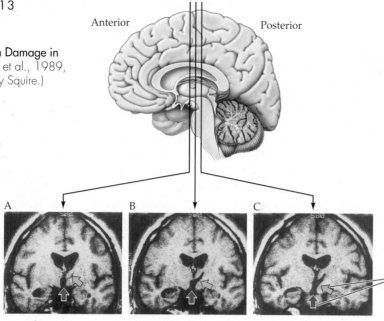

MRI scans show a prominent lesion of the dorsomedial thalamus on the left side of the brain (yellow arrows), as well as a lesion on the floor of the third ventricle where the mammillary bodies should be (red arrows).

patient N.A. A still-living patient who is unable to encode new declarative memories, because of damage to the dorsomedial thalamus and the mammillary bodies.

dorsomedial thalamus A limbic system structure that is connected to the hippocampus.

mammillary body One of a pair of limbic system structures that are connected to the hippocampus.

Korsakoff's syndrome A memory disorder, caused by thiamine deficiency, that is generally associated with chronic alcoholism.

confabulate To fill in a gap in memory with a falsification. Confabulation is often seen in Korsakoff's syndrome.

Damage to the mammillary bodies can also cause amnesia

In 1960, a young man now known to brain researchers as **patient N.A.** had a bizarre accident in which a miniature sword entered his nostril and injured his brain. Like Henry, N.A. has shown profound anterograde amnesia ever since his accident, primarily for verbal material (Squire and Moore, 1979), and he can give little information about events since his accident, although his memory for earlier events is near normal (Kaushall et al., 1981). MRI study of N.A. (**FIGURE 13.7**) shows damage to several limbic system structures that have connections to the hippocampus: the **dorsomedial thalamus** and the **mammillary bodies** (so called because they are shaped like a pair of breasts—see Figure 13.1A). Like Henry Molaison, N.A. shows normal short-term memory and can gain new nondeclarative/procedural memories, but he is impaired in forming declarative long-term memories. The similarity in symptoms suggests that the medial temporal lobe damaged in Henry's brain and these midline regions damaged in N.A. are parts of a larger memory system.

That idea is reinforced by studies of people with **Korsakoff's syndrome**, a degenerative disease in which damage is found in some parts of the limbic system, especially the mammillary bodies and dorsomedial thalamus (Mair et al., 1979), but not in temporal lobe structures like the hippocampus. The mammillary bodies may serve as a processing system connecting the medial temporal lobes (which were removed from Henry Molaison) to the thalamus and, from there, to other cortical sites (Vann and Aggleton, 2004). People with Korsakoff's syndrome often fail to recognize or sense any familiarity with some items, even when presented repeatedly, yet frequently they deny that anything is wrong with them. They often **confabulate**—that is, fill a gap in memory with a falsification that they seem to accept as true. Damage to the frontal cortex, also found in patients suffering from Korsakoff's syndrome, probably causes the denial and confabulation that differentiates them from other patients who have amnesia, such as Henry.

The main cause of Korsakoff's syndrome is lack of the vitamin thiamine. Alcoholics who obtain most of their calories from alcohol and neglect their diet often suffer this deficiency. Treating them with thiamine can prevent further deterioration of memory functions but will not reverse the damage already done. If alcoholic beverages were supplemented with thiamine, then many new cases of Korsakoff's syndrome might be prevented (Centerwall and Criqui, 1978), but political groups oppose this measure for fear that it might encourage drinking.

These studies make it clear that a brain circuit that includes the hippocampus, the mammillary bodies, and the dorsomedial thalamus is needed to *form* new declarative memories. But these case studies also clearly show that established declarative memories, formed before brain damage, are not *stored* in these structures. If the memories had been stored there, they would have been lost when the structures were damaged. So where are memories stored? We'll see next that a leading candidate is the cerebral cortex.

Brain damage can destroy autobiographical memories while sparing general memories

One striking case study suggests that at least some declarative memories are stored in the cortex, and it also illustrates an important distinction between two *subtypes* of declarative memory. Kent Cochrane, known to the world as **patient K.C.**, suffered brain damage in a motorcycle accident at age 30. He could no longer retrieve any *personal memory* of his past, although his general knowledge remained good. He conversed easily and played a good game of chess but could not remember where he learned to play chess or who taught him the game. Detailed autobiographical declarative memory of this sort is known as **episodic memory**; you show episodic memory when you recall a specific *episode* in your life or relate an event to a particular time and place. In contrast, **semantic memory** is generalized declarative memory, such as knowing the meaning of a word without knowing where or when you learned that word (Tulving, 1972). If care was taken to space out the trials, Kent could acquire new semantic knowledge (Tulving et al., 1991). But even with this method, Kent could not acquire new *episodic* knowledge—he wouldn't remember where he learned that new material.

Brain scans of Kent revealed extensive damage to the left frontoparietal and the right parieto-occipital cerebral cortex, as well as severe shrinkage of both right and left hippocampus and nearby cortex (Rosenbaum et al., 2005). As with Henry, the bilateral hippocampal damage probably accounts for Kent's anterograde declarative amnesia. But that damage cannot account for Kent's selective loss of nearly all his autobiographical memory, because other patients with damage restricted to the medial temporal lobe lack this symptom. Kent's inability to recall any autobiographical details of his life from many years before his accident may instead be a consequence of injuries to frontal and parietal cortex (Tulving, 1989). (Unlike dramatic portrayals of retrograde amnesia in fiction, Kent knew his name and recognized his family, although he couldn't remember any particular events with those people.)

Brain-imaging studies in volunteers confirm the distinction between semantic and episodic declarative memories, exemplified by Kent. The volunteers show greater activation of right frontal and temporal lobe regions when listening to passages about themselves than when listening to other people's stories (**FIGURE 13.8**). Thus, autobiographical memories and semantic memories appear to be processed in different locations and are probably stored in different parts of the cortex.

Patient K. C. Brain damage from a severe motorcycle accident left Kent Cochrane unable to retrieve episodic memories. He died in 2014 at age 62 (Branswell, 2014). (Photograph from Tulving, 2002.)

patient K.C. The late Kent Cochrane, a patient who sustained damage to the cortex that rendered him unable to form and retrieve episodic memories.

episodic memory Also called *autobiographical memory*. Memory of a particular incident or a particular time and place.

semantic memory Generalized declarative memory, such as knowing the meaning of a word.

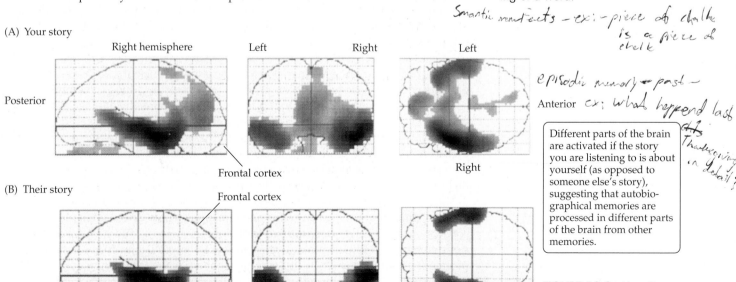

(A) Your story

Right hemisphere   Left   Right   Left

Posterior

Frontal cortex

Anterior

Right

(B) Their story

Frontal cortex

Sagittal section   Coronal section   Horizontal section

Different parts of the brain are activated if the story you are listening to is about yourself (as opposed to someone else's story), suggesting that autobiographical memories are processed in different parts of the brain from other memories.

FIGURE 13.8 Your Story versus Their Story (From G. R. Fink et al., 1996.)

Skill learning—① bat ready ② bat begging motion ③ bat mid motion ④ bat end motion

Priming—exposure to a stimulus
~~for~~ to be picked up and
~~learned~~ learned
ex: music lyrics

how's it going ❓

1. What are the two main types of amnesia, and which type did Henry Molaison exhibit?
2. What are the two main types of memory, and which type was affected in Henry?
3. How did research with animals help pin down the brain regions required for forming new memories that we can declare to others?
4. What are the two main types of declarative memory?

Different Forms of Nondeclarative Memory Involve Different Brain Regions

So far, we've seen that there are two different kinds of declarative memory: semantic and episodic. There are also different types of nondeclarative (procedural) memory, and we'll see that different brain regions are involved in these different forms.

Different types of nondeclarative memory serve varying functions

Skill learning is the process of learning how to perform a challenging task simply by repeating it over and over. Improving at the mirror-tracing task performed by Henry Molaison (see Figure 13.2) or learning to read mirror-reversed text (see Figure 13.3) are examples of skill learning. So, too, is the acquisition of everyday skills like learning to ride a bike or to juggle (well, okay, maybe juggling isn't an "everyday" skill, but you get the idea).

Imaging studies have investigated learning and memory for different kinds of skills, including *sensorimotor skills* (e.g., mirror tracing), *perceptual skills* (e.g., reading mirror-reversed text), and *cognitive skills* (tasks involving planning and problem solving, common in puzzles like the Tower of Hanoi problem, which you can play in Activity 13.2, on the website). All three kinds of skill learning are impaired in people with damage to the **basal ganglia** (see Figure 2.14A). Damage to other brain regions, especially the motor cortex and cerebellum, also affects aspects of some skills. Neuroimaging studies confirm that the basal ganglia, cerebellum, and motor cortex are important for sensorimotor skill learning (Grafton et al., 1992; Hazeltine et al., 1997).

Priming (or *repetition priming*) is a change in the way you process a stimulus, usually a word or a picture, because you've seen it, or something similar, previously. For example, if a person is shown the word *stamp* in a list and later is asked to complete the word stem *STA-*, then she is more likely to reply "stamp" than, say, "start." Priming does not require declarative memory of the stimulus—Henry Molaison and other patients with amnesia have shown priming for words they don't remember having seen. In contrast with skill learning, priming is not impaired by damage to the basal ganglia. In functional-imaging studies, perceptual priming (priming based on the visual *form* of words) is related to reduced activity in bilateral occipitotemporal cortex (Schacter et al., 1996), while conceptual priming (priming based on word *meaning*) is associated with reduced activation of the left frontal cortex (A. D. Wagner et al., 1997). So priming appears to be at least partly a function of the cortex.

Learning that involves relations between events—for example, between two or more stimuli, between a stimulus and a response, or between a response and its consequence—is called **associative learning**. In the best-studied form (**classical conditioning**), an initially neutral stimulus comes to predict an event. In famous experiments, Ivan Pavlov (1849–1936) found that a dog would learn to salivate when presented with an auditory or visual stimulus if the stimulus came to predict the

---

skill learning   The process of learning to perform a challenging task simply by repeating it over and over.

basal ganglia   A group of forebrain nuclei, including the caudate nucleus, globus pallidus, and putamen, found deep within the cerebral hemispheres. They are crucial for skill learning.

priming   Also called *repetition priming*. The phenomenon by which exposure to a stimulus facilitates subsequent responses to the same or a similar stimulus.

associative learning   A type of learning in which an association is formed between two stimuli or between a stimulus and a response. It includes both classical and instrumental conditioning.

classical conditioning   Also called *Pavlovian conditioning*. A type of associative learning in which an originally neutral stimulus acquires the power to elicit a conditioned response when presented alone.

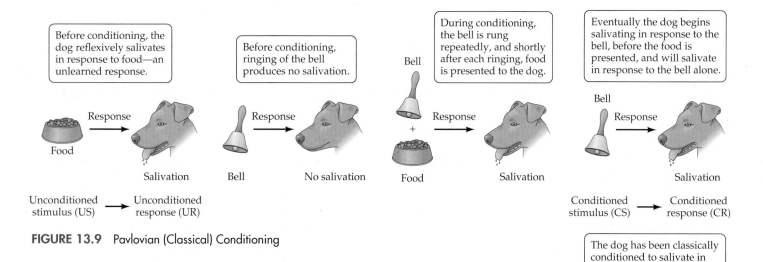

Before conditioning, the dog reflexively salivates in response to food—an unlearned response.

Food → Response → Salivation

Unconditioned stimulus (US) → Unconditioned response (UR)

Before conditioning, ringing of the bell produces no salivation.

Bell → Response → No salivation

During conditioning, the bell is rung repeatedly, and shortly after each ringing, food is presented to the dog.

Bell + Food → Response → Salivation

Eventually the dog begins salivating in response to the bell, before the food is presented, and will salivate in response to the bell alone.

Bell → Response → Salivation

Conditioned stimulus (CS) → Conditioned response (CR)

FIGURE 13.9 Pavlovian (Classical) Conditioning

The dog has been classically conditioned to salivate in response to the bell.

presentation of food. So, repeatedly ringing a bell before putting meat powder in a dog's mouth will eventually cause the dog to start salivating when it hears the bell alone. In this case the meat powder in the mouth is called the *unconditioned stimulus* (*US*), which already evokes an *unconditioned response* (*UR*; salivation in the example). The sound of the bell is called the *conditioned stimulus* (*CS*), and the learned response to the CS alone (salivation in response to the bell) is called the *conditioned response* (*CR*) (**FIGURE 13.9**).

Experimental evidence in lab animals shows that circuits in the **cerebellum** are crucial for simple eye-blink conditioning, in which a tone or other stimulus is associated with eye blinking in response to a puff of air. A PET study in humans confirmed this idea by showing a progressive increase in activity in the cerebellum during eye-blink conditioning (Logan and Grafton, 1995). Patients with hippocampal lesions can acquire the conditioned eye-blink response, but patients with damage to the cerebellum on one side can acquire a conditioned eye-blink response *only on the side where the cerebellum is intact* (Papka et al., 1994).

In **instrumental conditioning** (also called *operant conditioning*), an association is formed between the animal's behavior and the consequence(s) of that behavior. An example of an apparatus designed to study instrumental conditioning is called the *Skinner box*, named for its originator, B. F. Skinner (**FIGURE 13.10**). In a common setup, the animal learns that performing a certain action (e.g., pressing a bar) is followed by a reward (such as a food pellet). Research in animals has not pinpointed the brain regions that are crucial for instrumental conditioning, perhaps because this behavior taps so many different aspects of behavior.

Animal research confirms the various brain regions involved in different attributes of memory

The caricature of the white-coated biopsychologist watching rats run in mazes, a staple of cartoonists to this day, has its origins in the intensive memory research of the early twentieth century. The early work indicated that rats and other animals don't just learn a series of turns but instead form a **cognitive map** (an understanding of the relative spatial organization of objects and information) in order to solve a maze (Tolman, 1949). Animals apparently learn at least some of these details of their spatial environment simply by moving through it (Tolman and Honzik, 1930).

cerebellum A structure located at the back of the brain, dorsal to the pons, that is involved in the central regulation of movement, and in some forms of learning.

instrumental conditioning Also called *operant conditioning*. A form of associative learning in which the likelihood that an act (instrumental response) will be performed depends on the consequences (reinforcing stimuli) that follow it.

cognitive map A mental representation of the relative spatial organization of objects and information.

operant conditioning - act will occur depending on the consequence.

Inside this Skinner box, a rat presses a bar for food.

FIGURE 13.10 A Skinner Box

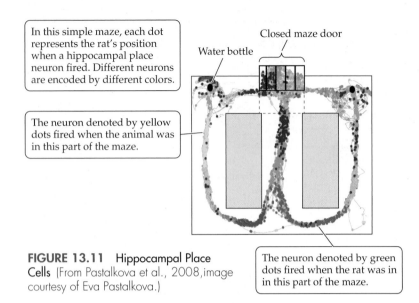

In this simple maze, each dot represents the rat's position when a hippocampal place neuron fired. Different neurons are encoded by different colors.

The neuron denoted by yellow dots fired when the animal was in this part of the maze.

Water bottle

Closed maze door

The neuron denoted by green dots fired when the rat was in in this part of the maze.

FIGURE 13.11 Hippocampal Place Cells (From Pastalkova et al., 2008, image courtesy of Eva Pastalkova.)

place cell A neuron in the hippocampus that selectively fires when the animal is in a particular location.

Place cells in the hippocampus that only fire in a particular location

We now know that, in parallel with its role in other types of declarative memory, the hippocampus is crucial for spatial learning. The rat hippocampus contains many neurons that selectively encode spatial location (O'Keefe and Dostrovsky, 1971; Leutgeb et al., 2005). These **place cells** become active when the animal is in—or moving toward—a particular location (**FIGURE 13.11**). If the animal is moved to a new environment, place cell activity indicates that the hippocampus remaps to the new locations (Moita et al., 2004). The Nobel Prize in Physiology or Medicine for 2014 was awarded to John O'Keefe and the wife-and-husband team of May-Britt and Edvard Moser for their work on understanding hippocampal place cells.

Bird species that hide food in many locations have a larger hippocampus than other birds have (Krebs et al., 1989), indicating that natural selection favors enlargement of the hippocampus to enhance spatial learning, as we discuss in **A STEP FURTHER 13.1**, on the website.

Brain regions involved in learning and memory: A summary

FIGURE 13.12 updates and summarizes the classification of long-term memory that we've been discussing. Several major conclusions should be apparent by now, especially (1) that many regions of the brain are involved in learning and memory; (2) that different forms of memory rely on at least partly different brain mechanisms, which may include several different regions of the brain; and (3) that the same brain structure can be a part of the circuitry for several different forms of learning. Next we'll discuss the stages by which memories, of any sort, can be preserved for a lifetime.

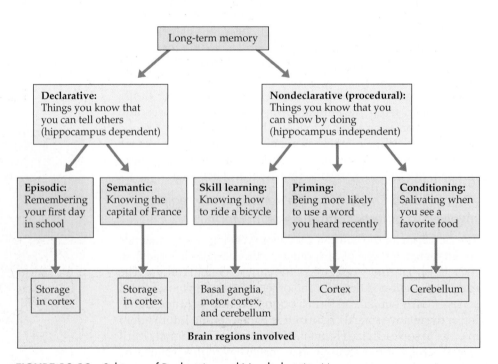

FIGURE 13.12 Subtypes of Declarative and Nondeclarative Memory

how's it going ❓

1. Name three different types of nondeclarative memories, and give an example of each. What different parts of the brain have been implicated in each type?
2. What is a cognitive map?
3. What are hippocampal place cells, and why do they suggest a role for the hippocampus in spatial learning?

Successive Processes Capture, Store, and Retrieve Information in the Brain

The span of time over which a piece of information is retained in the brain varies. There are at least three different stages of memory. The briefest memories are called **sensory buffers** (for the visual stimuli, they are sometimes called *iconic memories*); an example is the fleeting impression of a glimpsed scene that vanishes from memory seconds later. These brief memories are thought to be residual activity in sensory neurons.

Somewhat longer than sensory buffers are **short-term memories** (**STMs**). If you look up a phone number and keep it in mind (perhaps through rehearsal) just until you make the phone call, you are using STM. In the absence of rehearsal, STMs last only about 30 seconds (J. Brown, 1958; L. R. Peterson and Peterson, 1959). With rehearsal, you may be able to retain an STM until you turn to a new task a few minutes later; but when the STM is gone, it's gone for good. Many researchers now refer to this form of memory as *working memory*, in recognition of the way we use it. This is where we hold information while we're working with it to solve a problem or are otherwise actively manipulating the information. Eventually, some memories become really long-lasting—the address of your childhood home, how to ride a bike, your first love—and are called **long-term memories** (**LTMs**).

As shown in **FIGURE 13.13**, the memory system consists of at least three processes: (1) **encoding** of raw information from sensory channels into STM, (2) **consolidation** of the volatile STM into more-durable LTM, and (3) eventual **retrieval** of the stored

sensory buffer   A very brief type of memory that stores the sensory impression of a scene. In vision, it is sometimes called *iconic memory*.

short-term memory (STM)   Also called *working memory*. A form of memory that usually lasts only seconds, or as long as rehearsal continues.

long-term memory (LTM)   An enduring form of memory that lasts days, weeks, months, or years and has a very large capacity.

encoding   The first process in the memory system, in which the information entering sensory channels is passed into short-term memory.

consolidation   The second process in the memory system, in which information in short-term memory is transferred to long-term memory.

retrieval   The third process of the memory system, in which a stored memory is used by an organism.

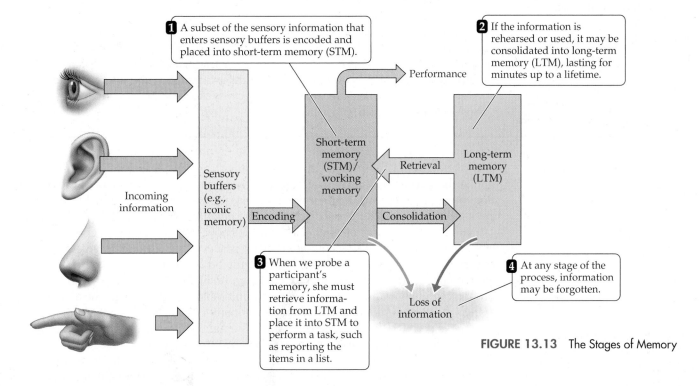

1 A subset of the sensory information that enters sensory buffers is encoded and placed into short-term memory (STM).

2 If the information is rehearsed or used, it may be consolidated into long-term memory (LTM), lasting for minutes up to a lifetime.

3 When we probe a participant's memory, she must retrieve information from LTM and place it into STM to perform a task, such as reporting the items in a list.

4 At any stage of the process, information may be forgotten.

Incoming information

Sensory buffers (e.g., iconic memory)

Encoding

Short-term memory (STM)/ working memory

Performance

Retrieval

Long-term memory (LTM)

Consolidation

Loss of information

FIGURE 13.13   The Stages of Memory

BOX 13.1 Emotions and Memory

Almost everyone knows from personal experience that strong emotions can affect memory formation and retrieval. Examples of memories enhanced in this way might include a strong association between special music and a first kiss, or uncomfortably vivid recollection of the morning of September 11, 2001. A large-scale research effort in many labs has identified a suite of biochemical agents that participate in the emotional enhancement of memory.

Epinephrine (adrenaline), released from the adrenal glands in large quantities during times of stress and strong emotion, appears to affect memory formation by influencing the amygdala, a brain region that is critical for fear conditioning (see Chapter 11). Electrical stimulation or lesions of the amygdala potently alter the memory-enhancing effects of epinephrine injections (Cahill and McGaugh, 1991), and tiny doses of epinephrine injected directly into the amygdala enhance memory formation in the same way that systemic injections do. This treatment appears to cause the release of norepinephrine within the amygdala, as do emotional experiences. Injecting propranolol, a blocker of beta-adrenergic receptors, into the amygdala blocks the memory-enhancing effects. In humans, the same drug can ease fears that have been conditioned in the lab (Kindt et al., 2009).

Can we develop pharmacological treatments to weaken or erase unwanted memories outside the lab? Some disorders would benefit greatly from such treatments. For example, people who have had life-threatening or other catastrophic

Flashbulb Memories Many people have vivid, detailed memories of where they were when they learned of the September 11, 2001, terrorist attack on the World Trade Center towers in New York City.

experiences often develop **posttraumatic stress disorder** (**PTSD**) (see Chapter 12), characterized as "reliving experiences such as intrusive thoughts, nightmares, dissociative flashbacks to elements of the original traumatic event, and … preoccupation with that event" (Keane, 1998, p. 398). In PTSD, each recurrence of the strong emotions and memories of the traumatic event may reactivate memories that, when reconsolidated in the presence of stress signals like epinephrine, become even stronger. Therefore, one strategy to prevent PTSD formation could be to block the effects of epinephrine in the amygdala by treating victims with antiadrenergic drugs either shortly before a traumatic

experience (e.g., in rescue workers) or as quickly as possible after it (in the case of victims of violence, for example) (Cahill, 1997). This treatment would not delete memories of the event but might diminish the traumatic aspects, and it might also be useful for weakening existing traumatic memories.

Perhaps one day it will be possible to selectively interfere with other neurotransmitters at work in the amygdala to provide more-specific and more-complete relief from traumatic memories. Whatever that treatment might be, it will probably have to be administered soon after the accident to effectively dull the painful memories.

information from LTM for use. A problem at any stage can cause us to lose information. Although not depicted in the figure, this model suggests that the flow of information into and out of working memory is supervised by another part of the brain, an *executive function*, which we will discuss in more detail in Chapter 14.

Not all memories are created equal. We all know from firsthand experience that emotion can powerfully affect our memory for past events. For example, an emotionally arousing story is remembered significantly better than a closely matched but emotionally neutral story (Reisberg and Heuer, 1995). But if people are treated with propranolol (a beta-adrenergic antagonist, or beta-blocker, that blocks the effects of epinephrine), this emotional enhancement of memory vanishes. It's not that treated volunteers perceive the story as being any less emotional; in fact, they rate the emotional content of the stories just the same as untreated people do. Instead, the drug seems to directly interfere with the ability of adrenal stress hormones to act on the brain to enhance memory (Cahill et al., 1994). **BOX 13.1** delves further into this topic.

postraumatic stress disorder A disorder in which memories of an unpleasant episode repeatedly plague the victim.

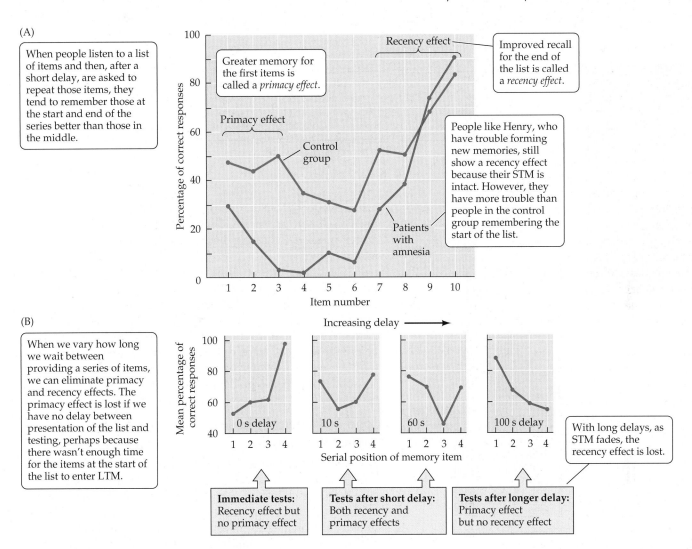

(A)

When people listen to a list of items and then, after a short delay, are asked to repeat those items, they tend to remember those at the start and end of the series better than those in the middle.

Greater memory for the first items is called a *primacy effect*.

Recency effect

Improved recall for the end of the list is called a *recency effect*.

Primacy effect

Control group

People like Henry, who have trouble forming new memories, still show a recency effect because their STM is intact. However, they have more trouble than people in the control group remembering the start of the list.

Patients with amnesia

Percentage of correct responses

Item number

(B)

When we vary how long we wait between providing a series of items, we can eliminate primacy and recency effects. The primacy effect is lost if we have no delay between presentation of the list and testing, perhaps because there wasn't enough time for the items at the start of the list to enter LTM.

Increasing delay ⟶

Mean percentage of correct responses

0 s delay   10 s   60 s   100 s delay

Serial position of memory item

With long delays, as STM fades, the recency effect is lost.

Immediate tests: Recency effect but no primacy effect

Tests after short delay: Both recency and primacy effects

Tests after longer delay: Primacy effect but no recency effect

FIGURE 13.14 Serial Position Curves (Part A after Baddeley and Warrington, 1970; B after A. A. Wright et al., 1985.)

STM and LTM appear to be different processes

Henry Molaison's case and the research it inspired have already told us several ways in which STM and LTM differ from one another. While the medial temporal lobe is not needed to encode sensory information into STM, or to retrieve that information from STM (Henry could repeat back to you a list of words or numbers), it is crucial for moving information from STM into LTM. In terms of the model, an intact hippocampus is required to *consolidate* declarative STMs into LTMs, indicating that the information is somehow transformed into a different format, one that may make it available for a lifetime.

Another indication that STM and LTM rely on different processes to store information comes from a classic demonstration of learning lists of words or numbers. If you hear a list of ten words and then try to repeat them back after a 30-second delay, you will probably do especially well with the earliest few words (this effect is termed a **primacy effect** because these words came first) and with the last few words (this effect is termed a **recency effect** because you heard these words most recently), compared with words in the middle of the list. The blue U-shaped curve in **FIGURE 13.14A**, called a *serial position* curve, shows typical results from such an experiment. If the delay prior to recall is extended from a few seconds to a few minutes, the recency effect goes away, while the primacy effect is still seen (**FIGURE 13.14B**). Because the recency effect is short-lived, lost in just a few minutes, it is attributed to STM. Conversely, the primacy effect is seen only if there is a delay in responding (see Figure 13.14B), suggesting that time is needed to get the early items into LTM. Similarly, patients with amnesia caused by impairment of the hippocampus (the red curve in Figure 13.14A) show a reduced

primacy effect The superior performance seen in a memory task for items at the start of a list. It is usually attributed to long-term memory.

recency effect The superior performance seen in a memory task for items at the end of a list. It is usually attributed to short-term memory.

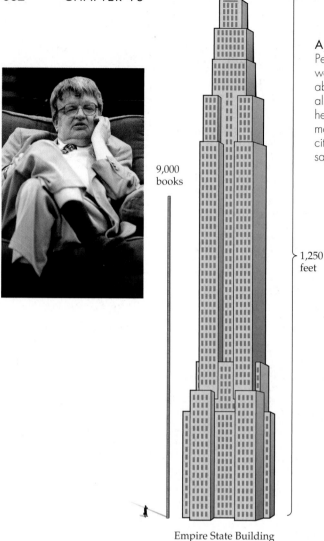

9,000 books

1,250 feet

Empire State Building

A Prodigious Memory The model for the fictional film *Rain Man*, the late Kim Peek, was a *savant* (from the French for "knowing"), a person with an unusually well-developed ability or skill. Born with several brain deformities, including an absence of the corpus callosum (Treffert and Christensen, 2005), Kim eventually memorized about 9,000 books, each taking about an hour. For example, he read the novel *The Hunt for Red October* in 75 minutes, and when asked, 4 months later, to name a minor character, not only did Kim know the name, but he cited the page number where the character appeared and quoted several passages on the page verbatim!

primacy effect, despite retaining the recency effect. Other animals also show primacy and recency effects, but only if shorter delays are used, indicating that STM lasts longer in humans than in other species.

The fact that brain trauma often causes retrograde amnesia (the inability to retrieve memories of events just before the accident) without affecting older memories suggests that STMs are more vulnerable to trauma—more likely to be lost—than are LTMs. When the retrograde amnesia stretches to several days before the accident, some recently acquired LTMs have been lost too. These cases suggest that, even after information has been consolidated into LTM, older memories are more "secure"—less likely to be lost—than freshly consolidated LTMs.

Long-term memory has vast capacity but is subject to distortion

How much information can be stored in LTM? There must be a limit, but no one has been able to come up with a way to measure it. In one classic experiment, people viewed long sequences of color photos of various scenes; several days later, they were shown pairs of images—in each case a new image plus one from the previous session—and asked to identify the images seen previously. Astonishingly, participants performed with a high degree of accuracy for series of up to 10,000 different stimuli (Standing, 1973)! For all practical purposes, there seems to be no upper bound to LTM capacity (Brady et al., 2014). Pigeons have a similarly impressive visual memory (Vaughan and Greene, 1984).

We take this capacity for granted and barely notice, for example, that knowledge of a language involves remembering at least 100,000 pieces of information. Most of us also store a huge assortment of information about faces, tunes, odors, skills, stories, and so on. Case studies of individuals with exceptional memory indicate that without the usual process of pruning out unimportant memories, continual perfect recall can become uncontrollable, distracting, and exhausting (Luria, 1987; Parker et al., 2006).

Despite the vast capacity of LTM, forgetting is a normal aspect of memory, helping to filter out unimportant information and freeing up needed cognitive resources (Kuhl et al., 2007). Interestingly, research indicates that the **memory trace** (the record laid down in memory by a learning experience) doesn't simply deteriorate from disuse and the passage of time; instead, memories tend to suffer interference from events before or after their formation.

For example, the process of retrieving information from LTM causes the memories to become temporarily unstable and susceptible to disruption or alteration before undergoing **reconsolidation** and returning to stable status (Nader and Hardt, 2009). Thus we can create *false memories* when we use leading questions to have people retrieve memories. Asking "Did you see the broken headlight?" rather than "Was the headlight broken?" can incorporate the false detail as the memory

memory trace A persistent change in the brain that reflects the storage of memory.

reconsolidation The return of a memory trace to stable long-term storage after it has been temporarily made changeable during the process of recall.

is reconsolidated (Loftus, 2003). This possibility of planting false memories clouds the issue of "recovered memories" of childhood sexual or physical abuse, because controversial therapeutic methods such as hypnosis or guided imagery (in which the patient is encouraged to imagine hypothetical abuse scenarios) can inadvertently plant false memories.

On the other hand, you can use the power of reconsolidation when studying—as long as you're careful to check your facts. One of the best ways to improve learning is simply repeated retrieval (and thus, repeated reconsolidation) of the stored information with feedback to let you know what you got right or wrong (Karpicke and Roediger, 2008). In other words, test yourself repeatedly, as with the "How's it going?" questions in this book. For your next exam, try making up some additional practice tests for yourself, or have a friend quiz you, instead of simply "cramming."

We still haven't talked about the nitty-gritty of memory in the brain—what exactly changes in the brain when we learn? That's what we'll take up in Part II.

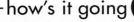 how's it going ❓

1. What are the stages of memory, and what do we call the processes by which information moves from one stage to the next?
2. What are the serial position effects of memory?
3. Explain how reconsolidation makes us vulnerable to distorted memories.

PART II
Neural Mechanisms of Memory

What are the basic cellular, synaptic, and molecular events that store information in the nervous system? In this next part of the chapter we will look at some of the ways in which new learning involves changes in the strength of existing synapses, and the biochemical signals that may produce those changes. We'll consider how the formation of memories may require the formation of new synapses, or even the birth of new neurons. This **neuroplasticity** (or *neural plasticity*) is found in virtually all animals, indicating that it is an ancient and vital product of evolution.

Memory Storage Requires Neuronal Remodeling

In introducing the term *synapse*, Charles Sherrington (1897) speculated that synaptic alterations might be the basis of learning. Sherrington's speculation anticipated an area of research that remains one of the most intensive efforts in all of neuroscience, since most theories of learning focus on plasticity of the structure and function of synapses.

Plastic changes at synapses can be physiological or structural

Synaptic changes that may store information can be measured physiologically. The changes can be presynaptic, postsynaptic, or both. They can include changes in the amount of neurotransmitter released and/or changes in the number or sensitivity of the postsynaptic receptors, resulting in larger (or smaller) postsynaptic potentials. Inhibiting inactivation of the transmitter (by altering reuptake or enzymatic degradation) can produce a similar effect (**FIGURE 13.15A**). Synaptic activity can also be influenced by inputs from other neurons, causing extra depolarization or hyperpolarization of the axon terminals and therefore changes in the amount of neurotransmitter released (**FIGURE 13.15B**).

Long-term memories may require changes in the nervous system so substantial that they can be directly observed (with the aid of a microscope, of course). After all, structural changes resulting from use are apparent in other parts of the body, as when exercise tones and shapes muscle. In a similar way, new synapses can form (or old synapses may die back) as a result of use (**FIGURE 13.15C**).

neuroplasticity Also called *neural plasticity*. The ability of the nervous system to change in response to experience or the environment.

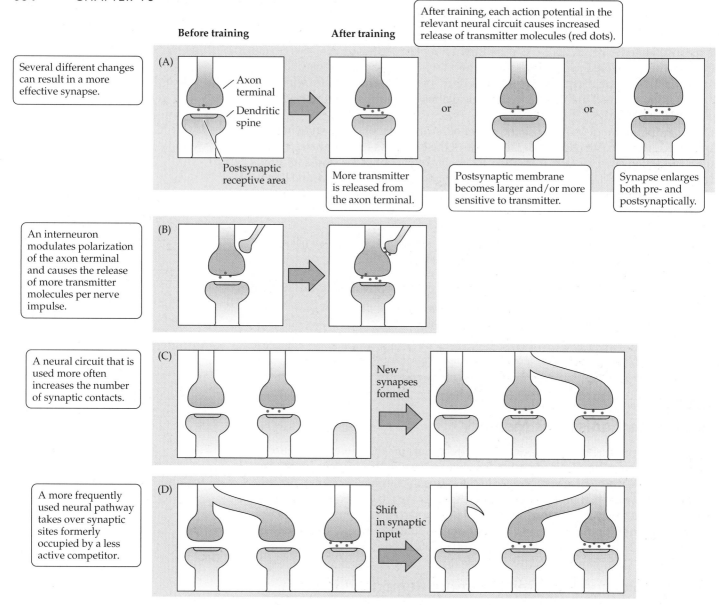

Several different changes can result in a more effective synapse.

Before training

After training

After training, each action potential in the relevant neural circuit causes increased release of transmitter molecules (red dots).

(A) Axon terminal

Dendritic spine

Postsynaptic receptive area

More transmitter is released from the axon terminal.

or

Postsynaptic membrane becomes larger and/or more sensitive to transmitter.

or

Synapse enlarges both pre- and postsynaptically.

(B) An interneuron modulates polarization of the axon terminal and causes the release of more transmitter molecules per nerve impulse.

(C) A neural circuit that is used more often increases the number of synaptic contacts.

New synapses formed

(D) A more frequently used neural pathway takes over synaptic sites formerly occupied by a less active competitor.

Shift in synaptic input

FIGURE 13.15 Synaptic Changes That May Store Memories

standard condition (SC) The usual environment for laboratory rodents, with a few animals in a cage and adequate food and water, but no complex stimulation.

impoverished condition (IC) Also called *isolated condition.* An environment for laboratory rodents in which each animal is housed singly in a small cage without complex stimuli.

enriched condition (EC) Also called *complex environment.* An environment for laboratory rodents in which animals are group-housed with a wide variety of stimulus objects.

Training can also lead to the reorganization of synaptic connections. For example, it can cause a more active pathway to take over sites formerly occupied by a less active competitor (**FIGURE 13.15D**).

Varied experiences and learning cause the brain to change and grow

The remarkable plasticity of the brain is not all that difficult to demonstrate. Simply living in a complex environment, with its many opportunities for new learning, produces pronounced biochemical and anatomical changes in the brains of rats (Renner and Rosenzweig, 1987).

In standard studies of environmental enrichment, rats are randomly assigned to one of three housing conditions:

1. **Standard condition (SC)** Animals are housed in small groups in standard lab cages (**FIGURE 13.16A**). This is the typical environment for laboratory animals.

2. **Impoverished condition (IC)** Animals are housed individually in standard lab cages (**FIGURE 13.16B**).

3. **Enriched condition (EC)** Animals are housed in large social groups in special cages containing various toys and other interesting features (**FIGURE 13.16C**). This condition provides enhanced opportunities for learning perceptual and motor skills, social learning, and so on.

(A) Standard condition

(B) Impoverished condition

(C) Enriched condition

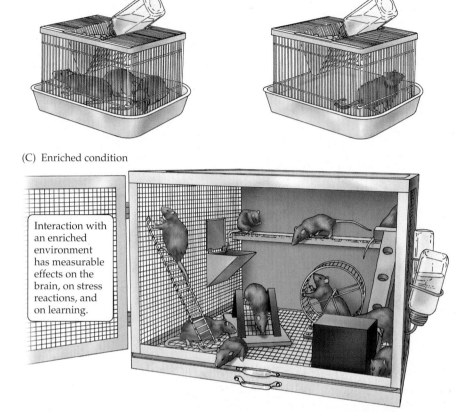

Interaction with an enriched environment has measurable effects on the brain, on stress reactions, and on learning.

FIGURE 13.16 Experimental Environments to Test the Effects of Enrichment on Learning and Brain Measures

In dozens of studies over several decades, a variety of changes in the brain were linked to environmental enrichment. For example, compared with IC animals:

- EC animals have a heavier, thicker cortex, especially in somatosensory and visual cortical areas (M. C. Diamond, 1967).

- EC animals show enhanced cholinergic activity throughout the cortex (Rosenzweig et al., 1961).

- EC animals have more dendritic branches on cortical neurons, and many more dendritic spines on those branches (**FIGURE 13.17**) (Greenough, 1976).

- EC animals have *larger* cortical synapses (M. C. Diamond et al., 1975), consistent with the storage of long-term memory in cortical areas through changes in synapses and circuits.

- EC animals have more neurons in the hippocampus because newly generated neurons (a topic of Part III of this chapter) live longer (Kempermann et al., 1997).

- EC animals show enhanced recovery from brain damage (Will et al., 2004).

These cerebral effects of experience, which were surprising when first reported for rats in the early 1960s, are now seen to occur widely in the animal kingdom—from flies to philosophers (Mohammed, 2001). But how can we study the physiology of learning when the mammalian cortex has many billions of neurons, organized in vast networks, and upwards of a billion synapses per cubic centimeter (Merchán-Pérez et al., 2009)? Researchers made progress by studying simple learning circuits, in various species including mammals, uncovering basic cellular principles of memory formation that may generalize to neurons throughout the brain.

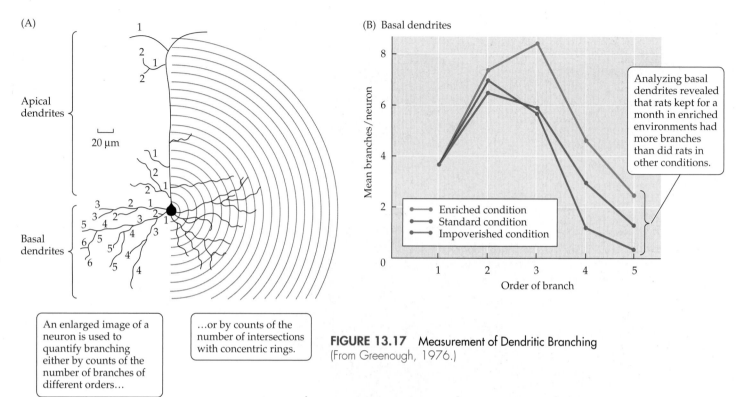

(A)

Apical dendrites

20 µm

Basal dendrites

(B) Basal dendrites

Mean branches/neuron

Order of branch

- Enriched condition
- Standard condition
- Impoverished condition

Analyzing basal dendrites revealed that rats kept for a month in enriched environments had more branches than did rats in other conditions.

An enlarged image of a neuron is used to quantify branching either by counts of the number of branches of different orders…

…or by counts of the number of intersections with concentric rings.

FIGURE 13.17 Measurement of Dendritic Branching (From Greenough, 1976.)

nonassociative learning A type of learning in which presentation of a particular stimulus alters the strength or probability of a response. It includes habituation.

habituation A form of nonassociative learning in which an organism becomes less responsive following repeated presentations of a stimulus.

Invertebrate nervous systems show synaptic plasticity

As we've discussed, neuroplasticity and the ability to learn are ancient adaptations found throughout the animal kingdom. At the neuronal level, even species that are only remotely related likely share the same basic cellular processes for information storage. Indeed, one fruitful research strategy has been to focus on memory mechanisms in the very simple nervous systems of certain invertebrates. Invertebrate nervous systems have relatively few neurons (on the order of hundreds to tens of thousands). Because these neurons are arranged identically in different individuals, it is possible to construct detailed neural circuit diagrams for particular behaviors and study the same few neurons in multiple individuals.

Even in these "simple" organisms, the search for memory mechanisms began with the simplest types of learning. Earlier we discussed one of the most basic forms of learning—associative learning about two stimuli, such as the case of a dog learning to associate the sound of a bell with food. Even simpler than associative learning are the types of learning that involve only *one* stimulus, called **nonassociative learning**. Perhaps the simplest form of nonassociative learning is **habituation**—a decrease in response to a stimulus as it is repeated. To be true habituation, the decreased response cannot be due to failure of the sensory system to detect the stimulus or due to an inability of the motor system to respond. Sitting in a café, you may stop noticing the door chime when someone enters. Your ears still detect the chime, and your body is perfectly capable of looking up to see what happened, but you've habituated to the sound.

Scientists uncovered how the sea slug *Aplysia* learns to habituate to a stimulus (Kandel, 2009). If you squirt water at the slug's siphon—a tube through which it draws water—the animal protectively retracts its delicate gill (**FIGURE 13.18**). But with repeated stimulation the animal retracts the gill less and less, as it learns that the stimulation represents no danger to the gill. Eric Kandel and associates demonstrated that this *short-term habituation* is caused by changes in the synapse between the sensory cell that detects the squirt of water and the motoneuron that retracts the gill. As less and less transmitter is released at this synapse, the gill withdrawal in response to the stimulation slowly fades (**FIGURE 13.19A**) (M. Klein et al., 1980).

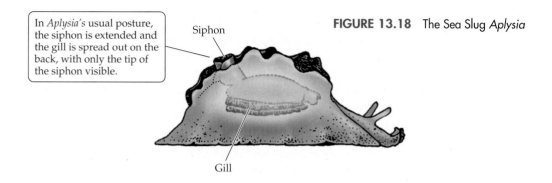

In *Aplysia's* usual posture, the siphon is extended and the gill is spread out on the back, with only the tip of the siphon visible.

Siphon

Gill

FIGURE 13.18   The Sea Slug *Aplysia*

The number and size of synapses can also vary with training in *Aplysia*. For example, if an *Aplysia* is tested in the habituation paradigm over a series of days, each successive day the animal habituates faster than it did the day before. This phenomenon represents *long-term habituation* (as opposed to the short-term habituation that we just described), and in this case there is a reduction in the *number of synapses* between the sensory cell and the motoneuron (**FIGURE 13.19B**) (C. H. Bailey and Chen, 1983).

A similar research program aimed at understanding simple learning in much more complicated species—mammals— also found that learning alters the strength of synaptic connections, as we'll see next.

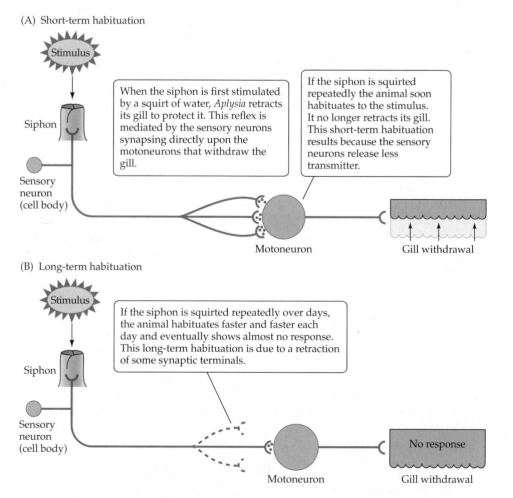

(A) Short-term habituation

Stimulus

Siphon

Sensory neuron (cell body)

When the siphon is first stimulated by a squirt of water, *Aplysia* retracts its gill to protect it. This reflex is mediated by the sensory neurons synapsing directly upon the motoneurons that withdraw the gill.

If the siphon is squirted repeatedly the animal soon habituates to the stimulus. It no longer retracts its gill. This short-term habituation results because the sensory neurons release less transmitter.

Motoneuron

Gill withdrawal

(B) Long-term habituation

Stimulus

Siphon

Sensory neuron (cell body)

If the siphon is squirted repeatedly over days, the animal habituates faster and faster each day and eventually shows almost no response. This long-term habituation is due to a retraction of some synaptic terminals.

Motoneuron

No response

Gill withdrawal

FIGURE 13.19   Synaptic Plasticity Underlying Habituation in *Aplysia*

Classical conditioning relies on circuits in the mammalian cerebellum

Success at studying the more complicated mammalian brain came when researchers probed simple associative learning: classical conditioning of the eye-blink reflex (R. F. Thompson, 1990; Lavond et al., 1993; R. F. Thompson and Steinmetz, 2009).

When a puff of air is aimed at the cornea of a rabbit, the animal reflexively blinks. The eye-blink reflex can be classically conditioned. Over several trials, if the air puff (US) immediately follows an acoustic tone (CS), a simple conditioned response (CR) develops rapidly: the rabbit comes to blink when the tone is sounded (to review these terms and the basics of conditioning, see Figure 13.9). The neural circuit of the eye-blink reflex is also simple, involving cranial nerves and some interneurons that connect their nuclei (**FIGURE 13.20A**). Sensory fibers from the cornea run along cranial nerve V (the trigeminal nerve) to its nucleus in the brainstem. From there, some interneurons send axons to synapse on other cranial nerve motor nuclei (VI and VII), which in turn activate the muscles of the eyelids, causing the blink.

Early studies showed that destruction of the hippocampus and the rest of the medial temporal lobe has little effect on the conditioned eye-blink response in rabbits (Lockhart and Moore, 1975). Instead, as we mentioned earlier, researchers found that a *cerebellar* circuit is both necessary and sufficient for eye-blink conditioning.

The trigeminal (cranial nerve V) pathway that carries information about the corneal stimulation (the US) to the cranial motor nuclei also sends axons to the brainstem. These brainstem neurons, in turn, send axons called *climbing fibers* to synapse on cerebellar neurons. The same cerebellar cells also receive information about the auditory CS by a pathway through the auditory nuclei (**FIGURE 13.20B**). So information about the US and CS converges in the cerebellum. After conditioning, the occurrence of the

FIGURE 13.20 The Neural Circuit for Conditioning the Eye-Blink Reflex (After R. F. Thompson and Krupa, 1994.)

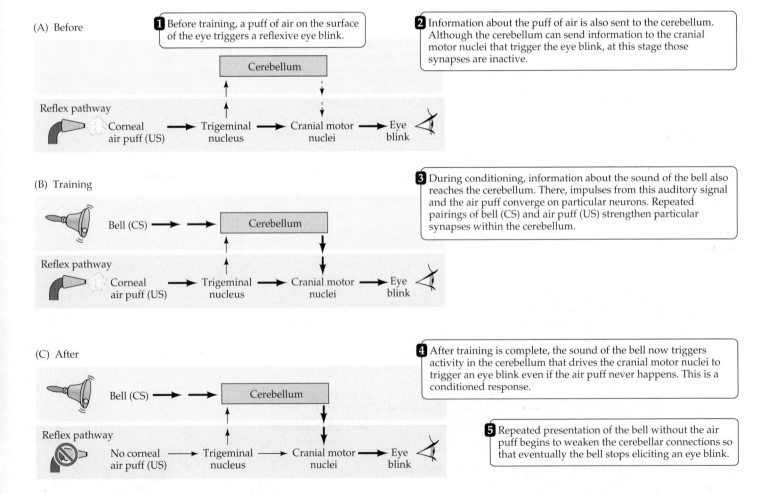

(A) Before

1 Before training, a puff of air on the surface of the eye triggers a reflexive eye blink.

2 Information about the puff of air is also sent to the cerebellum. Although the cerebellum can send information to the cranial motor nuclei that trigger the eye blink, at this stage those synapses are inactive.

Cerebellum

Reflex pathway

Corneal air puff (US) → Trigeminal nucleus → Cranial motor nuclei → Eye blink

(B) Training

3 During conditioning, information about the sound of the bell also reaches the cerebellum. There, impulses from this auditory signal and the air puff converge on particular neurons. Repeated pairings of bell (CS) and air puff (US) strengthen particular synapses within the cerebellum.

Bell (CS) → → Cerebellum

Reflex pathway

Corneal air puff (US) → Trigeminal nucleus → Cranial motor nuclei → Eye blink

(C) After

4 After training is complete, the sound of the bell now triggers activity in the cerebellum that drives the cranial motor nuclei to trigger an eye blink even if the air puff never happens. This is a conditioned response.

Bell (CS) → → Cerebellum

Reflex pathway

No corneal air puff (US) → Trigeminal nucleus → Cranial motor nuclei → Eye blink

5 Repeated presentation of the bell without the air puff begins to weaken the cerebellar connections so that eventually the bell stops eliciting an eye blink.

CS—the tone—has an enhanced effect on the cerebellar neurons, so they now trigger eye blink even in the absence of an air puff (**FIGURE 13.20C**). Imaging studies confirm that the cerebellum is important for conditioning of the eye-blink reflex and other simple conditioning in humans (Timmann et al., 2010). How does the training change the strength of those synapses, so that now the tone triggers a blink? At least some of these plastic changes in the cerebellar neurons rely on a special synaptic mechanism that has been best studied in the hippocampus (Mao and Evinger, 2001), and that's where we turn our attention next.

how's it going ?

1. What are some of the ways that learning could alter synaptic structure or function?
2. Describe the effects of different environments on the brains of rats.
3. How does the circuitry in *Aplysia* change in the course of short-term and long-term habituation?
4. Which brain region is crucial for classical conditioning in mammals, and how does it play its role?

Donald O. Hebb (1904–1985) "When an axon of cell A is near enough to excite cell B and repeatedly or persistently takes part in firing it, some growth process or metabolic change takes place in one or both cells such that A's efficiency, as one of the cells firing B, is increased" (Hebb, 1949). (Photo courtesy of Ms. Mary Ellen Hebb and Dr. Richard Brown.)

Synaptic Plasticity Can Be Measured in Simple Hippocampal Circuits

Modern ideas about synaptic plasticity have their origins in the theories of Donald Hebb, who proposed that when a presynaptic and a postsynaptic neuron were repeatedly activated together, the synaptic connection between them would become stronger and more stable (the phrase "cells that fire together wire together" captures the basic idea). These **Hebbian synapses** could then act together to store memory traces (Hebb, 1949).

This idea was eventually confirmed in various brain tissues. Researchers in the 1970s discovered an impressive form of neuroplasticity in the hippocampus that appeared to confirm Hebb's theories about synaptic changes (Bliss and Lømo, 1973; Schwartz-kroin and Wester, 1975). In experiments like theirs, electrodes are placed within the rat hippocampus, positioned so that the researchers can stimulate a group of *presynaptic* axons and immediately record the electrical response of a group of *postsynaptic* neurons. Normal, low-level activation of the presynaptic cells produces stable and predictable excitatory postsynaptic potentials (EPSPs) (see Chapter 3), as expected. But when a brief high-frequency burst of electrical stimuli (called a **tetanus**) is applied to the presynaptic neurons, causing them to produce a high rate of action potentials, the response of the postsynaptic neurons changes. Now the postsynaptic cells produce much larger EPSPs; in other words, the synapses appear to have become stronger, more effective. This stable and long-lasting enhancement of synaptic transmission, termed **long-term potentiation** (**LTP**; *potentiation* means a "strengthening"), is illustrated in **FIGURE 13.21**.

Hebbian synapse   A synapse that is strengthened when it successfully drives the postsynaptic cell.

tetanus   An intense volley of action potentials.

long-term potentiation (LTP)   A stable and enduring increase in the effectiveness of synapses following repeated strong stimulation.

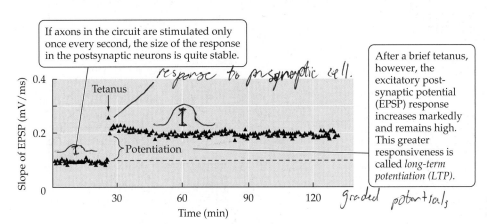

If axons in the circuit are stimulated only once every second, the size of the response in the postsynaptic neurons is quite stable.

After a brief tetanus, however, the excitatory post-synaptic potential (EPSP) response increases markedly and remains high. This greater responsiveness is called *long-term potentiation (LTP)*.

FIGURE 13.21   Long-Term Potentiation Occurs in the Hippocampus

dentate gyrus   A strip of gray matter in the hippocampal formation.

glutamate   An amino acid transmitter; the most common excitatory transmitter.

NMDA receptor   A glutamate receptor that also binds the glutamate agonist NMDA (N-methyl-D-aspartate) and that is both ligand-gated and voltage-sensitive.

AMPA receptor   A fast-acting ionotropic glutamate receptor that also binds the glutamate agonist AMPA.

We now know that LTP can be generated in conscious and freely behaving animals, in anesthetized animals, and in tissue slices and that LTP is evident in a variety of invertebrate and vertebrate species. LTP can also last for weeks or more (Bliss and Gardner-Medwin, 1973). So, at least superficially, LTP appears to have the hallmarks of a cellular mechanism of memory. This hint at a cellular origin prompted research into the molecular and physiological mechanisms underlying LTP.

NMDA receptors and AMPA receptors collaborate in LTP

The region called the *hippocampal formation* consists of two interlocking C-shaped structures: the hippocampus itself and the **dentate gyrus**. At least three different pathways in the hippocampal formation display LTP, and it is seen in other brain regions too (Malenka and Bear, 2004). The most studied form of LTP occurs at synapses that use the excitatory neurotransmitter **glutamate**, and it is critically dependent on a glutamate receptor subtype called the **NMDA receptor** (after its selective ligand, *N*-methyl-D-*a*spartate). Treatment with drugs that selectively block NMDA receptors completely prevents new LTP in this region, but it does not affect synaptic changes that have already been established. As you might expect, these postsynaptic NMDA receptors—working in conjunction with other glutamate receptors called **AMPA receptors**—have some unique characteristics, which are responsible for LTP.

During normal, low-level activity, the release of glutamate at the synapse activates only the AMPA receptors. The NMDA receptors cannot respond to the glutamate, because magnesium ions (Mg^{2+}) block the NMDA receptor's calcium ion (Ca^{2+}) channel (**FIGURE 13.22A**); thus, few Ca^{2+} ions can enter the neuron. The situation changes,

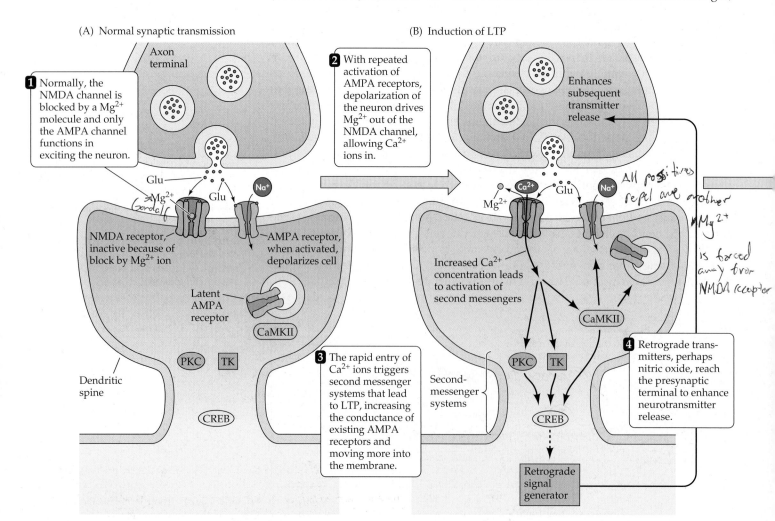

however, if larger quantities of glutamate are released—say, in response to a barrage of action potentials caused by a tetanus. That stronger stimulation of the AMPA receptors depolarizes the postsynaptic membrane so much that the Mg^{2+} plug is driven from the central channels of the NMDA receptors (**FIGURE 13.22B**). Now the NMDA receptors are also able to respond to glutamate, admitting large amounts of Ca^{2+} into the postsynaptic neuron. Thus, NMDA receptors are fully active only when "gated" by a combination of strong depolarization (via AMPA receptors) and the ligand (glutamate).

The large influx of Ca^{2+} at NMDA receptors activates a variety of intracellular enzymes that affect AMPA receptors in several important ways (**FIGURE 13.22C**) (Lisman et al., 2002; Kessels and Malinow, 2009). First, the enzymes cause existing nearby AMPA receptors to move to the active synapse (T. Takahashi et al., 2003), and they modify the AMPA receptors to increase their conductance of Na^+ and K^+ ions (Sanderson et al., 2008). In addition, more AMPA receptors are produced and inserted into the postsynaptic membrane. Thus, after the tetanus there are more AMPA receptors, and those receptors are more effective, so the synaptic response to glutamate is strengthened (see Figure 13.22B).

There are *presynaptic* changes in LTP too. When the postsynaptic cell is strongly stimulated and its NMDA receptors become active and admit Ca^{2+}, an intracellular process causes the postsynaptic cell to release a **retrograde transmitter**—often a diffusible gas—that travels back across the synapse and alters the functioning of the presynaptic neuron (see Figure 13.22B). The retrograde transmitter induces the presynaptic terminal to release more glutamate than previously, thereby strengthening the synapse some more. So, LTP involves active changes on both sides of the synapse.

retrograde transmitter   A neurotransmitter that is released by the postsynaptic region, diffuses back across the synapse, and alters the functioning of the presynaptic neuron.

To see the animation
AMPA and NMDA Receptors,
go to

2e.mindsmachine.com/13.3

(C) Enhanced synapse, after induction of LTP

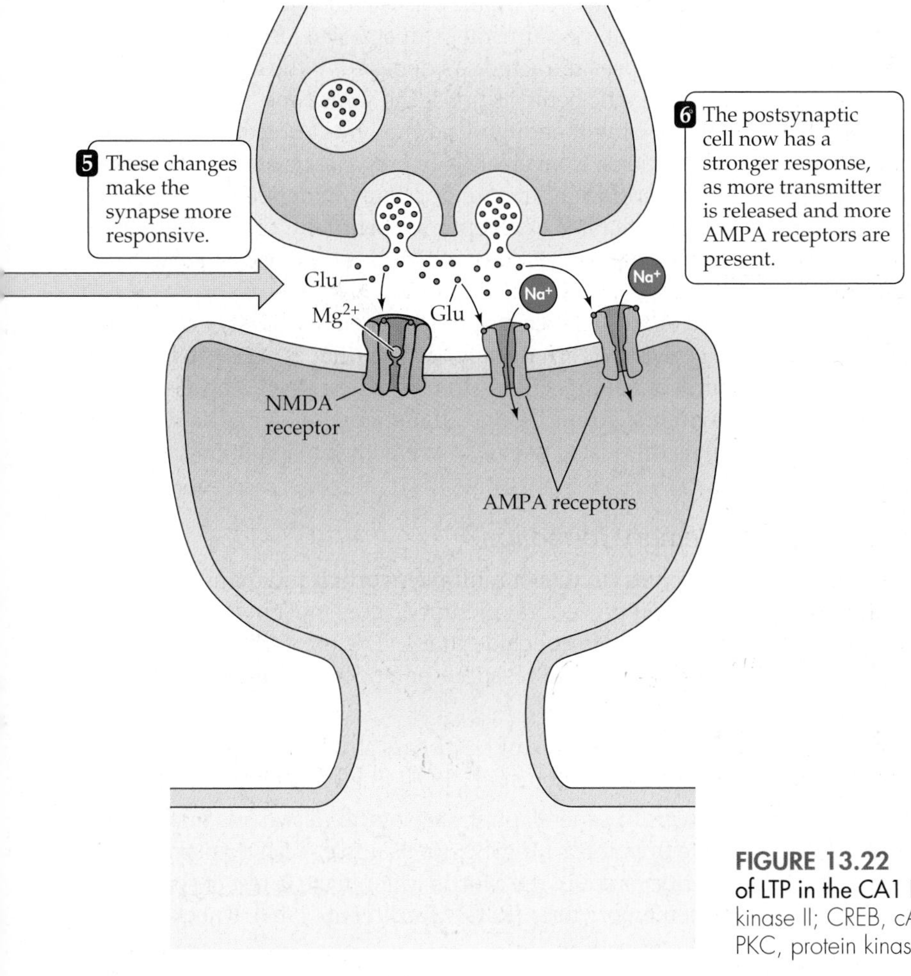

5 These changes make the synapse more responsive.

6 The postsynaptic cell now has a stronger response, as more transmitter is released and more AMPA receptors are present.

Glu

Mg^{2+} Glu   Na⁺   Na⁺

NMDA receptor

AMPA receptors

FIGURE 13.22 Roles of the NMDA and AMPA Receptors in the Induction of LTP in the CA1 Region CaMKII, calcium/calmodulin-dependent protein kinase II; CREB, cAMP responsive element–binding protein; Glu, glutamate; PKC, protein kinase C; TK, tyrosine kinase.

FIGURE 13.23   At Hebbian Synapses, Neurons That Fire Together Wire Together

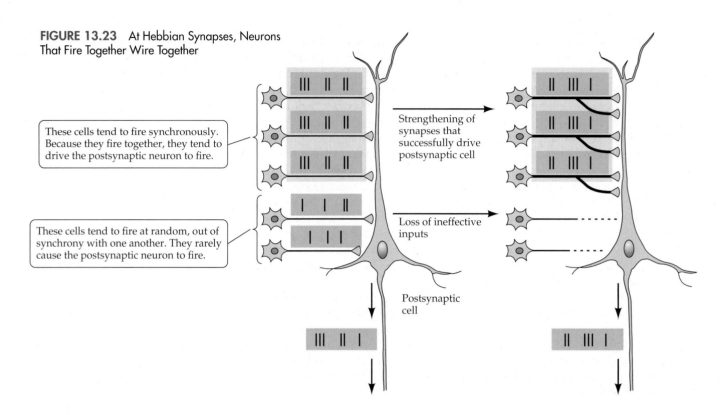

These cells tend to fire synchronously. Because they fire together, they tend to drive the postsynaptic neuron to fire.

These cells tend to fire at random, out of synchrony with one another. They rarely cause the postsynaptic neuron to fire.

Strengthening of synapses that successfully drive postsynaptic cell

Loss of ineffective inputs

Postsynaptic cell

So far, we've talked about how activity can make existing Hebbian synapses stronger. However, evidence suggests that the same mechanisms can affect whether new synapses are formed and old synapses retracted. As depicted in these systems it appears that when a group of presynaptic neurons fire at the same time, they "gang up" on the postsynaptic cell, depolarizing it enough that the NMDA receptors are activated to strengthen those connections. Conversely, any presynaptic neurons that tend to fire out of synchrony with the other inputs are not likely to depolarize the postsynaptic neurons enough to activate NMDA receptors. Eventually, the strengthened inputs seem to sprout new, additional connections, while the weakened synapses fade away (**FIGURE 13.23**).

Many scientists are excited about LTP because this momentary burst of neural activity, the tetanus, can change synaptic strength for a long time. It's easy to imagine how another momentary burst of neural activity, in this case triggered by a learning experience, could change synaptic strength, and that change in synaptic strength might be a memory trace. But is this just a case of an overactive imagination, or is LTP truly involved in learning?

Is LTP a mechanism of memory formation?

Even the simplest learning involves circuits of multiple neurons and many synapses, and more-complex declarative and procedural memory traces must involve vast networks of neurons, so we are unlikely to conclude that LTP is the *only* mechanism of learning. However, evidence from several research perspectives implicates LTP in memory:

1. *Correlational observations* The time course of LTP bears strong similarity to the time course of memory formation (Lynch et al., 1991; Staubli, 1995).

2. *Somatic intervention experiments* In general, pharmacological treatments that interfere with LTP also tend to impair learning. So, for example, NMDA receptor blockade interferes with performance in the Morris water maze (a test of spatial memory) and other types of memory tests (R. G. Morris et al., 1989). Knockout

mice that lack functional NMDA receptors only in CA1 appear normal in many respects, but their hippocampi are incapable of LTP and their declarative memory is impaired (Rampon et al., 2000). On the other hand, mice engineered to *overexpress* NMDA receptors in the hippocampus have enhanced LTP and better-than-normal long-term memory (Y. P. Tang et al., 2001). (For the full story of these mice, see **A STEP FURTHER 13.2**, on the website.)

3. *Behavioral intervention experiments* In principle, the most convincing evidence for a link between LTP and learning would be "behavioral LTP": a demonstration that training an animal in a memory task induces LTP somewhere in the brain. Such research is difficult because of uncertainty about exactly where to put the recording electrodes in order to detect any induced LTP. Nevertheless, several examples of successful behavioral LTP have been reported (McKernan and Shinnick-Gallagher, 1997; Rogan et al., 1997; Whitlock et al., 2006).

Taken together, these findings support the idea that LTP is a kind of synaptic plasticity that underlies (or is very similar to) certain forms of learning and memory.

Thus it seems that the cause of Henry's tragic amnesia may have been the loss of hippocampal mechanisms like LTP to consolidate short-term memories into long-term memories. It's strange to think that such microscopic changes in synapses in a particular brain region could be so crucial for living a full human life. Eventually Henry seemed to stop being shocked when he saw his reflection, and he learned he was no longer in his mid twenties. But it's not clear whether he understood, for long, that his parents had passed away. As a final act of generosity to a field of science that he helped launch, Henry arranged to donate his brain for further study after he died. Through webcasting technology, the dissection of Henry's brain was viewed live by thousands of people (see http://thebrainobservatory.ucsd.edu), and a series of more than 2,000 brain sections will eventually be made available.

To the end, although Henry could remember so little of his entire adult life, he was courteous and concerned about other people. Henry remembered the surgeon he had met several times before his operation: "He did medical research on people… What he learned about me helped others too, and I'm glad about that" (Corkin, 2002, p. 158). Henry never knew how famous he was, or how much his dreadful condition taught us about learning and memory; yet despite being deprived of one of the most important characteristics of a human being, he held fast to his humanity.

To see the video
Morris Water Maze,
go to

2e.mindsmachine.com/13.4

how's it going ❓

1. Describe how LTP is measured in the hippocampus.
2. What happens to AMPA receptors and NMDA receptors during LTP?
3. What evidence suggests that LTP may underlie some forms of learning and memory?

PART III
Development of the Brain

Age puts its stamp on us all. Although the rate, progression, and orderliness of changes in the brain are especially prominent early in life, change is a feature of the entire life span. In this final part of the chapter we describe brains in terms of their progress through life from the womb to the tomb. The fertilization of an egg leads to a body with a brain that contains billions of neurons with an incredible number of connections. The pace of this process is extraordinary: during the height of prenatal growth of the human brain, more than 250,000 neurons are added per minute! We will describe the emergence of nerve cells, the formation of their connections, and the role of genes in shaping the nervous system. But we'll see that experience, gained through behavioral interactions with the environment, also sculpts the developing brain.

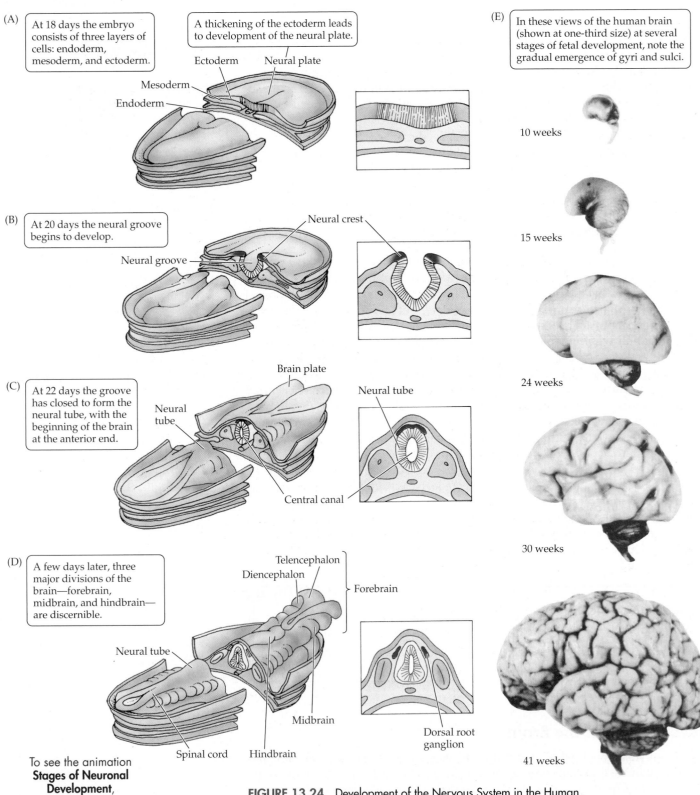

(A) At 18 days the embryo consists of three layers of cells: endoderm, mesoderm, and ectoderm.

A thickening of the ectoderm leads to development of the neural plate.

Ectoderm
Neural plate
Mesoderm
Endoderm

(B) At 20 days the neural groove begins to develop.

Neural crest
Neural groove

(C) At 22 days the groove has closed to form the neural tube, with the beginning of the brain at the anterior end.

Brain plate
Neural tube
Neural tube
Central canal

(D) A few days later, three major divisions of the brain—forebrain, midbrain, and hindbrain—are discernible.

Telencephalon
Diencephalon
Forebrain
Neural tube
Midbrain
Spinal cord
Hindbrain
Dorsal root ganglion

(E) In these views of the human brain (shown at one-third size) at several stages of fetal development, note the gradual emergence of gyri and sulci.

10 weeks

15 weeks

24 weeks

30 weeks

41 weeks

To see the animation **Stages of Neuronal Development**, go to

2e.mindsmachine.com/13.5

FIGURE 13.24 Development of the Nervous System in the Human Embryo and Fetus (Part E from Larroche, 1977.)

Growth and Development of the Brain Are Orderly Processes

Picture, if you can, the number of neurons in the mature human brain—nearly 100 billion (Herculano-Houzel, 2012). There are many types of neurons, each forming a vast array of hundreds or thousands of connections. Overall, the brain contains more than 100 trillion connections. If each synapse were represented by a grain of sand, the collected grains would form a cube longer than a football field on each edge! Yet each of us began as a single microscopic cell, the fertilized egg. How can one cell divide and grow to form the most complicated machines on Earth, perhaps in the universe?

Within 12 hours after a human egg is fertilized, the single cell begins dividing, forming a small mass of homogeneous cells, like a cluster of grapes, that is a mere 200 micrometers in diameter. Within a week the emerging human embryo shows three distinct cell layers (**FIGURE 13.24A**)—the beginnings of all the tissues of the body. The nervous system develops from the outer layer, called the **ectoderm** (from the Greek *ektos*, "out," and *derma*, "skin"). As the cell layers thicken, they grow to form a groove that will become the midline. At the head end of this neural groove, a thickened collection of cells forms (**FIGURE 13.24B**).

The crests of the neural groove come together to form the **neural tube** (**FIGURE 13.24C**). At the anterior part of the neural tube, three subdivisions become apparent. These subdivisions correspond to the future **forebrain** (cortical regions, thalamus, and hypothalamus), **midbrain**, and **hindbrain** (cerebellum, pons, and medulla) (**FIGURE 13.24D**), which were discussed in Chapter 2. The interior of the neural tube becomes the fluid-filled cerebral ventricles of the brain, the central canal of the spinal cord, and the passages that connect them.

By the end of the eighth week, the human embryo shows the rudimentary beginnings of most body organs. The rapid development of the brain is reflected in the fact that by this time the head is half the total size of the embryo! (Note that the developing human is called an **embryo** during the first 10 weeks after fertilization; thereafter it is called a **fetus**.) **FIGURE 13.24E** shows the prenatal development of the human brain from 10 weeks to 41 weeks. Even after birth, there are dramatic local changes as some brain regions grow more than others, as we'll see later.

Development of the Nervous System Can Be Divided into Six Distinct Stages

From a cellular viewpoint it is useful to consider brain development as a sequence of six distinct stages, most of which occur during prenatal life:

1. *Neurogenesis*, the mitotic division of nonneuronal cells to produce neurons
2. *Cell migration*, the massive movements of nerve cells or their precursors to establish distinct nerve cell populations (nuclei in the CNS, layers of the cerebral cortex, and so on)
3. *Cell differentiation*, the refining of cells into distinctive types of neurons or glial cells
4. *Synaptogenesis*, the establishment of synaptic connections as axons and dendrites grow
5. *Neuronal cell death*, the selective death of many nerve cells
6. *Synapse rearrangement*, the loss of some synapses and the development of others, to refine synaptic connections

The six stages proceed at different rates and times in different parts of the nervous system. Some of the stages may overlap even within a region. In the discussion that follows, we will take up each stage in succession. This sequence is portrayed in **FIGURE 13.25**.

ectoderm   The outer cellular layer of the developing embryo, giving rise to the skin and the nervous system.

neural tube   An embryonic structure with subdivisions that correspond to the future forebrain, midbrain, and hindbrain.

forebrain   The front division of the brain, which in the mature vertebrate contains the cerebral hemispheres, the thalamus, and the hypothalamus.

midbrain   The middle division of the brain.

hindbrain   The rear division of the brain, which in the mature vertebrate contains the cerebellum, pons, and medulla.

embryo   The earliest stage in a developing animal.

fetus   A developing individual after the embryo stage.

To see the video
Migration of a Neuron along a Radial Glial Cell,
go to

2e.mindsmachine.com/13.6

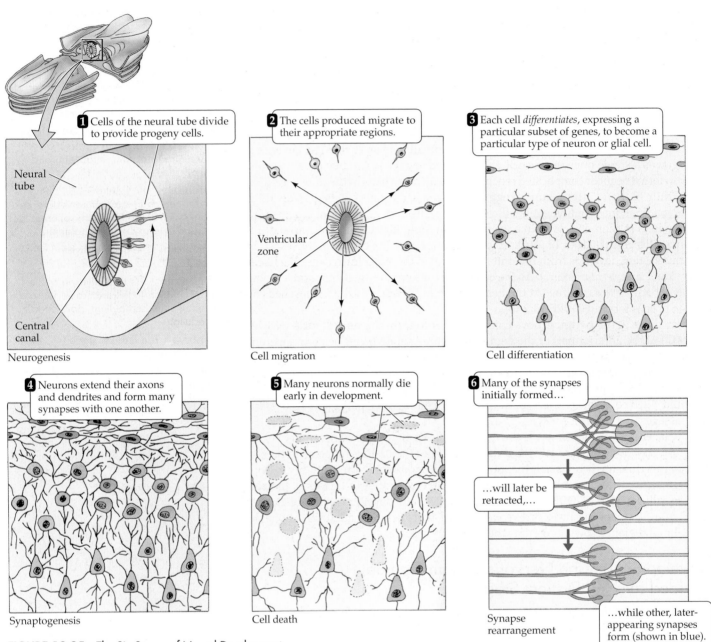

1 Cells of the neural tube divide to provide progeny cells.

Neural tube

Central canal

Neurogenesis

2 The cells produced migrate to their appropriate regions.

Ventricular zone

Cell migration

3 Each cell *differentiates*, expressing a particular subset of genes, to become a particular type of neuron or glial cell.

Cell differentiation

4 Neurons extend their axons and dendrites and form many synapses with one another.

Synaptogenesis

5 Many neurons normally die early in development.

Cell death

6 Many of the synapses initially formed…

…will later be retracted,…

Synapse rearrangement

…while other, later-appearing synapses form (shown in blue).

FIGURE 13.25 The Six Stages of Neural Development

Cell proliferation produces cells that become neurons or glial cells

The production of neurons is called **neurogenesis**. Nerve cells themselves do not divide, but the cells that will give rise to neurons begin as a single layer of cells along the inner surface of the neural tube. These cells divide (in a process called **mitosis**) within the **ventricular zone** inside the neural tube (see Figure 13.25, step 1). Eventually, some cells leave the ventricular zone and begin transforming into either neurons or glial cells. As the nervous system grows, **cell migration** follows, as the cells move over relatively long distances to fill out the brain (Figure 13.25, step 2).

Newly arrived cells in the brain bear no more resemblance to mature nerve cells than they do to the cells of other organs. Once they reach their destinations, however, the cells begin to use, or **express**, particular genes. This means that each type of cell makes use of a particular subset of genes to make the particular proteins that type needs. This process of **cell differentiation** enables cells to acquire the distinctive appearance and functions of neurons characteristic of their particular regions (Figure

neurogenesis The mitotic division of nonneuronal cells to produce neurons.

mitosis The process of division of somatic cells that involves duplication of DNA.

ventricular zone Also called *ependymal layer*. A region lining the cerebral ventricles that displays mitosis, providing neurons early in development and glial cells throughout life.

cell migration The movement of cells from site of origin to final location.

13.25, step 3). Once they take on the characteristics of neurons, they begin making synaptic connections with one another, in the process of **synaptogenesis** (Figure 13.25, step 4).

The particular fate of a differentiating cell depends on where in the brain the cell happens to be and what the cell's neighbors are doing. Cells in the developing brain are constantly sending chemical signals to one another, each shaping the development of the other. This is the hallmark of vertebrate development: cells sort themselves out via **cell-cell interactions**, taking on fates that are appropriate in the context of what neighboring cells are doing. When the negotiations are all over, if things go properly, a new person is formed with all the types of cells in the brain that she or he needs to live.

This system of cell-cell interactions determining how brain cells develop has an important consequence: If cells that have not yet differentiated extensively can be obtained and placed in a particular brain region, they can differentiate in an appropriate way and become properly integrated. Such undifferentiated cells, called **stem cells**, are present throughout embryonic tissues, so they can be gathered from umbilical cord blood, miscarried embryos, or unused embryos produced during in vitro fertilization. It may even be possible someday to take cells from adult tissue and, by treating them with various factors in a dish, transform them into stem cells (Hou et al., 2013). It is hoped that placing stem cells in areas of brain degeneration, such as loss of myelin in multiple sclerosis or loss of dopaminergic neurons in Parkinson's disease (see Chapter 5), might reverse such degeneration as the implanted cells differentiate to fill in for the missing components (Ross and Akimov, 2014).

In the adult brain, newly born neurons aid learning

At birth, mammals have already produced most of the neurons they will ever have. The postnatal increase of human brain weight is due primarily to growth in the size of neurons, branching of dendrites, elaboration of synapses, increase in myelin, and addition of glial cells. But research in the last decade or so has shown that we are also capable of **adult neurogenesis**, the generation of new neurons in adulthood, especially in the dentate gyrus of the hippocampal formation (**FIGURE 13.26**) (Gould, Reeves, et al., 1999; Magavi et al., 2000). The new neurons integrate into the functional circuitry of the hippocampus and adjacent cortex (Bruel-Jungerman et al., 2007), which we've seen to play a role in forming new memories.

Indeed, although the new neurons acquired in adulthood represent just a tiny minority of the total, there's reason to think they are important. In experimental animals, the birth and/or survival of new neurons is enhanced by factors like exercise,

gene expression The process by which a cell makes an mRNA transcript of a particular gene.

cell differentiation The developmental stage in which cells acquire distinctive characteristics, such as those of neurons, as a result of expressing particular genes.

synaptogenesis The establishment of synaptic connections as axons and dendrites grow.

cell-cell interaction The general process during development in which one cell affects the differentiation of other, usually neighboring, cells.

stem cell A cell that is undifferentiated and therefore can take on the fate of any cell that a donor organism can produce.

adult neurogenesis The creation of new neurons in the brain of an adult.

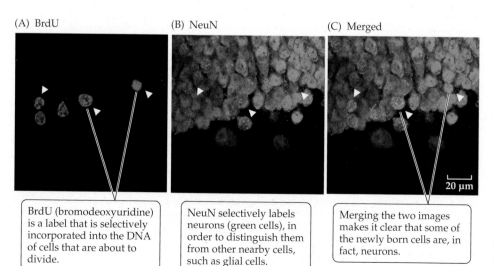

(A) BrdU   (B) NeuN   (C) Merged

BrdU (bromodeoxyuridine) is a label that is selectively incorporated into the DNA of cells that are about to divide.

NeuN selectively labels neurons (green cells), in order to distinguish them from other nearby cells, such as glial cells.

Merging the two images makes it clear that some of the newly born cells are, in fact, neurons.

FIGURE 13.26 Neurogenesis in the Dentate Gyrus (From Bruel-Jungerman et al., 2007, courtesy of Elodie Bruel-Jungerman and Serge Laroche.)

cell death Also called *apoptosis*. The developmental process during which "surplus" cells die.

death gene A gene that is expressed only when a cell becomes committed to natural cell death (apoptosis).

neurotrophic factor Also called simply *trophic factor*. A target-derived chemical that acts as if it "feeds" certain neurons to help them survive.

environmental enrichment, and training (Opendak and Gould, 2015; Waddell and Shors, 2008). Neurogenesis appears to enhance various forms of hippocampus-dependent learning, such as spatial memory and fear conditioning in some (but not all) studies (Saxe et al., 2006; Winocur et al., 2006; Kee et al., 2007). Mice with a genetic manipulation that turns off neurogenesis in the brains of adults showed a marked impairment in spatial learning with little effect on other behaviors (C. L. Zhang et al., 2008).

So by studying this chapter, you may be giving your brain a few more neurons to use on exam day! Physical exercise also boosts neurogenesis in rats—an effect that can be blocked by stressors such as social isolation (Stranahan et al., 2006)—so invest in exercise and a network of friends too.

The death of many neurons is a normal part of development

As strange as it may seem, cell death is a crucial phase of brain development (see Figure 13.25, step 5). This developmental stage is not unique to the nervous system. Naturally occurring **cell death**, also called *apoptosis* (from the Greek *apo*, "away from," and *ptosis*, "act of falling"), is evident as a kind of sculpting process in the emergence of other tissues in both animals and plants.

The number of neurons that die during early development is quite large. In some regions of the brain and spinal cord, *most* of the young nerve cells die during prenatal development. In 1958, Viktor Hamburger (1900–2001) first described naturally occurring neuronal cell death in chicks, in which nearly half the originally produced spinal motoneurons die before the chick hatches. A similar loss of spinal motoneurons was later reported in developing humans (**FIGURE 13.27**) (Forger and Breedlove, 1987).

These cells are not dying because of a defect. Rather, these cells die as a consequence of complex interactions with surrounding cells, and they are actively "committing suicide." Your chromosomes carry **death genes**—genes that are expressed only when a cell undergoes apoptosis (Peter et al., 1997). Genetically interfering with death genes in fetal mice causes them to grow brains that are too large to fit in the skull (Depaepe et al., 2005), so we can see how vital it is that some cells die.

Neurons compete for connections to target structures (other nerve cells or end organs, such as muscle). Cells that make adequate synapses remain; those without a place to form synaptic connections die. Apparently the cells compete not just for synaptic sites, but for a chemical that the target structure makes and releases (**FIGURE 13.28**). Neurons that receive enough of the chemical survive; those that do not, die. Such target-derived chemicals are called **neurotrophic factors** (or simply *trophic factors*) because they act as if they "feed" the neurons to help them survive (in Greek, *trophe* means "nourishment").

(A)

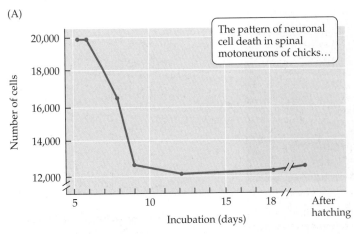

(B)

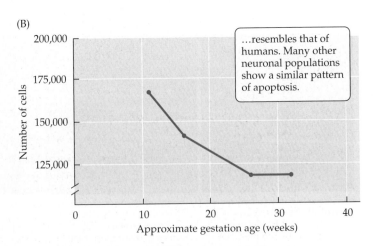

FIGURE 13.27 Many Neurons Die during Normal Early Development (Part A after Hamburger, 1975; B after Forger and Breedlove, 1987.)

FIGURE 13.28   A Model for the Action of Neurotrophic Factors

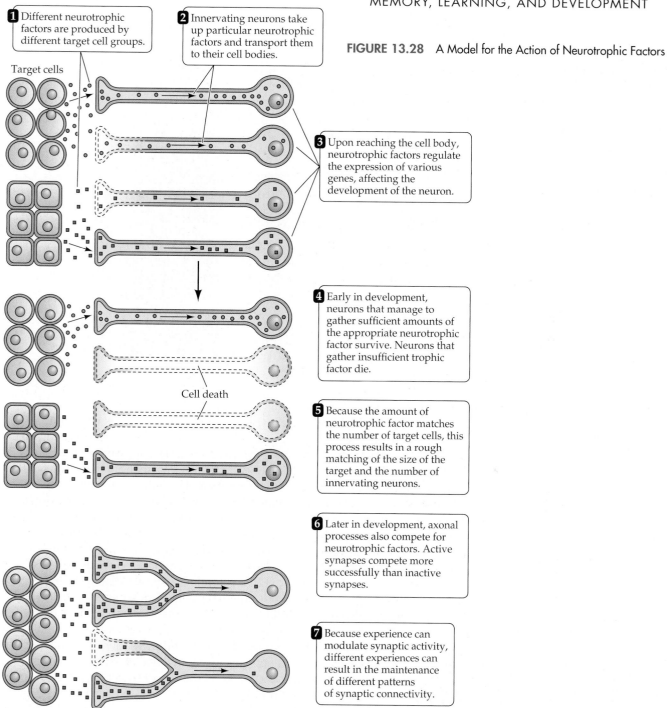

1 Different neurotrophic factors are produced by different target cell groups.

Target cells

2 Innervating neurons take up particular neurotrophic factors and transport them to their cell bodies.

3 Upon reaching the cell body, neurotrophic factors regulate the expression of various genes, affecting the development of the neuron.

4 Early in development, neurons that manage to gather sufficient amounts of the appropriate neurotrophic factor survive. Neurons that gather insufficient trophic factor die.

Cell death

5 Because the amount of neurotrophic factor matches the number of target cells, this process results in a rough matching of the size of the target and the number of innervating neurons.

6 Later in development, axonal processes also compete for neurotrophic factors. Active synapses compete more successfully than inactive synapses.

7 Because experience can modulate synaptic activity, different experiences can result in the maintenance of different patterns of synaptic connectivity.

An explosion of synapse formation is followed by synapse rearrangement

Before birth and after, neurons in the human cortex grow ever longer and more elaborate dendrites, each jammed with synapses. As we noted earlier, this massive increase in dendrites and synapses is responsible for most of the increase in brain size after birth (**FIGURE 13.29**). But just as not all the neurons produced by a developing individual are kept into adulthood, some of the synapses formed early in development are later retracted. Some original synapses are lost, and many, many new synapses are formed (see Figure 13.25, step 6). This **synapse rearrangement**, or *synaptic remodeling*, typically takes place after the period of cell death.

synapse rearrangement   Also called *synaptic remodeling*. The loss of some synapses and the development of others.

FIGURE 13.29 Cerebral Cortex Tissue in the Early Development of Humans (From Conel, 1939, 1947, 1959.)

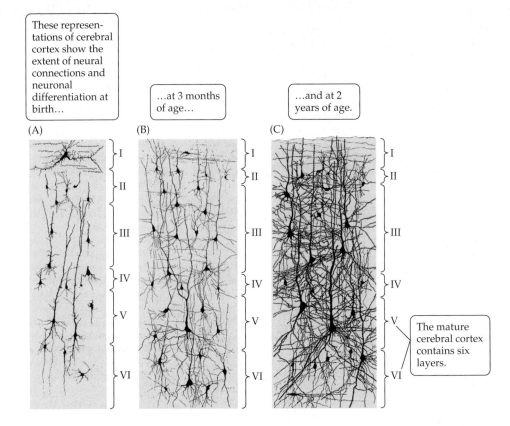

These representations of cerebral cortex show the extent of neural connections and neuronal differentiation at birth…

…at 3 months of age…

…and at 2 years of age.

The mature cerebral cortex contains six layers.

For example, as we learned already, about half of the spinal motoneurons that form die later (see Figure 13.27). By the end of the cell death period, each surviving motoneuron innervates many muscle fibers, and every muscle fiber is innervated by several motoneurons. But later the surviving motoneurons retract many of their axon collaterals, until each muscle fiber comes to be innervated by only one motoneuron. Again, which synaptic connections are retained, and which new connections are formed, is thought to depend on competition for trophic factors during development (see Figure 13.28) and/or competition between Hebbian synapses (see Figure 13.23).

Similar events have been documented in several neural regions, including the cerebellum (Mariani and Changeaux, 1981), the brainstem (Jackson and Parks, 1982), the visual cortex (Hubel et al., 1977), and the autonomic nervous system (Lichtman and Purves, 1980). In human cerebral cortex there is a net loss of synapses from late childhood until mid adolescence (**FIGURE 13.30**). This synaptic remodeling is evident in thinning of the cortical gray matter as pruning of dendrites and axon terminals progresses. The thinning process continues in a caudal–rostral (posterior–anterior) direction during maturation (**FIGURE 13.31**), so prefrontal cortex is affected last (Gogtay et al., 2004). Since prefrontal cortex is important for inhibiting behavior (see Chapter 14), this delayed brain maturation may contribute to teenagers' impulsivity and lack of control (Paus et al., 2008).

What determines which synapses are kept and which are lost? Although we don't know all the factors, one important influence is neural activity. One theory is that active synapses take up some neurotrophic factor that maintains the synapse, while inactive synapses get too little trophic factor to remain stable (see Figure 13.28). Intellectual stimulation probably contributes, as suggested by the fact that teenagers with the highest IQ show an especially prolonged period of cortical thinning (P. Shaw et al., 2006). Another stage of brain development, the formation of myelin sheaths for axons, is discussed in **A STEP FURTHER 13.3**, on the website.

This influence of experience on the developing brain brings us to the question of how both genes and the environment affect human intelligence.

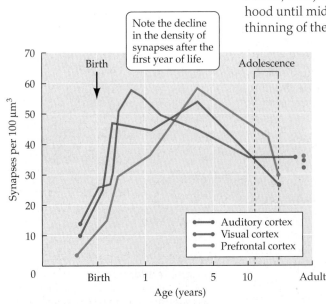

Note the decline in the density of synapses after the first year of life.

Birth

Adolescence

- Auditory cortex
- Visual cortex
- Prefrontal cortex

FIGURE 13.30 The Postnatal Development of Synapses in Human Cortex (From Huttenlocher et al., 1982.)

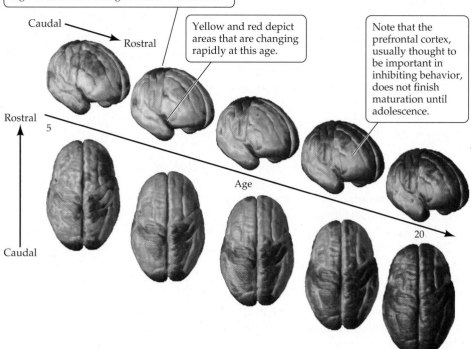

The layer of gray matter on the surface of the cortex gets thinner throughout development, as synapses are retracted. Purple and blue depict regions with little change in cortical thickness.

Yellow and red depict areas that are changing rapidly at this age.

Note that the prefrontal cortex, usually thought to be important in inhibiting behavior, does not finish maturation until adolescence.

Caudal

Rostral

Rostral

5

Caudal

Age

20

FIGURE 13.31 Synapse Rearrangement in the Developing Human Brain (From Gogtay et al., 2004, courtesy of Nitin Gogtay.)

how's it going ❓

1. What six stages of cellular processes take place in the developing brain?
2. What is cell differentiation, and what guides this process in each cell in the developing brain?
3. What two classes of structures undergo loss during development?
4. Speculate about why these "regressive" events might be important.

Genes Interact with Experience to Guide Brain Development

Many factors influence the emergence of the form, arrangements, and connections of the developing brain. One influence is genes, which direct the production of every protein the cell can make. An individual who has inherited an altered gene will make an altered protein, which will affect any cell structure that includes that protein. Thus, every neuronal structure, and therefore every behavior, can be altered by changes in the appropriate gene(s). It is useful to think of genes as *intrinsic* factors—that is, factors that originate within the developing cell itself. All other influences we can consider *extrinsic*—originating outside of the developing cell.

Genotype is fixed at birth, but phenotype changes throughout life

Two terms help illustrate how these intrinsic and extrinsic factors interact. The sum of all the intrinsic, genetic information that an individual has is its **genotype**. The sum of all the physical characteristics that make up an individual is its **phenotype**. Your genotype was determined at the moment of fertilization and remains the same throughout your life. But your phenotype changes constantly, as you grow up and grow old and even, in a tiny way, as you take each breath. In other words, phenotype is determined by the interaction of genotype and extrinsic factors, including experience. Thus, as we'll see, individuals who have identical genotypes do not have identical phenotypes, because they have not received identical extrinsic influences. And since their nervous system phenotypes are somewhat different, they do not behave exactly the same.

genotype All the genetic information that one specific individual has inherited.

phenotype The sum of an individual's physical characteristics at one particular time.

Be Careful What You Eat Millie Lonergan, who has phenylketonuria, eats fruit and protein-free rice and pasta (without cheese) for a diet low in phenylalanine.

phenylketonuria (PKU) An inherited disorder of protein metabolism in which the absence of an enzyme leads to a toxic buildup of certain compounds, causing intellectual disability.

clones Asexually produced organisms that are genetically identical.

epigenetics The study of factors that affect gene expression without making any changes in the nucleotide sequence of the genes themselves.

methylation A chemical modification of DNA that does not affect the nucleotide sequence of a gene but makes that gene less likely to be expressed.

Several hundred different genetic disorders affect the metabolism of proteins, carbohydrates, or lipids, having a profound impact on the developing brain. Characteristically, the genetic defect is the absence of a particular enzyme that controls a critical biochemical step in the synthesis or breakdown of a vital body product.

An example is **phenylketonuria** (**PKU**), a recessive hereditary disorder of protein metabolism that at one time resulted in many people with intellectual disability. About one out of 100 persons is a carrier; one in 10,000 births produces an affected victim. The basic defect is the absence of an enzyme necessary to metabolize phenylalanine, an amino acid that is present in many foods. As a result, the brain is damaged by an enormous buildup of phenylalanine, which becomes toxic.

The discovery of PKU marked the first time that an inborn error of metabolism was associated with intellectual disability. These days, the level of phenylalanine in the blood is measured in children a few days after birth. Early detection is important because brain impairment can be prevented simply by reducing phenylalanine in the diet. Such dietary control of PKU is critical during the early years of life, especially before age 2; after that, diet can be relaxed somewhat. Note this important example of the interaction of genes and the environment in PKU: the dysfunctional gene causes intellectual disability *only* in the presence of phenylalanine. Reducing phenylalanine consumption reduces or prevents this effect of the gene.

PKU illustrates one reason why, despite the importance of genes for nervous system development, understanding the genotype alone could never enable an understanding of the developing brain. Knowing that a baby is born with PKU doesn't tell you anything about how that child's brain will develop *unless* you also know something about the child's diet. Another reason why genes alone cannot tell the whole story is that experience can affect the activity of genes, as we discuss next.

Experience regulates gene expression in the developing and mature brain

Genetically identical animals, called **clones**, used to be known mainly in science fiction and horror films. But life imitates art. In grasshopper clones, many neurons show differences in neural connections despite the identical genotypes (Goodman, 1979). Likewise, genetically identical cloned pigs show as much variation in behavior and temperament as do normal siblings (G. S. Archer et al., 2003), and genetically identical mice raised in different laboratories behave very differently on a variety of tests (Finch and Kirkwood, 2000). If genes are so important to the developing nervous system, how can genetically identical individuals differ in their behavior?

Recall that although nearly all of the cells in your body have a complete copy of your genotype, each cell uses only a small subset of those genes at any one time. We mentioned earlier that when a cell uses a particular gene to make a particular protein, we say the cell has *expressed* that gene. **Epigenetics** is the study of factors that affect gene expression without making any changes in the nucleotide sequence of the genes.

An important epigenetic factor affecting the developing brain in mice is the mothering they receive. If genetically identical embryos of one mouse strain are implanted into the womb of a foster mother of either their own strain or another strain, their behavior is affected (Francis et al., 2003). Strain Black6 males carried and raised by mothers from the albino strain show significant differences in several behaviors, including maze running and measures of anxiety (**FIGURE 13.32**). Since the various males are *genetically identical* to one another, their different behaviors must be due to the effect of different prenatal environments and postnatal experiences on how those genes are expressed.

One particular influence of mothering on gene expression has been well documented. **Methylation** is a chemical modification of DNA that does not affect the nucleotide sequence of a gene but makes that gene less likely to be expressed. Michael Meaney and colleagues demonstrated that rodent pups provided with inattentive mothers, or subjected to interruptions in maternal care, secrete more glucocorticoids

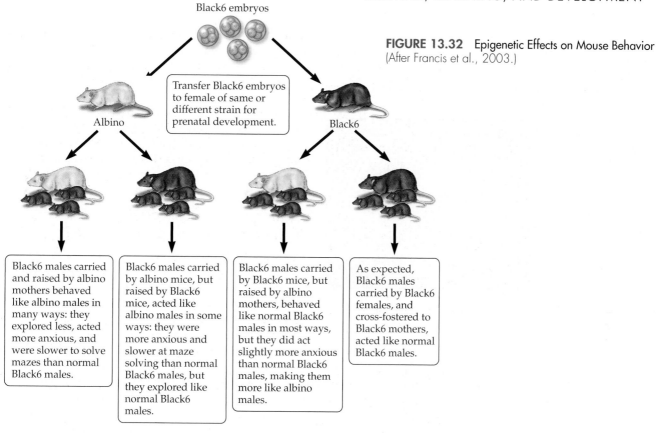

Black6 embryos

Transfer Black6 embryos to female of same or different strain for prenatal development.

Albino

Black6

Black6 males carried and raised by albino mothers behaved like albino males in many ways: they explored less, acted more anxious, and were slower to solve mazes than normal Black6 males.

Black6 males carried by albino mice, but raised by Black6 mice, acted like albino males in some ways: they were more anxious and slower at maze solving than normal Black6 males, but they explored like normal Black6 males.

Black6 males carried by Black6 mice, but raised by albino mothers, behaved like normal Black6 males in most ways, but they did act slightly more anxious than normal Black6 males, making them more like albino males.

As expected, Black6 males carried by Black6 females, and cross-fostered to Black6 mothers, acted like normal Black6 males.

FIGURE 13.32 Epigenetic Effects on Mouse Behavior (After Francis et al., 2003.)

in response to stress as adults (T. Y. Zhang and Meaney, 2010). Poor maternal care programs this heightened stress hormone response by inducing methylation of the glucocorticoid receptor gene in the brain, making the pups hyperresponsive to stress for the rest of their lives.

A similar mechanism may apply to humans, because this same gene is also more likely to be methylated in the postmortem brains of suicide victims than of controls, *but only in those victims who were subjected to childhood abuse.* Suicide victims who did not suffer childhood abuse were no more likely to have the gene methylated than were controls (McGowan et al., 2009). These results suggest that methylation of the gene in abused children may make them hyperresponsive to stress as adults—a condition that may lead them to take their own lives. This is a powerful demonstration of epigenetic influences on behavior. Other developmental disorders are also influenced by both genes and the environment, as we discuss in **A STEP FURTHER 13.4**, on the website.

Taken together, these studies lead us to the conclusion that the incredible intelligence of the adult human is due not only to the inheritance of genes provided us by natural selection, but also to the effect of the environment and experience that determines where and when those genes are expressed in the brain, especially in development. We wish we could tell you that once your brain has been sharpened by experience (including what you gain by reading this book), you will remain brilliant forever. Sadly, development continues relentlessly toward old age. Just as our faces and bodies weaken and fade, the brain also declines, the depressing topic that concludes this chapter.

─────────────────────────────── how's it going ❓

1. Compare changes in genotype and phenotype in an individual during development and aging.

2. Describe two examples of epigenetic effects on development.

3. How does PKU illustrate an interaction between genes and the environment?

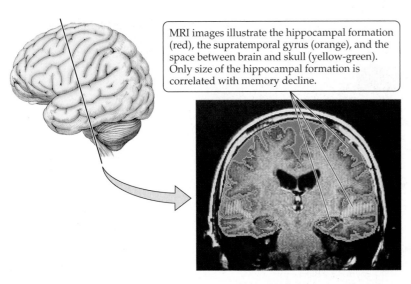

MRI images illustrate the hippocampal formation (red), the supratemporal gyrus (orange), and the space between brain and skull (yellow-green). Only size of the hippocampal formation is correlated with memory decline.

FIGURE 13.33 Hippocampal Shrinkage Correlates with Memory Decline in Aging (From Golomb et al., 1994, MRI courtesy of James Golomb.)

The Brain Continues to Change as We Grow Older

The passage of time brings us an accumulation of joys and sorrows—perhaps riches and fame—and a progressive decline in many of our abilities. Although slower responses seem inevitable with aging, many of our cognitive abilities show little change during the adult years, until we reach an advanced age. What happens to brain structure from adolescence to the day when we all become a little forgetful and walk more hesitantly?

Memory impairment correlates with hippocampal shrinkage during aging

In a study of healthy and cognitively normal people age 55–87, investigators asked whether mild impairment in memory is specifically related to reduction in size of the hippocampal formation or is better explained by generalized shrinkage of brain tissue (Golomb et al., 1994). Volunteers took a series of memory tests and were scored for both immediate recall and delayed recall, and their brains were measured from MRI images (**FIGURE 13.33**). When effects of sex, age, IQ, and overall brain atrophy were eliminated statistically, hippocampal formation volume was the only brain measure that correlated significantly with the delayed recall score.

PET scans of elderly people reveal that cerebral metabolism normally remains almost constant as we age. This stability is in marked contrast to the dramatic decline of cerebral metabolism in Alzheimer's disease, which we will consider next.

Alzheimer's disease is associated with a decline in cerebral metabolism

The population of elderly people in the United States is increasing dramatically. Most people reaching an advanced age lead happy, productive lives, although at a slower pace than they did in their earlier years. In a growing number of elderly people, however, age has brought a particular agony: the disorder called **Alzheimer's disease**, named after Alois Alzheimer (1864–1915), the neurologist who first described a type of **dementia** (drastic failure of cognitive ability, including memory failure and loss of orientation) appearing before the age of 65.

Over 4 million Americans suffer from Alzheimer's disease, and the progressive aging of our population means that these ranks will continue to swell. This disorder is found worldwide with almost no geographic differences. The frequency of Alzheimer's increases with aging up to age 85–90 (Rocca et al., 1991), but people who reach that age *without* symptoms become increasingly *less* likely ever to develop them (Breitner et al., 1999). This last finding indicates that Alzheimer's is in fact a disease, and not simply the result of wear and tear in the brain. The fact that remaining physically and mentally active reduces the risk of developing Alzheimer's disease (Smyth et al., 2004) also refutes the notion that brains simply "wear out" with age. Extensive use of the brain makes Alzheimer's *less* likely.

Alzheimer's disease begins as a loss of memory of recent events. Eventually this memory impairment becomes all-encompassing, so extensive that patients with Alzheimer's cannot maintain any form of conversation, because both context and prior information are rapidly lost. They cannot answer simple questions such as What year is it? Who is the president of the United States? or Where are you now? Cognitive decline is progressive and relentless. In time, patients become disoriented and easily lose themselves even in familiar surroundings.

Observations of the brains of patients with Alzheimer's reveal striking cortical atrophy, especially in the frontal, temporal, and parietal areas. PET scans show marked reduction of metabolism in posterior parietal cortex and some portions of the temporal

Alzheimer's disease A form of dementia that may appear in middle age but is more frequent among the aged.

dementia Drastic failure of cognitive ability, including memory failure and disorientation.

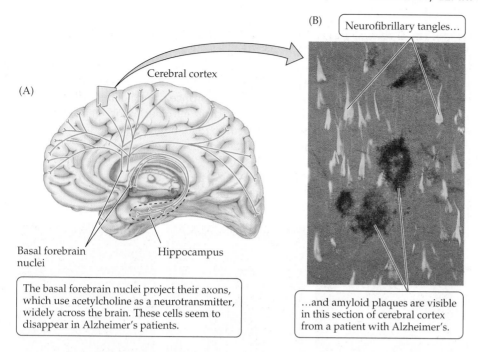

(A)

Cerebral cortex

(B)

Neurofibrillary tangles...

Basal forebrain
nuclei

Hippocampus

The basal forebrain nuclei project their axons,
which use acetylcholine as a neurotransmitter,
widely across the brain. These cells seem to
disappear in Alzheimer's patients.

...and amyloid plaques are visible
in this section of cerebral cortex
from a patient with Alzheimer's.

FIGURE 13.34   Patients with Alzheimer's
Show Structural Changes in the Brain (After
Roses, 1995; micrograph courtesy of Gary
W. Van Hoesen.)

lobe (N. L. Foster et al., 1984). The brains of individuals suffering from Alzheimer's
also reveal progressive changes at the cellular level (**FIGURE 13.34**):

- Strange patches termed **amyloid plaques** appear in cortex, the hippocampus, and
  associated limbic system sites. The plaques are formed by the buildup of a substance
  called **beta-amyloid** (Selkoe, 1991), which is how amyloid plaques got their name.

- Some cells show abnormalities called **neurofibrillary tangles**, which are abnormal
  whorls of neurofilaments that form a tangled array inside the cell. The number of
  neurofibrillary tangles is directly related to the magnitude of cognitive impairment,
  and they are probably a secondary response to amyloid plaques.

- Patients with Alzheimer's gradually lose many neurons in the basal forebrain,
  which make the transmitter acetylcholine. Drugs that boost ACh signaling may reduce some of the symptoms of Alzheimer's for a time.

One hypothesis about how these processes are related to each other is offered in
A STEP FURTHER 13.5, on the website.

amyloid plaque   Also called *senile
plaque*. A small area of the brain that
has abnormal cellular and chemical
patterns. Amyloid plaques correlate
with dementia.

beta-amyloid   A protein that accumulates in amyloid plaques in Alzheimer's
disease.

neurofibrillary tangle   An abnormal
whorl of neurofilaments within nerve
cells that is seen in Alzheimer's disease.

SIGNS & SYMPTOMS

Imaging Alzheimer's Plaques

At present, the only surefire diagnosis for Alzheimer's is postmortem examination
of the brain revealing amyloid plaques and neurofibrillary tangles. But one innovative approach is to inject a dye called Pittsburgh Blue (PiB), which has an affinity for beta-amyloid. Then a PET scan determines whether the dye accumulates in
the brain (Wolk et al., 2009). The brain of virtually every patient diagnosed with
Alzheimer's accumulates the dye, as do the brains of many elderly people showing mild cognitive impairment (**FIGURE 13.35**). A recent meta-analysis of findings
from thousands of participants confirmed that levels of amyloid, as revealed by
PiB imaging, indeed correlated with who would develop Alzheimer's (Ossenkoppele et al., 2015). One important implication of this finding is that now it will be
easier to track the effectiveness of various therapies for Alzheimer's.

One treatment strategy is to develop drugs that interfere with enzymes that
favor beta-amyloid production (O. Singer et al., 2005; Yu et al., 2015). In the
meantime, and in keeping with the repeated theme of this chapter—that genes

continued

SIGNS & SYMPTOMS *continued*

and experience interact—there is good evidence that physical activity (Ngandu et al., 2015), mental activity (Gates and Sachdev, 2014), and adequate sleep (Gelber et al., 2015) can postpone the appearance of Alzheimer's disease. So unless you want to pin your hopes on medical miracles in the future, the best way to avoid suffering from Alzheimer's is to remain physically and mentally active. Perhaps you should consider a career in neuroscience research...

FIGURE 13.35　Imaging Amyloid Plaques in the Brain (From Wolk et al., 2009, courtesy of the University of Pittsburgh Amyloid Imaging Group.)

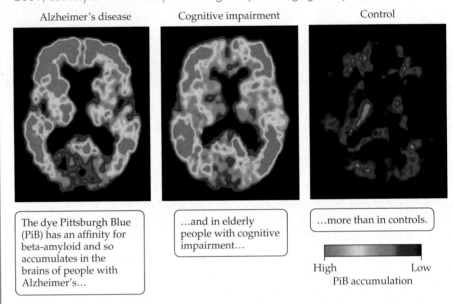

Alzheimer's disease　　Cognitive impairment　　Control

The dye Pittsburgh Blue (PiB) has an affinity for beta-amyloid and so accumulates in the brains of people with Alzheimer's...

...and in elderly people with cognitive impairment...

...more than in controls.

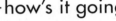

High　　　　Low
PiB accumulation

how's it going ❓

1. What is Alzheimer's disease, and how is it diagnosed?
2. Although genes clearly influence the risk of Alzheimer's, what environmental factors can postpone its onset?

Recommended Reading

Baddeley, A. D., Eysenck, M., and Anderson, M. C. (2009). *Memory.* London, England: Psychology Press.

Gilbert, S. F. (2014). *Developmental Biology* (10th ed.). Sunderland, MA: Sinauer.

Kesner, R. P., and Martinez, J. L. (Eds.). (2007). *The Neurobiology of Learning and Memory* (2nd ed.). San Diego, CA: Elsevier.

Marcus, G. (2008). *The Birth of the Mind: How a Tiny Number of Genes Creates the Complexities of Human Thought.* New York, NY: Basic Books.

Rudy, J. W. (2014). *The Neurobiology of Learning and Memory* (2nd ed.). Sunderland, MA: Sinauer.

Sanes, D. H., Reh, T. A., and Harris, W. A. (2012). *Development of the Nervous System* (3rd ed.). San Diego, CA: Academic Press.

Squire, L. R., and Kandel, E. R. (2008). *Memory: From Mind to Molecules.* Greenwood Village, CO: Roberts.

You should be able to relate each summary to the adjacent illustration, including structures and processes. Go to the online version of this summary (scan the QR code above) for links to figures, animations, and activities that will help you consolidate the material.

1 The **hippocampus**, **mammillary bodies**, and **dorsomedial thalamus** are part of a network that must be intact to form new **declarative memories**—memories that we can declare to others. Damage to these regions can cause **amnesia**, the impairment of memory. Review Figure 13.1, Animation 13.2

2 Removal of the hippocampus and nearby cortex left **patient H.M.** unable to form declarative memories that lasted more than a few minutes. He could, however, learn new skills, such as mirror tracing, showing that his **nondeclarative** (or procedural) **memory** for perceptual and motor behaviors was intact. Review Figures 13.2–13.4

3 Declarative memory consists of **semantic memory** of facts as well as **episodic memory** of particular incidents in the past. Brain damage can remove one type of declarative memory without affecting the other, indicating that the two kinds of memories are stored separately. Review Figures 13.4 and 13.12, Activity 13.1

4 Nondeclarative memory, which includes **skill learning**, **priming**, and conditioning, is demonstrated through performance. In nonhuman animals, most tests of memory measure nondeclarative memory, an exception being the **delayed non-matching-to-sample task** in monkeys. Review Figures 13.3–13.6, Activity 13.2

5 **Nonassociative learning** includes **habituation**, while **associative learning** includes **classical conditioning** (or Pavlovian conditioning) and **instrumental conditioning** (or operant conditioning). In the slug *Aplysia*, habituation is due to a weakening of the synapse between the sensory neuron and the motoneuron. In mammals, classical conditioning happens through changes in synaptic strength in the **cerebellum**. Review Figure 13.9, 13.18–13.20

6 Different kinds of learning depend on different brain regions. Spatial learning requires an intact hippocampus, while skill learning relies on the **basal ganglia**, and recognition relies on the cortex. Review Figures 13.11 and 13.12

7 Memories are classified by how long they last. The **sensory buffer** is a very brief recollection of sensations. **Short-term memory (STM)**, sometimes called *working memory*, lasts only a few minutes. Then the memory is either lost or transferred to **long-term memory (LTM)**, which may last a lifetime. The successive processes transferring information from one store to the other are **encoding**, **consolidation**, and **retrieval**. Memories are subject to distortion during recall and reconsolidation. Strong emotion can affect the strength of memories, as in **posttraumatic stress disorder (PTSD)**. Review Figures 13.12–13.14, Box 13.1

8 **Long-term potentiation (LTP)** is a lasting increase in amplitude of the response of neurons caused by brief high-frequency stimulation of their afferents (**tetanus**). In the hippocampus, LTP depends on the activation of **NMDA receptors**, which induces an increase in the number of postsynaptic **AMPA receptors** and greater neurotransmitter release. These are examples of changes in **Hebbian synapses**, which become stronger if they successfully drive the postsynaptic cell, and weaker if they are unsuccessful. Review Figures 13.21–13.23, Animation 13.3, Video 13.4

9 The brain develops in six stages: (1) **neurogenesis**, (2) **cell migration**, (3) **cell differentiation**, (4) **synaptogenesis**, (5) **cell death** (or apoptosis), and (6) **synapse rearrangement**. In adulthood, synapse rearrangement continues throughout the brain, and experience guides this process. Some neurogenesis also occurs in adults. Review Figures 13.24–13.31, Animation 13.5, Video 13.6, Activity 13.3

10 While genes play an important role in brain development, the environment and experience, such as mothering, can affect **gene expression**. That is, genetically identical individuals may behave very differently from one another because development is an **epigenetic** process. For example, the severe mental impairment that may accompany **phenylketonuria (PKU)** is avoided by controlling diet. While genes affect the buildup of **beta-amyloid** in **amyloid plaques** and therefore increase the risk of **Alzheimer's disease**, mental and physical activity postpone the onset of the disease. Review Figures 13.32–13.35

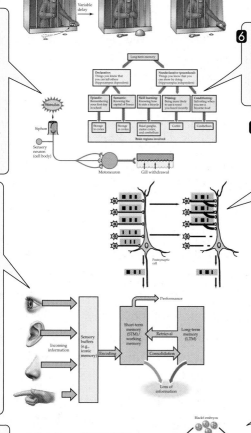

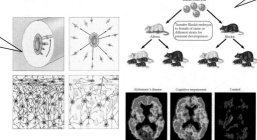

chapter

14

Attention and Consciousness

Attention to Details

Everyone agreed that Parminder and her family were bighearted—generous and friendly to a fault—but they were also "bad" hearted, in one sad sense. Many of Parminder's relatives had suffered early heart attacks and strokes, and now, in her 68th year, Parminder shared that unhappy fate. In February, and then again in September, blood clots that originated in Parminder's heart found their way into the complex of arteries in her brain, cutting off blood flow to the surrounding brain tissue. The two strokes that resulted were exceptional, however, because they were exact mirror images—they damaged identical regions of the left and right parietal lobes.

Parminder's unlikely lesions produced equally unlikely symptoms. A few weeks after her second stroke, Parminder had regained many of her intellectual powers—she could converse normally and remember things. Her visual fields were apparently normal too, but her visual *perception* was anything but normal. Parminder had lost the ability to perceive more than one thing at a time. For example, she could see her husband's face just fine, but she couldn't judge whether he had glasses on or not. It turned out that she could see the glasses *or* she could see the face, but she couldn't perceive them *both* at the same time. When shown a drawing of several overlapping items, she could perceive and name only one at a time. Furthermore, she couldn't understand where the objects she saw were located. It was as if Parminder was lost in space, able to focus on only one object or detail at a time, apparently alone in a world of its own. What could explain Parminder's symptoms?

To see the video
Attention and Perception,
go to

2e.mindsmachine.com/14.1

Although she may delight in pretending otherwise, the average 5-year-old knows exactly what it means when her exasperated parents shout, "Pay attention!" We all share an intuitive understanding of the term *attention*, but it is tricky to formally define. A definition provided in the nineteenth century by the great American psychologist William James is a good place to start:

> *Everyone knows what attention is. It is the taking possession by the mind, in clear and vivid form, of one out of what seem several simultaneously possible objects or trains of thought. Focalization, concentration, of consciousness are of its essence. It implies withdrawal from some things in order to deal effectively with others, and is a condition which has a real opposite in the confused, dazed, scatterbrained state. (W. James, 1890, pp. 403–404)*

Clearly, James understood that attention can be effortful, improves perception, and acts as a filter. This continual shifting of our focus from one interesting stimulus to the next lies at the heart of our innermost conscious experiences and our awareness of the world around us and our place in it. So, we open this chapter by exploring the behavioral and neural dimensions of attention before turning our focus to consciousness, the second major topic of the chapter.

FIGURE 14.1   Covert Attention

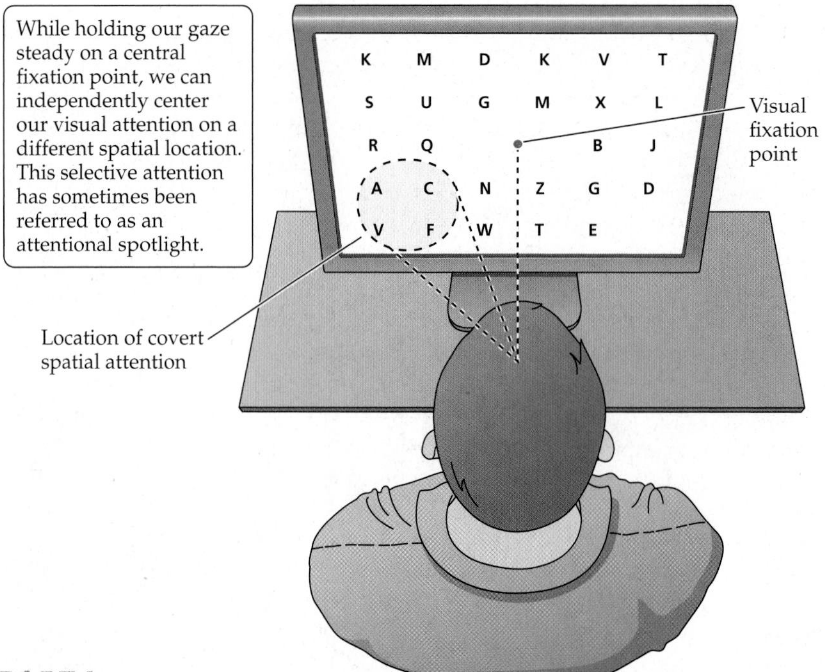

While holding our gaze steady on a central fixation point, we can independently center our visual attention on a different spatial location. This selective attention has sometimes been referred to as an attentional spotlight.

Visual fixation point

Location of covert spatial attention

To view the
Brain Explorer,
go to

2e.mindsmachine.com/14.2

attention   Also called *selective attention*. A state or condition of selective awareness or perceptual receptivity, by which specific stimuli are selected for enhanced processing.

overt attention   Attention in which the focus coincides with sensory orientation (e.g., you're attending to the same thing you're looking at).

covert attention   Attention in which the focus can be directed independently of sensory orientation (e.g., you're attending to one sensory stimulus while looking at another).

cocktail party effect   The selective enhancement of attention in order to filter out distracters, as you might do while listening to one person talking in the midst of a noisy party.

shadowing   A task in which the participant is asked to focus attention on one ear or the other while stimuli are being presented separately to both ears, and to repeat aloud the material presented to the attended ear.

PART I
Attention

Attention Focuses Cognitive Processing on Specific Objects

In general, **attention** (or *selective attention*) is the process by which we select or focus on one or more specific stimuli—either external phenomena or internal processes—for enhanced study and analysis. Most of the time we direct our eyes and our attention to the same target, a process known as **overt attention**. For example, as you read this sentence, it is both the center of your visual gaze and (we hope) the main item that your brain has selected for attention. But if we choose to, we can also shift the focus of our visual attention without changing the direction of visual gaze, keeping our eyes fixed on one location while "secretly" scrutinizing a different location in peripheral vision (Helmholtz, 1962; original work published in 1894). Remember that teacher who, even when looking out the window, somehow knew instantly when someone passed a note? That's an example of what is known as **covert attention** (**FIGURE 14.1**).

Selective attention isn't restricted to visual stimuli. Imagine yourself chatting with an old friend at a noisy party. Despite the background noise, you would probably find it relatively easy to focus on what your friend was saying, even if she was speaking quietly, because attention aids your sensory perception—paying close attention to your friend enhances your processing of her speech and helps filter out distracters. This phenomenon, known as the **cocktail party effect**, nicely illustrates how attention acts to *focus* cognitive processing resources on a particular target. If your attention drifts to a different stimulus—for example, if you start eavesdropping on a more interesting conversation nearby—it becomes almost impossible to simultaneously follow what your friend is saying.

There are limits on attention

The powers of attention that help you to focus on a conversation with a friend in a noisy room normally use several types of sensory cues, such as where her speech sounds are coming from, how her face moves as she speaks, and what unique sounds her voice makes. But what if we restrict our attention to just one type of stimulus?

In **shadowing** experiments, participants must focus their attention on just one out of two or more simultaneous streams of stimuli. In a classic example of this technique, Cherry (1953) presented different streams of speech simultaneously to people's left and right ears via headphones—the technique is called *dichotic presentation*—and asked them to focus their attention on one ear or the other and report what they heard. Participants were able to accurately report what they heard in the attended ear, but they reported very little about what was said in the *nonattended* ear, aside from simple characteristics, such as the sex of the speaker. In fact, if a shadowing task is difficult enough, people may fail to detect even their own names in the unattended ear about two-thirds of the time (N. Wood and Cowan, 1995)!

Similar restrictions of attention can be seen in other sensory modalities, such as musical notes (Zendel and Alain, 2009) and visual stimuli. Participants closely attending to one complex visual event against a background of other moving stimuli—dancers weaving through a basketball game, for example—may show **inattentional blindness**: a surprising failure to perceive nonattended stimuli. And the unperceived stimuli can be things that you might think impossible to miss, like a gorilla strolling across a movie screen out of the blue (Simons and Chabris, 1999; Simons and Jensen, 2009). You can see an example on the website.

In general, **divided-attention tasks**—in which the participant is asked to process two or more simultaneous stimuli—confirm that attention is a limited resource and that it's very difficult to attend to more than one thing at a time, particularly if the stimuli are spatially separated (Bonnel and Prinzmetal, 1998). So, our limited selective attention generally acts like an **attentional spotlight**,

Gorillas in the Midst Who could miss the gorilla here? Most people do if they are concentrating on another task, such as counting the number of times people in white shirts touch the ball. (From Simon and Chabris, 1999.)

shifting around the environment, highlighting stimuli for enhanced processing. It's an adaptation that we share with many other species because, like us, they are confronted with the problem of extracting important signals from a noisy background (Bee and Micheyl, 2008). Birds, for example, must isolate the vocalizations of specific individuals from a cacophony of calls and other noises in the environment—an avian version of the cocktail party problem (Benney and Braaten, 2000). Having a single attentional spotlight helps us focus cognitive resources and behavioral responses toward the most important things in the environment at any given moment (the smell of smoke, the voice of a potential mate, a glimpse of a big spotted cat), while filtering out extraneous information.

In a sense, attention acts as a filter, blocking unimportant stimuli and focusing cognitive resources on only the most important events, thereby protecting the brain from being overwhelmed by the world. But the details of this **attentional bottleneck** have been elusive. Initial research gave evidence of an *early-selection model* of attention, in which unattended information is filtered out right away, at the level of the initial sensory input, as in the shadowing experiments (Broadbent, 1958). But other researchers noted that important but unattended stimuli (such as your name) may undergo substantial unconscious processing, right up to the level of semantic meaning and awareness, before suddenly capturing attention (N. Wood and Cowan, 1995), thus illustrating a *late-selection model* of attention. Many contemporary models of attention now combine both early- and late-selection mechanisms (e.g., Wolfe, 1994), and vigorous debate continues over their relative importance (for a classic demonstration, see **A STEP FURTHER 14.1**, on the website). A possible resolution to this debate involves **perceptual load**—the immediate processing demands presented by a stimulus. According to this view, when we focus on a very complex stimulus, the load on our perceptual processing resources is so great that there is nothing left over. We are thus unable to process competing unattended items, so those extra stimuli are excluded from the outset (N. Lavie, 1995; N. Lavie et al., 2004). But when we focus on stimuli that are easier to process, we may have enough perceptual resources left over to simultaneously process additional stimuli, right up to the level of semantic meaning and awareness. In this case the result is a late selection of stimuli to attend to (N. Lavie et al., 2009). So, according to this account, attention is continually rebalanced between early and late selection, depending on the difficulty and specific features of the task at hand.

inattentional blindness The failure to perceive nonattended stimuli that seem so obvious as to be impossible to miss.

divided-attention task A task in which the participant is asked to focus attention on two or more stimuli simultaneously.

attentional spotlight The shifting of our limited selective attention around the environment to highlight stimuli for enhanced processing.

attentional bottleneck A filter created by the limits intrinsic to our attentional processes, whose effect is that only the most important stimuli are selected for special processing.

perceptual load The immediate processing demands presented by a stimulus.

To see the video
Inattentional Blindness,
go to

2e.mindsmachine.com/14.3

how's it going ❓

1. How do you define *attention*?
2. Distinguish between overt and covert attention, giving examples of each. What is the attentional spotlight?
3. What is inattentional blindness, and under what circumstances might it occur?
4. How do early-selection effects of attention differ from late-selection effects? What single aspect of a stimulus may determine whether early or late selection occurs?

sustained-attention task A task in which a single stimulus source or location must be held in the attentional spotlight for a protracted period.

endogenous attention Also called *voluntary attention*. The voluntary direction of attention toward specific aspects of the environment, in accordance with our interests and goals.

symbolic cuing A technique for testing endogenous attention in which a visual stimulus is presented and participants are asked to respond as soon as the stimulus appears on a screen. Each trial is preceded by a meaningful symbol used as a cue to hint at where the stimulus will appear.

Attention May Be Endogenous or Exogenous

We've now seen that through an act of willpower we can direct our attention to specific stimuli without moving our eyes or otherwise reorienting. Early experiments on this phenomenon employed **sustained-attention tasks**, like the one depicted in Figure 14.1, where a single stimulus location must be held in the attentional spotlight for an extended period. Although these tasks are useful for studying basic phenomena, several key questions about attention require another approach. For example, how do we shift our attention around? How does attention enhance the processing of stimuli, and which brain regions are involved? To answer these questions, researchers devised clever tasks that employ stimulus cuing to control attention, which revealed two general categories of attention: endogenous and exogenous. Let's consider these different forms of attention.

researchers at work

We can choose which stimuli we will attend to

So far, our discussion of attention has focused mostly on what researchers call **endogenous** (or *voluntary*) **attention**. As its name (from the Greek *endon*, "within," and *genes*, "born") implies, endogenous attention comes from within; it is the voluntary, or *top-down*, direction of attention toward specific aspects of the environment, according to our interests and goals. **FIGURE 14.2** features the *symbolic cuing task*, developed by Michael Posner and used extensively to study endogenous attention. Studies using cuing tasks have confirmed that consciously directing your attention to the correct location or stimulus improves processing speed and accuracy. Conversely, directing your attention to an *incorrect* location or stimulus impairs processing efficiency.

How much does it help to shift your attention to a location before a stimulus occurs there? Posner's (1980) **symbolic cuing** task allows us to quantify how endogenous attention benefits processing. In a symbolic cuing task, participants stare at a point in the center of a computer screen and must press a key as soon as a specific target appears on the screen (this technique thus measures reaction time, as described in **BOX 14.1**). The stimulus is preceded by a cue that briefly flashes on the screen, hinting where the stimulus will appear. Most of the time, as in **FIGURE 14.2A**, the participant is provided with a *valid cue*; for example a rightward arrow flashes on the screen moments before the stimulus appears on the right side of the screen. In a few trials, like the one in **FIGURE 14.2B**, the arrow points the wrong way and thus provides an *invalid cue*. And in "neutral" control trials (**FIGURE 14.2C**), the cue doesn't provide any hint at all. Both the cue and the stimulus are on the screen so briefly that the participant doesn't have time to shift his gaze to them (and in any case, he has been told to stare at the fixation point).

Averaged over many trials, the reaction-time data (**FIGURE 14.2D**) clearly showed that people swiftly learn to use cues to predict stimulus location, shifting their attention *without shifting their gaze*. Compared with neutral trials, processing is significantly faster for validly cued trials, and participants pay a price for misdirecting their attention on those few trials in which the cue is invalid, pointing to the wrong side of the display. Many variants of the symbolic cuing paradigm—changing the timing of the stimuli, altering their complexity, requiring a choice between different responses (as in Box 14.1), and so on—have been used to study the neurophysiological mechanisms of attention (R. D. Wright and Ward, 2008), as we'll discuss a little later in the chapter.

FIGURE 14.2 Measuring the Effects of Endogenous Shifts of Attention

■ **Question**
Does shifting your attention to a new location, without shifting your gaze, improve the processing of stimuli that appear at the new location?

■ **Hypothesis**
Cued attention will improve subsequent processing—as measured by reaction time—if the stimulus appears at the cued location, but not if the stimulus appears elsewhere.

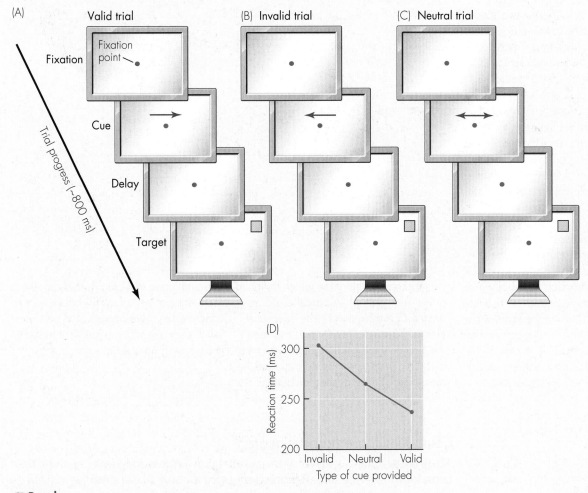

■ **Result**
Processing is improved for stimuli appearing at the cued location, and impaired for stimuli appearing at an uncued location.

■ **Conclusion**
Endogenous attention enhances processing independent of gaze.

BOX 14.1 Reaction-Time Responses, from Input to Output

Reaction-time measures are a mainstay of cognitive neuroscience research. In tests of *simple reaction time*, participants make a single response—for example, pressing a button—in response to experimental stimuli (the appearance of a target, the solution to a problem, a tone, or whatever the experiment is testing). In tests of *choice reaction time*, the situation is slightly more complicated: a person is presented with alternatives and has to choose among them (e.g., right versus wrong, same versus different) by pressing one of two or more buttons.

Reaction times in an uncomplicated choice-reaction-time test, in which the participant indicates whether two stimuli are the same or different, average about 300–350 milliseconds (ms). The delay between stimulus and response varies depending on the amount of neural processing required between input and output. The neural systems involved in an example task, and the timing of events in the response circuit, are illustrated in the figure. Brain activity proceeds from the primary visual cortex through a ventral visual object identification pathway (see Chapter 7) to prefrontal cortex, and then through premotor and primary motor cortex, down to the spinal motoneurons and out to the finger muscles. In the sequence shown in the figure—proceeding from the presentation of visual stimuli to a discrimination response—notice that it takes about 110 ms for the sensory system to recognize the stimulus (somewhere in the inferior temporal lobe), about 35 ms more for that information to reach the prefrontal cortex, and then about

A Reaction-Time Circuit in the Brain The sequence and timing of brain events that determine reaction time. LGN, lateral geniculate nucleus; V1, primary visual cortex; V2 and V4, extrastriate visual areas. (Timings based on Thorpe and Fabre-Thorpe, 2001.)

30 ms more to determine which button to push. After that, it takes another 75 ms or so for the movement to be executed (that is, 75 ms of time elapses between the moment the signal from the prefrontal cortex arrives in premotor cortex and the moment the finger pushes the button). It is

fascinating to think that something like this sequence of neural events happens over and over in more-complicated behaviors, such as recognizing a long-lost friend or composing an opera.

To see the animation **From Input to Output,** go to

2e.mindsmachine.com/14.4

exogenous attention Also called *reflexive attention*. The involuntary reorienting of attention toward a specific stimulus source, cued by an unexpected object or event.

Some stimuli are hard to ignore

There is a second way in which we pay attention to the world, involving more than just consciously steering our attentional spotlight around. Flashes, bangs, sudden movements—important changes generally—can instantly snatch our attention away from whatever we're doing, unless we are very focused. If you accidentally drop your glass in a restaurant, every conversation stops, and every head in the place swivels toward the source of the sound (you, embarrassingly). This sort of involuntary reorientation toward a sudden or important event is an example of **exogenous** (or *reflexive*) **attention**. It is considered to be a *bottom-up* process, because attention is being seized by sensory inputs from lower levels of the nervous system, rather than being directed by voluntary, conscious *top-down* processes of the forebrain.

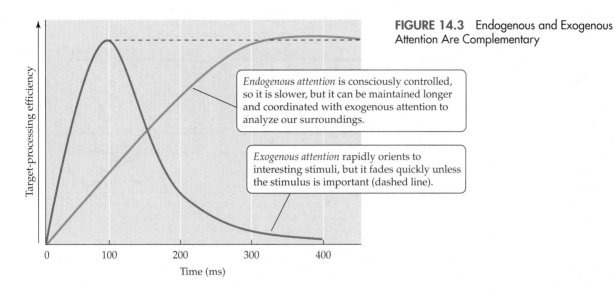

FIGURE 14.3 Endogenous and Exogenous Attention Are Complementary

Endogenous attention is consciously controlled, so it is slower, but it can be maintained longer and coordinated with exogenous attention to analyze our surroundings.

Exogenous attention rapidly orients to interesting stimuli, but it fades quickly unless the stimulus is important (dashed line).

Researchers study exogenous attention using a different kind of cuing task, called **peripheral spatial cuing**. In this task, instead of a meaningful symbol like an arrow, the cue that is presented is a simple sensory stimulus, such as a flash of light, occurring *in the location to which attention is to be drawn*. Research with this type of simple cuing confirmed that valid exogenous cues enhance the processing of subsequent stimuli at the same location, but only when the target stimulus closely follows the cue. At longer intervals between the cue and target, starting at about 200 milliseconds, a curious phenomenon is observed: detection of stimuli at the location where the valid cue occurred is increasingly *impaired* (Posner and Cohen, 1984; R. M. Klein, 2000). It's as though attention has moved on from where the cue occurred and is reluctant to return to that location. This **inhibition of return** probably evolved because it prevented exogenously controlled attention from settling on unimportant stimuli for more than an instant.

Normally, exogenous and endogenous attention work together to control cognitive activities (**FIGURE 14.3**), probably relying on somewhat overlapping neural

peripheral spatial cuing A technique for testing exogenous attention in which a visual stimulus is preceded by a simple task-irrelevant sensory stimulus either in the location where the stimulus will appear or in an incorrect location.

inhibition of return The phenomenon, observed in peripheral spatial cuing tasks when the interval between cue and target stimulus is 200 milliseconds or more, in which the detection of stimuli at the former location of the cue is increasingly impaired.

Where's Waldo? Puzzles like the "Where's Waldo?" series are classic examples of conjunction searches: You can find Waldo only if you search for the right combination of striped sweater, hat, glasses, and slightly goofy expression. Imagine how much easier it would be to find Waldo if everyone else on the beach were wearing green! In that case, finding Waldo would be a feature search (the only person not dressed in green) and he would "pop out" in the picture... but that wouldn't be any fun. (From *Where's Waldo*, © Martin Handford 2005.)

(A) Feature search

Find a green object   Find a square

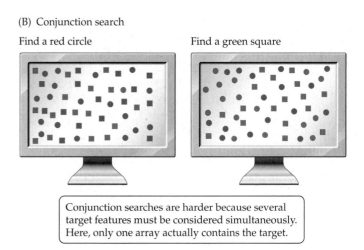

In these two feature search arrays, targets "pop out" because they differ from all other stimuli on one key feature, such as color or shape.

(B) Conjunction search

Find a red circle   Find a green square

Conjunction searches are harder because several target features must be considered simultaneously. Here, only one array actually contains the target.

FIGURE 14.4 **Visual Search** (After A. M. Treisman and Gelade, 1980.)

(C)

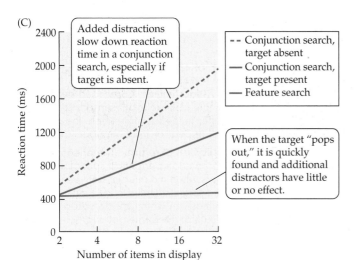

Added distractions slow down reaction time in a conjunction search, especially if target is absent.

-- Conjunction search, target absent
— Conjunction search, target present
— Feature search

When the target "pops out," it is quickly found and additional distractors have little or no effect.

Reaction time (ms)

Number of items in display

mechanisms. Anyone who has watched a squirrel at work has seen that twitchy interplay. When it comes to single-mindedly searching for tasty tidbits (an example of endogenous attention), a squirrel has few rivals. But even slight noises and movements (examples of exogenous attention cues) cause the squirrel to stop and scan its surroundings—a sensible precaution if, like a squirrel, you are yourself a tasty tidbit. Exogenous attentional cues can cross the boundaries between different sensory modalities in order to aid the processing of stimuli; a sudden sound coming from a particular location, for example, can improve the *visual* processing of a stimulus that appears there (Hillyard et al., 2015), especially if the cue is associated with reward (Pooresmaeili et al., 2014).

We use visual search to make sense of a cluttered world

Another familiar way that we use attention is to scan the world to locate specific objects among many—your car in a parking lot, for example, or your friend's face in a crowd. If the sought-after item varies in just one key attribute, the task can be pretty easy—searching for your red car among a bunch of silver and black ones, for example. In a simple **feature search** like this (**FIGURE 14.4A**), the sought-after item "pops out" immediately, no matter how many distracters are present (Joseph et al., 1997); effortful voluntary attention isn't needed. But that's something of a luxury.

More commonly we must use a **conjunction search**—searching for an item on the basis of a *combination* of two or more features, such as size and color (**FIGURE 14.4B**). This can become very difficult when, for example, you must simultaneously consider the hair, nose, eyes, and smile of your friend's face in a crowd—and the bigger the crowd grows, the harder the task becomes (unless your friend waves, in which case you can fall back on using a feature search—phew!).

Experimental results (**FIGURE 14.4C**) confirm what you probably already intuitively know: conjunction searches can be relatively slow and laborious, involving a large cognitive effort. That's because your brain has to deal with what is known as the **binding problem** (A. M. Treisman, 1996), which is this: How do we know which different features, processed by different regions of the brain, are bound together in a single object? And if those objects appear only infrequently—say, weapons in luggage or tumors in X-rays—they may go undetected with alarmingly high frequency, even by highly trained screeners (Wolfe et al., 2005). Our innate tendency is to let our attention wander between the various sources of stimuli in our environment.

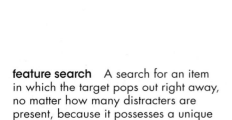

feature search   A search for an item in which the target pops out right away, no matter how many distracters are present, because it possesses a unique attribute.

conjunction search   A search for an item that is based on two or more features (e.g., size and color) that together distinguish the target from distracters that may share some of the same attributes.

binding problem   The question of how the brain understands which individual attributes blend together into a single object, when these different features are processed by different regions in the brain.

To uncover finer details of attentional mechanisms, neuroscientists use two general experimental strategies. First, we can look at *consequences of attention* in the brain, asking how selective attention to stimuli modifies brain activity and thus enhances stimulus processing. To do this, we need techniques with excellent **temporal resolution**—the ability to track changes in the brain that occur very quickly, in fractions of a second. Electrophysiological measures are commonly used for this purpose because they can track rapid oscillations in brain areas involved in stimulus processing. A second class of questions is concerned with the *mechanisms of attention*, the brain regions that produce and control attention, shifting it between different stimuli in different sensory modalities. For this approach we need the excellent **spatial resolution** of anatomical and brain-imaging studies in order to observe the detailed structure of the brain. We discuss the two types of experimental approaches individually in the following sections.

temporal resolution   The ability to track changes in the brain that occur very quickly.

spatial resolution   The ability to observe the detailed structure of the brain.

event-related potential (ERP)   Also called *evoked potential*. Averaged EEG recordings measuring brain responses to repeated presentations of a stimulus. Components of the ERP tend to be reliable because the background noise of the cortex has been averaged out.

──────────────── how's it going ❓

1. Summarize Posner's symbolic cuing task. What did this task reveal?

2. Compare and contrast endogenous attention and exogenous attention, and identify the principal ways in which they differ. What is inhibition of return, and does it relate to endogenous attention or to exogenous attention?

3. While conducting a visual search for something, we sometimes experience "pop-out." What is it? Is pop-out more closely associated with feature search or with conjunction search, and how do those differ?

4. Distinguish between temporal resolution and spatial resolution as they apply to brain-imaging techniques. How are they related?

The Electrical Activity of the Brain Provides Clues about Mechanisms of Attention

When many cortical neurons work together on a specific task, their activity becomes synchronized to some degree. You might think this would be easy to see in a standard EEG recording (that is, an *electroencephalogram*, the recording of the brain's electrical activity that we described in Chapter 3), but it isn't. Because of variation in the firing of the neurons, not to mention regional differences in the timing of brain activity, a real-time EEG recorded during an attention task looks surprisingly random. So instead, researchers have participants do the same task over and over again, and they *average* all the EEGs recorded during the repeated trials (**FIGURE 14.5A**). Over enough trials, the random variation averages out, and what's left is the overall electrical activity specifically associated with task performance (**FIGURE 14.5B**). This averaged activity, called the **event-related potential** (**ERP**) (Luck, 2005), tracks regional changes in brain activity much faster than brain-imaging techniques like fMRI do. For this reason, ERP has become the favorite tool of neuroscientists studying moment-to-moment consequences of attention in the brain, as we'll see.

Distinctive patterns of brain electrical activity mark voluntary shifts of attention

Consciously directing your attention to a particular auditory stimulus—for example, shadowing one ear, as we described earlier—has a predictable effect on the ERP. Between about 100 and 150 milliseconds after the onset of a sound stimulus, two large waves are seen in the ERP from the auditory cortex: an initial positive-going wave called *P1*, immediately followed by a larger negative-going wave called *N1* (see Figure 14.5B). The N1 wave reflects an important aspect of auditory attention: it is much larger following a stimulus that is being attended to than it is for the very same stimulus presented at the same ear but *not* attended to (Hillyard et al., 1973). Because the only thing that changes between conditions is the participants' attention to the

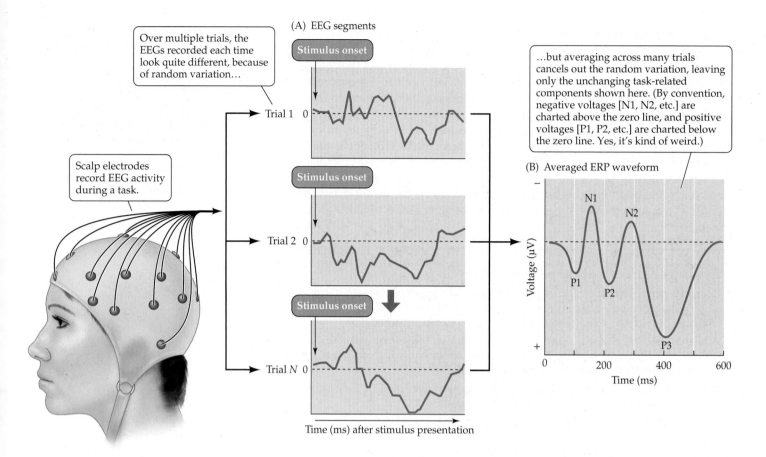

(A) EEG segments

Over multiple trials, the EEGs recorded each time look quite different, because of random variation…

Stimulus onset

Trial 1 0

…but averaging across many trials cancels out the random variation, leaving only the unchanging task-related components shown here. (By convention, negative voltages [N1, N2, etc.] are charted above the zero line, and positive voltages [P1, P2, etc.] are charted below the zero line. Yes, it's kind of weird.)

Scalp electrodes record EEG activity during a task.

Stimulus onset

Trial 2 0

(B) Averaged ERP waveform

Stimulus onset

Trial N 0

Time (ms) after stimulus presentation

FIGURE 14.5 Event-Related Potentials

auditory N1 effect A negative deflection of the event-related potential, occurring about 100 milliseconds after stimulus presentation, that is enhanced for selectively attended auditory input compared with ignored input.

P20–50 effect A positive deflection of the event-related potential, occurring about 20–50 milliseconds after stimulus presentation, that is enhanced for selectively attended auditory input compared with ignored input.

P3 effect A positive deflection of the event-related potential, occurring about 300 milliseconds after stimulus presentation, that is associated with higher-order auditory stimulus processing and late attentional selection.

stimuli, this **auditory N1 effect** must be a result of selective attention somehow acting on neural mechanisms to enhance processing of the sound.

The N1 wave isn't the only ERP feature affected by endogenous auditory attention. Using extra sensitive ERP techniques, researchers have found that just 20–50 milliseconds after an attended stimulus occurs, an enhanced positive-going wave is seen (Woldorff and Hillyard, 1991). Called the **P20–50 effect**, this component occurs so soon after stimulus onset that researchers think it must represent an unconscious, early-selection aspect of attention. There simply hasn't been enough time for the auditory stimuli to have been processed to the level of speech cues or meaning. Conversely, auditory attention also enhances much later ERP components, especially a wave called *P3* (or *auditory P300*) (see Figure 14.5). P3 is believed to reflect higher-order cognitive processing of the stimulus (Herrmann and Knight, 2001), so the **P3 effect** is an example of a late-selection effect of attention. Abnormal P3 responses are a reliable finding in people with schizophrenia, consistent with their difficulty in filtering out distracting information (Ford, 1999).

What about effects of attention on ERPs from *visual* stimuli? Because the neural systems involved in visual perception are different from those involved in audition, endogenous visual attention causes its own distinctive changes in the ERP. We can study these visual effects by collecting ERP data over occipital cortex—the primary visual area of the brain—while a participant performs a symbolic cuing task. **FIGURE 14.6** depicts this sort of experiment. On valid trials (remember, this is when the target appears as expected, in the location indicated by the cue, as in **FIGURE 14.6A**), electrodes over occipital cortex show a substantial enhancement of the ERP component P1, the positive wave that we saw occur over auditory cortex. It occurs about 70–100 milliseconds after stimulus onset, often carrying over into an enhancement of the N1 component immediately afterward (**FIGURE 14.6C**). Notice that the enhancement

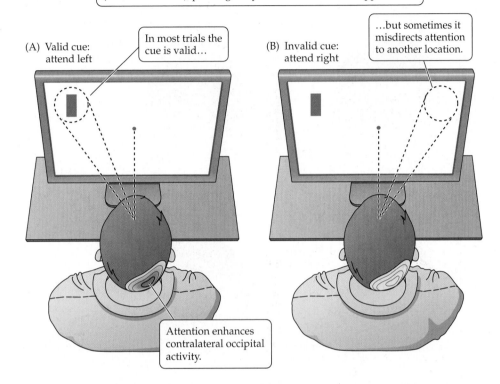

The individual fixates on the red point and covertly orients attention (dashed circle) in the direction indicated by a prior symbolic cue (such as an arrow), pressing a key as soon as a stimulus appears.

(A) Valid cue: attend left

In most trials the cue is valid…

(B) Invalid cue: attend right

…but sometimes it misdirects attention to another location.

Attention enhances contralateral occipital activity.

Correct direction of attention enhances neural processing, resulting in larger P1 and N1 components in the corresponding ERP. Note that neither the stimulus nor the individual's gaze differs between conditions; the change in *attention* is solely responsible for the ERP effect.

(C) Right occipital ERP

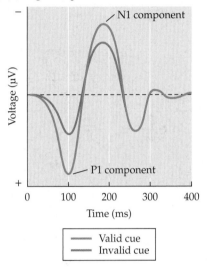

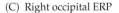

N1 component

P1 component

Voltage (μV)

Time (ms)

— Valid cue
— Invalid cue

FIGURE 14.6 ERP Changes in Endogenous Visual Attention

occurs over the *contralateral* occipital cortex; this is because information from the left visual field is processed in right occipital cortex, and vice versa, as we discussed in Chapter 7.

For invalid trials (**FIGURE 14.6B**), where attention is being directed elsewhere, the **visual P1 effect** isn't evident, even though the visual stimulus is identical and in the same location as in the valid trials. Interestingly, the P1 effect is evident only in visual tasks involving manipulations of *spatial* attention (*where* is the target?)—not other features, like color, orientation, or more complex properties that would be characteristic of late-selection tasks. These later-occurring aspects of endogenous visual attention may be evident in other types of tasks, affecting the later-occurring ERP component P3, as we discussed earlier (see Figure 14.5). Evidence suggests that extensive experience with action video games, which heavily rely on visual attention, is associated with plastic changes in neural mechanisms of attention and corresponding enhancements of longer-latency ERP components (Mishra et al., 2011). There may be a trade-off, however, in the form of impairments in social and emotional processing (yes, really: K. Bailey and West, 2013). And of course, some people are drawn to gaming because they already have good visuospatial abilities (Boot et al., 2008).

Reflexive visual attention has its own electrophysiological signature

When an external event captures our attention—that is, when we experience an exogenous shift of attention—what brain regions are involved, and how do they differ from those involved in voluntary shifts of attention? Determining whether exogenous attention involves the same brain mechanisms as endogenous attention is an interesting topic (to cognitive neuroscientists, anyway) that can be addressed using ERP.

In exogenous spatial cuing tasks, a visual P1 effect is seen in the contralateral ERP (Hopfinger and Mangun, 1998, 2001). This augmented P1 is just the same as what

visual P1 effect A positive deflection of the event-related potential, occurring 70–100 milliseconds after stimulus presentation, that is enhanced for selectively attended visual input compared with ignored input.

FIGURE 14.7 ERP Maps in Exogenous Visual Attention (Courtesy of Joe Hopfinger.)

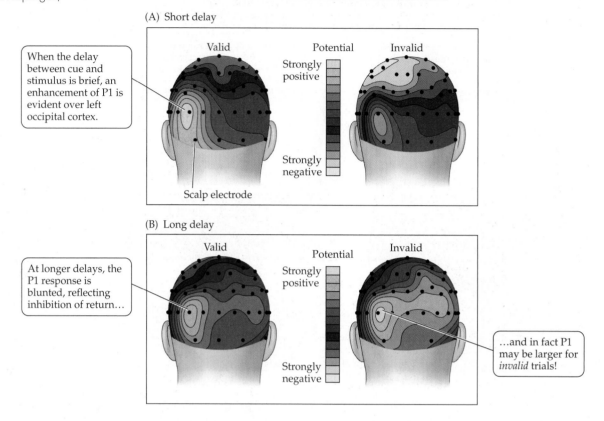

The same right-sided stimulus was presented in each case, but in some trials the stimulus was preceded by an unexpected flash in the location where the stimulus then appeared (valid trials). In other trials the flash drew attention to a different location (invalid trials).

(A) Short delay

When the delay between cue and stimulus is brief, an enhancement of P1 is evident over left occipital cortex.

Valid   Potential   Invalid
Strongly positive
Strongly negative

Scalp electrode

(B) Long delay

At longer delays, the P1 response is blunted, reflecting inhibition of return…

Valid   Potential   Invalid
Strongly positive
Strongly negative

…and in fact P1 may be larger for *invalid* trials!

we saw for endogenous visual attention in the previous section, and it is associated with enhanced processing by the visual cortex. At least, that's what happens when the target stimulus follows closely after the sensory cue. But as the delay between the sensory cue and the target lengthens, the P1 enhancement is reduced and even becomes inverted, as **FIGURE 14.7** shows; this is the electrophysiological manifestation of inhibition of return, which we discussed earlier. In contrast, the experience of "pop-out" in visual search tasks is associated with a unique ERP component called the *N2pc wave*, an enhancement of N2 (Luck and Hillyard, 1994).

Attention affects the activity of individual neurons

PET and fMRI operate too slowly to track the rapid changes in brain activity that occur in reaction-time tests. Instead, researchers have used "sustained-attention tasks" to confirm that attention enhances activity in brain regions that process key aspects of the target stimulus. So, for example, when participants are presented with stimuli made up of faces overlaid on pictures of houses, and asked to focus and sustain their attention only on the faces, fMRI reveals enhanced activation of a part of the brain (called the *fusiform face area*) that is specialized for face processing (O'Craven et al., 1999). A different brain area (the *parahippocampal place area*) is activated if attention is focused on the houses in the stimuli, instead of the faces. Results like these suggest that attention somehow acts directly on neurons, boosting the activity of those brain regions that process whichever stimulus characteristic has been targeted.

In Chapter 7 we discussed the distinctive receptive fields of visual neurons and how stimuli falling within these fields can excite or inhibit the neurons, causing them to produce more or fewer action potentials. In an important early study (**FIGURE 14.8**), Moran and Desimone (1985) recorded the activity of neurons in visual cortex while attention was shifted *within the cell's receptive field*. Using a system of rewards, the researchers trained monkeys to covertly attend to one spatial location or another while recordings

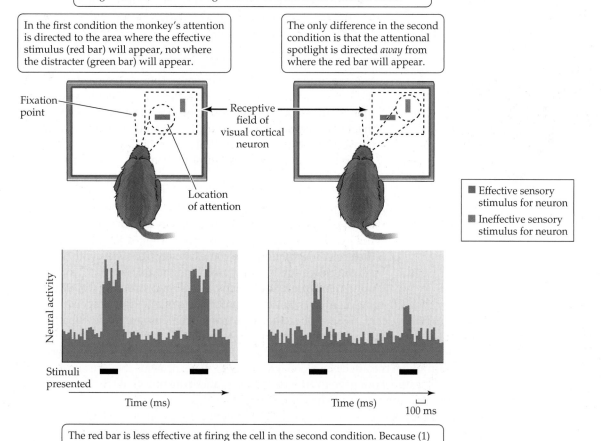

Here, a monkey has been trained to maintain central fixation while moving covert attention. Within the receptive field for the particular cortical neuron being recorded, the area being attended to is shown as a dashed circle.

FIGURE 14.8 Effect of Selective Attention on the Activity of Single Visual Neurons (After Moran and Desimone, 1985.)

In the first condition the monkey's attention is directed to the area where the effective stimulus (red bar) will appear, not where the distracter (green bar) will appear.

The only difference in the second condition is that the attentional spotlight is directed *away* from where the red bar will appear.

Fixation point

Receptive field of visual cortical neuron

Location of attention

■ Effective sensory stimulus for neuron
■ Ineffective sensory stimulus for neuron

Neural activity

Stimuli presented

Time (ms)

Time (ms)

100 ms

The red bar is less effective at firing the cell in the second condition. Because (1) the stimuli are identical in both conditions, (2) the fixation point hasn't changed, and (3) the same cell is being recorded in both conditions, attentional mechanisms must have directly altered the individual neuron's responsiveness.

were made from single neurons in visual cortex. A display was presented that included the cell's most preferred stimulus, as well as an ineffective stimulus (one that, by itself, did not affect the cell's firing) a short distance away but still within the cell's receptive field. As long as attention was covertly directed at the preferred stimulus, the cell responded by producing a high volume of action potentials. But when the monkey's attention was shifted elsewhere within the cell's receptive field, even though the animal's gaze had not shifted, that same stimulus provoked far fewer action potentials from the neuron. Only the shift in attention could account for this sort of modulation of the cell's excitability. Subsequent work has confirmed that attention can remold the receptive fields of neurons in a variety of ways (Mounts 2000; Womelsdorf et al., 2008).

how's it going ❓

1. Define *EEG* and *ERP*, and explain how ERPs are measured. Why is the ERP a favored technique in cognitive neuroscience?

2. Match each of the following ERP phenomena—N1, P20–50, P1, P3, N2pc— with one of these terms: *pop-out, early selection, auditory attention, late selection, visual attention.*

3. Describe an experimental procedure that can demonstrate the effects of selective attention on the activity of an individual neuron.

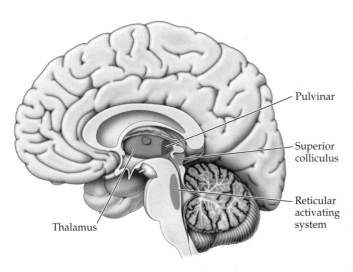

Pulvinar

Superior
colliculus

Reticular
activating
system

Thalamus

FIGURE 14.9 Subcortical Sites Implicated in Visual Attention

superior colliculus Paired gray matter structures of the dorsal midbrain that process visual information and are involved in direction of visual gaze and visual attention to intended stimuli.

pulvinar In humans, the posterior portion of the thalamus. It is heavily involved in visual processing and direction of attention.

lateral intraparietal area (LIP) A region in the monkey parietal lobe, homologous to the human intraparietal sulcus, that is especially involved in voluntary, top-down control of attention.

Many Brain Regions Are Involved in Processes of Attention

The research that we've discussed so far confirms that attention alters neural processing, and it also indicates that endogenous and exogenous attention may share much the same kinds of effects on sensory analysis mechanisms. That doesn't mean that the *sources* of the different forms of attention are identical, however, or even that they are similar—just that their *consequences* are somewhat comparable. So let's turn to some of the details of the brain mechanisms that are the *source* of attention. Our moment-to-moment engagement with the world may seem effortless, but it requires a complex network of specialized subcortical and cortical systems, working together to extract a few important stimuli from a chaotic background.

Two subcortical systems guide shifts of attention

Subcortical structures can be difficult to study because, deep in the center of the brain and skull, their activity is harder to measure with EEG/ERP and other noninvasive techniques. Our knowledge of their roles in attention thus comes mostly from work with animals.

Single-cell recordings from individual neurons have implicated the **superior colliculus**, a midbrain structure (**FIGURE 14.9**), in controlling the movement of the eyes toward objects of attention, especially overt attention (Wurtz et al., 1982). When the same eye movements are made but attention is directed elsewhere, a lower rate of firing by the collicular neurons is recorded. And in patients with lesions in one superior colliculus, inhibition of return was reduced for visual stimuli on the affected side (Sapir et al., 1999). So it seems that the superior colliculus helps direct our gaze to attended objects, and it ensures that we don't return to them too soon after our gaze has moved on.

Making up the posterior quarter of the thalamus in humans (see Figure 14.9), the **pulvinar** is heavily involved in visual processing, with widespread interconnections between lower visual pathways, the superior colliculus, and many cortical areas. The pulvinar is important for the orienting and shifting of attention. In monkeys, direct treatment of the pulvinar with a drug that inhibits its activity causes the animals to have great difficulty orienting *covert* attention toward visual targets (D. L. Robinson and Petersen, 1992). The pulvinar is also needed to filter out and ignore distracting stimuli while we're engaged in covert attention tasks. In humans, attention tasks with larger numbers of distracters induce greater activation of the pulvinar (M. S. Buchsbaum et al., 2006), indicating that this nucleus is also important for human attention.

Several cortical areas are crucial for generating and directing attention

The extensive connections between subcortical mechanisms of attention and the parietal lobes, along with observations from clinical cases that we will discuss shortly, point to a special role of the parietal lobes for attention control. Evidence is mounting that two integrated networks—dorsal frontoparietal and right temporoparietal—work together to continually select and shift between objects of interest, in coordination with subcortical mechanisms of attention.

A DORSAL FRONTOPARIETAL SYSTEM PROVIDES CONSCIOUS CONTROL OF ATTENTION In monkeys, recordings from single cells show that a region called the **lateral intraparietal area**, or just **LIP**, is crucial for endogenous attention. LIP neurons increase their firing rate when attention—not gaze—is directed to particular locations, and it doesn't matter whether the endogenous attention is being directed toward visual or auditory targets (Bisley and Goldberg, 2003; Gottlieb, 2007). So it's the top-down steering of the attentional spotlight that is important to LIP neurons, not the sensory characteristics of the stimuli.

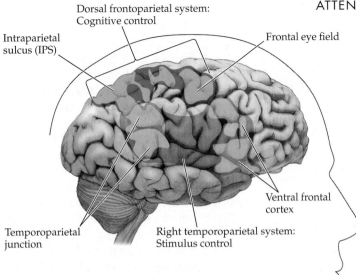

Dorsal frontoparietal system:
Cognitive control

Intraparietal
sulcus (IPS)

Frontal eye field

Ventral frontal
cortex

Temporoparietal
junction

Right temporoparietal system:
Stimulus control

FIGURE 14.10 Cortical Regions Implicated in the Top-Level Control of Attention

The human equivalent of this system is a region around the **intraparietal sulcus** (**IPS**) (**FIGURE 14.10**) that behaves much like the monkey LIP. For example, on tasks designed so that covert attention can be sustained long enough to make fMRI images, enhanced activity in the IPS is associated with the control of attention (Corbetta and Shulman, 1998). And when researchers used transcranial magnetic stimulation (see Chapter 2) to interfere with the IPS in research participants, the participants found it difficult to voluntarily shift attention (Koch et al., 2005).

People with damage to a frontal lobe region called the **frontal eye field** (**FEF**) (see Figure 14.10) struggle to prevent their gaze from being drawn away toward peripheral distracters while they're performing an endogenous attention task (Paus et al., 1991). Neurons of the FEF appear to be crucial for ensuring that our gaze is directed among stimuli according to cognitive goals rather than being guided by any characteristics of stimuli. In effect, the FEF ensures that cognitively controlled top-down attention gets priority. It's no surprise, then, that the FEF is closely connected to the superior colliculus, which, as we discussed earlier, is important for planned eye movements.

A modified form of fMRI that links rapid behavioral events to changes in activity of selected brain regions (called, unsurprisingly, *event-related fMRI*, or *ER-fMRI*) reveals network activities during top-down attentional processing (Corbetta et al., 2000; Hopfinger et al., 2000). **FIGURE 14.11** shows patterns of activation while endogenous attention is shifting in response to a symbolic cue. Enhanced activity is evident in the vicinity of the frontal eye fields (dorsolateral frontal cortex) and, simultaneously, in the IPS.

A RIGHT TEMPOROPARIETAL SYSTEM FACILITATES REFLEXIVE SHIFTS OF ATTENTION A second attention system, located at the border of the temporal and parietal lobes of the right hemisphere—and named, a little unimaginatively, the **temporoparietal junction** (**TPJ**) (see Figure 14.10)—is involved in steering toward novel or unexpected

intraparietal sulcus (IPS) A region in the human parietal lobe, homologous to the monkey lateral intraparietal area, that is especially involved in voluntary, top-down control of attention.

frontal eye field (FEF) An area in the frontal lobe of the brain that contains neurons important for establishing gaze in accordance with cognitive goals (top-down processes) rather than with any characteristics of stimuli (bottom-up processes).

temporoparietal junction (TPJ) The point in the brain where the temporal and parietal lobes meet. It plays a role in shifting attention to a new location after target onset.

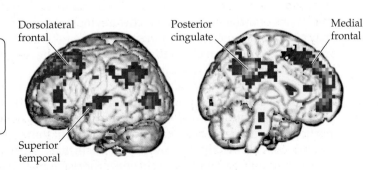

During conscious shifts of covert attention, ER-fMRI shows activation of the frontal eye field and the intraparietal sulcus (along with some activation of the temporal lobe).

Dorsolateral
frontal

Posterior
cingulate

Medial
frontal

Superior
temporal

FIGURE 14.11 The Frontoparietal Attention Network (Image courtesy of Joe Hopfinger and George Mangun.)

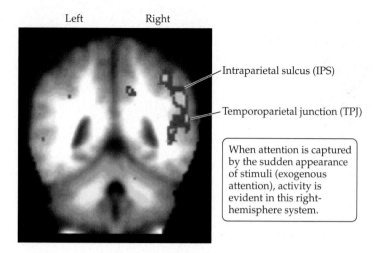

Left   Right

Intraparietal sulcus (IPS)

Temporoparietal junction (TPJ)

When attention is captured by the sudden appearance of stimuli (exogenous attention), activity is evident in this right-hemisphere system.

FIGURE 14.12 The Right Temporoparietal System and Exogenous Attention (Image courtesy of Maurizio Corbetta.)

hemispatial neglect Failure to pay any attention to objects presented to one side of the body.

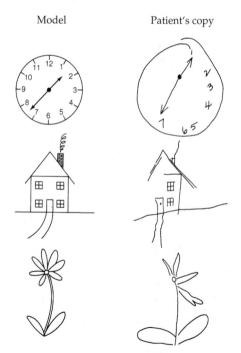

Model   Patient's copy

Diagnostic Test for Hemispatial Neglect
When asked to duplicate drawings of common symmetrical objects, patients suffering from hemispatial neglect ignore the left side of the model that they're copying.

stimuli (flashes, color changes, and so on). ER-fMRI studies (**FIGURE 14.12**) confirm that there's a spike in right-hemisphere TPJ activity if a relevant stimulus suddenly appears in an unexpected location (Corbetta et al., 2000; Corbetta and Shulman, 2002), and people with TPJ damage may fail to react to unanticipated targets (Friedrich et al., 1998). Interestingly, the TPJ system receives direct input from the visual cortex, presumably providing direct access for information about visual stimuli. The TPJ also has strong connections with the ventral frontal cortex (VFC), a region that is involved in working memory (see Chapter 13). Because working memory tracks sensory inputs over short time frames, the VFC's contribution may be to analyze *novelty* by comparing present stimuli with those of the recent past. Overall, the ventral TPJ system seems to act as an alerting signal, or "circuit breaker," overriding our current attentional priority if something new and unexpected happens.

Ultimately, the dorsal and ventral attention-control networks have to function as a single interactive system. According to one influential model (Corbetta and Shulman, 2002), the more dorsal stream of processing is responsible for *endogenous* attention, enhancing neural processing of stimuli and interacting with the pulvinar and superior colliculus to steer the attentional spotlight around. At the same time, the right-sided ventral system scans for novel salient stimuli (drawing *exogenous* attention), rapidly reassigning attention as interesting stimuli pop up. This basic model seems to apply across sensory modalities (Kanwisher and Wojciulik, 2000), including visual and auditory stimuli (Brunetti et al., 2008; Walther et al., 2010).

Brain disorders can cause specific impairments of attention

One way to learn about attention systems in the brain is to note the behavioral consequences of damage to specific regions of the brain. Research with neurological patients shows that damage of cortical or subcortical attention mechanisms can dramatically alter our ability to understand and interact with the environment.

RIGHT-HEMISPHERE LESIONS   We've discussed evidence that the right hemisphere normally plays a special role in attention (see Figure 14.12). Unfortunately, it is not uncommon for people to suffer strokes or other types of brain damage that particularly affect this part of the brain. The result—**hemispatial neglect**—is an extraordinary attention syndrome in which the patient tends to completely disregard the left side of the world. People and objects to the left of the person's midline may be completely ignored, as if unseen, even though the person's vision is otherwise normal (Rafal, 1994). The patient may fail to dress the left side of her body, will not notice visitors if they approach from the left, and may fail to eat the food on the left side of her dinner plate. If touched lightly on both hands at the same moment, the patient may notice only the right-hand touch—a symptom called *simultaneous extinction*. She may even deny ownership of her left arm or leg—"My sister must've left that arm in my bed; wasn't that an awful thing to do?!"—despite normal sensory function and otherwise intact intellectual capabilities.

It is as if the normally balanced competition for attention between the two sides has become skewed and now the input from the right side overrules or extinguishes the input from the left. Lesions in patients with hemispatial neglect (**FIGURE 14.13A**) neatly overlap the frontoparietal attention network that we discussed earlier (shown again in **FIGURE 14.13B**). This overlap suggests that hemispatial neglect is a disorder of attention itself, and not a problem with processing spatial relationships, as was once thought (Mesulam, 1985; Bartolomeo, 2007).

BILATERAL LESIONS   Parminder, the stroke victim we met at the beginning of the chapter, had *bilateral* lesions of the parietal lobe regions that are implicated in the attention network. While rare, biparietal damage can result in a dramatic disorder called **Balint's syndrome**, made up of three principal symptoms. First, people with Balint's syndrome have great difficulty steering their visual gaze appropriately (a symptom called *oculomotor apraxia*). Second, they are unable to accurately reach for objects using visual guidance (*optic ataxia*). And third—the most striking symptom—people with Balint's show a profound restriction of attention, to the point that only one object or feature can be consciously observed at any moment. This problem, called **simultagnosia**, can be likened to an extreme narrowing of the attentional spotlight, to the point that it can't encompass more than one object at a time. Hold up a comb or a pencil, and Parminder has no trouble identifying the object. But hold up both the comb *and* the pencil, and she can identify only one or the other. It's as though she is simply unable to consciously experience more than one external feature at a time, despite having little or no loss of vision. Balint's syndrome thus illustrates the coordination of attention and awareness with mechanisms that orient us within our environment.

SUBCORTICAL DAMAGE   A rare, degenerative disease of the brain called **progressive supranuclear palsy** (**PSP**) begins with marked impairment of gaze control; specifically, the superior colliculi are damaged. In the early phases of the disease, before more-widespread mental deterioration becomes evident, patients with PSP have trouble with visual tasks, such as moving the eyes voluntarily, and with converging the two eyes to view close-up objects. But patients with PSP also experience problems with covert attention; they find it increasingly difficult to switch between different *attentional* targets, even without eye movements. So, damage to the human superior colliculus appears to impair a mechanism for shifting the attentional spotlight (R. D. Wright and Ward, 2008), in agreement with work in monkeys that indicates that the superior colliculus plays a role in shifting visual attention.

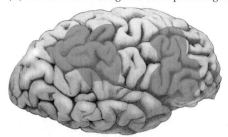

(A) Critical areas damaged in hemispatial neglect

This brain map is an average of the lesions of several people with hemispatial neglect.

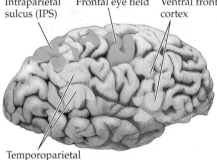

(B) Model of cortical attention control network

Intraparietal sulcus (IPS)   Frontal eye field   Ventral frontal cortex

Temporoparietal junction

Note how the hemispatial-neglect lesions overlap with the proposed attentional networks discussed in the text and shown here.

FIGURE 14.13   Brain Damage in Hemispatial Neglect

SIGNS **&** SYMPTOMS

Difficulty with Sustained Attention Can Sometimes Be Relieved with Stimulants

About 5% of all children are diagnosed with **attention deficit hyperactivity disorder** (**ADHD**), and three-fourths of these are male. Children with ADHD have trouble paying attention, and they are more impulsive than other children of the same age. Estimating the prevalence of ADHD (**FIGURE 14.14**) is somewhat complicated and controversial; for example, there is significant variation in ADHD diagnosis and medication between different (sometimes neighboring) states within the USA, raising questions about the reliability of current diagnostic practices (Fulton et al., 2009). Nevertheless, evidence has accumulated indicating that there can be some neurological changes in the disorder. Affected children tend to have slightly reduced overall brain volumes (about 3–4% smaller than in unaffected children), with reductions especially evident in the cerebellum and the frontal lobes (Arnsten, 2006). As we discuss elsewhere in the chapter, frontal lobe function is important for myriad complex cognitive processes, including the inhibition of impulsive behavior.

In addition to structural changes, ADHD has been associated with abnormalities in connectivity between brain regions, such as within the default mode network (discussed in the next section) (Cao et al., 2014). Children with ADHD

continued

Balint's syndrome   A disorder, caused by damage to both parietal lobes, that is characterized by difficulty in steering visual gaze (oculomotor apraxia), in accurately reaching for objects using visual guidance (optic ataxia), and in directing attention to more than one object or feature at a time (simultagnosia).

simultagnosia   A profound restriction of attention, often limited to a single item or feature.

progressive supranuclear palsy (PSP)   A rare, degenerative disease of the brain that begins with marked, persistent visual symptoms and leads to more widespread intellectual deterioration.

attention deficit hyperactivity disorder (ADHD)   A syndrome characterized by distractibility, impulsiveness, and hyperactivity that, in children, interferes with school performance.

SIGNS **&** SYMPTOMS *continued*

FIGURE 14.14 Prevalence of ADHD in the United States (From CDC, 2011.)

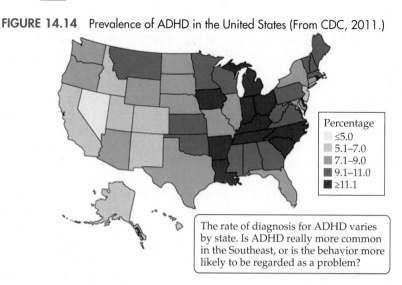

Percentage
≤5.0
5.1–7.0
7.1–9.0
9.1–11.0
≥11.1

The rate of diagnosis for ADHD varies by state. Is ADHD really more common in the Southeast, or is the behavior more likely to be regarded as a problem?

may have abnormal activity levels in some brain locations, including the system that signals the rewarding aspects of activities (Volkow et al., 2009). For this reason, some researchers advocate treating these children with stimulant drugs like methylphenidate (Ritalin) or norepinephrine reuptake inhibitors like atomexetine (Strattera) (Schwartz and Correll, 2014). Stimulant treatment often improves the focus and performance of ADHD children in traditional school settings, but this treatment remains controversial because of the significant risk of side effects and because some researchers view ADHD as being simply one extreme on a continuum of normal behavior. In fact, simply allowing kids diagnosed with ADHD to fidget and engage in more intense physical activity effectively reduces their symptoms and improves task performance (Hartanto et al., 2015; Sarver et al., 2015). You can read more about ADHD and another developmental disorder—autism spectrum disorder—in **A STEP FURTHER 13.4** on the website.

how's it going ?

1. Identify two subcortical structures that are implicated in the control of attention. What functions do they perform? Which one is damaged in PSP, and what are the symptoms of that disorder?
2. What is the general name for the cortical system responsible for conscious shifts of attention? What are its components, and what happens when those components are damaged?
3. Name the cortical system implicated in exogenous shifts of attention. Which specific regions are part of this system, and what happens if they are damaged?

PART II
Consciousness

Consciousness Is a Mysterious Product of the Brain

So far, this chapter has been mostly about attention, but when you think about it (no joke intended), consciousness is completely dependent on attention. Whenever we are conscious, we're attending to *something*, be it internal or external. William James (1890) tried to capture the relationship between attention and consciousness when he wrote, "My experience is what I agree to attend to. Only those items which I notice shape my mind—without selective interest, experience is an utter chaos." We all experience consciousness, so we know what it is, but that experience is so personal,

TABLE 14.1 Elements of Consciousness in Humans and Other Animals

Element	Definition	Other species
Theory of mind	Insight into the mental lives of others; understanding that other people act on their own unique beliefs, knowledge, and desires	Only chimpanzees, so far
Mirror recognition	Ability to recognize the self as depicted in a mirror	All great apes; dolphins; elephants
Imitation	Ability to copy the actions of others; thought to be a stepping-stone to awareness and empathy	Many species, including cephalopods like the octopus
Empathy and emotion	Possession of complex emotions and the ability to imagine the feelings of other individuals	Most mammals, ranging from primates and dolphins, to hippos and rodents; most vertebrates able to experience pleasure (and other basic emotions)
Tool use	Ability to employ found objects to achieve intermediate and/or ultimate goals	Chimps and other primates; other mammals such as elephants, otters, and dolphins; birds such as crows and gulls
Language	Use of a system of arbitrary symbols, with specific meanings and strict grammar, to convey concrete or abstract information to any other individual that has learned the same language	Generally considered to be an exclusively human ability, with controversy over the extent to which the great apes can acquire language skills
Metacognition	"Thinking about thinking": the ability to consider the contents of one's own thoughts and cognitions	Nonhuman primates; dolphins

so subjective, that it's difficult to come up with an objective definition. Perhaps a reasonable attempt is to say that **consciousness** is the state of being *aware* that we are conscious, coupled with our perception of what is going on in our minds. However, a lot of different activities are embedded in the phrase "perception of what is going on in our minds." It includes our perception of what's going on around us, our sense of time passing, our sense of being aware, our recollection of events that happened in the past, and our imaginings about what might happen in the future. Add to that our belief that we are free to direct our attention and make decisions, and we have a concept of immense scope. And while attention and awareness are clearly closely related to each other and central to consciousness, they are not synonymous: as we saw earlier, attention to a stimulus enhances activity in visual cortex, but *conscious awareness* of the stimulus doesn't add any extra activation (Watanabe et al., 2011). Some of the basic elements of human consciousness that researchers agree on are identified in **TABLE 14.1**, which also lists other species that may have comparable capacities and experiences.

Which brain regions are active when we are conscious?

Despite definitional complexities, consciousness is an active area of neuroscience research. So far, there are numerous competing theoretical models of consciousness, but it has been difficult to obtain hard physiological data. One approach is to look for patterns of synchronized activity in neural networks as people engage in conscious, inwardly focused thought. Using fMRI, researchers have identified a large circuit of brain regions—collectively called the **default mode network**, consisting of parts of the frontal, temporal, and parietal lobes—that seems to be selectively activated when we are at our most introspective and reflective, and relatively deactivated during behavior directed toward external goals (Raichle, 2015). In some ways you could think of it as a daydream network (and thereby also engage in metacognition: see Table 14.1). It has been proposed that dysfunction within the default mode network contributes to the symptoms of various neuropsychological disorders, such as attention deficit hyperactivity disorder, autism spectrum disorder, schizophrenia, and dementia, in both adults and childen (Whitfield-Gabrieli and Ford, 2012; Sato et al., 2015). Monkeys and lab rats have circuits that resemble the human default mode network on structural and functional grounds, raising the possibility that nonhuman species may likewise engage in self-reflection or other introspective mental activity (Mantini et al., 2011; Sierakowiak et al., 2015).

consciousness   The state of awareness of one's own existence, thoughts, emotions, and experiences.

default mode network   A circuit of brain regions that is active during quiet introspective thought.

FIGURE 14.15 The Unconscious Brain
(Courtesy of Ralph Adolphs.)

These fMRI studies range from the temporary unconsciousness we all experience (sleep) to the profound and long-lasting unconsciousness of a persistent vegetative state. They share reduced activity of a frontoparietal network that includes dorsolateral prefrontal cortex (F), medial frontal cortex (MF), posterior parietal cortex (P), and posterior cingulate (Pr).

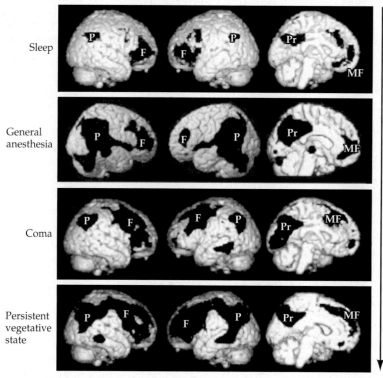

A practical alternative approach has been to study consciousness by focusing on people who lack it—people in comas or other states of reduced consciousness. Maps of brain activity in such people—or more precisely, maps of *deactivated* areas (Tsuchiya and Adolphs, 2007)—suggest that consciousness depends on a specific frontoparietal network (**FIGURE 14.15**) that includes much of the cortical attention network we've been discussing, along with regions of medial frontal and cingulate cortices.

But is clinical unconsciousness really the inverse of consciousness? Perhaps it's not that simple. For one thing, some patients in apparently deep and unresponsive coma can be instructed to use two different forms of mental imagery to create distinct "yes" and "no" patterns of activity on fMRI and then to use this mental activity to answer questions (**FIGURE 14.16**) (Monti et al., 2010). Although this ability is observed in only a small number of patients in comas—most give no evidence of responding to their environments—studies like this suggest that it isn't accurate to view a coma as an exact inverse of what we experience as consciousness. In any case, there seems to be more to consciousness than just being awake, aware, and attending. How can we identify and study the additional dimensions of consciousness?

Some aspects of consciousness are easier to study than others

Most of the activity of the central nervous system is unconscious. Scientists call these brain functions **cognitively impenetrable**: they involve basic neural processing operations that cannot be experienced through introspection. For example, we see whole objects and hear whole words and can't really imagine what the primitive sensory precursors of those perceptions would feel like. Sweet food tastes sweet, and we can't mentally break it down any further. But those simpler mechanisms, operating below the surface of awareness, are the foundation that conscious experiences are built on.

In principle, then, we might someday develop technology that would let us directly reconstruct people's conscious experience—read their minds—by decoding the primitive neural activity and assembling identifiable patterns from it. This is sometimes called the **easy problem of consciousness**: understanding how particular patterns

cognitively impenetrable Referring to basic neural processing operations that cannot be experienced through introspection—in other words, that are unconscious.

easy problem of consciousness Understanding how particular patterns of neural activity create specific conscious experiences by reading brain activity directly from people's brains as they're having particular experiences.

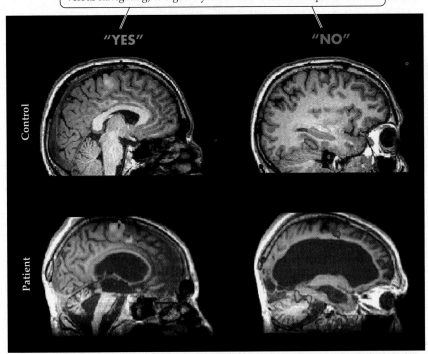

Functional MRI shows the brain activity of a healthy control participant asked to use two different mental images (playing tennis versus navigating) to signal "yes" or "no" answers to questions.

FIGURE 14.16 Communication in "Unconscious" Patients (Image courtesy of Adrian Owen.)

The strikingly similar brain activity of a patient in a persistent vegetative state raises questions about the definition of unconsciousness. The patient was able to correctly answer a variety of questions using this technique, despite being in a state of apparently deep unconsciousness due to profound brain damage (as evident in the grossly abnormal MRI scan) and lacking behavioral responses.

of neural activity create *specific* conscious experiences. Of course, it's almost a joke to call this problem "easy," but at least we can fairly say that, someday, the necessary technology and knowledge may be available to accomplish the task of eavesdropping on large networks of neurons, in real time.

Present-day technology offers a glimpse of that possible future. For example, if participants are repeatedly scanned while viewing several distinctive scenes, a computer can eventually learn to identify which of the scenes the participant is viewing on each trial, solely on the basis of the pattern of brain activation (**FIGURE 14.17A**) (Kay et al., 2008). Of course, this outcome relies on having the participants repeatedly view the same static images—hardly a normal state of consciousness. A much more difficult problem is to reconstruct conscious experience from neural activity during a person's initial exposure to a stimulus. So far, this has been accomplished for only relatively simple visual stimuli (**FIGURE 14.17B**) (Miyawaki et al., 2008) or brief video reconstructions (**BOX 14.2**). But it's a good start, and rapid progress seems likely as technological problems are solved.

Alas, there is also the **hard problem of consciousness**, and it may prove impossible to crack. How can we understand the brain processes that produce people's *subjective experiences* of their conscious perceptions? To use a simple example, everyone with normal vision will agree that a ripe tomato is "red." That's the label that children all learn to apply to the particular pattern of information, entering consciousness from the color-processing areas of visual cortex, that is provoked by looking at something like a tomato. But that doesn't mean that your friend's internal *personal* experience of "red" is the same as yours. These purely subjective experiences of perceptions are referred to as **qualia** (singular *quale*). Because they are subjective and impossible to communicate to others—how can your friend know if "redness" feels the same in your mind as it does in hers?—qualia may prove impossible to study (**FIGURE 14.17C**). At this point, anyway, we are unable to even conceive of a technology that would make it possible.

To see the video
Reconstructing Brain Activity,
go to

2e.mindsmachine.com/14.5

hard problem of consciousness
Understanding the brain processes that produce people's subjective experiences of their conscious perceptions—that is, their qualia.

quale A purely subjective experience of perception.

(A) Pattern identification

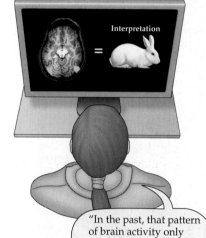

FIGURE 14.17 Easy and Hard Problems of Consciousness (Part A after Kay et al., 2008; B after Miyawaki et al., 2008.)

For a person in a brain scanner, repeatedly viewing the same image causes the same pattern of brain activity to occur each time. Over enough trials, researchers can develop computer models that identify which of about 20 images the participant is looking at.

"In the past, that pattern of brain activity only appeared when he was looking at the rabbit."

(B) Visual reconstruction

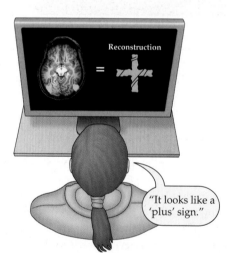

Scientists also know enough about how different parts of the brain are activated by light striking the retina that they can even predict what type of simple shapes a person is viewing.

"It looks like a 'plus' sign."

(C) Subjective experience

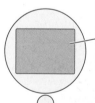

Both people instantly identify the color as red, but what differs is the way red *feels* in their minds.

Actual color

"Red"

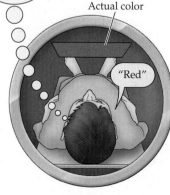

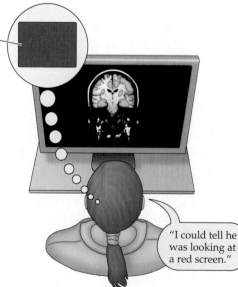

The "hard" problem is to go beyond predicting what a person is seeing to knowing what that person's *subjective experience* is like. Here, the participant and the researcher would both immediately identify the color being viewed as "red" but, as suggested by their respective thought bubbles, the participant's personal, internal, *subjective* experience may be quite different from that of the researcher. It is difficult to see how we could ever be sure that we share any subjective experiences of consciousness.

"I could tell he was looking at a red screen."

BOX 14.2 Building a Better Mind Reader

The idea of a machine that can record and play back a person's thoughts is a staple of science fiction, but is it plausible? By tracking the activity of the retinotopic map in primary visual cortex with fMRI, researchers have succeeded in detecting whether people are viewing crosses, circles, or other large and simple stimuli (Miyawaki et al., 2008; depicted in Figure 14.17B) or static pictures of faces or scenes (Naselaris et al., 2009). However, traditional fMRI is normally too slow to "read" anything more than relatively simple, static visual experiences. But of course visual perception is almost never static; it's a continuously rolling movie, directed by attention and powerfully shaping our consciousness.

So how might we be able to look in on people's internal movies? One clever approach involves building a huge library of predicted brain responses to visual scenes and using those to reconstruct visual experiences from recordings of brain activity. Researchers in Jack Gallant's lab at UC Berkeley (Nishimoto and Gallant, 2011) began by themselves spending many, many hours watching two sets of training videos—Hollywood movie trailers—while fMRI data were recorded from the occipitotemporal visual cortex. They couldn't justify asking anyone else to spend that much time in fMRI. These data were used to develop a mathematical model that predicts how shape and motion are encoded for every voxel in the visual cortex. (*Voxel* is short for **vo**lumetric pi**xel**, and just as pixels define the two dimensions of a computer screen, voxels make up the three-dimensional volume of an imaged brain.)

Next, the researchers cut up a random assortment of YouTube videos into 18 million clips, each 1 second long, none of which were from the original training videos. This library of short clips was fed into a computer program containing the voxel-by-voxel model of brain activity for each participant. The program created a prediction for the pattern of brain activity that each of the 18 million clips would produce.

Finally, the computer compared the actual fMRI activity recorded during the second set of test videos (again, movie trailers) with the *predicted* activity associated with the YouTube clips. For each 1-second "frame" of the test videos, a computer program selected the 100 clips whose predicted activity most closely matched the actual activity recorded when the participant viewed the original frame. These clips were then merged together into a movie reconstruction. The results of this process are shown in the figure. For each of three test video clips (A, B, and C), three individual frames are shown across the top row. The next five rows show the five best-fitting YouTube frames, as selected by the modeled brain activity of the participant. Mostly, they don't much resemble the original frames, but when averaged together, as shown in the bottom row, a blurry but pretty strong resemblance begins to emerge. The accuracy of the reconstruction is much more evident when viewing the original and reconstructed videos playing side by side, which you can see on the website for this text. It is almost unnerving. As technology develops, we can expect more progress on this, the "easy" problem of consciousness. The same cannot be said for the hard problem of consciousness, however: any progress in reading the personal and subjective feelings that accompany conscious experiences lies in a distant, and largely unimagined, future.

The Mind's Eye (Courtesy of Drs. Shinji Nishimoto and Jack Gallant, University of California, Berkeley.)

(A) (B) (C)

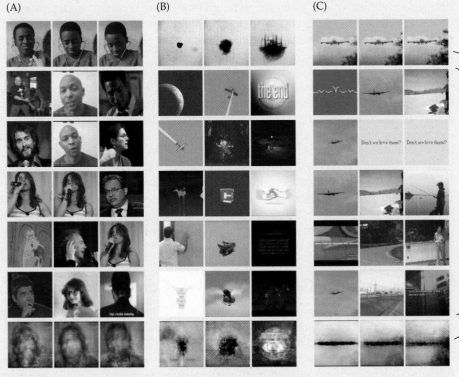

Three frames from each of three videos (A, B, and C) are shown here.

These rows show five 1-second YouTube clips identified by the computer model as being most likely to produce brain activity similar to that for the original videos.

This row shows the amalgamation of YouTube clips. Although blurry, these reconstructions bear obvious similarities to the viewed videos.

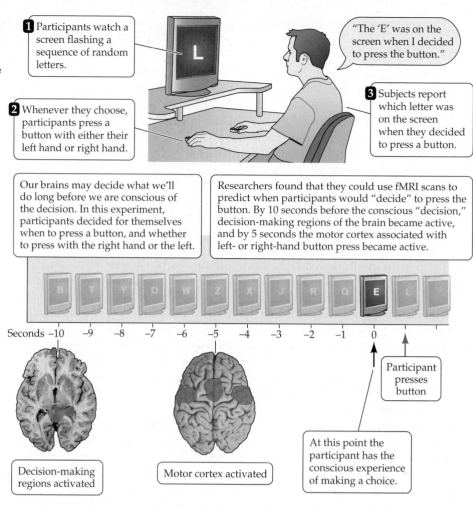

FIGURE 14.18   Reading the Future
(After Soon et al., 2008.)

① Participants watch a screen flashing a sequence of random letters.

"The 'E' was on the screen when I decided to press the button."

② Whenever they choose, participants press a button with either their left hand or right hand.

③ Subjects report which letter was on the screen when they decided to press a button.

Our brains may decide what we'll do long before we are conscious of the decision. In this experiment, participants decided for themselves when to press a button, and whether to press with the right hand or the left.

Researchers found that they could use fMRI scans to predict when participants would "decide" to press the button. By 10 seconds before the conscious "decision," decision-making regions of the brain became active, and by 5 seconds the motor cortex associated with left- or right-hand button press became active.

Seconds –10 –9 –8 –7 –6 –5 –4 –3 –2 –1 0

Decision-making regions activated

Motor cortex activated

Participant presses button

At this point the participant has the conscious experience of making a choice.

free will   The feeling that our conscious self is the author of our actions and decisions.

Our subjective experience of consciousness is closely tied up with the notion of **free will**: the belief that our conscious self is unconstrained in deciding our actions and decisions and that for any given moment, given exactly the same circumstances, we *could* have chosen to engage in a different behavior. After hundreds of years of argument, there's still no agreement on whether we actually have free will, but most people behave as though there are always options, and in any event there must be a neural substrate for the universal *feeling* of having free will. When participants *intend* to act (push a button, say), there is selective activation of the IPS (which we implicated earlier in top-down attention) and frontal regions including dorsal prefrontal cortex (Lau et al., 2004), suggesting that these regions are important for our feelings of control over our behavior. But consider this: brain activity associated with making a decision may be evident in fMRI scans as much as 5–10 seconds *before* participants are consciously aware of making a choice (**FIGURE 14.18**) (Soon et al., 2008)!

In any event, the earliest indications of the decision-making process are found in frontal cortex. Such involvement of prefrontal systems in most aspects of attention and consciousness, regardless of sensory modality or emotional tone, suggests that the frontal cortex is the main source of goal-driven behaviors (E. K. Miller and Cohen, 2001), as we discuss next.

The Frontal Lobes Govern Our Most Complex Behaviors

How do we decide what to do, and when to do it? Humans are distinctive among mammals for the comparatively large size of our prefrontal cortex—about a third of the entire cortex (**FIGURE 14.19A**) (Semendeferi et al., 2002). Perhaps it reflects a bit of vanity about our species, but size is one reason why researchers have long viewed the frontal cortex as the source of intelligence, abstract thinking, and self-awareness.

Adding to the mystery of frontal lobe function is the unusual assortment of behavioral changes that accompany damage to the frontal lobes.

The posterior portion of the frontal cortex includes motor and premotor regions (see Chapter 5). The anterior portion, usually referred to as **prefrontal cortex**, is immensely interconnected with the rest of the brain (Fuster, 1990; Mega and Cummings, 1994), and it is usually subdivided into *dorsolateral* and *orbitofrontal* regions (**FIGURE 14.19B**). The study of prefrontal cortical function began with Carlyle Jacobsen's work in the 1930s on delayed-response learning in chimpanzees. The animals were shown where food was hidden, but they had to wait before being allowed to reach for it. Chimps with prefrontal lesions were strikingly impaired on this simple task, compared with animals that sustained lesions in other brain regions. Jacobsen believed the impairment was caused by memory problems. But more recent observations, which we'll discuss next, suggest that the chimps' problems were with directing attention and formulating a plan of action.

Frontal lobe injury in humans leads to emotional, motor, and cognitive changes

Because they are large, the frontal lobes are vulnerable to injury; frontal lobe damage is a common consequence of strokes, tumors, and trauma. The exact symptoms induced by frontal lobe damage depend to some extent on the lesion location, but they include an unusual collection of emotional, motor, and cognitive changes. A common symptom in patients with frontal lobe damage is a persistent strange apathy, broken by bouts of euphoria (an exalted sense of well-being). Ordinary social conventions are readily cast aside by impulsive behavior. Concern for the past or the future is absent (Petrides and Milner, 1982; Duffy and Campbell, 1994). Forgetfulness is shown in many tasks requiring sustained attention. And yet, standard IQ test performance shows only slight changes after injury or stroke.

Clinical examination of patients with frontal lesions also reveals an array of strange impairments in their behavior, especially in the realm of **executive function**, the high-level control of other cognitive functions in order to attend to important stimuli and make suitable "plans" for action. For example, a patient with frontal damage who is given a simple set of errands may be unable to complete them without numerous false starts, backtracking, and confusion (Shallice and Burgess, 1991). Regions of dorsolateral prefrontal cortex, along with anterior portions of the cingulate cortex, are closely associated with executive control. Dorsolateral prefrontal cortex is also crucial for working memory, the ability to hold information in mind while processing it or using the information to solve problems.

People with frontal lobe damage often struggle with *task shifting* and tend to **perseverate** (continue beyond a reasonable degree) in any activity (B. Milner, 1963; Alvarez and Emory, 2006). Similarly, frontal lobe lesions may cause motor perseveration, repeating a simple movement over and over. However, the overall level of ordinary spontaneous motor activity is often quite diminished in people with frontal lesions. Along with movement of the head and eyes, facial expression of emotions may be greatly reduced.

One explanation for these disparate effects of prefrontal lesions is that this region of cortex may be important for organizing different aspects of goal-directed behavior, including the capacity for prolonged attention and sensitivity to potential rewards and punishments. Patients with prefrontal lesions often have an inability to plan future acts and use foresight, as in the famous case of Phineas Gage. Their social skills may decline, especially the ability to inhibit inappropriate behaviors, and they may be unable to stay focused on any but short-term projects. They may agonize over even simple decisions. Some of the symptom clusters associated with damage of the major subdivisions of prefrontal cortex are summarized in **TABLE 14.2**.

The evidence suggests that prefrontal cortex—especially the orbitofrontal region—controls goal-directed behaviors. For example, monkeys that must make decisions

> The human prefrontal cortex can be subdivided into a dorsolateral region (blue) and an orbitofrontal region (green). Lesions in these different areas of prefrontal cortex have different effects on behavior.

(A)

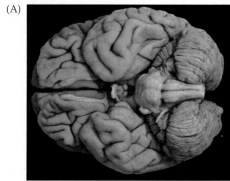

(B)

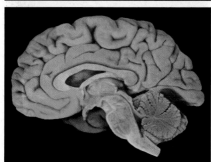

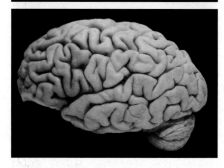

FIGURE 14.19 The Prefrontal Cortex

prefrontal cortex The anteriormost region of the frontal lobe.

executive function A neural and cognitive system that helps develop plans of action and organizes the activities of other high-level processing systems.

perseverate To continue to show a behavior repeatedly.

TABLE 14.2 Regional Prefrontal Syndromes

Prefrontal damage site	Syndrome type	Characteristics
Dorsolateral	Dysexecutive	Diminished judgment, planning, insight, and temporal organization; reduced cognitive focus; motor programming deficits (possibly including aphasia and apraxia); diminished self-care
Orbitofrontal	Disinhibited	Stimulus-driven behavior; diminished social insight; distractibility; emotional lability
Mediofrontal	Apathetic	Diminished spontaneity; diminished verbal output; diminished motor behavior; urinary incontinence; lower-extremity weakness and sensory loss; diminished spontaneous prosody; increased response latency

that could lead to rewards show increased orbitofrontal activation (Matsumoto et al., 2003); in general, orbitofrontal cortex seems to link pleasant experiences (e.g., eating a delicious meal) with reward signals (dopamine release in the reward pathway) generated elsewhere in the brain (Kringelbach, 2005). In fact, some researchers believe that orbitofrontal cortex is actually more important for signaling expected outcomes (which is likewise related to reward) than for learning (Schoenbaum et al., 2009). In humans performing tasks in which some stimuli have more reward value than others, the level of activation in prefrontal cortex correlates with how rewarding the stimulus is (Gottfried et al., 2003). This relationship seems to be a significant factor in gambling behavior and, more generally, is important for our decision-making processes. In fact, progress in understanding the decision-making process has spawned a new field, *neuroeconomics*, discussed in **BOX 14.3**.

We may never have a full understanding of the deepest secrets of consciousness, or an answer to the question of whether we possess free will. But that doesn't prevent us from marveling that our consciousness has become so self-aware that it can study itself to a high degree. Perhaps it's best to allow ourselves at least one or two mysteries, if only for the sake of art. Would life seem as rich if we could predict other people's behavior, or even our own, with perfect accuracy?

Phineas Gage Phineas P. Gage was a sober, polite, and capable member of a rail-laying crew, responsible for placing the charges used to blast rock from new rail beds. That's Gage on the left, holding a meter-long steel tamping rod. Perhaps the image on the right can help you guess why there appears to be something wrong with the left side of his face. In a horrific accident in 1848, a premature detonation blew that rod right through Gage's skull, on the trajectory shown in red, severely damaging both frontal lobes, especially in the orbitofrontal regions. Amazingly, Gage could speak shortly after the accident, and he walked to the doctor's office, although no one expected him to live (Macmillan, 2000). In fact, Gage survived another 12 years, but he was definitely a changed man, so rude and aimless, and his powers of attention so badly impaired, "that his friends and acquaintances said that he was 'no longer Gage.'" The historical account of Gage's case, now a neuroscience classic, was eventually found to closely agree with the symptoms of modern patients with frontal damage (H. Damasio et al., 1994; Wallis, 2007).

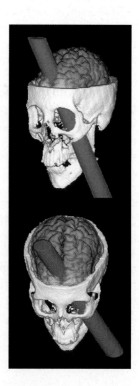

The waiter has brought over the dessert trolley, and it's decision time: do you go with the certain delight of the chocolate cake, or do you succumb to the glistening allure of the sticky toffee pudding? Or, do you allow yourself only a cup of black coffee, for the sake of your waistline? What happens in your brain when you finally make your difficult choice?

In the lab, researchers usually evaluate decision making by using monetary rewards (instead of desserts, darn it) because money is convenient: you can vary how much money is at stake, how great a reward is offered, and so on, to accurately gauge how we really make economic decisions. These studies show that most of us are very averse to loss and risk: we are more sensitive to losing a certain amount of money than we are to gaining that amount. In other words, losing $20 makes us feel a lot worse than gaining $20 makes us feel good. From a strictly logical point of view, the value of money, whether lost or gained, should be exactly the same. Our tendency to overemphasize loss is just one of several ways in which people fail to act rationally in the marketplace.

Neuroeconomics is the study of brain mechanisms at work during economic decision making, and our attention to environmental factors and evaluation of rewards has a tremendous impact on these decisions. In general, findings from human and animal research suggest that two main systems underlie decision processes (Kable and Glimcher, 2009). The first, consisting of the ventromedial prefrontal cortex (including the anterior cingulate cortex) plus the dopamine-based reward system of the brain (see Chapter 4), is a *valuation system*, a network that ranks choices on the basis of their perceived worth and potential reward. The second system, involving mostly dorsolateral prefrontal cortex and parietal regions (like the LIP or IPS discussed in the text), is thought to be a *choice system*, sifting through the valuated alternatives and producing the conscious decision.

Neuroeconomics research is also confirming that the prefrontal cortex normally inhibits impulsive decision making as a way to avoid loss (Tom et al., 2007; Muhlert and Lawrence, 2015). As people are faced with more and more uncertainty, the prefrontal cortex becomes more and more active (Hsu et al., 2005; Huettel et al., 2006). Likewise, when

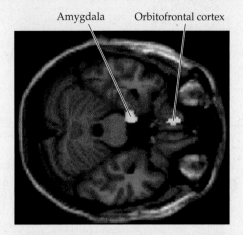

Amygdala   Orbitofrontal cortex

A Poor Choice A costly decision is associated with activation of the amygdala and orbitofrontal cortex, signaling diminished reward and aversion to loss. (From Coricelli et al., 2005; courtesy of Angela Sirigu.)

people have made wrong, costly decisions that they regret, activity increases in the amygdala and in the orbitofrontal aspect of the prefrontal cortex, as shown in the figure (Coricelli et al., 2005), probably reflecting the participant's perception of diminished reward and increasing aversion to loss.

---

how's it going

1. Define *consciousness* (or at least try!).
2. Discuss unconsciousness in states like deep coma. Is this unconsciousness the inverse of consciousness?
3. Contrast the easy and hard problems of consciousness, giving examples of each. What are *qualia*?
4. Name the main subdivisons of the prefrontal cortex. How do they differ in function?
5. What are some of the main symptoms of frontal lobe lesions?
6. What are the two main neural systems that are thought to operate in the process of decision making, as identified in neuroeconomics research?

neuroeconomics The study of brain mechanisms at work during economic decision making.

Recommended Reading

Glimcher, P. W., Camerer, C., Poldrack, R. A., and Fehr, E. (2008). *Neuroeconomics: Decision Making and the Brain.* San Diego, CA: Academic Press.

Koch, C. (2012). *Consciousness: Confessions of a Romantic Reductionist.* Cambridge, MA: MIT Press.

Laureys, S., and Tononi, G. (Eds.). (2008). *The Neurology of Consciousness: Cognitive Neuroscience and Neuropathology.* New York, NY: Academic Press.

Posner, M. I. (Ed.). (2012). *Cognitive Neuroscience of Attention* (2nd ed.). New York, NY: Guilford Press.

Stuss, D. T., and Knight, R. T. (2012). *Principles of Frontal Lobe Function* (2nd ed.). Oxford, England: Oxford University Press.

Wright, R. D., and Ward, L. M. (2008). *Orienting of Attention.* Oxford, England: Oxford University Press.

VISUAL SUMMARY

14

You should be able to relate each summary to the adjacent illustration, including structures and processes. Go to the online version of this summary (scan the QR code above) for links to figures, animations, and activities that will help you consolidate the material.

1 Although we pay mostly **overt attention** to stimuli, we can also pay **covert attention** to stimuli or locations of our choosing. **Attention** has been likened to a spotlight, helping us to distinguish stimuli from distracters. Review **Figure 14.1**, Animation 14.2 and Video 14.3

2 In **endogenous attention** we voluntarily select objects to attend to; it is studied using **symbolic cuing** tasks. **Exogenous attention** is the involuntary capture of attention by stimuli, studied using **peripheral spatial cuing** tasks. Review **Figures 14.2** and **14.3**, Animation 14.4

3 In visual search, targets may pop out if they are distinctive for a particular feature (**feature search**). More often, though, we use **conjunction searches**, identifying target stimuli on the basis of two or more features. Review **Figure 14.4**

4 **Event-related potentials (ERPs)** are created by the averaging of many EEG recordings from repeated experimental trials. ERPs can track neural operations with excellent **temporal resolution**. Review **Figure 14.5**

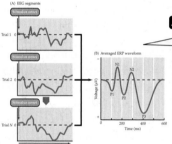

5 Endogenous and exogenous attention enhance various ERP components. Which specific ERP components are affected depends on the sensory modality (e.g., audition versus vision) and the task employed. Review **Figures 14.5–14.7**

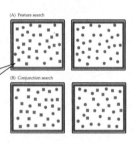

6 Single-cell recordings in lab animals confirm that attention affects the responses of individual neurons. Review **Figure 14.8**

7 Subcortical mechanisms involving the **superior colliculi** and the **pulvinar** are crucial for shifting visual attention and gaze between important objects of attention. Review **Figure 14.9**, Activity 14.1

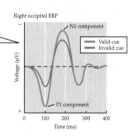

8 A dorsal frontoparietal system directs voluntary, endogenous attention. A right-sided temporoparietal attention system is responsible for detecting and shifting attention to novel stimuli. The two networks interact extensively. Review **Figures 14.10–14.13**, Activity 14.2

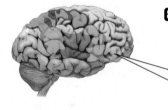

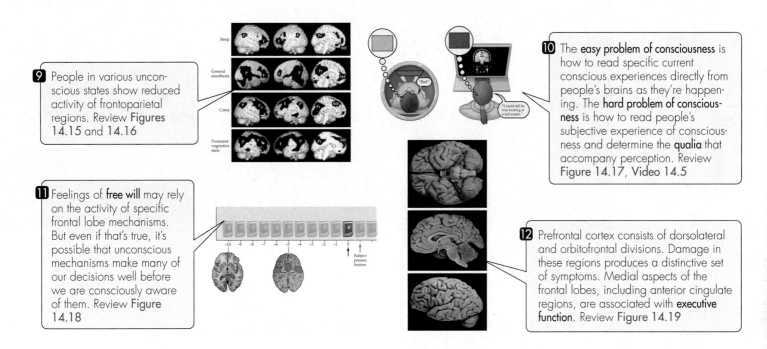

9 People in various unconscious states show reduced activity of frontoparietal regions. Review **Figures 14.15 and 14.16**

10 The **easy problem of consciousness** is how to read specific current conscious experiences directly from people's brains as they're happening. The **hard problem of consciousness** is how to read people's subjective experience of consciousness and determine the **qualia** that accompany perception. Review **Figure 14.17, Video 14.5**

11 Feelings of **free will** may rely on the activity of specific frontal lobe mechanisms. But even if that's true, it's possible that unconscious mechanisms make many of our decisions well before we are consciously aware of them. Review **Figure 14.18**

12 Prefrontal cortex consists of dorsolateral and orbitofrontal divisions. Damage in these regions produces a distinctive set of symptoms. Medial aspects of the frontal lobes, including anterior cingulate regions, are associated with **executive function**. Review **Figure 14.19**

Go to **2e.mindsmachine.com** for study questions, quizzes, flashcards, and other resources.

Brain Asymmetry, Spatial Cognition, and Language

Putting a Name to a Face

Humpty Dumpty, the famously grouchy egg in Lewis Carroll's *Through the Looking-Glass, and What Alice Found There*, was grumbling to Alice about his problem distinguishing between people:

"The face is what one goes by, generally," Alice remarked in a thoughtful tone.

"That's just what I complain of," said Humpty Dumpty. "Your face is the same as everybody has—the two eyes, so—" (marking their places in the air with his thumb) "nose in the middle, mouth under. It's always the same. Now, if you had the two eyes on the same side of the nose, for instance—or the mouth at the top—that would be of some help."

Louis can relate. Throughout his 54 years of life, Louis has had great difficulty distin-

guishing between people. He regularly fails to recognize his seven children, his wife, or even himself reflected in a mirror. He relies on name badges to identify colleagues he has worked with for years. Yet it's clear that Louis can make out other people's features perfectly well, because he can describe the color of the eyes, the shape of the nose and mouth, and so on, and is otherwise just like anyone else.

Intriguingly, sometimes people who suffer a stroke, especially in the right hemisphere, have difficulty recognizing faces, but there's no evidence that Louis ever had a stroke. What might account for Louis's—and Humpty's—difficulty with putting a name to a face?

To see the video
Face Blindness,
go to

2e.mindsmachine.com/15.1

Faces, coloration, smells, sounds—many species use physical and behavioral sig-nals to engage in **communication**, the transmission of information between in-dividuals. But we humans may be alone in our use of **language**, the highly specialized form of communication in which arbitrary symbols or behaviors are assembled and re-assembled in almost infinite variety and associated with a vast range of things, actions, and concepts. Because the speakers of a language all understand the same strict set of rules, or **grammar**, language allows us to share information on topics ranging from the mundane to the sublime, and from the concrete (like, say, concrete) to the abstract (the meaning of life, the feeling of love, why chocolate tastes soooo good).

In almost everyone, verbal abilities are especially associated with the left hemi-sphere of the brain, while the right hemisphere plays a special role in **spatial cogni-tion**, our ability to navigate and to understand the spatial relationships between ob-jects. In this chapter we survey the neuroscience of asymmetry in cerebral function, with a special emphasis on the acquisition and use of language. Because much of what we know about human brain organization comes from studies of people who have suffered strokes and other forms of brain injury, we also consider evidence that some recovery of function is possible even after severe brain damage.

communication Information transfer between two individuals.

language Communication in which arbitrary sounds or symbols are arranged according to a grammar in order to convey an almost limitless variety of concepts.

grammar All of the rules for usage of a particular language.

spatial cognition The ability to navigate and to understand the spatial relationship between objects.

corpus callosum The main band of axons that connects the two cerebral hemispheres.

split-brain individual An individual whose corpus callosum has been severed, halting communication between the right and left hemispheres.

contralateral In anatomy, pertaining to a location on the opposite side of the body.

To view the
Brain Explorer,
go to

2e.mindsmachine.com/15.2

PART I
Cerebral Asymmetry

One of the major advances of twentieth-century biological psychology was the realization that in controlling ongoing behavior, the two cerebral hemispheres are not necessarily doing the same things. Instead, prompted by the discovery that most language functions are contained within the left hemisphere, researchers started to think in terms of hemispheric specialization, or *lateralization* (Finger, 1994). While we'll see plenty of evidence that the two sides of the brain differ in how readily they handle various tasks, keep in mind that normally the two halves work together. Claims that some people are especially "right-brained" (supposedly more random, intuitive, and creative) while others are more "left-brained" (supposedly logical, sequential, and analytical) have little scientific basis.

The Left and Right Brains Are Different

Lateralization of function is not really such a surprising idea; other body organs also show considerable asymmetry between the right and left sides. But when we study the *behavior* of people, cerebral lateralization of function is masked by the rich neural connections between the hemispheres: they communicate with each other so quickly and so thoroughly that they seem to act as one. So researchers had to develop clever techniques to study the functioning of each hemisphere in isolation from the other.

Disconnection of the cerebral hemispheres reveals their individual processing specializations

Some rare and unfortunate people develop epilepsy that becomes very difficult to control with medication. They experience frequent seizures that start in one hemisphere and then spread through the **corpus callosum**—the huge axon pathway connecting the two hemispheres—to involve the whole brain. One last-ditch approach to controlling unmanageable epilepsy is to cut the corpus callosum, preventing the spread of the seizure discharges from one hemisphere to the other. This operation, developed in the 1960s, greatly reduced the severity of seizures in patients, but it also had a significant impact on behavior—and thus created a research opportunity. Because the hemispheres were now isolated from one another, detailed behavioral testing of these **split-brain individuals** allowed researchers to document cerebral hemispheric specialization in cognitive, perceptual, emotional, and motor activities.

In Nobel Prize–winning research, Roger Sperry and his collaborators began using behavioral techniques perfected in split-brain cats to study split-brain humans. The researchers realized that by exploiting the organization of the sensory systems, stimuli could be directed exclusively to one hemisphere or the other. For example, objects felt with the left hand, or seen only in the left visual field, are first processed in the sensory cortex of the **contralateral** (opposite side) hemisphere (in this case, the right hemisphere). Normally, information about the object would be passed to the other hemisphere immediately, via the corpus callosum. But in split-brain individuals, the sensory information remains trapped within the receiving hemisphere, so the patient's response to the stimuli reflects the processing specializations of that hemisphere in isolation.

In some of these studies, words were projected visually to either the left or the right hemisphere. The results were dramatic. Split-brain participants could easily read and verbally report words projected to the left hemisphere (via the right visual field) but not words directed to the right hemisphere (**FIGURE 15.1**). Subsequent work (Zaidel, 1976) showed that the right hemisphere does have a limited amount of linguistic ability; for example, it can recognize simple words and the emotional content of verbal material. But in most people, vocabulary and grammar are the exclusive domain of the left hemisphere.

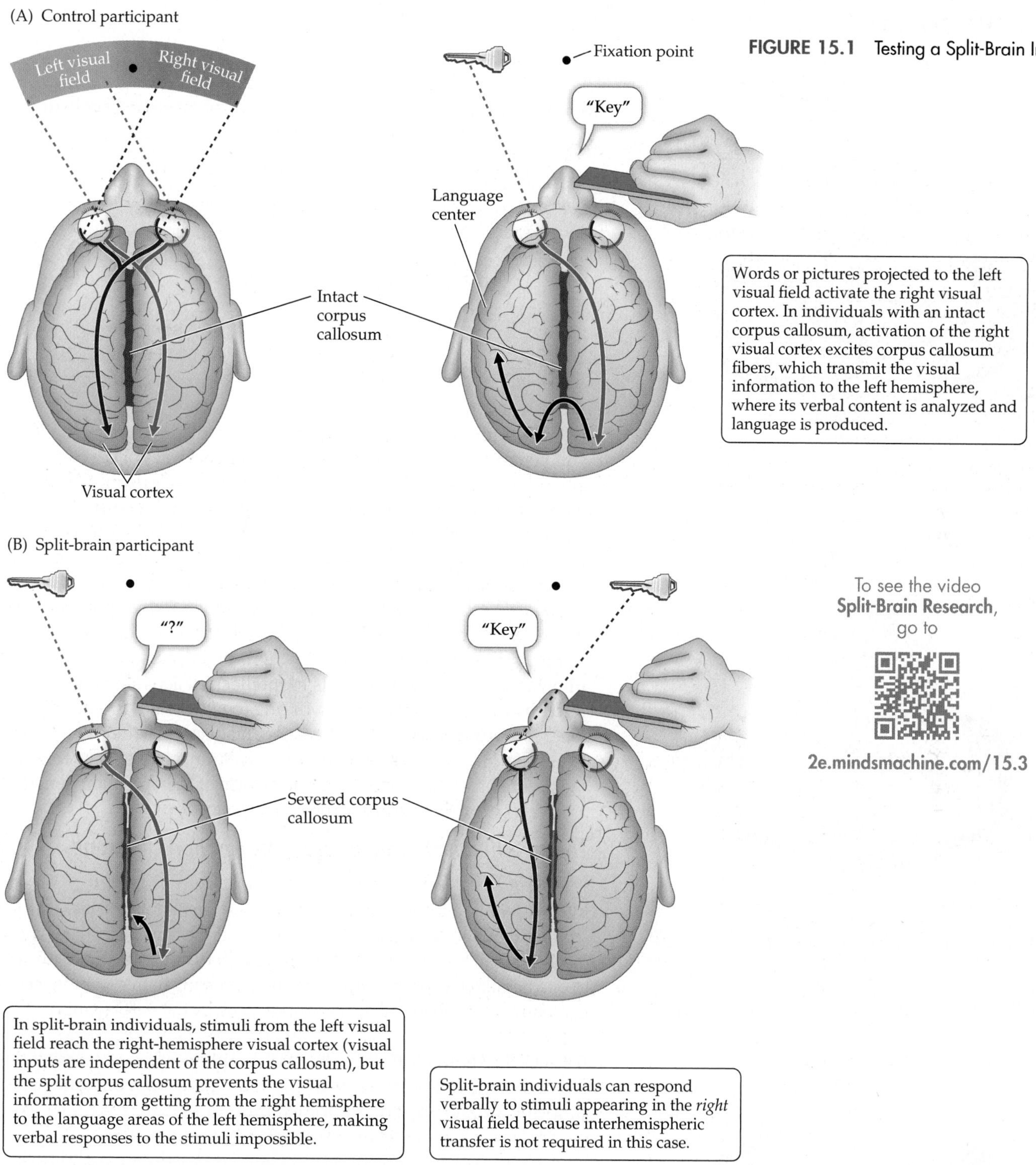

(A) Control participant

Left visual field • Right visual field

Fixation point

"Key"

FIGURE 15.1 Testing a Split-Brain Individual

Language center

Intact corpus callosum

Words or pictures projected to the left visual field activate the right visual cortex. In individuals with an intact corpus callosum, activation of the right visual cortex excites corpus callosum fibers, which transmit the visual information to the left hemisphere, where its verbal content is analyzed and language is produced.

Visual cortex

(B) Split-brain participant

"?"

"Key"

Severed corpus callosum

To see the video **Split-Brain Research**, go to

2e.mindsmachine.com/15.3

In split-brain individuals, stimuli from the left visual field reach the right-hemisphere visual cortex (visual inputs are independent of the corpus callosum), but the split corpus callosum prevents the visual information from getting from the right hemisphere to the language areas of the left hemisphere, making verbal responses to the stimuli impossible.

Split-brain individuals can respond verbally to stimuli appearing in the *right* visual field because interhemispheric transfer is not required in this case.

The capabilities of the "mute" right hemisphere had to be tested by nonverbal means. For example, a picture of a key might be projected to the left visual field and so reach only the right visual cortex. The participant would then be asked to touch several different objects that she could not see and hold up the correct one. Such a task could be performed correctly by the left hand (controlled by the right hemisphere) but not by the right hand (controlled by the left hemisphere). So in this case, the left hemisphere literally does not know what the left hand is doing! In general, these and other studies with split-brain individuals provided evidence that, in most people, the right hemisphere is specialized for processing spatial information. Right-hemisphere mechanisms are also crucial for face perception, for processing emotional aspects of language, and for controlling attention (as we discuss in Chapter 14).

(A)

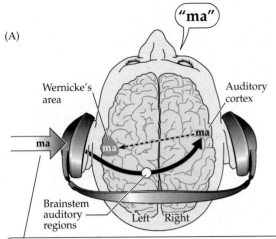

Information to the left ear goes to right auditory cortex and then to Wernicke's area in the left hemisphere. Subject repeats word.

(B)

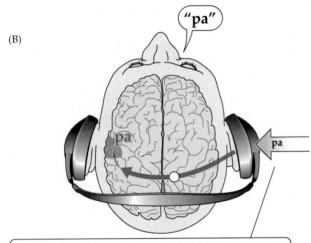

Information to the right ear goes to left auditory cortex and then to Wernicke's area. Subject repeats word.

(C)

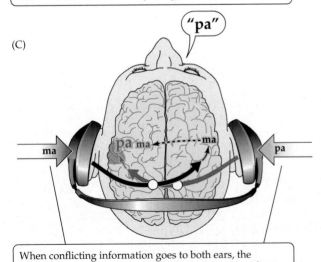

When conflicting information goes to both ears, the information to the right ear reaches Wernicke's area first. Subject repeats only the right-ear information.

FIGURE 15.2 The Right-Ear Advantage in Dichotic Presentation (After Kimura, 1973.)

TABLE 15.1 Proposed Cognitive Modes of the Two Cerebral Hemispheres in Humans

Left hemisphere	Right hemisphere
Phonetic	Nonlinguistic
Sequential	Holistic
Analytical	Synthetic
Propositional	Gestalt
Discrete temporal analysis	Form perception
Linguistic	Spatial

The dramatic findings of split-brain research prompted a lot of commentary and speculation about the different roles of the two hemispheres, often promoting the view that the left hemisphere provides analytical and sequential processing, while the right hemisphere offers a more holistic, general analysis of information (**TABLE 15.1**). Many untested or inconsistent ideas about laterality have received wide circulation, in various formats like self-help books and instructional materials. For example, the notion that the two hemispheres are so different that they need separate instruction lacks any scientific basis, as does the idea that people can have "left-brain" or "right-brain" personalities. Almost all behavior, from manual skills to intellectual activity, is performed better by the two hemispheres working together than by either hemisphere working alone. For this reason, and because better medical options have also been developed, the split-brain surgery has remained a very rare procedure, and the modern form of the operation generally transects only about a third of the corpus callosum.

The two hemispheres process information differently in the human brain

Most research on brain asymmetry in healthy people focuses on two sensory modalities: hearing and vision. That's because researchers have devised clever procedures for directing auditory or visual stimuli mostly to one hemisphere or the other, and then inferring hemispheric specializations from the behavioral responses made by the participants.

THE RIGHT-EAR ADVANTAGE Through earphones, we can present different sounds to the two ears at the same time—a technique called **dichotic presentation**. So, for example, a participant may hear a particular speech sound in one ear and, at the same time, a different vowel, consonant, or word in the other ear. The participant is asked to try to identify or recall both sounds. Although this procedure may seem designed to produce confusion, in general, right-handed people identify verbal stimuli delivered to the right ear more accurately than the stimuli simultaneously presented to the left ear. This result is described as a right-ear "advantage" for verbal information. In contrast, up to 50% of left-handed individuals may show a reduced or reversed pattern, with either no difference between the ears or a clear left-ear advantage.

As a consequence of the preferential connections between the right ear and the left hemisphere, the right-ear advantage for verbal stimuli confirms the idea that the left hemisphere is specialized for language (**FIGURE 15.2**). Although we can normally use either ear for processing speech sounds, speech presented to the right ear in dichotic presentation tests exerts stronger control over language mechanisms in the left hemisphere than does speech simultaneously presented to the left ear (Kimura, 1973). The competition between the left- and right-ear inputs is the key; presentation of speech stimuli to one ear at a time (*monaural* presentation) does not produce a right-ear advantage.

VISUAL PERCEPTION OF LINGUISTIC STIMULI   Another way to study hemispheric specialization is to use a **tachistoscope test** to pit the two hemispheres against one another, using a device (called a *tachistoscope*, surprisingly) that very briefly presents visual stimuli to the left or right visual half field (see Figures 7.10 and 15.1). If the stimulus exposure lasts less than 150 milliseconds or so, input is restricted to one hemisphere because there is not enough time for the eyes to shift their direction. In humans with intact brains, of course, further processing may involve the transmission of information through the corpus callosum to the other hemisphere.

Most tachistoscopic studies confirm the general verbal-spatial division of labor between the hemispheres. Verbal stimuli (words and letters) presented to the right visual field / left hemisphere are better recognized than the same input presented to the left visual field / right hemisphere. On the other hand, nonverbal visual stimuli (such as faces or geometric forms) presented to the left visual field / right hemisphere are better recognized than the same stimuli presented to the other side. Simpler kinds of visual processing, such as the detection of light, hue, or simple patterns, are equivalent in the two hemispheres.

Does the left hemisphere hear words and the right hemisphere hear music?

Anatomical studies of primary auditory cortex in the left and right hemispheres are consistent with the view that the two play different roles in auditory perception. In one early postmortem study of auditory cortex, the **planum temporale**—an auditory region on the superior surface of the temporal lobe—was found to be larger in the left hemisphere than in the right in 65% of brains studied (Geschwind and Levitsky, 1968) (**FIGURE 15.3A**). In only 11% of adults was the right side larger. The planum temporale includes part of a posterior cortical region (called *Wernicke's area*), which we'll see a little later is important for language, so it seems likely that the larger left planum

dichotic presentation   The simultaneous delivery of different stimuli to both the right and the left ears at the same time.

tachistoscope test   A test in which stimuli are very briefly presented to either the left or right visual half field.

planum temporale   An auditory region of superior temporal cortex.

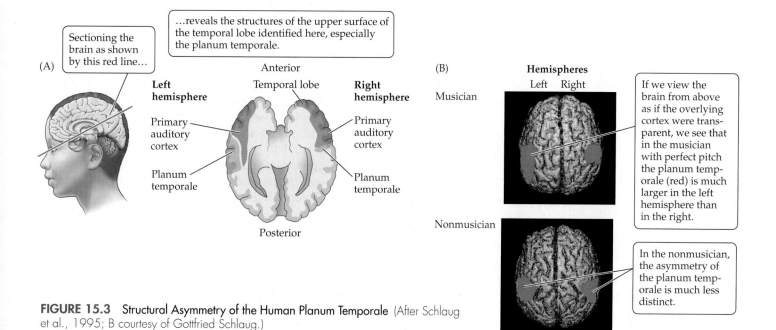

FIGURE 15.3   Structural Asymmetry of the Human Planum Temporale (After Schlaug et al., 1995; B courtesy of Gottfried Schlaug.)

prosody The perception of emotional tone-of-voice aspects of language.

temporale is related to the left hemisphere's specialization for language. Direct evidence of this relationship, however, remains somewhat elusive: in one MRI study, for example, asymmetry of the planum temporale did not correlate with direct measures of language lateralization (Dorsaint-Pierre et al., 2006).

As in adults, the left planum temporale is larger than the right in the brains of infants (Wada et al., 1975). The observation that this difference is present in our brains before we begin to speak therefore bolsters the view that we have an inborn neural mechanism for language. In fact, by just 12–14 weeks of gestation, a variety of genes show asymmetrical expression in the fetal human brain (Sun et al., 2005). The planum temporale likewise tends to be larger on the left than on the right in chimpanzees (Gannon et al., 1998); an anterior zone implicated in speech in humans (Broca's area, which we'll discuss shortly) is also larger on the left in chimps, bonobos, and gorillas, as it is in humans (Cantalupo and Hopkins, 2001). Furthermore, just as our left hemisphere is more activated than the right when we hear speech rather than other sounds, the left hemisphere of monkeys is more activated than the right when they hear monkey vocalizations rather than human speech (Poremba et al., 2004). These results suggest that other primates also possess a left-hemisphere specialization for communication (but not necessarily *language*, as we discuss a little later).

In contrast, the auditory areas of the *right* hemisphere play a major role in the perception of music. Musical perception is especially impaired after damage to the right hemisphere (Samson and Zatorre, 1994), and many aspects of music activate the right hemisphere more than the left (Zatorre et al., 1994). In musicians, however, perfect pitch (the ability to accurately name a musical note just by listening to it) appears to strongly rely on left-hemisphere mechanisms (**FIGURE 15.3B**) (Schlaug et al., 1995), perhaps owing to its verbal aspects.

Despite these data, we cannot assign the perception of speech and pitch entirely to the left hemisphere and the perception of music entirely to the right hemisphere. We have seen that the right hemisphere can play a role in speech perception even in people in whom the left hemisphere is speech-dominant. In addition, the perception of emotional tone-of-voice aspects of language, termed **prosody**, is a *right*-hemisphere specialization. Furthermore, although damage to the right hemisphere can impair the perception of music, it does not abolish it. Damage to *both* sides of the brain can completely wipe out musical perception (Samson and Zatorre, 1991). Thus, even though each hemisphere plays a greater role than the other in different kinds of auditory perception, the two hemispheres appear to collaborate in these functions, as well as in many others.

How does handedness relate to brain asymmetry?

About 10% of the population is left-handed, although the estimated prevalence seems to vary somewhat through history and across geographic regions (Leask and Beaton, 2007). Surveys of left-handed writing in American college populations indicate an incidence of 14% in that group (Spiegler and Yeni-Komshian, 1983). This is nothing new. Artifacts from prehistoric human societies suggest that a predominance of right-handedness is probably an ancient human characteristic, possibly in concert with hemispheric specializations for throwing and speech (Watson, 2001). Some researchers even argue that hemispheric asymmetry and limb preferences reflect a left-right division of labor that, as a matter of processing efficiency, arose in the first vertebrates, hundreds of millions of years ago (MacNeilage et al., 2009). This view is bolstered by reports that on both physiological and behavioral grounds, left-hemisphere neurons of mice differ from their right-hemisphere counterparts (Kohl et al., 2011; Shipton et al., 2014). Perhaps cerebral asymmetry is an ancient and ubiquitous adaptation. (For other examples of the evolution of asymmetry, see **A STEP FURTHER 15.1**, on the website.)

Handedness may have a genetic component, but if so, it is not a simple single-gene effect. Genes that play a role in asymmetry throughout the body, ranging from the shape of internal organs to the pattern of hair whorls on the scalp, may be involved.

But in any case, just as in right-handers, language is vested in the *left* hemisphere in the great majority of left-handers. Only in those very rare cases where the right hemisphere is dominant for language is the person more likely to be left-handed than right-handed.

In a tiny minority of cases, early brain injuries may be responsible for a shift to left-handedness (Silva and Satz, 1979). Simply because most people are right-handed, early one-sided brain injury is more likely to cause a change from right-handedness to left-handedness than the reverse. However, studies of achievement, ability, and cognitive function in thousands of school-age children showed that left-handed children do not differ from right-handed children on any measure of cognitive performance (Hardyck et al., 1976).

There are some rare circumstances in which a change in handedness can occur in adults. Two right-handed people who received double hand transplants after losing their own hands in accidents reportedly both became left-handers after the surgery (Vargas et al., 2009). One speculation is that the right hemisphere required less reorganization in order to take control of its contralateral hand, whereas reconnection of the left hemisphere to the new right hand was initially hampered by the strong preexisting cortical representation of the former right hand. After getting an early start, perhaps the right hemisphere–left hand combo simply outcompeted the left hemisphere–right hand combo. Of course, no one yet knows whether this process bears any relationship whatsoever to the normal establishment of hand preference, but it suggests a surprising degree of flexibility in handedness.

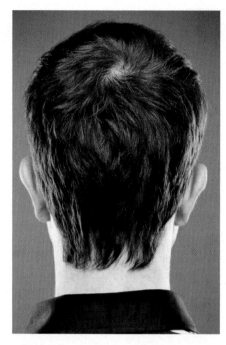

Whorls Apart Because of the relationship between handedness and features such as the whorl of hair at the crown of the scalp—the whorl is clockwise in 93% of right-handers but randomly clockwise or counterclockwise in non-right-handers—it has been suggested that a single gene has a major (but not absolute) influence on asymmetries throughout the body (Klar, 2003). Other aspects of development presumably account for the remaining variability in hand preference.

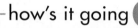

how's it going ?

1. Summarize the cardinal differences between the left and right hemispheres of the human brain. How were these functional differences first identified?

2. Describe the dichotic listening task and tachistoscope testing. How do these tasks reveal lateralization of function?

3. Discuss the representation of music in the human brain. What is the planum temporale, and how does its structure relate to musical experience?

4. Provide a survey of the neural underpinnings of left-handedness. Do left-handers generally show reversed asymmetry, with language vested in the *right* hemisphere?

Right-Hemisphere Damage Impairs Spatial Cognition

When studied with procedures such as tachistoscope tests, people reliably demonstrate a right-hemisphere advantage for processing spatial stimuli. Geometric shapes and their relations, direction sense and navigation, face processing, imagined three-dimensional rotation of objects held in the mind's eye—these are a few examples of the kinds of spatial processing that preferentially rely on the right hemisphere. It is therefore no surprise that right-hemisphere lesions—especially more-posterior lesions that involve the temporal and parietal lobes—tend to produce a variety of striking impairments of spatial cognition, such as inability to recognize faces, spatial disorientation, inability to recognize objects by touch, or the complete neglect of one side of the body that we discussed in Chapter 14.

The diversity of behavioral changes following injury to the parietal lobe is related partly to the large expanse of this lobe and its critical position, abutting all three of the other major lobes of the brain. The anterior end of the parietal region includes the postcentral gyrus, which is the primary cortical receiving area for somatic sensation. Brain injury in this area can produce impairments beyond simple numbness, instead causing changes in complex sensory processing. In one example, objects placed in the hand opposite the injured somatosensory area can be *felt* but cannot be identified by touch and active manipulation. This deficit is called **astereognosis** (from the Greek *a-*, "not"; *stereos*, "solid"; and *gnosis*, "knowledge").

astereognosis The inability to recognize objects by touching and feeling them.

BOX 15.1 The Wada Test

Scientists estimate that 90–95% of humans have a left-hemisphere specialization for language. But how can we be certain which hemisphere is dominant for language (or other functions, such as spatial memory or music perception) in people who haven't had a stroke? In some cases, prior to brain surgery for example, it can be crucial to understand an individual's specific lateralization.

By injecting a short-acting anesthetic (amobarbital) into the carotid artery—first on one side (as shown in the figure) and then, several minutes later, on the other—it is possible to simulate a massive stroke, shutting down the entire hemisphere for a few minutes (Wada and Rasmussen, 1960). This is just long enough to use behavioral measures to document the specializations of each hemisphere.

This **Wada test** confirms that most people have left-hemisphere specialization for language, regardless of handedness. The reverse pattern (right-hemisphere dominance for language) is very rare, but when it occurs, it is more

common in left-handed people. Similar results can be obtained in healthy people using transcranial magnetic stimulation (TMS) (Knecht et al., 2002) or fMRI.

2 ...temporarily shuts down the cerebral hemisphere on the same side, thereby revealing the functions performed by that hemisphere.

1 Injection of the anesthetic amobarbital into the carotid artery, via a catheter...

"I can still talk just fine, but I can't seem to hold my left arm up"

A Simulated Stroke The Wada test uses anesthetic to shut down one hemisphere (in this example, the right hemisphere).

prosopagnosia Also called *face blindness*. A condition characterized by the inability to recognize faces.

Wada test A test in which a short-lasting anesthetic is delivered into one carotid artery to determine which cerebral hemisphere principally mediates language.

More-extensive injuries in the parietal cortex, beyond the primary somatosensory cortex located in the postcentral gyrus, affect interactions between or among sensory modalities, such as visual or tactile matching tasks, which require the participant to visually identify an object that is touched or to reach for an object that is identified visually.

In prosopagnosia, faces are unrecognizable

Suppose that one day you look in the mirror, and someone totally unfamiliar is looking back at you. As incredible as this scenario might seem, it's a fate that some individuals suffer following very specific brain damage. In this rare syndrome, called **prosopagnosia** (from the Greek *prosop-*, "face"; *a-*, "not"; and *gnosis*, "knowledge") or *face blindness*, afflicted individuals fail to recognize not only their own faces but also the faces of relatives and friends. No amount of remedial training restores their ability to recognize anyone's face. In contrast, the ability to visually recognize *objects* may be retained, and the patient may have no difficulty identifying people by their voices.

To people with prosopagnosia, faces simply lack meaning. No disorientation or confusion accompanies this condition, nor is there evidence of diminished intellectual abilities or significant visual impairment. Most research indicates that the right hemisphere is especially important for recognizing faces. For example, shutting down the right hemisphere during the Wada test (**BOX 15.1**) can cause difficulty in recognizing faces, whereas anesthetizing the left hemisphere has less effect on facial recognition (**FIGURE 15.4**).

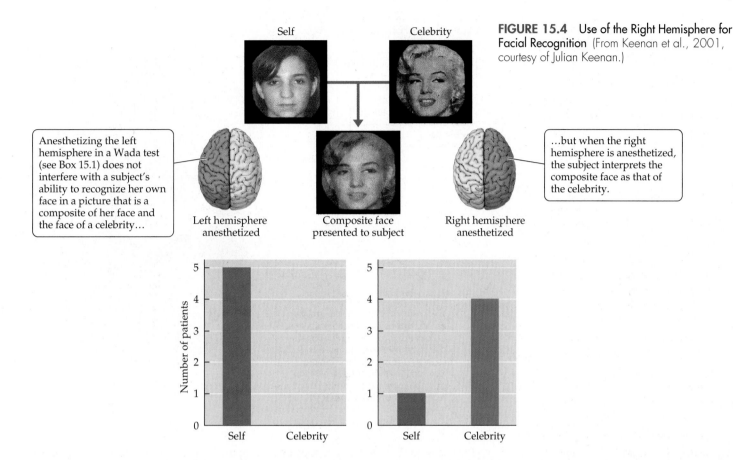

Self   Celebrity

FIGURE 15.4 Use of the Right Hemisphere for Facial Recognition (From Keenan et al., 2001, courtesy of Julian Keenan.)

Anesthetizing the left hemisphere in a Wada test (see Box 15.1) does not interfere with a subject's ability to recognize her own face in a picture that is a composite of her face and the face of a celebrity…

Left hemisphere anesthetized

Composite face presented to subject

Right hemisphere anesthetized

…but when the right hemisphere is anesthetized, the subject interprets the composite face as that of the celebrity.

Similarly, split-brain patients do a better job of recognizing faces that are presented to the right hemisphere than to the left (Gazzaniga and Smylie, 1983). Still, data from split-brain patients and functional imaging studies make it clear that both hemispheres have *some* capacity for recognizing faces. Thus, although damage restricted to the right hemisphere can impair face processing, the most complete cases of prosopagnosia are caused by *bilateral* damage. The **fusiform gyrus**, a region of cortex on the inferior surface of the brain where the occipital and temporal cortices meet (**FIGURE 15.5**), is crucial for face recognition, and cases of prosopagnosia following brain damage almost always involve damage here. Symptoms may be exacerbated when lesions also include a nearby occipital area that performs initial face-specific visual processing, as well as when they include the superior temporal sulcus, which has been implicated in connecting faces with speech and facial expressions.

Until recently, it was believed that prosopagnosia occurred solely as a result of brain damage (*acquired prosopagnosia*), but a number of cases of *developmental* (or *congenital*, which means "present at birth") *prosopagnosia* have now been identified. This is apparently the problem that has dogged Louis, whom we met at the beginning of the chapter, and the discovery that he's not alone in this problem has been a huge relief to him (we can only guess how Humpty Dumpty would feel about it).

Unexpectedly, surveys have revealed that about 2.5% of the general population is sufficiently impaired in processing faces to meet the criteria for congenital prosopagnosia (Kennerknecht et al., 2006; Duchaine et al., 2007). The developmental form of prosopagnosia appears to run in families, indicating a genetic aspect to the disorder (Grüter et al., 2008). Developmental prosopagnosia is associated with reduced activation of the fusiform gyrus, in keeping with the anatomical findings in acquired prosopagnosia that we've already discussed. (You can learn more, and test yourself for prosopagnosia, online at www.faceblind.org/facetests.)

fusiform gyrus A region on the inferior surface of the cortex, at the junction of the temporal and occipital lobes, that has been associated with recognition of faces.

FIGURE 15.5 The Fusiform Gyrus

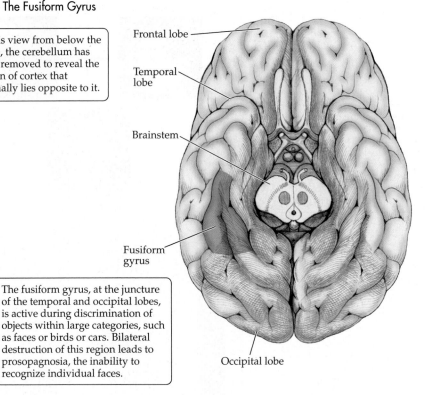

In this view from below the brain, the cerebellum has been removed to reveal the region of cortex that normally lies opposite to it.

Frontal lobe

Temporal lobe

Brainstem

Fusiform gyrus

The fusiform gyrus, at the juncture of the temporal and occipital lobes, is active during discrimination of objects within large categories, such as faces or birds or cars. Bilateral destruction of this region leads to prosopagnosia, the inability to recognize individual faces.

Occipital lobe

agnosia The inability to recognize objects, despite being able to describe them in terms of form and color. Agnosia may occur after localized brain damage.

aphasia An impairment in language understanding and/or production that is caused by brain injury.

paraphasia A symptom of aphasia that is distinguished by the substitution of a word by a sound, an incorrect word, an unintended word, or a neologism (a meaningless word).

agraphia The inability to write.

alexia The inability to read.

Prosopagnosia of either type may be accompanied by additional forms of **agnosia**, an inability to identify individual items—makes of cars, tools, bird species, sounds, and so on—in the absence of any specific sensory deficits or memory problems (Gauthier et al., 1999). Functional-MRI studies of healthy people show that, as with face recognition, the fusiform gyrus is important for identifying individual members of large categories (e.g., faces or birds or cars) in which all members have many things in common (Gauthier et al., 2000).

how's it going ?

1. In general, what are the behavioral consequences of right-hemisphere damage in humans?

2. Define, compare, and contrast two of the most striking symptoms of right-hemisphere damage: astereognosis and prosopagnosia. Which areas of the brain appear to be involved in each?

3. Is prosopagnosia always associated with brain damage? What other behavioral abnormalities may co-occur with prosopagnosia?

Language Disorders Result from Region-Specific Brain Damage

For thousands of years, dating back to the earliest medical records and beyond, people have known that a selective impairment of language abilities, known as **aphasia**, could result from injury to specific regions of the brain, especially the left hemisphere (Finger, 1994). Left-hemisphere damage can impair language to varying degrees, depending on the location and extent of the injury. In cases of severe damage, people may lose the ability to produce any speech whatsoever. In less severe cases, patients may exhibit speech with **paraphasia**—insertion of incorrect sounds or words—along with labored, effortful speech production.

Most patients with aphasia also show some impairment in writing (**agraphia**) and disturbances in reading (**alexia**). Brain damage that produces aphasia also produces a

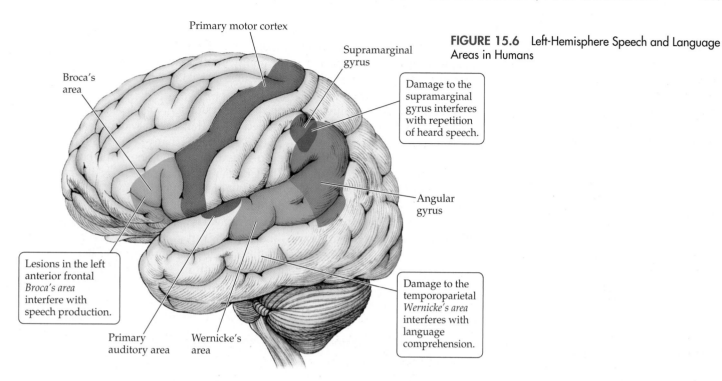

Primary motor cortex

Broca's area

Supramarginal gyrus

FIGURE 15.6 Left-Hemisphere Speech and Language Areas in Humans

Damage to the supramarginal gyrus interferes with repetition of heard speech.

Angular gyrus

Lesions in the left anterior frontal *Broca's area* interfere with speech production.

Damage to the temporoparietal *Wernicke's area* interferes with language comprehension.

Primary auditory area

Wernicke's area

distinctive motor impairment called **apraxia** (see Chapter 5), characterized by great difficulty in making precise sequences of movements, despite the absence of weakness or paralysis. In fact, as we discuss later, some theorists view aphasia as primarily a disorder of motor control. Research has distinguished several major categories of aphasia, differing from one another in the patterns of symptoms that occur.

Linking these distinctive language disorders to specific regions of brain damage has uncovered details of a left-hemisphere language network (**FIGURE 15.6**). The specific language deficit experienced depends on which particular components are compromised, as we discuss in the following sections.

Damage to a left anterior speech zone causes nonfluent (or Broca's) aphasia

In the mid-1800s, French neurologist Paul Broca (1824–1880) examined a man who had lost the ability to utter more than the single syllable "tan" (**FIGURE 15.7**). Following postmortem analysis, Broca reported that this patient, and other patients with similar severe impairments of speech production, had suffered damage to a left inferior frontal region that now bears his name—**Broca's area** (see Figure 15.6). Anterior lesions that include Broca's area often produce a type of aphasia known as **nonfluent aphasia**, also called *Broca's aphasia*. Patients with nonfluent aphasia have a lot

apraxia   An impairment in the ability to carry out complex sequential movements, even though there is no muscle paralysis.

Broca's area   A region of the frontal lobe of the brain that is involved in the production of speech.

nonfluent aphasia   Also called *Broca's aphasia*. A language impairment characterized by difficulty with speech production but not with language comprehension. It is related to damage in Broca's area.

Study of this brain and similar cases led Paul Broca to identify a region in the anterior left hemisphere that is specialized for speech.

Damage in what is now known as Broca's area is clearly evident in these horizontal sections, corresponding to levels 2 and 3 on the orientation figure.

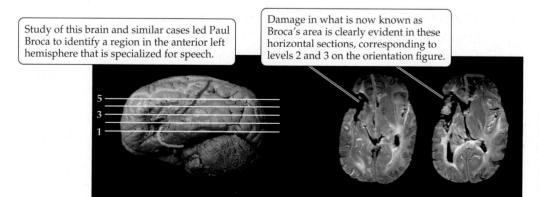

5
3
1

FIGURE 15.7   The Brain of "Tan" Photo and MRI image of the preserved brain of M. Leborgne, who could only utter the syllable "tan" after his brain injury. (From Dronkers et al., 2007.)

hemiplegia Paralysis of one side of the body.

hemiparesis Weakness of one side of the body.

Wernicke's area A region of temporoparietal cortex in the brain that is involved in the perception and production of speech.

fluent aphasia Also called *Wernicke's aphasia*. A language impairment characterized by fluent, meaningless speech and little language comprehension. It is related to damage in Wernicke's area.

anomia The inability to name persons or objects readily.

word deafness A selective inability to understand spoken words.

word blindness A selective inability to understand written words.

global aphasia The total loss of ability to understand language, or to speak, read, or write.

of difficulty producing speech, talking only in a labored and hesitant manner. Reading and writing are also impaired. The ability to utter automatic speech, however, is often preserved. Such speech includes greetings ("Hello"); short, common expressions ("Oh, my God!"); and swear words.

Compared with their difficulty with speech production, *comprehension* of language is relatively good in nonfluent aphasics. Because of the proximity of primary and supplementary motor cortex to Broca's area (see Chapter 5), many people who suffer from nonfluent aphasia also have **hemiplegia**—paralysis of one side of the body (usually the right side, which is controlled by the left hemisphere). Sometimes there is unilateral weakness, termed **hemiparesis**, rather than full paralysis.

The CT scans in **FIGURE 15.8A** and maps of lesion sites in **FIGURE 15.8B** are from several patients with nonfluent aphasia. Seven years after a stroke, one such patient still spoke slowly, used mainly nouns and very few verbs or function words (a selective loss of action words, called *averbia*, sometimes occurs in nonfluent aphasia), and spoke only with great effort. When asked to repeat the phrase "Go ahead and do it if possible," she could say only, "Go to do it," with a pause between each word. People who suffer brain lesions as extensive as hers show little recovery of speech functions with the passing of time, but people with milder cases can show significant recovery.

Damage to a left posterior speech zone causes fluent (or Wernicke's) aphasia

Not long after Broca's groundbreaking discovery of the left anterior speech zone, German neurologist Carl Wernicke (pronounced "VER-nih-keh") (1848–1905) described a different form of aphasia, resulting from damage to a more posterior region of the left hemisphere, centered on a region of the superior temporal cortex that is now known as **Wernicke's area** (see Figure 15.6). Unlike the nonfluent aphasics, people who have this **fluent aphasia** (or *Wernicke's aphasia*) produce plenty of verbal output, but their utterances, although speechlike, tend to contain many paraphasias, such as sound substitutions (e.g., "girl" becomes "curl") and/or word substitutions (e.g., *bread* becomes *cake*). Some fluent aphasias are marked by a particular difficulty in naming persons or objects—an impairment referred to as **anomia**. The ability to repeat words and sentences is impaired. For this reason, people with fluent aphasia are believed to have difficulty *understanding* what they read or hear.

Lesions that produce fluent aphasia usually include the posterior parts of the superior left temporal lobe, extending into adjacent regions of parietal cortex. Examples of lesions causing fluent aphasia are shown in **FIGURE 15.8C**. Sometimes stroke victims experience **word deafness** (the inability to understand spoken words), which usually means that auditory regions of the temporal lobe are particularly affected. Other patients may instead experience **word blindness** (the inability to understand written words), usually indicating damage that includes connections to visual regions (see Figure 15.6). And because the typical lesion is posterior, postcentral somatosensory cortex is more likely to be damaged than precentral motor cortex, so patients are more likely to have a right-sided numbness than a right-sided weakness.

Widespread left-hemisphere damage can obliterate language capabilities

In some patients, brain injury or disease results in total loss of the ability to understand or produce language, called **global aphasia**. Patients suffering from global aphasia may retain some ability to make speechlike sounds, especially emotional exclamations. But they can utter very few words, and no semblance of syntax remains. Global aphasia generally results from very large left-hemisphere lesions that encompass both anterior and posterior language zones. Frontal, temporal, and parietal cortex—including Broca's area, Wernicke's area, and the supramarginal gyrus (see Figure 15.6)—are usually affected (**FIGURE 15.8D**). The prognosis for language recovery in these patients is quite poor, and because of the extent of their lesions, the aphasia is generally accompanied by other debilitating neurological impairments.

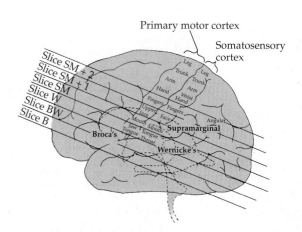

Primary motor cortex
Somatosensory cortex

FIGURE 15.8 Brain Lesions That Produce Aphasia (After Naeser and Hayward, 1978, CT scans courtesy of Margaret Naeser.)

To make the maps shown below, brain CT scan "slices" were labeled according to the brain language regions shown in each slice: B, Broca's area; SM, supramarginal gyrus; W, Wernicke's area.

(A) This 51-year-old patient had nonfluent (Broca's) aphasia as a result of a stroke that included Broca's area.

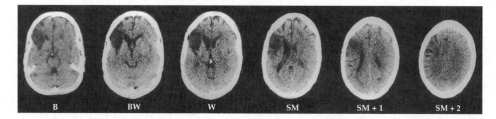

B   BW   W   SM   SM + 1   SM + 2

(B) Here the lesions (in blue) of four patients with nonfluent (Broca's) aphasia are drawn as overlapping maps.

(C) Lesion sites for four cases of fluent (Wernicke's) aphasia are mapped on these drawings.

(D) These drawings show lesion sites for five cases of global aphasia. Large lesions are present in every language area.

Slice B   Slice BW   Slice W   Slice SM   Slice SM + 1   Slice SM + 2

(A) Speaking a *heard* word

1 Information about the sound is analyzed by primary auditory cortex and transmitted to Wernicke's area.

2 Wernicke's area analyzes the sound information to determine the word that was said.

3 Under the connectionist model this information is transmitted via the arcuate fasciculus. (Note, however, that anatomical research casts doubt on this projection.)

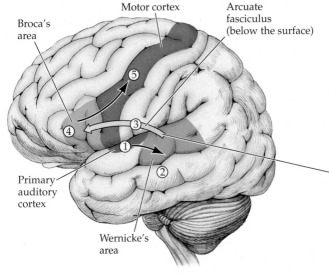

4 Broca's area forms a motor plan to repeat the word and sends that information to motor cortex.

5 Motor cortex implements the plan, manipulating the larynx and related structures to say the word.

Lesions of the arcuate fasciculus disrupt the transfer from Wernicke's area to Broca's area, so the patient has difficulty repeating spoken words (so-called conduction aphasia), but may retain comprehension of spoken language (because of intact Wernicke's area) and may still be able to speak spontaneously (because of intact Broca's area).

(B) Speaking a *written* word

1 Visual cortex analyzes the image and transmits the information about the image to the angular gyrus.

2 The angular gyrus decodes the image information to recognize the word and associate this visual form with the spoken form in Wernicke's area.

3 Information about the word is transmitted via the arcuate fasciculus to Broca's area.

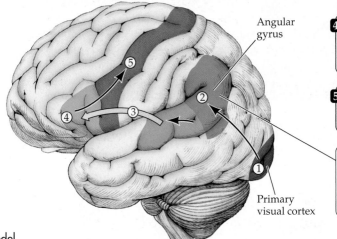

4 Broca's area formulates a motor plan to say the appropriate word and transmits that plan to motor cortex for implementation.

5 Motor cortex implements the plan, manipulating the larynx and related structures to say the word.

A lesion of the angular gyrus disrupts the flow of information from visual cortex, so the person has difficulty saying words he has seen but not words he has heard.

FIGURE 15.9 The Connectionist Model of Aphasia (After Geschwind, 1976.)

connectionist model of aphasia Also called *Wernicke-Geschwind model*. A theory proposing that left-hemisphere language deficits result from disconnection between the brain regions in a language network, each of which serves a particular linguistic function.

arcuate fasciculus A fiber tract classically viewed as a connection between Wernicke's speech area and Broca's speech area.

conduction aphasia An impairment in the ability to repeat words and sentences.

Disconnection of language regions may result in specific verbal problems

The traditional **connectionist model of aphasia**—also known as the *Wernicke-Geschwind model*—views language deficits as resulting from *disconnection* between the brain regions in a language network. Each of these regions is proposed to serve a particular feature of language analysis or production. So, when a word or sentence is heard, the auditory cortex transmits information about the sounds to a speech reception mechanism in Wernicke's area, where the sounds are analyzed to decode what they mean. For the word to be spoken, Wernicke's area must transmit this information, via a fiber pathway called the **arcuate fasciculus**, to the expressive mechanism in Broca's area, where a speech plan is activated. Broca's area then transmits this plan to adjacent motor cortex, which controls the muscles of the chest, throat, and mouth that are used for speech production (**FIGURE 15.9**). According to the model, patients with lesions that selectively disrupt the connection between the posterior speech reception zone (Wernicke's area) and the anterior speech production zone (Broca's area) will especially struggle with repetition of words and phrases that they hear, despite good speech comprehension and production—a condition termed **conduction aphasia**. **TABLE 15.2** summarizes the main features of the aphasias that we've discussed.

Critics of the connectionist model argue that it oversimplifies the neural mechanisms of language, and brain-imaging studies indicate that the left hemisphere contains several speech areas well outside of the classical Broca's and Wernicke's areas

TABLE 15.2 Language Symptomatology in Aphasia

Type of aphasia	Brain area affected	Spontaneous speech	Comprehension	Paraphasia	Repetition	Naming
Nonfluent (Broca's) aphasia		Nonfluent	Good	Uncommon	Poor	Poor
Fluent (Wernicke's) aphasia		Fluent	Poor	Common	Poor	Poor
Global aphasia		Nonfluent	Poor	Variable	Poor	Poor
Conduction aphasia	L R	Fluent	Good	Common	Poor	Poor

(Bates et al., 2003; Dronkers et al., 2004), implying greater complexity than was previously supposed. Furthermore, technological advances in the visualization of white matter pathways in the living brain have raised questions about the assumptions underlying the connectionist model (**BOX 15.2**). Detailed brain-imaging research suggests that so-called conduction aphasia actually results from a specific type of lesion of superior temporal cortex, rather than the disruption of white matter pathways as proposed under the connectionist model (B. R. Buchsbaum et al., 2011). So it remains to be seen exactly how the various mechanisms of the left hemisphere collaborate to give us the verbal abilities that seem so effortless.

diffusion tensor imaging (DTI)   A modified form of MRI imaging in which the diffusion of water in a confined space is exploited to produce images of axonal fiber tracts.

DTI tractography   Also called *fiber tracking*. Visualization of the orientation and terminations of white matter tracts in the living brain via diffusion tensor imaging.

BOX 15.2 Studying Connectivity in the Living Brain

A modern brain-imaging technology is allowing researchers to reexamine the traditional connectionist model of left hemisphere language functions. In **diffusion tensor imaging (DTI)**, MRI technology is used in a new way to specifically study white matter tracts—axon bundles—within the living brain.

As we discussed in Chapter 2, MRI images are created from the radio-frequency energy that is emitted by relaxing protons within water molecules. In DTI, the unique behavior of water molecules that are constrained within axons (known as *fractional anisotropy*) is exploited to create images of axonal fiber pathways between areas, a procedure called **DTI tractography**, or *fiber tracking* (Assaf and Pasternak, 2008). Although the technology doesn't have the resolution to portray individual axons, the origin, orientation, course, and termination of axonal projections can be effectively visualized, as

shown in the figure. One important discovery using this technique is that the arcuate fasciculus—long believed to connect Wernicke's area to Broca's area—appears to terminate in the precentral gyrus (motor cortex), just short of Broca's area, in most people (Bernal and Altman, 2010; E. C. Brown et al., 2014). Coupled with the clinical observation that people with purely cortical lesions sometimes have "conduction" aphasia, this finding is prompting researchers to reevaluate existing notions of language organization in the brain. The more nuanced picture of the language network emerging from this research will have implications for both fundamental understanding of brain organization and treatment of language disorders.

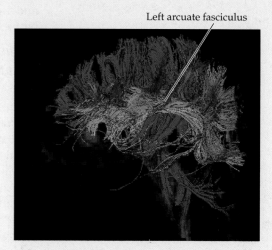

Left arcuate fasciculus

Fiber Tracking the Language Areas In DTI tractography, the orientation, origin, and termination of fiber tracts can be visualized. Here, the left arcuate fasciculus is visualized in light green. (From Vandermosten et al., 2012.)

motor theory of language The theory that speech is perceived using the same left-hemisphere mechanisms that are used to produce the complex movements that go into speech.

An alternative model of speech mechanisms—the **motor theory of language** (Lieberman, 1985; Kimura, 1993)—suggests that the anterior and posterior left-hemisphere language zones are both specialized for motor control. According to the motor theory, when we listen to speech, we simply process the speech sounds using the same neural systems that we would use to *make* the same sounds: an anterior system programs simple phonemic units, and a posterior system strings speech sounds together into long sequences of movements (Kimura and Watson, 1989). Interestingly, deaf people who use American Sign Language (ASL)—a language based entirely on movements—employ the same language-related regions of the left hemisphere as hearing people who use spoken language (Neville et al., 1998; Petitto et al., 2000), and they show comparable aphasia-like symptoms after focal left-hemisphere damage (Kimura, 1981; Bellugi et al., 1983).

how's it going

1. Distinguish among aphasia, agraphia, and alexia.
2. Identify the main types of aphasia, and summarize the distinctive symptoms of each. Which type of aphasia is most often associated with paralysis?
3. Provide a brief outline of the traditional "connectionist" model of aphasia. How does it get its name, and what shortcomings of the model have been identified by critics?
4. Briefly describe the motor theory of language. Can you think of ways that this theory may relate to the evolutionary origins of language?

Brain Mapping Helps Us Understand the Organization of Language in the Brain

In order to study the organization of the brain's mechanisms of language in people with intact language capabilities, researchers employ two general experimental approaches. In cortical stimulation studies, discrete cortical regions are directly activated, and the resultant changes in language function are measured. In functional brain-imaging studies, the approach is reversed: participants engage in language behaviors, and associated changes in brain activity are measured. Together, these techniques have extended our understanding of the neural bases of language.

Cortical stimulation mapping shows localized functions within language areas

Early studies of the organization of language areas in the brain employed electrical stimulation mapping, in which a surgeon used a handheld electrode to electrically stimulate discrete regions of cortex while the effect on behavior was observed. Most of these studies collected data from neurosurgical patients undergoing surgery to remove a tumor or epileptic tissue. Once the brain was exposed, small stimulating electrodes were touched to the surface, disrupting the normal functioning of neurons in the immediate vicinity. Because patients were given only local anesthesia, they were conscious and able to perform various cognitive tests (the main aim of the procedure was to identify tissue that could be removed without impairing language). Data from numerous patients were superimposed to create a map of language-related zones of the left hemisphere (**FIGURE 15.10A**) (Penfield and Roberts, 1959). Stimulation anywhere within a large anterior zone often stopped speech outright. Other forms of language interference, such as misnaming or impaired repetition of words, occurred with stimulation throughout the anterior and posterior cortical speech zones.

Later cortical stimulation studies revealed anatomical compartmentalization of linguistic systems such as naming, reading, speech production, and verbal memory (Calvin and Ojemann, 1994). An interesting example of the effects of cortical stimulation on naming is illustrated in **FIGURE 15.10B**, which shows the different places where

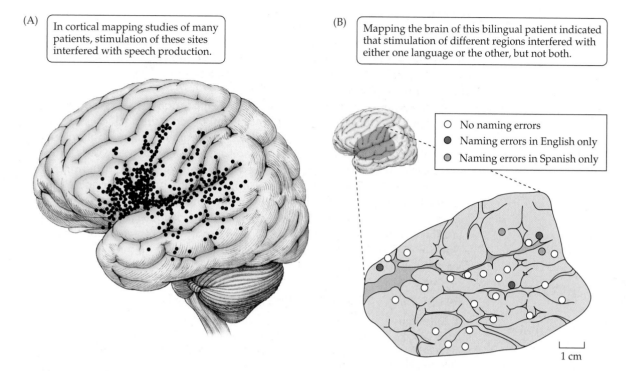

(A) In cortical mapping studies of many patients, stimulation of these sites interfered with speech production.

(B) Mapping the brain of this bilingual patient indicated that stimulation of different regions interfered with either one language or the other, but not both.

○ No naming errors
● Naming errors in English only
● Naming errors in Spanish only

1 cm

FIGURE 15.10 Electrical Stimulation of Some Brain Sites Can Interfere with Language (Part A after Penfield and Roberts, 1959; B after Ojemann and Mateer, 1979.)

stimulation caused naming errors in English and Spanish in a bilingual person. Note that this very fine-grained approach reveals different subregions that disrupt either English or Spanish function. People who are bilingual from an early age show a complete overlap of the English and Spanish language zones at the gross neuroanatomical level as provided by fMRI measures (Perani and Abutalebi, 2005). More recent studies employ a noninvasive cortical stimulation technology called *transcranial magnetic stimulation* to further probe the organization of language areas in healthy volunteers. As we discuss next, these studies have confirmed the general organization of the left-hemisphere language network and have revealed new details about the compartmentalization of functions within traditional speech areas.

researchers at work

Noninvasive stimulation mapping reveals details of the brain's language areas

Transcranial magnetic stimulation (TMS) is a relatively new technology that allows researchers to stimulate small clusters of cortical neurons with good precision, from outside the scalp (see Figure 2.21). By using MRI scans to select targets, TMS provides a noninvasive method for inducing a sort of temporary brain lesion, disrupting the activity of the selected brain region for up to an hour. Alternatively, TMS can be used along with PET or fMRI to precisely map regions of increased activity.

Using TMS mapping, researchers have generally replicated and extended the earlier findings regarding the cortical organization of language functions. For example, TMS mapping has revealed that speech production is associated with activation of not only face areas in motor cortex but also hand areas (Meister et al., 2003), confirming the linkage and possible evolutionary relationship of hand gestures and speech. Similarly, TMS mapping has been used to show that speech *perception*

continued

researchers at work *continued*

activates specific regions that TMS shows to be involved with speech *production* (Scott and Wise, 2004), providing support for the motor theory of language, which we discussed earlier.

Impressively, the TMS temporary-lesion approach has revealed previously unknown functional subregions *within* Broca's area (**FIGURE 15.11**). In these experiments (Gough et al., 2005), researchers found that anterior parts of Broca's area are involved in the semantic *meaning* of words, while a more posterior part of Broca's area is important for the patterning of speech *sounds*. Other research has shown that the posterior speech zone likewise contributes to both word meaning and sound-related aspects of language (Stoeckel et al., 2009). So the TMS procedure is providing new insights into the fine details of the cortical organization of language, especially in conjunction with traditional neuroimaging techniques (Devlin and Watkins, 2007).

FIGURE 15.11   Subregions of Broca's Area Revealed by TMS

■ **Question**

Does Broca's area perform a single language function (as proposed in classical models of language organization), or is Broca's a collection of structures with differing linguistic functions?

■ **Test**

Guided by precise structural MRI scans, apply transcranial magnetic stimulation (TMS) to activate discrete regions within Broca's area.

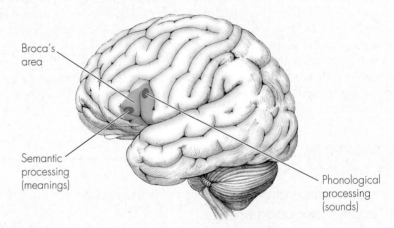

■ **Result**

Anterior regions of Broca's area appear to be important for semantic processing (word meanings), while more posterior regions within Broca's area specialize in phonological processing (sounds of words).

Functional neuroimaging technologies let us visualize activity in the brain's language zones during speech

Different aspects of language processing produce noticeably different patterns of brain activation, as shown in **FIGURE 15.12**. In this study, passive *viewing* of words activated a posterior area within the left hemisphere (**FIGURE 15.12A**), but passive *hearing* of words shifted the focus of brain activation to the temporal lobes (**FIGURE 15.12B**). Repeating words orally activated the motor cortex of both sides, along with supplementary motor cortex and some of the cerebellum (**FIGURE 15.12C**). During word repetition or reading aloud, little activity was seen in Broca's area. But when participants were

(A) Passively viewing words

Each series of PET scans shows the pattern of brain activity associated with the verbal task depicted at left.

Level = 40   20   0   −12   −20

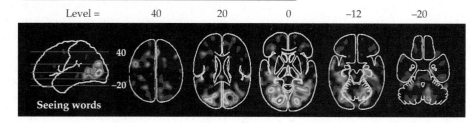

Seeing words

(B) Listening to words

"Scissors"

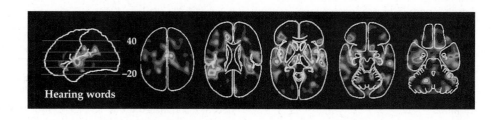

Hearing words

(C) Speaking words

"Scissors"

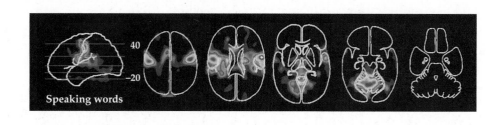

Speaking words

(D) Generating a verb associated with each noun shown

"Cut"

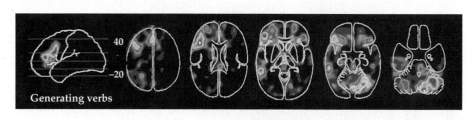

Generating verbs

FIGURE 15.12 PET Scans of Brain Activation in Progressively More Complex Language Tasks (After Posner and Raichle, 1994, PET scans courtesy of Marcus Raichle.)

required to generate an appropriate verb to go with a supplied noun, language-related regions in the left hemisphere, including Broca's area, suddenly became markedly activated (**FIGURE 15.12D**).

Even languages that sound very, very different tend to activate much the same brain regions in native speakers. Silbo Gomero is a very unusual whistled language of the Canary Islands, used by shepherds (known as *silbadores*) to communicate over long distances. In Silbo, whistled notes serve as the phonemes and morphemes of a stripped-down form of Spanish (for an audio sample of Silbo, with translation, see **A STEP FURTHER 15.2**, on the website). Functional MRI reveals that *silbadores* process Silbo using the same left-hemisphere mechanisms that they (and everyone else) use to process spoken language (Carreiras et al., 2005). Non-*silbadore* controls, in contrast, process the whistle sounds of Silbo using completely different regions of the brain; for these people, of course, the whistle sounds have no linguistic content. So the brain's left-hemisphere language systems appear to be sufficiently plastic to adapt to many types of signals, provided that they're heard early enough. Even 3-month-old infants, when exposed to speech, show more metabolic activity in the left hemisphere than in the right (Dehaene-Lambertz et al., 2002).

Event-related potentials (ERPs; see Chapters 3 and 14) also provide hints about the brain's language network, by revealing the time base for language processing by the brain. In one such study, participants were asked to read a sentence in which they encountered a word that was grammatically correct but, because of its meaning, didn't fit—for example, "The man started the car engine and stepped on the pancake." About 400 milliseconds after the participant read the word *pancake*, a negative wave was detectable on the scalp (Kutas and Hillyard, 1980). Such "N400" responses (*N* denotes "negative," and the number represents the response time in milliseconds; see Chapter 14) to word meanings seem to be centered over the temporal lobe (Neville et al., 1992) and therefore may originate from temporoparietal cortex (including Wernicke's area). But *grammatically* inappropriate words elicit a positive potential about 600 milliseconds after they are encountered (a "P600" response), indicating that detection of this level of error requires an extra 200 milliseconds of brain processing by other components of the language network (Osterhout, 1997).

stuttering The tendency of otherwise healthy people to produce speech sounds only haltingly, tripping over certain syllables or being unable to start vocalizing certain words.

Williams syndrome A disorder characterized by impairments of spatial cognition and IQ but superior linguistic abilities.

how's it going ❓

1. Summarize and discuss the organization of language systems in the human brain, as revealed by stimulation mapping. Include TMS mapping in this discussion.

2. Discuss the patterns of brain activity that are observed in various verbal tasks, using functional-imaging technologies and ERPs. How well do these results align with traditional models of the organization of language areas in the brain?

3. What can we learn about general principles of language localization in the brain by studying unusual languages like Silbo?

PART II
Speech and Language
Some Aspects of Language Are Innate, but Others Must Be Learned

A few years ago, researchers discovered a most unusual family in England. Across at least three generations, about half of the members of the KE family have suffered from a severe language disorder. Affected family members take a long time to learn to speak and have difficulty with particular language tasks, such as learning verb tenses (Lai et al., 2001). Brain activation during language tasks is altered in this family too (Liégeois et al., 2003).

By studying the KE family's pedigree, researchers soon identified a gene, called *FOXP2*, that must be important for the normal acquisition of human language, because affected members of the KE family all share a mutation in this gene (**FIGURE 15.13**). In fact, multiple variants of *FOXP2* tend to produce different abnormalities in language-associated areas of the brain (Pinel et al., 2012), probably because *FOXP2* is a regulatory gene that can alter a variety of other genes (Vernes et al., 2011). **Stuttering**—the tendency of otherwise healthy people to produce speech sounds only haltingly, tripping over certain

The King's Speech England's King George VI—shown here during a wartime radio broadcast—suffered from severe stuttering. His struggle and eventual success in coping with speech difficulties, as portrayed in the film *The King's Speech*, shows that despite the possible genetic bases of the condition, effective therapy is possible.

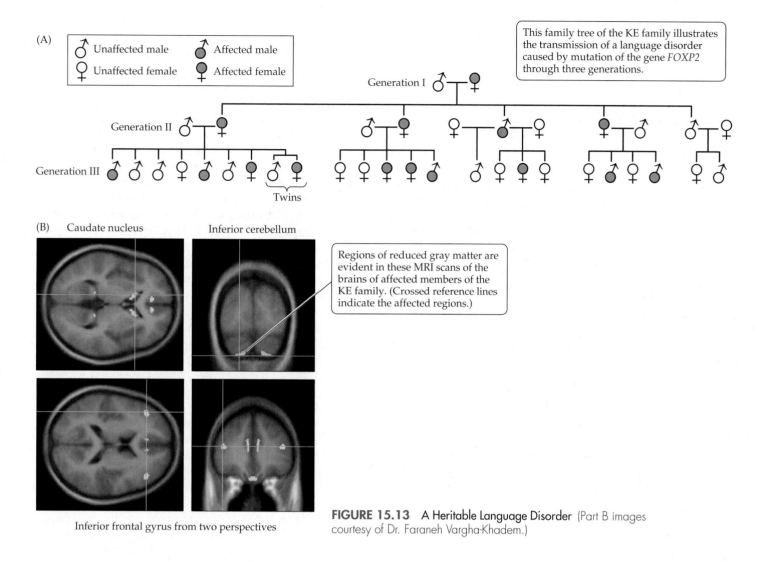

FIGURE 15.13 A Heritable Language Disorder (Part B images courtesy of Dr. Faraneh Vargha-Khadem.)

syllables or being unable to start vocalizing certain words—is likewise at least partly heritable (C. Kang et al., 2010).

On the flip side of the coin, children born with **Williams syndrome**—caused by the deletion of 28 genes from chromosome 7—have various intellectual deficits but excellent verbal skills. No one knows exactly what developmental mechanism results in this hyperverbal behavior (Paterson et al., 1999), but possession of *extra* copies of the identified genes on chromosome 7—rather than deletions of these genes—produces a syndrome of very poor expressive language that is, in many ways, the converse of Williams syndrome (Somerville et al., 2005).

These findings confirm that basic mechanisms of language are heritable components of the human brain, and the product of a long evolutionary history. Indeed, the use of language is one of the key adaptations of humans, with basic capabilities probably arising in an ancient ancestor of our species. For example, the Neandertals shared our version of *FOXP2* (Krause et al., 2007), so perhaps major differences in just a few genes like *FOXP2* and the stuttering genes are enough to explain why we write books and give speeches and chimps do not, despite our otherwise great genetic similarity.

The basic mechanisms of language may be universal, but the particular *forms* of language we use must be learned in childhood,

The Appearance of Williams Syndrome Children with Williams syndrome often have a characteristic facial shape, caused by the loss of a copy of the *elastin* gene. The loss of copies of other nearby genes is thought to cause mild mental disability paired with verbal fluency. (Photo courtesy of the Williams Syndrome Association.)

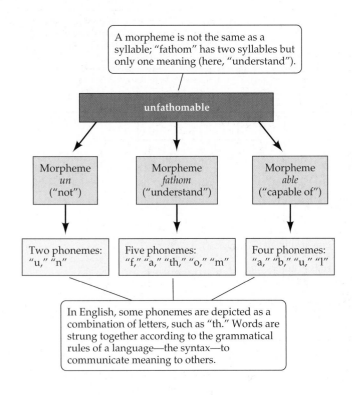

FIGURE 15.14   It's All in a Word

phoneme   A sound that is produced for language.

morpheme   The smallest grammatical unit of a language; a word or meaningful part of a word.

sensitive period   Also called *critical period*. The period during development in which an organism can be permanently altered by a particular experience or treatment.

and they are incredibly diverse: about 7,000 different languages, most of which have never been formally studied (Wuethrich, 2000). Nevertheless, all languages share certain basic features. Each language has basic speech sounds, or **phonemes**, that are assembled into simple units of meaning called **morphemes** (**FIGURE 15.14**). Morphemes are assembled into words (the word *unfathomable*, for example, consists of the morphemes *un*, *fathom*, and *able*), which have meaning (termed *semantics*). In turn, the words are assembled into meaningful strings (which may be complete sentences or just phrases) according to the language's *syntax* (rules for constructing phrases—an element of grammar). But how does each of us end up with the correct set of sounds and rules: the ones we need for our particular native language?

A child's brain is an incredible linguistic machine, rapidly acquiring the phonemes, vocabulary, and grammar of the local language without need of formal instruction. A human baby starts out babbling nearly all the phonemes known in all human languages, but she soon comes to use only the subset of phonemes that she hears in use around her. And the baby's developing language abilities are especially shaped by "parentese," the singsong speech of parents that helps babies to attach emotion and meaning to speech sounds (Falk, 2004). By 7 months of age, infants already have a sense of the grammar of the language used in their homes, and they react to exceptions (Marcus et al., 1999).

The human brain contains specialized mechanisms for language acquisition that show a clear-cut **sensitive period** (or *critical period*): a limited span of time during which exposure and practice with language must occur in order for language skills to develop normally. This sensitive period tapers down from the maximal sensitivity of early childhood to an eventual end of special sensitivity around puberty. Individuals who are exposed to language only late in the sensitive period or after it has ended show much-reduced language development (Curtiss, 1989). And as many of us know from first-hand experience, learning a second language is much more difficult in adulthood, after the sensitive period is over. In fact, people learning a second language later than age 11 apparently use different brain regions for each language (K. H. Kim et al., 1997), unlike those who learn multiple languages early in life. Nevertheless, bilingualism is clearly worth the effort, as evidence suggests a lifelong benefit with respect to preservation of brain volume and cognitive function (R. K. Olsen et al., 2015).

Speech mechanisms may have evolved from more-ancient systems controlling gestures of the face and hands (Hewes, 1973; Corballis, 2002), in agreement with the motor theory of language that we discussed earlier. Even today there is a close relationship between speaking and gesturing with the hands (Krauss, 1998); in fact, most people find it difficult *not* to gesture when speaking. An evolutionary origin of the brain's mechanisms of gesturing and speaking suggests that the rudiments of language—or at least some capacity to learn language-like behavior—may be shared with our closest primate relatives. If so, perhaps modern-day nonhuman primates could learn enough grammar to engage in linguistic behavior—an idea we take up next.

Can nonhuman primates acquire language with training?

Apes and monkeys employ a wide range of vocal behaviors for communication between individuals, particularly for relaying emotional information like alarm or territoriality (Ploog, 1992; Seyfarth and Cheney, 2003). The shrieking, purring, peeping, growling, and cackling sounds of squirrel monkeys, for example, can generally be related to specific social situations, and chimpanzees issue specific alarm calls to alert other members of their group to the presence of a viper (Crockford et al.,

2012). But despite their apparent adaptive importance, most nonhuman primate vocalizations seem to have a somewhat "preprogrammed" quality, being repeatedly produced in much the same fashion and order. Electrical stimulation of the brain indicates that vocal behavior in monkeys and apes relies primarily on subcortical systems, especially sites in the limbic system, rather than on cortex. The vocalizations elicited by subcortical stimulation are associated with strongly emotional behaviors such as defense, attack, feeding, and sex (**FIGURE 15.15**). Vocalizations are more common if subcortical stimulation is provided to the left hemisphere, indicating a special role of the left hemisphere in the communicative behavior of monkeys and apes (Meguerditchian and Vauclair, 2006; Taglialatela et al., 2006), mirroring what we've seen for human speech.

Nonhuman primates will never produce human speech, because they lack our vocal tracts and vocal repertoires. But can these animals be taught other forms of communication, with features similar to those of human language? Can they learn to represent objects with symbols and to manipulate those symbols according to rules of order? Can animals other than humans generate a novel string of symbols, such as a new sentence, in a grammatical form?

Chimpanzees, gorillas, and orangutans—the great apes—are our nearest primate relatives. Wild chimpanzees and other apes reportedly use a variety of hand gestures for communication (Hobaiter and Byrne, 2014, and apes are capable of learning hundreds of the hand gestures of American Sign Language. Given enough training, apes may learn to use ASL signs spontaneously, sometimes in novel sequences (R. A. Gardner and Gardner, 1969, 1984; F. Patterson and Linden, 1981). Through extensive practice, apes can also be trained to communicate by assembling abstract symbols, such as colored plastic chips or computerized symbols, into new sentences (Premack, 1971; Rumbaugh, 1977).

Some researchers conclude that apes are able to acquire words (or equivalent symbols) and then string them together into novel, meaningful chains; that is, they seem to employ a grammar to communicate ideas. Other researchers argue that these sequences may simply be subtle forms of imitation (Terrace, 1979), perhaps unconsciously cued by the experimenter who is providing the training. Native ASL users dispute the linguistic validity of the signs generated by apes, and Pinker (1994) concludes that "even putting aside vocabulary, phonology, morphology, and syntax, what impresses one the most about chimpanzee signing is that fundamentally, deep down, chimps just don't get it" (p. 349). Others, such as linguist Noam Chomsky, argue that teaching primates to emulate a quintessentially human behavior in which they do not naturally engage can tell us little about the behavior, other than the obvious conclusion that ape evolution did not favor the use of language.

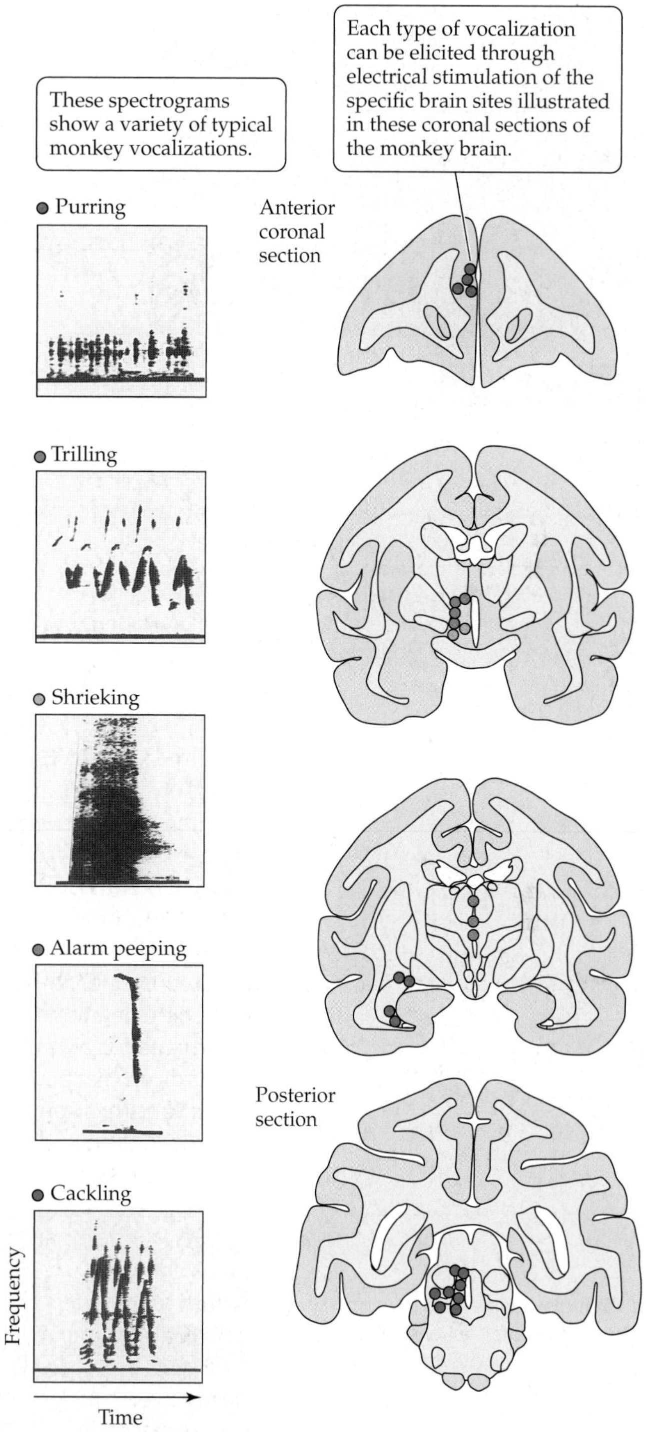

FIGURE 15.15 Electrical Stimulation of the Monkey Brain Elicits Vocalizations (After Ploog, 1992; spectrograms courtesy of Uwe Jürgens.)

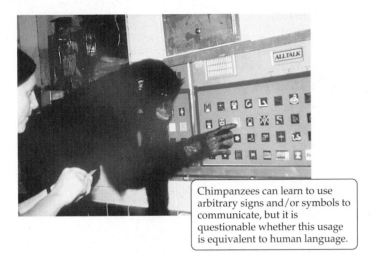

Chimpanzees can learn to use arbitrary signs and/or symbols to communicate, but it is questionable whether this usage is equivalent to human language.

Chimpanzee Using Symbols Some researchers argue that chimpanzees are able to string arbitrary symbols together into meaningful, sentence-like statements. Others believe that while chimps can learn subtle associations between symbols and meanings, their communicative behavior lacks the sense of grammar that is the hallmark of human language. (Photo courtesy of Sue Savage-Rumbaugh.)

Nevertheless, considering that apes can comprehend spoken words, produce novel combinations of words, and respond appropriately to sentences arranged according to a syntactic rule, it seems likely that the capabilities of apes have been underestimated (Savage-Rumbaugh, 1993). For example, a bonobo (pygmy chimpanzee) named Kanzi, the focus of a long-term research program (Savage-Rumbaugh and Lewin, 1994), reportedly learned numerous symbols and ways to assemble them in novel combinations, entirely through observational learning rather than the usual intensive training. And in natural settings, monkeys combine certain vocalizations into higher-order, more complex calls, suggesting the presence of both syntax and semantic meaning, at least on a rudimentary level (K. Arnold and Zuberbühler, 2006; Ouattara et al., 2009).

Vocal behavior is a feature of many different species

In addition to primates, many other animal species use vocalizations—chirps, barks, meows, songs, and more—to communicate important information to members of their own or other species. Although these communication sounds do not constitute language, in that they lack the features of language that we've discussed, such sounds nevertheless broadcast crucial signals about readiness to mate, danger, territorial defense, emotional state, and so on. Whales sing and may imitate songs that they've heard in distant oceans (Noad et al., 2000), and some seal mothers recognize their pups' vocalizations even after 4 years of separation (Insley, 2000). In fact, many species—from elephants to bats to birds to dolphins—are capable of vocal learning and use their vocalizations to help form social bonds and identify individuals (Tyack, 2003; Poole et al., 2005).

In the lab, measurement with special instruments reveals that rats and mice produce complex ultrasonic vocalizations that they use to communicate emotional information (Panksepp, 2005; Burgdorf et al., 2011). These ultrasonic vocalizations are impaired in mice with mutations in the *FoxP2* gene (Shu et al., 2005), providing an intriguing parallel to the situation in humans with *FOXP2* mutations that we discussed earlier.

Birds are particularly vocal animals. Many bird species produce only simple vocalizations, but songbirds like canaries and zebra finches produce rich and melodious vocalizations that are critical components for reproductive behavior. Although

Talk to the Animals People and animals can clearly communicate: anyone who has watched a sheepdog at work has to acknowledge that the dog's human handler is transmitting lots of information to his highly intelligent companion. Instilling *language* in a nonhuman is a different matter, however. Every day, you utter sentences that you've never said before, yet the meaning is clear to both you and your listener because you both understand the rules. Animals generally are incapable of similar feats, instead requiring extensive training with each specific utterance (e.g., each voice command to the sheepdog) in order for communication to occur at all. In other words, most animals appear to lack grammar.

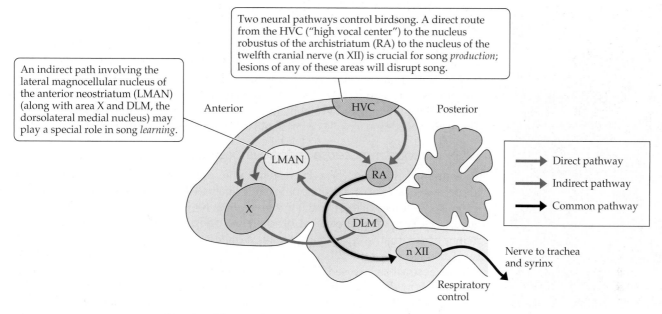

FIGURE 15.16 Song Control Nuclei of the Songbird Brain (After A. P. Arnold, 1980.)

birdsong has evolved quite independently of human speech, there are some interesting parallels between the two (Marler, 1970). For example, the songbird brain contains a specialized neural system for the control of vocal behavior (**FIGURE 15.16**), and as in humans, damage to the left rather than the right hemisphere has a greater effect on vocal behavior. What's more, the juvenile bird must be exposed to the songs of adults during a distinct critical period in order for its own singing behavior to develop normally (DeVoogd, 1994)—a requirement that also parallels human language. And yes, as with human speech, the *FoxP2* gene is implicated in the learning and production of birdsong (Bolhuis et al., 2010). When *FoxP2* expression is blocked in parts of the song control system, young males fail to properly learn and recite their tutor's song, producing errors that in some ways resemble those in the humans with abnormal *FOXP2* that we described earlier (Haesler et al., 2007).

In many important ways, the birdsong system is also dramatically different from the human speech system. For one thing, because it is a sexually selected trait, most birdsong is produced by males. In humans, of course, both sexes use language (in fact, women outperform men on some verbal tasks; see Chapter 8). Also in stark contrast to the human system, the birdsong system crystallizes after the sensitive period ends; from then on, the bird's singing becomes highly stereotyped, and he fixedly produces the same unchanging song for the rest of his life (not something that humans do, although sometimes it can seem that way!). (For more on birdsong and its social roles, see **A STEP FURTHER 15.3**, on the website.)

how's it going ?

1. Why do researchers think that a capacity for language is inborn in the human brain? Briefly describe some pertinent research findings that support this position, including genetic evidence.

2. What are the basic linguistic components of verbal language?

3. Discuss the process of language acquisition in infants. What is the significance of the term *sensitive period* in this regard?

4. Summarize some of the ways in which studies of nonhuman animals help us understand the neural mechanisms of language.

Overcoming Dyslexia Many highly intelligent, highly successful people—like self-made billionaire Richard Branson, pictured here—have coped with dyslexia on their way to fame and fortune.

Reading Skills Are Difficult to Acquire and Frequently Impaired

Why is it so much harder to learn to read and write than to speak? Compared with speech, the written word is a relatively new development, so we haven't had enough time to evolve brain mechanisms for reading and writing of the sort we have for speech. Therefore, learning the written form of a language is a slow and laborious chore of childhood that is vulnerable to developmental disruptions that result in **dyslexia** (from the Greek *dys*, "bad," and *lexis*, "word"), a mild to severe difficulty with reading.

Brain damage may cause specific impairments in reading

Sometimes people who learned to read just fine as children suddenly become dyslexic in adulthood as a result of disease or injury, usually to the left hemisphere. This *acquired dyslexia* (sometimes called *alexia*) offers hints about how the brain processes written language. One type of acquired dyslexia, known as **deep dyslexia**, is characterized by semantic errors (that is, errors related to the *meanings* of words); for example, the printed word *cow* is read as *horse*. Patients with deep dyslexia are also unable to read aloud words that are abstract as opposed to concrete, and they make frequent errors in which they seem to fail to see small differences in words. It's as though they grasp words whole, without noting the details of the letters, so they have a hard time sounding out nonsense words.

In another form of acquired dyslexia, **surface dyslexia**, the patient makes different types of errors when reading. These patients can read nonsense words without problems, indicating that they understand which letters make which sounds. But they find it difficult to recognize words in which the letter-to-sound rules are irregular. *The Tough Coughs as He Ploughs the Dough* by Dr. Seuss (1987), for example, would utterly confound them. In contrast to patients with deep dyslexia, those with surface dyslexia have difficulties that are restricted to the details and sounds of letters. Interestingly, surface dyslexia doesn't occur in native speakers of languages that are perfectly phonetic (such as Italian, where every letter is pronounced). This finding indicates that what's lost in speakers of nonphonetic languages, like English, is purely a learned aspect of language. In contrast, deep dyslexia probably involves inborn language mechanisms that are important for all languages.

Other kinds of brain damage can impair reading. For example, patients with hemispatial neglect following right parietal lobe damage (discussed in Chapter 14) disregard the left half of the world, despite having otherwise normal vision. Such people thus also fail to notice the left halves of the words that they see, necessarily resulting in poor reading. Some severe cases of acquired dyslexia exhibit *letter-by-letter reading*, a striking impairment in which the person laboriously spells out each word to herself (aloud or silently). In these cases, it seems that conscious attention to the spelling of each word is the only way by which words can be identified, so reading is dramatically slower.

Some people struggle throughout their lives to read

Some children seem to take forever to learn to read, and not even extended practice can make their reading easy and accurate. Affecting about 5% of children—especially boys and left-handers—this *developmental dyslexia* is a problem unique to written language, and not a general cognitive deficit. (Indeed, children with dyslexia can have high IQs [B. Morris, 2002]; several notable examples have grown up to amass billion-dollar fortunes.) Instead, the problem seems to lie in connecting reading with the more ancient brain mechanisms for speech.

Developmental dyslexia has been associated with several types of neurological abnormalities (**FIGURE 15.17**). In both postmortem investigations and anatomical studies using MRI, the brains of dyslexic people have been found to have aberrant layering of the neurons of the cerebral cortex, along with excessive cortical folding and clusters of extra neurons in unexpected locations (Galaburda, 1994; Chang et al., 2005). Cortical abnormalities are especially evident in the frontal and temporal lobes, possibly because of defective migration of newborn neurons during fetal development

dyslexia Also called *alexia*. A reading disorder attributed to brain impairment.

deep dyslexia Acquired dyslexia in which the patient reads a word as another word that is semantically related.

surface dyslexia Acquired dyslexia in which the patient seems to attend only to the fine details of reading.

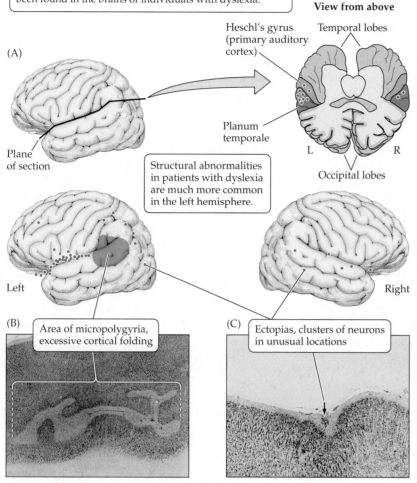

In these drawings of the left and right upper temporal lobe from the brain of a dyslexic person, the planum temporale is nearly symmetrical; in most people the left planum temporale is considerably larger. The dots and the shaded area represent regions where cellular disorganization has been found in the brains of individuals with dyslexia.

FIGURE 15.17 Neural Disorganization in Dyslexia (After Galaburda, 1994, micrographs courtesy of Albert Galaburda.)

View from above

Heschl's gyrus (primary auditory cortex)   Temporal lobes

(A)

Planum temporale

Plane of section

L   R

Occipital lobes

Structural abnormalities in patients with dyslexia are much more common in the left hemisphere.

Left   Right

(B) Area of micropolygyria, excessive cortical folding

(C) Ectopias, clusters of neurons in unusual locations

(Galaburda et al., 2006). Studies using fMRI confirm that people with dyslexia show impaired neural activity in left posterior speech zones (Pugh et al., 2000; Shaywitz et al., 1998, 2003) while displaying a relative overactivation of anterior regions. Abnormality in the nearby temporoparietal region has been linked to the phonological (phoneme-processing) aspects of dyslexia (Hoeft et al., 2006). And it looks like some of these abnormalities have genetic bases; for example, disruption of genes involved in brain development and the migration of neurons into adult positions are associated with developmental dyslexia (Cope et al., 2005; Meng et al., 2005; Harold et al., 2006).

Taken together, imaging and behavioral studies indicate that our brains rely on two different language systems during reading: one focused on the sounds of letters, the other on the meanings of whole words (McCarthy and Warrington, 1990). Presumably, these systems are shaped by training: as with learning any highly skilled behavior, our brains mold themselves to accommodate our acquired expertise with written language. For example, remedial training in people with dyslexia induces changes in the left-hemisphere systems that are used for reading (Temple et al., 2003). And there are also positive findings: for example, evidence is accumulating that people with dyslexia actually outperform unaffected individuals in certain learning domains, such as aspects of spatial learning (Schneps et al., 2012). Findings like these may have important implications for developing new educational strategies in dyslexia. Perhaps, then, coupling education interventions with early genetic screening for dyslexia will help affected people overcome their trouble with words completely.

how's it going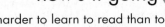

1. Why do researchers believe that it is so much harder to learn to read than to learn to speak?
2. What are the two principal types of acquired dyslexia? Summarize their respective features.
3. What is developmental dyslexia, and what are some of the neurological findings associated with this form of reading disorder?

recovery of function   The recovery of behavioral capacity following brain damage from stroke or injury.

concussion   A form of closed head injury caused by a jarring blow to the head, resulting in damage to the tissue of the brain with short- or long-term consequences for cognitive function.

chronic traumatic encephalopathy (CTE)   A form of dementia that may develop following multiple concussions, such as in athletes engaged in contact sports.

embryonic stem cell   A cell, derived from an embryo, that has the capacity to form any type of tissue.

PART III
Recovery of Function

Because of the brain's extreme complexity, high metabolic demands, delicate structure, and vulnerable location, disabling brain injuries are all too common. In the United States alone, according to the 2012 National Health Interview Survey (Blackwell et al., 2014), almost 7 million adults are survivors of stroke, and many more are living with the consequences of other disease processes, such as tumors and brain trauma. In addition to diseases, motor vehicle accidents and mishaps during horseback riding, diving, and contact sports such as boxing are major causes of injuries to the brain and spinal cord (**BOX 15.3**). So it's no surprise that discovering the mechanisms mediating **recovery of function** after brain damage, and developing new treatments to exploit those mechanisms, are topics of intense research activity. But in spite of encouraging recent developments, prevention is clearly better than treatment.

BOX 15.3 Contact Sports Can Be Costly

Jarring blows to the body are common in a number of sports—football, hockey, and wrestling, for example—and when the part of the body receiving the impact is the head, a brain injury called a **concussion** may result. Even one concussion, and certainly a number of concussions, places the athlete at risk of permanent brain damage. Although uncomplicated concussions generally clear up with time, up to 25% of concussions may cause persistent cognitive symptoms (Ponsford, 2005), some of which may not become evident until later in an athlete's life (Thornton et al., 2008), and repeated concussions greatly elevate the risk of lasting cognitive impairments. A large-scale CT study of 338 active boxers—athletes whose whole goal is to rain blows upon the head of an opponent—showed that scans were abnormal in 7% (showing brain atrophy) and borderline in 12% (B. D. Jordan et al., 1992).

A marked cognitive impairment resulting from too many concussions, once called *dementia pugilistica* (the Latin *pugil* means "boxer") or *punch-drunk syndrome*

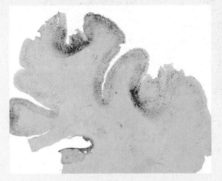
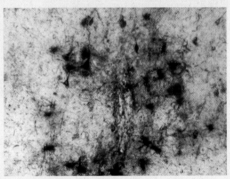

Tau Protein in the Brain of a Boxer Suffering from CTE (*Left*) Unmagnified section of cortex. (*Right*) Cortical gray matter magnified 350×. (Courtesy of Ann McKee.)

(Erlanger et al., 1999), is known today as **chronic traumatic encephalopathy (CTE)**. Neuropathological evidence indicates that, like Alzheimer's disease, CTE in athletes is a type of tauopathy, in which excess tau protein within neurons interferes with their functioning (McKee et al., 2009). The photos in the figure show the brain of a former boxer who, by his mid thirties, was experiencing symptoms including memory loss, confusion, and a tendency to fall. As in other cases of

CTE, an excessive amount of tau (brown in the photos) is evident in the brain, and it is found forming tangles within many neurons. CTE is a real and serious risk in contact sports, particularly where heavy blows to the head are a regular occurrence—no small reason why many believe that the rules of some of these sports, especially where youths are participating, are in serious need of revision. And some sports, such as boxing, should perhaps retire from the ring altogether.

Stabilization and Reorganization Are Crucial for Recovery of Function

In the months following a brain injury, patients often show conspicuous improvements in neural function as the injury site stabilizes, unaffected tissue reorganizes, and compensation occurs. We now know that the nervous system has much more potential for plasticity and recovery than was previously believed. For example, damaged neurons can regrow their connections under some circumstances, through a process called *collateral sprouting*. (For more information on collateral sprouting, see **A STEP FURTHER 15.4**, on the website.) And, as we've discussed in several places in this book, it has become evident that the adult brain is capable of producing new neurons; although this neurogenesis normally plays a minimal role in recovery from brain damage, perhaps we will learn how to bend these new neurons to our will and use them to replace damaged brain tissue.

One of the most exciting prospects for brain repair following stroke or injury or in many other neurological conditions is the use of **embryonic stem cells** (Lindvall and Kokaia, 2006). Derived from embryos, these cells have not yet differentiated into specific roles and therefore are able to develop, under the control of local chemical cues, into the type of cell needed. Controversy surrounds the use of human embryos as cell donors, so researchers hope to find ways of creating stem cells from other sources, such as skin, that can then be induced to form into new neurons.

Several factors determine how thoroughly a person will recover from a brain injury. One of these is simply the passage of time. Immediate medical treatment at the onset of a stroke can greatly limit the extent of damage, reducing cell death and inflammation. (Mechanisms of brain damage, and the mitigation of brain injuries, are discussed in **A STEP FURTHER 15.5**, on the website.) However, it takes months for the extent of recovery to become evident. For patients with aphasia following a stroke, most recovery occurs during the first 3 months following brain damage, but steady recovery continues for 1–1.5 years (**FIGURE 15.18**) (Kertesz et al., 1979). New rehabilitation therapies may extend the recovery period; for example, using alternate communication channels, such as singing, can help in some cases (Racette et al., 2006). Age is also a factor. As in adults, left-hemisphere damage can cause aphasia in children, but they often show rapid and full recovery. In fact, children can recover their language function even if they lose the *entire*

Perseverance and Plasticity Despite suffering a severe gunshot wound to the left cerebral hemisphere during an assassination attempt, former U.S. Representative Gabrielle Giffords has made striking progress in regaining her language and cognitive functions, thanks to intensive rehabilitation therapy and strategies for compensating for the damage.

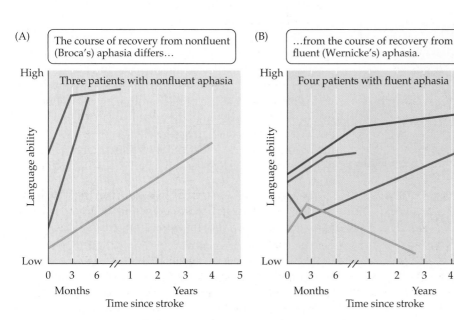

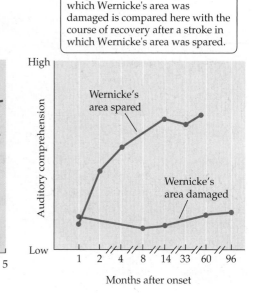

(C) The recovery of auditory speech comprehension after a stroke in which Wernicke's area was damaged is compared here with the course of recovery after a stroke in which Wernicke's area was spared.

(A) The course of recovery from nonfluent (Broca's) aphasia differs…

(B) …from the course of recovery from fluent (Wernicke's) aphasia.

FIGURE 15.18 Courses of Recovery for Patients with Aphasia (Parts A and B after Kertesz et al., 1979; C after Naeser et al., 1990.)

lesion momentum   The phenomenon in which the brain is impaired more by a lesion that develops quickly than by a lesion that develops slowly.

constraint-induced movement therapy (CIMT)   A therapy for recovery of movement after stroke or injury in which the person's unaffected limb is constrained while he is required to perform tasks with the affected limb.

left hemisphere, as we discuss in Signs & Symptoms at the end of this chapter, powerfully demonstrating that the right hemisphere *can* take over the language function if the left-hemisphere damage occurs early in life. Unfortunately, as we age, our capacity for this sort of plastic reorganization of the brain diminishes. Reversing this age-related decline in neuroplasticity is an important focus of neuroscience research.

Statistically, there are additional considerations that affect the degree of impairment a brain injury causes and the patient's extent of recovery. For example, in neurological disorders such as stroke and tumors, **lesion momentum** is important: patients experience less impairment when the lesion develops more gradually than when it occurs suddenly. Consequently, two or three sequential strokes may have less impact than a single stroke even if they damage as much or more cortex, probably because there is enough time for compensatory reorganization to occur in between the strokes. Recovery also tends to be better when the brain injury is due to trauma, such as a blow to the head, rather than a stroke; this difference may be due to the generation of cell-damaging signaling chemicals in cells that have been starved of oxygen during a stroke. Patients with more-severe initial loss of language recover less completely. And left-handed people show better recovery than those who are right-handed, perhaps because of reduced lateralization of function in left-handers.

Rehabilitation and Retraining Can Help Recovery from Brain and Spinal Cord Injury

Cognitive and/or perceptual handicaps that develop from brain impairments can be modified by training; similarly, intensive training may restore some measure of walking ability after certain spinal cord injuries (Barbeau et al., 1998). But it is important to distinguish restoration of function from compensation. Practice can significantly reduce the impact of brain injury by fostering compensatory behavior. For example, vigorous eye movements can make up for the scotomas (blind spots) that commonly result from strokes or other injuries that affect the visual cortex. Indeed, people can develop a wide variety of behavioral strategies after a brain injury to enable successful performance on a variety of tasks.

As with recovery of language functions after stroke, a growing body of evidence indicates that people can substantially regain the use of limbs that have been paralyzed following brain injury, especially if they are *forced* to use the limbs repeatedly. **Constraint-induced movement therapy (CIMT)** persuades patients to use a stroke-affected arm by simply tying the "good" arm to a splint for up to 90% of waking hours (Taub et al., 2002) in conjunction with daily rehabilitation therapy, including practice moving the affected limb repeatedly (**FIGURE 15.19**). Some patients who receive this treatment reportedly regain up to 75% of normal use of the paralyzed arm after only 2 weeks of therapy. Although the underlying mechanisms are not well understood—perhaps involving a remapping of the motor cortex—it's clear that CIMT offers substantial benefits that persist over the long term (Liepert et al., 2000; Kwakkel et al., 2015).

Another surprising use of experience for rehabilitation involves a simple mirror. Altschuler et al. (1999) treated patients who had reduced use of one arm after a stroke by placing them before a mirror with only their "good" arm visible. To the patients, it looked as though they were seeing the entire body, but what they saw were mirror images of the good arm. The patients were told to make symmetrical fluid motions with both arms. In the mirror, the motions looked perfectly symmetrical (of course); surprisingly, even

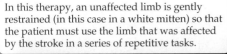

In this therapy, an unaffected limb is gently restrained (in this case in a white mitten) so that the patient must use the limb that was affected by the stroke in a series of repetitive tasks.

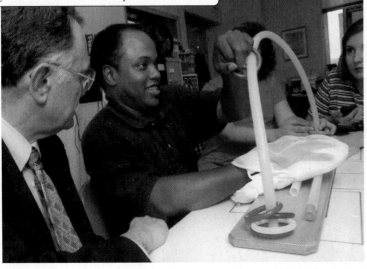

FIGURE 15.19   Constraint-Induced Movement Therapy (Photo courtesy of Edward Taub.)

though their real arm movements weren't perfect, most of the patients soon learned to use the "weak" arm more extensively. It was as though the visible feedback, indicating that the weak arm was moving perfectly, overcame the brain's reluctance to use that arm. It is likely that the mirror neurons of the brain, discussed in Chapter 5, mediate some of the beneficial effects of this sort of rehearsal (Buccino et al., 2006).

Brain damage will remain a serious problem for the foreseeable future—one that will affect most of us in some way as our friends and relatives go through their lives. Perhaps the most important and encouraging message to convey to victims of stroke and other nervous system damage is that, with effort and perseverance, they can help their remarkably plastic brains to regain a significant amount of the lost behavioral capacity.

SIGNS & SYMPTOMS

The Amazing Resilience of a Child's Brain

Complications in pregnancy and fetal development occasionally result in children who are born with severe epilepsy that can't be controlled by medication. These children may suffer paralysis on one side of the body, along with al- most continual seizures. Some- times, in these extreme cases, the only way to save the life of the child is to remove an entire brain hemisphere, as shown in the scan of a girl whose left hemisphere was completely re- moved when she was 7 years old (**FIGURE 15.20**).

> The brain can compensate even for the loss of a whole hemisphere, provided that this radical surgery is performed early enough in the patient's development.

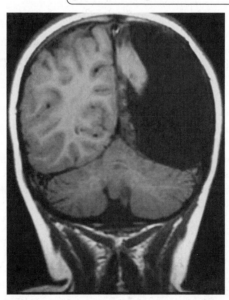

Hemispherectomy surgery removes the malfunctioning brain tissue and saves the life of the child (Griessenauer et al., 2015), but it also produces severe symptoms of its own, such as complete paralysis of one side, speech loss, or visual impairments. However, pro- vided the surgery occurs early enough, the child may show almost complete recovery of function over a long period of time. The girl whose brain scan is shown in Figure 15.20 expe- rienced paralysis and complete

FIGURE 15.20 Hemispherectomy (From Borgstein and Grootendorst, 2002.)

speech loss after her surgery, but at follow-up a few years later she had recovered normal language function (including her bilingual capacity!), and her paralysis had resolved almost completely.

Provided that the remaining hemisphere is healthy, children who have hemispherectomy surgery tend to show stable or improving intellectual func- tioning in the years following surgery (Boshuisen et al., 2010; Thomas et al., 2010) and may have normal IQs in adulthood. So, although extensive hemi- spheric damage in an adult usually results in drastic and largely permanent functional and intellectual impairments, cases of childhood hemispherectomy vividly illustrate the amazing plasticity of the young brain.

how's it going ?

1. Just how prevalent is brain damage due to stroke, and to other diseases? Once the brain is damaged, is recovery more or less complete in a matter of hours, days, or months?

2. Discuss some of the factors that determine how thoroughly a person will recover from brain injury.

3. Explain the concept of lesion momentum.

4. Is recovery better when brain damage is caused by trauma, or when the damage is caused by stroke? What is believed to be responsible for the difference?

5. Discuss some of the types of therapy that appear to help maximize recovery following brain damage.

Recommended Reading

Bradbury, J. W., and Vehrencamp, S. L. (2011). *Principles of Animal Communication* (2nd ed.). Sunderland, MA: Sinauer.

Ellard, C. (2009). *You Are Here: Why We Can Find Our Way to the Moon, but Get Lost in the Mall.* New York, NY: Doubleday.

Fitch, W. T. (2010). *The Evolution of Language.* Cambridge, England: Cambridge University Press.

Gazzaniga, M. S. (2008). *Human: The Science behind What Makes Us Unique.* New York, NY: HarperCollins.

Harrison, D. W. (2015) *Brain Asymmetry and Neural Systems: Foundations in Clinical Neuroscience and Neuropsychology.* New York, NY: Springer.

Kolb, B., and Whishaw, I. Q. (2008). *Fundamentals of Human Neuropsychology* (6th ed.). New York, NY: Worth.

McManus, I. C. (2003). *Right Hand, Left Hand: The Origins of Asymmetry in Brains, Bodies, Atoms, and Cultures.* Cambridge, MA: Harvard University Press.

Meyer, J. (2015). *Whistled Languages: A Worldwide Enquiry on Human Whistled Speech.* New York, NY: Springer.

Patel, A. (2007). *Music, Language, and the Brain.* Oxford, England: Oxford University Press.

Purves, D., Cabeza, R., Huettel, S. A., et al. (2012). *Principles of Cognitive Neuroscience* (2nd ed.). Sunderland, MA: Sinauer.

Tomasello, M. (2010). *Origins of Human Communication.* Cambridge, MA: Bradford Books/MIT Press.

Zeigler, H. P., and Marler, P. (2008). *Neuroscience of Birdsong.* Cambridge, England: Cambridge University Press.

2e.mindsmachine.com/vs15

VISUAL SUMMARY

15

You should be able to relate each summary to the adjacent illustration, including structures and processes. Go to the online version of this summary (scan the QR code above) for links to figures, animations, and activities that will help you consolidate the material.

1 Split-brain individuals show striking examples of hemispheric specialization (lateralization). Most words projected only to the right hemisphere, for example, cannot be read, but the same stimuli directed to the left hemisphere can be read. Spatial tasks, however, are performed better by the right hemisphere than by the left. Review **Figure 15.1, Animation 15.2, Video 15.3**

Visual cortex

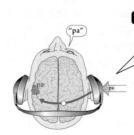

2 Most humans show many forms of cognitive specialization of the cerebral hemispheres. For example, most people show an advantage for verbal stimuli presented to the right ear or right visual field. Review **Figure 15.2**

3 Anatomical asymmetry of the hemispheres is seen in some brain regions, such as the **planum temporale** (larger in the left hemisphere than in the right hemisphere of most right-handed individuals). Perfect pitch in musicians is also associated with a larger left planum temporale, but the right hemisphere is dominant for many other aspects of music processing. Review **Figure 15.3**

4 In most patients, parietal cortical injuries produce perceptual changes, including alterations in sensory and spatial processing. Damage affecting the **fusiform gyrus** can produce acquired **prosopagnosia**, a dramatic inability to recognize the faces of familiar people. A developmental form of prosopagnosia affects about 2.5% of the population. Review **Figures 15.4** and **15.5**

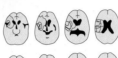

5 The left hemisphere of the human brain contains a network of structures specialized for processing speech and language. Damage in this network can cause the language impairment **aphasia**. Review **Figure 15.6, Activity 15.1**

6 Left inferior frontal lesions produce an impairment in speech production called **nonfluent** (or Broca's) **aphasia**. More-posterior lesions, involving the temporoparietal cortex, cause **fluent** (or Wernicke's) **aphasia**. Extensive destruction of the left hemisphere causes a more complete loss of language called **global aphasia**. Review **Figure 15.8**

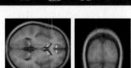

7 The **connectionist model of aphasia** involves a loop from a posterior speech reception zone to an anterior expressive zone that, when severed, may produce a **conduction aphasia**. However, the **motor theory of language** suggests that the entire circuit serves motor control and is used for both production and perception. Review **Figure 15.9, Activity 15.2**

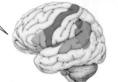

8 Noninvasive stimulation has confirmed the left-hemisphere organization for language functions. Stimulation of anterior regions causes speech arrest, while stimulation in other locations causes misnaming and other speech errors. Transcranial magnetic stimulation indicates that the anterior speech zone contains further subdivisions. Review **Figures 15.10** and **15.11**

9 Studies using PET and fMRI reveal that distinct regions of the left hemisphere are active during viewing, hearing, repeating, or assembling of verbal material. ERP studies have documented differential left-hemisphere processing of speech sounds versus speech meanings. Review **Figure 15.12**

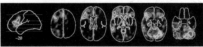

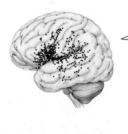

10 Languages are made up of speech sounds, called **phonemes** and **morphemes**, that are assembled into words and sentences according to syntax. Humans acquire language during an early **sensitive period**, involving genes like *FOXP2*, suggesting that mechanisms of language acquisition are innate. Review **Figures 15.13**, and **15.14**

11 Monkeys produce emotional vocalizations but cannot produce speech. Certain species of apes can learn American Sign Language, but it is not clear that this use constitutes language. Review **Figure 15.15**

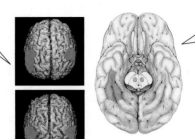

12 Many species use vocalizations for **communication** (but not **language**); for example, birdsong is controlled by a lateralized system in the brains of some species of songbirds, and early experience is essential for proper song development. Review **Figure 15.16, Activity 15.3**

13 Acquired dyslexia is a difficulty with reading due to brain damage; in **deep dyslexia** there is a disturbance in reading whole words; **surface dyslexia** involves a difficulty with the sounds of words. Developmental dyslexia is a congenital difficulty with reading that is associated with brain abnormalities. Review **Figure 15.17**

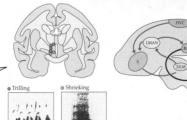

14 The brain can show at least partial **recovery of function** after injury, especially during the first year or so, as the damaged brain stabilizes and reorganizes. Retraining is a significant part of functional recovery and may involve both compensation, by establishing new solutions to adaptive demands, and reorganization of surviving networks. Review **Figure 15.18**

Molecular Biology
Basic Concepts and Important Techniques

Genes Carry Information That Encodes Proteins

The most important thing about **genes** is that they are pieces of information, inherited from parents, that affect the development and function of our cells. Information carried by the genes is a very specific sort: each gene carries the code for putting together a specific string of amino acids to form a particular **protein** molecule. This is *all* that genes do; they do not *directly* encode intelligence, or memories, or any other sort of complex behavior. The various proteins, each encoded by its own gene, make up the physical structure and most of the constituents of cells, such as **enzymes**, which are proteins that enable chemical reactions in our cells. All these proteins make complex behavior possible, and in that context they are also the targets upon which the forces of evolution act.

Proteins are specific. For example, only cells that have liver-typical proteins will look like liver cells and be able to perform liver functions. Neurons, on the other hand, are cells that make neuron-typical proteins so that they can look and act like neurons. The genetic information for making these various proteins is crucial for an animal to live and for a nervous system to work properly.

One thing we hope this book will help you understand is that everyday experience can affect whether and when particular genetic recipes for making various proteins are used. To aid in that understanding, let's review how genetic information is stored and how proteins are made. Our discussion will be brief, but many online tutorials can provide you with more detailed information (see **A STEP FURTHER A.1**, on the website).

Genetic information is stored in molecules of DNA

The information for making all of our proteins could, in theory, be stored in any sort of format—on sheets of paper, magnetic tape, a DVD, an iPod—but creatures on this planet store their genetic information in a chemical called **deoxyribonucleic acid**, or **DNA**. Each molecule of DNA consists of a long strand of chemicals called **nucleotides** strung one after the other. DNA has only four nucleotides: guanine, cytosine, thymine, and adenine (abbreviated G, C, T, and A). The particular sequence of nucleotides (e.g., GCTTACC or TGGTCC or TGA) holds the information that will eventually make a protein. Because many millions of these nucleotides can be joined one after the other, a tremendous amount of information can be stored in very little space—on a single molecule of DNA.

gene   A length of DNA that encodes the information for constructing a particular protein.

protein   A long string of amino acids. Proteins are the basic building material of organisms.

enzyme   A complicated protein whose action increases the probability of a specific chemical reaction.

deoxyribonucleic acid (DNA)   A nucleic acid that is present in the chromosomes of cells and codes hereditary information.

nucleotide   A portion of a DNA or RNA molecule that is composed of a single base and the adjoining sugar-phosphate unit of the strand.

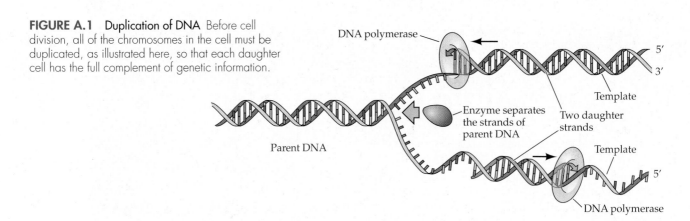

FIGURE A.1 Duplication of DNA Before cell division, all of the chromosomes in the cell must be duplicated, as illustrated here, so that each daughter cell has the full complement of genetic information.

hybridization   The process by which one string of nucleotides becomes linked to a complementary series of nucleotides.

chromosome   A complex of condensed strands of DNA and associated protein molecules. Chromosomes are found in the nucleus of cells.

eukaryote   Any organism whose cells have the genetic material contained within a nuclear envelope.

cell nucleus   The spherical central structure of a cell that contains the chromosomes.

ribonucleic acid (RNA)   A nucleic acid that implements information found in DNA.

transcription   The process during which mRNA forms bases complementary to a strand of DNA. The resulting message (called a *transcript*) is then used to translate the DNA code into protein molecules.

messenger RNA (mRNA)   Also called *transcript* or *message*. A strand of RNA that carries the code of a section of a DNA strand to the cytoplasm.

ribosome   An organelle in the cell body where genetic information is translated to produce proteins.

translation   The process by which amino acids are linked together (directed by an mRNA molecule) to form protein molecules.

codon   A set of three nucleotides that encodes one particular amino acid.

peptide   A short string of amino acids. Longer strings of amino acids are called *proteins*.

genome   Also called *genotype*. All the genetic information that one specific individual has inherited.

A set of nucleotides that has been strung together can snuggle tightly against another string of nucleotides if it has the proper sequence: T nucleotides preferentially link with A nucleotides, and G nucleotides link with C nucleotides. Thus, T and A are said to be complementary nucleotides, and C and G are complementary nucleotides. In fact, most of the time, our DNA consists not of a single strand of nucleotides, but of two complementary strands of nucleotides wrapped around one another.

The two strands of nucleotides are said to **hybridize** with (link to) one another, coiling slightly to form the famous double helix. The double-stranded DNA twists and coils further, becoming visible in microscopes as **chromosomes**, which resemble twisted lengths of yarn. Humans and the many other organisms known as **eukaryotes** store our chromosomes in a membranous sphere called a **nucleus** (plural *nuclei*) inside each cell. The ability of DNA to exist as two complementary strands of nucleotides is crucial for the duplication of the chromosomes (**FIGURE A.1**), but the details of that story will not concern us. Just remember that, with very few exceptions, every cell in your body has a faithful copy of all the DNA you received from your parents.

DNA is transcribed to produce messenger RNA

The information from DNA is used to assemble another molecule—**ribonucleic acid**, or **RNA**—that serves as a template for later steps in protein synthesis. Like DNA, RNA is made up of a long string of four types of nucleotides. For RNA, those nucleotides are G and C (which, recall, are complementary to each other), and A and U (uracil), which are also complementary to each other. Note that the T nucleotide is found only in DNA, and the U nucleotide is found only in RNA.

When a particular gene becomes active, the double strand of DNA unwinds enough so that one strand becomes free of the other and becomes available to special cellular machinery (including an enzyme called *transcriptase*) that begins **transcription**—the construction of a specific string of RNA nucleotides that are complementary to the exposed strand of DNA (**FIGURE A.2**). This length of RNA goes by several names: **messenger RNA (mRNA)**, *transcript*, or sometimes *message*. Each DNA nucleotide encodes a specific RNA nucleotide (an RNA G for every DNA C, an RNA C for every DNA G, an RNA U for every DNA A, and an RNA A for every DNA T). This transcript is made in the nucleus where the DNA resides; then the mRNA molecule moves to the cytoplasm, where protein molecules are assembled.

RNA molecules direct the formation of protein molecules

In the cytoplasm are special organelles, called **ribosomes**, that attach themselves to a molecule of RNA, "read" the sequence of RNA nucleotides, and using that information, begin linking together amino acids to form a protein molecule. The structure and function of a protein molecule depend on which particular amino acids are put together and in what order. The decoding of an RNA transcript to manufacture a particular protein is called **translation** (see Figure A.2), as distinct from *transcription*, the construction of the mRNA molecule.

Each trio of RNA nucleotides, or **codon**, encodes one of 20 or so different amino acids. Special molecules associated with the ribosome recognize the codon and bring a molecule of the appropriate amino acid so that the ribosome can fuse that amino acid to the previous one. If the resulting string of amino acids is short (say, 50 amino acids or less), it is called a **peptide**; if it is long, it is called a *protein*. Thus the ribosome assembles a very particular sequence of amino acids at the behest of a very particular sequence of RNA nucleotides, which were themselves encoded in the DNA inherited from our parents. In short, the biological secret of life is that DNA makes RNA, and RNA makes protein.

There are fascinating additions to this short story. Often the information from separate stretches of DNA is spliced together to make a single transcript; this so-called alternative splicing can create different transcripts from the same gene. Sometimes a protein is modified extensively after translation ends; special chemical processes can cleave long proteins to create one or several active peptides.

Keep in mind that each cell has the complete library of genetic information (collectively known as the **genome**) but makes only a fraction of all the proteins encoded in that DNA. In modern biology we say that each cell **expresses** only some genes; that is, the cell transcribes certain genes and makes the corresponding gene products (protein molecules). Thus, each cell must come to express all the genes needed to perform its function. Modern biologists refer to the expression of a particular subset of the genome as **cell differentiation**: the process by which different types of cells acquire their unique appearance and function. During development, individual cells appear to become more and more specialized, expressing progressively fewer genes. Many molecular biologists are striving to understand which cellular and molecular mechanisms "turn on" or "turn off" gene expression, in order to understand development and pathologies such as cancer, or to provide crucial proteins to afflicted organs in a variety of diseases.

Molecular Biologists Have Craftily Enslaved Microorganisms and Enzymes

Many basic methods of molecular biology are not explicitly discussed in the text, so we will not describe them in detail here. However, you should understand what some of the terms *mean*, even if you don't know exactly how the methods are performed.

Molecular biologists have found ways to incorporate DNA from other species into the DNA of microorganisms such as bacteria and viruses. After the foreign DNA is incorporated, the microorganisms are allowed to reproduce rapidly, producing more and more copies of the (foreign) gene of interest. At this point the gene is said to be **cloned** because researchers can make as many copies as they want. To ensure that the right gene is being cloned, the researcher generally clones many, many different genes—each into different bacteria—and then "screens" the bacteria rapidly to find the rare one that has incorporated the gene of interest.

When enough copies of the DNA have been made, the microorganisms are ground up and the DNA extracted. If sufficient DNA has been generated, chemical steps can then determine the exact sequence of nucleotides found in that stretch of DNA—a process known as **DNA sequencing**. Once the sequence of nucleotides has been determined, the sequence of complementary nucleotides in the messenger RNA for that

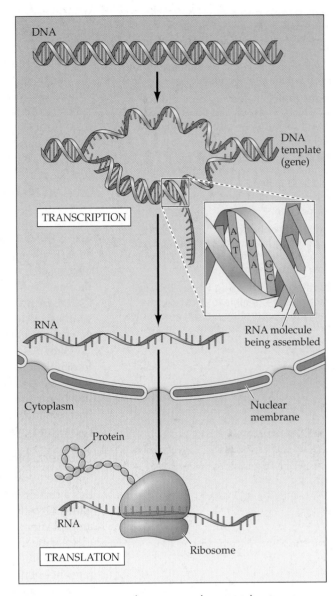

FIGURE A.2   DNA Makes RNA, and RNA Makes Protein

gene expression   The process by which a cell makes an mRNA transcript of a particular gene.

cell differentiation   The developmental stage in which cells acquire distinctive characteristics, such as those of neurons, as the result of expressing particular genes.

clones   Asexually produced organisms that are genetically identical.

DNA sequencing   The process by which the order of nucleotides in a gene is identified.

polymerase chain reaction (PCR) Also called *gene amplification*. A method for reproducing a particular RNA or DNA sequence manyfold, allowing amplification for sequencing or manipulating the sequence.

transgenic Referring to an animal in which a new or altered gene has been deliberately introduced into the genome.

probe In molecular biology, a manufactured sequence of DNA that is made to include a label (a colorful or radioactive molecule) that lets us track its location.

gel electrophoresis A method of separating molecules of differing size or electrical charge by forcing them to flow through a gel.

blotting Transferring DNA, RNA, or protein fragments to nitrocellulose following separation via gel electrophoresis. The blotted substance can then be labeled.

gene can be inferred. The sequence of mRNA nucleotides tells the investigator the sequence of amino acids that will be made from that transcript because biologists know which amino acid is encoded by each trio of DNA nucleotides. For example, scientists discovered the amino acid sequence of neurotransmitter receptors by this process.

The business of obtaining many copies of DNA has been boosted by a technique called the **polymerase chain reaction**, or **PCR**. This technique exploits a special type of polymerase enzyme that, like other such enzymes, induces the formation of a DNA molecule that is complementary to an existing single strand of DNA (see Figure A.1). Because this particular polymerase enzyme (called *Taq polymerase*) evolved in bacteria that inhabit geothermal hot springs, it can function in a broad range of temperatures. By heating double-stranded DNA, we can cause the two strands to separate, making each strand available to polymerase enzymes that, when the temperature has decreased enough, construct a new "mate" for each strand so that they are both double-stranded again. The first PCR yields only double the original number of DNA molecules; repeating the process results in 4 times as many molecules as at first. Repeatedly heating and cooling the DNA of interest in the presence of this heat-resistant polymerase enzyme soon yields millions of copies of the original DNA molecule, which is why this process is also referred to as *gene amplification*. In practice, PCR usually requires the investigator to provide primers—short nucleotide sequences synthesized to hybridize on either side of the gene of interest to amplify that particular gene more than others.

With PCR, sufficient quantities of DNA are produced for chemical analysis or other manipulations, such as introducing DNA into cells. For example, we might inject some of the DNA encoding a protein of interest into a fertilized mouse egg (a zygote) and then return the zygote to a pregnant mouse to grow. Occasionally the injected DNA becomes incorporated into the zygote's genome, resulting in a **transgenic** mouse that carries and expresses the foreign gene.

Southern blots identify particular genes

Suppose we want to know whether a particular individual or a particular species carries a certain gene. Because all cells contain a complete copy of the genome, we can gather DNA from just about any kind of cell population: blood, skin, or muscle, for example. After the cells are ground up, a chemical extraction procedure isolates the DNA (discarding the RNA and protein). Finding a particular gene in that DNA boils down just to finding a particular sequence of DNA nucleotides. To do that, we can exploit the tendency of nucleic acids to hybridize with one another.

If we were looking for the DNA sequence GCT, for example, we could manufacture the sequence CGA (there are machines to do that), which would then stick to (hybridize with) any DNA sequence of GCT. The manufactured sequence CGA is called a **probe** because it is made to include a label (a colorful or radioactive molecule) that lets us track its location. Of course, such a short length of nucleotides will be found in many genes. In order for a probe to recognize one particular gene, it has to be about 15 nucleotides long.

When we extract DNA from an individual, it's convenient to let enzymes cut up the very long stretches of DNA into more manageable pieces of 1,000–20,000 nucleotides each. A process called **gel electrophoresis** uses electrical current to separate these millions of pieces more or less by size (**FIGURE A.3**). Large pieces move slowly through a tube of gelatin-like material, and small pieces move rapidly. The tube of gel is then sliced and placed on top of a sheet of paperlike material called *nitrocellulose*. When fluid is allowed to flow through the gel and nitrocellulose, DNA molecules are pulled out of the gel and deposited on the waiting nitrocellulose. This process of making a "sandwich" of gel and nitrocellulose and using fluid to move molecules from the former to the latter is called **blotting** (see Figure A.3).

If the gene we're looking for is among those millions of DNA fragments sitting on the nitrocellulose, our labeled probe should recognize and hybridize to the sequence.

To see the animation **Gel Electrophoresis**, go to

2e.mindsmachine.com/A.1

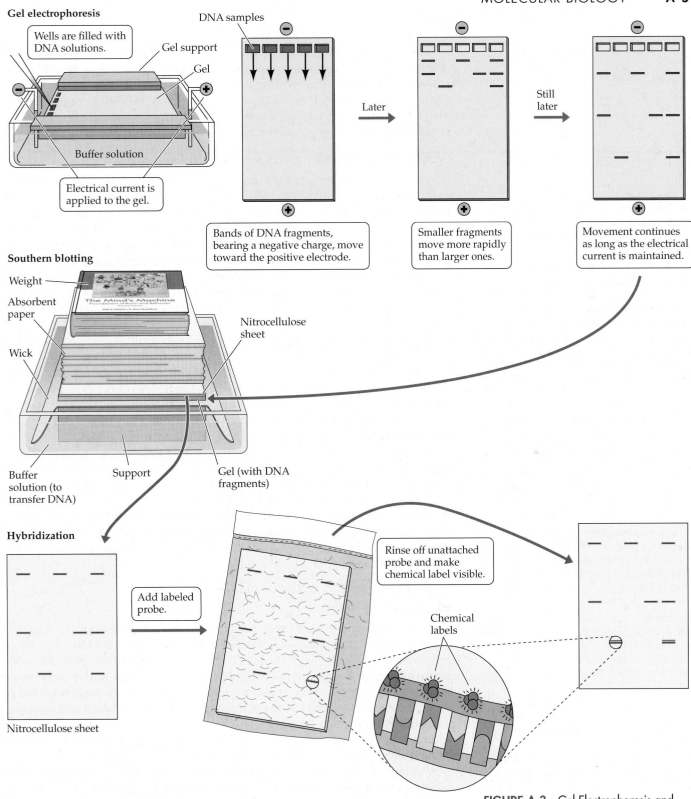

Gel electrophoresis

Wells are filled with DNA solutions.

Gel support

Gel

Electrical current is applied to the gel.

Buffer solution

DNA samples

Bands of DNA fragments, bearing a negative charge, move toward the positive electrode.

Later

Smaller fragments move more rapidly than larger ones.

Still later

Movement continues as long as the electrical current is maintained.

Southern blotting

Weight

Absorbent paper

Wick

Nitrocellulose sheet

Buffer solution (to transfer DNA)

Support

Gel (with DNA fragments)

Hybridization

Add labeled probe.

Rinse off unattached probe and make chemical label visible.

Chemical labels

Nitrocellulose sheet

FIGURE A.3 Gel Electrophoresis and Southern Blotting

The nitrocellulose sheet is soaked in a solution containing our labeled probe; we wait for the probe to find and hybridize with the gene of interest (if it is present), and we rinse the sheet to remove probe molecules that did not find the gene. Then we *visualize* the probe, either by causing the label to show its color or, if radioactive, by letting the probe expose photographic film to identify the locations where the probe has accumulated. In either case, if the probe found the gene, a labeled band will be evident, corresponding to the size of DNA fragment that contained the gene (see Figure A.3).

Southern blot A method of detecting a particular DNA sequence in the genome of an organism by separating DNA with gel electrophoresis, blotting the separated DNA molecules onto nitrocellulose, and then using a nucleotide probe to hybridize with, and highlight, the gene of interest.

Northern blot A method of detecting a particular RNA transcript in a tissue or organ by separating RNA from that source with gel electrophoresis, blotting the separated RNA molecules onto nitrocellulose, and then using a nucleotide probe to hybridize with, and highlight, the transcript of interest.

in situ hybridization A method for detecting particular RNA transcripts in tissue sections by providing a nucleotide probe that is complementary to, and will therefore hybridize with, the transcript of interest.

antibody Also called *immunoglobulin*. A large protein that recognizes and permanently binds to particular shapes, normally as part of the immune system attack on foreign particles.

Western blot A method of detecting a particular protein molecule in a tissue or organ by separating proteins from that source with gel electrophoresis, blotting the separated proteins onto nitrocellulose, and then using an antibody that binds, and highlights, the protein of interest.

immunocytochemistry (ICC) A method for detecting a particular protein in tissues in which an antibody recognizes and binds to the protein and then chemical methods are used to leave a visible reaction product around each antibody.

This process of looking for a particular sequence of DNA is called a **Southern blot**, named after the man who developed the technique, Edward Southern. Southern blots are useful for determining whether related individuals share a particular gene or for assessing the evolutionary relatedness of different species. The developed blots, with their lanes of labeled bands (see Figure A.3), are often seen in popular-media accounts of "DNA fingerprinting" of individuals.

Northern blots identify particular mRNA transcripts

A method more relevant for our discussions is the **Northern blot** (whimsically named as the opposite of the Southern blot). A Northern blot can identify which tissues are making a particular RNA transcript. If liver cells are making a particular protein, for example, then some transcripts for the gene that encodes that protein should be present. So we can take the liver, grind it up, and use chemical processes to isolate most of the RNA (discarding the DNA and protein). The resulting mixture consists of RNA molecules of many different sizes: long, medium, and short transcripts. Gel electrophoresis will separate the transcripts by size, and we can blot the size-sorted mRNA molecules onto nitrocellulose sheets; the process is very similar to the Southern blot procedure.

To see whether the particular transcript we're looking for is among the mRNAs, we construct a labeled probe (of either DNA nucleotides or RNA nucleotides) that is complementary to the mRNA transcript of interest and long enough that it will hybridize only to that particular transcript. We incubate the nitrocellulose in the probe, allow time for the probe to hybridize with the targeted transcript (if present), rinse off any unused probe molecules, and then visualize the probe as before. If the transcript of interest is present, we should see a band on the film (see Figure A.3). The presence of several bands indicates that the probe has hybridized to more than one transcript and we may need to make a more specific probe or alter chemical conditions to make the probe less likely to bind similar transcripts.

Because different gene transcripts have different lengths, the transcript of interest should have reached a particular point in the electrophoresis gel: small transcripts should have moved far; large transcripts should have moved only a little. If our probe has found the right transcript, the single band of labeling should be at the point that is appropriate for a transcript of that length.

In situ hybridization localizes mRNA transcripts within specific cells

Northern blots can tell us whether a particular *organ* has transcripts for a particular gene product. For example, Northern blot analyses have indicated that thousands of genes are transcribed only in the brain. Presumably the proteins encoded by these genes are used exclusively in the brain. But such results alone are not very informative, because the brain consists of so many different kinds of glial and neuronal cells. We can refine Northern blot analyses somewhat, by dissecting out a particular part of the brain—say, the hippocampus—to isolate mRNAs. Sometimes, though, it is important to know *exactly which cells* are making the transcript. In that case we use **in situ hybridization**.

With in situ hybridization we use the same sort of labeled probe, constructed of nucleotides that are complementary to (and will therefore hybridize with) the targeted transcript, as in Northern blots. Instead of using the probe to find and hybridize with the transcript on a sheet of nitrocellulose, however, we use the probe to find the transcripts "in place" (*in situ* in Latin)—that is, on a section of tissue. After rinsing off the probe molecules that didn't find a match, we visualize the probe right in the tissue section. Any cells in the section that were transcribing the gene of interest will have transcripts in the cytoplasm that should have hybridized with our labeled probe. In situ hybridization therefore can tell us exactly which cells are expressing a particular gene (**FIGURE A.4**; see also Box 2.1).

Western blots identify particular proteins

Sometimes we wish to study a particular protein rather than its transcript. In such cases we can use antibodies. **Antibodies** are large, complicated molecules (proteins, in fact) that our immune system adds to the bloodstream to identify and fight invading microbes, thereby arresting and preventing disease. But if we inject a rabbit or mouse with a sample of a protein of interest, we can induce the animal to create antibodies that recognize and attach to that particular protein, just as if it were an invader.

Once these antibodies have been purified and chemically labeled, we can use them to search for the target protein. We grind up an organ, isolate the proteins (discarding the DNA and RNA), and separate them by means of gel electrophoresis. Then we blot these proteins out of the gel and onto nitrocellulose. Next we use the antibodies to tell us whether the targeted protein is among those made by that organ. If the antibodies identify only the protein we care about, there should be a single band of labeling (if there are two or more, then the antibodies recognize more than one protein). Because proteins come in different sizes, the single band of label should be at the position corresponding to the size of the protein that we're studying. Such blots are called **Western blots**.

To review, Southern blots identify particular DNA pieces (genes), Northern blots identify particular RNA pieces (transcripts), and Western blots identify particular proteins (sometimes called *products*).

Antibodies can also tell us which cells possess a particular protein

If we need to know which particular cells within an organ such as the brain are making a particular protein, we can use the same sorts of antibodies that we use in Western blots, but in this case directed at that protein in tissue sections. We slice up the brain, expose the sections to the antibodies, allow time for them to find and attach to the protein, rinse off unattached antibodies, and use chemical treatments to visualize the antibodies. Cells that were making the protein will be labeled from the chemical treatments (see Box 2.1).

Because antibodies from the *immune* system are used to identify *cells* with the aid of *chemical* treatment, this method is called **immunocytochemistry**, or **ICC**. This technique can even tell us where, within the cell, the protein is found. Such information can provide important clues about the function of the protein. For example, if the protein is found in axon terminals, it may be a neurotransmitter.

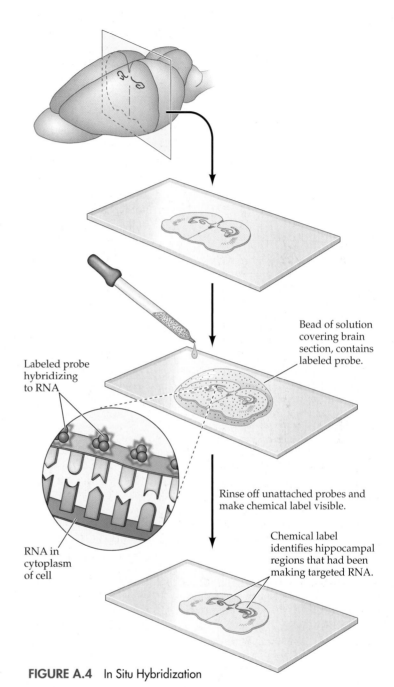

Labeled probe hybridizing to RNA

RNA in cytoplasm of cell

Bead of solution covering brain section, contains labeled probe.

Rinse off unattached probes and make chemical label visible.

Chemical label identifies hippocampal regions that had been making targeted RNA.

FIGURE A.4 In Situ Hybridization

Glossary

Numbers in brackets refer to the chapter(s), Appendix (App), or A Step Further (ASF) where the term is introduced.

5-alpha-reductase An enzyme that converts testosterone into dihydrotestosterone. [8]

5-HT See *serotonin*. [4]

17-beta-estradiol See *estradiol*. [8]

A

A delta (Aδ) fiber A moderately large, myelinated, and therefore fast-conducting, axon, usually transmitting pain information. Compare *C fiber*. [5]

absence attack See *petit mal seizure*. [3]

absolute refractory phase A brief period of complete insensitivity to stimuli. Compare *relative refractory phase*. [3]

accommodation The process by which the ciliary muscles adjust the lens to focus a sharp image on the retina. [7]

acetylcholine (ACh) A neurotransmitter that is produced and released by the autonomic nervous system, by motoneurons, and by neurons throughout the brain. See Figure 4.3; Table 4.1. [3–5]

acetylcholinesterase (AChE) An enzyme that inactivates the transmitter acetylcholine. [3, ASF 4.1]

ACh See *acetylcholine*. [3–5]

AChE See *acetylcholinesterase*. [3, ASF 4.1]

acid See *LSD*. [4]

act Complex behavior, as distinct from a simple movement. [5]

ACTH See *adrenocorticotropic hormone*. [ASF 8.3]

action potential Also called *spike*. A rapid reversal of the membrane potential that momentarily makes the inside of the membrane positive with respect to the outside. See Figures 3.6, 3.7. [3]

activational effect A temporary change in behavior resulting from the administration of a hormone to an adult animal. Compare *organizational effect*. [8]

acupuncture The insertion of needles at designated points on the skin to alleviate pain or neurological malfunction. [5]

adaptation 1. See *sensory adaptation*. [5] 2. In the context of evolution, a trait that increases the probability that an individual will leave offspring in subsequent generations.

ADH See *antidiuretic hormone*. [8, 9]

ADHD See *attention deficit hyperactivity disorder*. [14, ASF 13.4]

adipose tissue Commonly called *fat tissue*. Tissue made up of fat cells. [9]

adrenal cortex The steroid-secreting outer rind of the adrenal gland. See Figure 8.1. Compare *adrenal medulla*. [11]

adrenal gland An endocrine gland atop the kidney. See Figure 8.1.

adrenal medulla The inner core of the adrenal gland. The adrenal medulla secretes epinephrine and norepinephrine. See Figure 8.1. Compare *adrenal cortex*. [11]

adrenal steroid hormone A steroid hormone that is secreted by the adrenal cortex. [11]

adrenaline See *epinephrine*. [11]

adrenocorticotropic hormone (ACTH) A tropic hormone, secreted by the anterior pituitary gland, that controls the production and release of hormones of the adrenal cortex. [ASF 8.3]

adult neurogenesis The creation of new neurons in the brain of an adult. [1, 13]

afferent Carrying action potentials toward the brain, or toward one region of inter-

est from another region of interest See Box 2.2. Compare *efferent*. [2]

affinity See *binding affinity*. [4]

afterpotential The positive or negative change in membrane potential that may follow an action potential. [3]

aggression Behavior that is intended to cause pain or harm to others. [11]

agnosia The inability to recognize objects, despite being able to describe them in terms of form and color. Agnosia may occur after localized brain damage. [15]

agonist A substance that mimics or potentiates the actions of a transmitter or other signaling molecule. Compare *antagonist* (definition 1). [3, 4]

agraphia The inability to write. Compare *alexia*. [15]

AIS See androgen insensitivity syndrome. [8]

aldosterone A mineralocorticoid hormone, secreted by the adrenal cortex, that promotes the conservation of sodium by the kidneys. [9]

alexia The inability to read. See *dyslexia*. Compare *agraphia*. [15]

all-or-none property Referring to the fact that the size (amplitude) of the action potential is independent of the size of the stimulus. See Table 3.2. Compare *postsynaptic potential*. [3]

allele Any particular version of a gene.

allomone A chemical signal that is released outside the body by one species and affects the behavior of other species. See Figure 8.3. Compare *pheromone*. [8]

allostasis The combined set of behavioral and physiological adjustments that an individual makes in response to current and predicted behavioral and environmental stressors. [9]

alpha-fetoprotein A protein found in the plasma of fetuses. In rodents, alpha-fetoprotein binds estrogens and prevents them from entering the brain. [ASF 8.6]

alpha rhythm A brain potential of 8–12 hertz that occurs during relaxed wakefulness. See Figure 10.9. Compare *desynchronized EEG*. [10]

alpha-synuclein A protein that has been implicated in Parkinson's disease. [ASF 5.5]

ALS See *amyotrophic lateral sclerosis*. [ASF 5.5]

Alzheimer's disease A form of dementia that may appear in middle age but is more frequent among the aged. [13]

amacrine cell A specialized retinal cell that contacts both bipolar cells and ganglion cells and is especially significant in inhibitory interactions within the retina. Compare *horizontal cell*. [7]

amblyopia Reduced visual acuity that is not caused by optical or retinal impairments. [7, ASF 13.4]

AMH See *anti-müllerian hormone*. [8]

amine hormone Also called *monoamine hormone*. A hormone composed of a single amino acid that has been modified into a related molecule, such as melatonin or epinephrine. Compare *peptide hormone* and *steroid hormone*. [8]

amine neurotransmitter A neurotransmitter based on modifications of a single amino acid nucleus. Examples include acetylcholine, serotonin, and dopamine. See Table 4.1. Compare *amino acid neurotransmitter, gas neurotransmitter,* and *peptide neurotransmitter*. [4]

amino acid neurotransmitter A neurotransmitter that is itself an amino acid. Examples include GABA, glycine, and glutamate. See Table 4.1. Compare *amine neurotransmitter, gas neurotransmitter,* and *peptide neurotransmitter*. [4]

amnesia Severe impairment of memory. [13]

AMPA receptor A fast-acting ionotropic glutamate receptor that also binds the glutamate agonist AMPA. Compare *NMDA receptor*. [ASF 4.2, 13]

amphetamine A molecule that resembles the structure of the catecholamine transmitters and enhances their activity. [4]

amphetamine psychosis A delusional and psychotic state, closely resembling acute schizophrenia, that is brought on by repeated use of high doses of amphetamine. [12]

amplitude Also called *intensity*. The force that sound exerts per unit area, usually measured as dynes per square centimeter. In practical terms, amplitude corresponds to the volume of a sound. See Box 6.1. [6]

ampulla (pl. ampullae) An enlarged region of each semicircular canal that contains the receptor cells (hair cells) of the vestibular system. See Figure 6.14. [6]

amusia A disorder characterized by the inability to discern tunes accurately or to sing. [6]

amygdala A group of nuclei in the medial anterior part of the temporal lobe. See Figure 2.14. [2, 11]

amyloid plaque Also called *senile plaque*. A small area of the brain that has abnormal cellular and chemical patterns. Amyloid plaques correlate with dementia. See Figure 13.34. [13]

amyloid precursor protein (APP) A protein that, when cleaved by several enzymes, produces beta-amyloid, which can lead to Alzheimer's disease. [ASF 13.5]

amyotrophic lateral sclerosis (ALS) Also called *Lou Gehrig's disease*. A disease in which motoneurons and their target muscles waste away. [ASF 5.5]

analgesia Absence of or reduction in pain. [5]

analgesic Having painkilling properties. [4]

anandamide An endogenous substance that binds the cannabinoid receptor molecule. [4]

androgen Any of a class of hormones that includes testosterone and other male hormones. See Figure 8.13. [8]

androgen insensitivity syndrome (AIS) A syndrome caused by a mutation of the androgen receptor gene that renders tissues insensitive to androgenic hormones like testosterone. Affected XY individuals are phenotypic females, but they have internal testes and regressed internal genital structures. See Figure 8.27. [8]

angel dust See *phencyclidine*. [12]

angiotensin II A hormone that is produced in the blood by the action of renin and that may play a role in the control of thirst. [9]

angular gyrus A brain region in which strokes can lead to word blindness.

anion A negatively charged ion, such as a protein or a chloride ion. Compare *cation*. [3]

anomia The inability to name persons or objects readily. [15]

anorexia nervosa A syndrome in which individuals severely deprive themselves of food. [9]

anosmia The inability to detect odors. [6]

ANP See *atrial natriuretic peptide*. [9]

antagonist 1. A substance that blocks or attenuates the actions of a transmitter or other signaling molecule. Compare *agonist*. [3, 4] 2. A muscle that counteracts the effect of another muscle. Compare *synergist*. [5]

anterior Also called *rostral*. In anatomy, toward the head end of an organism. See Box 2.2. Compare *posterior*. [2]

anterior cerebral artery Either of two large arteries, arising from the carotid arteries, that provide blood to the anterior poles and medial surfaces of the cerebral hemispheres. Compare *middle cerebral artery* and *posterior cerebral artery*. [ASF 2.1]

anterior pituitary The front division of the pituitary gland. It secretes tropic hormones. See Figures 8.1, 8.12. Compare *posterior pituitary*. [8]

anterograde amnesia Difficulty in forming new memories beginning with the onset of a disorder. Compare *retrograde amnesia*. [13]

anterolateral system Also called *spinothalamic system*. A somatosensory system that carries most of the pain information from the body to the brain. See Figure 5.12. Compare *dorsal column system*. [5]

antibody Also called *immunoglobulin*. A large protein that recognizes and permanently binds to particular shapes, normally as part of the immune system attack on foreign particles. [ASF 11.4, App]

antidepressant A drug that relieves the symptoms of depression. Major categories include monoamine oxidase inhibitors, tricyclics, and selective serotonin reuptake inhibitors. [4]

antidiuretic hormone (ADH) See *vasopressin*.

anti-müllerian hormone (AMH) Also called *müllerian regression hormone*. A protein hormone secreted by the fetal testes that inhibits müllerian duct development. [8]

antipsychotic Also called *neuroleptic*. Any of a class of antipsychotic drugs that alleviate symptoms of schizophrenia, typically by blocking dopamine receptors. [4, 12]

anxiety disorder Any of a class of psychological disorders that includes recurrent panic states, generalized persistent anxiety disorders, and posttraumatic stress disorder. [12]

anxiolytic A substance that is used to combat anxiety. Examples include alcohol, opiates, barbiturates, and the benzodiazepines. [4, 12]

aphasia An impairment in language understanding and/or production that is caused by brain injury. [15]

ApoE See *apolipoprotein E*. [ASF 13.5]

apolipoprotein E (ApoE) A protein that may help break down beta-amyloid. Individuals carrying the *ApoE4* allele are more likely to develop Alzheimer's disease. [ASF 13.5]

apoptosis See *cell death*. [13]

APP See *amyloid precursor protein*. [ASF 13.5]

appetitive behavior The second stage of mating behavior. It helps establish or maintain sexual interaction. See Figure 8.16. [8]

apraxia An impairment in the ability to carry out complex sequential movements, even though there is no muscle paralysis. [5, 15]

arachnoid The thin covering (one of the three meninges) of the brain that lies between the dura mater and the pia mater. See Figure 2.8. [2]

arcuate fasciculus A fiber tract classically viewed as a connection between Wernicke's speech area and Broca's speech area. See Figure 15.9. [15]

arcuate nucleus An arc-shaped hypothalamic nucleus implicated in appetite control. See Figure 9.15 . [9]

area 17 See *primary visual cortex*.

arginine vasopressin (AVP) See *vasopressin*.

aromatase An enzyme that converts many androgens into estrogens. [ASF 8.6]

aromatization The chemical reaction that converts testosterone to estradiol, and other androgens to other estrogens. [ASF 8.6]

aromatization hypothesis The hypothesis that testicular androgens enter the brain and are converted there into estrogens to masculinize the developing nervous system of some rodents. [ASF 8.6]

arousal The global, nonselective level of alertness of an individual.

Asperger's syndrome Also called *high-functioning autism*. A syndrome characterized by difficulties in social cognitive processing. It is usually accompanied by strong language skills. Compare *autism*. [ASF 13.4]

associative learning A type of learning in which an association is formed between two stimuli or between a stimulus and a response. It includes both classical and instrumental conditioning. Compare *nonassociative learning*. [13]

astereognosis The inability to recognize objects by touching and feeling them. [15]

astrocyte A star-shaped glial cell with numerous processes (extensions) that run in all directions. See Figure 2.5. [2]

ataxia A loss of movement coordination, often caused by disease of the cerebellum. [5]

atrial natriuretic peptide (ANP) A hormone, secreted by the heart, that normally reduces blood pressure, inhibits drinking, and promotes the excretion of water and salt at the kidneys. [9]

attention Also called *selective attention*. A state or condition of selective awareness or perceptual receptivity, by which specific stimuli are selected for enhanced processing. [14]

attention deficit hyperactivity disorder (ADHD) A syndrome characterized by distractibility, impulsiveness, and hyperactivity that, in children, interferes with school performance. [14, ASF 13.4]

attentional blink The reduced ability of subjects to detect a target stimulus if it follows another target stimulus by about 200–450 milliseconds. [14]

attentional bottleneck A filter created by the limits intrinsic to our attentional processes, whose effect is that only the most important stimuli are selected for special processing. [14]

attentional spotlight The shifting of our limited selective attention around the environment to highlight stimuli for enhanced processing. [14]

atypical antipsychotic Also called *atypical neuroleptic*. An antipsychotic drug that has actions other than or in addition to the dopamine D_2 receptor antagonism that characterizes the typical antipsychotics. **[4, 12]**

auditory canal See *ear canal*. [6]

auditory N1 effect A negative deflection of the event-related potential, occurring about 100 milliseconds after stimulus presentation, that is enhanced for selectively attended auditory input compared with ignored input. Compare *visual P1 effect*. [14]

auditory P300 See *P3 effect*. [14]

aura In epilepsy, the unusual sensations or premonition that may precede the beginning of a seizure. [3]

autism spectrum disorder (ASD) A disorder, arising during childhood, that is characterized by social withdrawal and perseverative behavior. Compare *Asperger's syndrome*. [ASF 13.4]

autobiographical memory See *episodic memory*. [13]

autocrine Referring to a signal that is secreted by a cell into its environment and that feeds back to the same cell.

autoimmune disorder A disorder caused when the immune system mistakenly

attacks a person's own body, thereby interfering with normal functioning. [ASF 5.5]

autonomic ganglion A collection of nerve cell bodies, belonging to the autonomic division of the peripheral nervous system, that is found in any of various locations and contributes to the innervation of major organs.

autonomic nervous system A part of the peripheral nervous system that provides the main neural connections to glands and to smooth muscles of internal organs. Its two divisions (sympathetic and parasympathetic) act in opposite fashion. See Figure 2.9. [2]

autoradiography A staining technique that shows the distribution of radioactive chemicals in tissues. See Boxes 2.1, 8.1. [2, 8]

autoreceptor A receptor for a synaptic transmitter that is located in the presynaptic membrane and tells the axon terminal how much transmitter has been released. [4]

AVP See *arginine vasopressin*. [8]

axo-axonic synapse A synapse at which a presynaptic axon terminal synapses onto the axon terminal of another neuron. Compare *axo-dendritic synapse*, *axo-somatic synapse*, and *dendro-dendritic synapse*. [3]

axo-dendritic synapse A synapse at which a presynaptic axon terminal synapses onto a dendrite of the postsynaptic neuron, either via a dendritic spine or directly onto the dendrite itself. Compare *axo-axonic synapse*, *axo-somatic synapse*, and *dendro-dendritic synapse*. [3]

axo-somatic synapse A synapse at which a presynaptic axon terminal synapses onto the cell body (soma) of the postsynaptic neuron. Compare *axo-axonic synapse*, *axo-dendritic synapse*, and *dendro-dendritic synapse*. [3]

axon Also called *nerve fiber*. A single extension from the nerve cell that carries action potentials from the cell body toward the axon terminals. Functionally, the axon is the conduction zone of the neuron. See Figures 2.1, 2.3. [2]

axon collateral A branch of an axon. [2]

axon hillock The cone-shaped area on the cell body from which the axon originates. See Figure 2.4. [2, 3]

axon terminal Also called *synaptic bouton*. The end of an axon or axon collateral, which forms a synapse on a neuron or other target cell. Functionally, the axon terminals are the output zone of the neuron. See Figures 2.1, 2.3 [2]

axonal transport The transportation of materials from the neuronal cell body toward the axon terminals, and from the axon terminals back toward the cell body. [2]

B

B cell See *B lymphocyte*. [ASF 11.4]

B lymphocyte Also called *B cell*. An immune system cell, formed in the bone marrow (hence the *B*), that mediates humoral immunity. Compare *T lymphocyte*. [ASF 11.4]

Balint's syndrome A disorder, caused by damage to both parietal lobes, that is characterized by difficulty in steering visual gaze (oculomotor ataxia), in accurately reaching for objects using visual guidance (optic ataxia), and in directing attention to more than one object or feature at a time (simultagnosia). [14]

bar detector See *simple cortical cell*. [7]

barbiturate An early anxiolytic drug and sleep aid that has depressant activity in the nervous system. [4]

bariatrics The branch of medicine that deals with the causes, prevention, and treatment of obesity. [9]

baroreceptor A pressure receptor in the heart or a major artery that detects a change in blood pressure. [9]

basal "Toward the base" or "toward the bottom" of a structure. See Box 2.2. [2]

basal forebrain A region, ventral to the basal ganglia, that is the major source of acetylcholine in the brain and has been implicated in sleep. [4, 10]

basal ganglia A group of forebrain nuclei, including the caudate nucleus, globus pallidus, and putamen, found deep within the cerebral hemispheres. See Figures 2.14, 2.15, 5.27. [2, 5, 13]

basal metabolism The consumption of energy to fuel processes such as heat production, maintenance of membrane potentials, and all the other basic life-sustaining functions of the body. [9]

basilar artery An artery, formed by the fusion of the vertebral arteries, that supplies blood to the brainstem and to the posterior cerebral arteries. [ASF 2.1]

basilar membrane A membrane in the cochlea that contains the principal structures involved in auditory transduction. See Figures 6.1, 6.2. [6]

behavioral intervention An approach to finding relations between body variables and behavioral variables that involves intervening in the behavior of an organism and looking for resultant changes in body structure or function. See Figure 1.9. Compare *somatic intervention*. [1]

behavioral medicine See *health psychology*. [11]

behavioral neuroscience See *biological psychology*. [1]

benzodiazepine Any of a class of antianxiety drugs that are agonists of GABA$_A$ receptors in the central nervous system. One example is diazepam (Valium). [4, 12]

beta activity See *desynchronized EEG*. [10]

beta-amyloid A protein that accumulates in amyloid plaques in Alzheimer's disease. [13]

beta-secretase An enzyme that cleaves amyloid precursor protein, forming beta-amyloid, which can lead to Alzheimer's disease. See also *presenilin*. [ASF 13.5]

between-participants experiment An experiment in which an experimental group of individuals is compared with a control group of individuals that have been treated identically in every way except that they haven't received the experimental manipulation. Compare *within-participants experiment*. [1]

binaural Pertaining to two ears. Compare *monaural*.

binding affinity Also called simply *affinity*. The propensity of molecules of a drug (or other ligand) to bind to receptors. Drugs with high affinity for their receptors are effective even at low doses. [4]

binding problem The question of how the brain understands which individual attributes blend together into a single object, when these different features are processed by different regions in the brain. [14]

binge eating The rapid intake of large quantities of food, often poor in nutritional value and high in calories. [9]

binocular Referring to two-eyed processes. [7]

binocular deprivation Depriving both eyes of form vision, as by sealing the eyelids. Compare *monocular deprivation*. [ASF 13.4]

bioavailable Referring to a substance, usually a drug, that is present in the body in a form that is able to interact with physiological mechanisms. [4]

biological psychology Also called *behavioral neuroscience*, *brain and behavior*, and *physiological psychology*. The study of the biological bases of psychological processes and behavior. [1]

biological rhythm A regular periodic fluctuation in any living process. [10]

biotransformation The process in which enzymes convert a drug into a metabolite that is itself active, possibly in ways that are substantially different from the actions of the original substance. [4]

bipolar cell An interneuron in the retina that receives information from rods and cones and passes the information to retinal ganglion cells. See Figure 7.3.

Compare *amacrine cell* and *horizontal cell*. [7]

bipolar disorder A psychiatric disorder characterized by periods of depression that alternate with excessive, expansive moods. [12]

bipolar neuron A nerve cell that has a single dendrite at one end and a single axon at the other end. See Figure 2.3. Compare *unipolar neuron* and *multipolar neuron*. [2]

blind spot The portion of the visual field from which light falls on the optic disc. [7]

blindsight The paradoxical phenomenon whereby, within a scotoma, a person cannot *consciously* perceive visual cues but may still be able to make some visual discrimination. [7]

blood-brain barrier The protective property of cerebral blood vessels that impedes the movement of some harmful substances from the blood stream into the brain. [2, 4]

blotting Transferring DNA, RNA, or protein fragments to nitrocellulose following separation via gel electrophoresis. The blotted substance can then be labeled. See Appendix Figure A.3. [App]

brain and behavior See *biological psychology*. [1]

brain self-stimulation The process in which animals will work to provide electrical stimulation to particular brain sites, presumably because the experience is very rewarding. [11]

brainstem The region of the brain that consists of the midbrain, the pons, and the medulla. [2]

brightness One of three basic dimensions of light perception, varying from dark to light. Compare *hue* and *saturation*. [7]

Broca's aphasia See *nonfluent aphasia*. [15]

Broca's area A region of the frontal lobe of the brain that is involved in the production of speech. See Figures 15.6, 15.8, 15.9. Compare *Wernicke's area*. [15]

brown fat Also called *brown adipose tissue*. A specialized type of fat tissue that generates heat through intense metabolism. [ASF 9.1]

bulimia Also called *bulimia nervosa*. A syndrome in which individuals periodically gorge themselves, usually with "junk food," and then either vomit or take laxatives to avoid weight gain. [9]

bungarotoxin A neurotoxin, isolated from the venom of the many-banded krait, that selectively blocks acetylcholine receptors. [3]

C

C fiber A small, unmyelinated axon that conducts pain information slowly and adapts slowly. [5]

c-fos An immediate early gene commonly used to identify activated neurons. See Box 2.1.

CAH See *congenital adrenal hyperplasia*. [8]

calcium ion (Ca²⁺) A calcium atom that carries a double positive charge. [3]

carotid artery Either of the two major arteries that ascend the left and right sides of the neck to the brain, supplying blood to the anterior and middle cerebral arteries. The branch that enters the brain is called the internal carotid artery. [ASF 2.1]

castration Removal of the gonads, usually the testes. [8]

CAT See *computerized axial tomography*. [2]

cataplexy Sudden loss of muscle tone, leading to collapse of the body without loss of consciousness. Cataplexy is sometimes a component of narcoleptic attacks. [10]

cation A positively charged ion, such as a potassium or sodium ion. Compare *anion*. [3]

cauda equina Literally, "horse's tail" (in Latin). The caudalmost spinal nerves, which extend beyond the spinal cord proper to exit the spinal column.

caudal See *posterior*. See Box 2.2. [2]

caudate nucleus One of the basal ganglia. It has a long extension or tail. See Figure 2.14. [2]

causality The relation of cause and effect, such that we can conclude that an experimental manipulation has specifically caused an observed result. [1]

CBT See *cognitive behavioral therapy*. [12]

CCK See *cholecystokinin*. [9]

cell body Also called *soma*. The region of a neuron that is defined by the presence of the cell nucleus. Functionally, the cell body is the integration zone of the neuron. See Figures 2.1, 2.3. [2]

cell-cell interaction The general process during development in which one cell affects the differentiation of other, usually neighboring, cells. [13]

cell death Also called *apoptosis*. The developmental process during which "surplus" cells die. See Figure 13.25. [13]

cell differentiation The developmental stage in which cells acquire distinctive characteristics, such as those of neurons, as a result of expressing particular genes. See Figure 13.25. [13, App]

cell membrane The lipid bilayer that ensheathes a cell. [3]

cell migration The movement of cells from site of origin to final location. See Figure 13.25. [13]

cell nucleus The spherical central structure of a cell that contains the chromosomes. [App]

central deafness A hearing impairment in which the auditory areas of the brain fail to process and interpret action potentials from sound stimuli in meaningful ways, usually as a consequence of damage in auditory brain areas. See Figure 6.11. Compare *conduction deafness* and *sensorineural deafness*. [6]

central modulation of sensory information The process in which higher brain centers, such as the cortex and thalamus, suppress some sources of sensory information and amplify others. [5]

central nervous system (CNS) The portion of the nervous system that includes the brain and the spinal cord. See Figures 2.6, 2.12. Compare *peripheral nervous system*. [2]

central sulcus A fissure that divides the frontal lobe from the parietal lobe. See Figure 2.10. [2]

cerebellum A structure located at the back of the brain, dorsal to the pons, that is involved in the central regulation of movement and in some forms of learning. See Figures 2.10, 2.12, 2.16, 5.27. [2, 5, 13]

cerebral arteries The three pairs of large arteries within the skull that supply blood to the cerebral cortex. [2]

cerebral cortex Also called simply *cortex*. The thick and convoluted outermost rind of the cerebral hemispheres, consisting largely of nerve cell bodies and their branches. See Figure 2.13. [2]

cerebral hemisphere One of the two halves—right or left—of the forebrain. See Figure 2.12. [2]

cerebrocerebellum The lowermost part of the cerebellum, consisting especially of the lateral parts of each cerebellar hemisphere. It is implicated in planning complex movements. Compare *spinocerebellum* and *vestibulocerebellum*. [ASF 5.4]

cerebrospinal fluid (CSF) The fluid that fills the cerebral ventricles. See Figure 2.17. [2]

cerveau isolé See *isolated forebrain*. [10]

cervical Referring to the topmost eight segments of the spinal cord, in the neck region. See Figures 2.8, 2.9. Compare *thoracic, lumbar, sacral,* and *coccygeal*. [2]

change blindness A failure to notice changes in comparisons of two alternating static visual scenes.

ChAT See *choline acetyltransferase*. [ASF 4.1]

chemical transmitter See *neurotransmitter*. [2, 3]

chloride ion (Cl⁻) A chlorine atom that carries a negative charge. [3]

chlorpromazine An early antipsychotic drug that revolutionized the treatment of schizophrenia. [12]

cholecystokinin (CCK) A peptide hormone that is released by the gut after ingestion of food that is high in protein and/or fat. [9]

choline acetyltransferase (ChAT) An important enzyme involved in the synthesis of the neurotransmitter acetylcholine. [ASF 4.1]

cholinergic Referring to cells that use acetylcholine as their synaptic transmitter. [3, 4]

choroid plexus A specialized membrane lining the ventricles that produces cerebrospinal fluid by filtering blood. [2]

chromosome A complex of condensed strands of DNA and associated protein molecules. Chromosomes are found in the nucleus of cells. [App]

chronic traumatic encephalopathy (CTE) Also called *dementia pugilistica* or *punch-drunk syndrome*. A form of dementia that may develop following multiple concussions, such as in athletes engaged in contact sports. See Box 15.3. [15]

ciliary muscle One of the muscles that control the shape of the lens inside the eye, focusing an image on the retina. See Figure 7.1. [7]

CIMT See *constraint-induced movement therapy*. [15]

cingulate cortex Also called *cingulum*. A region of medial cerebral cortex that lies dorsal to the corpus callosum. [5]

cingulate gyrus Also called *cingulate cortex* or *cingulum*. A strip of cortex, found in the frontal and parietal midline, that is part of the limbic system and is implicated in many cognitive functions. See Figures 2.14, 2.16. [2]

circadian rhythm A pattern of behavioral, biochemical, or physiological fluctuation that has a 24-hour period. [10]

circle of Willis A structure at the base of the brain that is formed by the joining of the carotid and basilar arteries. [ASF 2.1]

circumventricular organ Any of multiple distinct sites that lie in the wall of a cerebral ventricle and monitor the composition of the cerebrospinal fluid. See Figure 9.8. [9]

classical conditioning Also called *Pavlovian conditioning*. A type of associative learning in which an originally neutral stimulus (the conditioned stimulus, or CS)—through pairing with another stimulus (the unconditioned stimulus, or US) that elicits a particular response—acquires the power to elicit that response when presented alone. A response elicited by the US is called an unconditioned response (UR); a response elicited by the CS alone is called a conditioned response (CR). See Figure 13.9. Compare *instrumental conditioning*. [13]

clitoris The female phallus. Compare *penis*. [8]

cloacal exstrophy A rare medical condition in which XY individuals are born completely lacking a penis. [8]

clones Asexually produced organisms that are genetically identical. [13, App]

clozapine An atypical antipsychotic. [12]

CNS See *central nervous system*. [2]

cocaine A drug of abuse, derived from the coca plant, that acts by enhancing catecholamine neurotransmission. [4]

coccygeal Referring to the lowest spinal vertebra (the coccyx, also known as the "tailbone"). See Figure 2.9. Compare *cervical*, *thoracic*, *lumbar*, and *sacral*. [2]

cochlea A snail-shaped structure in the inner ear that contains the primary receptor cells for hearing. See Figure 6.1. [6]

cochlear implant An electromechanical device that detects sounds and selectively stimulates nerves in different regions of the cochlea via surgically implanted electrodes. [6]

cochlear nucleus Either of two brainstem nuclei—left and right—that receive input from auditory hair cells and send output to the superior olivary nuclei. See Figure 6.6. [6]

cocktail party effect The selective enhancement of attention in order to filter out distracters, as you might do while listening to one person talking in the midst of a noisy party. [14]

codon A set of three nucleotides that encodes one particular amino acid. A series of codons determines the structure of a peptide or protein. [App]

cognitive behavioral therapy (CBT) Psychotherapy aimed at correcting negative thinking and consciously changing behaviors as a way of changing feelings. [12]

cognitive map A mental representation of the relative spatial organization of objects and information. [13]

cognitively impenetrable Referring to basic neural processing operations that cannot be experienced through introspection—in other words, that are unconscious. [14]

coitus See *copulation*. [8]

collateral sprouting The formation of a new branch on an axon, usually in response to the uncovering of unoccupied postsynaptic sites. [ASF 15.4]

communication Information transfer between two individuals. [15]

complex cortical cell A cell in the visual cortex that responds best to a bar of a particular size and orientation anywhere within a particular area of the visual field and that needs movement to make it respond actively. Compare *simple cortical cell*. [7]

complex environment See *enriched condition*. [13]

complex partial seizure In epilepsy, a type of seizure that doesn't involve the entire brain and therefore can cause a wide variety of symptoms. [3]

computerized axial tomography (CAT or CT) A noninvasive technique for examining brain structure through computer analysis of X-ray absorption at several positions around the head. See Figure 2.19. Compare *magnetic resonance imaging*. [2]

concordance Sharing of a characteristic by both individuals of a pair of twins. [12]

concussion A form of closed head injury caused by a jarring blow to the head, resulting in damage to the tissue of the brain with short- or long-term consequences for cognitive function. [15]

conduction aphasia An impairment in the ability to repeat words and sentences. [15]

conduction deafness A hearing impairment in which the ears fail to convert sound vibrations in air into waves of fluid in the cochlea. It is associated with defects of the external ear or middle ear. See Figure 6.11. Compare *central deafness* and *sensorineural deafness*. [6]

conduction velocity The speed at which an action potential is propagated along the length of an axon. [3]

conduction zone The part of a neuron—typically the axon—over which the action potential is actively propagated. See Figures 2.1, 2.3. Compare *input zone*, *integration zone*, and *output zone*. [2]

cone Any of several classes of photoreceptor cells in the retina that are responsible for color vision. See Figure 7.3. Compare *rod*. [7]

confabulate To fill in a gap in memory with a falsification. Confabulation is often seen in Korsakoff's syndrome. [13]

congenital adrenal hyperplasia (CAH) Any of several genetic mutations that can cause a female fetus to be exposed to adrenal androgens, resulting in partial masculinization at birth. [8]

conjunction search A search for an item that is based on two or more features (e.g., size and color) that together distinguish the target from distracters that may share some of the same attributes. Compare *feature search*. [14]

connectionist model of aphasia Also called *Wernicke-Geschwind model*. A theory proposing that left-hemisphere language deficits result from disconnection between the brain regions in a language network, each of which serves a particular linguistic function. Compare *motor theory of language*. [15]

consciousness The state of awareness of one's own existence, thoughts, emotions, and experiences. [1, 14]

conserved In the context of evolution, referring to a trait that is passed on from a common ancestor to two or more descendant species. [1]

consolidation The second process in the memory system, in which information in short-term memory is transferred to long-term memory. See Figure 13.13. Compare *encoding* and *retrieval*. [13]

constraint-induced movement therapy (CIMT) A therapy for recovery of movement after stroke or injury in which the person's unaffected limb is constrained while he is required to perform tasks with the affected limb. [15]

contralateral In anatomy, pertaining to a location on the opposite side of the body. See Box 2.2. Compare *ipsilateral*. [2, 15]

control group In research, a group of individuals that are identical to those in an experimental (or test) group in every way except that they do not receive the experimental treatment or manipulation. The experimental group is then compared with the control group to assess the effect of the treatment. [1]

convergence The phenomenon of neural connections in which many cells send signals to a single cell. Compare *divergence*. [ASF 3.2, 7]

Coolidge effect The propensity of an animal that appears sexually satiated with a current partner to resume sexual activity when provided with a new partner. [8]

copulation Also called *coitus*. The sexual act. [8]

cornea The transparent outer layer of the eye, whose curvature is fixed. The cornea bends light rays and is primarily responsible for forming the image on the retina. See Figure 7.1. [7]

coronal plane Also called *frontal plane* or *transverse plane*. The plane that divides the body or brain into front and back parts. See Box 2.2. Compare *horizontal plane* and *sagittal plane*. [2]

corpus callosum The main band of axons that connects the two cerebral hemispheres. See Figures 2.11, 2.16. [2, 15]

corpus luteum (pl. corpora lutea) The structure that forms from the collapsed ovarian follicle after ovulation. The corpora lutea are a major source of progesterone. [8, ASF 8.3]

correlation The tendency of two measures to vary in concert, such that a change in one measure is matched by a change in the other. [1]

cortex (pl. cortices) The outer layer of a structure. See also *cerebral cortex* and *neocortex*. [2]

cortical column One of the vertical columns that constitute the basic organization of the cerebral cortex. [2]

cortical deafness A form of central deafness, caused by damage to both sides of the auditory cortex, that is characterized by difficulty in recognizing all complex sounds, whether verbal or nonverbal. [6]

corticospinal system See *pyramidal system*. [5]

cortisol A glucocorticoid stress hormone of the adrenal cortex. [11]

covert attention Attention in which the focus can be directed independently of sensory orientation (e.g., you're attending to one sensory stimulus while looking at another). Compare *overt attention*. [14]

cranial nerve One of the 12 pairs of nerves that arise directly from the brain rather than the spinal cord, supplying senory and motor connections to the head and neck. See Figure 2.7. Compare *spinal nerve*. [2]

crib death See *sudden infant death syndrome*. [10]

critical period See *sensitive period*. [15]

cross-tolerance A condition in which the development of tolerance for one drug causes an individual to develop tolerance for another drug. [4]

crystallization The final stage of birdsong formation, in which fully formed adult song is achieved. [ASF 15.3]

CSF See *cerebrospinal fluid*. [2]

CT See *computerized axial tomography*. [2]

CTE See *chronic traumatic encephalopathy*. [15]

curare A neurotoxin that causes paralysis by blocking acetylcholine receptors in muscle. [3]

Cushing's syndrome A condition in which levels of adrenal glucocorticoids are abnormally high. [ASF 12.2]

cytokine A protein that induces the proliferation of other cells, as in the immune system. Examples include interleukins and interferons. [ASF 11.4]

cytoplasm See *intracellular fluid*. [3]

D

DA See *dopamine*. [4]

dB See *decibel*. [6]

DBS See *deep brain stimulation*. [12]

death gene A gene that is expressed only when a cell becomes committed to natural cell death (apoptosis). [13]

decibel (dB) A measure of sound intensity, perceived as loudness. See Box 6.1. [6]

declarative memory A memory that can be stated or described. See Figures 13.4, 13.12. Compare *nondeclarative memory*. [13]

decomposition of movement Difficulty of movement in which gestures are broken up into individual segments instead of being executed smoothly; it is a symptom of cerebellar lesions. [5]

decorticate rage Also called *sham rage*. Sudden intense rage characterized by actions (such as snarling and biting in dogs) that lack clear direction. [11]

deep brain stimulation (DBS) Mild electrical stimulation through an electrode that is surgically implanted deep in the brain. [12]

deep dyslexia Acquired dyslexia in which the patient reads a word as another word that is semantically related. Compare *surface dyslexia*. [15]

default mode network A circuit of brain regions that is active during quiet introspective thought. [14]

degradation The chemical breakdown of a neurotransmitter into inactive metabolites. [3]

dehydration Excessive loss of water.

delayed non-matching-to-sample task A test in which the subject must respond to the unfamiliar stimulus in a pair of stimuli. See Figure 13.5. [13]

delta-9-tetrahydrocannabinol (THC) The major active ingredient in marijuana. [4]

delta wave The slowest type of EEG wave, characteristic of stage 3 sleep. See Figure 10.10. [10]

delusion A false belief strongly held in spite of contrary evidence. [12]

dementia Drastic failure of cognitive ability, including memory failure and loss of orientation. [13]

dementia pugilistica See *chronic traumatic encephalopathy*. [15]

dendrite An extension of the cell body that receives information from other neurons. Functionally, the dendrites are the input zone of the neuron. See Figures 2.1, 2.3. [2]

dendritic spine An outgrowth along the dendrite of a neuron. See Figure 2.4. [2]

dendro-dendritic synapse A synapse at which a synaptic connection forms between the dendrites of two neurons. Compare *axo-axonic synapse, axo-dendritic synapse*, and *axo-somatic synapse*. [3]

dentate gyrus A strip of gray matter in the hippocampal formation. [13]

deoxyribonucleic acid (DNA) A nucleic acid that is present in the chromosomes of cells and codes hereditary information. Compare *ribonucleic acid*. [App]

dependent variable The factor that an experimenter measures to monitor a change in response to changes in an independent variable. [1]

depolarization A decrease in membrane potential (the interior of the neuron becomes less negative). See Figure 3.5. Compare *hyperpolarization*. [3]

depressant A drug that reduces the excitability of neurons. Compare *stimulant*. [4]

depression A psychiatric condition characterized by such symptoms as an unhappy mood; loss of interests, energy, and appetite; and difficulty concentrating. See also *bipolar disorder*. [12]

dermatome A strip of skin innervated by a particular spinal nerve. [5]

desynchronized EEG Also called *beta activity*. A pattern of EEG activity comprising a mix of many different high frequencies with low amplitude. Compare *alpha rhythm*. [10]

dexamethasone suppression test A test of pituitary-adrenal function in which the subject is given dexamethasone, a synthetic glucocorticoid hormone, which should cause a decline in the production of adrenal corticosteroids. [ASF 12.2]

DHT See *dihydrotestosterone*. [8]

diabetes mellitus A condition, characterized by excessive glucose in the blood and urine and by reduced glucose utilization by body cells, that is caused by the failure of insulin to induce glucose absorption. [9]

dichotic presentation The simultaneous delivery of different stimuli to both the right and the left ears at the same time. See Figure 15.2. [15]

diencephalon The posterior part of the fetal forebrain, which will become the thalamus and hypothalamus in the adult brain. See Figures 2.12, 13.24. Compare *telencephalon*. [2]

differentiation See *cell differentiation*. [13, App]

diffusion The spontaneous spread of molecules from an area of high concentration to an area of low concentration until a uniform concentration is achieved. See Figure 3.3. Compare *osmosis*. [3, 9]

diffusion tensor imaging (DTI) A modified form of MRI imaging in which the diffusion of water in a confined space is exploited to produce images of axonal fiber tracts. [15]

digestion The process by which food is broken down to provide energy and nutrients. [ASF 9.3]

dihydrotestosterone (DHT) The 5-alpha-reduced metabolite of testosterone. DHT is a potent androgen that is principally responsible for the masculinization of the external genitalia in mammals. [8]

distal In anatomy, toward the periphery of an organism or toward the end of a limb. See Box 2.2. Compare *proximal*. [2]

diurnal Active during the light periods of the daily cycle. Compare *nocturnal*. [10]

divergence The phenomenon of neural connections in which one cell sends signals to many other cells. Compare *convergence*. [ASF 3.2]

divided-attention task A task in which the subject is asked to focus attention on two or more stimuli simultaneously. Compare *sustained-attention task*. [14]

dizygotic Referring to twins derived from separate eggs (fraternal twins). Compare *monozygotic*. [12]

DNA See *deoxyribonucleic acid*. [App]

DNA sequencing The process by which the order of nucleotides in a gene is identified. [App]

dopamine (DA) A monoamine transmitter found in the midbrain—especially the substantia nigra—and in the basal forebrain. See Figure 4.4; Table 4.1. [4]

dopamine hypothesis The idea that schizophrenia results from either excessive levels of synaptic dopamine or excessive postsynaptic sensitivity to dopamine. [12]

dopaminergic Referring to cells that use dopamine as their synaptic transmitter. [4]

dorsal In anatomy, toward the back of the body or the top of the brain. See Box 2.2. Compare *ventral*. [2]

dorsal column system A somatosensory system that delivers most touch stimuli via the dorsal columns of spinal white matter to the brain. See Figure 5.7. Compare *anterolateral system*. [5]

dorsomedial thalamus A limbic system structure that is connected to the hippocampus. [13]

dose-response curve (DRC) A formal graph of a drug's effects (on the y-axis) versus the dose given (on the x-axis). See Figure 4.6. [4]

down-regulation A compensatory decrease in receptor availability at the synapses of a neuron. Compare *up-regulation*. [4]

DRC See *dose-response curve*. [4]

drug tolerance Also called simply *tolerance*. A condition in which, with repeated exposure to a drug, an individual becomes less responsive to a constant dose. [4]

DTI See *diffusion tensor imaging*. [15]

DTI tractography Also called *fiber tracking*. Visualization of the orientation and terminations of white matter tracts in the living brain via diffusion tensor imaging. [15]

dualism The notion, promoted by René Descartes, that the mind has an immaterial aspect that is distinct from the material body and brain. [1]

duplex theory A theory that we localize sound by combining information about intensity differences and latency differences between the two ears.

dura mater The outermost of the three meninges that surround the brain and spinal cord. See also *pia mater* and *arachnoid*. See Figure 2.8. [2]

dynorphin One of the three kinds of endogenous opioids. Compare *endorphin* and *enkephalin*. See Table 4.1. [4]

dyskinesia Difficulty or distortion in voluntary movement. See Box 12.1.[12]

dyslexia Also called *alexia*. A reading disorder attributed to brain impairment. [15]

dysphoria Unpleasant feelings; the opposite of euphoria. [4]

dystrophin A protein that is needed for normal muscle function. Dystrophin is defective in some forms of muscular dystrophy. [ASF 5.5]

E

ear canal Also called *auditory canal*. The tube leading from the pinna to the tympanic membrane. [6]

eardrum See *tympanic membrane*. [6]

easy problem of consciousness Understanding how particular patterns of neural activity create specific conscious experiences by reading brain activity directly from people's brains as they're having particular experiences. Compare *hard problem of consciousness*. [14]

EC See *enriched condition*. [13]

ecological niche The unique assortment of environmental opportunities and challenges to which each organism is adapted. [10]

Ecstasy See *MDMA*. [4]

ECT See electroconvulsive shock therapy. [12]

ectoderm The outer cellular layer of the developing embryo, giving rise to the skin and the nervous system. [13]

ectotherm An animal whose body temperature is regulated by, and whose heat comes mainly from, the environment. Examples include snakes and bees. Compare *endotherm*. [9]

ED$_{50}$ Effective dose 50%; the dose of a drug that is required to produce half of its maximal effect. See Figure 4.6. Compare LD_{50}.

edema The swelling of tissue in response to injury. [2]

edge detector See *simple cortical cell*. [7]

EEG See *electroencephalography*. [3, 10]

efferent Carrying action potentials away from the brain, or away from one region of interest toward another region of interest. See Box 2.2. Compare *afferent*. [2]

efficacy Also called *intrinsic activity*. The extent to which a drug activates a response when it binds to a receptor. Receptor antagonist drugs have low efficacy; receptor agonists have high efficacy. See Figure 4.6. [4]

egg See *ovum*. [8]

ejaculation The forceful expulsion of semen from the penis. [8]

electrical synapse Also called *gap junction*. The region between neurons where the presynaptic and postsynaptic membranes are so close that the action potential can jump to the postsynaptic membrane without first being translated into a chemical message. [ASF 3.1]

electroconvulsive shock therapy (ECT) A last-resort treatment for unmanageable depression, in which a strong electrical current is passed through the brain, causing a seizure. [12]

electroencephalography (EEG) The recording of gross electrical activity of the brain via large electrodes placed on the scalp. The abbreviation EEG may refer either to the process of encephalography or to its product, the encephalogram. See Figures 3.15, 10.9. [3, 10]

electromyography (EMG) The electrical recording of muscle activity. See Figure 5.15. [5]

electrostatic pressure The propensity of charged molecules or ions to move toward areas with the opposite charge. [3]

embryo The earliest stage in a developing animal. Humans are considered to be embryos until 8–10 weeks after conception. Compare *fetus*. [13]

embryonic stem cell A cell, derived from an embryo, that has the capacity to form any type of tissue. [15]

EMG See *electromyography*. [5]

emotion A subjective mental state that is usually accompanied by distinctive behaviors as well as involuntary physiological changes. [11]

encéphale isolé See *isolated brain*. [10]

encoding The first process in the memory system, in which the information entering sensory channels is passed into short-term memory. See Figure 13.13. Compare *consolidation* and *retrieval*. [13]

endocannabinoid An endogenous ligand of cannabinoid receptors, thus a marijuana analog that is produced by the brain.[4, 9]

endocrine Referring to glands that release chemicals to the interior of the body. These glands secrete the principal hormones used by the body. See Figure 8.3. [8]

endocrine gland A gland that secretes hormones into the bloodstream to act on distant targets. See Figure 8.1. [8]

endogenous Produced inside the body. Compare *exogenous*. [4]

endogenous attention Also called *voluntary attention*. The voluntary direction of attention toward specific aspects of the environment, in accordance with our interests and goals. Compare *exogenous attention*. [14]

endogenous ligand Any substance, produced within the body, that selectively binds to the type of receptor that is under study. Compare *exogenous ligand*. [4]

endogenous opioid Any of a class of opium-like peptide transmitters that have been referred to as the body's own narcotics. The three kinds are enkephalins, endorphins, and dynorphins. See Table 4.1. [4]

endorphin One of the three kinds of endogenous opioids. Compare *dynorphin* and *enkephalin*. See Table 4.1. [4, 5]

endotherm An animal whose body temperature is regulated chiefly by internal metabolic processes. Examples include mammals and birds. Compare *ectotherm*. [9]

enkephalin One of the three kinds of endogenous opioids. Compare *dynorphin* and *endorphin*. See Table 4.1. [4]

enriched condition (EC) Also called *complex environment*. An environment for laboratory rodents in which animals are group-housed with a wide variety of stimulus objects. See Figure 13.16. Compare *impoverished condition* and *standard condition*. [13]

entrainment The process of synchronizing a biological rhythm to an environmental stimulus. See Figure 10.1. [10]

enzyme A complicated protein whose action increases the probability of a specific chemical reaction. [App]

ependymal layer See *ventricular zone*. [13]

epigenetic regulation Changes in gene expression that are due to environmental effects rather than to changes in the nucleotide sequence of the gene. [11]

epigenetic transmission The passage from one individual to another of changes in the expression of targeted genes, without modifications to the genes themselves. [9]

epigenetics The study of factors that affect gene expression without making any changes in the nucleotide sequence of the genes themselves. [1, 13]

epilepsy A brain disorder marked by major, sudden changes in the electrophysiological state of the brain that are referred to as seizures. See Figure 3.16. [3]

epinephrine Also called *adrenaline*. A compound that acts both as a hormone (secreted by the adrenal medulla under the control of the sympathetic nervous system) and as a synaptic transmitter. See Tables 4.1, 8.1. [11]

episodic memory Also called *autobiographical memory*. Memory of a particular incident or a particular time and place. Compare *semantic memory*. [13]

EPSP See *excitatory postsynaptic potential*. [3]

equilibrium potential The point at which the movement of ions across the cell membrane is balanced, as the electrostatic pressure pulling ions in one direction is offset by the diffusion force pushing them in the opposite direction. [3]

ERP See *event-related potential*. [3, 14]

estradiol The primary type of estrogen that is secreted by the ovary. Its formal name is *17-beta-estradiol*. [8]

estrogen Any of a class of steroid hormones, including estradiol, produced by female gonads. See Figure 8.13. [8]

estrus The period during which female animals are sexually receptive. [8]

eukaryote Any organism whose cells have the genetic material contained within a nuclear envelope. [App]

event-related potential (ERP) Also called *evoked potential*. Averaged EEG recordings measuring brain responses to repeated presentations of a stimulus. Components of the ERP tend to be reliable because the background noise of the cortex has been averaged out. See Figures 3.15, 14.6. [3, 14]

evoked potential See *event-related potential*. [3, 14]

evolution by natural selection The Darwinian theory that evolution proceeds by differential success in reproduction.

evolutionary psychology A field of study devoted to asking how natural selection has shaped behavior in humans and other animals. [1]

excitatory postsynaptic potential (EPSP) A depolarizing potential in the postsynaptic neuron that is normally caused by synaptic excitation. EPSPs increase the probability that the postsynaptic neuron will fire an action potential. See Figure

3.9. Compare *inhibitory postsynaptic potential*. [3]

excitotoxicity The property by which neurons die when overstimulated, as with large amounts of glutamate. [ASF 15.5]

executive function A neural and cognitive system that helps develop plans of action and organizes the activities of other high-level processing systems. [14]

exocytosis A cellular process that results in the release of a substance into the extracellular space. [4]

exogenous Arising from outside the body. Compare *endogenous*. [4]

exogenous attention Also called *reflexive attention*. The involuntary reorienting of attention toward a specific stimulus source, cued by an unexpected object or event. Compare *endogenous attention*. [14]

exogenous ligand Any substance, originating from outside the body, that selectively binds to the type of receptor that is under study. Compare *endogenous ligand*. [4]

expression See *gene expression*. [1, 13, App]

external ear The part of the ear that we readily see (the pinna) and the canal that leads to the eardrum. See Figure 6.1.

extracellular compartment The fluid space of the body that exists outside the cells. See Figure 9.7. Compare *intracellular compartment*. [9]

extracellular fluid The fluid in the spaces between cells (interstitial fluid) Compare *intracellular fluid*. [3]

extraocular muscle One of the muscles attached to the eyeball that control its position and movements. [7]

extrapyramidal system A motor system that includes the basal ganglia and some closely related brainstem structures. Axons of this system pass into the spinal cord outside the pyramids of the medulla. Compare *pyramidal system*. [5]

extrastriate cortex Visual cortex outside of the primary visual (striate) cortex. [7]

F

face blindness See *prosopagnosia*. [15]

facial feedback hypothesis The idea that sensory feedback from our facial expressions can affect our mood. [11]

fat See *lipid*. [9]

fat tissue See *adipose tissue*. [9]

fatal familial insomnia An inherited disorder in which humans sleep normally at the beginning of their life but in midlife stop sleeping and, 7–24 months later, die. See Box 10.1. [10]

fear conditioning A form of classical conditioning in which fear comes to be associated with previously neutral stimuli. [11, 12]

feature search A search for an item in which the target pops out right away, no matter how many distracters are present, because it possesses a unique attribute. Compare *conjunction search*. [14]

FEF See *frontal eye field*. [14]

fetal alcohol syndrome A disorder, including intellectual disability and characteristic facial abnormalities, that affects children exposed to too much alcohol (through maternal ingestion) during fetal development. [4]

fetus A developing individual after the embryo stage. Humans are considered to be fetuses from 10 weeks after fertilization until birth. Compare *embryo*. [13]

fiber tracking See *DTI tractography*. [15]

final common pathway The motoneurons of the brain and spinal cord, so called because they receive and integrate all motor signals from the brain to direct movement. [5]

flavor The sense of taste combined with the sense of smell. Compare *taste*. [6]

fluent aphasia Also called *Wernicke's aphasia*. A language impairment characterized by fluent, meaningless speech and little language comprehension. It is related to damage in Wernicke's area. See Figure 15.8. Compare *nonfluent aphasia*. [15]

fMRI See *functional MRI*. [2]

follicle The structure of the ovary that contains an immature ovum (egg). [8, ASF 8.3]

follicle-stimulating hormone (FSH) A gonadotropin, named for its actions on ovarian follicles. See Figure 8.13. [8, ASF 8.3]

forebrain The frontmost division of the brain, which in the mature vertebrate contains the cerebral hemispheres, the thalamus, and the hypothalamus. See Figures 2.12, 13.24. Compare *hindbrain* and *midbrain*. [2, 13]

fornix A fiber tract that extends from the hippocampus to the mammillary body. See Figures 2.14, 2.16. [2]

Fourier analysis The mathematical decomposition of a complex pattern into a sum of sine waves. [6, ASF 7.2]

fourth ventricle The passageway within the pons that receives cerebrospinal fluid from the third ventricle and releases it to surround the brain and spinal cord. See Figure 2.17. Compare *lateral ventricle* and *third ventricle*. [2]

fovea The central portion of the retina, which is packed with the highest den-

sity of photoreceptors and is the center of our gaze. See Figure 7.1. [7]

fraternal birth order effect A phenomenon in human populations, such that the more older biological brothers a boy has, the more likely he is to develop a homosexual orientation. [8]

free nerve ending An axon that terminates in the skin and has no specialized cell associated with it. Free nerve endings detect pain and/or changes in temperature. See Figure 5.3. [5]

free-running Referring to a rhythm of behavior shown by an animal deprived of external cues about time of day. See Figure 10.1. [10]

free will The feeling that our conscious self is the author of our actions and decisions. [14]

frequency The number of cycles per second in a sound wave, measured in hertz. See Box 6.1. [6]

frontal eye field (FEF) An area in the frontal lobe of the brain that contains neurons important for establishing gaze in accordance with cognitive goals (top-down processes) rather than with any characteristics of stimuli (bottom-up processes). [14]

frontal lobe The most anterior portion of the cerebral cortex. See Figure 2.10. Compare *occipital lobe, parietal lobe,* and *temporal lobe*. [2]

frontal plane See *coronal plane*. [2]

FSH See *follicle-stimulating hormone*. [8, ASF 8.3]

functional MRI (fMRI) Magnetic resonance imaging that detects changes in blood flow and therefore identifies regions of the brain that are particularly active during a given task. See Figure 2.19. Compare *positron emission tomography*. [2]

functional tolerance The form of drug tolerance that arises when repeated exposure to the drug causes receptors to be up-regulated or down-regulated. Compare *metabolic tolerance*. [4]

fundamental The predominant frequency of an auditory tone. Compare *harmonic*. See Box 6.1. [6]

fusiform gyrus A region on the inferior surface of the cortex, at the junction of the temporal and occipital lobes, that has been associated with recognition of faces. See Figure 15.5. [15]

G

GABA See *gamma-aminobutyric acid*. [4]

gamete A sex cell (sperm or ovum) that contains only unpaired chromosomes and therefore has only half of the usual number of chromosomes. [8]

gamma-aminobutyric acid (GABA) A widely distributed amino acid transmitter, the main inhibitory transmitter in the mammalian nervous system. See Table 4.1. [4]

ganglion (pl. ganglia) A collection of nerve cell bodies outside the central nervous system. Compare *nucleus* (definition 1).

ganglion cell Any of a class of cells in the retina whose axons form the optic nerve. See Figure 7.15. Compare *amacrine cell, bipolar cell,* and *horizontal cell*. [7]

gap junction See *electrical synapse*. [ASF 3.1]

gas neurotransmitter A neurotransmitter that is a soluble gas. Examples include nitric oxide and carbon monoxide. Usually gas neurotransmitters act, in a retrograde fashion, on presynaptic neurons. See Table 4.1. Compare *amine neurotransmitter, amino acid neurotransmitter,* and *peptide neurotransmitter*. [4]

gel electrophoresis A method of separating molecules of differing size or electrical charge by forcing them to flow through a gel. See Appendix Figure A.3. [App]

gene A length of DNA that encodes the information for constructing a particular protein. [App]

gene amplification See *polymerase chain reaction*. [App]

gene expression The turning on or off of specific genes. The process by which a cell makes an mRNA transcript of a particular gene. [1, App]

general anesthetic A drug that renders an individual unconscious. [10]

generator potential A local change in the resting potential of a receptor cell in response to stimuli, which may initiate an action potential. [5]

genital tubercle In the early fetus, a "bump" between the legs that can develop into either a clitoris or a penis. [8]

genome See *genotype*. [App]

genotype Also called *genome*. All the genetic information that one specific individual has inherited. Compare *phenotype*. [13, App]

GH See *growth hormone*. [8, ASF 8.3]

ghrelin A peptide hormone produced and released by the gut. See Figure 9.15. Compare PYY_{3-36}. [9]

glial cells Also called *glia*. Nonneuronal brain cells that provide structural, nutritional, and other types of support to the brain. See Figure 2.5. [2]

global aphasia The total loss of ability to understand language, or to speak, read, or write. See Figure 15.8. [15]

globus pallidus One of the basal ganglia. See Figure 2.14. [2]

glomerulus (pl. glomeruli) A complex arbor of dendrites from a group of olfactory cells. [6]

glucagon A pancreatic hormone that converts glycogen to glucose and thus increases blood glucose. Compare *insulin*. [9]

glucocorticoid Any of a class of steroid hormones, released by the adrenal cortex, that affect carbohydrate metabolism and inflammation.

glucodetector A specialized type of liver cell that detects and informs the nervous system about levels of circulating glucose. [9]

glucose An important sugar molecule used by the body and brain for energy. [9]

glutamate An amino acid transmitter, the most common excitatory transmitter. See Table 4.1. [4, 13]

glutamate hypothesis The idea that schizophrenia may be caused, in part, by understimulation of glutamate receptors. [12]

glutamatergic Referring to cells that use glutamate as their synaptic transmitter.

glycine An amino acid transmitter, often inhibitory. See Table 4.1.

glycogen A complex carbohydrate made by the combining of glucose molecules for a short-term store of energy. [9]

GnRH See *gonadotropin-releasing hormone*. [8]

Golgi stain A tissue stain that completely fills a small proportion of neurons with a dark, silver-based precipitate. See Box 2.1. [2]

Golgi tendon organ A type of receptor found within tendons that sends impulses to the central nervous system when a muscle contracts. See Figure 5.20. Compare *muscle spindle*. [5]

gonad Any of the sexual organs (ovaries in females, testes in males) that produce gametes for reproduction. See Figure 8.1. [8]

gonadotropin An anterior pituitary hormone that selectively stimulates the cells of the gonads to produce sex steroids and gametes. See *luteinizing hormone* and *follicle-stimulating hormone*. [8, ASF 8.3]

gonadotropin-releasing hormone (GnRH) A hypothalamic hormone that controls the release of luteinizing hormone and follicle-stimulating hormone from the pituitary. See Figure 8.13. [8]

grammar All of the rules for usage of a particular language. [15]

grand mal seizure A type of generalized epileptic seizure in which nerve cells fire in high-frequency bursts, usually accompanied by involuntary rhythmic contractions of the body. See Figure 3.16. Compare *petit mal seizure*. [3]

gray matter Areas of the brain that are dominated by cell bodies and are devoid of myelin. Gray matter mostly receives and processes information. See Figure 2.11. Compare *white matter*. [2]

gross neuroanatomy Anatomical features of the nervous system that are apparent to the naked eye. [2]

growth hormone (GH) Also called *somatotropin* or *somatotropic hormone*. A tropic hormone, secreted by the anterior pituitary, that promotes the growth of cells and tissues. [8, ASF 8.3]

guevedoces Literally "eggs at 12" (in Spanish). A nickname for individuals who are raised as girls but at puberty change appearance and begin behaving as boys. [8]

gustatory system The sensory system that detects taste. See Figure 6.18. [6]

gyrus (pl. gyri) A ridged or raised portion of a convoluted brain surface. Compare *sulcus*. [2]

H

habituation A form of nonassociative learning in which an organism becomes less responsive following repeated presentations of a stimulus. See Figure 13.19. [13]

hair cell One of the receptor cells for hearing in the cochlea, named for the stereocilia that protrude from the top of the cell and transduce vibrational energy in the cochlea into neural activity. See Figures 6.1, 6.3. [6]

hallucinogen A drug that alters sensory perception and produces peculiar experiences. Compare *dissociative*. [4]

hard problem of consciousness Understanding the brain processes that produce people's subjective experiences of their conscious perceptions—that is, their qualia. Compare *easy problem of consciousness*. [14]

harmonic A multiple of a particular frequency called the *fundamental*. See Box 6.1. [6]

health psychology Also called *behavioral medicine*. A field of study that focuses on the influence of psychological influences on health and disease-related processes. [11]

Hebbian synapse A synapse that is strengthened when it successfully drives the postsynaptic cell. [13, ASF 13.4]

hemiparesis Weakness of one side of the body. Compare *hemiplegia*. [15]

hemiplegia Paralysis of one side of the body. Compare *hemiparesis*. [15]

hemispatial neglect Failure to pay any attention to objects presented to one side of the body. [14]

hermaphrodite An individual possessing the reproductive organs of both sexes, either simultaneously or at different points in time. [ASF 8.5]

heroin Diacetylmorphine, an artificially modified, very potent form of morphine. [4]

hertz (Hz) Cycles per second, as of an auditory stimulus. Hertz is a measure of frequency. See Box 6.1. [6]

hindbrain The rear division of the brain, which in the mature vertebrate contains the cerebellum, pons, and medulla. See Figures 2.12, 13.24. Compare *forebrain* and *midbrain*. [2, 13]

hippocampus (pl. hippocampi) A medial temporal lobe structure that is important for spatial cognition, learning and memory. See Figures 2.14, 13.1, 13.21. [2, 13]

histology The study of tissue structure.

homeostasis The active process of keeping a particular physiological parameter relatively constant. [9]

horizontal cell A specialized retinal cell that contacts both photoreceptors and bipolar cells. Compare *amacrine cell* and *ganglion cell*. [7]

horizontal plane The plane that divides the body or brain into upper and lower parts. See Box 2.2. Compare *coronal plane* and *sagittal plane*. [2]

hormone A chemical, usually secreted by an endocrine gland, that is conveyed by the bloodstream and regulates target organs or tissues. See Table 8.1. [8]

horseradish peroxidase (HRP) An enzymatic label, originating in horseradish and other plants, that is used to determine the cells of origin of a particular set of axons. See Box 2.1. [2]

HRP See *horseradish peroxidase*. [2]

hue One of three basic dimensions of light perception, varying through the spectrum from blue to red. Compare *brightness* and *saturation*. [7]

hunger The internal state of an animal seeking food. Compare *satiety*. [9]

huntingtin A protein produced by a gene (called *HTT*) that, when containing too many trinucleotide repeats, results in Huntington's disease in a carrier. [ASF 5.5]

Huntington's disease A genetic disorder, with onset in middle age, in which the destruction of basal ganglia results in a syndrome of abrupt, involuntary writhing movements and changes in mental functioning. Compare *Parkinson's disease*. [5]

hybridization The process by which one string of nucleotides becomes linked to a complementary series of nucleotides. [App]

hyperpolarization An increase in membrane potential (the interior of the neuron becomes even more negative). See Figure 3.5. Compare *depolarization*. [3]

hypertonic Referring to a solution with a higher concentration of salt than that found in interstitial fluid and blood plasma (more than about 0.9% salt). Compare *hypotonic* and *isotonic*. [9]

hypocretin See *orexin*. [9, 10]

hypofrontality hypothesis The idea that schizophrenia may reflect underactivation of the frontal lobes. [12]

hypothalamic-pituitary portal system An elaborate bed of blood vessels leading from the hypothalamus to the anterior pituitary. [8]

hypothalamus Part of the diencephalon, lying ventral to the thalamus. See Figures 2.12, 2.14, 2.16. [2]

hypotonic Referring to a solution with a lower concentration of salt than that found in interstitial fluid and blood plasma (less than about 0.9% salt). Compare *hypertonic* and *isotonic*. [9]

hypovolemic thirst A desire to ingest fluids that is stimulated by a reduction in volume of the extracellular fluid. Compare *osmotic thirst*. [9]

Hz See *hertz*. [6]

I

IC See impoverished condition. [13]

ICC See *immunocytochemistry*. [2, 8, App]

iconic memory See *sensory buffer*. [13]

IEG See *immediate early gene*. [2]

IHC See *inner hair cell*. [6]

immediate early gene (IEG) A gene that shows rapid but temporary increases in expression in cells that have become activated. See Box 2.1. [2]

immunocytochemistry (ICC) A method for detecting a particular protein in tissues in which an antibody recognizes and binds to the protein and then chemical methods are used to leave a visible reaction product around each antibody. See Boxes 2.1, 8.1. [2, 8, App]

immunoglobulin See *antibody*. [ASF 11.4, App]

impoverished condition (IC) Also called *isolated condition*. An environment for laboratory rodents in which each animal is housed singly in a small cage without complex stimuli. See Figure 13.16. Compare *enriched condition* and *standard condition*. [13]

in situ hybridization A method for detecting particular RNA transcripts in tissue sections by providing a nucleotide probe that is complementary to, and will therefore hybridize with, the transcript of interest. See Boxes 2.1, 8.1; Appendix Figure A.4. [2, 8, App]

in vitro Literally "in glass" (in Latin). Usually, in a laboratory dish; outside the body.

inattentional blindness The failure to perceive nonattended stimuli that seem so obvious as to be impossible to miss. [14]

incus Latin for "anvil." A middle-ear bone situated between the malleus (attached to the tympanic membrane) and the stapes (attached to the cochlea). It is one of the three ossicles that conduct sound across the middle ear. See Figure 6.1. [6]

independent variable The factor that is manipulated by an experimenter. Compare *dependent variable*. [1]

indifferent gonads The undifferentiated gonads of the early mammalian fetus, which will eventually develop into either testes or ovaries. See Figure 8.24. See also *gonad*. [8]

indoleamines A class of monoamines that serve as neurotransmitters, including serotonin and melatonin. See Table 4.1.

inferior In anatomy, below. See Box 2.2. Compare *superior*. [2]

inferior colliculi (sing. colliculus) Paired gray matter structures of the dorsal midbrain that process auditory information. See Figure 2.16. Compare *superior colliculi*. [2, 6]

infradian Referring to a rhythmic biological event with a period longer than a day. Compare *ultradian*. [10]

infrasound Very low frequency sound; in general, below the threshold for human hearing, at about 20 Hz. Compare *ultrasound*. [6]

infundibulum See *pituitary stalk*. [8]

inhibition of return The phenomenon, observed in peripheral spatial cuing tasks when the interval between cue and target stimulus is 200 milliseconds or more, in which the detection of stimuli at the former location of the cue is increasingly impaired. [14]

inhibitory postsynaptic potential (IPSP) A hyperpolarizing potential in the postsynaptic neuron. IPSPs decrease the probability that the postsynaptic neuron will fire an action potential. See Figure 3.9. Compare *excitatory postsynaptic potential*. [3]

inner ear The cochlea and vestibular apparatus. See Figure 6.1. [6]

inner hair cell (IHC) One of the two types of receptor cells for hearing in the co-

chlea. Compared with *outer hair cells*, IHCs are positioned closer to the central axis of the coiled cochlea. See Figure 6.1. [6]

innervate To provide neural input to. [2]

input zone The part of a neuron that receives information from other neurons or from specialized sensory structures. This zone usually corresponds to the cell's dendrites. See Figures 2.1, 2.3. Compare *conduction zone, integration zone*, and *output zone*. [2]

instrumental conditioning Also called *operant conditioning*. A form of associative learning in which the likelihood that an act (instrumental response) will be performed depends on the consequences (reinforcing stimuli) that follow it. Compare *classical conditioning*. [13]

insula A region of cortex lying below the surface, within the lateral sulcus, of the frontal, temporal, and parietal lobes. [4]

insulin A pancreatic hormone that lowers blood glucose, promotes energy storage, and facilitates glucose utilization by cells. Compare *glucagon*. [9]

integration zone The part of a neuron that initiates neural electrical activity. This zone usually corresponds to the neuron's cell body. See Figures 2.1, 2.3. Compare *conduction zone, input zone*, and *output zone*. [2]

intensity See *amplitude*. [6]

intensity difference A perceived difference in loudness between the two ears, which the nervous system can use to localize a sound source. Compare *latency difference*. [6]

intermale aggression Aggression between males of the same species. [11]

interneuron A nerve cell that is neither a sensory neuron nor a motoneuron; interneurons receive input from and send output to other neurons. Compare *motoneuron* and *sensory neuron*. [2]

intersex Referring to an individual with atypical genital development and sexual differentiation that generally resembles a form intermediate between typical male and typical female genitalia. [8]

intracellular compartment The fluid space of the body that is contained within cells. See Figure 9.7. Compare *extracellular compartment*. [9]

intracellular fluid Also called *cytoplasm*. The watery solution found within cells. Compare *extracellullar fluid*. [3]

intrafusal fiber Any of the small muscle fibers that lie within each muscle spindle. See Figure 5.20. [5]

intraparietal sulcus (IPS) A region in the human parietal lobe, homologous to the monkey lateral intraparietal area, that is especially involved in voluntary, top-down control of attention. [14]

intrinsic activity See *efficacy*. [4]

intromission Insertion of the penis into the vagina during copulation. [8]

inverse agonist A substance that binds to a receptor and causes it to do the opposite of what the naturally occurring transmitter does.

ion An atom or molecule that has acquired an electrical charge by gaining or losing one or more electrons. [3]

ion channel A pore in the cell membrane that permits the passage of certain ions through the membrane when the channels are open. See Figure 3.2. [3]

ionotropic receptor Also called *ligand-gated ion channel*. A receptor protein containing an ion channel that opens when the receptor is bound by an agonist. See Figure 4.2. Compare *metabotropic receptor*. [4]

IPS See *intraparietal sulcus*. [14]

ipsilateral In anatomy, pertaining to a location on the same side of the body. See Box 2.2. Compare *contralateral*. [2]

IPSP See *inhibitory postsynaptic potential*. [3]

iris (pl. irides) The circular structure of the eye that provides an opening to form the pupil. See Figure 7.1. [7]

isocortex See *neocortex*.

isolated brain Also called *encéphale isolé*. An experimental preparation in which an animal's brainstem has been separated from the spinal cord by a cut below the medulla. See Figure 10.17. Compare *isolated forebrain*. [10]

isolated condition See *impoverished condition*. [13]

isolated forebrain Also called *cerveau isolé*. An experimental preparation in which an animal's nervous system has been cut in the upper midbrain, dividing the forebrain from the brainstem. See Figure 10.17. Compare *isolated brain*. [10]

isotonic Referring to a solution with a concentration of salt that is the same as that found in interstitial fluid and blood plasma (about 0.9% salt). Compare *hypertonic* and *hypotonic*. [9]

K

K complex A sharp, negative EEG potential that is seen in stage 2 sleep. [10]

kcal See *kilocalorie*. [ASF 9.1]

ketamine A dissociative anesthetic drug, similar to PCP, that acts as an NMDA receptor antagonist. [12]

ketone A compound, liberated by the breakdown of body fats and proteins, that is a metabolic fuel source. [9]

kilocalorie (kcal) A measure of energy commonly applied to food; formally defined as the quantity of heat required to raise the temperature of 1 kilogram of water by 1°C. [ASF 9.1]

Klüver-Bucy syndrome A condition, brought about by bilateral amygdala damage, that is characterized by dramatic emotional changes including reduction in fear and anxiety. [11]

knee jerk reflex A variant of the stretch reflex in which stretching of the tendon beneath the knee leads to an upward kick of the leg. See Figure 3.14. [3]

knockout organism An individual in which a particular gene has been disabled by an experimenter. See Box 8.1. [8]

Korsakoff's syndrome A memory disorder, caused by thiamine deficiency, that is generally associated with chronic alcoholism. [13]

L

L-dopa The immediate precursor of the transmitter dopamine. It is known to markedly reduce symptoms in patients with Parkinson's, decreasing tremors and increasing the speed of movements. [ASF 5.5]

labeled lines The concept that each nerve input to the brain reports only a particular type of information. [5]

lamellated corpuscle See *Pacinian corpuscle*. [5]

language Communication in which arbitrary sounds or symbols are arranged according to a grammar in order to convey an almost limitless variety of concepts. [15]

latency difference A difference between the two ears in the time of arrival of a sound, which the nervous system can use to localize a sound source. Compare *intensity difference*. [6]

lateral In anatomy, toward one side. See Box 2.2. Compare *medial*. [2]

lateral geniculate nucleus (LGN) The part of the thalamus that receives information from the optic tract and sends it to visual areas in the occipital cortex. [7]

lateral hypothalamus (LH) A hypothalamic region involved in the control of appetite and other functions. See Figure 9.12. [9]

lateral inhibition The phenomenon by which interconnected neurons inhibit their neighbors, producing contrast at the edges of regions. See Figure 7.15. [7]

lateral intraparietal area (LIP) A region in the monkey parietal lobe, homologous to the human intraparietal sulcus, that is especially involved in voluntary, top-down control of attention. [14]

lateral sulcus See *Sylvian fissure*. [2]

lateral tegmental area A brainstem region that provides some of the norepinephrine-containing projections of the brain. [4]

lateral ventricle A complex C-shaped lateral portion of the ventricular system within each hemisphere of the brain. See Figures 2.15, 2.17. Compare *fourth ventricle* and *third ventricle*. [2]

lateralization The tendency for the right and left halves of a system to differ from one another.

LD$_{50}$ Lethal dose 50%; the dose of a drug at which half the treated animals will die. See Figure 4.6. Compare *ED$_{50}$*.

learned helplessness A learning paradigm in which individuals are subjected to inescapable, unpleasant conditions. [12]

learning The process of acquiring new and relatively enduring information, behavior patterns, or abilities, characterized by modifications of behavior as a result of practice, study, or experience. [13]

lens A structure in the eye that helps focus an image on the retina. See Figure 7.1. [7]

leptin A peptide hormone released by fat cells. [9]

lesion momentum The phenomenon in which the brain is impaired more by a lesion that develops quickly than by a lesion that develops slowly. [15]

levels of analysis The scope of experimental approaches. A scientist may try to understand behavior by monitoring molecules, nerve cells, brain regions, or social environments or using some combination of these levels of analysis. [1]

LGN See *lateral geniculate nucleus*. [7]

LH 1. See *lateral hypothalamus*. [9] 2. See *luteinizing hormone*. [8, ASF 8.3]

lie detector See *polygraph*. [11]

ligand A substance that binds to receptor molecules, such as a neurotransmitter or drug that binds postsynaptic receptors. [3, 4]

ligand-gated ion channel See *ionotropic receptor*. [4]

limbic system A loosely defined, widespread group of brain nuclei that innervate each other and form a network. These nuclei are implicated in emotions. See Figure 2.14. [2, 11]

LIP See *lateral intraparietal area*. [14]

lipid A large molecule (commonly called a *fat*) that consists of fatty acids and glycerol. Lipids are insoluble in water. [9]

liposuction The surgical removal of fat tissue. [9]

lithium An element that, administered to patients, often relieves the symptoms of bipolar disorder. [12]

lobotomy The surgical separation of a portion of the frontal lobes from the rest of the brain, once used as a treatment for schizophrenia and many other ailments. [12]

local anesthetic A drug, such as procaine or lidocaine, that blocks sodium channels to stop neural transmission in pain fibers.

local potential An electrical potential that is initiated by stimulation at a specific site, which is a graded response that spreads passively across the cell membrane, decreasing in strength with time and distance.[3]

localization of function The concept that different brain regions specialize in specific behaviors. [1]

locus coeruleus A small nucleus in the brainstem whose neurons produce norepinephrine and modulate large areas of the forebrain. Compare *substantia nigra*. [4, 10]

long-term memory (LTM) An enduring form of memory that lasts days, weeks, months, or years and has a very large capacity. Compare *sensory buffer* and *short-term memory*. See Figure 13.13. [13]

long-term potentiation (LTP) A stable and enduring increase in the effectiveness of synapses following repeated strong stimulation. See Figures 13.21, 13.22. [13]

lordosis A female receptive posture in four-legged animals in which the hindquarters are raised and the tail is turned to one side, facilitating intromission by the male. See Figures 8.17, 8.20. [8]

Lou Gehrig's disease See *amyotrophic lateral sclerosis*. [ASF 5.5]

LSD Also called *acid*. Lysergic acid diethylamide, a hallucinogenic drug. [4]

LTM See *long-term memory*. [13]

LTP See **long-term potentiation**. [13]

lumbar Referring to the five spinal segments that make up the upper part of the lower back. See Figures 2.8, 2.9. Compare *cervical*, *thoracic*, *sacral*, and *coccygeal*. [2]

luteinizing hormone (LH) A gonadotropin, named for its stimulatory effects on the ovarian corpora lutea. See Figure 8.13. [8, ASF 8.3]

lysergic acid diethylamide See *LSD*. [4]

M

M1 See *primary motor cortex*. [5]

magnetic resonance imaging (MRI) A noninvasive brain imaging technology that uses magnetism and radio-frequency energy to create images of the gross structure of the living brain. See Figures 1.5, 2.19. Compare *computerized axial tomography*. [2]

magnetoencephalography (MEG) A noninvasive brain-imaging technology that creates maps of brain activity during cognitive tasks by measuring the tiny magnetic fields produced by active neurons. Compare *transcranial magnetic stimulation*. [2]

malleus Latin for "hammer." A middle-ear bone that is connected to the tympanic membrane. It is one of the three ossicles that conduct sound across the middle ear. See Figure 6.1. Compare *incus* and *stapes*. [6]

mammillary body One of a pair of limbic system structures that are connected to the hippocampus. See Figure 2.14. [13]

MAO See *monoamine oxidase*. [4, ASF 4.1, 12]

maternal aggression Aggression of a mother defending her nest or offspring. [11]

MD See *muscular dystrophy*. [ASF 5.5]

MDMA Also called *Ecstasy*. 3,4-Methylenedioxymethamphetamine, a drug of abuse. [4]

medial In anatomy, toward the middle. See Box 2.2. Compare *lateral*. [2]

medial amygdala A portion of the amygdala that receives olfactory and pheromonal information. [8, 11]

medial forebrain bundle A collection of axons traveling in the midline region of the forebrain. See Figure 4.17. [11]

medial geniculate nucleus Either of two nuclei—left and right—in the thalamus that receive input from the inferior colliculi and send output to the auditory cortex. See Figure 6.6. [6]

medial preoptic area (mPOA) A region of the anterior hypothalamus implicated in the control of many behaviors, including sexual behavior, gonadotropin secretion, and thermoregulation. [8]

median eminence A midline feature on the base of the brain that marks the point at which the pituitary stalk exits the hypothalamus to connect to the pituitary. The median eminence contains elements of the hypothalamic-pituitary portal system. See Figure 8.12. [8]

medulla The posterior part of the hindbrain, continuous with the spinal cord. See Figures 2.12, 2.16. [2]

MEG See *magnetoencephalography*. [2]

Meissner's corpuscle Also called *tactile corpuscle*. A skin receptor cell type that detects light touch, responding especially to changes in stimuli. See Figures 5.3, 5.4. Compare *Merkel's disc*, *Pacinian corpuscle*, and *Ruffini corpuscle*. [5]

melanopsin A photopigment found in those retinal ganglion cells that project to the suprachiasmatic nucleus. See Figure 10.5. [10]

melatonin An amine hormone that is secreted by the pineal gland at night, thereby signaling day length to the brain. See Table 4.1. [10, ASF 8.1]

memory 1. The capability to learn and neurally encode information, consolidate the information for longer term storage, and retrieve or reactivate the consolidated memory at a later time. 2. The specific information that is stored in the brain. [13]

memory trace A persistent change in the brain that reflects the storage of memory. [13]

meninges The three protective membranes—dura mater, pia mater, and arachnoid—that surround the brain and spinal cord. See Figure 2.8. [2]

meningioma A noninvasive tumor of the meninges. [2]

meningitis An acute inflammation of the meninges, usually caused by a viral or bacterial infection. [2]

Merkel's disc A skin receptor cell type that detects light touch, responding especially to edges and isolated points on a surface. See Figures 5.3, 5.4. Compare *Meissner's corpuscle*, *Pacinian corpuscle*, and *Ruffini corpuscle*. [5]

message See *messenger RNA*. [App]

messenger RNA (mRNA) Also called *transcript* or *message*. A strand of RNA that carries the code of a section of a DNA strand to the cytoplasm. [App]

meta-analysis A type of quantitative review of a field of research, in which the results of multiple previous studies are combined in order to identify overall patterns that are consistent across studies. [12]

metabolic tolerance The form of drug tolerance that arises when repeated exposure to the drug causes the metabolic machinery of the body to become more efficient at clearing the drug. Compare *functional tolerance*. [4]

metabolism The breakdown of complex molecules into smaller molecules. [ASF 9.1]

metabotropic receptor A receptor protein that does not contain ion channels but may, when activated, use a second-messenger system to open nearby ion channels or to produce other cellular effects. See Figure 4.2. Compare *ionotropic receptor*. [4]

methylation A chemical modification of DNA that does not affect the nucleotide sequence of a gene but makes that gene less likely to be expressed. [13]

microelectrode An especially small electrode used to record electrical potentials in living cells. [3]

microglial cells Also called *microglia*. Extremely small, motile glial cells that remove cellular debris from injured or dead cells. [2]

midbrain The middle division of the brain. See Figures 2.12, 13.24. Compare *forebrain* and *hindbrain*. [2, 13]

middle canal See *scala media*. See Figure 6.1.

middle cerebral artery Either of two large arteries, arising from the carotid arteries, that provide blood to most of the forebrain. Compare *anterior cerebral artery* and *posterior cerebral artery*. [ASF 2.1]

middle ear The cavity between the tympanic membrane and the cochlea. See Figure 6.1. [6]

milk letdown reflex The reflexive release of milk by the mammary glands of a nursing female in response to suckling or to stimuli associated with suckling. See Figure 8.9. [8]

millivolt (mV) A thousandth of a volt. [3]

mirror neuron A neuron that is active both when an individual makes a particular movement and when that individual sees another individual make the same movement. [5]

mitochondrion (pl. mitochondria) A cellular organelle that provides metabolic energy for the cell's processes. See Figure 2.4. [2]

mitosis The process of division of somatic cells that involves duplication of DNA. [13]

monaural Pertaining to one ear. Compare *binaural*.

monoamine hormone See *amine hormone*. [8]

monoamine oxidase (MAO) An enzyme that breaks down monoamine neurotransmitters, thereby inactivating them. [4, 12, ASF 4.1]

monocular deprivation Depriving one eye of form vision. Compare *binocular deprivation*. [ASF 13.4]

monopolar neuron See *unipolar neuron*. [2]

monozygotic Referring to twins derived from a single fertilized egg (identical twins). Such individuals share an identical set of genes. Compare *dizygotic*. [12]

morpheme The smallest grammatical unit of a language; a word or meaningful part of a word. [15]

morphine An opiate compound derived from the poppy flower. [4]

motion sickness The experience of nausea brought on by unnatural passive movement, as may occur in a car or boat. [6]

motivation The psychological process that induces or sustains a particular behavior. [9]

motoneuron Also called *motor neuron*. A neuron that transmits neural messages to muscles (or glands). See Figure 5.19. Compare *interneuron* and *sensory neuron*. [2, 5]

motor nerve A nerve that transmits information from the central nervous system to the muscles and glands. Compare *sensory nerve*. [2]

motor neuron See *motoneuron*. [2]

motor plan Also called *motor program*. A plan for a series of muscular contractions, established in the nervous system prior to its execution. [5]

motor theory of language The theory that speech is perceived using the same left-hemisphere mechanisms as are used to produce the complex movements that go into speech. Compare *connectionist model of aphasia*. [15]

motor unit A single motor axon and all the muscle fibers that it innervates. [5]

movement A single relocation of a body part, usually resulting from a brief muscle contraction. It is less complex than an act. [5]

mPOA See *medial preoptic area*. [8]

MRI See *magnetic resonance imaging*. [2]

mRNA See *messenger RNA*. [App]

müllerian duct A duct system in the embryo that will develop into female reproductive structures (oviducts, uterus, and upper vagina) if androgens are not present. See Figure 8.24. Compare *wolffian duct*. [8]

müllerian regression hormone See *anti-müllerian hormone*. [8]

multiple sclerosis (MS) Literally "many scars." A disorder characterized by the widespread degeneration of myelin. [2, 3, ASF 13.3]

multipolar neuron A nerve cell that has many dendrites and a single axon. See Figure 2.3. Compare *bipolar neuron* and *unipolar neuron*. [2]

multisensory See *polymodal*. [ASF 6.3]

muscarinic Referring to cholinergic receptors that respond to the chemical muscarine as well as to acetylcholine. Muscarinic receptors mediate chiefly the inhibitory activities of acetylcholine. Compare *nicotinic*.

muscle fiber Large, cylindrical cells, making up most of a muscle, that can contract in response to neurotransmitter released from a motoneuron. See Figures 5.19, 5.20. See *intrafusal fiber*.

muscle spindle A muscle receptor that lies parallel to a muscle and sends impulses to the central nervous system when the muscle is stretched. See Figure 5.19. Compare *Golgi tendon organ*. [5]

muscular dystrophy (MD) A disease that leads to degeneration of and functional changes in muscles. [ASF 5.5]

musth An annual period of heightened aggressiveness and sexual activity in male elephants. [ASF 8.6]

mV See *millivolt*. [3]

myasthenia gravis A disorder characterized by a profound weakness of skeletal muscles. It is caused by a loss of acetylcholine receptors. [ASF 5.5]

myelin The fatty insulation around an axon, formed by glial cells. This sheath boosts the speed at which action potentials are conducted. See Figures 2.5, 3.8. [2, 3]

myelination The process by which myelin sheaths develop around axons. See Figure 2.5. [ASF 13.3]

myopia Nearsightedness; the inability to focus the retinal image of objects that are far away. [7]

N

N1 effect See *auditory N1 effect*. [14]

naloxone A potent antagonist of opiates that is often administered to people who have taken drug overdoses. It binds to receptors for endogenous opioids. [5]

narcolepsy A disorder that involves frequent, intense episodes of sleep, which last from 5 to 30 minutes and can occur anytime during the usual waking hours. [10]

NE See *norepinephrine*. [4, 11]

negative correlation A covariation of two measures in which one of the two usually goes up when the other goes down (and vice versa). Compare *positive correlation*. [1]

negative feedback The property by which some of the output of a system feeds back to reduce the effect of input signals. [8, 9]

negative symptom In psychiatry, an abnormality that reflects insufficient functioning. Examples include emotional and social withdrawal, and blunted affect. Compare *positive symptom*. [12]

neocortex Also called *isocortex* or simply *cortex*. Cerebral cortex that is made up of six distinct layers.

neonatal Referring to newborns. [8]

nerve A collection of axons bundled together outside of the central nervous system. See Figures 2.6, 2.7. Compare *tract*. [2]

nerve cell See *neuron*. [1, 2]

nerve fiber See *axon*. [2]

neural chain A simple kind of neural circuit in which neurons are attached linearly, end-to-end. [ASF 3.2]

neural groove In the developing embryo, the groove between the neural folds. See Figure 13.24.

neural plasticity See *neuroplasticity*. [1, 2, 13]

neural tube An embryonic structure with subdivisions that correspond to the future forebrain, midbrain, and hindbrain. The cavity of this tube will include the cerebral ventricles and the passages that connect them. See Figure 13.24. [2, 13]

neuroeconomics The study of brain mechanisms at work during economic decision making. [1, 14]

neuroendocrine cell A neuron that releases hormones into local or systemic circulation. [8]

neurofibrillary tangle An abnormal whorl of neurofilaments within nerve cells that is seen in Alzheimer's disease. See Figure 13.34. [13]

neurogenesis The mitotic division of non-neuronal cells to produce neurons. See Figure 13.25. [13]

neuroleptic See *antipsychotic*. [4, 12]

neuromuscular junction The region where the motoneuron terminal and the adjoining muscle fiber meet. It is the point where the nerve transmits its message to the muscle fiber. [5]

neuron Also called *nerve cell*. The basic unit of the nervous system, each composed of receptive extensions called *dendrites*, an integrating cell body, a conducting axon, and a transmitting axon terminal. See Figures 2.2, 2.3. [1, 2]

neuron doctrine The hypothesis that the brain is composed of separate cells that are distinct structurally, metabolically, and functionally.

neuropathic pain Pain that persists long after the injury that started it has healed. [5]

neuropeptide See peptide neurotransmitter. [4]

neurophysiology The study of the life processes of neurons. [3]

neuroplasticity Also called *neural plasticity*. The ability of the nervous system to change in response to experience or the environment. [1, 2, 13]

neuroscience The scientific study of the nervous system. [1]

neurotransmitter Also called simply *transmitter, synaptic transmitter,* or *chemical transmitter*. A signaling chemical, released by a presynaptic neuron, that diffuses across the synaptic cleft to alter the functioning of the postsynaptic

neuron. See Figure 3.11; Table 4.1. [2, 3, 4]

neurotransmitter receptor Also called simply *receptor*. A specialized protein, often embedded in the cell membrane, that selectively senses and reacts to molecules of a corresponding neurotransmitter or hormone. [2, 3, 4]

neurotrophic factor Also called simply *trophic factor*. A target-derived chemical that acts as if it "feeds" certain neurons to help them survive. See Figure 13.28. [13]

nicotine A compound found in plants, including tobacco, that acts as an agonist on a large class of cholinergic receptors. [4]

nicotinic Referring to cholinergic receptors that respond to nicotine as well as to acetylcholine. Nicotinic receptors mediate chiefly the excitatory activities of acetylcholine, including at the neuromuscular junction. Compare *muscarinic*.

night terror A sudden arousal from stage 3 sleep that is marked by intense fear and autonomic activation. Compare *nightmare*. [10]

nightmare A long, frightening dream that awakens the sleeper from REM sleep. Compare *night terror*. [10]

Nissl stain A tissue stain that outlines all cell bodies because the dyes are attracted to RNA, which encircles the nucleus. See Box 2.1. [2]

NMDA receptor A glutamate receptor that also binds the glutamate agonist NMDA (N-methyl-D-aspartate) and that is both ligand-gated and voltage-sensitive. Compare *AMPA receptor*. [13, ASF 4.2]

nociceptor A receptor that responds to stimuli that produce tissue damage or pose the threat of damage. [5]

nocturnal Active during the dark periods of the daily cycle. Compare *diurnal*. [10]

node of Ranvier A gap between successive segments of the myelin sheath where the axon membrane is exposed. See Figures 2.5, 3.8. [2, 3]

nonassociative learning A type of learning in which presentation of a particular stimulus alters the strength or probability of a response. It includes habituation and sensitization. See Figure 13.19. Compare *associative learning*. [13]

noncompetitive ligand A drug that affects a transmitter receptor while binding at a site other than that bound by the endogenous ligand.

nondeclarative memory Also called *procedural memory*. A memory that is shown by performance rather than by conscious recollection. See Figures 13.4, 13.12. Compare *declarative memory*. [13]

nonfluent aphasia Also called *Broca's aphasia*. A language impairment characterized by difficulty with speech production but not with language comprehension. It is related to damage in Broca's area. See Figure 15.8. Compare *fluent aphasia*. [15]

nonprimary motor cortex Frontal lobe regions adjacent to the primary motor cortex that contribute to motor control and modulate the activity of the primary motor cortex. See Figure 5.25. [5]

nonprimary sensory cortex Also called *secondary sensory cortex*. For a given sensory modality, the cortical regions receiving direct projections from primary sensory cortex for that modality. Compare *primary sensory cortex*. [5]

non-REM sleep Sleep, divided into stages 1–3, that is defined by the presence of distinctive EEG activity that differs from that seen in REM sleep. [10]

noradrenaline See *norepinephrine*. [4, 11]

noradrenergic Referring to cells using norepinephrine (noradrenaline) as a transmitter. [4]

norepinephrine (NE) Also called *noradrenaline*. A neurotransmitter that is produced and released by sympathetic postganglionic neurons to accelerate organ activity. It is also produced in the brainstem and found in projections throughout the brain. See Figure 4.4; Table 4.1. [4, 11]

Northern blot A method of detecting a particular RNA transcript in a tissue or organ by separating RNA from that source with gel electrophoresis, blotting the separated RNA molecules onto nitrocellulose, and then using a nucleotide probe to hybridize with, and highlight, the transcript of interest. Compare *Southern blot* and *Western blot*. [App]

NPY neuron A neuron, involved in the hypothalamic appetite control system, that produces both neuropeptide Y and agouti-related peptide. Compare *POMC neuron*. [9]

NST See *nucleus of the solitary tract*. [9]

nucleotide A portion of a DNA or RNA molecule that is composed of a single base and the adjoining sugar-phosphate unit of the strand. [App]

nucleus (pl. nuclei) 1. A collection of neuronal cell bodies within the central nervous system (e.g., the caudate nucleus). Compare *ganglion*. [2] 2. See *cell nucleus*. [App]

nucleus accumbens A region of the forebrain that receives dopaminergic innervation from the ventral tegmental area, often associated with reward and pleasurable sensations. [4, 11]

nucleus of the solitary tract (NST) A complicated brainstem nucleus that receives visceral and taste information via several cranial nerves. See Figure 9.15. [9]

nutrient A chemical that is needed for growth, maintenance, and repair of the body but is not used as a source of energy. [9]

O

obsessive-compulsive disorder (OCD) An anxiety disorder in which the affected individual experiences recurrent unwanted thoughts and engages in repetitive behaviors without reason or the ability to stop. [12]

occipital cortex Also called *visual cortex*. Cortex of the occipital lobe of the brain, corresponding to the primary visual area of the cortex. See Figure 7.10. [7]

occipital lobe A large region of cortex that covers much of the posterior part of each cerebral hemisphere. See Figure 2.10. Compare *frontal lobe, parietal lobe,* and *temporal lobe.* [2]

OCD See *obsessive-compulsive disorder.* [12]

ocular dominance histogram A graph that plots how strongly a brain neuron responds to stimuli presented to either the left eye or the right eye. Ocular dominance histograms are used to determine the effects of manipulating visual experience. [ASF 13.4]

odor The sensation of smell. [6]

off-center bipolar cell A retinal bipolar cell that is inhibited by light in the center of its receptive field. See Figures 7.13, 7.14. Compare *on-center bipolar cell.* [7]

off-center ganglion cell A retinal ganglion cell that is activated when light is presented to the periphery, rather than the center, of the cell's receptive field. See Figures 7.13, 7.14. Compare *on-center ganglion cell.* [7]

off-center/on-surround Referring to a concentric receptive field in which stimulation of the center inhibits the cell of interest while stimulation of the surround excites it. See Figure 7.14. Compare *on-center/off-surround.* [7]

OHC See *outer hair cell.* [6]

olfaction The sensory system that detects smell; the act of smelling. [6]

olfactory bulb An anterior projection of the brain that terminates in the upper nasal passages and, through small openings in the skull, provides receptors for smell. See Figures 2.10, 2.16, 6.19. [2, 6]

olfactory epithelium (pl. epithelia) A sheet of cells, including olfactory receptors, that lines the dorsal portion of the nasal cavities and adjacent regions. See Figures 6.18, 6.19. [6]

oligodendrocyte A type of glial cell that forms myelin in the central nervous system. See Figure 2.5. Compare *Schwann cell.* [2]

on-center bipolar cell A retinal bipolar cell that is excited by light in the center of its receptive field. See Figures 7.13, 7.14. Compare *off-center bipolar cell.* [7]

on-center ganglion cell A retinal ganglion cell that is activated when light is presented to the center, rather than the periphery, of the cell's receptive field. See Figures 7.13, 7.14. Compare *off-center ganglion cell.* [7]

on-center/off-surround Referring to a concentric receptive field in which stimulation of the center excites the cell of interest while stimulation of the surround inhibits it. See Figure 7.14. Compare *off-center/on-surround.* [7]

ontogeny The process by which an individual changes in the course of its lifetime—that is, grows up and grows old. [1]

Onuf's nucleus The human homolog of the spinal nucleus of the bulbocavernosus (SNB) in rats. [8]

operant conditioning See *instrumental conditioning.* [13]

opiate Any of a class of compounds that exert an effect like that of opium, including reduced pain sensitivity. [4]

opioid peptide A type of endogenous peptide that mimics the effects of morphine in binding to opioid receptors and producing marked analgesia and reward. See Table 4.1. [4]

opioid receptor A receptor that responds to endogenous opioids and/or exogenous opiates. [4]

opium An extract of the seedpod juice of the opium poppy, *Papaver somniferum.* Drugs based on opium are potent painkillers. [4]

opponent-process hypothesis A hypothesis of color perception stating that different systems produce opposite responses to light of different wavelengths. See Figures 7.23, 7.24. [7]

optic ataxia Spatial disorientation in which the patient is unable to accurately reach for objects using visual guidance. [7]

optic chiasm The point at which parts of the two optic nerves cross the midline. See Figure 7.10. [7]

optic disc The region of the retina that is devoid of receptor cells because ganglion cell axons and blood vessels exit the eyeball there. See Figure 7.7. [7]

optic nerve Cranial nerve II; the collection of ganglion cell axons that extend from the retina to the brain. See Figures 2.7, 7.10. [7]

optic radiation Axons from the lateral geniculate nucleus that terminate in the primary visual areas of the occipital cortex. See Figure 7.10. [7]

optic tract The axons of retinal ganglion cells after they have passed the optic chiasm. Most of these axons terminate in the lateral geniculate nucleus. See Figure 7.10. [7]

optogenetics The use of lasers to excite or inhibit neurons expressing light-sensitive membrane channels, typically in transgenic mice. [11]

oral contraceptive A birth control pill, typically consisting of steroid hormones to prevent ovulation. [8]

orexin Also called *hypocretin.* A neuropeptide produced in the hypothalamus that is involved in switching between sleep states, in narcolepsy, and in the control of appetite. [9, 10]

organ of Corti A structure in the inner ear that lies on the basilar membrane of the cochlea and contains the hair cells and terminations of the auditory nerve. See Figure 6.1. [6]

organizational effect A permanent alteration of the nervous system, and thus permanent change in behavior, resulting from the action of a steroid hormone on an animal early in its development. Compare *activational effect.* [8]

orgasm The climax of sexual behavior, marked by extremely pleasurable sensations. [8]

osmosensory neuron A specialized neuron that monitors the concentration of the extracellular fluid by measuring the movement of water into and out of the intracellular compartment. See Figures 9.7, 9.9. [9]

osmosis The passive movement of a solvent, usually water, through a semipermeable membrane until a uniform concentration of solute (often salt) is achieved on both sides of the membrane. See Figure 9.6. Compare *diffusion.* [9]

osmotic pressure The tendency of a solvent to move across a membrane in order to equalize the concentration of solute on both sides of the membrane. [9]

osmotic thirst A desire to ingest fluids that is stimulated by a high concentration of solute (like salt) in the extracellular compartment. Compare *hypovolemic thirst.* [9]

ossicles Three small bones (incus, malleus, and stapes) that transmit vibrations across the middle ear, from the tympanic membrane to the oval window. See Figure 6.1. [6]

otolith A small crystal on the gelatinous membrane in the vestibular system.

outer hair cell (OHC) One of the two types of receptor cells for hearing in the cochlea. Compared with inner hair cells, OHCs are positioned farther from the central axis of the coiled cochlea. See Figure 6.1. [6]

output zone The part of a neuron at which the cell sends information to another cell. This zone usually corresponds to the axon terminals. See Figures 2.1, 2.3. Compare *conduction zone*, *input zone*, and *integration zone*. [2]

oval window The opening from the middle ear to the inner ear. See Figure 6.1. [6]

ovaries The female gonads, which produce eggs (ova) for reproduction. See Figure 8.1. Compare *testes*. [8]

overt attention Attention in which the focus coincides with sensory orientation (e.g., you're attending to the same thing you're looking at). Compare *covert attention*. [14]

ovulation The production and release of an egg (ovum). [8]

ovulatory cycle The periodic occurrence of ovulation in females. See Figure 8.19. [8]

ovum (pl. ova) An egg, the female gamete. Compare *sperm*. [8]

oxytocin A hormone, released from the posterior pituitary, that triggers milk letdown in the nursing female but is also associated with a variety of complex behaviors. See Figures 8.8, 8.9. [8]

P

P1 effect See *visual P1 effect*. [14]

P3 effect Also called *auditory P300*. A positive deflection of the event-related potential, occurring about 300 milliseconds after stimulus presentation, that is associated with higher-order auditory stimulus processing and late attentional selection. [14]

P20–50 effect A positive deflection of the event-related potential, occurring about 20–50 milliseconds after stimulus presentation, that is enhanced for selectively attended auditory input compared with ignored input. [14]

Pacinian corpuscle Also called *lamellated corpuscle*. A skin receptor cell type that detects vibration and pressure. See Figures 5.3, 5.4. Compare *Meissner's corpuscle*, *Merkel's disc*, and *Ruffini corpuscle*. [5]

pain The discomfort normally associated with tissue damage. [5]

pair-bond A durable and exclusive relationship between two individuals. [8]

Papez circuit A group of brain regions within the limbic system.

papilla (pl. papillae) A small bump that projects from the surface of the tongue. Papillae contain most of the taste receptor cells. See Figures 6.15, 6.16. [6]

parabiotic Referring to a surgical preparation that joins two animals to share a single blood supply. [8]

paradoxical sleep See *rapid-eye-movement sleep (REM)* sleep. [10]

paraphasia A symptom of aphasia that is distinguished by the substitution of a word by a sound, an incorrect word, an unintended word, or a neologism (a meaningless word). [15]

parasympathetic nervous system The part of the autonomic nervous system that generally prepares the body to relax and recuperate. See Figure 2.9. Compare *sympathetic nervous system*. [2, 11]

parental behavior Behavior of adult animals that has the goal of enhancing the well-being of their own offspring, often at some cost to the parents. [8]

paresis Muscular weakness, often the result of damage to motor cortex. Compare *plegia*. [5]

parietal lobe The large region of cortex lying between the frontal and occipital lobes in each cerebral hemisphere. See Figure 2.10. Compare also *temporal lobe*. [2]

parkin A protein that has been implicated in Parkinson's disease. [ASF 5.5]

Parkinson's disease A degenerative neurological disorder, characterized by tremors at rest, muscular rigidity, and reduction in voluntary movement, caused by loss of the dopaminergic neurons of the substantia nigra. Compare *Huntington's disease*. [5]

partial agonist A drug that, when bound to a receptor, has less effect than the endogenous ligand would. The term *partial antagonist* is equivalent. [4]

patient H.M. The late Henry Molaison, a patient who was unable to encode new declarative memories because of surgical removal of medial temporal lobe structures.. See Figure 13.1. [13]

patient K.C. The late Kent Cochrane, a patient who sustained damage to cortex that rendered him unable to form and retrieve episodic memories. [13]

patient N.A. A still-living patient who is unable to encode new declarative memories, because of damage to the dorsomedial thalamus and the mammillary bodies. [13]

Pavlovian conditioning See *classical conditioning*. [13]

PCP See *phencyclidine*. [12]

PCR See *polymerase chain reaction*. [App]

penis The male phallus. Compare *clitoris*. [8]

peptide A short string of amino acids. Longer strings of amino acids are called *proteins*. [App]

peptide hormone Also called *protein hormone*. A hormone that consists of a string of amino acids. [8]

peptide neurotransmitter Also called *neuropeptide*. A neurotransmitter consisting of a short chain of amino acids. See Table 4.1. Compare *amine neurotransmitter*, *amino acid neurotransmitter*, and *gas neurotransmitter*. [4]

perceptual load The immediate processing demands presented by a stimulus. [14]

periaqueductal gray A midbrain region involved in pain perception. [2, 4, 8]

period The interval of time between two similar points of successive cycles, such as sunset to sunset. [10]

peripheral nervous system The portion of the nervous system that includes all the nerves and neurons outside the brain and spinal cord. See Figures 2.6, 2.12. Compare *central nervous system*. [2]

peripheral spatial cuing A technique for testing exogenous attention in which a visual stimulus is preceded by a simple task-irrelevant sensory stimulus either in the location where the stimulus will appear or in an incorrect location. Compare *symbolic cuing*. [14]

perseverate To continue to show a behavior repeatedly. [14, ASF 13.4]

PET See positron emission tomography. [2]

petit mal seizure Also called *absence attack*. A seizure that is characterized by a spike-and-wave EEG and often involves a loss of awareness and inability to recall events surrounding the seizure. See Figure 3.16. Compare *grand mal seizure*. [3]

phagocyte An immune system cell that engulfs invading molecules or microbes. [ASF 11.4]

phallus The clitoris or penis. [8]

pharmacokinetics Collective name for all the factors that affect the movement of a drug into, through, and out of the body. [4]

phase shift A shift in the activity of a biological rhythm, typically provided by a synchronizing environmental stimulus. [10]

phasic receptor A receptor in which the frequency of action potentials drops rapidly as stimulation is maintained. Compare *tonic receptor*. [5]

phencyclidine (PCP) Also called *angel dust*. An anesthetic agent that is also a psychedelic drug. PCP makes many people feel dissociated from themselves and their environment. [12]

phenotype The sum of an individual's physical characteristics at one particular time. Compare *genotype*. [13]

phenylketonuria (PKU) An inherited disorder of protein metabolism in which the absence of an enzyme leads to a toxic buildup of certain compounds, causing intellectual disability. [13]

pheromone A chemical signal that is released outside the body of an animal and affects other members of the same species. See Figure 8.3. Compare *allomone*. [6, 8]

phoneme A sound that is produced for language. [15]

photopic system A system in the retina that operates at high levels of light, shows sensitivity to color, and involves the cones. See Table 7.1. Compare *scotopic system*. [7]

photoreceptor A neural cell in the retina that responds to light. [7]

photoreceptor adaptation The tendency of rods and cones to adjust their light sensitivity to match current levels of illumination. [7]

phrenology The belief that bumps on the skull reflect enlargements of brain regions responsible for certain behavioral faculties. See Figure 1.4. [1]

physiological psychology See *biological psychology*. [1]

pia mater The innermost of the three meninges that surround the brain and spinal cord. See also *dura mater* and *arachnoid*. See Figure 2.8. [2]

pineal gland A secretory gland in the brain midline that is the source of melatonin release. See Figure 8.1. [8, ASF 8.1]

pinna (pl. pinnae) The external part of the ear. [6]

pitch A dimension of auditory experience in which sounds vary from low to high. See Box 6.1. **[6]**

pituitary gland A small, complex endocrine gland located in a socket at the base of the skull. See Figures 2.16, 8.8, 8.12. [8]

pituitary stalk Also called *infundibulum*. A thin piece of tissue that connects the pituitary gland to the hypothalamus. [8]

PKU See *phenylketonuria*. [13]

place cell A neuron in the hippocampus that selectively fires when the animal is in a particular location. [13]

place coding Frequency discrimination in which the pitch of a sound is determined by the location of activated hair cells along the length of the basilar membrane. Compare *temporal coding*. [6]

placebo effect Relief of a symptom, such as pain, that results following a treatment that is known to be ineffective or inert. [5]

planum temporale An auditory region of superior temporal cortex. See Figure 15.3. [15]

plegia Paralysis, the loss of the ability to move. Compare *paresis*. [5]

polarized Exhibiting a difference in electrical charge between the inside and outside of the cell. [3]

poliovirus A virus that destroys motoneurons of the spinal cord and brainstem, causing permanent paralysis. [ASF 5.5]

polygraph Popularly known as a *lie detector*. A device that measures several bodily responses, such as heart rate and blood pressure. See Box 11.1. [11]

polymerase chain reaction (PCR) Also called *gene amplification*. A method for reproducing a particular RNA or DNA sequence manyfold, allowing amplification for sequencing or manipulating the sequence. [App]

polymodal Also called *multisensory*. Involving several sensory modalities. [ASF 6.3]

polymodal neuron A neuron upon which information from more than one sensory system converges. [5]

POMC neuron A neuron, involved in the hypothalamic appetite control system, that produces both pro-opiomelanocortin and cocaine- and amphetamine-related transcript. Compare *NPY neuron*. [9]

pons The portion of the brainstem that connects the midbrain to the medulla. See Figures 2.12, 2.16. [2]

positive correlation A covariation of two measures in which they both usually increase together, or decrease together. Compare *negative correlation*. [1]

positive symptom In psychiatry, an abnormal behavioral state. Examples include hallucinations, delusions, and excited motor behavior. Compare *negative symptom*. [12]

positron emission tomography (PET) A brain imaging technology that tracks the metabolism of injected radioactive substances in the brain, in order to map brain activity.. See Figure 2.19. Compare *functional MRI*. [2]

postcentral gyrus The strip of parietal cortex, just behind the central sulcus, that receives somatosensory information from the entire body. See Figure 2.10. Compare *precentral gyrus*. [2]

postcopulatory behavior The final stage in mating behavior. Species-specific postcopulatory behaviors include rolling (in the cat) and grooming (in the rat). See Figure 8.16. [8]

posterior Also called *caudal*. In anatomy, toward the tail end of an organism. See Box 2.2. Compare *anterior*. [2]

posterior cerebral artery Either of two large arteries, arising from the basilar artery, that provide blood to posterior aspects of the cerebral hemispheres, cerebellum, and brainstem. Compare *anterior cerebral artery* and *middle cerebral artery*. [ASF 2.1]

posterior pituitary The rear division of the pituitary gland. See Figures 8.1, 8.8. Compare *anterior pituitary*. [8]

postpartum depression A bout of depression that afflicts a woman either immediately before or after giving birth. [12]

postsynaptic Referring to the region of a synapse that receives and responds to neurotransmitter. See Figure 2.4. Compare *presynaptic*. [2, 3, 4]

postsynaptic membrane The specialized membrane on the surface of a neuron that receives information by responding to neurotransmitter from a presynaptic neuron. See Figure 2.4. Compare *presynaptic membrane*. [2]

postsynaptic potential A local potential that is initiated by stimulation at a synapse, which can vary in amplitude, and spreads passively across the cell membrane, decreasing in strength with time and distance. Compare *all-or-none property*. [3]

posttraumatic stress disorder (PTSD) A disorder in which memories of an unpleasant episode repeatedly plague the victim. See Box 13.1. [12, 13]

potassium ion (K⁺) A potassium atom that carries a positive charge. [3]

precentral gyrus The strip of frontal cortex, just in front of the central sulcus, that is crucial for motor control. See Figure 2.10. Compare *postcentral gyrus*. [2, 5]

prefrontal cortex The anteriormost region of the frontal lobe. [14]

premotor cortex A region of nonprimary motor cortex just anterior to the primary motor cortex. See Figure 5.25. [5]

presenilin An enzyme that cleaves amyloid precursor protein, forming beta-amyloid, which can lead to Alzheimer's disease. See *beta-secretase*. [ASF 13.5]

presynaptic Located on the "transmitting" side of a synapse. See Figure 2.4. Compare *postsynaptic*. [2, 3, 4]

presynaptic membrane The specialized membrane on the axon terminal of a nerve cell that transmits information by releasing neurotransmitter. See Figure 2.4. [2]

primacy effect The superior performance seen in a memory task for items at the start of a list. It is usually attributed to long-term memory. See Figure 13.14. Compare *recency effect*. [13]

primary auditory cortex Also called *A1*. The cortical region, located on the superior surface of the temporal lobe, that processes complex sounds transmitted from lower auditory pathways. [6]

primary motor cortex (M1) The apparent executive region for the initiation of movement. It is primarily the precentral gyrus. Compare *nonprimary motor cortex*. [5]

primary sensory cortex For a given sensory modality, the region of cortex that receives most of the information about that modality from the thalamus (or, in the case of olfaction, directly from the secondary sensory neurons). Compare *nonprimary sensory cortex*. [5]

primary somatosensory cortex Also called *somatosensory 1* or *S1*. Primarily the post-central gyrus of the parietal lobe, , where sensory inputs from the body surface are mapped. See Figures 5.7, 5.9. [5]

primary visual cortex (V1) Also called *striate cortex* or *area 17*. The region of the occipital cortex where most visual information first arrives. See Figures 7.10, 7.11, 7.20. [7]

priming Also called *repetition priming*. In memory, the phenomenon by which exposure to a stimulus facilitates subsequent responses to the same or a similar stimulus. [13]

probe In molecular biology, a manufactured sequence of DNA that is made to include a label (a colorful or radioactive molecule) that lets us track its location. [App]

procedural memory See *nondeclarative memory*. [13]

proceptive Referring to a state in which a female advertises its readiness to mate through species-typical behaviors. [8]

progesterone The primary type of progestin secreted by the ovary. See Figure 8.13. [8]

progestin Any of a major class of steroid hormones that are produced by the ovary, including progesterone. See Figure 8.13. [8]

progressive supranuclear palsy (PSP) A rare, degenerative disease of the brain that begins with marked, persistent visual symptoms and leads to more widespread intellectual deterioration. [14]

prolactin A protein hormone, produced by the anterior pituitary, that promotes mammary development for lactation in female mammals. [ASF 8.3]

proprioception Body sense; information about the position and movement of the body. [5]

prosody The perception of emotional tone-of-voice aspects of language. [15]

prosopagnosia Also called *face blindness*. A condition characterized by the inability to recognize faces. [15]

protein A long string of amino acids. Proteins are the basic building material of organisms. Compare *peptide*. [App]

protein hormone See *peptide hormone*.

proximal In anatomy, near the trunk or center of an organism. See Box 2.2. Compare *distal*. [2]

PSP See *progressive supranuclear palsy*. [14]

psychoneuroimmunology The study of the immune system and its interaction with the nervous system and behavior. [11]

psychopath An individual incapable of experiencing remorse. [11]

psychosocial dwarfism A syndrome of stunted growth in children subjected to social stress, such as abusive caregivers. See Figure 8.36. [8]

psychosomatic medicine A field of study that emphasizes the role of psychological factors in disease. [11]

psychosurgery Surgery in which brain lesions are produced to modify severe psychiatric disorders.

psychotomimetic A drug that induces a state resembling schizophrenia. [12]

PTSD See *posttraumatic stress disorder*. [12, 13]

pulvinar In humans, the posterior portion of the thalamus. It is heavily involved in visual processing and direction of attention. [14]

punch-drunk syndrome See *chronic traumatic encephalopathy*. [15]

pupil The opening, formed by the iris, that allows light to enter the eye. See Figures 7.1, 7.6. [7]

pure tone A tone with a single frequency of vibration. See Box 6.1. [6]

Purkinje cell A type of large nerve cell in the cerebellar cortex.

putamen One of the basal ganglia. See Figure 2.14. [2]

pyramidal cell A type of large nerve cell that has a roughly pyramid-shaped cell body and is found in the cerebral cortex. See Figure 2.13. [2]

pyramidal system Also called *corticospinal system*. The motor system that includes neurons within the cerebral cortex and their axons, which form the pyramidal tract. See Figure 5.22. Compare *extrapyramidal system*. [5]

PYY$_{3-36}$ A peptide hormone, secreted by the intestines, that probably acts on hypothalamic appetite control mechanisms to suppress appetite. Compare *ghrelin*. [9]

Q

quale (pl. qualia) A purely subjective experience of perception. [14]

R

range fractionation The means by which sensory systems cover a wide range of intensity values, as each sensory receptor cell specializes in just one part of the overall range of intensities. [7]

raphe nuclei A string of nuclei in the midline of the midbrain and brainstem that contain most of the serotonergic neurons of the brain. [4]

rapid-eye-movement (REM) sleep Also called *paradoxical sleep*. A stage of sleep characterized by small-amplitude, fast-EEG waves, flaccid muscles, and rapid eye movements. *REM* rhymes with "gem." See Figure 10.9. Compare *stage 3 sleep*. [10]

RBD See *REM behavior disorder*. [10]

recency effect The superior performance seen in a memory task for items at the end of a list. It is usually attributed to short-term memory. See Figure 13.14. Compare *primacy effect*. [13]

receptive field The stimulus region and features that affect the activity of a cell in a sensory system. See Figures 5.5, 7.14, 7.17. [5, 7]

receptor See *neurotransmitter receptor*. [2, 3, 4]

receptor cell A specialized cell that responds to a particular energy or substance in the internal or external environment and converts this energy into a change in the electrical potential across its membrane. [5]

receptor subtype Any type of receptor having functional characteristics that distinguish it from other types of receptors for the same neurotransmitter. For example, at least 15 different subtypes of receptor molecules respond to serotonin.

reconsolidation The return of a memory trace to stable long-term storage after it has been temporarily made changeable during the process of recall. [13]

recovery of function The recovery of behavioral capacity following brain damage from stroke or injury. [15]

reductionism The scientific strategy of breaking a system down into increasingly smaller parts in order to understand it. [1]

reflex A simple, highly stereotyped, and unlearned response to a particular stimulus (e.g., an eye blink in response to a puff of air). See Figures 3.14, 5.21. [5]

reflexive attention See *exogenous attention*. [14]

refraction The bending of light rays by a change in the density of a medium, such as the cornea and the lens of the eyes. [7]

refractory Temporarily unresponsive or inactivated. [3]

refractory phase 1. A period during and after an action potential in which the responsiveness of the axonal membrane is reduced. A brief period of complete insensitivity to stimuli (absolute refractory phase) is followed by a longer period of reduced sensitivity (relative refractory phase) during which only strong stimulation produces an action potential. [3] 2. A period following copulation during which an individual does not recommence copulation. See Figure 8.23. [8]

relative refractory phase A period of reduced sensitivity during which only strong stimulation produces an action potential. Compare *absolute refractory phase*. [3]

releasing hormone Any of a class of hormones, produced in the hypothalamus, that traverse the hypothalamic-pituitary portal system to control the pituitary's release of tropic hormones. See Figure 8.12. [8]

REM behavior disorder (RBD) A sleep disorder in which a person physically acts out a dream. [10]

REM sleep See *rapid-eye-movement (REM) sleep*. [10]

repetition priming See *priming*. [13]

repetitive transcranial magnetic stimulation (rTMS) A noninvasive treatment in which repeated pulses of focused magnetic energy are used to stimulate the cortex through the scalp. [12]

resting potential The difference in electrical potential across the membrane of a nerve cell at rest. See Figures 3.1, 3.5. [3]

reticular formation Also called *reticular activating system*. An extensive region of the brainstem (extending from the medulla through the thalamus) that is involved in arousal (waking). See Figure 10.18. [2, 10]

retina The receptive surface inside the eye that contains photoreceptors and other neurons. See Figures 7.1, 7.3. [7]

retinohypothalamic pathway The route by which retinal ganglion cells send their axons to the suprachiasmatic nuclei. [10]

retrieval The third process of the memory system, in which a stored memory is used by an organism. See Figure 13.13. Compare *encoding* and *consolidation*. [13]

retrograde amnesia Difficulty in retrieving memories formed before the onset of amnesia. Compare *anterograde amnesia*. [13]

retrograde transmitter A neurotransmitter that is released by the postsynaptic region, diffuses back across the synapse, and alters the functioning of the presynaptic neuron. [4, 13]

reuptake The process by which released synaptic transmitter molecules are taken up and reused by the presynaptic neuron, thus stopping synaptic activity. [3, 4]

rhodopsin The photopigment in rods that responds to light. [7]

ribonucleic acid (RNA) A nucleic acid that implements information found in DNA. Compare *deoxyribonucleic acid*. [App]

ribosome An organelle in the cell body where genetic information is translated to produce proteins. See Appendix Figure A.2. [App]

RNA See **ribonucleic acid (RNA)**. [App]

rod A photoreceptor cell in the retina that is most active at low levels of light. See Figure 7.3. Compare *cone*. [7]

root Either of two distinct branches of a spinal nerve, each of which serves a separate function. The dorsal root enters the dorsal horn of the spinal cord and carries sensory information, the ventral root arises from the ventral horn of the spinal cord and carries motor messages. See Figure 2.8. [2]

rostral See *anterior*. [2]

round window A membrane separating the tympanic canal from the middle ear. See Figure 6.1. [6]

Ruffini corpuscle A skin receptor cell type that detects stretching of the skin. See Figure 5.3. Compare *Meissner's corpuscle*, *Merkel's disc*, and *Pacinian corpuscle*. [5]

S

S1 See *primary somatosensory cortex*. [5]

sacral Referring to the five spinal segments that make up the lower part of the lower back. See Figures 2.8, 2.9. Compare *cervical*, *thoracic*, *lumbar*, and *coccygeal*. [2]

SAD See *seasonal affective disorder*. [ASF 12.3]

sagittal plane The plane that divides the body or brain into right and left portions. See Box 2.2. Compare *coronal plane* and *horizontal plane*. [2]

saltatory conduction The form of conduction that is characteristic of myelinated axons, in which the action potential jumps from one node of Ranvier to the next. [3]

satiety A feeling of fulfillment or satisfaction. Compare *hunger*. [9]

saturation One of three basic dimensions of light perception, varying from rich to pale. Compare *brightness* and *hue*. [7]

SC See *standard condition*. [13]

scala media Also called *middle canal*. The central of the three spiraling canals inside the cochlea, situated between the *scala vestibuli* and the *scala tympani*. See Figure 6.1. [6]

scala tympani Also called *tympanic canal*. One of three principal canals running along the length of the cochlea. Compare *scala media* and *scala vestibuli*. See Figure 6.1. [6]

scala vestibuli Also called *vestibular canal*. One of three principal canals running along the length of the cochlea. Compare *scala media* and *scala tympani*. See Figure 6.1. [6]

schizophrenia A severe psychopathological disorder characterized by negative symptoms such as emotional withdrawal and flat affect and by positive symptoms such as hallucinations and delusions. [12]

Schwann cell A type of glial cell that forms myelin in the peripheral nervous system. Compare *oligodendrocyte*. [2]

SCN See *suprachiasmatic nucleus*. [10]

scotoma A region of blindness within the visual fields, caused by injury to the visual pathway or brain. [7]

scotopic system A system in the retina that operates at low levels of light and involves the rods. See Table 7.1. Compare *photopic system*. [7]

SDN-POA See *sexually dimorphic nucleus of the preoptic area*. [8]

seasonal affective disorder (SAD) A depression putatively brought about by the short days of winter. [ASF 12.3]

second messenger A slow-acting substance in a target cell that amplifies the effects of synaptic or hormonal activity and regulates activity within the target cell. [8]

secondary sensory cortex See *nonprimary sensory cortex*. [5]

seizure A wave of abnormally synchronous electrical activity in the brain. See Figure 3.16. [3]

selective attention See *attention*. [14]

selective permeability The property of a membrane that allows some substances to pass through, but not others. [3]

selective serotonin reuptake inhibitor (SSRI) An antidepressant drug that blocks the reuptake of transmitter at serotonergic synapses. [4, 12]

semantic memory Generalized declarative memory, such as knowing the meaning of a word. Compare *episodic memory*. [13]

semantics The meanings or interpretation of words and sentences in a language.

semen A mixture of fluid and sperm that is released during ejaculation. [8]

semicircular canal Any one of the three fluid-filled tubes in the inner ear that are part of the vestibular system. Each of the tubes, which are at right angles to each other, detects angular acceleration in a particular direction. See Figure 6.14. [6]

senile dementia A neurological disorder of the aged that is characterized by progressive behavioral deterioration, including personality change and profound intellectual decline. It includes, but is not limited to, Alzheimer's disease.

senile plaque See *amyloid plaque*. [13]

sensitive period Also called *critical period*. The period during development in which an organism can be permanently altered by a particular experience or treatment. [8, 15, ASF 13.4]

sensorineural deafness A hearing impairment most often caused by the permanent damage or destruction of hair cells, or by interruption of the vestibulocochlear nerve that carries auditory information to the brain. See Figure 6.11. Compare *central deafness* and *conduction deafness*. [6]

sensory adaptation The progressive loss of receptor response as stimulation is maintained. See Figure 5.6. [5]

sensory buffer A very brief type of memory that stores the sensory impression of a scene. In vision, it is sometimes called *iconic memory*. See Figure 13.13. [13]

sensory nerve A nerve that conveys information from the body to the central nervous system. Compare *motor nerve*. [2]

sensory neuron A nerve cell that is directly affected by changes in the environment, such as light, odor, or touch. Compare *interneuron* and *motoneuron*. [2]

sensory transduction The process in which a receptor cell converts the energy in a stimulus into a change in the electrical potential across its membrane. [5]

serotonergic Referring to cells that use serotonin as their synaptic transmitter. [4]

serotonin (5-HT) A synaptic transmitter that is produced in the raphe nuclei and is active in structures throughout the central nervous system. See Figure 4.4; Table 4.1. [4, 11]

set point The point of reference in a feedback system. An example is the temperature at which a thermostat is set. Compare *set zone*. [9]

set zone The optimal range of a variable that a feedback system tries to maintain. Compare *set point*. [9]

sex determination The process that normally establishes whether a fetus will develop as a male or a female. [8]

sexual attraction The first step in the mating behavior of many animals, in which animals emit stimuli that attract members of the opposite sex. See Figure 8.16. [8]

sexual differentiation The process by which individuals develop either male-like or femalelike bodies and behavior. See Figure 8.24. [8]

sexual dimorphism The condition in which males and females of the same species show pronounced sex differences in appearance. [8]

sexually dimorphic nucleus of the preoptic area (SDN-POA) A region of the preoptic area that is 5 to 6 times larger in volume in male than in females rats. See Figure 8.30. [8]

sexually receptive Referring to the state in which an individual (in mammals, typically the female) is willing to copulate. [8]

shadowing A task in which the subject is asked to focus attention on one ear or the other while stimuli are being presented separately to both ears, and to repeat aloud the material presented to the attended ear. [14]

sham rage See *decorticate rage*. [11]

shivering Rapid involuntary muscle contractions that generate heat in hypothermic animals. [ASF 9.1]

short-term memory (STM) Also called *working memory*. A form of memory that usually lasts only seconds, or as long as rehearsal continues. See Figure 13.13. Compare *sensory buffer* and *long-term memory*. [13]

SIDS See *sudden infant death syndrome*. [10]

simple cortical cell Also called *bar detector* or *edge detector*. A cell in the visual cortex that responds best to an edge or a bar that has a particular width, as well as a particular orientation and location in the visual field. Compare *complex cortical cell*. [7]

simultagnosia A profound restriction of attention, often limited to a single item or feature. Simultagnosia is one of the three primary symptoms of Balint's syndrome. [14]

skeletal muscle A muscle that is used for movement of the skeleton, typically under our conscious control. [5]

skill learning The process of learning to perform a challenging task simply by repeating it over and over. [13]

sleep apnea A sleep disorder in which respiration slows or stops periodically, waking the patient. Excessive daytime sleepiness results from the frequent nocturnal awakening. [10]

sleep cycle A period of slow-wave sleep followed by a period of REM sleep. In humans, a sleep cycle lasts 90–110 minutes.

sleep deprivation The partial or total prevention of sleep. [10]

sleep enuresis Bed-wetting. [10]

sleep-maintenance insomnia Difficulty in staying asleep. Compare *sleep-onset insomnia*. [10]

sleep-onset insomnia Difficulty in falling sleep. Compare *sleep-maintenance insomnia*. [10]

sleep paralysis A state, during the transition to or from sleep, in which the ability to move or talk is temporarily lost. [10]

sleep recovery The process of sleeping more than normally after a period of sleep deprivation, as though in compensation. [10]

sleep spindle A characteristic 14- to 18-hertz wave in the EEG of a person said to be in stage 2 sleep. See Figure 10.9. [10]

sleep state misperception Commonly, a person's perception that he has not been asleep when in fact he has. It typically occurs at the start of a sleep episode. [10]

slow wave sleep (SWS) Also called *non-REM sleep*. Sleep, divided into stages 1–3, that is defined by the presence of slow-wave EEG activity. See Figure 10.9. Compare **rapid-eye-movement (REM) sleep**. [10]

SMA See *supplementary motor area*. [5]

SNB See *spinal nucleus of the bulbocavernosus*. [8]

social neuroscience A field of study that uses the tools of neuroscience to discover both the biological bases of social behavior and the effects of social circumstances on brain activity. [1]

sodium ion (Na+) A sodium atom that carries a positive charge. [3]

sodium-potassium pump The energetically expensive mechanism that pushes sodium ions out of a cell, and potassium ions in. [3]

solute A solid compound that is dissolved in a liquid. Compare *solvent*.

solvent The liquid (often water) in which a compound is dissolved. Compare *solute*.

soma (pl. somata) See *cell body*. [2]

somatic intervention An approach to finding relations between body variables and behavioral variables that involves manipulating body structure or function and looking for resultant changes in behavior. See Figure 1.9. Compare *behavioral intervention*. [1]

somatic nerve See *spinal nerve*. [2]

somatic nervous system A part of the peripheral nervous system that supplies neural connections mostly to the skeletal muscles and sensory systems of the body. It consists of cranial nerves and spinal nerves. [2]

somatosensory 1 (S1) See *primary somatosensory cortex*. [5]

somatosensory system A set of specialized receptors and neural mechanisms responsible for body sensations such as touch and pain. [5]

somatotropic hormone Also called **somatotropin**. See *growth hormone*. [8]

somnambulism Sleepwalking. [10]

Southern blot A method of detecting a particular DNA sequence in the genome of an organism by separating DNA with gel electrophoresis, blotting the separated DNA molecules onto nitrocellulose, and then using a nucleotide probe to hybridize with, and highlight, the gene of interest. See Appendix Figure A.3. Compare *Northern blot* and *Western blot*. [App]

spasticity Markedly increased rigidity in response to forced movement of the limbs.

spatial cognition The ability to navigate and to understand the spatial relationship between objects. [15]

spatial-frequency model A model of vision that emphasizes the analysis of different spatial frequencies, of various orientations and in various parts of the visual field, as the basis of visual perception of form. [7, ASF 7.2]

spatial resolution The ability to observe the detailed structure of the brain. Compare *temporal resolution*. [14]

spatial summation The summation of postsynaptic potentials that reach the axon hillock from different locations across the cell body. If this summation reaches threshold, an action potential is triggered. See Figure 3.10. Compare *temporal summation*. [3]

spectral filtering The process by which the hills and valleys of the external ear alter the amplitude of some, but not all, frequencies in a sound. [6]

spectrally opponent cell Also called *color-opponent cell*. A visual receptor neuron that has opposite firing responses to different regions of the spectrum. See Figures 7.26, 7.27. [7]

sperm The gamete produced by males for the fertilization of eggs (ova). [8]

spike See *action potential*. [3]

spinal nerve Also called *somatic nerve*. A nerve that emerges from the spinal cord. There are 31 pairs of spinal nerves.

Compare *cranial nerve*. See Figure 2.8. [2]

spinal nucleus of the bulbocavernosus (SNB) A group of motoneurons in the spinal cord of rats that innervate muscles controlling the penis. See Figure 8.32. See also *Onuf's nucleus*. [8]

spinocerebellum The uppermost part of the cerebellum, consisting mostly of the vermis and the anterior lobe. It receives sensory information about the current spatial location of the parts of the body and anticipates subsequent movement. Compare *cerebrocerebellum* and *vestibulocerebellum*. [ASF 5.4]

spinothalamic system See *anterolateral system*. [5]

split-brain individual An individual whose corpus callosum has been severed, halting communication between the right and left hemispheres. [15]

SRY gene A gene on the Y chromosome that directs the developing gonads to become testes. The name *SRY* stands for *s*ex-determining *r*egion on the *Y* chromosome. [8]

SSRI See *selective serotonin reuptake inhibitor*. [4, 12]

stage 1 sleep The initial stage of non-REM sleep, which is characterized by small-amplitude EEG waves of irregular frequency, slow heart rate, and reduced muscle tension. See Figure 10.9. [10]

stage 2 sleep A stage of sleep that is defined by bursts of regular 14- to 18-hertz EEG waves called *sleep spindles*. See Figure 10.9. [10]

stage 3 sleep Also called *slow wave sleep* (*SWS*). A stage of non-REM sleep that is defined by the presence of large-amplitude, slow delta waves. See Figure 10.9. [10]

standard condition (SC) The usual environment for laboratory rodents, with a few animals in a cage and adequate food and water, but no complex stimulation. See Figure 13.16. Compare *enriched condition* and *impoverished condition*. [13]

stapedius A middle-ear muscle that is attached to the stapes. See Figure 6.1.

stapes Latin for "stirrup." A middle-ear bone that is connected to the oval window. It is one of the three ossicles that conduct sounds across the middle ear. See Figure 6.1. [6]

stem cell A cell that is undifferentiated and therefore can take on the fate of any cell that a donor organism can produce. [13]

stereocilium (pl. stereocilia) A tiny bristle that protrudes from a hair cell in the auditory or vestibular system. See Figure 6.1. [6]

steroid hormone Any of a class of hormones, each of which is composed of four interconnected rings of carbon atoms. Compare *amine hormone* and *peptide hormone*. [8]

stimulant A drug that enhances the excitability of neurons. Compare *depressant*. [4]

stimulus (pl. stimuli) A physical event that triggers a sensory response. [5]

stimulus cuing A technique for testing reaction time to sensory stimuli, in which a cue to where the stimulus will be presented is provided before the stimulus itself.

STM See *short-term memory*. [13]

stress Any circumstance that upsets homeostatic balance. [11]

stress immunization The concept that mild stress early in life makes an individual better able to handle stress later in life. The benefits seem to be due to effective comforting after stressful events, not the stressful events themselves. [11]

stretch reflex The contraction of a muscle in response to stretch of that muscle. See Figure 5.21. [5]

striate cortex See *primary visual cortex*. [7]

striate muscle A type of muscle that has a striped appearance. It is generally under voluntary control. [5]

striatum The caudate nucleus and putamen together.

stroke Damage to a region of brain tissue that results from the blockage or rupture of vessels that supply blood to that region. [2]

stuttering The tendency of otherwise healthy people to produce speech sounds only haltingly, tripping over certain syllables or being unable to start vocalizing certain words. [15]

substance P A peptide transmitter that is involved in pain transmission. [5]

substantia nigra A brainstem structure that innervates the basal ganglia and is a major source of dopaminergic projections. Compare *locus coeruleus*. [2, 4, 5]

sudden infant death syndrome (SIDS) Also called *crib death*. The sudden, unexpected death of an apparently healthy human infant who simply stops breathing, usually during sleep. [10]

sulcus (pl. sulci) A crevice or valley of a convoluted brain surface. Compare *gyrus*. [2]

superior In anatomy, above. See Box 2.2. Compare *inferior*. [2]

superior colliculi (sing. superior colliculus) Paired gray matter structures of the dorsal midbrain that process visual information and are involved in direction of visual gaze and visual attention to intended stimuli. See Figures 2.16, 7.10. Compare *inferior colliculi*. [2, 7, 14]

superior olivary nucleus Either of two brainstem nuclei—left and right—that receive input from both right and left cochlear nuclei, and provide the first binaural analysis of auditory information. See Figure 6.6. [6]

supersensitivity psychosis An exaggerated psychosis that may emerge when doses of antipsychotic medication are reduced, probably as a consequence of the up-regulation of receptors that occurred during drug treatment. See Box 12.1. [12]

supplementary motor area (SMA) A region of nonprimary motor cortex that receives input from the basal ganglia and modulates the activity of the primary motor cortex. See Figure 5.25. [5]

suprachiasmatic nucleus (SCN) A small region of the hypothalamus above the optic chiasm that is the location of a circadian clock. [10]

surface dyslexia Acquired dyslexia in which the patient seems to attend only to the fine details of reading. Compare *deep dyslexia*. [15]

sustained-attention task A task in which a single stimulus source or location must be held in the attentional spotlight for a protracted period. Compare *divided-attention task*. [14]

SWS See *stage 3 sleep*. [10]

Sylvian fissure Also called *lateral sulcus*. A deep fissure that demarcates the temporal lobe. See Figure 2.10. [2]

symbolic cuing A technique for testing endogenous attention in which a visual stimulus is presented and subjects are asked to respond as soon as the stimulus appears on a screen. Each trial is preceded by a meaningful symbol used as a cue to hint at where the stimulus will appear. Compare *peripheral spatial cuing*. [14]

sympathetic nervous system The part of the autonomic nervous system that acts as the "fight or flight" system, generally activating the body for action. See Figure 2.9. Compare *parasympathetic nervous system*. [2, 11]

synapse The cellular location at which information is transmitted from a neuron to another cell. See Figure 2.4. [2, 4, 8]

synapse rearrangement Also called *synaptic remodeling*. The loss of some synapses and the development of others; a refinement of synaptic connections that is often seen in development. See Figure 13.25. [13]

synaptic bouton See *axon terminal*. [2]

synaptic cleft The space between the presynaptic and postsynaptic neurons at a synapse. This gap measures about 20–40 nm. See Figures 2.2, 2.4, 3.11. [2, 3]

synaptic delay The brief delay between the arrival of an action potential at the axon terminal and the creation of a postsynaptic potential. [3]

synaptic remodeling See *synapse rearrangement*. [13]

synaptic transmitter See *neurotransmitter*. [2, 3]

synaptic vesicle A small, spherical structure that contains molecules of neurotransmitter. See Figure 2.4. [2, 3]

synaptogenesis The establishment of synaptic connections as axons and dendrites grow. See Figure 13.25. [13]

synergist A muscle that acts together with another muscle. Compare *antagonist* (definition 2). [5]

synesthesia A condition in which stimuli in one modality evoke the involuntary experience of an additional sensation in another modality. [5]

syrinx The vocal organ in birds. [ASF 15.3]

T

T cell See *T lymphocyte*. [ASF 11.4]

T lymphocyte Also called *T cell*. An immune system cell, formed in the thymus (hence the *T*), that attacks foreign microbes or tissue; "killer cell." Compare *B lymphocyte*. [ASF 11.4]

T1R A family of taste receptor proteins that, when particular members bind together, form taste receptors for sweet flavors and umami flavors. Compare *T2R*. [6]

T2R A family of bitter taste receptors. Compare *T1R*. [6]

TAAR See *trace amine–associated receptor*. [6]

tachistoscope test A test in which stimuli are very briefly presented to either the left or right visual half field. [15]

tactile Referring to touch.

tactile corpuscle See *Meissner's corpuscle*. [5]

tailbone See *coccygeal*. [2]

tardive dyskinesia A disorder characterized by involuntary movements, especially involving the face, mouth, lips, and tongue. It is related to prolonged use of antipsychotic drugs, such as chlorpromazine. See Box 12.1. [12]

taste Any of the five basic sensations detected by the tongue: sweet, salty, sour, bitter, and umami. Compare *flavor*. [6]

taste bud A cluster of 50–150 cells that detects tastes. Taste buds are found in papillae. See Figures 6.15, 6.16. [6]

tau A protein associated with neurofibrillary tangles in Alzheimer's disease. [13]

tectorial membrane A membrane that sits atop the organ of Corti in the cochlear duct. See Figure 6.1. [6]

tectum The dorsal portion of the midbrain consisting of the inferior and superior colliculi. [2]

tegmentum The main body of the midbrain, containing the substantia nigra, periaqueductal gray, part of the reticular formation, and multiple fiber tracts. [2]

telencephalon The anterior part of the fetal forebrain, which will become the cerebral hemispheres in the adult brain. See Figure 2.12, 13.24. Compare *diencephalon*. [2, 13]

temporal coding Frequency discrimination in which the pitch of a sound is determined by the rate of firing of auditory neurons. Compare *place coding*. [6]

temporal lobe The large lateral region of cortex in each cerebral hemisphere. It is continuous with the parietal lobe posteriorly and separated from the frontal lobe by the Sylvian fissure. See Figure 2.10. Compare *occipital lobe*. [2]

temporal resolution The ability to track changes in the brain that occur very quickly. Compare *spatial resolution*. [14]

temporal summation The summation of postsynaptic potentials that reach the axon hillock at different times. The closer in time the potentials occur, the more complete the summation is. See Figure 3.10. Compare *spatial summation*. [3]

temporoparietal junction (TPJ) The point in the brain where the temporal and parietal lobes meet. It plays a role in shifting attention to a new location after target onset. [14]

TENS See *transcutaneous electrical nerve stimulation*. [5]

testes (sing. testis) The male gonads, which produce sperm and androgenic steroid hormones. See Figure 8.1. Compare *ovaries*. [8]

testosterone A hormone, produced by male gonads, that controls a variety of bodily changes that become visible at puberty. It is one of a class of hormones called *androgens*. See Figure 8.13. [8, 11]

tetanus An intense volley of action potentials. [13]

tetrahydrocannabinol (THC) See *delta-9-tetrahydrocannabinol*. [4]

thalamus (pl. thalami) The brain regions at the top of the brainstem that trade information with the cortex. See Figures 2.14, 2.15, 2.16. [2, 5]

THC See *delta-9-tetrahydrocannabinol*. [4]

therapeutic index The margin of safety for a given drug, expressed as the distance between effective doses and toxic doses. See Figure 4.6. [4]

thermoregulation The active process of maintaining a constant internal temperature through behavioral and physiological adjustments. [9]

third ventricle The midline ventricle that conducts cerebrospinal fluid from the lateral ventricles to the fourth ventricle. See Figure 2.17. Compare *fourth ventricle* and *lateral ventricle*. [2]

thoracic Referring to the 12 spinal segments below the cervical (neck) portion of the spinal cord, corresponding to the chest. See Figures 2.8, 2.9. Compare *cervical, lumbar, sacral,* and *coccygeal*. [2]

threshold The stimulus intensity that is just adequate to trigger an action potential in an axon. [3, 5]

thrombolytic A substance that is used to unblock blood vessels and restore circulation. [ASF 15.5]

thyroid-stimulating hormone (TSH) A tropic hormone, released by the anterior pituitary gland, that signals the thyroid gland to secrete its hormones. [ASF 8.3]

timbre The characteristic sound quality of a musical instrument, as determined by the relative intensities of its various harmonics. See Box 6.1. [6]

tinnitus A sensation of noises or ringing in the ears not caused by external sound. [6]

tip link A fine, threadlike fiber that runs along and connects the tips of stereocilia. See Figure 6.3. [6]

TMS See *transcranial magnetic stimulation.* [2]

tolerance See *drug tolerance.* [4]

tonic receptor A receptor in which the frequency of action potentials declines slowly or not at all as stimulation is maintained. Compare *phasic receptor.* [5]

tonotopic organization A major organizational feature in auditory systems, in which neurons are arranged as an orderly map of stimulus frequency, with cells responsive to high frequencies located at a distance from those responsive to low frequencies. [6]

topographic projection A mapping that preserves the point-to-point correspondence between neighboring parts of space. For example, the retina extends a topographic projection onto the cortex. [7]

Tourette's syndrome A disorder that is characterized by heightened sensitivity to tactile, auditory, and visual stimuli that may be accompanied by the build-up of an urge to emit verbal or phonic tics. See Box 12.2. [12]

TPJ See *temporoparietal junction.* [14]

trace amine–associated receptor (TAAR) Any one of a family of probable pheromone receptors produced by neurons in the main olfactory epithelium. TAARs are candidate pheromone receptors, despite being situated outside the vomeronasal organ. [6]

tract A bundle of axons found within the central nervous system. Compare *nerve.* [2]

transcranial magnetic stimulation (TMS) A noninvasive technique for examining brain function that applies strong magnetic fields to stimulate cortical neurons, in order to identify discrete areas of the brain that are particularly active during specific behaviors. Compare *magnetoencephalography.* [2]

transcript See *messenger RNA.* [App]

transcription The process during which mRNA forms bases complementary to a strand of DNA. The resulting message (called a *transcript*) is then used to translate the DNA code into protein molecules. See Appendix Figure A.2. Compare *translation.* [App]

transcutaneous electrical nerve stimulation (TENS) The delivery of electrical pulses through electrodes attached to the skin, which excite nerves that supply the region to which pain is referred. [5]

transduction The conversion of one form of energy to another, as converting light into neuronal activity. [6, 7]

transgenic Referring to an animal in which a new or altered gene has been deliberately introduced into the genome. [App]

transient ischemic attack (TIA) A temporary blood restriction to part of the brain that causes strokelike symptoms that quickly resolve, serving as a warning of elevated stroke risk. [2]

transient receptor potential 2 (TRP2) A receptor, found in some free nerve endings, that opens its channel in response to rising temperatures. [5]

translation The process by which amino acids are linked together (directed by an mRNA molecule) to form protein molecules. See Appendix Figure A.2. Compare *transcription.* [App]

transmitter See *neurotransmitter.* [2, 3, 4]

transporter A specialized membrane component that returns transmitter molecules to the presynaptic neuron for reuse. [3, 4]

transverse plane See *coronal plane.* [2]

trichromatic hypothesis A hypothesis of color perception stating that there are three different types of cones, each excited by a different region of the spectrum and each having a separate pathway to the brain. [7]

tricyclic antidepressant An antidepressant that acts by increasing the synaptic accumulation of serotonin and norepinephrine. [4]

trinucleotide repeat Repetition of the same three nucleotides within a gene, which can lead to dysfunction, as in Huntington's disease. [ASF 5.5]

trophic factor See *neurotrophic factor.* [13]

tropic hormone Any of a class of anterior pituitary hormones that affect the secretion of hormones by other endocrine glands. See Figure 8.12. [8]

TRP2 See *transient receptor potential 2.* [5]

TSH See *thyroid-stimulating hormone.* [ASF 8.3]

tuberomammillary nucleus A region of the basal hypothalamus, near the pituitary stalk, that plays a role in generating stage 3 (slow wave) sleep. [10]

tuning curve A graph of the responses of a single auditory nerve fiber or neuron to sounds that vary in frequency and intensity. See Figure 6.5. [6]

Turner's syndrome A condition, seen in individuals carrying a single X chromosome but no other sex chromosome, in which an apparent female has underdeveloped but recognizable ovaries. [8]

tympanic canal See *scala tympani.* [6]

tympanic membrane Also called *eardrum.* The partition between the external ear and the middle ear. See Figure 6.1. [6]

typical antipsychotic Also called *typical neuroleptic.* An antischizophrenic drug that shows antagonist activity at dopamine D_2 receptors. Compare *atypical antipsychotic.* [12]

U

ultradian Referring to a rhythmic biological event with a period shorter than a day, usually from several minutes to several hours long. Compare *infradian.* [10]

ultrasound High-frequency sound; in general, above the threshold for human hearing, at about 20,000 Hz. Compare *infrasound.* [6]

umami One of the five basic tastes—the meaty, savory flavor. (The other four tastes are salty, sour, sweet, and bitter.) [6]

unipolar neuron Also called *monopolar neuron.* A nerve cell with a single branch that leaves the cell body and then extends in two directions; one end is the input zone, and the other end is the output zone. See Figure 2.3. Compare *bipolar neuron* and *multipolar neuron.* [2]

up-regulation A compensatory increase in receptor availability at the synapses of a neuron. Compare *down-regulation.* [4]

V

V1 See *primary visual cortex*. [7]

vagina The opening from the outside of the body to the cervix and uterus in females. [8]

vagus nerve Cranial nerve X, which provides extensive innervation of the viscera (organs). The vagus both regulates visceral activity and transmits signals from the viscera to the brain. See Figures 2.7, 9.15. [2, 9]

vasopressin Also called *arginine vasopressin* or *antidiuretic hormone*. A peptide hormone from the posterior pituitary that promotes water conservation and increases blood pressure. [8, 9]

ventral In anatomy, toward the belly or front of the body, or the bottom of the brain. See Box 2.2. Compare *dorsal*. [2]

ventral tegmental area (VTA) A portion of the midbrain that projects dopaminergic fibers to the nucleus accumbens. [4]

ventricular system A system of fluid-filled cavities inside the brain. See Figure 2.17. [2]

ventricular zone Also called *ependymal layer*. A region lining the cerebral ventricles that displays mitosis, providing neurons early in development and glial cells throughout life. See Figure 13.25. [13]

ventromedial hypothalamus (VMH) A hypothalamic region involved in sexual behaviors, eating, and aggression. See Figures 8.20, 9.12. [8, 9, 11]

vertex spike A sharp-wave EEG pattern that is seen during stage 1 sleep. See Figure 10.9. [10]

vestibular canal See *scala vestibuli*. See Figure 6.1. [6]

vestibular nucleus A brainstem nucleus that receives information from the vestibular organs through cranial nerve VIII (the vestibulocochlear nerve). [6]

vestibular system The sensory system that detects balance. It consists of several small inner-ear structures that adjoin the cochlea. [6]

vestibulocerebellum The middle portion of the cerebellum, sandwiched between the spinocerebellum and the cerebrocerebellum and consisting of the nodule and the flocculus. It helps the motor systems to maintain posture and appropriate orientation toward the external world. [ASF 5.4]

vestibulocochlear nerve Cranial nerve VIII, which runs from the cochlea to the brainstem auditory nuclei. See Figures 2.7, 6.1. [6]

visual acuity Sharpness of vision. [7]

visual cortex See *occipital cortex*. [7]

visual field The whole area that you can see without moving your head or eyes. [7]

visual P1 effect A positive deflection of the event-related potential, occurring 70–100 milliseconds after stimulus presentation, that is enhanced for selectively attended visual input compared with ignored input. Compare *auditory N1 effect*. [14]

VMH See *ventromedial hypothalamus*. [8, 9, 11]

VNO See *vomeronasal organ*. [6, 8]

voltage-gated Na$^+$ channel A Na$^+$-selective channel that opens or closes in response to changes in the voltage of the local membrane potential. It mediates the action potential. Compare *ionotropic receptor*. [3]

voluntary attention See *endogenous attention*. [14]

vomeronasal organ (VNO) A collection of specialized receptor cells, near to but separate from the olfactory epithelium, that detect pheromones and send electrical signals to the accessory olfactory bulb in the brain. [6, 8]

vomeronasal system A specialized sensory system that detects pheromones and transmits information to the brain. [6]

VTA See *ventral tegmental area*. [4]

W

Wada test A test in which a short-lasting anesthetic is delivered into one carotid artery to determine which cerebral hemisphere principally mediates language. See Box 15.1. [15]

wavelength The length between two peaks in a repeated stimulus such as a wave, light, or sound. See Figure 7.23. [7]

Wernicke-Geschwind model See *connectionist model of aphasia*. [15]

Wernicke's aphasia See *fluent aphasia*. [15]

Wernicke's area A region of temporoparietal cortex in the brain that is involved in the perception and production of speech. See Figures 15.5, 15.6, 15.7. Compare *Broca's area*. [15]

Western blot A method of detecting a particular protein molecule in a tissue or organ by separating proteins from that source with gel electrophoresis, blotting the separated proteins onto nitrocellulose, and then using an antibody that binds, and highlights, the protein of interest. Compare *Northern blot* and *Southern blot*. [App]

white matter A light-colored layer of tissue, consisting mostly of myelin-sheathed axons, that lies underneath the gray matter of the cortex. White matter mostly transmits information. See Figures 2.8, 2.11. Compare *gray matter*. [2]

Williams syndrome A disorder characterized by impairments of spatial cognition and IQ but superior linguistic abilities. [15]

withdrawal symptom An uncomfortable symptom that arises when a person stops taking a drug that he or she has used frequently, especially at high doses. [4]

within-participants experiment An experiment in which the same set of individuals is compared before and after an experimental manipulation. The experimental group thus serves as its own control group. Compare *between-participants experiment*. [1]

wolffian duct A duct system in the embryo that will develop into male reproductive structures (epididymis, vas deferens, and seminal vesicle) if androgens are present. See Figure 8.24. Compare *müllerian duct*. [8]

word blindness A selective inability to understand written words. [15]

word deafness 1. A form of central deafness that is characterized by the specific inability to hear words, although other sounds can be detected. [6] 2. A selective inability to understand spoken words. [15]

working memory See *short-term memory*. [13]

Z

zeitgeber Literally "time-giver" (in German). The stimulus (usually the light-dark cycle) that entrains circadian rhythms. [10]

zygote The fertilized egg. [8]

Illustration Credits

Chapter 1
1.2: Reproduced with gracious permission of Her Majesty Queen Elizabeth II, copyright reserved. 1.3: Bettmann/Corbis. 1.4A: © The Print Collector/Alamy. 1.8: © Dwayne Godwin 2011. p. 5 *bottom*: © Scala/Art Resource, NY. p. 13: © Roman Sigaev/istock. p. 18: © Caters News/Zuma Press.

Chapter 2
2.6A: From *Gray's Anatomy*, 35th ed., Figure 2.9, page 807. Reprinted with permission of the publisher, Churchill Livingstone. Dissection by M. C. E. Hutchinson, photograph by Kevin Fitzpatrick, Guy's Hospital Medical School, London. 2.19A: © Living Art Enterprises, LLC/Science Source. 2.19B: © Guy Croft SciTech/Alamy. 2.19D: Courtesy of Jamie Eberling. 2.19E: © Shutterstock. 2.22: Courtesy of Dr. Mario Liotti and Anthony Herdman, Simon Fraser University, and Down Syndrome Research Foundation.

Chapter 3
3.15A: Courtesy of Neuroscan Labs, a division of Neurosoft, Inc. 3.16C: Courtesy of Hal Blumenfeld, Rik Stokking, Susan Spencer, and George Zubal, Yale School of Medicine. p. 52: Photographs of events and activities documenting Yale University, Manuscripts and Archives, Yale University Library.

Chapter 4
4.10: Courtesy of E. Riley. p. 95: © Alfred Eisenstaedt/Time & Life Pictures/Getty Images. p. 96: © South West Images Scotland/Alamy. p. 98: © Blaine Harrington III/Corbis. p. 101: Art by Wes Black, courtesy of www.blotterart.com. p. 102: © ejwhite/Shutterstock.

Chapter 5
5.28: © Associated Press.

Chapter 6
6.15: © Astrid & Hanns-Frieder Michler/Science Source. p. 147 *opossums*: © Stuart Elflett/ShutterStock. p. 147 *rabbit*: © Heather Craig/istock. p. 147 *elephant*: © Gerry Ellis/Digital Vision. p. 147 *chimp*: © Holger Ehlers/ShutterStock. p. 149: © Bradley Smith/Corbis.

Chapter 7
7.7A: © Paul Parker/SPL/Science Source. 7.27: Courtesy of Thomas Eisner. Box 7.1A: Photo by David McIntyre; simulation created using software from Vischeck (www.vischeck.com). Box 7.1B: © Brand X Pictures/Alamy. p. 204: Courtesy of Patch Pals, www.PatchPals.com.

Chapter 8
8.26: © The Wellcome Photo Library. p. 219: Image by Vicky Tobin and Mike Ludwig, Centre for Integrative Physiology, University of Edinburgh. p. 228: © Jane Burton/naturepl.com. p. 234: © Bettmann/Corbis. p. 236: David McIntyre. p. 237: © Biophoto Associates/Science Source. p. 248: © Murat Sarica/istock. p. 251: © Bettmann/Corbis.

Chapter 9
9.14: © John Sholtis/Rockefeller University. p. 257: © Karen Hadley/Shutterstock. p. 266A: © Dmitry Lobanov/Shutterstock. p. 266B: © Ace Stock Limited/Alamy. p. 274 *left*: © Ash Knotek/Zuma Press. p. 274 *right*: Kunsthistorisches Museum, Vienna.

Chapter 10
p. 289: © Hank Morgan/Science Source. p. 292: © Robin Chittenden/Alamy. p. 296: © San Diego History Center. p. 297: © Hoberman Collection/Alamy. p. 198: Courtesy of Ray Meddis. p. 305: Courtesy of Joanne Delphia. p. 308 *sleeping kitten*: © Ira Bachinskaya/istock.

Chapter 11

11.4: © Sinauer Associates. 11.6: © Ken Cedeno/Corbis. 11.8 *left*: © Richard Green/Commercial/Alamy. 11.8 *right*: Courtesy of Jennifer Basil-Whitaker. 11.14: © Lucy Nicholson/Reuters/Corbis. p. 312: © blickwinkel/Alamy. p. 327: © David Osborn/Alamy. p. 329: © Bettmann/Corbis. p. 334: © Richard Garvey-Williams/Alamy.

Chapter 12

Box 12.1: Courtesy of Steven J Frucht. p. 340: Courtesy of Howard Dully. p. 348: © Bettmann/Corbis. p. 356: © MBI/Alamy. p. 359: © AF archive/Alamy. p. 361: © Jeff Cook/Quad-City Times/Alamy Live News.

Chapter 13

13.1: From Corkin et al., 1997; courtesy of Suzanne Corkin. 13.10: Courtesy of Med Associates. Box 13.1: © Sara K. Schwittek/Reuters/Corbis. p. 382: Kathleen Turley/Tri-Valley Herald/Zuma Press. p. 402: © Noah Goodrich/Caters News/Zuma Press.

Chapter 14

p. 411: Figure provided by Daniel Simons. p. 415: From *Where's Waldo?* © Martin Handford 2005. p. 434 *left*: From the collection of Jack and Beverly Wilgus. p. 434 *right*: Courtesy of Hanna Damasio.

Chapter 15

p. 445: © Marta Benavides/istock. p. 458: © Hulton-Deutsch Collection/Corbis. p. 462 *bottom*: © Paul White–UK Rural Communites/Alamy. p. 464: © Everett Collection Inc/Alamy. p. 467: © UPPA/Zuma Press.

References

A

Abbott, S. M., and Videnovic, A. (2014). Sleep disorders in atypical parkinsonism. *Movement Disorders Clinical Practice* (Hoboken), *1*, 89–96.

Abe, N., Suzuki, M., Mori, E., Itoh, M., et al. (2007). Deceiving others: Distinct neural responses of the prefrontal cortex and amygdala in simple fabrication and deception with social interactions. *Journal of Cognitive Neuroscience, 19*, 287–295.

Abelson, J. F., Kwan, K. Y., O'Roak, B. J., Baek, D. Y., et al. (2005). Sequence variants in *SLITRK1* are associated with Tourette's syndrome. *Science, 310*, 317–320.

Absil, P., Pinxten, R., Balthazart, J., and Eens, M. (2003). Effect of age and testosterone on autumnal neurogenesis in male European starlings (*Sturnus vulgaris*). *Behavioural Brain Research, 143*, 15–30.

Adams, C. S., Korytko, A. I., and Blank, J. L. (2001). A novel mechanism of body mass regulation. *Journal of Experimental Biology, 204*, 1729–1734.

Ader, R. (2001). Psychoneuroimmunology. *Current Directions in Psychological Science, 10*(3), 94–98.

Adler, E., Hoon, M. A., Mueller, K. L., Chandrashekar, J., et al. (2000). A novel family of mammalian taste receptors. *Cell, 100*, 693–702.

Adolphs, R., Gosselin, F., Buchanan, T. W., Tranel, D., et al. (2005). A mechanism for impaired fear recognition after amygdala damage. *Nature, 433*, 68–72.

Agarwal, N., Pacher, P., Tegeder, I., Amay, F., et al. (2007). Cannabinoids mediate analgesia largely via peripheral type cannabinoid receptors in nociceptors. *Nature Neuroscience, 10*, 870–878.

Agmon-Snir, H., Carr, C. E., and Rinzel, J. (1998). The role of dendrites in auditory coincidence detection. *Nature, 393*, 268–272.

Ahmed, E. I., Zehr, J. L., Schulz, K. M., Lorenz, B. H., et al. (2008). Pubertal hormones modulate the addition of new cells to sexually dimorphic brain regions. *Nature Neuroscience, 11*, 995–997.

Ahn, S., and Phillips, A. G. (2002). Modulation by central and basolateral amygdalar nuclei of dopaminergic correlates of feeding to satiety in the rat nucleus accumbens and medial prefrontal cortex. *Journal of Neuroscience, 22*, 10958–10965.

Albanese, A., Hamill, G., Jones, J., Skuse, D., et al. (1994). Reversibility of physiological growth hormone secretion in children with psychosocial dwarfism. *Clinical Endocrinology (Oxford), 40*, 687–692.

Al-Barazanji, K. A., Buckingham, R. E., Arch, J. R., Haynes, A., et al. (1997). Effects of intracerebroventricular infusion of leptin in obese Zucker rats. *Obesity Research, 5*, 387–394.

Alberts, J. R. (1978). Huddling by rat pups: Multisensory control of contact behavior. *Journal of Comparative and Physiological Psychology, 92*, 220–230.

Aldrich, M. A. (1993). The neurobiology of narcolepsy-cataplexy. *Progress in Neurobiology, 41*, 533–541.

Almer, G., Hainfellner, J. A., Brücke, T., Jellinger, K., et al. (1999). Fatal familial insomnia: A new Austrian family. *Brain, 122*, 5–16.

Altschuler, E. L., Wisdom, S. B., Stone, L., Foster, C., et al. (1999). Rehabilitation of hemiparesis after stroke with a mirror. *Lancet, 353*, 2035–2036.

Alvarez, J. A., and Emory, E. (2006). Executive function and the frontal lobes: A meta-analytic review. *Neuropsychology Review, 16*, 17–42.

American Psychiatric Association. (2013). *Diagnostic and statistical manual of mental disorders: DSM-5*. Washington, DC: American Psychiatric Association.

Amici, R., Bastianini, S., Berteotti, C., Cerri, M., et al. (2014). Sleep and bodily functions: The physiological interplay between body homeostasis and sleep homeostasis. *Archives Italiennes de Biologie, 152*, 66–78.

Amunts, K., Schlaug, G., Jaencke, L., Steinmetz, H., et al. (1997). Motor cortex and hand motor skills: Structural compliance in the human brain. *Human Brain Mapping, 5*, 206–215.

Anand, B. K., and Brobeck, J. R. (1951). Localization of a "feeding center" in the hypothalamus of the rat. *Proceedings of the Society for Experimental Biology and Medicine, 77*, 323–324.

Andersen, B. B., Korbo, L., and Pakkenberg, B. (1992). A quantitative study of the human cerebellum with unbiased stereological techniques. *Journal of Comparative Neurology, 326*, 549–560.

Andersen, P. M., Nilsson, P., Ala-Hurula, V., Keranen, M. L., et al. (1995). Amyotrophic lateral sclerosis associated with homozygosity for an Asp90Ala mutation in CuZn-superoxide dismutase. *Nature Genetics, 10*, 61–66.

Andreasen, N. C. (1991). Assessment issues and the cost of schizophrenia. *Schizophrenia Bulletin, 17*, 475–481.

Andreasen, N., Nassrallah, H. A., Dunn, V., Olson, S. C., et al. (1986). Structural abnormalities in the frontal system in schizophrenia. *Archives of General Psychiatry, 43*, 136–144.

Angier, N. (2004, April 13). No time for bullies: Baboons retool their culture. *The New York Times*.

Apkarian, A. V., Sosa, Y., Sonty, S., Levy, R. M., et al. (2004). Chronic back pain is associated with decreased prefrontal and thalamic gray matter density. *Journal of Neuroscience, 24*, 10410–10415.

Archer, G. S., Friend, T. H., Piedrahita, J., Nevill, C. H., et al. (2003). Behavioral variation among cloned pigs. *Applied Animal Behaviour Science, 82*, 151–161.

Archer, J. (2006). Testosterone and human aggression: An evaluation of the challenge hypothesis. *Neuroscience and Biobehavioral Reviews, 30*, 319–345.

Argyll-Robertson, D. M. C. L. (1869). On an interesting series of eye symptoms in a case of spinal disease, with remarks on the action of belladonna on the iris. *Edinburgh Medical Journal, 14*, 696–708.

Arnold, A. P. (1980). Sexual differences in the brain. *American Scientist, 68*, 165–173.

Arnold, A. P., and Schlinger, B. A. (1993). Sexual differentiation of brain and behavior: The zebra finch is not just a flying rat. *Brain, Behavior and Evolution, 42*, 231–241.

Arnold, K., and Zuberbühler, K. (2006). Language evolution: Semantic combinations in primate calls. *Nature, 441*, 303.

Arnone, D., Cavanagh, J., Gerber, D., Lawrie, S. M., et al. (2009). Magnetic resonance imaging studies in bipolar disorder and schizophrenia: Meta-analysis. *British Journal of Psychiatry, 195*, 194–201.

Arnsten, A. F. (2006). Fundamentals of attention-deficit/hyperactivity disorder: Circuits and pathways. *Journal of Clinical Psychiatry, 67*, 7–12.

Asberg, M., Nordstrom, P., and Traskman-Bendz, L. (1986). Cerebrospinal fluid studies in suicide. An overview. *Annals of the New York Academy of Sciences, 487*, 243–255.

Aschner, M., and Ceccatelli, S. (2010). Are neuropathological conditions relevant to ethylmercury exposure? *Neurotoxicity Research, 18*, 59–68.

Aserinsky, E., and Kleitman, N. (1953). Regularly occurring periods of eye motility, and concomitant phenomena, during sleep. *Science, 118*, 273–274.

Ashmore, J. F. (1994). The cellular machinery of the cochlea. *Experimental Physiology, 79*, 113–134.

Ashtari, M., Kumra, S., Bhaskar, S. L., Clarke, T., et al. (2005). Attention-deficit/hyperactivity disorder: A preliminary diffusion tensor imaging study. *Biological Psychiatry, 57*, 448–455.

Assaf, Y., and Pasternak, O. (2008). Diffusion tensor imaging (DTI)-based white matter mapping in brain research: A review. *Journal of Molecular Neuroscience, 34*, 51–61.

Audero, E., Coppi, E., Mlinar, B., Rossetti, T., et al. (2008). Sporadic autonomic dysregulation and death associated with excessive serotonin autoinhibition. *Science, 321*, 130–133.

Aungst, J. L., Heyward, P. M., Puche, A. C., Karnup, S. V., et al. (2003). Centre-surround inhibition among olfactory bulb glomeruli. *Nature, 426*, 623–629.

Avila, M. T., Hong, L. E., Moates, A., Turano, K. A., et al. (2006). Role of anticipation in schizophrenia-related pursuit initiation deficits. *Journal of Neurophysiology, 95*, 593–601.

B

Baddeley, A. D., and Warrington, E. K. (1970). Amnesia and the distinction between long- and short-term memory. *Journal of Verbal Learning and Verbal Behavior, 9*, 176–189.

Bagemihl, B. (1999). *Biological exuberance: Animal homosexuality and natural diversity.* New York, NY: St. Martin's Press.

Bailey, C. H., and Chen, M. (1983). Morphological basis of long-term habituation and sensitization in *Aplysia. Science, 220*, 91–93.

Bailey, J. M., Pillard, R. C., Neale, M. C., and Agyei, Y. (1993). Heritable factors influence sexual orientation in women. *Archives of General Psychiatry, 50*, 217–223.

Bailey, K., and West, R. (2013). The effects of an action video game on visual and affective information processing. *Brain Research, 1504*, 35–46.

Bakker, J., De Mees, C., Douhard, Q., Balthazart, J., et al. (2006). Alpha-fetoprotein protects the developing female mouse brain from masculinization and defeminization by estrogens. *Nature Neuroscience, 9*, 220–226.

Baldwin, M. W., Toda, Y., Nakagita, T., O'Connell, M. J., et al. (2014). Sensory biology: Evolution of sweet taste perception in hummingbirds by transformation of the ancestral umami receptor. *Science, 345*, 929–933.

Ball, G. F., and Hulse, S. H. (1998). Birdsong. *American Psychologist, 53*, 37–58.

Bancaud, J., Brunet-Bourgin, F., Chauvel, P., and Halgren, E. (1994). Anatomical origin of deja vu and vivid "memories" in human temporal lobe. *Brain, 117*, 71–90.

Bao, S., Chan, V. T., and Merzenich, M. M. (2001). Cortical remodelling induced by activity of ventral tegmental dopamine neurons. *Nature, 412*, 79–83.

Baptista, L. F. (1996). Nature and its nurturing in avian vocal development. In D. E. Kroodsma and E. H. Miller (Eds.), *Ecology and evolution of acoustic communication in birds* (pp. 39–60). Ithaca, NY: Cornell University Press.

Baptista, L., and Petrinovich, L. (1986). Song development in the white-crowned sparrow: Social factors and sex differences. *Animal Behaviour, 34*, 1359–1371.

Barbeau, H., Norman, K., Fung, J., Visintin, M., et al. (1998). Does neurorehabilitation play a role in the recovery of walking in neurological populations? *Annals of the New York Academy of Sciences, 860*, 377–392.

Bark, N. (2002). Did schizophrenia change the course of English history? The mental illness of Henry VI. *Medical Hypotheses, 59*, 416–421.

Barkow, J. H., Cosmides, L., and Tooby, J. (1992). *The adapted mind: Evolutionary psychology and the generation of culture.* New York, NY: Oxford University Press.

Barnea, G., O'Donnell, S., Mancia, F., Sun, X., et al. (2004). Odorant receptors on axon termini in the brain. *Science, 304*, 1468.

Baron-Cohen, S. (2003). *The essential difference: Men, women and the extreme male brain.* London, England: Allen Lane.

Bartels, A., and Zeki, S. (2000). The neural basis of romantic love. *NeuroReport, 11*, 3829–3834.

Bartfai, T. (2001). Telling the brain about pain. *Nature, 410*, 425–426.

Bartolomeo, P. (2007). Visual neglect. *Current Opinion in Neurology, 20*, 381–386.

Bartoshuk, L. M. (1993). Genetic and pathological taste variation: What can we learn from animal models and human disease? In D. Chadwick, J. Marsh, and J. Goode (Eds.), *The molecular basis of smell and taste transduction* (pp. 251–267). New York, NY: Wiley.

Bartoshuk, L. M., and Beauchamp, G. K. (1994). Chemical senses. *Annual Review of Psychology, 45*, 419–449.

Basson, R. (2001). Human sex-response cycles. *Journal of Sex & Marital Therapy, 27*, 33–43.

Basson, R. (2008).Women's sexual function and dysfunction: Current uncertainties, future directions. *International Journal of Impotence Research, 20*, 466–478.

Bates, E., Wilson, S. M., Saygin, A. P., Dick, F., et al. (2003). Voxel-based lesion-symptom mapping. *Nature Neuroscience, 6*, 448–450.

Batterham, R. L., and Bloom, S. R. (2003). The gut hormone peptide YY regulates appetite. *Annals of the New York Academy of Sciences, 994*, 162–168.

Baum, A., and Posluszny, D. M. (1999). Health psychology: Mapping biobehavioral contributions to health and illness. *Annual Review of Psychology, 50*, 137–163.

Bautista, D. M., Siemens, J., Glazer, J. M., Tsuruda, P. R., et al. (2007). The menthol receptor TRPM8 is the principal detector of environmental cold. *Nature, 448*, 204–208.

Baynes, K. C., Dhillo, W. S., and Bloom, S. R. (2006). Regulation of food intake by gastrointestinal hormones. *Current Opinion in Gastroenterology, 22*, 626–631.

Beach, F. A. (1977). *Human sexuality in four perspectives.* Baltimore, MD: Johns Hopkins University Press.

Bedrosian, T. A., Vaughn, C. A., Galan, A., Daye, G., et al. (2013). Nocturnal light exposure impairs affective responses in a wavelength-dependent manner. *Journal of Neuroscience, 33*, 13081–13087.

Bee, M. A., and Micheyl, C. (2008). The cocktail party problem: What is it? How can it be solved? And why should animal behaviorists study it? *Journal of Comparative Psychology, 122*, 235–251.

Beeli, G., Esslen, M., and Jäncke, L. (2005). When coloured sounds taste sweet. *Nature, 434*, 38.

Beggs, S., Trang, T., and Salter, M. W. (2012). P2X4R+ microglia drive neuropathic pain. *Nature Neuroscience, 15*, 1068–1073.

Beggs, W. D., and Foreman, D. L. (1980). Sound localization and early binaural experience in the deaf. *British Journal of Audiology, 14*, 41–48.

Bellinger, D. L., Ackerman, K. D., Felten, S. Y., and Felten, D. L. (1992). A longitudinal study of age-related loss of noradrenergic nerves and lymphoid cells in the rat spleen. *Experimental Neurology, 116*, 295–311.

Bellugi, U., Poizner, H., and Klima, E. S. (1983). Brain organization for language: Clues from sign aphasia. *Human Neurobiology, 2*, 155–171.

Bennett, A. F., and Ruben, J. A. (1979). Endothermy and activity in vertebrates. *Science, 206*, 649–654.

Bennett, M. V. (2000). Electrical synapses, a personal perspective (or history). *Brain Research Reviews, 32*, 16–28.

Bennett, W. (1983). The nicotine fix. *Rhode Island Medical Journal, 66*, 455–458.

Benney, K. S., and Braaten, R. F. (2000). Auditory scene analysis in estrildid finches (*Taeniopygia guttata* and *Lonchura striata domestica*): A species advantage for detection of conspecific song. *Journal of Comparative Psychology, 114*, 174–182.

Berenbaum, S. A. (2001). Cognitive function in congenital adrenal hyperplasia. *Endocrinology and Metabolism Clinics of North America, 30*, 173–192.

Bernal, B., and Altman, N. (2010). The connectivity of the superior longitudinal fasciculus: A tractography DTI study. *Magnetic Resonance Imaging, 28*, 217–225.

Bernhardt, P. C. (1997). Influences of serotonin and testosterone in aggression and dominance: Convergence with social psychology. *Current Directions in Psychological Science, 2*(6), 44–48.

Bernhardt, P. C., Dabbs, J. M., Jr., Fielden, J. A., and Lutter, C. D. (1998). Testosterone changes during vicarious experiences of winning and losing among fans at sporting events. *Physiology & Behavior, 65*, 59–62.

Bernstein, I. S., and Gordon, T. P. (1974). The function of aggression in primate societies. *American Scientist, 62*, 304–311.

Bernstein, L. E., Auer, E. T., Jr., Moore, J. K., Ponton, C. W., et al. (2002). Visual speech perception without primary auditory cortex activation. *NeuroReport, 13*, 311–315.

Bernstein-Goral, H., and Bregman, B. S. (1993). Spinal cord transplants support the regeneration of axotomized neurons after spinal cord lesions at birth: A quantitative double-labeling study. *Experimental Neurology, 123*, 118–132.

Berthold, A. (1849). Transplantation der Hoden. *Archiv für Anatomie, Physiologie und Wissenschaftliche Medicin, 16*, 42–46.

Bertram, L., and Tanzi, R. E. (2008). Thirty years of Alzheimer's disease genetics: The implications of systematic meta-analyses. *Nature Reviews Neuroscience, 9*, 768–778.

Besedovsky, H. O., and del Rey, A. (1992). Immune-neuroendocrine circuits: Integrative role of cytokines. *Frontiers of Neuroendocrinology, 13*, 61–94.

Beurg, M., Fettiplace, R., Nam, J.-H., and Ricci, A. J. (2009). Localization of inner hair cell mechanotransducer channels using high-speed calcium imaging. *Nature Neuroscience, 12*, 553–558.

Binder, J. R., Rao, S. M., Hammeke, T. A., Yetkin, F. Z., et al. (1994). Functional magnetic resonance imaging of human auditory cortex. *Annals of Neurology, 35*, 662–672.

Birch, L. L., Fisher, J. O., and Davison, K. K. (2003). Learning to overeat: Maternal use of restrictive feeding practices promotes girls' eating in the absence of hunger. *American Journal of Clinical Nutrition, 78*, 215–220.

Bisley, J. W., and Goldberg, M. E. (2003). Neuronal activity in the lateral intraparietal area and spatial attention. *Science, 299*, 81–86.

Blackwell, D. L, Lucas, J. W, and Clarke, T. C. (2014). *Summary health statistics for U.S. adults: National Health Interview Survey, 2012.* Atlanta, GA: Centers for Disease Control and Prevention, National Center for Health Statistics. Vital and Health Statistics 10(260).

Blake, D. J., Weir, A., Newey, S. E., and Davies, K. E. (2002). Function and genetics of dystrophin and dystrophin-related proteins in muscle. *Physiology Review, 82*, 291–329.

Blake, D. T., Heiser, M. A., Caywood, M., and Merzenich, M. M. (2006). Experience-dependent adult cortical plasticity requires cognitive association between sensation and reward. *Neuron, 52*, 371–381.

Blakemore, C., and Campbell, F. W. (1969). On the existence of neurones in the human visual system selectively sensitive to the orientation and size of retinal images. *Journal of Physiology (London), 203*, 237–260.

Blanchard, R. (2012). A possible second type of maternal-fetal immune interaction involved in both male and female homosexuality. *Archives of Sexual Behavior, 41*, 1507–1511.

Blanchard, R., Cantor, J. M., Bogaert, A. F., Breedlove, S. M., et al. (2006). Interaction of fraternal birth order and handedness in the development of male homosexuality. *Hormones and Behavior, 49*, 405–414.

Blehar, M. C., and Rosenthal, N. E. (1989). Seasonal affective disorders and phototherapy. Report of a National Institute of Mental Health-sponsored workshop. *Archives of General Psychiatry, 46*, 469–474.

Bleuler, E. (1950). *Dementia praecox; or, The group of schizophrenias* (J. Zinkin, Trans.). New York, NY: International Universities Press.

Bliss, T. V. P., and Gardner-Medwin, A. R. (1973). Long-lasting potentiation of synaptic transmission in the dentate area of the unanaesthetized rabbit following stimulation of the perforant path. *Journal of Physiology (London), 232*, 357–374.

Bliss, T. V. P., and Lømo, T. (1973). Long-lasting potentiation of synaptic transmission in the dentate area of the anaesthetized rabbit following stimulation of the perforant path. *Journal of Physiology (London), 232*, 331–356.

Bliwise, D. L. (1989). Neuropsychological function and sleep. *Clinics in Geriatric Medicine, 5*, 381–394.

Blumberg, M. S., Sokoloff, G., and Kirby, R. F. (1997). Brown fat thermogenesis and cardiac rate regulation during cold challenge in infant rats. *American Journal of Physiology, 272*, R1308–R1313.

Bogaert, A. F. (2006). Biological versus nonbiological older brothers and men's sexual orientation. *Proceedings of the National Academy of Sciences, USA, 103*, 10771–10774.

Bogaert, A. F. (2007). Extreme right-handedness, older brothers, and sexual orientation in men. *Neuropsychology, 21*, 141–148.

Bolhuis, J. J., and Gahr, M. (2006). Neural mechanisms of birdsong memory. *Nature Reviews Neuroscience, 7*, 347–357.

Bolhuis, J. J., Okanoya, K., and Scharff, C. (2010). Twitter evolution: Converging mechanisms in birdsong and human speech. *Nature Reviews Neuroscience, 11*, 747–759.

Bonnel, A. M., and Prinzmetal, W. (1998). Dividing attention between the color and the shape of objects. *Perception & Psychophysics, 60*, 113–124.

Boodman, S. G. (2006, March 21). Mood machine: Now there's a device to treat depression. If only there were solid evidence that it works. *Washington Post*, p. HE01.

Boolell, M., Gepi-Attee, S., Gingell, J. C., and Allen, M. J. (1996). Sildenafil, a novel effective oral therapy for male erectile dysfunction. *British Journal of Urology, 78*, 257–261.

Boot, W. R., Kramer, A. F., Simons, D. J., Fabiani, M., et al. (2008). The effects of video game playing on attention, memory, and executive control. *Acta Psychologica* (Amsterdam), *129*, 387–398.

Borgstein, J., and Grootendorst, C. (2002). Clinical picture: Half a brain. *Lancet, 359*, 473.

Borod, J. C., Haywood, C. S., and Koff, E. (1997). Neuropsychological aspects of facial asymmetry during emotional expression: A review of the normal adult literature. *Neuropsychology Review, 7*, 41–60.

Boshuisen, K., van Schooneveld, M. M., Leijten, F. S., de Kort, G. A., et al. (2010). Contralateral MRI abnormalities affect seizure and cognitive outcome after hemispherectomy. *Neurology, 75,* 1623–1630.

Boström, P., Wu, J., Jedrychowski, M. P., Korde, A., et al. (2012). A PGC1-α-dependent myokine that drives brown-fat-like development of white fat and thermogenesis. *Nature, 481,* 463–846.

Bottjer, S. W., Miesner, E. A., and Arnold, A. P. (1984). Forebrain lesions disrupt development but not maintenance of song in passerine birds. *Science, 224,* 901–903.

Bourke, C. H., Stowe, Z. N., and Owens, M. J. (2014). Prenatal antidepressant exposure: Clinical and preclinical findings. *Pharmacological Reviews, 66,* 435–465.

Bourque, C. W. (2008). Central mechanisms of osmosensation and systemic osmoregulation. *Nature Reviews Neuroscience, 9,* 519–531.

Bouwknecht, J. A., Hijzen, T. H., van der Gugten, J., Maes, R. A., et al. (2001). Absence of 5-HT(1B) receptors is associated with impaired impulse control in male 5-HT(1B) knockout mice. *Biological Psychiatry, 49,* 557–568.

Bower, B. (2006). Prescription for controversy: Medications for depressed kids spark scientific dispute. *Science News, 169,* 168–172.

Bowman, M. L. (1997). *Individual differences in posttraumatic response.* Mahway, NJ: Erlbaum.

Brady, T. F., Konkle, T., Alvarez, G. A., and Oliva, A. (2014). Visual long-term memory has a massive storage capacity for object details. *Proceedings of the National Academy of Sciences, USA, 105,* 14325–14329.

Brain, P. F. (1994). Neurotransmission, the individual and the alcohol/aggression link. Commentary on Miczek et al. "Neuropharmacological characteristics of individual differences in alcohol effects on aggression in rodents and primates." *Behavioural Pharmacology, 5,* 422–424.

Brainard, D. H., Roorda, A., Yamauchi, Y., Calderone, J. B., et al. (2000). Functional consequences of the relative numbers of L and M cones. *Journal of the Optical Society of America. Part A, Optics, Image Science and Vision, 17,* 607–614.

Brandon, N. J., and Sawa, A. (2011). Linking neurodevelopmental and synaptic theories of mental illness through DISC1. *Nature Reviews Neuroscience, 12,* 707–722.

Branswell, H. (2014, April 1). Toronto amnesiac whose case helped rewrite chapters of the book on memory dies. *Toronto Star* (www.thestar.com/news/gta/2014/04/01/toronto_amnesiac_whose_case_helped_rewrite_chapters_of_the_book_on_memory_dies.html).

Brasser, S. M., Mozhui, K., and Smith, D. V. (2005). Differential covariation in taste responsiveness to bitter stimuli in rats. *Chemical Senses, 30,* 793–799.

Bray, G. A. (1969). Effect of caloric restriction on energy expenditure in obese patients. *Lancet, 2,* 397–398.

Breitner, J. C., Wyse, B. W., Anthony, J. C., Welsh-Bohmer, K. A., et al. (1999). APOE-epsilon4 count predicts age when prevalence of AD increases, then declines: The Cache County Study. *Neurology, 53,* 321–331.

Bremer, F. (1938). L'activité électrique de l'écorce cérébrale. *Actualités Scientifiques et Industrielles, 658,* 3–46.

Bremner, J. D., Randall, P., Scott, T. M., Bronen, R. A., et al. (1995). MRI-based measurement of hippocampal volume in patients with combat-related posttraumatic stress disorder. *American Journal of Psychiatry, 152,* 973–981.

Bremner, J. D., Scott, T. M., Delaney, R. C., Southwick, S. M., et al. (1993). Deficits in short-term memory in posttraumatic stress disorder. *American Journal of Psychiatry, 150,* 1015–1019.

Brennan, P., Kaba, H., and Keverne, E. B. (1990). Olfactory recognition: A simple memory system. *Science, 250,* 1223–1226.

Brigande, J. V., and Heller, S. (2009). *Quo vadis,* hair cell regeneration? *Nature Neuroscience, 12,* 679–685.

Broadbent, D. A. (1958). *Perception and communication.* New York, NY: Pergamon Press.

Broberg, D. J., and Bernstein, I. L. (1989). Cephalic insulin release in anorexic women. *Physiology & Behavior, 45,* 871–874.

Broughton, R., Billings, R., Cartwright, R., Doucette, D., et al. (1994). Homicidal somnambulism: A case report. *Sleep, 17,* 253–264.

Brown, A. S. (2011). The environment and susceptibility to schizophrenia. *Progress in Neurobiology, 93,* 23–58.

Brown, C. (2003, February 2). The man who mistook his wife for a deer. *The New York Times,* Section 6, p. 32.

Brown, E. C., Jeong, J. W., Muzik, O., Rothermel, R., et al. (2014). Evaluating the arcuate fasciculus with combined diffusion-weighted MRI tractography and electrocorticography. *Human Brain Mapping, 35,* 2333–2347.

Brown, H. (2006, November 26). One spoonful at a time. *New York Times Magazine.*

Brown, J. (1958). Some tests of the decay theory of immediate memory. *Quarterly Journal of Experimental Psychology, 10,* 12–21.

Brownlee, S., and Schrof, J. M. (1997). The quality of mercy. Effective pain treatments already exist. Why aren't doctors using them? *U.S. News & World Report, 122,* 54–67.

Bruel-Jungerman, E., Rampon, C., and Laroche, S. (2007). Adult hippocampal neurogenesis, synaptic plasticity and memory: Facts and hypotheses. *Review in the Neurosciences, 18,* 93–114.

Brunetti, M., Della Penna, S., Ferretti, A., Del Gratta, C., et al. (2008). A frontoparietal network for spatial attention reorienting in the auditory domain: A human fMRI/MEG study of functional and temporal dynamics. *Cerebral Cortex, 18,* 1139–1147.

Bryant, P., Trinder, J., and Curtis, N. (2004). Sick and tired: Does sleep have a vital role in the immune system? *Nature Reviews Immunology, 4,* 457–467.

Bu, G. (2009). Apolipoprotein E and its receptors in Alzheimer's disease: Pathways, pathogenesis and therapy. *Nature Reviews Neuroscience, 10,* 333–344.

Buccino, G., Lui, F., Canessa, N., Patteri, I., et al. (2004). Neural circuits involved in the recognition of actions performed by nonconspecifics: An fMRI study. *Journal of Cognitive Neuroscience, 16,* 114–126.

Buccino, G., Solodkin, A., and Small, S. L. (2006). Functions of the mirror neuron system: Implications for neurorehabilitation. *Cognitive and Behavioral Neurology, 19,* 55–63.

Buchsbaum, B. R., Baldo, J., Okada, K., Berman, K. F., et al. (2011). Conduction aphasia, sensory-motor integration, and phonological short-term memory—An aggregate analysis of lesion and fMRI data. *Brain and Language, 119,* 119–128.

Buchsbaum, M. S., Buchsbaum, B. R., Chokron, S., Tang, C., et al. (2006). Thalamocortical circuits: fMRI assessment of the pulvinar and medial dorsal nucleus in normal volunteers. *Neuroscience Letters, 404,* 282–287.

Buchsbaum, M. S., Mirsky, A. F., DeLisi, L. E., Morihisa, J., et al. (1984). The Genain quadruplets: Electrophysiological, positron emission and X-ray tomographic studies. *Psychiatry Research, 13,* 95–108.

Buck, L., and Axel, R. (1991). A novel multigene family may encode odorant receptors: A molecular basis for odor recognition. *Cell, 65,* 175–187.

Burgdorf, J., Kroes, R. A., Moskal, J. R., Pfaus, J. G., et al. (2008). Ultrasonic vocalizations of rats (*Rattus norvegicus*) during mating, play, and aggression: Behavioral concomitants, relationship to reward, and self-administration of playback. *Journal of Comparative Psychology, 122,* 357–367.

Burgdorf, J., Panksepp, J., and Moskal, J. R. (2011). Frequency-modulated 50 kHz ultrasonic vocalizations: A tool for uncovering the molecular substrates of positive affect. *Neuroscience & Biobehavioral Reviews, 35,* 1831–1836.

Bushman, J. D., Ye, W., and Liman, E. R. (2015). A proton current associated with sour taste: Distribution and functional properties. *FASEB Journal, 29,* 3014–3026.

Buss, D. (2013). *Evolutionary psychology: The new science of the mind*. New York, NY: Psychology Press.

Butler, A. C., Chapman, J. E., Forman, E. M., and Beck, A. T. (2006). The empirical status of cognitive-behavioral therapy: A review of meta-analyses. *Clinical Psychology Review, 26*, 17–31.

Byne, W., Tobet, S., Mattiace, L. A., Lasco, M. S., et al. (2001). The interstitial nuclei of the human anterior hypothalamus: An investigation of variation with sex, sexual orientation, and HIV status. *Hormones and Behavior, 40*, 86–92.

C

Cahill, L. (1997). The neurobiology of emotionally influenced memory: Implications for understanding traumatic memory. In R. Yehuda and A. C. McFarlane (Eds.), *Annals of the New York Academy of Sciences: Vol. 41. Psychobiology of traumatic stress disorder* (pp. 238–246). New York, NY: New York Academy of Sciences.

Cahill, L., and McGaugh, J. L. (1991). NMDA-induced lesions of the amygdaloid complex block the retention-enhancing effect of posttraining epinephrine. *Psychobiology, 19*, 206–210.

Cahill, L., Prins, B., Weber, M., and McGaugh, J. L. (1994). Beta-adrenergic activation and memory for emotional events. *Nature, 371*, 702–704.

Calder, A. J., Keane, J., Manes, F., Antoun, N., et al. (2000). Impaired recognition and experience of disgust following brain injury. *Nature Neuroscience, 3*, 1077.

Calvert, G. A., Bullmore, E. T., Brammer, M. J., Campbell, R., et al. (1997). Activation of auditory cortex during silent lipreading. *Science, 276*, 593–596.

Calvin, W. H., and Ojemann, G. A. (1994). *Conversation's with Neil's brain: The neural nature of thought and language*. Reading, MA: Adison-Wesley.

Campbell, F. W., and Robson, J. G. (1968). Application of Fourier analysis to the visibility of gratings. *Journal of Physiology (London), 197*, 551–566.

Cannon, T. D., Chung, Y., He, G., Sun, D., et al. (2015). Progressive reduction in cortical thickness as psychosis develops: A multisite longitudinal neuroimaging study of youth at elevated clinical risk. *Biological Psychiatry, 77*, 147–157.

Cannon, W. B. (1929). *Bodily changes in pain, hunger, fear and rage*. New York, NY: Appleton.

Cantalupo, C., and Hopkins, W. D. (2001). Asymmetric Broca's area in great apes. *Nature, 414*, 505.

Cantor, J. M., Blanchard, R., Paterson, A. D., and Bogaert, A. F. (2002). How many gay men owe their sexual orientation to fraternal birth order? *Archives of Sexual Behavior, 31*, 63–71.

Cao, M., Shu, N., Cao, Q., Wang, Y., et al. (2014). Imaging functional and structural brain connectomics in attention-deficit/hyperactivity disorder. *Molecular Neurobiology, 50*, 1111–1123.

Cao, Y. Q., Mantyh, P. W., Carlson, E. J., Gillespie, A. M., et al. (1998). Primary afferent tachykinins are required to experience moderate to intense pain. *Nature, 392*, 390–394.

Caramazza, A., Anzellotti, S., Strnad, L., and Lingnau, A. (2014). Embodied cognition and mirror neurons: A critical assessment. *Annual Review of Neuroscience, 37*, 1–15.

Cardno, A. G., and Gottesman, I. I. (2000). Twin studies of schizophrenia: From bow-and-arrow concordances to star wars Mx and functional genomics. *American Journal of Medical Genetics, 97*, 12–17.

Carhart-Harris, R. L., Erritzoe, D., Williams, T., Stone, J. M., et al. (2012). Neural correlates of the psychedelic state as determined by fMRI studies with psilocybin. *Proceedings of the National Academy of Sciences, USA, 109*, 2138–2143.

Carmichael, M. S., Warburton, V. L., Dixen, J., and Davidson, J. M. (1994). Relationships among cardiovascular, muscular, and oxytocin responses during human sexual activity. *Archives of Sexual Behavior, 23*, 59–79.

Carpen, J. D., Archer, S. N., Skene, D. J., Smits, M., et al. (2005). A single-nucleotide polymorphism in the 59-untranslated region of the *hPER2* gene is associated with diurnal preference. *Journal of Sleep Research, 14*, 293–297.

Carreiras, M., Lopez, J., Rivero, F., and Corina, D. (2005). Linguistic perception: Neural processing of a whistled language. *Nature, 433*, 31–32.

Carroll, J., McMahon, C., Neitz, M., and Neitz, J. (2000). Flicker-photometric electroretinogram estimates of L:M cone photoreceptor ratio in men with photopigment spectra derived from genetics. *Journal of the Optical Society of America. Part A, Optics, Image Science, and Vision, 17*, 499–509.

Carter, C. S. (1992). Oxytocin and sexual behavior. *Neuroscience and Biobehavioral Reviews, 16*, 131–144.

Cartwright, R. D. (1979). The nature and function of repetitive dreams: A survey and speculation. *Psychiatry, 42*, 131–137.

Casey, D. E. (1989). Clozapine: Neuroleptic-induced EPS and tardive dyskinesia. *Psychopharmacology (Berlin), 99*, S47–S53.

Cassens, G., Wolfe, L., and Zola, M. (1990). The neuropsychology of depressions. *Journal of Neuropsychiatry and Clinical Neurosciences, 2*, 202–213.

Castellanos, F. X., Lee, P. P., Sharp, W., Jeffries, N. O., et al. (2002). Developmental trajectories of brain volume abnormalities in children and adolescents with attention-deficit/hyperactivity disorder. *Journal of the American Medical Association, 288*, 1740–1748.

Caterina, M. J., Leffler, A., Malmberg, A. B., Martin, W. J., et al. (2000). Impaired nociception and pain sensation in mice lacking the capsaicin receptor. *Science, 288*, 306–313.

Caterina, M. J., Schumacher, M. A., Tominaga, M., Rosen, T. A., et al. (1997). The capsaicin receptor: A heat-activated ion channel in the pain pathway. *Nature, 389*, 816–824.

CDC (Centers for Disease Control and Prevention). (2010). Current depression among adults—United States, 2006 and 2008. *Morbidity and Mortality Weekly Report, 59*, 1229–1235 (www.cdc.gov/mmwr/preview/mmwrhtml/mm5938a2.htm).

Centerwall, B. S., and Criqui, M. H. (1978). Prevention of the Wernicke-Korsakoff syndrome: A cost-benefit analysis. *New England Journal of Medicine, 299*, 285–289.

Chamberlain, S. R., Menzies, L., Hampshire, A., Suckling, J., et al. (2008). Orbitofrontal dysfunction in patients with obsessive-compulsive disorder and their unaffected relatives. *Science, 321*, 421–422.

Champagne, F., Diorio, J., Sharma, S., and Meaney, M. J. (2001). Naturally occurring variations in maternal behavior in the rat are associated with differences in estrogen-inducible central oxytocin receptors. *Proceedings of the National Academy of Sciences, USA, 98*, 12736–12741.

Chandrashekar, J., Hoon, M. A., Ryba, N. J., and Zuker, C. S. (2006). The receptors and cells for mammalian taste. *Nature, 444*, 288–294.

Chandrashekar, J., Mueller, K. L., Hoon, M. A., Adler, E., et al. (2000). T2Rs function as bitter taste receptors. *Cell, 100*, 703–711.

Chandrashekar, J., Yarmolinsky, D., von Buchholtz, L., Oka, Y., et al. (2009). The taste of carbonation. *Science, 326*, 443–445.

Chang, B. S., Ly, J., Appignani, B., Bodell, A., et al. (2005). Reading impairment in the neuronal migration disorder of periventricular nodular heterotopia. *Neurology, 64*, 799–803.

Chapman, C. D., Dono, L. M., French, M. C., Weinberg, Z. Y., et al. (2012). Paraventricular nucleus anandamide signaling alters eating and substrate oxidation. *NeuroReport, 23*, 425–429.

Charney, D. S., Deutch, A. Y., Krystal, J. H., Southwick, S. M., et al. (1993). Psychobiologic mechanisms of posttraumatic stress disorder. *Archives of General Psychiatry, 50*, 295–305.

Chaudhari, N., Landin, A. M., and Roper, S. D. (2000). A metabotropic glutamate receptor variant functions as a taste receptor. *Nature Neuroscience, 3*, 113–119.

Chelikani, P. K., Haver, A. C., and Reidelberger, R. D. (2005). Intravenous infusion of peptide YY(3-36) potently inhibits food intake in rats. *Endocrinology, 146*, 879–888.

Chemelli, R. M., Willie, J. T., Sinton, C. M., Elmquist, J. K., et al. (1999). Narcolepsy in orexin knockout mice: Molecular genetics of sleep regulation. *Cell, 98,* 437–451.

Chen, L., and Feany, M. B. (2005). α-Synuclein phosphorylation controls neurotoxicity and inclusion formation in a *Drosophila* model of Parkinson disease. *Nature Neuroscience, 8,* 657–663.

Cherry, E. C. (1953). Some experiments on the recognition of speech, with one and with two ears. *Journal of the Acoustical Society of America, 25,* 975–979.

Cheyne, J. A. (2002). Situational factors affecting sleep paralysis and associated hallucinations: Position and timing effects. *Journal of Sleep Research, 11,* 169–177.

Ciccocioppo, R., Martin-Fardon, R., and Weiss, F. (2004). Stimuli associated with a single cocaine experience elicit long-lasting cocaine-seeking. *Nature Neuroscience, 7,* 495–496.

Clapham, J. C., Arch, J. R. S., Chapman, H., Haynes, A., et al. (2000). Mice overexpressing human uncoupling protein-3 in skeletal muscle are hyperphagic and lean. *Nature, 406,* 415–418.

Clarke, M. C., Tanskanen, A., Huttunen, M., Leon, D. A., et al. (2011). Increased risk of schizophrenia from additive interaction between infant motor developmental delay and obstetric complications: Evidence from a population-based longitudinal study. *American Journal of Psychiatry, 168,* 1295–1302.

Classen, J., Liepert, J., Wise, S. P., Hallett, M., et al. (1998). Rapid plasticity of human cortical movement representation induced by practice. *Journal of Neurophysiology, 79,* 1117–1123.

Clemente, C. D., and Sterman, M. B. (1967). Limbic and other forebrain mechanisms in sleep induction and behavioral inhibition. *Progress in Brain Research, 27,* 34–37.

Coghill, R. C., McHaffie, J. G., and Yen, Y.-F. (2003). Neural correlates of interindividual differences in the subjective experience of pain. *Proceedings of the National Academy of Sciences, USA, 100,* 8538–8542.

Cohen, A. H., Baker, M. T., and Dobrov, T. A. (1989). Evidence for functional regeneration in the adult lamprey spinal cord following transection. *Brain Research, 496,* 368–372.

Cohen, A. J., and Leckman, J. F. (1992). Sensory phenomena associated with Gilles de la Tourette's syndrome. *Journal of Clinical Psychiatry, 53,* 319–323.

Cohen, S., Alper, C. M., Doyle, W. H., Treanor, J. J., et al. (2006). Positive emotional style predicts resistance to illness after experimental exposure to rhinovirus or influenza A virus. *Psychosomatic Medicine, 68,* 809–815.

Cohen, S., Doyle, W. J., Alper, C. M., Janicki-Deverts, D., et al. (2009). Sleep habits and susceptibility to the common cold. *Archive of Internal Medicine, 169,* 62–67.

Cohen, S., Lichtenstein, E., Prochaska, J. O., Rossi, J. S., et al. (1989). Debunking myths about quitting: Evidence from 10 perspective studies of persons who attempt to quit smoking by themselves. *American Psychologist, 44,* 1355–1365.

Cole, J. (1995). *Pride and a daily marathon.* Cambridge, MA: MIT Press.

Colman, R. J., Beasley, T. M., Kemnitz, J. W., Johnson, S. C., et al. (2014). Caloric restriction reduces age-related and all-cause mortality in rhesus monkeys. *Nature Communications, 5,* 3557.

Conel, J. L. (1939). *The postnatal development of the human cerebral cortex: Vol. 1. The cortex of the newborn.* Cambridge, MA: Harvard University Press.

Conel, J. L. (1947). *The postnatal development of the human cerebral cortex: Vol. 3. The cortex of the three-month infant.* Cambridge, MA: Harvard University Press.

Conel, J. L. (1959). *The postnatal development of the human cerebral cortex: Vol. 6. The cortex of the twenty-four-month infant.* Cambridge, MA: Harvard University Press.

Conrad, A. J., Abebe, T., Austin, R., Forsythe, S., et al. (1991). Hippocampal pyramidal cell disarray in schizophrenia as a bilateral phenomenon. *Archives of General Psychiatry, 48,* 413–417.

Cooke, B. M., Breedlove, S. M., and Jordan, C. L. (2003). Both estrogen receptors and androgen receptors contribute to testosterone-induced changes in the morphology of the medial amygdala and sexual arousal in male rats. *Hormones and Behavior, 43,* 336–346.

Cooke, B. M., Chowanadisai, W., and Breedlove, S. M. (2000). Post-weaning social isolation of male rats reduces the volume of the medial amygdala and leads to deficits in adult sexual behavior. *Behavioural Brain Research, 117,* 107–113.

Cope, N., Harold, D., Hill, G., Moskvina, V., et al. (2005). Strong evidence that KIAA0319 on chromosome 6p is a susceptibility gene for developmental dyslexia. *American Journal of Human Genetics, 76,* 581–591.

Corballis, M. C. (2002). *From hand to mouth: The origins of language.* Princeton, NJ: Princeton University Press.

Corbetta, M., Kincade, J. M., Ollinger, J. M., McAvoy, M. P., et al. (2000). Voluntary orienting is dissociated from target detection in human posterior parietal cortex. *Nature Neuroscience, 3,* 292–297.

Corbetta, M., and Shulman, G. L. (1998). Human cortical mechanisms of visual attention during orienting and search. *Philosophical Transactions of the Royal Society of London. Series B: Biological Sciences, 353,* 1353–1362.

Corbetta, M., and Shulman, G. L. (2002). Control of goal-directed and stimulus-driven attention in the brain. *Nature Reviews Neuroscience, 3,* 201–215.

Corcoran, A. J., Barber, J. R., and Conner, W. E. (2009). Tiger moth jams bat sonar. *Science, 325,* 325–327.

Coricelli, G., Critchley, H. D., Joffily, M., O'Doherty, J. P., et al. (2005). Regret and its avoidance: A neuroimaging study of choice behavior. *Nature Neuroscience, 8,* 1255–1262.

Corkin, S. (2002). What's new with the amnesic patient H.M.? *Neuroscience, 3,* 153–159.

Corkin, S., Amaral., D. G., Gonzalez, R. G., Johnson, K. A., et al. (1997). H.M.'s medial temporal lobe lesion: Findings from magnetic resonance imaging. *Journal of Neuroscience, 17,* 3964–3979.

Coryell, W., Noyes, R., Jr., and House, J. D. (1986). Mortality among outpatients with anxiety disorders. *American Journal of Psychiatry, 143,* 508–510.

Costanzo, R. M. (1991). Regeneration of olfactory receptor cells. *CIBA Foundation Symposium, 160,* 233–242.

Counotte, D. S., Goriounova, N. A., Li, K. W., Loos, M., et al. (2011). Lasting synaptic changes underlie attention deficits caused by nicotine exposure during adolescence. *Nature Neuroscience, 14,* 417–419.

Courtney, S., Ungerleider, L., Keil, K., and Haxby, J. (1996). Object and spatial visual working memory activate separate neural systems in human cortex. *Cerebral Cortex, 6,* 39–49.

Cox, J. J., Reimann, F., Nicholas, A. K., Thornton, G., et al. (2006). An SCN9A channelopathy causes congenital inability to experience pain. *Nature, 444,* 894–898.

Cox, S. S., Speaker, K. J., Beninson, L. A., Craig, W. C., et al. (2014). Adrenergic and glucocorticoid modulation of the sterile inflammatory response. *Brain, Behavior, and Immunity, 36,* 183–192.

Coyle, J. T., Tsai, G., and Goff, D. (2003). Converging evidence of NMDA receptor hypofunction in the pathophysiology of schizophrenia. *Annals of the New York Academy of Sciences, 1003,* 318–327.

Cragg, B. G. (1975). The development of synapses in the visual system of the cat. *Journal of Comparative Neurology, 160,* 147–166.

Crasto, C., Singer, M. S., and Shepherd, G. M. (2001). The olfactory receptor family album. *Genome Biology, 2,* reviews1027.1–reviews1027.4.

Crockford, C., Wittig, R. M., Mundry, R., and Zuberbühler, K. (2012). Wild chimpanzees inform ignorant group members of danger. *Current Biology, 22,* 142–146.

Cruce, J. A. F., Greenwood, M. R. C., Johnson, P. R., and Quartermain, D. (1974). Genetic versus hypothalamic obesity: Studies of intake and dietary manipulation in rats. *Journal of Comparative and Physiological Psychology, 87,* 295–301.

Cryns, K., and Van Camp, G. (2004). Deafness genes and their diagnostic applications. *Audiology & Neuro-Otology, 9,* 2–22.

Cummings, D. E. (2006). Ghrelin and the short- and long-term regulation of appetite and body weight. *Physiology & Behavior, 89,* 71–84.

Cummings, J. L. (1995). Dementia: The failing brain. *Lancet, 345,* 1481–1484.

Curcio, C. A., Sloan, K. R., Packer, O., Hendrickson, A. E., et al. (1987). Distribution of cones in human and monkey retina: Individual variability and radial asymmetry. *Science, 236,* 579–582.

Curtis, V., Aunger, R., and Rabie, T. (2004). Evidence that disgust evolved to protect from risk of disease. *Proceedings of the Royal Society of London. Series B: Biological Sciences, 271*(Suppl. 4), S131–S133.

Curtiss, S. (1989). The independence and task-specificity of language. In M. H. Bornstein and J. S. Bruner (Eds.), *Interaction in human development* (pp. 105–137). Hillsdale, NJ: Erlbaum.

Cytowic, R. E., and Eagleman, D. M. (2009). *Wednesday is indigo blue: Discovering the brain of synesthesia.* Cambridge, MA: MIT Press.

D

Dabbs, J. M., Jr., and Hargrove, M. F. (1997). Age, testosterone, and behavior among female prison inmates. *Psychosomatic Medicine, 59,* 477–480.

Dabbs, J. M., and Morris, R. (1990). Testosterone, social class, and antisocial behavior in a sample of 4,462 men. *Psychological Science, 1,* 209–211.

Dabbs, J. M., Ruback, R. B., Frady, R. L., Hopper, C. H., et al. (1988). Saliva testosterone and criminal violence among women. *Personality and Individual Differences, 9,* 269–275.

Dale, R. C., Heyman, I., Giovannoni, G., and Church, A. W. (2005). Incidence of anti-brain antibodies in children with obsessive-compulsive disorder. *British Journal of Psychiatry, 187,* 314–319.

Dalton, K. M., Nacewicz, B. M., Johnstone, T., Schaefer, H. S., et al. (2005). Gaze fixation and the neural circuitry of face processing in autism. *Nature Neuroscience, 8,* 519–526.

Damasio, A. R., Grabowski, T. J., Bechara, A., Damasio, H., et al. (2000). Subcortical and cortical brain activity during the feeling of self-generated emotions. *Nature Neuroscience, 3,* 1049–1056.

Damasio, H., Grabowski, T., Frank, R., Galaburda, A. M., et al. (1994). The return of Phineas Gage: Clues about the brain from the skull of a famous patient. *Science, 264,* 1102–1105.

Damassa, D. A., Smith, E. R., Tennent, B., and Davidson, J. M. (1977). The relationship between circulating testosterone levels and male sexual behavior in rats. *Hormones and Behavior, 8,* 275–286.

Daniele, C. A., and MacDermott, A. B. (2009). Low-threshold primary afferent drive onto GABAergic interneurons in the superficial dorsal horn of the mouse. *Journal of Neuroscience, 29,* 686–695.

Dantz, B., Edgar, D. M., and Dement, W. C. (1994). Circadian rhythms in narcolepsy: Studies on a 90 minute day. *Electroencephalography and Clinical Neurophysiology, 90,* 24–35.

Dantzer, R., O'Connor, J. C., Freund, G. G., Johnson, R. W., et al. (2008). From inflammation to sickness and depression: When the immune system subjugates the brain. *Nature Reviews Neuroscience, 9,* 46–56.

Dapretto, M., Davies, M. S., Pfeifer, J. H., Scott, A. A., et al. (2006). Understanding emotions in others: Mirror neuron dysfunction in children with autism spectrum disorders. *Nature Neuroscience, 9,* 28–30.

D'Ardenne, K., McClure, S. M., Nystrom, L. E., and Cohen, J. D. (2008). BOLD responses reflecting dopaminergic signals in the human ventral tegmental area. *Science, 319,* 1264–1267.

Darwin, C. (1872). *The expression of the emotions in man and animals.* London, England: J. Murray.

Davalos, D., Grutzendler, J., Yang, G., Kim, J. V., et al. (2005). ATP mediates rapid microglial response to local brain injury in vivo. *Nature Neuroscience, 8,* 752–758.

Davey-Smith, G., Frankel, S., and Yarnell, J. (1997). Sex and death: Are they related? Findings from the Caerphilly Cohort Study. *British Medical Journal (Clinical Research Edition), 315,* 1641–1644.

Davidson, J. M., Camargo, C. A., and Smith, E. R. (1979). Effects of androgen on sexual behavior in hypogonadal men. *Journal of Clinical Endocrinology and Metabolism, 48,* 955–958.

Davis, J. I., Senghas, A., Brandt, F., and Ochsner, K. N. (2010). The effects of BOTOX injections on emotional experience. *Emotion, 10,* 433–440.

Davis, J. I., Senghas, A., and Ochsner, K. N. (2009). How does facial feedback modulate emotional experience? *Journal of Research in Personality, 43,* 822–829.

Dawson, T. M., and Dawson, V. L. (2003). Molecular pathways of neurodegeneration in Parkinson's disease. *Science, 302,* 819–822.

Dearborn, G. V. N. (1932). A case of congenital general pure analgesia. *Journal of Nervous and Mental Disease, 75,* 612–615.

De Felipe, C., Herrero, J. F., O'Brien, J. A., Palmer, J. A., et al. (1998). Altered nociception, analgesia and aggression in mice lacking the receptor for substance P. *Nature, 392,* 394–397.

De Gelder, B., Tamietto, M., van Boxtel, G., Goebel, R., et al. (2008). Intact navigation skills after bilateral loss of striate cortex. *Current Biology, 18,* R1128–R1129.

De Groot, C. M., Janus, M. D., and Bornstein, R. A. (1995). Clinical predictors of psychopathology in children and adolescents with Tourette syndrome. *Journal of Psychiatric Research, 29,* 59–70.

Dehaene-Lambertz, G., Dehaene, S., and Hertz-Pannier, L. (2002). Functional neuroimaging of speech perception in infants. *Science, 298,* 2013–2015.

Delgado, J. M. R. (1969). *Physical control of the mind: Toward a psychocivilized society.* New York, NY: Harper & Row.

Dennis, S. G., and Melzack, R. (1983). Perspectives on phylogenetic evolution of pain expression. In R. L. Kitchell, H. H. Erickson, E. Carstens, and L. E. Davis (Eds.), *Animal pain* (pp. 151–161). Bethesda, MD: American Physiological Society.

Denton, D., Shade, R., Zamarippa, F., Egan, G., et al. (1999). Neuroimaging of genesis and satiation of thirst and an interoceptor-driven theory of origins of primary consciousness. *Proceedings of the National Academy of Sciences, USA, 96,* 5304–5309.

Depaepe, V., Suarez-Gonzalez, N., Dufour, A., Passante, L., et al. (2005). Ephrin signalling controls brain size by regulating apoptosis of neural progenitors. *Nature, 435,* 1244–1250.

DeRubeis, R. J., Siegle, G. J., and Hollon, S. D. (2008). Cognitive therapy versus medication for depression: Treatment outcomes and neural mechanisms. *Nature, 9,* 788–796.

Desimone, R., Albright, T. D., Gross, C. G., and Bruce, C. (1984). Stimulus-selective properties of inferior temporal neurons in the macaque. *Journal of Neuroscience, 4,* 2051–2062.

De Valois, K. K., De Valois, R. L., and Yund, E. W. (1979). Responses of striate cortex cells to grating and checkerboard patterns. *Journal of Physiology (London), 291,* 483–505.

De Valois, R. L., and De Valois, K. K. (1980). Spatial vision. *Annual Review of Psychology, 31,* 309–341.

De Valois, R. L., and De Valois, K. K. (1988). *Spatial vision.* New York, NY: Oxford University Press.

De Valois, R. L., and De Valois, K. K. (1993). A multi-stage color model. *Vision Research, 33,* 1053–1065.

Devane, W. A., Dysarz, F. A., Johnson, M. R., Melvin, L. S., et al. (1988). Determination and characterization of a cannabinoid receptor in rat brain. *Molecular Pharmacology, 34,* 605–613.

Devane, W. A., Hanus, L., Breuer, A., Pertwee, R. G., et al. (1992). Isolation and structure of a brain constituent that binds the cannabinoid receptor. *Science, 258,* 1946–1949.

Devlin, J. T., and Watkins, K. E. (2007). Stimulating language: Insights from TMS. *Brain, 130*(Pt. 3), 610–622.

DeVoogd, T. J. (1994). Interactions between endocrinology and learning in the avian song system. *Annals of the New York Academy of Sciences, 743,* 19–41.

De Waal, F. B. M. (2003). Darwin's legacy and the study of primate visual communication. *Annals of the New York Academy of Sciences, 1000,* 7–31.

De Win, M. M., Jager, G., Booij, J., Reneman, L., et al. (2008). Sustained effects of ecstasy on the human brain: A prospective neuroimaging study in novel users. *Brain, 131*(Pt. 11), 2936–2945.

Dewsbury, D. A. (1972). Patterns of copulatory behavior in male mammals. *Quarterly Review of Biology, 47,* 1–33.

Diamond, J., Cooper, E., Turner, C., and Macintyre, L. (1976). Trophic regulation of nerve sprouting. *Science, 193,* 371–377.

Diamond, M. C. (1967). Extensive cortical depth measurements and neuron size increases in the cortex of environmentally enriched rats. *Journal of Comparative Neurology, 131,* 357–364.

Diamond, M. C., Lindner, B., Johnson, R., Bennett, E. L., et al. (1975). Differences in occipital cortical synapses from environmentally enriched, impoverished, and standard colony rats. *Journal of Neuroscience Research, 1,* 109–119.

Dichgans, J. (1984). Clinical symptoms of cerebellar dysfunction and their topodiagnostical significance. *Human Neurobiology, 2,* 269–279.

Di Chiara, G., Tanda, G., Bassareo, V., Pontieri, F., et al. (1999). Drug addiction as a disorder of associative learning. Role of nucleus accumbens shell/extended amygdala dopamine. *Annals of the New York Academy of Sciences, 877,* 461–485.

Dietz, P. M., Williams, S. B., Callaghan, W. M., Bachman, D. J., et al. (2007). Clinically identified maternal depression before, during, and after pregnancies ending in live births. *American Journal of Psychiatry, 164,* 1457–1459.

Di Marzo, V., and Matias, I. (2005). Endocannabinoid control of food intake and energy balance. *Nature Neuroscience, 8,* 585–589.

Dittmann, R. W., Kappes, M. E., and Kappes, M. H. (1992). Sexual behavior in adolescent and adult females with congenital adrenal hyperplasia. *Psychoneuroendocrinology, 17,* 153–170.

Do, M. T. H., Kang, S. H., Zue, T., Zhong, H., et al. (2009). Photon capture and signalling by melanopsin retinal ganglion cells. *Nature, 457,* 281–287.

Dohanich, G. (2003). Ovarian steroids and cognitive function. *Current Directions in Psychological Science, 12,* 57–61.

Dohrenwend, B. P., Turner, J. B., Turse, N. A., Adams, B. G., et al. (2006). The psychological risks of Vietnam for U.S. veterans: A revisit with new data and methods. *Science, 313,* 979–982.

Dolan, R. J. (2002). Emotion, cognition, and behavior. *Science, 298,* 1191–1194.

Dolder, C. R., and Nelson, M. H. (2008). Hypnosedative-induced complex behaviours: Incidence, mechanisms and management. *CNS Drugs, 22,* 1021–1036.

Domjan, M., and Purdy, J. E. (1995). Animal research in psychology: More than meets the eye of the general psychology student. *American Psychologist, 50,* 496–503.

Donaldson, Z. R., and Young, L. J. (2008). Oxytocin, vasopressin, and the neurogenetics of sociality. *Science, 322,* 900–903.

Dorsaint-Pierre, R., Penhune, V. B., Watkins, K. E., Neelin, P., et al. (2006). Asymmetries of the planum temporale and Heschl's gyrus: Relationship to language lateralization. *Brain, 129,* 1164–1176.

Drea, C. M., Weldele, M. L., Forger, N. G., Coscia, E. M., et al. (1998). Androgens and masculinization of genitalia in the spotted hyaena (*Crocuta crocuta*). 2. Effects of prenatal anti-androgens. *Journal of Reproduction and Fertility, 113,* 117–127.

Drevets, W. C. (1998). Functional neuroimaging studies of depression: The anatomy of melancholia. *Annual Review of Medicine, 49,* 341–361.

Drickamer, L. C. (1992). Behavioral selection of odor cues by young female mice affects age of puberty. *Developmental Psychobiology, 25,* 461–470.

Dronkers, N. F., Plaisant, O., Iba-Zizen, M. T., and Cabanis, E. A. (2007). Paul Broca's historic cases: High resolution MR imaging of the brains of Leborgne and Lelong. *Brain, 130* (Pt. 5), 1432–1441.

Dronkers, N. F., Wilkins, D. P., Van Valin, R. D., Jr., Redfern, B. B., et al. (2004). Lesion analysis of the brain areas involved in language comprehension. *Cognition, 92,* 145–177.

Druckman, D., and Bjork, R. A. (1994). *Learning, remembering, believing: Enhancing human performance.* Washington, DC: National Academy Press.

Du, J. L., and Poo, M. M. (2004). Rapid BDNF-induced retrograde synaptic modification in a developing retinotectal system. *Nature, 429,* 878–882.

Du, L., Bakish, D., Lapierre, Y. D., Ravindran, A. V., et al. (2000). Association of polymorphism of serotonin 2A receptor gene with suicidal ideation in major depressive disorder. *American Journal of Medical Genetics, 96,* 56–60.

Duchaine, B., Germine, L., and Nakayama, K. (2007). Family resemblance: Ten family members with prosopagnosia and within-class object agnosia. *Cognitive Neuropsychology, 24,* 419–430.

Duchamp-Viret, P., Chaput, M. A., and Duchamp, A. (1999). Odor response properties of rat olfactory receptor neurons. *Science, 284,* 2171–2174.

Duffy, J. D., and Campbell, J. J. (1994). The regional prefrontal syndromes: A theoretical and clinical overview. *Journal of Neuropsychiatry and Clinical Neurosciences, 6,* 379–387.

Dulac, C., and Torello, A. T. (2003). Molecular detection of pheromone signals in mammals, from genes to behaviour. *Nature Reviews Neuroscience, 4,* 551–562.

Dully, H., and Fleming, C. (2007). *My lobotomy.* New York, NY: Crown.

Dunah, A. W., Hyunkyung, J., Griffin, A., Kim, Y.-M., et al. (2002). Sp1 and TAFII130 transcriptional activity disrupted in early Huntington's disease. *Science, 296,* 2238–2242.

E

Earnest, D. J., Liang, F. Q., Ratcliff, M., and Cassone, V. M. (1999). Immortal time: Circadian clock properties of rat suprachiasmatic cell lines. *Science, 283,* 693–695.

Ebbinghaus, H. (1908). *Psychology: An elementary textbook.* Boston: Heath.

Edelsohn, G. A. (2006). Hallucinations in children and adolescents: Considerations in the emergency setting. *American Journal of Psychiatry, 163,* 781–785.

Edwards, R. R., Grace, E., Peterson, S., Klick, B., et al. (2009). Sleep continuity and architecture: Associations with pain-inhibitory processes in patients with temporomandibular joint disorder. *European Journal of Pain, 13,* 1043–1047.

Egaas, B., Courchesne, E., and Saitoh, O. (1995). Reduced size of corpus callosum in autism. *Archives of Neurology, 52,* 794–801.

Elbert, T., Pantev, C., Wienbruch, C., Rockstroh, B., et al. (1995). Increased cortical representation of the fingers of the left hand in string players. *Science, 270,* 305–307.

Ellenbogen, J. M., Hu, P. T., Payne, J. D., Titone, D., et al. (2007). Human relational memory requires time and sleep. *Proceedings of the National Academy of Sciences, USA, 104,* 7317–7318.

Emery, N. J., Capitanio, J. P., Mason, W. A., Machado, C. J., et al. (2001). The effects of bilateral lesions of the amygdala on dyadic social interactions in rhesus monkeys (*Macaca mulatta*). *Behavioral Neuroscience, 115,* 515–544.

Engel, J., Jr. (1992). Recent advances in surgical treatment of temporal lobe epilepsy. *Acta Neurologica Scandinavica. Supplementum, 140,* 71–80.

English, P. J., Ghatei, M. A., Malik, I. A., Bloom, S. R., et al. (2002). Food fails to suppress ghrelin levels in obese humans. *Journal of Clinical Endocrinology and Metabolism, 87,* 2984–2987.

Enoch, M. A., Kaye, W. H., Rotondo, A., Greenberg, B. D., et al. (1998). 5-HT2A promoter polymorphism -1438G/A, anorexia nervosa, and obsessive-compulsive disorder. *Lancet, 351,* 1785–1786.

Epstein, A. N., Fitzsimons, J. T., and Rolls, B. J. (1970). Drinking induced by injection of angiotensin into the brain of the rat. *Journal of Physiology (London), 210,* 457–474.

Eriksson, A., and Lacerda, F. (2007). Charlantry in forensic speech science: A problem to be taken seriously. *International Journal of Speech Language and the Law, 14,* 169–193.

Erlanger, D. M., Kutner, K. C., Barth, J. T., and Barnes, R. (1999). Neuropsychology of sports-related head injury: Dementia pugilistica to post concussion syndrome. *Clinical Neuropsychologist, 13,* 193–209.

Ernst, T., Chang, L., Leonido-Yee, M., and Speck, O. (2000). Evidence for long-term neurotoxicity associated with methamphetamine abuse: A 1H MRS study. *Neurology, 54,* 1344–1349.

Erren, T. C., Morfeld, P., Stork, J., Knauth, P., et al. (2009). Shift work, chronodisruption and cancer?—The IARC 2007 challenge for research and prevention and 10 theses from the Cologne Colloquium 2008. *Scandinavian Journal of Work, Environment & Health, 35,* 74–79.

Everitt, B. J., and Stacey, P. (1987). Studies of instrumental behavior with sexual reinforcement in male rats (*Rattus norvegicus*): II. Effects of preoptic area lesions, castration, and testosterone. *Journal of Comparative Psychology, 101,* 407–419.

Everson, C. A. (1993). Sustained sleep deprivation impairs host defense. *American Journal of Physiology, 265,* R1148–R1154.

Everson, C. A., Bergmann, B. M., and Rechtschaffen, A. (1989). Sleep deprivation in the rat: III. Total sleep deprivation. *Sleep, 12,* 13–21.

Eybalin, M. (1993). Neurotransmitters and neuromodulators of the mammalian cochlea. *Physiological Reviews, 73,* 309–373.

F

Falk, D. (2004). Prelinguistic evolution in early hominins: Whence motherese? *Behavioral and Brain Sciences, 27,* 491–503.

Falkner, A. L., and Lin, D. (2014). Recent advances in understanding the role of the hypothalamic circuit during aggression. *Frontiers in Systems Neuroscience, 8,* 168. doi:10.3389/fnsys.2014.00168

Faraone, S. V., Glatt, S. J., Su, J., and Tsuang, M. T. (2004). Three potential susceptibility loci shown by a genome-wide scan for regions influencing the age at onset of mania. *American Journal of Psychiatry, 161,* 625–630.

Farbman, A. I. (1994). The cellular basis of olfaction. *Endeavour, 18,* 2–8.

Feder, H. H., and Whalen, R. E. (1965). Feminine behavior in neonatally castrated and estrogen-treated male rats. *Science, 147,* 306–307.

Feinstein, J. S., Adolphs, R., Damasio, A., and Tranel, D. (2011). The human amygdala and the induction and experience of fear. *Current Biology, 21,* 34–38.

Feinstein, J. S., Buzza, C., Hurelmann, R., Follmer, R. L., et al. (2013). Fear and panic in humans with bilateral amygdala damage. *Nature Neuroscience, 16,* 270–272. doi:10.1038/nn.3323

Fenstemaker, S. B., Zup, S. L., Frank, L. G., Glickman, S. E., et al. (1999). A sex difference in the hypothalamus of the spotted hyena. *Nature Neuroscience, 2,* 943–945.

Ferguson, J. N., Young, L. J., Hearn, E. F., Matzuk, M. M., et al. (2000). Social amnesia in mice lacking the oxytocin gene. *Nature Genetics, 25,* 284–288.

Fields, R. D., and Stevens-Graham, B. (2002). New insights into neuron-glia communication. *Science, 298,* 556–562.

Fields, S. (1990). Pheromone response in yeast. *Trends in Biochemical Sciences, 15,* 270–273.

Finch, C. E., and Kirkwood, T. B. L. (2000). *Chance, development, and aging.* New York, NY: Oxford University Press.

Finger, S. (1994). *Origins of neuroscience: A history of explorations into brain function.* New York, NY: Oxford University Press.

Fink, G. R., Markowitsch, H. J., Reinkemeier, M., Bruckbauer, T., et al. (1996). Cerebral representation of one's own past: Neural networks involved in autobiographical memory. *Journal of Neuroscience, 16,* 4275–4282.

Fink, H., Rex, A., Voits, M., and Voigt, J. P. (1998). Major biological actions of CCK—A critical evaluation of research findings. *Experimental Brain Research, 123,* 77–83.

Fink, M., and Taylor, M. A. (2007). Electroconvulsive therapy: Evidence and challenges. *JAMA, 298,* 330–332.

Fishman, R. B., Chism, L., Firestone, G. L., and Breedlove, S. M. (1990). Evidence for androgen receptors in sexually dimorphic perineal muscles of neonatal male rats. Absence of androgen accumulation by the perineal motoneurons. *Journal of Neurobiology, 21,* 694–704.

Fitzsimmons, J. T. (1998). Angiotensin, thirst, and sodium appetite. *Physiological Reviews, 78,* 583–686.

Flegal, K. M., Carroll, M. D., Ogden, C. L., and Johnson, C. L. (2002). Prevalence and trends in obesity among US adults, 1999–2000. *JAMA, 288,* 1723–1727.

Fleming, A. S., Kraemer, G. W., Gonzalez, A., Lovic, V., et al. (2002). Mothering begets mothering: The transmission of behavior and its neurobiology across generations. *Pharmacology, Biochemistry, and Behavior, 73,* 61–75.

Fleming, A. S., Ruble, D., Krieger, H., and Wong, P. Y. (1997). Hormonal and experiential correlates of maternal responsiveness during pregnancy and the puerperium in human mothers. *Hormones and Behavior, 31,* 145–158.

Florence, S. L., Taub, H. B., and Kaas, J. H. (1998). Large-scale sprouting of cortical connections after peripheral injury in adult macaque monkeys. *Science, 282,* 1117–1121.

Foerster, O., and Penfield, W. (1930). The structural basis of traumatic epilepsy and results of radical operation. *Brain, 53,* 8–119.

Ford, J. M. (1999). Schizophrenia: The broken P300 and beyond. *Psychophysiology, 36,* 667–682.

Forger, N. G., and Breedlove, S. M. (1986). Sexual dimorphism in human and canine spinal cord: Role of early androgen. *Proceedings of the National Academy of Sciences, USA, 83,* 7527–7531.

Forger, N. G., and Breedlove, S. M. (1987). Seasonal variation in mammalian striated muscle mass and motoneuron morphology. *Journal of Neurobiology, 18,* 155–165.

Forger, N. G., Frank, L. G., Breedlove, S. M., and Glickman, S. E. (1996). Sexual dimorphism of perineal muscles and motoneurons in spotted hyenas. *Journal of Comparative Neurology, 375,* 333–343.

Foster, G. D., Wyatt, H. R., Hill, J. O., McGuckin, B. G., et al. (2003). A randomized trial of a low-carbohydrate diet for obesity. *New England Journal of Medicine, 348,* 2082–2090.

Foster, N. L., Cahse, T. N., Mansi, L., Brooks, R., et al. (1984). Cortical abnormalities in Alzheimer's disease. *Annals of Neurology, 16,* 649–654.

Fournier, J. C., DeRubeis, R. J., Hollon, S. D., Dimidjian, S., et al. (2010). Antidepressant drug effects and depression severity: A patient-level meta-analysis. *JAMA, 303,* 47–53.

Francis, D. D., Szegda, K., Campbell, G., Martin, W. D., et al. (2003). Epigenetic sources of behavioral differences in mice. *Nature Neuroscience, 6,* 445–446.

Frank, L. G., Glickman, S. E., and Licht, P. (1991). Fatal sibling aggression, precocial development, and androgens in neonatal spotted hyenas. *Science, 252,* 702–704.

Frank, M. J., Samanta, J., Moustafa, A. A., and Sherman, S. J. (2007). Hold your horses: Impulsivity, deep brain stimulation, and medication in parkinsonism. *Science, 318,* 1309–1312.

Frankenhaeuser, M. (1978). Psychoneuroendocrine approaches to the study of emotion as related to stress and coping. *Nebraska Symposium on Motivation, 26,* 123–162.

Franklin, T. R., Acton, P. D., Maldjian, J. A., Gray, J. D., et al. (2002). Decreased gray matter concentration in the insular, orbitofrontal, cingulate, and temporal cortices of cocaine patients. *Biological Psychiatry, 51,* 134–142.

Franks, N. P. (2008). General anaesthesia: From molecular targets to neuronal pathways of sleep and arousal. *Nature, 9,* 370–386.

Franssen, C. L., Bardi, M., Shea, E. A., Hampton, J. E., et al. (2011). Fatherhood alters behavioural and neural responsiveness in a spatial task. *Journal of Neuroendocrinology, 23*, 1177–1187.

Freed, C. R., Greene, P. E., Breeze, R. E., Tsai, W.-Y., et al. (2001). Transplantation of embryonic dopamine neurons for severe Parkinson's disease. *New England Journal of Medicine, 344*, 710–719.

Freedman, M. S., Lucas, R. J., Soni, B., von Schantz, M., et al. (1999). Regulation of mammalian circadian behavior by non-rod, non-cone, ocular photoreceptors. *Science, 284*, 502–504.

Freitag, J., Ludwig, G., Andreini, P., Roessler, P., et al. (1998). Olfactory receptors in aquatic and terrestrial vertebrates. *Journal of Comparative Physiology, 183*, 635–650.

Freiwald, W. A., Tsao, D. Y., and Livingstone, M. S. (2009). A face feature space in the macaque temporal lobe. *Nature Neuroscience, 12*, 1187–1196.

Frey, S. H., Bogdanov, S., Smith, J. C., Watrous, S., et al. (2008). Chronically deafferented sensory cortex recovers a grossly typical organization after allogenic hand transplantation. *Current Biology, 18*, 1530–1534.

Fried, I., Wilson, C. L., MacDonald, K. A., and Behnke, E. J. (1998). Electric current stimulates laughter. *Nature, 391*, 650.

Friedman, L., and Jones, B. E. (1984). Study of sleep-wakefulness states by computer graphics and cluster analysis before and after lesions of the pontine tegmentum in the cat. *Electroencephalography and Clinical Neurophysiology, 57*, 43–56.

Friedrich, F. J., Egly, R., Rafal, R. D., and Beck, D. (1998). Spatial attention deficits in humans: A comparison of superior parietal and temporal-parietal junction lesions. *Neuropsychology, 12*, 193–207.

Fritz, J., Shamma, S., Elhilali, M., and Klein, D. (2003). Rapid task-related plasticity of spectrotemporal receptive fields in primary auditory cortex. *Nature Neuroscience, 6*, 1216–1223.

Fukuda, K., Ogilvie, R. D., Chilcott, L., Vendittelli, A.-M., et al. (1998). The prevalence of sleep paralysis among Canadian and Japanese college students. *Dreaming: Journal of the Association for the Study of Dreams, 8*(2), 59–66.

Fulton, B. D., Scheffler, R. M., Hinshaw, S. P., Levine, P., et al. (2009). National variation of ADHD diagnostic prevalence and medication use: Health care providers and education policies. *Psychiatric Services, 60*, 1075–1083.

Fuster, J. M. (1990). Prefrontal cortex and the bridging of temporal gaps in the perception-action cycle. *Annals of the New York Academy of Sciences, 608*, 318–336.

G

Galaburda, A. M. (1994). Developmental dyslexia and animal studies: At the interface between cognition and neurology. *Cognition, 56*, 833–839.

Galaburda, A. M., LoTurco, J., Ramus, F., Fitch, R. H., et al. (2006). From genes to behavior in developmental dyslexia. *Nature Neuroscience, 9*, 1213–1217.

Gallant, J. L., Braun, J., and Van Essen, D. C. (1993). Selectivity for polar, hyperbolic, and Cartesian gratings in macaque visual cortex. *Science, 259*, 100–103.

Gallese, V., and Sinigaglia, C. (2011). What is so special about embodied simulation? *Trends in Cognitive Sciences, 15*, 512–519.

Gallopin, T., Fort, P., Eggermann, E., Cauli, B., et al. (2000). Identification of sleep-promoting neurons *in vitro*. *Nature, 404*, 992–995.

Gangwisch, J. E., Heymsfield, S. B., Boden-Albala, B., Buijs, R. M., et al. (2007). Sleep duration as a risk factor for diabetes incidence in a large U.S. sample. *Sleep, 30*, 1667–1673.

Gannon, P. J., Holloway, R. L., Broadfield, D. C., and Braun, A. R. (1998). Asymmetry of chimpanzee planum temporale: Human-like pattern of brain language area homolog. *Science, 279*, 220–222.

Gaoni, Y., and Mechoulam, R. (1964). Isolation, structure, and partial synthesis of an active constituent of hashish. *Journal of the American Chemical Society, 86*, 1646–1647.

Gardner, L. I. (1972). Deprivation dwarfism. *Scientific American, 227*(1), 76–82.

Gardner, R. A., and Gardner, B. T. (1969). Teaching sign language to a chimpanzee. *Science, 165*, 664–672.

Gardner, R. A., and Gardner, B. T. (1984). A vocabulary test for chimpanzees (*Pan troglodytes*). *Journal of Comparative Psychology, 98*, 381–404.

Gardner, T. J., Naef, F., and Nottebohm, F. (2005). Freedom and rules: The acquisition and reprogramming of a bird's learned song. *Science, 308*, 1046–1049.

Garver, D. L., Holcomb, J. A., and Christensen, J. D. (2000). Heterogeneity of response to antipsychotics from multiple disorders in the schizophrenia spectrum. *Journal of Clinical Psychiatry, 61*, 964–972.

Gates, N. J, and Sachdev, P. (2014). Is cognitive training an effective treatment for preclinical and early Alzheimer's disease? *Journal of Alzheimer's Disease, 42*, S551–S559.

Gaulin, S. J. C, and Fitzgerald, R. W. (1989). Sexual selection for spatial-learning ability. *Animal Behaviour, 37*, 322–331.

Gauthier, I., Behrmann, M., and Tarr, M. J. (1999). Can face recognition really be dissociated from object recognition? *Journal of Cognitive Neuroscience, 11*, 349–370.

Gauthier, I., Skudlarski, P., Gore, J. C., and Anderson, A. W. (2000). Expertise for cars and birds recruits brain areas involved in face recognition. *Nature Neuroscience, 3*, 191–197.

Gazzaniga, M. S., and Smylie, C. S. (1983). Facial recognition and brain asymmetries: Clues to underlying mechanisms. *Annals of Neurology, 13*, 536–540.

Gelber, R. P., Redline, S., Ross, G. W., Petrovitch, H., et al. (2015). Associations of brain lesions at autopsy with polysomnography features before death. *Neurology, 84*, 296–303.

Gelstein, S., Yeshurun, Y., Rozenkrantz, L., Shushan, S., et al. (2011). Human tears contain a chemosignal. *Science, 331*, 226–230.

Georgiadis, J. R., Reinders, A. A., Paans, A. M., Renken, R., et al. (2009). Men versus women on sexual brain function: Prominent differences during tactile genital stimulation, but not during orgasm. *Human Brain Mapping, 10*, 3089–3101.

Georgopoulos, A. P., Kalaska, J. F., Caminiti, R., and Massey, J. T. (1982). On the relations between the direction of two-dimensional arm movements and cell discharge in primate motor cortex. *Journal of Neuroscience, 2*, 1527–1537.

Georgopoulos, A. P., Taira, M., and Lukashin, A. (1993). Cognitive neurophysiology of the motor cortex. *Science, 260*, 47–52.

Gerashchenko, D., Kohls, M. D., Greco, M. A., Waleh, N. S., et al. (2001). Hypocretin-2-saporin lesions of the lateral hypothalamus produce narcoleptic-like sleep behavior in the rat. *Neuroscience, 21*, 7273–7283.

Geschwind, N. (1976). Language and cerebral dominance. In T. N. Chase (Ed.), *Nervous system: Vol. 2. The clinical neurosciences* (pp. 433–439). New York, NY: Raven Press.

Geschwind, N., and Levitsky, W. (1968). Human brain: Left-right asymmetries in temporal speech region. *Science, 161*, 186–187.

Gibbons, R. D., Hur, K., Brown, C. H., Davis, J. M., et al. (2012). Benefits from antidepressants: Synthesis of 6-week patient-level outcomes from double-blind placebo-controlled randomized trials of fluoxetine and venlafaxine. *Archives of General Psychiatry, 69*, 572–579.

Gilbert, A. N., and Wysocki, C. J. (1987). The smell survey results. *National Geographic, 172*, 514–525.

Gilbertson, M. W., Shenton, M. E., Ciszewski, A., Kasai, K., et al. (2002). Smaller hippocampal volume predicts pathologic vulnerability to psychological trauma. *Nature Neuroscience, 5*, 1242–1247.

Gill, R. E., Tibbitts, T. L., Douglas, D. C., Hanel, C. M., et al. (2009). Extreme endurance flights by landbirds crossing the Pacific Ocean: Ecological corridor rather than barrier? *Proceedings of the Royal Society of*

London. Series B: Biological Sciences, 276, 447–457.

Gillin, J. C., Duncan, W. C., Murphy, D. L., Post, R. M., et al. (1981). Age-related changes in sleep in depressed and normal subjects. Psychiatry Research, 4, 73–78.

Glantz, M., and Pickens, R. (1992). Vulnerability to drug abuse. Washington, DC: American Psychological Association.

Glaser, R., Rice, J., Speicher, C. E., Stout, J. C., et al. (1986). Stress depresses interferon production by leukocytes concomitant with a decrease in natural killer cell activity. Behavioral Neuroscience, 100, 675–678.

Glickman, S. E. (1977). Comparative psychology. In P. Mussen and M. R. Rosenzweig (Eds.), Psychology: An introduction (2nd ed., pp. 625–703). Lexington, MA: Heath.

Glickman, S. E., Frank, L. G., Davidson, J. M., Smith, E. R., et al. (1987). Androstenedione may organize or activate sex-reversed traits in female spotted hyenas. Proceedings of the National Academy of Sciences, USA, 84, 344–347.

Glusman, G., Yanai, I., Rubin, I., and Lancet, D. (2001). The complete human olfactory subgenome. Genome Research, 11, 685–702.

Goel, V., and Dolan, R. J. (2001). The functional anatomy of humor: Segregating cognitive and affective components. Nature Neuroscience, 4, 237–238.

Gogtay, N., Giedd, J. N., Lusk, L., Hayashi, K. M., et al. (2004). Dynamic mapping of human cortical development during childhood through early adulthood. Proceedings of the National Academy of Sciences, USA, 101, 8174–8179.

Gold, S. M., and Voskuhl, R. R. (2006). Testosterone replacement therapy for the treatment of neurological and neuropsychiatric disorders. Current Opinion in Investigational Drugs, 7, 625–630.

Golden, R. N., Gaynes, B. N., Ekstrom, R. D., Hamer, R. M., et al. (2005). The efficacy of light therapy in the treatment of mood disorders: A review and meta-analysis of the evidence. American Journal of Psychiatry, 162, 656–662.

Goldstein, J. M., Seidman, L. J., Horton, N. J., Makris, N., et al. (2001). Normal sexual dimorphism of the adult human brain assessed by in vivo magnetic resonance imaging. Cerebral Cortex, 11, 490–497.

Golomb, J., de Leon, M. J., George, A. E., Kluger, A., et al. (1994). Hippocampal atrophy correlates with severe cognitive impairment in elderly patients with suspected normal pressure hydrocephalus. Journal of Neurology, Neurosurgery and Psychiatry, 57, 590–593.

Gonçalves, T. C., Londe, A. K., Albano, R. I., et al. (2014). Cannabidiol and endogenous opioid peptide-mediated mechanisms modulate antinociception induced by transcutaneous electrostimulation of the peripheral nervous system. Journal of Neurological Sciences, 347, 82–89.

Goodale, M. A., Milner, A. D., Jakobson, L. S., and Carey, D. P. (1991). A neurological dissociation between perceiving objects and grasping them. Nature, 349, 154–156.

Goodman, C. (1979). Isogenic grasshoppers: Genetic variability and development of identified neurons. In X. O. Breakefeld (Ed.), Neurogenetics (pp. 101–151). New York, NY: Elsevier.

Gooley, J. J., Rajaratnam, S. M., Brainard, G. C., Kronauer, R. E., et al. (2010). Spectral responses of the human circadian system depend on the irradiance and duration of exposure to light. Science Translational Medicine, 2, 31ra33.

Gordon, N. S., Burke, S., Akil, H., Watson, S. J., et al. (2003). Socially-induced brain "fertilization": Play promotes brain derived neurotrophic factor transcription in the amygdala and dorsolateral frontal cortex in juvenile rats. Neuroscience Letters, 341, 17–20.

Gorski, R. A., Gordon, J. H., Shryne, J. E., and Southam, A. M. (1978). Evidence for a morphological sex difference within the medial preoptic area of the rat brain. Brain Research, 148, 333–346.

Gorzalka, B. B., Mendelson, S. D., and Watson, N. V. (1990). Serotonin receptor subtypes and sexual behavior. Annals of the New York Academy of Sciences, 600, 435–444.

Gottesman, I. I. (1991). Schizophrenia genesis: The origins of madness. New York, NY: Freeman.

Gottfried, J. A., O'Doherty, J., and Dolan, R. J. (2003). Encoding predictive reward value in human amygdala and orbitofrontal cortex. Science, 301, 1104–1107.

Gottlieb, J. (2007). From thought to action: The parietal cortex as a bridge between perception, action, and cognition. Neuron, 53, 9–16.

Gough, P. M., Nobre, A. C., and Devlin, J. T. (2005). Dissociating linguistic processes in the left inferior frontal cortex with transcranial magnetic stimulation. Journal of Neuroscience, 25, 8010–8016.

Gould, E., Reeves, A. J., Graziano, M. S., and Gross, C. G. (1999). Neurogenesis in the neocortex of adult primates. Science, 286, 548–552.

Grados, M. A., Walkup, J., and Walford, S. (2003). Genetics of obsessive-compulsive disorders: New findings and challenges. Brain & Development, 25(Suppl. 1), S55–S61.

Graeber, M. B. (2010). Changing face of microglia. Science, 330, 783–788.

Grafton, S. T., Mazziotta, J. C., Presty, S., Friston, K. J., et al. (1992). Functional anatomy of human procedural learning determined with regional cerebral blood flow and PET. Journal of Neuroscience, 12, 2542–2548.

Graham, R. K., Deng, Y., Slow, E. J., Haigh, B., et al. (2006). Cleavage at the caspase-6 site is required for neuronal dysfunction and degeneration due to mutant huntingtin. Cell, 125, 1179–1191.

Gray, N. S., MacCulloch, M. J., Smith, J., Morris, M., et al. (2003). Violence viewed by psychopathic murderers. Nature, 423, 497.

Graybiel, A. M., Aosaki, T., Flaherty, A. W., and Kimura, M. (1994). The basal ganglia and adaptive motor control. Science, 265, 1826–1831.

Graziano, M. (2006). The organization of behavioral repertoire in motor cortex. Annual Review of Neuroscience, 29, 105–134.

Graziano, M. S., and Aflalo, T. N. (2007). Mapping behavioral repertoire onto the cortex. Neuron, 56, 239–251.

Green, C. S., and Bavelier, D. (2003). Action video game modifies visual selective attention. Nature, 423, 534–537.

Green, W. H., Campbell, M., and David, R. (1984). Psychosocial dwarfism: A critical review of the evidence. Journal of the American Academy of Child Psychiatry, 23, 39–48.

Greenewalt, C. H. (1968). Bird song: Acoustics and physiology. Washington, DC: Smithsonian Institution Press.

Greenough, W. T. (1976). Enduring brain effects of differential experience and training. In M. R. Rosenzweig and E. L. Bennett (Eds.), Neural mechanisms of learning and memory (pp. 255–278). Cambridge, MA: MIT Press.

Grieco-Calub, T. M., Saffran, J. R., and Litovsky, R. Y. (2009). Spoken word recognition in toddlers who use cochlear implants. Journal of Speech, Language, and Hearing Research, 52, 1390–1400.

Griessenauer, C. J., Salam, S., Hendrix, P., Patel, D. M., et al. (2015). Hemispherectomy for treatment of refractory epilepsy in the pediatric age group: A systematic review. Journal of Neurosurgery: Pediatrics, 15, 34–44.

Grimbos, T., Dawood, K., Burriss, R. P., Zucker, K. J., et al. (2010). Sexual orientation and the second to fourth finger length ratio: A meta-analysis in men and women. Behavioral Neuroscience, 124, 278–287.

Grimm, S., and Bajbouj, M. (2010). Efficacy of vagus nerve stimulation in the treatment of depression. Expert Review of Neurotherapeutics, 19, 87–92.

Grothe, B. (2003). New roles for synaptic inhibition in sound localization. Nature Reviews Neuroscience, 4, 540–550.

Grover, G. J., Mellstrom, K., Ye, L., Malm, J., et al. (2003). Selective thyroid hormone receptor-β activation: A strategy for reduction of weight, cholesterol, and lipoprotein (a) with reduced cardiovascular liability. Proceedings of the National Academy of Sciences, USA, 100, 10067–10072.

Grumbach, M. M., and Auchus, R. J. (1999). Estrogen: Consequences and implications of human mutations in synthesis and action. *Journal of Clinical Endocrinology and Metabolism, 84,* 4677–4694.

Grunt, J. A., and Young, W. C. (1953). Consistency of sexual behavior patterns in individual male guinea pigs following castration and androgen therapy. *Journal of Comparative and Physiological Psychology, 46,* 138–144.

Grüter, T., Grüter, M., and Carbon, C. C. (2008). Neural and genetic foundations of face recognition and prosopagnosia. *Journal of Neuropsychology, 2,* 79–97.

Gulevich, G., Dement, W., and Johnson, L. (1966). Psychiatric and EEG observations on a case of prolonged (264 hours) wakefulness. *Archives of General Psychiatry, 15,* 29–35.

Gurney, M. E., Pu, H., Chiu, A. Y., Dal Canto, M. C., et al. (1994). Motor neuron degeneration in mice that express a human Cu,Zn superoxide dismutase mutation. *Science, 264,* 1772–1775.

Gusella, J. F., and MacDonald, M. E. (1993). Hunting for Huntington's disease. *Molecular Genetic Medicine, 3,* 139–158.

H

Haesler, S., Rochefort, C., Georgi, B., Licznerski, P., et al. (2007). Incomplete and inaccurate vocal imitation after knockdown of FoxP2 in songbird basal ganglia nucleus Area X. *PLOS Biology, 5,* e321.

Haesler, S., Wada, K., Nshdejan, A., Morrisey, E. E., et al. (2004). FoxP2 expression in avian vocal learners and non-learners. *Journal of Neuroscience, 24,* 3164–3175.

Hagstrom, S. A., Neitz, J., and Neitz, M. (1998). Variations in cone populations for red-green color vision examined by analysis of mRNA. *NeuroReport, 9,* 1963–1967.

Hall, D., Dhilla, A., Charalambous, A., Gogos, J. A., et al. (2003). Sequence variants of the brain-derived neurotrophic factor (*BDNF*) gene are strongly associated with obsessive-compulsive disorder. *American Journal of Human Genetics, 73,* 370–376.

Halsband, U., Matsuzaka, Y., and Tanji, J. (1994). Neuronal activity in the primate supplementary, pre-supplementary and premotor cortex during externally and internally instructed sequential movements. *Neuroscience Research, 20,* 149–155.

Hamburger, V. (1975). Cell death in the development of the lateral motor column of the chick embryo. *Journal of Comparative Neurology, 160,* 535–546.

Hamer, D. H., Hu, S., Magnuson, V. L., Hu, N., et al. (1993). A linkage between DNA markers on the X chromosome and male sexual orientation. *Science, 261,* 321–327.

Hamson, D. K., Csupity, A. S., Ali, F. M., and Watson, N. V. (2009). Partner preference and mount latency are masculinized in androgen insensitive rats. *Physiology & Behavior, 98,* 25–30.

Hamson, D. K., and Watson, N. V. (2004). Regional brainstem expression of Fos associated with sexual behavior in male rats. *Brain Research, 1006,* 233–240.

Hardyck, C., Petrinovich, L. F., and Goldman, R. D. (1976). Left-handedness and cognitive deficit. *Cortex, 12,* 266–279.

Hare, R. D., Harpur, T. J., Hakstian, A. R., Forth, A. E., et al. (1990). The revised psychopathy checklist: Descriptive statistics, reliability, and factor structure. *Psychological Assessment, 2,* 338–341.

Harel, N. Y., and Strittmatter, S. M. (2006). Can generating axons recapitulate developmental guidance during recovery from spinal cord injury? *Nature Reviews Neuroscience, 7,* 603–616.

Harold, D., Paracchini, S., Scerri, T., Dennis, M., et al. (2006). Further evidence that the *KIAA0319* gene confers susceptibility to developmental dyslexia. *Molecular Psychiatry, 11,* 1085–1091, 1061.

Hart, B. L. (1988). Biological basis of the behavior of sick animals. *Neuroscience and Biobehavioral Reviews, 12,* 123–137.

Hartanto, T. A., Krafft, C. E., Iosif, A. M., and Schweitzer, J. B. (2015). A trial-by-trial analysis reveals more intense physical activity is associated with better cognitive control performance in attention-deficit/hyperactivity disorder. *Child Neuropsychology,* doi: 10.1080/09297049.2015.1044511

Hartmann, E. (1978). *The sleeping pill.* New Haven, CT: Yale University Press.

Hartse, K. M. (2011). The phylogeny of sleep. *Handbook of Clinical Neurology, 98,* 97–109.

Haynes, K. F., Gemeno, C., Yeargan, K. V., Millar, J. G., et al. (2002). Aggressive chemical mimicry of moth pheromones by a bolas spider: How does this specialist predator attract more than one species of prey? *Chemoecology, 12,* 99–105.

Hazeltine, E., Grafton, S. T., and Ivry, R. (1997). Attention and stimulus characteristics determine the locus of motor-sequence encoding. A PET study. *Brain, 120,* 123–140.

Heath, R. G. (1972). Pleasure and brain activity in man. *Journal of Nervous and Mental Diseases, 154,* 3–18.

Hebb, D. O. (1949). *The organization of behavior.* New York, NY: Wiley.

Heckert, J. (2012, November 8). The hazards of growing up painlessly. *New York Times Sunday Magazine,* p. MM26.

Heidenreich, M., Lechner, S. G., Vardanyan, V., Wetzel, C., et al. (2011). KCNQ4 K(+) channels tune mechanoreceptors for normal touch sensation in mouse and man. *Nature Neuroscience, 15,* 138–145.

Heinrichs, R. W. (2003). Historical origins of schizophrenia: Two early madmen and their illness. *Journal for the History of Behavioral Sciences, 39,* 349–363.

Heit, S., Owens, M. J., Plotsky, P., and Nemeroff, C. B. (1997). Corticotropin-releasing factor, stress, and depression. *Neuroscientist, 3,* 186–194.

Helmholtz, H. von. (1962). *Treatise on physiological optics* (J. P. C. Southall, Trans.). New York, NY: Dover. (Original work published 1894)

Hemmingsen, A. M. (1960). Energy metabolism as related to body size and respiratory surfaces, and its evolution. *Reports of Steno Memorial Hospital, Copenhagen, 9,* 1–110.

Hendriks, W. T., Ruitenberg, M. J., Blits, B., Boer, G. J., et al. (2004). Viral vector-mediated gene transfer of neurotrophins to promote regeneration of the injured spinal cord. *Progress in Brain Research, 146,* 451–476.

Henry, J. F., and Sherwin, B. B. (2012). Hormones and cognitive functioning during late pregnancy and postpartum: A longitudinal study. *Behavioral Neuroscience, 126,* 73–85.

Herculano-Houzel, S. (2012). The remarkable, yet not extraordinary, human brain as a scaled-up primate brain and its associated cost. *Proceedings of the National Academy of Sciences, USA, 109* (Suppl. 1), 10661–10668.

Herek, G. M., and McLemore, K. A. (2013). Sexual prejudice. *Annual Review of Psychology, 64,* 309–333.

Heres, S., Davis, J., Maino, K., Jetzinger, E., et al. (2006). Why olanzapine beats risperidone, risperidone beats quetiapine, and quetiapine beats olanzapine: An exploratory analysis of head-to-head comparison studies of second-generation antipsychotics. *American Journal of Psychiatry, 163,* 185–194.

Herrmann, C., and Knight, R. (2001). Mechanisms of human attention: Event-related potentials and oscillations. *Neuroscience and Biobehavioral Reviews, 25,* 465–476.

Hertel, P., Fagerquist, M. V., and Svensson, T. H. (1999). Enhanced cortical dopamine output and antipsychotic-like effects of raclopride by alpha-2 adrenoceptor blockade. *Science, 286,* 105–107.

Hetherington, A. W., and Ranson, S. W. (1940). Hypothalamic lesions and adiposity in the rat. *Anatomical Record, 78,* 149–172.

Hewes, G. (1973). Primate communication and the gestural origin of language. *Current Anthropology, 14,* 5–24.

Hickmott, P. W., and Steen, P. A. (2005). Large-scale changes in dendritic structure during reorganization of adult somatosensory cortex. *Nature Neuroscience, 8,* 140–142.

Hicks, M. J., De, B. P., Rosenberg, J. B., Davidson, J. T., et al. (2011). Cocaine analog coupled to disrupted adenovirus: A vaccine strategy to evoke high-titer immunity against addictive drugs. *Molecular Therapy, 19,* 612–619.

Highley, J. R., Walker, M. A., Esiri, M. M., McDonald, B., et al. (2001). Schizophrenia and the frontal lobes: Post-mortem stereological study of tissue volume. *British Journal of Psychiatry, 178,* 337–343.

Higley, J. D., Mehlman, P. T., Taub, D. M., Higley, S. B., et al. (1992). Cerebrospinal fluid monoamine and adrenal correlates of aggression in free-ranging rhesus monkeys. *Archives of General Psychiatry, 49,* 436–441.

Hillyard, S. A., Hink, R. F., Schwent, V. L., and Picton, T. W. (1973). Electrical signs of selective attention in the human brain. *Science, 182,* 177–180.

Hillyard, S. A., Störmer, V. S., Feng, W., Martinez, A., et al. (2015, June 11). Cross-modal orienting of visual attention. *Neuropsychologia,* pii: S0028-3932(15)30051-8. doi: 10.1016/j.neuropsychologia.2015.06.003

Himle, M. B., Woods, D. W., Piacentini, J. C., and Walkup, J. T. (2006). Brief review of habit reversal training for Tourette syndrome. *Journal of Child Neurology, 21,* 719–725.

Hirsch, E., Moye, D., and Dimon, J. H. (1995). Congenital indifference to pain: Long-term follow-up of two cases. *Southern Medical Journal, 88,* 851–857.

Hirtz, D., Thurman, D. J., Gwinn-Hardy, K., Mohamed, M., et al. (2007). How common are the "common" neurologic disorders? *Neurology, 68,* 326–337.

Hitt, E. (2007). Careers in neuroscience: From protons to poetry. *Science, 318,* 661–665.

Hobaiter, C., and Byrne, R. W. (2014). The meanings of chimpanzee gestures. *Current Biology, 24,* 1596–1600.

Hobson, J. A., and Friston, K. J. (2012). Waking and dreaming consciousness: Neurobiological and functional considerations. *Progress in Neurobiology, 98,* 82–98.

Hochberg, L. R., Serruya, M. D., Friehs, G. M., Mukand, J. A., et al. (2006). Neuronal ensemble control of prosthetic devices by a human with tetraplegia. *Nature, 442,* 164–171.

Hodgkin, A. L., and Katz, B. (1949). The effect of sodium ions on the electrical activity of the giant axon of the squid. *Journal of Physiology (London), 108,* 37–77.

Hoeft, F., Hernandez, A., McMillon, G., Taylor-Hill, H., et al. (2006). Neural basis of dyslexia: A comparison between dyslexic and nondyslexic children equated for reading ability. *Journal of Neuroscience, 26,* 10700–10708.

Hohmann, A. G., Suplita, R. L., Bolton, N. M., Neely, M. H., et al. (2005). An endocannobinoid mechanism for stress-induced analgesia. *Nature, 435,* 1108–1112.

Hohmann, G. W. (1966). Some effects of spinal cord lesions on experienced emotional feelings. *Psychophysiology, 3,* 143–156.

Hollander, E. (1999). Managing aggressive behavior in patients with obsessive-compulsive disorder and borderline personality disorder. *Journal of Clinical Psychiatry, 60*(Suppl.), 38–44.

Hollon, S. D., Thase, M. E., and Markowitz, J. C. (2002). Treatment and prevention of depression. *Psychological Science in the Public Interest, 3,* 39–77.

Holmes, C., Boche, D., Wilkinson, D., Yadegarfar, G., et al. (2008). Long-term effects of Abeta42 immunisation in Alzheimer's disease: Follow-up of a randomized, placebo-controlled phase I trial. *Lancet, 372,* 216–223.

Honey, G. D., Bullmore, E. T., Soni, W., Varatheesan, M., et al. (1999). Differences in frontal cortical activation by a working memory task after substitution of risperidone for typical antipsychotic drugs in patients with schizophrenia. *Proceedings of the National Academy of Sciences, USA, 96,* 13432–13437.

Hopfinger, J. B., Buonocore, M. H., and Mangun, G. R. (2000). The neural mechanisms of top-down attentional control. *Nature Neuroscience, 3,* 284–291.

Hopfinger, J., and Mangun, G. (1998). Reflexive attention modulates processing of visual stimuli in human extrastriate cortex. *Psychological Science, 6,* 441–447.

Hopfinger, J., and Mangun, G. (2001). Tracking the influence of reflexive attention on sensory and cognitive processing. *Cognitive, Affective & Behavioral Neuroscience, 1,* 56–65.

Horton, J. C., and Adams, D. L. (2005). The cortical column: A structure without a function. *Philosophical Transactions of the Royal Society of London. Series B: Biological Sciences, 360,* 837–862.

Hou, P., Li, Y., Zhang, X., Liu, C., et al. (2013). Pluripotent stem cells induced from mouse somatic cells by small-molecule compounds. *Science, 341,* 651–654.

Hsu, M., Bhatt, M., Adolphs, R., Tranel, D., et al. (2005). Neural systems responding to degrees of uncertainty in human decision-making. *Science, 310,* 1680–1683.

Huang, A. L., Chen, X., Hoon, M. A., Chandrashekar, J., et al. (2006). The cells and logic for mammalian sour taste detection. *Nature, 442,* 934–938.

Hubel, D. H., and Wiesel, T. N. (1959). Receptive fields of single neurones in the cat's striate cortex. *Journal of Physiology (London), 148,* 573–591.

Hubel, D. H., and Wiesel, T. N. (1962). Receptive fields, binocular interaction and functional architecture in the cat's visual cortex. *Journal of Physiology (London), 160,* 106–154.

Hubel, D. H., and Wiesel, T. N. (1965). Binocular interaction in striate cortex of kittens reared with artificial squint. *Journal of Neurophysiology, 28,* 1041–1059.

Hubel, D. H., Wiesel, T. N., and LeVay, S. (1977). Plasticity of ocular dominance in monkey striate cortex. *Philosophical Transactions of the Royal Society of London. Series B: Biological Sciences, 278,* 377–409.

Hudspeth, A. J. (1997). How hearing happens. *Neuron, 19,* 947–950.

Hudspeth, A. J. (2014). Integrating the active process of hair cells with cochlear function. *Nature Reviews Neuroscience, 15,* 600–614.

Hudspeth, A. J., Choe, Y., Mehta, A. D., and Martin, P. (2000). Putting ion channels to work: Mechanoelectrical transduction, adaptation, and amplification by hair cells. *Proceedings of the National Academy of Sciences, USA, 97,* 11765–11772.

Huettel, S. A., Stowe, C. J., Gordon, E. M., Warner, B. T., et al. (2006). Neural signatures of economic preferences for risk and ambiguity. *Neuron, 49,* 765–775.

Hughes, J., Smith, T. W., Kosterlitz, H. W., Fothergill, L. A., et al. (1975). Identification of two related pentapeptides from the brain with potent opiate agonist activity. *Nature, 258,* 577–580.

Huntington, G. (1872). On chorea. *Medical and Surgical Reporter, 26,* 317–321.

Hurtado, M. D., Sergeyev, V. G., Acosta, A., Spegele, M., et al. (2013). Salivary peptide tyrosine-tyrosine 3-36 modulates ingestive behavior without inducing taste aversion. *Journal of Neuroscience, 33,* 18368–18380.

Huttenlocher, P. R., deCourten, C., Garey, L. J., and Van der Loos, H. (1982). Synaptogenesis in human visual cortex—Evidence for synapse elimination during normal development. *Neuroscience Letters, 33,* 247–252.

Hyde, K. L., and Peretz I. (2004). Brains that are out of tune but in time. *Psychological Science, 15,* 356–360.

Hyde, K. L., Zatorre, R. J., Griffiths, T. D., Lerch, J. P., et al. (2006). Morphometry of the amusic brain: A two-site study. *Brain, 129,* 2562–2570.

Hyde, T. M., and Weinberger, D. R. (1990). The brain in schizophrenia. *Seminars in Neurology, 10,* 276–286.

Hyde, T. M., Aaronson, B. A., Randolph, C., Rickler, K. C., et al. (1992). Relationship of birth weight to the phenotypic expression of Gilles de la Tourette's syndrome in monozygotic twins. *Neurology, 42,* 652–658.

I

Imai, T., Yamazaki, T., Kobayakawa, R., Kobayakawa, K., et al. (2009). Pre-target axon sorting establishes the neural map topography. *Science, 325,* 585–590.

Imeri, L., and Opp, M. R. (2009). How (and why) the immune system makes us sleep. *Nature Reviews Neuroscience, 10,* 199–210.

Imperato-McGinley, J., Guerrero, L., Gautier, T., and Peterson, R. E. (1974). Steroid 5α-reductase deficiency in man: An inherited form of male pseudohermaphroditism. *Science, 86,* 1213–1215.

Insley, S. J. (2000). Long-term vocal recognition in the northern fur seal. *Nature, 406,* 404–405.

Institute of Medicine. (1990). *Broadening the base of treatment for alcohol problems.* Washington, DC: National Academy Press.

Institute of Medicine. (2010). *Gulf War and health: Vol. 8. Update of health effects of serving in the Gulf War.* Washington, DC: National Academies Press.

Isacson, O., Bjorklund, L., and Sanchez Pernaute, R. (2001). Parkinson's disease: Interpretations of transplantation study erroneous. *Nature Neuroscience, 4,* 533.

Isles, A. R., Baum, M. J., Ma, D., Keverne, E. B., et al. (2001). Urinary odour preferences in mice. *Nature, 409,* 783–784.

Izumikawa, M., Minoda, R., Kawamoto, K., Abrashkin, K. A., et al. (2005). Auditory hair cell replacement and hearing improvement by *Atoh1* gene therapy in deaf mammals. *Nature Medicine, 11,* 271–276.

J

Jackson, H., and Parks, T. N. (1982). Functional synapse elimination in the developing avian cochlear nucleus with simultaneous reduction in cochlear nerve axon branching. *Journal of Neuroscience, 2,* 1736–1743.

Jacobs, G. H. (1993). The distribution and nature of colour vision among the mammals. *Biological Reviews of the Cambridge Philosophical Society, 68,* 413–471.

Jacobs, G. H., Williams, G. A., Cahill, H., and Nathans, J. (2007). Emergence of novel color vision in mice engineered to express a human cone photopigment. *Science, 315,* 1723–1725.

Jacobs, L. F., Gaulin, S. J., Sherry, D. F., and Hoffman, G. E. (1990). Evolution of spatial cognition: Sex-specific patterns of spatial behavior predict hippocampal size. *Proceedings of the National Academy of Sciences, USA, 87,* 6349–6352.

Jacobs, L. F., and Spencer, W. D. (1994). Natural space-use patterns and hippocampal size in kangaroo rats. *Brain, Behavior and Evolution, 44,* 125–132.

James, T. W., Culham, J., Humphery, G. K., Milner, A. D., et al. (2003). Ventral occipital lesions impair object recognition but not object-directed grasping: An fMRI study. *Brain, 126,* 2464–2475.

James, W. (1890). *Principles of psychology.* New York, NY: Holt.

Jamieson, D., and Roberts, A. (2000). Responses of young *Xenopus laevis* tadpoles to light dimming: Possible roles for the pineal eye. *Journal of Experimental Biology, 203,* 1857–1867.

Jaskiw, G. E., and Popli, A. P. (2004). A meta-analysis of the response to chronic l-dopa in patients with schizophrenia: Therapeutic and heuristic implications. *Psychopharmacology (Berlin), 171,* 365–374.

Jasper, H., and Penfield, W. (1954). *Epilepsy and the functional anatomy of the human brain* (2nd ed.). New York, NY: Little, Brown.

Jeffress, L. A. (1948). A place theory of sound localization. *Journal of Comparative and Physiological Psychology, 41,* 35–39.

Jenkins, J., and Dallenbach, K. (1924). Oblivescence during sleep and waking. *American Journal of Psychology, 35,* 605–612.

Jentsch, J. D., Redmond, D. E., Jr., Elsworth, J. D., Taylor, J. R., et al. (1997). Enduring cognitive deficits and cortical dopamine dysfunction in monkeys after long-term administration of phencyclidine. *Science, 277,* 953–955.

Johansson, R. S., and Flanagan, J. R. (2009). Coding and use of tactile signals from the fingertips in object manipulation tasks. *Nature Reviews Neuroscience, 10,* 345–358.

Johnson, L. C. (1969). Psychological and physiological changes following total sleep deprivation. In A. Kales (Ed.), *Sleep: Physiology & pathology; a symposium* (pp. 206–220). Philadelphia, PA: Lippincott.

Jones, P. B., Barnes, T. R. E., Davies, L., Dunn, G., et al. (2006). Randomized controlled trial of the effect on Quality of Life of second- vs first-generation antipsychotic drugs in schizophrenia. *Archives of General Psychiatry, 39,* 1079–1087.

Jordan, B. D., Jahre, C., Hauser, W. A., Zimmerman, R. D., et al. (1992). CT of 338 active professional boxers. *Radiology, 185,* 509–512.

Jordan, C. L., Breedlove, S. M., and Arnold, A. P. (1991). Ontogeny of steroid accumulation in spinal lumbar motoneurons of the rat: Implications for androgen's site of action during synapse elimination. *Journal of Comparative Neurology, 313,* 441–448.

Jordt, S.-E., Bautista, D. M., Chuang, H., McKemy, D. D., et al. (2004). Mustard oils and cannabinoids excite sensory nerve fibres through the TRP channel ANKTM1. *Nature, 427,* 260–265.

Joseph, J. S., Chun, M. M., and Nakayama, K. (1997). Attentional requirements in a "preattentive" feature search task. *Nature, 387,* 805–807.

Julian, T., and McKenry, P. C. (1979). Relationship of testosterone to men's family functioning at mid-life: A research note. *Aggressive Behavior, 15,* 281–289.

K

Kaar, G. F., and Fraher, J. P. (1985). The development of alpha and gamma motoneuron fibres in the rat. I. A comparative ultrastructural study of their central and peripheral axon growth. *Journal of Anatomy, 141,* 77–88.

Kaas, J. H., Nelson, R. J., Sur, M., Lin, C. S., et al. (1979). Multiple representations of the body within the primary somatosensory cortex of primates. *Science, 204,* 521–523.

Kable, J. W., and Glimcher, P. W. (2009). The neurobiology of decision: Consensus and controversy. *Neuron, 63,* 733–745.

Kaiser, D. (2013). Infralow frequencies and ultradian rhythms. *Seminars in Pediatric Neurology, 20,* 242–245.

Kajimura, S., and Saito, M. (2014). A new era in brown adipose tissue biology: Molecular control of brown fat development and energy homeostasis. *Annual Review of Physiology, 76,* 225–249.

Kales, A., and Kales, J. (1970). Evaluation, diagnosis and treatment of clinical conditions related to sleep. *JAMA, 213,* 2229–2235.

Kales, A., and Kales, J. D. (1974). Sleep disorders. Recent findings in the diagnosis and treatment of disturbed sleep. *New England Journal of Medicine, 290,* 487–499.

Kandel, E. R. (2009). The biology of memory: A forty-year perspective. *Journal of Neuroscience, 29,* 12748–12756.

Kandler, K., Clause, A., and Noh, J. (2009). Tonotopic reorganization of developing auditory brainstem circuits. *Nature Neuroscience, 12,* 711–716.

Kang, C., Riazuddin, S., Mundorff, J., Krasnewich, D., et al. (2010). Mutation in the lysosomal enzyme–targeting pathway and persistent stuttering. *New England Journal of Medicine, 362,* 677–685.

Kang, J.-E., Lim, M. M., Bateman, R. J., Lee, J. J., et al. (2009). Amyloid-β dynamics are regulated by orexin and the sleep-wake cycle. *Science, 326,* 1005–1007.

Kanwisher, N., and Wojciulik, E. (2000). Visual attention: Insights from brain imaging. *Nature Reviews Neuroscience, 1,* 91–100.

Karch, S. B. (2006). *Drug abuse handbook* (2nd ed.). Boca Raton, FL: CRC Press.

Karlin, A. (2002). Emerging structure of the nicotinic acetylcholine receptors. *Nature Reviews Neuroscience, 3,* 102–114.

Karlsgodt, K. H., Sun, D., and Cannon, T. D. (2010). Structural and functional brain abnormalities in schizophrenia. *Current Directions in Psychological Science, 19,* 226–231.

Karni, A., Tanne, D., Rubenstein, B. S., Askenasy, J. J., et al. (1994). Dependence on REM sleep of overnight improvement of a perceptual skill. *Science, 265,* 679–682.

Karpicke, J. D., and Roediger, H. L., III. (2008). The critical importance of retrieval for learning. *Science, 319,* 966–968.

Karra, E., Chandarana, K., and Batterham, R. L. (2009). The role of peptide YY in appetite regulation and obesity. *Journal of Physiology, 587,* 19–25.

Kass, A. E., Kolko, R. P., and Wilfley, D. E. (2013). Psychological treatments for eating disorders. *Current Opinion in Psychiatry, 26,* 549–555.

Katz, D. B., and Steinmetz, J. E. (2002). Psychological functions of the cerebellum. *Behavioral Cognitive Neuroscience Review, 1,* 229–241.

Katzenberg, D., Young, T., Finn, L., Lin, L., et al. (1998). A CLOCK polymorphism associated with human diurnal preference. *Sleep, 21*, 569–576.

Kaushall, P. I., Zetin, M., and Squire, L. R. (1981). A psychosocial study of chronic, circumscribed amnesia. *Journal of Nervous and Mental Disease, 169*, 383–389.

Kay, K. N., Naselaris, T., Prenger, R. J., and Gallant, J. L. (2008). Identifying natural images from human brain activity. *Nature, 452*, 352–355.

Kaye, W. H., Fudge, J. L., and Paulus, M. (2009). New insights into symptoms and neurocircuit function of anorexia nervosa. *Nature Reviews Neuroscience, 10*, 573–584.

Keane, T. M. (1998). Psychological and behavioral treatments of post-traumatic stress disorder. In P. E. Nathan and J. M. Gorman (Eds.), *A guide to treatments that work* (pp. 398–407). New York, NY: Oxford University Press.

Kee, N., Teixeira, C. M., Wang, A. H., and Frankland, P. W. (2007). Preferential incorporation of adult-generated granule cells into spatial memory networks in the dentate gyrus. *Nature Neuroscience, 10*, 355–362.

Keenan, J. P., Nelson, A., O'Connor, M., and Pascual-Leone, A. (2001). Self-recognition and the right hemisphere. *Nature, 409*, 305.

Keesey, R. E. (1980). A set-point analysis of the regulation of body weight. In A. J. Stunkard (Ed.), *Obesity* (pp. 144–165). Philadelphia, PA: Saunders.

Keesey, R. E., and Corbett, S. W. (1984). Metabolic defense of the body weight set-point. *Research Publications—Association for Research in Nervous and Mental Disease, 62*, 87–96.

Keesey, R. E., and Powley, T. L. (1986). The regulation of body weight. *Annual Review of Psychology, 37*, 109–133.

Keltner, D., and Ekman, P. (2000). Facial expression of emotion. In M. Lewis and J. M. Haviland-Jones (Eds.), *Handbook of emotions* (2nd ed., pp. 236–250). New York, NY: Guilford Press.

Keltner, D., and Ekman, P. (2015, July 3). The science of "Inside Out." *The New York Times*, p. SR10.

Kemp, J. A., and McKernan, R. M. (2002). NMDA receptor pathways as drug targets. *Nature Neuroscience, 5*(Suppl.), 1039–1042.

Kempermann, G., Kuhn, H. G., and Gage, F. H. (1997). More hippocampal neurons in adult mice living in an enriched environment. *Nature, 386*, 493–495.

Kendler, K. S., Gardner, C. O., and Prescott, C. A. (1999). Clinical characteristics of major depression that predict risk of depression in relatives. *Archives of General Psychiatry, 56*, 322–327.

Kenis, G., and Maes, M. (2002). Effects of antidepressants on the production of cytokines. *International Journal of Neuropsychopharmacology, 5*, 401–412.

Kennedy, J. L., Farrer, L. A., Andreasen, N. C., Mayeux, R., et al. (2003). The genetics of adult-onset neuropsychiatric disease: Complexities and conundra? *Science, 302*, 822–826.

Kennerknecht, I., Grueter, T., Welling, B., Wentzek, S., et al. (2006). First report of prevalence of non-syndromic hereditary prosopagnosia (HPA). *American Journal of Medical Genetics Part A, 140*, 1617–1622.

Kertesz, A., Harlock, W., and Coates, R. (1979). Computer tomographic localization, lesion size, and prognosis in aphasia and nonverbal impairment. *Brain and Language, 8*, 34–50.

Kessels, H. W., and Malinow, R. (2009). Synaptic AMPA receptor plasticity and behavior. *Neuron, 61*, 340–350.

Kessler, R. C., Angermeyer, M., Anthony, J. C., de Graaf, R., et al. (2007). Lifetime prevalence and age-of-onset distributions of mental disorders in the World Health Organization's World Mental Health Survey Initiative. *World Psychiatry, 6*, 168–176.

Kessler, R. C., Berglund, P., Demler, O., Jin, R., et al. (2005). Lifetime prevalence and age-of-onset distributions of DSM-IV disorders in the National Comorbidity Survey Replication. *Archives of General Psychiatry, 62*, 593–602.

Kety, S., Rosenthal, D., Wender, P. H., Schulsinger, F., et al. (1975). Mental illness in the biological and adoptive families of adopted individuals who have become schizophrenic: A preliminary report based on psychiatric interviews. *Proceedings of the Annual Meeting of the American Psychopathological Association, 63*, 147–165.

Kety, S. S., Wender, P. H., Jacobsen, B., Ingraham, L. J., et al. (1994). Mental illness in the biological and adoptive relatives of schizophrenic adoptees. Replication of the Copenhagen Study in the rest of Denmark. *Archives of General Psychiatry, 51*, 442–455.

Kheirbek, M. A., Klemenhagen, K. C., Sahay, A., and Hen, R. (2012). Neurogenesis and generalization: A new approach to stratify and treat anxiety disorders. *Nature Neuroscience, 15*, 1613–1620.

Kiang, N. Y. S. (1965). *Discharge patterns of single fibers in the cat's auditory nerve.* Cambridge, MA: MIT Press.

Kim, D. R., Pesiridou, A., and O'Reardon, J. P. (2009). Transcranial magnetic stimulation in the treatment of psychiatric disorders. *Current Psychiatry Reports, 11*, 447–452.

Kim, K. H., Relkin, N. R., Lee, K. M., and Hirsch, J. (1997). Distinct cortical areas associated with native and second languages. *Nature, 388*, 171–174.

Kimura, D. (1973). The asymmetry of the human brain. *Scientific American, 228*(3), 70–78.

Kimura, D. (1981). Neural mechanisms in manual signing. *Sign Language Studies, 33*, 291–312.

Kimura, D. (1993). *Neuromotor mechanisms in human communication.* Oxford, England: Oxford University Press.

Kimura, D., and Watson, N. V. (1989). The relation between oral movement control and speech. *Brain and Language, 37*, 565–590.

Kindt, M., Soeter, M., and Vervliet, B. (2009). Beyond extinction: Erasing human fear responses and preventing the return of fear. *Nature Neuroscience, 12*, 256–258.

King, S., St-Hilaire, A., and Heidkamp, D. (2010). Prenatal factors in schizophrenia. *Current Directions in Psychological Science, 19*, 209–213.

Kingsbury, S. J., and Garver, D. L. (1998). Lithium and psychosis revisited. *Progress in Neuro-Psychopharmacology & Biological Psychiatry, 22*, 249–263.

Kinney, H. C. (2009). Brainstem mechanisms underlying the sudden infant death syndrome: Evidence from human pathologic studies. *Developmental Psychobiology, 51*, 223–233.

Kinsey, A. C., Pomeroy, W. B., and Martin, C. E. (1948). *Sexual behavior in the human male.* Philadelphia, PA: Saunders.

Kinsey, A. C., Pomeroy, W. B., Martin, C. E., and Gebhard, P. H. (1953). *Sexual behavior in the human female.* Philadelphia, PA: Saunders.

Kinsley, C. H., and Lambert, K. G. (2006). The maternal brain. *Scientific American, 294*, 72–79.

Kirik, D., Georgievska, B., and Björklund, A. (2004). Localized striatal delivery of GDNF as a treatment for Parkinson disease. *Nature Neuroscience, 7*, 105–110.

Klar, A. J. (2003). Human handedness and scalp hair-whorl direction develop from a common genetic mechanism. *Genetics, 165*, 269–276.

Kleiber, M. (1947). Body size and metabolic rate. *Physiological Reviews, 15*, 511–541.

Klein, B. A. (2003). Signatures of sleep in a paper wasp. *Sleep, 26*, A115–A116.

Klein, M., Shapiro, K. M., and Kandel, E. R. (1980). Synaptic plasticity and the modulation of the Ca2+ current. *Journal of Experimental Biology, 89*, 117–157.

Klein, R. M. (2000). Inhibition of return. *Trends in Cognitive Science, 4*, 138–147.

Kleitman, N., and Engelmann, T. (1953). Sleep characteristics of infants. *Journal of Applied Physiology, 6*, 269–282.

Kluger, M. J. (1978). The evolution and adaptive value of fever. *American Scientist, 66*, 38–43.

Klüver, H., and Bucy, P. C. (1938). An analysis of certain effects of bilateral temporal lobectomy in the rhesus monkey, with special reference to "psychic blindness." *Journal of Psychology, 5*, 33–54.

Knecht, S., Flöel, A., Dräger, B., Breitenstein, C., et al. (2002). Degree of language lateralization determines susceptibility to unilateral brain lesions. *Nature Neuroscience, 5,* 695–699.

Knibestol, M., and Valbo, A. B. (1970). Single unit analysis of mechanoreceptor activity from the human glabrous skin. *Acta Physiologica Scandinavica, 80,* 178–195.

Knudsen, E. I. (1982). Auditory and visual maps of space in the optic tectum of the owl. *Journal of Neuroscience, 2,* 1177–1194.

Knudsen, E. I. (1984). The role of auditory experience in the development and maintenance of sound localization. *Trends in Neurosciences, 7,* 326–330.

Knudsen, E. I. (1998). Capacity for plasticity in the adult owl auditory system expanded by juvenile experience. *Science, 279,* 1531–1533.

Knudsen, E., and Knudsen, P. (1985). Vision guides adjustment of auditory localization in young barn owls. *Science, 230,* 545–548.

Knudsen, E. I., Knudsen, P. F., and Esterly, S. D. (1984). A critical period for the recovery of sound localization accuracy following monaural occlusion in the barn owl. *Journal of Neuroscience, 4,* 1012–1020.

Knudsen, E. I., and Konishi, M. (1978). A neural map of auditory space in the owl. *Science, 200,* 795–797.

Kobatake, E., and Tanaka, K. (1994). Neuronal selectivities to complex object features in the ventral visual pathway of the macaque cerebral cortex. *Journal of Neurophysiology, 71,* 856–867.

Koch, G., Oliveri, M., Torriero, S., and Caltagirone, C. (2005). Modulation of excitatory and inhibitory circuits for visual awareness in the human right parietal cortex. *Experimental Brain Research, 160,* 510–516.

Kodama, T., Lai, Y. Y., and Siegel, J. M. (2003). Changes in inhibitory amino acid release linked to pontine-induced atonia: An *in vivo* microdialysis study. *Journal of Neuroscience, 23,* 1548–1554.

Koehler, K. R., Mikosz, A. M., Molosh, A. I., Patel, D., et al. (2013). Generation of inner ear sensory epithelia from pluripotent stem cells in 3D culture. *Nature, 500,* 217–221.

Koh, K., Joiner, W. J., Wu, M. N., Yue, Z., et al. (2008). Identification of SLEEPLESS, a sleep-promoting factor. *Science, 321,* 372–376.

Kohl, M. M., Shipton, O. A., Deacon, R. M., Rawlins, J. N., et al. (2011). Hemisphere-specific optogenetic stimulation reveals left-right asymmetry of hippocampal plasticity. *Nature Neuroscience, 14,* 1413–1415. doi:10.1038/nn.2915. (Erratum in 2011 *Nature Neuroscience, 14,* 1617.)

Kojima, M., Hosoda, H., Date, Y., Nakazato, M., et al. (1999). Ghrelin is a growth-hormone-releasing acylated peptide from stomach. *Nature, 402,* 656–660.

Kokrashvili, Z., Mosinger, B., and Margolskee, R. F. (2009). T1r3 and alpha-gustducin in gut regulate secretion of glucagon-like peptide-1. *Annals of the New York Academy of Sciences, 1170,* 91–94.

Kondo, Y., Sachs, B. D., and Sakuma, Y. (1997). Importance of the medial amygdala in rat penile erection evoked by remote stimuli from estrous females. *Behavioural Brain Research, 88,* 153–160.

Konishi, M. (1985). Birdsong: From behavior to neuron. *Annual Review of Neuroscience, 8,* 125–170.

Konopka, R. J., and Benzer, S. (1971). Clock mutants of *Drosophila melanogaster*. *Proceedings of the National Academy of Sciences, USA, 68,* 2112–2116.

Koob, G. F. (1995). Animal models of drug addiction. In F. E. Bloom and D. J. Kupfer (Eds.), *Psychopharmacology: The fourth generation of progress* (pp. 759–772). New York, NY: Raven Press.

Kopell, B. H., Machado, A. G., and Rezai, A. R. (2005). Not your father's lobotomy: Psychiatric surgery revisited. *Clinical Neurosurgery, 52,* 315–330.

Kordower, J. H., Chu, Y., Hauser, R. A., Freeman, T. B., et al. (2008). Lewy body-like pathology in long-term embryonic nigral transplants in Parkinson's disease. *Nature Medicine, 14,* 504–506.

Korman, M., Doyon, J., Doljansky, J., Carrier, J., et al. (2007). Daytime sleep condenses the time course of motor memory consolidation. *Nature Neuroscience, 10,* 1206–1213.

Kovelman, J. A., and Scheibel, A. B. (1984). A neurohistological correlate of schizophrenia. *Biological Psychiatry, 19,* 1601.

Krause, J., Lalueza-Fox, C., Orlando, L., Enard, W., et al. (2007). The derived FOXP2 variant of modern humans was shared with Neandertals. *Current Biology, 17,* 1908–1912.

Krauss, R. M. (1998). Why do we gesture when we speak? *Current Directions in Psychological Science, 7(2),* 54–60.

Krebs, J. R., Sherry, D. F., Healy, S. D., Perry, V. H., et al. (1989). Hippocampal specialization of food-storing birds. *Proceedings of the National Academy of Sciences, USA, 86,* 1388–1392.

Kril, J., Halliday, G., Svoboda, M., and Cartwright, H. (1997). The cerebral cortex is damaged in chronic alcoholics. *Neuroscience, 79,* 983–998.

Kringelbach, M. L. (2005). The human orbitofrontal cortex: Linking reward to hedonic experience. *Nature Reviews Neuroscience, 6,* 691–702.

Kringelbach, M. L., Jenkinson, N., Owen, S. L. F., and Aziz, T. Z. (2007). Translational principles of deep brain stimulation. *Nature Reviews Neuroscience, 8,* 623–634.

Kripke, D. F., Garfinkel, L., Wingard, D. L., Klauber, M. R., et al. (2002). Mortality associated with sleep duration and insomnia. *Archives of General Psychiatry, 59,* 131–136.

Krystal, A., Krishnan, K. R., Raitiere, M., Poland, R., et al. (1990). Differential diagnosis and pathophysiology of Cushing's syndrome and primary affective disorder. *Journal of Neuropsychiatry and Clinical Neurosciences, 2,* 34–43.

Kuhl, B. A., Dudukovic, N. M., Kahn, I., and Wagner, A. D. (2007). Decreased demands on cognitive control reveal the neural processing benefits of forgetting. *Nature Neuroscience, 10,* 908–914.

Kulkarni, A., and Colburn, H. S. (1998). Role of spectral detail in sound-source localization. *Nature, 396,* 747–749.

Kupfer, D. J., Reynolds, C. F., Ulrich, R. F., Shaw, D. H., et al. (1982). EEG sleep, depression, and aging. *Neurobiology of Aging, 3,* 351–360.

Kutas, M., and Hillyard, S. A. (1980). Reading senseless sentences: Brain potentials reflect semantic incongruity. *Science, 207,* 203–205.

Kwakkel, G., Veerbeek, J. M., van Wegen, E. E., and Wolf, S. L. (2015). Constraint-induced movement therapy after stroke. *Lancet Neurology, 14,* 224–234. doi:10.1016/S1474-4422(14)70160-7

L

LaBar, K. S., Gatenby, J. C., Gore, J. C., LeDoux, J. E., et al. (1998). Human amygdala activation during conditioned fear acquisition and extinction: A mixed-trial fMRI study. *Neuron, 20,* 937–945.

LaFerla, F. M., Green, K. N., and Oddo, S. (2007). Intracellular amyloid-β in Alzheimer's disease. (2007). *Nature Reviews Neuroscience, 8,* 499–508.

Lai, C. S. L., Fisher, S. E., Hurst, J. A., Vargha-Khadem, F., et al. (2001). A forkhead-domain gene is mutated in a severe speech and language disorder. *Nature, 413,* 519–523.

Landau, B., and Levy, R. M. (1993). Neuromodulation techniques for medically refractory chronic pain. *Annual Review of Medicine, 44,* 279–287.

Langleben, D. D., Schroeder, L., Maldjian, J. A., Gur, R. C., et al. (2002). Brain activity during simulated deception: An event-related functional magnetic resonance study. *NeuroImage, 15,* 727–732.

Larroche, J.-C. (1977). *Developmental pathology of the neonate.* Amsterdam, Netherlands: Excerpta Medica.

Larsson, J., Gulyas, B., and Roland, P. E. (1996). Cortical representation of self-paced finger movement. *NeuroReport, 7,* 463–468.

Larsson, M., and Willander, J. (2009). Autobiographical odor memory. *Annals of the New York Academy of Sciences, 1170,* 318–323.

Lau, H. C., Rogers, R. D., Haggard, P., and Passingham, R. E. (2004). Attention to intention. *Science, 303,* 1208–1210.

Laverty, P. H., Leskovar, A., Breur, G. J., Coates, J. R., et al. (2004). A preliminary study of intravenous surfactants in paraplegic dogs: Polymer therapy in canine clinical SCI. *Journal of Neurotrauma, 21,* 1767–1777.

Lavie, N. (1995). Perceptual load as a necessary condition for selective attention. *Journal of Experimental Psychology: Human Perception and Performance, 21,* 451–468.

Lavie, N., Hirst, A., de Fockert, J. W., and Viding, E. (2004). Load theory of selective attention and cognitive control. *Journal of Experimental Psychology: General, 133,* 339–354.

Lavie, N., Lin, Z., Zokaei, N., and Thoma, V. (2009). The role of perceptual load in object recognition. *Journal of Experimental Psychology: Human Perception and Performance, 35,* 1346–1358.

Lavie, P. (1996). *The enchanted world of sleep* (A. Berris, Trans.). New Haven, CT: Yale University Press.

Lavond, D. G., Kim, J. J., and Thompson, R. F. (1993). Mammalian brain substrates of aversive classical conditioning. *Annual Review of Psychology, 44,* 317–342.

Leask, S. J., and Beaton, A. A. (2007). Handedness in Great Britain. *Laterality, 12,* 559–572.

LeDoux, J. E. (1994). Emotion, memory and the brain. *Scientific American, 270*(6), 50–57.

LeDoux, J. E. (1995). Emotion: Clues from the brain. *Annual Review of Psychology, 46,* 209–235.

LeDoux, J. E. (1996). *The emotional brain: The mysterious underpinnings of emotional life.* London, England: Simon & Schuster.

Lee, H., Kim, D. W., Remedios, R., Anthony, T. E., et al. (2014). Scalable control of mounting and attack by Esr1+ neurons in the ventromedial hypothalamus. *Nature, 509,* 627–632.

Le Grange, D. (2005). The Maudsley family-based treatment for adolescent anorexia nervosa. *World Psychiatry, 4,* 142–146.

Leinders-Zufall, T., Lane, A. P., Puche, A. C., Ma, W., et al. (2000). Ultrasensitive pheromone detection by mammalian vomeronasal neurons. *Nature, 405,* 792–796.

Lepage, J. F., and Theoret, H. (2006). EEG evidence for the presence of an action observation-execution matching system in children. *European Journal of Neuroscience, 23,* 2505–2510.

Lereya, S. T., Copeland, W. E., Costello, E. J., and Wolke, D. (2015). Adult mental health consequences of peer bullying and maltreatment in childhood: Two cohorts in two countries. *Lancet Psychiatry, 2,* 524–531. doi:http://dx.doi.org/10.1016/S2215-0366(15)00165-0

Leroi, I., Sheppard, J. M., and Lyketsos, C. G. (2002). Cognitive function after 11.5 years of alcohol use: Relation to alcohol use. *American Journal of Epidemiology, 156,* 747–752.

Lesku, J. A., Roth, T. C., II, Rattenborg, N. C., Amlaner, C. J., et al. (2009). History and future of comparative analyses in sleep research. *Neuroscience and Biobehavioral Reviews, 33,* 1024–1036.

Lesné, S., Koh, M. T., Kotilinek, L., Kayed, R., et al. (2006). A specific amyloid-β protein assembly in the brain impairs memory. *Nature, 440,* 352–357.

Leuner, B., Glasper, E. R., and Gould, E. (2010). Sexual experience promotes adult neurogenesis in the hippocampus despite an initial elevation in stress hormones. *PLOS ONE, 5,* e11597.

Leung, C. T., Coulombe, P. A., and Reed, R. R. (2007). Contribution of olfactory neural stem cells to tissue maintenance and regeneration. *Nature Neuroscience, 10,* 720–726.

Leutgeb, S., Leutgeb, J. K., Barnes, C. A., Moser, E. I., et al. (2005). Independent codes for spatial and episodic memory in hippocampal neuronal ensembles. *Science, 309,* 619–623.

LeVay, S. (1991). A difference in hypothalamic structure between heterosexual and homosexual men. *Science, 253,* 1034–1037.

Levine, J. D., Gordon, N. C., and Fields, H. L. (1978). The mechanism of placebo analgesia. *Lancet, 2,* 654–657.

Levine, S., Haltmeyer, G. C., and Karas, G. G. (1967). Physiological and behavioral effects of infantile stimulation. *Physiology & Behavior, 2,* 55–59.

Lewis, D. O. (1990). Neuropsychiatric and experiential correlates of violent juvenile delinquency. *Neuropsychology Review, 1,* 125–136.

Lewis, D. O., Shankok, S. S., and Pincus, J. (1979). Juvenile male sexual assaulters. *American Journal of Psychiatry, 136,* 1194–1195.

Lewy, A. J., Bauer, V. K., Cutler, N. L., Sack, R. L., et al. (1998). Morning vs evening light treatment of patients with winter depression. *Archives of General Psychiatry, 55,* 890–896.

Lewy, A. J., Rough, J. N., Songer, J. B., Mishra, N., et al. (2007). The phase shift hypothesis for the circadian component of winter depression. *Dialogues in Clinical Neuroscience, 9,* 291–300.

Li, B., Piriz, J., Mirrione, M., Chung, C., et al. (2011). Synaptic potentiation onto habenula neurons in the learned helplessness model of depression. *Nature, 470,* 535–539.

Li, J. Y., Englund, E., Holton, J. L., Soulet, D., et al. (2008). Lewy bodies in grafted neurons in subjects with Parkinson's disease suggest host-to-graft disease propagation. *Nature Medicine, 14,* 501–503.

Li, X., Glaser, D., Li, W., Johnson, W. E., et al. (2009). Analyses of sweet receptor gene (Tas1r2) and preference for sweet stimuli in species of Carnivora. *Journal of Heredity, 100*(Suppl. 1), S90–S100.

Liberles, S. D. (2009). Trace amine-associated receptors are olfactory receptors in vertebrates. *Annals of the New York Academy of Science, 1170,* 168–172.

Liberles, S. D., and Buck, L. B. (2006). A second class of chemosensory receptors in the olfactory epithelium. *Nature, 442,* 645–650.

Licht, P., Frank, L. G., Pavgi, S., Yalcinkaya, T. M., et al. (1992). Hormonal correlates of "masculinization" in female spotted hyenas (*Crocuta crocuta*). 2. Maternal and fetal steroids. *Journal of Reproduction and Fertility, 95,* 463–474.

Lichtenstein, P., Yip, B. H., Björk, C., Pawitan, Y., et al. (2009). Common genetic determinants of schizophrenia and bipolar disorder in Swedish families: A population-based study. *Lancet, 373,* 234–239.

Lichtman, J. W., and Purves, D. (1980). The elimination of redundant preganglionic innervation to hamster sympathetic ganglion cells in early post-natal life. *Journal of Physiology (London), 301,* 213–228.

Lieberman, P. (1985). On the evolution of human syntactic ability: Its pre-adaptive bases—motor control and speech. *Journal of Human Evolution, 14,* 657–668.

Liégeois, F., Baldeweg, T., Connelly, A., Gadian, D. G., et al. (2003). Language fMRI abnormalities associated with FOXP2 gene mutation. *Nature Neuroscience, 6,* 1230–1237.

Liepert, J., Bauder, H., Wolfgang, H. R., Miltner, W. H., et al. (2000). Treatment-induced cortical reorganization after stroke in humans. *Stroke, 31,* 1210–1216.

Lim, M. M., Wang, Z., Olazabal, D. E., Ren, X., et al. (2004). Enhanced partner preference in a promiscuous species by manipulating the expression of a single gene. *Nature, 429,* 754–757.

Lim, M. M., and Young, L. J. (2006). Neuropeptidergic regulation of affiliative behavior and social bonding in animals. *Hormones and Behavior, 50,* 506–557.

Lin, F. H., Witzel, T., Hamalainen, M. S., Dale, A. M., et al. (2004). Spectral spatiotemporal imaging of cortical oscillations and interactions in the human brain. *NeuroImage, 23,* 582–595.

Lin, L., Faraco, J., Li, R., Kadotani, H., et al. (1999). The sleep disorder canine narcolepsy is caused by a mutation in the hypocretin (orexin) receptor 2 gene. *Cell, 98,* 365–376.

Linde, K., Allais, G., Brinkhaus, B., Manheimer, E., et al. (2009). Acupuncture for tension-type headache. *Cochrane Database of Systematic Reviews, 1,* CD007587.

Lindemann, B. (1995). Sweet and salty: Transduction in taste. *News in Physiological Sciences, 10,* 166–170.

Lindvall, O., and Kokaia, Z. (2006). Stem cells for the treatment of neurological disorders. *Nature, 441,* 1094–1096.

Lindvall, O., Sawle, G., Widner, H., Rothwell, J. C., et al. (1994). Evidence for long-term survival and function of dopaminergic grafts in progressive Parkinson's disease. *Annals of Neurology, 35,* 172–180.

Lisk, R. D. (1962). Diencephalic placement of estradiol and sexual receptivity in the female rat. *American Journal of Physiology, 203,* 493–496.

Lisman, J., Schulman, H., and Cline, H. (2002). The molecular basis of CAMKII function in synaptic and behavioural memory. *Nature Reviews Neuroscience, 3,* 175–190.

Liu, D., Diorio, J., Tannenbaum, B., Caldji, C., et al. (1997). Maternal care, hippocampal glucocorticoid receptors, and hypothalamic-pituitary-adrenal responses to stress. *Science, 277,* 1659–1662.

Liu, J., Lillo, C., Jonsson, P. A., Vande Velde, C., et al. (2004). Toxicity of familial ALS-linked SOD1 mutants from selective recruitment to spinal mitochondria. *Neuron, 42,* 5–17.

Liu, Y., Gao, J. H., Liotti, M., Pu, Y., et al. (1999). Temporal dissociation of parallel processing in the human subcortical outputs. *Nature, 400,* 364–367.

Liu, Y., Gao, J.-H., Liu, H.-L., and Fox, P. T. (2000). The temporal response of the brain after eating revealed by functional MRI. *Nature, 405,* 1058–1062.

Livingstone, M. S. (2000). Is it warm? Is it real? Or just low spatial frequency? *Science, 290,* 1299.

Llewellyn, S., and Hobson, J. A. (2015). Not only … but also: REM sleep creates and NREM Stage 2 instantiates landmark junctions in cortical memory networks. *Neurobiology of Learning and Memory, 122,* 69–87.

Lloyd, J. A. (1971). Weights of testes, thymi, and accessory reproductive glands in relation to rank in paired and grouped house mice (*Mus musculus*). *Proceedings of the Society for Experimental Biology and Medicine, 137,* 19–22.

Lo, E. H., Dalkara, T., and Moskowitz, M. A. (2003). Mechanisms, challenges and opportunities in stroke. *Nature Reviews Neuroscience, 4,* 399–415.

Lockhart, M., and Moore, J. W. (1975). Classical differential and operant conditioning in rabbits (*Oryctolagus cuniculus*) with septal lesions. *Journal of Comparative and Physiological Psychology, 88,* 147–154.

Loconto, J., Papes, F., Chang, E., Stowers, L., et al. (2003). Functional expression of murine V2R pheromone receptors involves selective association with the M10 and M1 families of MHC class Ib molecules. *Cell, 112,* 607–618.

Loeb, G. E. (1990). Cochlear prosthetics. *Annual Review of Neuroscience, 13,* 357–371.

Loewenstein, W. R. (Ed.). (1971). Mechano-electric transduction in the Pacinian corpuscle. Initiation of sensory impulses in mechanoreception. In *Handbook of sensory physiology: Vol. 1. Principles of receptor physiology* (pp. 269–290). Berlin, Germany: Springer-Verlag.

Loftus, E. F. (2003). Make-believe memories. *American Psychologist, 58,* 867–873.

Logan, C. G., and Grafton, S. T. (1995). Functional anatomy of human eyeblink conditioning determined with regional cerebral glucose metabolism and positron emission tomography. *Proceedings of the National Academy of Sciences, USA, 92,* 7500–7504.

Long, M. A., Jutras, M. J., Connors, B. W., and Burwell, R. D. (2005). Electrical synapses coordinate activity in the suprachiasmatic nucleus. *Nature Neuroscience, 8,* 61–66.

Loui, P., Alsop, D., and Schlaug, G. (2009). Tone deafness: A new disconnection syndrome? *Journal of Neuroscience, 29,* 10215–10220.

Lu, X. Y. (2007). The leptin hypothesis of depression: A potential link between mood disorders and obesity? *Current Opinion in Pharmacology, 7,* 648–652.

Lübke, K. T., and Pause, B. M. (2015). Always follow your nose: The functional significance of social chemosignals in human reproduction and survival. *Hormones and Behavior, 68,* 134–144.

Luck, S. J. (2005). *An introduction to the event-related potential technique.* Cambridge, MA: MIT Press.

Luck, S. J., and Hillyard, S. A. (1994). Electrophysiological correlates of feature analysis during visual search. *Psychophysiology, 31,* 291–308.

Lucking, C. B., Durr, A., Bonifati, V., Vaughan, J., et al. (2000). Association between early-onset Parkinson's disease and mutations in the parkin gene. *New England Journal of Medicine, 342,* 1560–1567.

Luria, A. R. (1987). *The mind of a mnemonist.* Cambridge, MA: Harvard University Press.

Lush, I. E. (1989). The genetics of tasting in mice. VI. Saccharin, acesulfame, dulcin and sucrose. *Genetical Research, 53,* 95–99.

Lyamin, O., Pryaslova, J., Lance, V., and Siegel, J. (2005). Continuous activity in cetaceans after birth: The exceptional wakefulness of newborn whales and dolphins has no ill-effect on their development. *Nature, 435,* 1177.

Lynch, G., Larson, J., Staubli, U., and Granger, R. (1991). Variants of synaptic potentiation and different types of memory operations in hippocampus and related structures. In L. R. Squire, N. M. Weinberger, G. Lynch, and J. L. McGaugh (Eds.), *Memory: Organization and locus of change* (pp. 330–363). New York, NY: Oxford University Press.

M

Machado-Vieira, R., Salvador, G., Diazgranados, N., and Zarate, C. A., Jr. (2009). Ketamine and the next generation of antidepressants with a rapid onset of action. *Pharmacology & Therapeutics, 123,* 143–150.

MacLean, P. D. (1949). Psychosomatic disease and the "visceral brain": Recent developments bearing on the Papez theory of emotion. *Psychosomatic Medicine, 11,* 338–353.

MacLeod, C. M. (1991). Half a century of research on the Stroop effect: An integrative review. *Psychological Bulletin, 109,* 163–203.

Macmillan, M. (2000). *An odd kind of fame: Stories of Phineas Gage.* Cambridge, MA: MIT Press.

MacNeilage, P. F., Rogers, L. J., and Vallortigara, G. (2009). Origins of the left & right brain. *Scientific American, 301*(1), 60–67.

Maes, M., Bosmans, E., Suy, E., Vandervorst, C., et al. (1991). Depression-related disturbances in mitogen-induced lymphocyte responses and interleukin-1 beta and soluble interleukin-2 receptor production. *Acta Psychiatrica Scandinavica, 84,* 379–386.

Magavi, S. S., Leavitt, B. R., and Macklis, J. D. (2000). Induction of neurogenesis in the neocortex of adult mice. *Nature, 405,* 951–955.

Maggioncalda, A. N., and Sapolsky, R. M. (2002). Disturbing behaviors of the orangutan. *Scientific American, 286*(6), 60–65.

Magnusson, A., and Stefansson, J. G. (1993). Prevalence of seasonal affective disorder in Iceland. *Archives of General Psychiatry, 50,* 941–946.

Mahowald, M. W., and Schenck, C. H. (2005). Insights from studying human sleep disorders. *Nature, 437,* 1279–1285.

Mair, W. G. P., Warrington, E. K., and Wieskrantz, L. (1979). Memory disorder in Korsakoff's psychosis. *Brain, 102,* 749–783.

Mak, G. K., Enwere, E. K., Gregg, C., Pakarainen, T., et al. (2007). Male pheromone-stimulated neurogenesis in the adult female brain: Possible role in mating behavior. *Nature Neuroscience, 10,* 1003–1011.

Maki, P. M., and Resnick, S. M. (2000). Longitudinal effects of estrogen replacement therapy on pet cerebral blood flow and cognition. *Neurobiology of Aging, 21,* 373–383.

Malenka, R. C., and Bear, M. F. (2004). LTP and LTD: An embarrassment of riches. *Neuron, 44,* 5–21.

Mancuso, K., Hauswirth, W. W., Li, Q., Connor, T. B., et al. (2009). Gene therapy for red-green colour blindness in adult primates. *Nature, 461,* 784–788.

Mani, S. K., Fienberg, A. A., O'Callaghan, J. P., Snyder, G. L., et al. (2000). Requirement for DARPP-32 in progesterone-facilitated sexual receptivity in female rats and mice. *Science, 287,* 1053–1056.

Manova, M. G., and Kostadinova, I. I. (2000). Some aspects of the immunotherapy of multiple sclerosis. *Folia Medica, 42*(1), 5–9.

Mantini, D., Gerits, A., Nelissen, K., Durand, J. B., et al. (2011). Default mode of brain function in monkeys. *Journal of Neuroscience, 31,* 12954–12962.

Mantyh, P. W., Rogers, S. D., Honore, P., Allen, B. J., et al. (1997). Inhibition of hyperalgesia by ablation of lamina I spinal neurons expressing the substance P receptor. *Science, 278,* 275–279.

Mao, J. B., and Evinger, C. (2001). Long-term potentiation of the human blink reflex. *Journal of Neuroscience, 21,* RC151.

March, J., Silva, S., Petrychi, S., Curry, J., et al. (2004). Fluoxetine, cognitive-behavioral therapy, and their combination for adolescents with depression: Treatment for Adolescents with Depression Study (TADS) randomized controlled trial. *JAMA, 292,* 807–820.

Marcus, G. F., Vijayan, S., Bandi Rao, S., and Vishton, P. M. (1999). Rule learning by seven-month-old infants. *Science, 283,* 77–80.

Maren, S., and Quirk, G. J. (2004). Neuronal signalling of fear memory. *Nature, 5,* 844–852.

Mariani, J., and Changeaux, J.-P. (1981). Ontogenesis of olivocerebellar relationships. I. Studies by intracellular recordings of the multiple innervation of Purkinje cells by climbing fibers in the developing rat cerebellum. *Journal of Neuroscience, 1,* 696–702.

Mark, V. H., and Ervin, F. R. (1970). *Violence and the brain.* New York, NY: Harper & Row.

Marler, P. (1970). Birdsong and speech development: Could there be parallels? *American Scientist, 58,* 669–673.

Marler, P. (1991). Song-learning behavior: The interface with neuroethology. *Trends in Neurosciences, 14,* 199–206.

Marler, P., and Peters, S. (1982). Developmental overproduction and selective attrition: New processes in the epigenesis of birdsong. *Developmental Psychobiology, 15,* 369–378.

Marler, P., and Sherman, V. (1983). Song structure without auditory feedback: Emendations of the auditory template hypothesis. *Journal of Neuroscience, 3,* 517–531.

Marler, P., and Sherman, V. (1985). Innate differences in singing behaviour of sparrows reared in isolation from adult conspecific song. *Animal Behavior, 33,* 57–71.

Marshall, L., Helgadóttir, H., Mölle, M., and Born, J. (2006). Boosting slow oscillations during sleep potentiates memory. *Nature, 444,* 610–613.

Marsicano, G., Wotjak, C. T., Azad, S. C., Bisogno, T., et al. (2002). The endogenous cannabinoid system controls extinction of aversive memories. *Nature, 418,* 530–532.

Martin, C. K., Heilbronn, L., de Jonge, L., Delany, J. P., et al. (2007). Effect of calorie restriction on resting metabolic rate and spontaneous physical activity. *Obesity (Silver Spring), 15,* 2964–2973.

Martin, J. T., and Nguyen, D. H. (2004). Anthropometric analysis of homosexuals and heterosexuals: Implications for early hormone exposure. *Hormones and Behavior, 45,* 31–39.

Martuza, R. L., Chiocca, E. A., Jenike, M. A., Giriunas, I. E., et al. (1990). Stereotactic radiofrequency thermal cingulotomy for obsessive compulsive disorder. *Journal of Neuropsychiatry and Clinical Neurosciences, 2,* 331–336.

Marucha, P. T., Kiecolt-Glaser, J. K., and Favagehi, M. (1998). Mucosal wound healing is impaired by examination stress. *Psychosomatic Medicine, 60,* 362–365.

Maruyama, Y., Pereira, E., Margolskee, R. F., Chaudhari, N., et al. (2006). Umami responses in mouse taste cells indicate more than one receptor. *Journal of Neuroscience, 26,* 2227–2234.

Marzani, D., and Wallman, J. (1997). Growth of the two layers of the chick sclera is modulated reciprocally by visual conditions. *Investigative Ophthalmology & Visual Science, 38,* 1726–1739.

Maskos, U., Molles, B. E., Pons, S., Besson, M., et al. (2005). Nicotine reinforcement and cognition restored by targeted expression of nicotinic receptors. *Nature, 436,* 103–107.

Masters, W. H., and Johnson, V. E. (1966). *Human sexual response.* Boston, MA: Little, Brown.

Masters, W. H., and Johnson, V. E. (1970). *Human sexual inadequacy.* Boston, MA: Little, Brown.

Masters, W. H., Johnson, V. E., and Kolodny, R. C. (1994). *Heterosexuality.* New York, NY: HarperCollins.

Masterton, R. B. (1997). Neurobehavioral studies of the central auditory system. *Annals of Otology Rhinology Laryngology, Supplement, 168,* 31–34.

Mastrianni, J. A., Nixon, R., Layzer, R., Telling, G. C., et al. (1999). Prion protein conformation in a patient with sporadic fatal insomnia. *New England Journal of Medicine, 340,* 1630–1638.

Mateo, J. M., and Johnston, R. E. (2000). Kin recognition and the "armpit effect": Evidence of self-referent phenotype matching. *Proceedings of the Royal Society of London. Series B: Biological Sciences, 267,* 695–700.

Matsumoto, K., Suzuki, W., and Tanaka, K. (2003). Neuronal correlates of goal-based motor selection in the prefrontal cortex. *Science, 301,* 229–232.

Mauch, D. H., Nägler, K., Schumacher, S., Göritz, C., et al. (2001). CNS synaptogenesis promoted by glia-derived cholesterol. *Science, 294,* 1354–1357.

Maurer, P., and Bachmann, M. F. (2007). Vaccination against nicotine: An emerging therapy for tobacco dependence. *Expert Opinion on Investigational Drugs, 16,* 1775–1783.

Mayberg, H. S., Lozano, A. M., Voon, V., McNeely, H. E., et al. (2005). Deep brain stimulation for treatment-resistant depression. *Neuron, 45,* 651–660.

Mazur, A., and Booth, A. (1998). Testosterone and dominance in men. *Behavioral and Brain Sciences, 21,* 353–363.

McAllister, A. K., Katz, L. C., and Lo, D. C. (1997). Opposing roles for endogenous BDNF and NT-3 in regulating cortical dendritic growth. *Neuron, 18,* 767–778.

McAlpine, D., Jiang, D., and Palmer, A. R. (2001). A neural code for low-frequency sound localization in mammals. *Nature Neuroscience, 4,* 396–401.

McBurney, D. H., Smith, D. V., and Shick, T. R. (1972). Gustatory cross adaptation: Sourness and bitterness. *Perception & Psychophysics, 11,* 2228–2232.

McCall, W. V., and Edinger, J. D. (1992). Subjective total insomnia: An example of sleep state misperception. *Sleep, 15,* 71–73.

McCarthy, R. A., and Warrington, E. K. (1990). *Cognitive neuropsychology: A clinical introduction.* San Diego, CA: Academic Press.

McCrae, C. S., Rowe, M. A., Tierney, C. G., Dautovich, N. D., et al. (2005). Sleep complaints, subjective and objective sleep patterns, health, psychological adjustment, and daytime functioning in community-dwelling older adults. *Journals of Gerontology. Series B, Psychological Sciences and Social Sciences, 60*(4), P182–P189.

McDonald, J. J., Teder-Sälejärvi, W. A., and Hillyard, S. A. (2000). Involuntary orienting to sound improves visual perception. *Nature, 407,* 906–908.

McEwen, B. S., and Wingfield, J. C. (2010). What is in a name? Integrating homeostasis, allostasis and stress. *Hormones and Behavior, 57,* 105–111.

McFadden, D., and Pasanen, E. (1998). Comparison of the auditory systems of heterosexuals and homosexuals: Click-evoked otoacoustic emissions. *Proceedings of the National Academy of Sciences, USA, 95,* 2709–2713.

McGinty, D. J., and Sterman, M. B. (1968). Sleep suppression after basal forebrain lesions in the cat. *Science, 160,* 1253–1255.

McGowan, P. O., Sasaki, A., D'Alessio, A. C., Dymov, S., et al. (2009). Epigenetic regulation of the glucocorticoid receptor in human brain associates with childhood abuse. *Nature Neuroscience, 12,* 342–348.

McKee, A. C., Cantu, R. C., Nowinski, C. J., Hedley-Whyte, E. T., et al. (2009). Chronic traumatic encephalopathy in athletes: Progressive tauopathy after repetitive head injury. *Journal of Neuropathology and Experimental Neurology, 68,* 709–735.

McKenna, K. (1999). The brain is the master organ in sexual function: Central nervous system control of male and female sexual function. *International Journal of Impotence Research, 11*(Suppl. 1), S48–S55.

McKernan, M. G., and Shinnick-Gallagher, P. (1997). Fear conditioning induces a lasting potentiation of synaptic currents in vitro. *Nature, 390,* 607–611.

McKim, W. A. (1991). *Drugs and behavior: An introduction to behavioral pharmacology* (2nd ed.). Englewood Cliffs, NJ: Prentice Hall.

McKinley, M. J., and Johnson, A. K. (2004). The physiological regulation of thirst and fluid intake. *News in Physiological Sciences, 19,* 1–6.

McKinney, T. D., and Desjardins, C. (1973). Postnatal development of the testis, fighting behavior, and fertility in house mice. *Biology of Reproduction, 9,* 279–294.

McLaughlin, S. K., McKinnon, P. J., Spickofsky, N., Danho, W., et al. (1994). Molecular cloning of G proteins and phosphodiesterases from rat taste cells. *Physiology & Behavior, 56,* 1157–1164.

Meddis, R. (1975). On the function of sleep. *Animal Behavior, 23,* 676–691.

Meddis, R. (1977). *The sleep instinct.* London, England: Routledge & Kegan Paul.

Mednick, S. A., Huttunen, M. O., and Machon, R. A. (1994). Prenatal influenza infections and adult schizophrenia. *Schizophrenia Bulletin, 20,* 263–267.

Medori, R., Montagna, P., Tritschler, H. J., LeBlanc, A., et al. (1992). Fatal familial insomnia: A second kindred with mutation of prion protein gene at codon 178. *Neurology, 42,* 669–670.

Mega, M. S., and Cummings, J. L. (1994). Frontal-subcortical circuits and neuropsychiatric disorders. *Journal of Neuropsychiatry and Clinical Neurosciences, 6,* 358–370.

Meguerditchian, A., and Vauclair, J. (2006). Baboons communicate with their right hand. *Behavioural Brain Research, 171,* 170–174.

Mei, L., and Xiong, W.-C. (2008). Neuregulin 1 in neural development, synaptic plasticity and schizophrenia. *Nature Reviews Neuroscience, 9,* 437–452.

Meier, M. H., Caspi, A., Ambler, A., Harrington, H., et al. (2012). Persistent cannabis users show neuropsychological decline from childhood to midlife. *Proceedings of the National Academy of Sciences, USA, 109,* E2657–E2664.

Meisel, R. L., and Luttrell, V. R. (1990). Estradiol increases the dendritic length of ventromedial hypothalamic neurons in female Syrian hamsters. *Brain Research Bulletin, 25,* 165–168.

Meisel, R. L., and Sachs, B. D. (1994). The physiology of male sexual behavior. In E. Knobil and J. D. Neill (Eds.), *The physiology of reproduction* (2nd ed., Vol. 1, pp. 3–105). New York, NY: Raven Press.

Mello, C. V., Vicario, D. S., and Clayton, D. F. (1992). Song presentation induces gene expression in the songbird forebrain. *Proceedings of the National Academy of Sciences, USA, 89,* 6818–6822.

Melzack, R. (1984). Neuropsychological basis of pain measurement. *Advances in Pain Research, 6,* 323–341.

Melzack, R. (1990). The tragedy of needless pain. *Scientific American, 262*(2), 27–33.

Melzack, R., and Wall, P. D. (1965). Pain mechanisms: A new history. *Science, 150,* 971–979.

Meng, H., Smith, S. D., Hager, K., Held, M., et al. (2005). DCDC2 is associated with reading disability and modulates neuronal development in the brain. *Proceedings of the National Academy of Sciences, USA, 102,* 17053–17058.

Menninger, W. C. (1948). Facts and statistics of significance for psychiatry. *Bulletin of the Menninger Clinic, 12,* 1–25.

Merchán-Pérez, A., Rodriguez, J. R., Alonso-Nanclares, L., Schertel, A., et al. (2009). Counting synapses using FIB/SEM microscopy: A true revolution for ultrastructural volume reconstruction. *Frontiers in Neuroanatomy, 3,* 18.

Mersch, P. P., Middendorp, H. M., Bouhuys, A. L., Beersma, D. G., et al. (1999). Seasonal affective disorder and latitude: A review of the literature. *Journal of Affective Disorders, 53,* 35–48.

Merzenich, M. M., Schreiner, C., Jenkins, W., and Wang, X. (1993). Neural mechanisms underlying temporal integration, segmentation, and input sequence representation: Some implications for the origin of learning disabilities. *Annals of the New York Academy of Sciences, 682,* 1–22.

Meshberger, F. L. (1990). An interpretation of Michelangelo's *Creation of Adam* based on neuroanatomy. *JAMA, 264,* 1837–1841.

Messias, E., Kirkpatrick, B., Bromet, E., Ross, D., et al. (2004). Summer birth and deficit schizophrenia: A pooled analysis from 6 countries. *Archives of General Psychiatry, 61,* 985–989.

Mesulam, M.-M. (1985). Attention, confusional states and neglect. In M.-M. Mesulam (Ed.), *Principles of behavioral neurology.* Philadelphia, PA: Davis.

Michael, N., and Erfurth, A. (2004). Treatment of bipolar mania with right prefrontal rapid transcranial magnetic stimulation. *Journal of Affective Disorders, 78,* 253–257.

Miczek, K. A., Fish, E. W., De Bold, J. F., and De Almeida, R. M. (2002). Social and neural determinants of aggressive behavior: Pharmacotherapeutic targets at serotonin, dopamine and gamma-aminobutyric acid systems. *Psychopharmacology (Berlin), 163,* 434–458.

Miles, L. E., and Dement, W. C. (1980). Sleep and aging. *Sleep, 3,* 1220.

Miller, E. K., and Cohen, J. D. (2001). An integrative theory of prefrontal cortex function. *Annual Review of Neuroscience, 24,* 167–202.

Miller, G. F. (2000). *The mating mind: How sexual choice shaped the evolution of human nature.* New York, NY: Doubleday.

Miller, J. M., and Spelman, F. A. (1990). *Cochlear implants: Models of the electrically stimulated ear.* New York, NY: Springer-Verlag.

Milner, A. D., Perrett, D. I., Johnston, R. S., Benson, P. J., et al. (1991). Perception and action in "visual form agnosia." *Brain, 114,* 405–428.

Milner, B. (1963). Effect of different brain lesions on card sorting. *Archives of Neurology, 9,* 90–100.

Milner, B. (1965). Memory disturbance after bilateral hippocampal lesions. In P. M. Milner and S. E. Glickman (Eds.), *Cognitive processes and the brain; an enduring problem in psychology* (pp. 97–111). Princeton, NJ: Van Nostrand.

Milner, B. (1970). Memory and the medial temporal regions of the brain. In D. H. Pribram and D. E. Broadbent (Eds.), *Biology of memory* (pp. 29–50). New York, NY: Academic Press.

Minzenberg, M. J., Laird, A. R., Thelen, S., Carter, C. S., et al. (2009). Meta-analysis of 41 functional neuroimaging studies of executive function in schizophrenia. *Archives of General Psychiatry, 66,* 811–822.

Miras, M., Serrano, M., Durán, C., Valiño, C., et al. (2015). Early experience with customized, meal-triggered gastric electrical stimulation in obese patients. *Obesity Surgery, 25,* 174–179.

Mirescu, C., Peters, J. D., and Gould, E. (2004). Early life experience alters response of adult neurogenesis to stress. *Nature Neuroscience, 7,* 841–846.

Mirsky, A. F., and Duncan, C. C. (1986). Etiology and expression of schizophrenia: Neurobiological and psychosocial factors. *Annual Review of Psychology, 37,* 291–321.

Mishkin, M., and Ungerleider, L. (1982). Contribution of striate inputs to the visuospatial functions of parieto-preoccipital cortex in monkeys. *Behavioural Brain Research, 6,* 57–77.

Mishra, J., Zinni, M., Bavelier, D., and Hillyard, S. A. (2011). Neural basis of superior performance of action videogame players in an attention-demanding task. *Journal of Neuroscience, 31,* 992–998.

Mithoefer, M. C., Wagner, M. T., Mithoefer, A. T., Jerome, L., et al. (2013). Durability of improvement in post-traumatic stress disorder symptoms and absence of harmful effects or drug dependency after 3,4-methylenedioxymethamphetamine-assisted psychotherapy: A prospective long-term follow-up study. *Journal of Psychopharmacology, 27,* 28–39.

Miyawaki, Y., Uchida, H., Yamashita, O., Sato, M. A., et al. (2008). Visual image reconstruction from human brain activity using a combination of multiscale local image decoders. *Neuron, 60*, 915–929.

Moghaddam, B., and Adams, B. W. (1998). Reversal of phencyclidine effects by a group II metabotropic glutamate receptor agonist in rats. *Science, 281*, 1349–1352.

Mohammed, A. (2001). *Enrichment and the brain. Plasticity in the adult brain: From genes to neurotherapy.* 22nd International Summer School of Brain Research, Amsterdam, Netherlands.

Moita, M. A., Rosis, S., Zhou, Y., LeDoux, J. E., et al. (2004). Putting fear in its place: Remapping of hippocampal place cells during fear conditioning. *Journal of Neuroscience, 24*, 7015–7023.

Money, J., and Ehrhardt, A. A. (1972). *Man and woman, boy and girl.* Baltimore, MD: Johns Hopkins University Press.

Monfils, M. H., Plautz, E. J., and Kleim, J. A. (2005). In search of the motor engram: Motor map plasticity as a mechanism for encoding motor experience. *Neuroscientist, 11*, 471–483.

Monks, D. A., and Watson N. V. (2001). N-cadherin expression in motoneurons is directly regulated by androgens: A genetic mosaic analysis in rats. *Brain Research, 895*, 73–79.

Montague, C. T., Farooqi, I. S., Whitehead, J. P., Soos, M. A., et al. (1997). Congenital leptin deficiency is associated with severe early-onset obesity in humans. *Nature, 387*, 903–908.

Monti, M. M., Vanhaudenhuyse, A., Coleman, M. R., Boly, M., et al. (2010). Willful modulation of brain activity in disorders of consciousness. *New England Journal of Medicine, 362*, 579–589.

Moore, C. L., Dou, H., and Juraska, J. M. (1992). Maternal stimulation affects the number of motor neurons in a sexually dimorphic nucleus of the lumbar spinal cord. *Brain Research, 572*, 52–56.

Moore, G. J., Bebchuk, J. M., Wilds, I. B., Chen, G., et al. (2000). Lithium-induced increase in human brain grey matter. *Lancet, 356*, 241–242.

Moore, R. Y. (1983). Organization and function of a central nervous system circadian oscillator: The suprachiasmatic nucleus. *Federation Proceedings, 42*, 2783–2789.

Moore, R. Y., and Eichler, V. B. (1972). Loss of circadian adrenal corticosterone rhythm following suprachiasmatic lesions in the rat. *Brain Research, 42*, 201–206.

Moorhead, T. W., McKirdy, J., Sussmann, J. E., Hall, J., et al. (2007). Progressive gray matter loss in patients with bipolar disorder. *Biological Psychiatry, 62*, 894–900.

Moran, J., and Desimone, R. (1985). Selective attention gates visual processing in the extrastriate cortex. *Science, 229*, 782–784.

Moreno, F. A., Wiegand, C. B., Taitano, E. K., and Delgado, P. L. (2006). Safety, tolerability, and efficacy of psilocybin in 9 patients with obsessive-compulsive disorder. *Journal of Clinical Psychiatry, 67*, 1735–1740.

Mori, K., Nagao, H., and Yoshihara, Y. (1999). The olfactory bulb: Coding and processing of odor molecule information. *Science, 286*, 711–715.

Morihisa, J., and McAnulty, G. B. (1985). Structure and function: Brain electrical activity mapping and computed tomography in schizophrenia. *Biological Psychiatry, 20*, 3–19.

Morris, B. (2002). Overcoming dyslexia. *Fortune, 145*(10), 1–7.

Morris, R. G., Halliwell, R. F., and Bowery, N. (1989). Synaptic plasticity and learning. II: Do different kinds of plasticity underlie different kinds of learning? *Neuropsychologia, 27*, 41–59.

Morrison, A. R. (1983). A window on the sleeping brain. *Scientific American, 248*(4), 94–102.

Morrison, A. R., Sanford, L. D., Ball, W. A., Mann, G. L., et al. (1995). Stimulus-elicited behavior in rapid eye movement sleep without atonia. *Behavioral Neuroscience, 109*, 972–979.

Morrison, R. G., and Nottebohm, F. (1993). Role of a telencephalic nucleus in the delayed song learning of socially isolated zebra finches. *Journal of Neurobiology, 24*, 1045–1064.

Moruzzi, G. (1972). The sleep-waking cycle. *Ergebnisse der Physiologie, biologischen Chemie und experimentellen Pharmakologie, 64*, 1–165.

Moruzzi, G., and Magoun, H. W. (1949). Brain stem reticular formation and activation of the EEG. *Clinical Neurophysiology, 1*, 455–473.

Moscovitch, A., Blashko, C. A., Eagles, J. M., Darcourt, G., et al. (2004). A placebo-controlled study of sertraline in the treatment of outpatients with seasonal affective disorder. *Psychopharmacology (Berlin), 171*, 390–397.

Mott, F. W. (1895). Experimental inquiry upon the afferent tracts of the central nervous system of the monkey. *Brain, 18*, 1–20.

Motta, S. C., Guimarães, C. C., Furigo, I. C., Sukikara, M. H., et al. (2013). Ventral premammillary nucleus as a critical sensory relay to the maternal aggression network. *Proceedings of the National Academy of Sciences, USA, 110*, 14438–14443.

Mountcastle, V. B. (1979). An organizing principle for cerebral function: The unit module and the distributed system. In F. O. Schmitt and F. G. Worden (Eds.), *The neurosciences: Fourth study program* (pp. 21–24). Cambridge, MA: MIT Press.

Mountcastle, V. B. (1984). Central nervous mechanisms in mechanoreceptive sensibility. In I. Darian-Smith (Ed.), *Handbook of physiology, Section 1: Vol. 3. Sensory processes* (pp. 789–878). Bethesda, MD: American Physiological Society.

Mounts, J. R. (2000). Attentional capture by abrupt onsets and feature singletons produces inhibitory surrounds. *Perception & Psychophysics, 62*, 1485–1493.

Mowry, B. J., Holmans, P. A., Pulver, A. E., Gejman, P. V., et al. (2004). Multicenter linkage study of schizophrenia loci on chromosome 22q. *Molecular Psychiatry, 9*, 784–795.

Mueller, H. T., Haroutunian, V., Davis, K. L., and Meador-Woodruff, J. H. (2004). Expression of the ionotropic glutamate receptor subunits and NMDA receptor-associated intracellular proteins in the substantia nigra in schizophrenia. *Brain Research. Molecular Brain Research, 121*, 60–69.

Muhlert, N., and Lawrence, A. D. (2015). Brain structure correlates of emotion-based rash impulsivity. *NeuroImage, 115*, 138–146.

Mukamal, K. J., Conigrave, K. M., Mittleman, M. A., Camargo, C. A., Jr., et al. (2003). Roles of drinking pattern and type of alcohol consumed in coronary heart disease in men. *New England Journal of Medicine, 348*, 109–118.

Mukhametov, L. M. (1984). Sleep in marine mammals. In A. Borbély and J. L. Valatx (Eds.), *Experimental Brain Research: Suppl. 8. Sleep mechanisms* (pp. 227–238). Berlin, Germany: Springer-Verlag.

Münte, T. F., Altenmüller, E., and Jäncke, L. (2002). The musician's brain as a model of neuroplasticity. *Nature Reviews Neuroscience, 3*, 473–478.

N

Nader, K., and Hardt, O. (2009). A single standard for memory: The case for reconsolidation. *Nature Reviews Neuroscience, 10*, 224–234.

Naeser, M., Gaddie, A., Palumbo, C., and Stiassny-Eder, D. (1990). Late recovery of auditory comprehension in global aphasia. Improved recovery observed with subcortical temporal isthmus lesion vs. Wernicke's cortical area lesion. *Archives of Neurology, 47*, 425–432.

Naeser, M., and Hayward, R. (1978). Lesion localization in aphasia with cranial computed tomography and the Boston Diagnostic Aphasia Exam. *Neurology, 28*, 545–551.

Nair, K. S., Rizza, R. A., O'Brien, P., Dhatariya, K., et al. (2006). DHEA in elderly women and DHEA or testosterone in elderly men. *New England Journal of Medicine, 355*, 1647–1659.

Nakazato, M., Murakami, N., Date, Y., Kojima, M., et al. (2001). A role for ghrelin in the central regulation of feeding. *Nature, 409*, 194–198.

Naqvi, N. H., Rudrauf, D., Damasio, H., and Bechara, A. (2007). Damage to the insula disrupts addiction to cigarette smoking. *Science, 315,* 531–534.

Naselaris, T., Prenger, R. J., Kay, K. N., Oliver, M., et al. (2009). Bayesian reconstruction of natural images from human brain activity. *Neuron, 63,* 902–915.

Nathans, J. (1987). Molecular biology of visual pigments. *Annual Review of Neuroscience, 10,* 163–194.

Nation, E. F. (1973). William Osler on penis captivus and other urologic topics. *Urology, 2,* 468–470.

National Academy of Sciences. (2003). *The polygraph and lie detection.* Washington, DC: National Academies Press (www.nap.edu/openbook.php?isbn=0309084369).

National Center for Health Statistics. (2007). *Health, United States, 2007, with chartbook on trends in the health of Americans.* Hyattsville, MD (www.cdc.gov/nchs/data/hus/hus07.pdf).

National Institute of Mental Health. (2012). *Serious mental illness (SMI) among U.S. adults.* Behtesda, MD (www.nimh.nih.gov/health/statistics/prevalence/serious-mental-illness-smi-among-us-adults.shtml).

Neary, M. T., and Batterham, R. L. (2009). Gut hormones: Implications for the treatment of obesity. *Pharmacology & Therapeutics, 124,* 44–56.

Neff, W. D., and Casseday, J. H. (1977). Effects of unilateral ablation of auditory cortex on monaural cat's ability to localize sound. *Journal of Neurophysiology, 40,* 44–52.

Neitz, M., Kraft, T.W., and Neitz, J. (1998). Expression of L cone pigment gene subtypes in females. *Vision Research, 38,* 3221–3225.

Nelson, G., Chandrashekar, J., Hoon, M. A., Feng, L., et al. (2002). An amino-acid taste receptor. *Nature, 416,* 199–202.

Nelson, G., Hoon, M. A., Chandrashekar, J., Zhang, Y., et al. (2001). Mammalian sweet taste receptors. *Cell, 106,* 381–390.

Nestler, E. J., and Hyman, S. E. (2010). Animal models of neuropsychiatric disorders. *Nature Neuroscience, 13,* 1161–1169.

Neumeister, A., Bain, E., Nugent, A. C., Carson, R. E., et al. (2004). Reduced serotonin type 1A receptor binding in panic disorder. *Journal of Neuroscience, 24,* 589–591.

Neville, H. J., Bavelier, D., Corina, D., Rauschecker, J., et al. (1998). Cerebral organization for language in deaf and hearing subjects: Biological constraints and effects of experience. *Proceedings of the National Academy of Sciences, USA, 95,* 922–929.

Neville, H. J., Mills, D. L., and Lawson, D. S. (1992). Fractionating language: Different neural subsystems with different sensitive periods. *Cerebral Cortex, 2,* 244–258.

Newsome, W. T., Wurtz, R. H., Dursteler, M. R., and Mikami, A. (1985). Deficits in visual motion processing following ibotenic acid lesions of the middle temporal visual area of the macaque monkey. *Journal of Neuroscience, 5,* 825–840.

Ng, S. F., Lin, R. C., Laybutt, D. R., Barres, R., et al. (2010). Chronic high-fat diet in fathers programs β-cell dysfunction in female rat offspring. *Nature, 467,* 963–966.

Ngandu, T., Lehtisalo, J., Solomon, A., Levälahti, E., et al. (2015). A 2 year multidomain intervention of diet, exercise, cognitive training, and vascular risk monitoring versus control to prevent cognitive decline in at-risk elderly people (FINGER): A randomised controlled trial. *Lancet,* pii: S0140-6736(15)60461-5. doi:10.1016/S0140-6736(15)60461-5

Nichols, M. J., and Newsome, W. T. (1999). The neurobiology of cognition. *Nature, 402,* C35–C38.

Nieto-Sampedro, M., and Cotman, C. W. (1985). Growth factor induction and temporal order in central nervous system repair. In C. W. Cotman (Ed.), *Synaptic plasticity* (pp. 407–457). New York, NY: Guilford Press.

Nietzel, M. T. (2000). Police psychology. In A. E. Kazdin (Ed.), *Encyclopedia of psychology* (Vol. 6, pp. 224–226). Washington, DC: American Psychological Association.

Nishida, M., and Walker, M. P. (2007). Daytime naps, motor memory consolidation and regionally specific sleep spindles. *PLOS ONE, 2,* e341.

Nishimoto, S., and Gallant, J. L. (2011). A three-dimensional spatiotemporal receptive field model explains responses of area MT neurons to naturalistic movies. *Journal of Neuroscience, 31,* 14551–14564.

Nixon, K., and Crews, F. T. (2002). Binge ethanol exposure decreases neurogenesis in adult rat hippocampus. *Journal of Neurochemistry, 83,* 1087–1093.

Noad, M. J., Cato, D. H., Bryden, M. M., Jenner, M.-N., et al. (2000). Cultural revolution in whale songs. *Nature, 408,* 537–538.

Noguchi, Y., Watanabe, E., and Sakai, K. L. (2003). An event-related optical topography study of cortical activation induced by single-pulse transcranial magnetic stimulation. *NeuroImage, 19,* 156–162.

Nordeen, E. J., Nordeen, K. W., Sengelaub, D. R., and Arnold, A. P. (1985). Androgens prevent normally occurring cell death in a sexually dimorphic spinal nucleus. *Science, 229,* 671–673.

Norman, A. W., and Henry, H. L. (2014). *Hormones* (3rd ed.). San Diego, CA: Academic Press.

Nottebohm, F. (1980). Brain pathways for vocal learning in birds: A review of the first 10 years. *Progress in Psychobiology and Physiological Psychology, 9,* 85–124.

Nottebohm, F. (1981). A brain for all seasons: Cyclical anatomical changes in song control nuclei of the canary brain. *Science, 214,* 1368–1370.

Numan, M., and Numan. M. J. (1991). Preoptic-brainstem connections and maternal behavior in rats. *Behavioral Neuroscience, 105,* 1010–1029.

Nutt, J. G., Rufener, S. L., Carter, J. H., Anderson, V. C., et al. (2001). Interactions between deep brain stimulation and levodopa in Parkinson's disease. *Neurology, 57,* 1835–1842.

O

Oades, R. D., and Halliday, G. M. (1987). Ventral tegmental (A10) system: Neurobiology. 1. Anatomy and connectivity. *Brain Research Reviews, 12,* 117–165.

Oberlander, T. F., Papsdorf, M., Brain, U. M., Misri, S., et al. (2010). Prenatal effects of selective serotonin reuptake inhibitor antidepressants, serotonin transporter promoter genotype (*SLC6A4*), and maternal mood on child behavior at 3 years of age. *Archives of Pediatrics and Adolescent Medicine, 164,* 444–451.

O'Connell-Rodwell, C. E. (2007). Keeping an "ear" to the ground: Seismic communication in elephants. *Physiology (Bethesda), 22,* 287–294.

O'Connor, D. B., Archer, J., and Wu, F. C. (2004). Effects of testosterone on mood, aggression, and sexual behavior in young men: A double-blind, placebo-controlled, cross-over study. *Journal of Clinical Endocrinology & Metabolism, 89,* 2837–2845.

O'Craven, K. M., Downing, P. E., and Kanwisher, N. (1999). fMRI evidence for objects as the units of attentional selection. *Nature, 401,* 584–587.

O'Donovan, A., Chao, L. L., Paulson, J., Samuelson, K. W., et al. (2015). Altered inflammatory activity associated with reduced hippocampal volume and more severe posttraumatic stress symptoms in Gulf War veterans. *Psychoneuroendocrinology, 51,* 557–566.

Ojemann, G., and Mateer, C. (1979). Human language cortex: Localization of memory, syntax, and sequential motor-phoneme identification systems. *Science, 205,* 1401–1403.

Okano, H., Ogawa, Y., Nakamura, M., Kaneko, S., et al. (2003). Transplantation of neural stem cells into the spinal cord after injury. *Seminars in Cell and Developmental Biology, 14,* 191–198.

O'Keefe, J., and Dostrovsky, J. (1971). The hippocampus as a spatial map. Preliminary evidence from unit activity in the freely-moving rat. *Brain Research, 34,* 171–175.

Olanow, C. W., Goetz, C. G., Kordower, J. H., Stoessl, A. J., et al. (2003). A double-blind controlled trial of bilateral fetal nigral transplantation in Parkinson's disease. *Annals of Neurology, 54,* 403–414.

Olatunji, B. O., Davis, M. L., Powers, M. B., and Smits, J. A. (2013). Cognitive-behavioral therapy for obsessive-compulsive disorder: A meta-analysis of treatment outcome and moderators. *Journal of Psychiatric Research, 47,* 33–41.

Olds, J., and Milner, P. (1954). Positive reinforcement produced by electrical stimulation of septal area and other regions of rat brain. *Journal of Comparative and Physiological Psychology, 47,* 419–427.

Oler, J. A., Fox, A. S., Shelton, S. E., Rogers, J., et al. (2010). Amygdalar and hippocampal substrates of anxious temperament differ in their heritability. *Nature, 466,* 864–868.

Olfson, M., Marcus, S. C., and Shaffer, D. (2006). Antidepressant drug therapy and suicide in severely depressed children and adults: A case-control study. *Archives of General Psychiatry, 63,* 865–872.

Olsen, K. L. (1979). Androgen-insensitive rats are defeminised by their testes. *Nature, 279,* 238–239.

Olsen, R. K., Pangelinan, M. M., Bogulski, C., Chakravarty, M. M., et al. (2015). The effect of lifelong bilingualism on regional grey and white matter volume. *Brain Research, 1612,* 128–139.

Olson, S. (2004). Making sense of Tourette's. *Science, 305,* 1390–1392.

Olsson, A., and Phelps, E. A. (2007). Social learning of fear. *Nature Neuroscience, 10,* 1095–1102.

Opendak, M., and Gould, E. (2015). Adult neurogenesis: A substrate for experience-dependent change. *Trends in Cognitive Science, 19,* 151–161.

Oppenheim, K. (2006, February 3). Life full of danger for little girl who can't feel pain. *CNN* (http://www.cnn.com/2006/HEALTH/conditions/02/03/btsc.oppenheim)

Osorio, D., and Vorobyev, M. (2008). A review of the evolution of animal colour vision and visual communication signals. *Vision Research, 48,* 2042–2051.

Ossenkoppele, R., Jansen, W. J., Rabinovici, G. D., Knol, D. L., et al. (2015). Prevalence of amyloid PET positivity in dementia syndromes: A meta-analysis. *JAMA, 313,* 1939–1949.

Osterhout, L. (1997). On the brain response to syntactic anomalies: Manipulations of word position and word class reveal individual differences. *Brain and Language, 59,* 494–522.

Ouattara, K., Lemasson, A., and Zuberbühler, K. (2009). Campbell's monkeys concatenate vocalizations into context-specific call sequences. *Proceedings of the National Academy of Sciences, USA, 22,* 22026–22031.

Overstreet, D. H. (1993). The Flinders sensitive line rats: A genetic animal model of depression. *Neuroscience and Biobehavioral Reviews, 17,* 51–68.

Ozelius, L. J., Senthil, G., Saunders-Pullman, R., Ohmann, E., et al. (2006). *LRRK2 G2019S* as a cause of Parkinson's disease in Ashkenazi Jews. *New England Journal of Medicine, 354,* 424–425.

P

Pack, A. I. (2003). Should a pharmaceutical be approved for the broad indication of excessive sleepiness? *American Journal of Respiratory and Critical Care Medicine, 167,* 109–111.

Pagel, J. F., and Helfter, P. (2003). Drug induced nightmares—An etiology based review. *Human Psychopharmacology, 18,* 59–67.

Palmer, S. M, Crowther, S. G, Carey, L. M, and the START Project Team. (2015). A meta-analysis of changes in brain activity in clinical depression. *Frontiers in Human Neuroscience, 8,* 1045.

Panksepp, J. (1998). *Affective neuroscience.* New York, NY: Oxford University Press.

Panksepp, J. (2000). Emotions as natural kinds within the mammalian brain. In M. Lewis and J. M. Haviland-Jones (Eds.), *Handbook of emotions* (2nd ed., pp. 137–156). New York, NY: Guilford Press.

Panksepp, J. (2005). Beyond a joke: From animal laughter to human joy? *Science, 308,* 62–63.

Panksepp, J. (2007). Neuroevolutionary sources of laughter and social joy: Modeling primal human laughter in laboratory rats. *Behavioural Brain Research, 182,* 231–244.

Panksepp, J. B., Yue, Z., Drerup, C., and Huber, R. (2003). Amine neurochemistry and aggression in crayfish. *Microscopy Research and Technique, 60,* 360–368.

Panov, A. V., Gutekunst, C.-A., Leavitt, B. R., Hayden, M. R., et al. (2002). Early mitochondrial calcium defects in Huntington's disease are a direct effect of polyglutamines. *Nature Neuroscience, 5,* 731–736.

Pantev, C., Oostenveld, R., Engelien, A., Ross, B., et al. (1998). Increased auditory cortical representation in musicians. *Nature, 392,* 811–814.

Pantle, A., and Sekuler, R. (1968). Size-detecting mechanisms in human vision. *Science, 162,* 1146–1148.

Papez, J. W. (1937). A proposed mechanism of emotion. *Archives of Neurology and Psychiatry, 38,* 725–745.

Papka, M., Ivry, R., and Woodruff-Pak, D. S. (1994). Eyeblink classical conditioning and time production in patients with cerebellar damage. *Society of Neuroscience Abstracts, 20,* 360.

Pare, M., Behets, C., and Cornu, O. (2003). Paucity of presumptive ruffini corpuscles in the index finger pad of humans. *Journal of Comparative Neurology, 456,* 260–266.

Park, S., Como, P. G., Cui, L., and Kurlan, R. (1993). The early course of the Tourette's syndrome clinical spectrum. *Neurology, 43,* 1712–1715.

Parker, G., Cahill, L., and McGaugh, J. L. (2006). A case of unusual autobiographical remembering. *Neurocase, 12,* 35–49.

Parkes, J. D. (1985). *Sleep and its disorders.* Philadelphia, PA: Saunders.

Parrott, A. C. (2013). Human psychobiology of MDMA or "Ecstasy": An overview of 25 years of empirical research. *Human Psychopharmacology, 28,* 289–307.

Parton, L. E., Ye, C. P., Coppari, R., Enriori, P. J., et al. (2007). Glucose sensing by POMC neurons regulates glucose homeostasis and is impaired in obesity. *Nature, 449,* 228–232.

Pastalkova, E., Itskov, V., Amarasingham, A., and Buzsáki, G. (2008). Internally generated cell assembly sequences in the rat hippocampus. *Science, 321,* 1322–1327.

Paterson, S. J., Brown, J. H., Gsödl, M. K., Johnson, M. H., et al. (1999). Cognitive modularity and genetic disorders. *Science, 286,* 2355–2358.

Patterson, F., and Linden, E. (1981). *The education of Koko.* New York, NY: Holt, Rinehart, and Winston.

Patterson, P. H. (2007). Maternal effects on schizophrenia risk. *Science, 318,* 576–578.

Paus, T., Kalina, M., Patocková, L., Angerová, Y., et al. (1991). Medial vs lateral frontal lobe lesions and differential impairment of central-gaze fixation maintenance in man. *Brain, 114,* 2051–2067.

Paus, T., Keshavan, M., and Giedd, J. N. (2008). Why do many psychiatric disorders emerge during adolescence? *Nature Reviews Neuroscience, 9,* 947–956.

Pedersen, C. B., and Mortensen, P. B. (2001). Evidence of a dose-response relationship between urbanicity during upbringing and schizophrenia risk. *Archives of General Psychiatry, 58,* 1039–1046.

Pediatric Eye Disease Investigator Group. (2005). Randomized trial of treatment of amblyopia in children aged 7 to 17 years. *Archives of Ophthalmology, 13,* 437–447.

Pedreira, C., Mormann, F., Kraskov, A., Cerf, M., et al. (2010). Responses of human medial temporal lobe neurons are modulated by stimulus repetition. *Journal of Neurophysiology, 103,* 97–107.

Peever, J., Luppi, P. H., and Montplaisir, J. (2014). Breakdown in REM sleep circuitry underlies REM sleep behavior disorder. *Trends in Neurosciences, 37,* 279–288.

Pegna, A. J., Khateb, A., Lazeyras, F., and Seghier, M. L. (2005). Discriminating emotional faces without primary visual cortices involves the right amygdala. *Nature Neuroscience, 8,* 24–25.

Peña, J. L., and Konishi, M. (2000). Cellular mechanisms for resolving phase ambiguity in the owl's inferior colliculus. *Proceedings of the National Academy of Sciences, USA, 97,* 11787–11792.

Penfield, W., and Rasmussen, T. (1950). *The cerebral cortex in man.* New York, NY: Macmillan.

Penfield, W., and Roberts, L. (1959). *Speech and brain-mechanisms.* Princeton, NJ: Princeton University Press.

Peplau, L. A. (2003). Human sexuality: How do men and women differ? *Current Directions in Psychological Science, 12,* 37–40.

Pepper, J., Hariz, M., and Zrinzo, L. (2015). Deep brain stimulation versus anterior capsulotomy for obsessive-compulsive disorder: A review of the literature. *Journal of Neurosurgery, 122,* 1028–1037.

Perani, D., and Abutalebi, J. (2005). The neural basis of first and second language processing. *Current Opinion in Neurobiology, 15,* 202–206.

Perenin, M. T., and Vighetto, A. (1988). Optic ataxia: A specific disruption in visuomotor mechanisms. I. Different aspects of the deficit in reaching for objects. *Brain, 111,* 643–674.

Pernía-Andrade, A. J., Kato, A., Witschi, R., Nyilas, R., et al. (2009). Spinal endocannabinoids and CB1 receptors mediate C-fiber–induced heterosynaptic pain sensitization. *Science, 325,* 760–764.

Peschanski, M., Defer, G., N'Guyen, J. P., Ricolfi, F., et al. (1994). Bilateral motor improvement and alteration of L-dopa effect in two patients with Parkinson's disease following intrastriatal transplantation of foetal ventral mesencephalon. *Brain, 117,* 487–499.

Peter, M. E., Medema, J. P., and Krammer, P. H. (1997). Does the *Caenorhabditis elegans* protein CED-4 contain a region of homology to the mammalian death effector domain? *Cell Death and Differentiation, 4,* 51–134.

Peterhans, E., and von der Heydt, R. (1989). Mechanisms of contour perception in monkey visual cortex. II. Contours bridging gaps. *Journal of Neuroscience, 9,* 1749–1763.

Peters, A., Palay, S. L., and Webster, H. deF. (1991). *The fine structure of the nervous system: Neurons and their supporting cells* (3rd ed.). New York, NY: Oxford University Press.

Peterson, B. S., Warner, V., Bansal, R., Zhu, H., et al. (2009). Cortical thinning in persons at increased familial risk for major depression. *Proceedings of the National Academy of Sciences, USA, 106,* 6273–6278.

Peterson, L. R., and Peterson, M. J. (1959). Short-term retention of individual verbal items. *Journal of Experimental Psychology, 58,* 193–198.

Petit, C., and Richardson, G. P. (2009). Linking genes underlying deafness to hair-bundle development and function. *Nature Neuroscience, 12,* 703–710.

Petitto, L. A., Zatorre, R. J., Gauna, K., Nikelski, E. J., et al. (2000). Speech-like cerebral activity in profoundly deaf people processing signed languages: Implications for the neural basis of human language. *Proceedings of the National Academy of Sciences, USA, 97,* 13961–13966.

Petrides, M., and Milner, B. (1982). Deficits on subject-ordered tasks after frontal- and temporal-lobe lesions in man. *Neuropsychologia, 20,* 249–262.

Petrone, A. B., Simpkins, J. W., and Barr, T. L. (2014). 17β-Estradiol and inflammation: Implications for ischemic stroke. *Aging and Disease, 5,* 340–345.

Petrovic, P., Kalso, E., Petersson, K. M., and Ingvar, M. (2002). Placebo and opioid analgesia imaging—A shared neuronal network. *Science, 295,* 1737–1740.

Pettit, H. O., and Justice, J. B., Jr. (1991). Effect of dose on cocaine self-administration behavior and dopamine levels in the nucleus accumbens. *Brain Research, 539,* 94–102.

Petty, F., Kramer, G., Wilson, L., and Jordan, S. (1994). In vivo serotonin release and learned helplessness. *Psychiatry Research, 52,* 285–293.

Pfaff, D. W. (1980). *Estrogens and brain function: Neural analysis of a hormone-controlled mammalian reproductive behavior.* New York, NY: Springer-Verlag.

Pfaff, D. W. (1997). Hormones, genes, and behavior. *Proceedings of the National Academy of Sciences, USA, 94,* 14213–14216.

Pfefferbaum, A., Sullivan, E. V., Mathalon, D. H., Shear, P. K., et al. (1995). Longitudinal changes in magnetic resonance imaging brain volumes in abstinent and relapsed alcoholics. *Alcoholism: Clinical and Experimental Research, 19,* 1177–1191.

Phillips, M. L., Young, A. W., Scott, S. K., Calder, A. J., et al. (1998). Neural responses to facial and vocal expressions of fear and disgust. *Proceedings of the Royal Society of London. Series B: Biological Sciences, 265,* 1809–1817.

Phoenix, C. H., Goy, R. W., Gerall, A. A., and Young, W. C. (1959). Organizing action of prenatally administered testosterone propionate on the tissues mediating mating behavior in the female guinea pig. *Endocrinology, 65,* 369–382.

Pickens, R., and Thompson, T. (1968). Drug use by U.S. Army enlisted men in Vietnam: A followup on their return home. *Journal of Pharmacology and Experimental Therapeutics, 161,* 122–129.

Pierce, K., Müller, R.-A., Ambrose, J., Allen, G., et al. (2001). Face processing occurs outside the fusiform "face area": Evidence from functional MRI. *Brain, 124,* 2059–2073.

Pierce, R. C., and Kumaresan, V. (2006). The mesolimbic dopamine system: The final common pathway for the reinforcing effect of drugs of abuse? *Neuroscience and Biobehavioral Reviews, 30,* 215–238.

Pinel, P., Fauchereau, F., Moreno, A., Barbot, A., et al. (2012). Genetic variants of *FOXP2* and *KIAA0319/TTRAP/THEM2* locus are associated with altered brain activation in distinct language-related regions. *Journal of Neuroscience, 32,* 817–825.

Pinker, S. (1994). *The language instinct.* New York, NY: Morrow.

Pletnikov, M. V., Ayhan, Y., Nikolskaia, O., Xu, Y., et al. (2008). Inducible expression of mutant human DISC1 in mice is associated with brain and behavioral abnormalities reminiscent of schizophrenia. *Molecular Psychiatry, 13,* 13–186.

Plihal, W., and Born, J. (1999). Effects of early and late nocturnal sleep on priming and spatial memory. *Psychophysiology, 36,* 571–582.

Ploog, D. W. (1992). Neuroethological perspectives on the human brain: From the expression of emotions to intentional signing and speech. In A. Harrington (Ed.), *So human a brain: Knowledge and values in the neurosciences* (pp. 3–13). Boston, MA: Birkhauser.

Plutchik, R. (1994). *The psychology and biology of emotion.* New York, NY: HarperCollins.

Polymeropoulos, M. H., Lavedan, C., Leroy, E., Ide, S. E., et al. (1997). Mutation in the alpha-synuclein gene identified in families with Parkinson's disease. *Science, 276,* 2045–2047.

Ponsford, J. (2005). Rehabilitation interventions after mild head injury. *Current Opinion in Neurology, 18,* 692–697.

Poole, J. H., Tyack, P. L., Stoeger-Horwath, A. S., and Watwood, S. (2005). Animal behaviour: Elephants are capable of vocal learning. *Nature, 434,* 455–456.

Pooresmaeili, A., FitzGerald, T. H., Bach, D. R., Toelch, U., et al. (2014). Cross-modal effects of value on perceptual acuity and stimulus encoding. *Proceedings of the National Academy of Sciences, USA, 111,* 15244–15249.

Pope, H. G., Jr., Kouri, E. M., and Hudson, J. I. (2000). Effects of supraphysiologic doses of testosterone on mood and aggression in normal men: A randomized controlled trial. *Archives of General Psychiatry, 57,* 133–140.

Poremba, A., Malloy, M., Saunders, R. C., Carson, R. E., et al. (2004). Species-specific calls evoke asymmetric activity in the monkey's temporal poles. *Nature, 427,* 448–451.

Porta, M., Brambilla, A., Cavanna, A. E., Servello, D., et al. (2009). Thalamic deep brain stimulation for treatment-refractory Tourette syndrome: Two-year outcome. *Neurology, 73,* 1375–1380.

Porter, J., Craven, B., Khan, R. H., Chang, S.-J., et al. (2007). Mechanisms of scent-tracking in humans. *Nature Neuroscience, 10,* 27–29.

Posner, M. I. (1980). Orienting of attention. *Quarterly Journal of Experimental Psychology, 32,* 3–25.

Posner, M. I., and Cohen, Y. (1984). Components of visual orienting. In H. Bouma and D. Bowhuis (Eds.), *Attention and performance: Vol 10. Control of language processes* (pp. 531–556). Hillsdale, NJ: Erlbaum.

Posner, M. I., and Raichle, M. E. (1994). *Images of mind.* New York, NY: Scientific American Library.

Poulet, J. F. A., and Petersen, C. C. H. (2008). Internal brain state regulates membrane potential synchrony in barrel cortex of behaving mice. *Nature, 454,* 881–885.

Powell, S. B. (2010). Models of neurodevelopmental abnormalities in schizophrenia. *Current Topics in Behavioral Neurosciences, 4,* 435–481.

Powley, T. L. (2000). Vagal circuitry mediating cephalic-phase responses to food. *Appetite, 34,* 184–188.

Pratt, L. A., Brody, D. J., and Gu, Q. (2011). *Antidepressant use in persons aged 12 and over: United States, 2005–2008* (NCHS Data Brief, No. 76). Hyattsville, MD: National Center for Health Statistics.

Premack, D. (1971). Language in a chimpanzee? *Science, 172,* 808–822.

Prentice, R. L. (2014). Postmenopausal hormone therapy and the risks of coronary heart disease, breast cancer, and stroke. *Seminars in Reproductive Medicine, 32,* 419–425.

Price, M. A., and Vandenbergh, J. G. (1992). Analysis of puberty-accelerating pheromones. *Journal of Experimental Zoology, 264,* 42–45.

Prudente, C. N., Stilla, R., Buetefisch, C. M., Singh, S., et al. (2015). Neural substrates for head movements in humans: A functional magnetic resonance imaging study. *Journal of Neuroscience, 35,* 9163–9172.

Pugh, K. R., Mencl, W. E., Shaywitz, B. A., Shaywitz, S. E., et al. (2000). The angular gyrus in developmental dyslexia: Task-specific differences in functional connectivity within posterior cortex. *Psychological Science, 11,* 51–56.

Pulak, L. M., and Jensen, L. (2014). Sleep in the intensive care unit: A review. *Journal of Intensive Care Medicine,* pii: 0885066614538749.

Purves, D., Augustine, G. J., Fitzpatrick, D., Katz, L., et al. (Eds.). (2001). *Neuroscience* (2nd ed.). Sunderland, MA: Sinauer.

Purves, D., and Lotto, B. (2011). *Why we see what we do redux: A wholly empirical theory of vision.* Sunderland, MA: Sinauer.

Putman, C. T., Xu, X., Gillies, E., MacLean, I. M., et al. (2004). Effects of strength, endurance and combined training on myosin heavy chain content and fibre-type distribution in humans. *European Journal of Applied Physiology, 92,* 376–384.

R

Racette, A., Bard, C., and Peretz, I. (2006). Making non-fluent aphasics speak: Sing along! *Brain, 129,* 2571–2584.

Racine, E., Bar-Ilan, O., and Illes, J. (2005). fMRI in the public eye. *Nature Reviews Neuroscience, 6,* 159–164.

Rafal, R. D. (1994). Neglect. *Current Opinion in Neurobiology, 4,* 231–236.

Rahman, Q. (2005). The neurodevelopment of human sexual orientation. *Neuroscience and Biobehavioral Reviews, 29,* 1057–1066.

Raichle, M. E. (2015). The brain's default mode network. *Annual Review of Neuroscience, 38,* 433–447.

Raine, A., Lencz, T., Bihrle, S., LaCasse, L., et al. (2000). Reduced prefrontal gray matter volume and reduced autonomic activity in antisocial personality disorder. *Archives of General Psychiatry, 57,* 119–127.

Raine, A., Meloy, J. R., Bihrle, S., Stoddard, J., et al. (1998). Reduced prefrontal and increased subcortical brain functioning assessed using positron emission tomography in predatory and affective murderers. *Behavioral Sciences & the Law, 16,* 319–332.

Rainville, P., Duncan, G. H., Price, D. D., Carrier, B., et al. (1997). Pain affect encoded in human anterior cingulate but not somatosensory cortex. *Science, 277,* 968–971.

Raisman, G. (1978). What hope for repair of the brain? *Annals of Neurology, 3,* 101–106.

Raisman, G., and Field, P. M. (1971). Sexual dimorphism in the preoptic area of the rat. *Science, 173,* 731–733.

Raisman, G., and Li, Y. (2007). Repair of neural pathways by olfactory ensheathing cells. *Nature Reviews Neuroscience, 8,* 312–319.

Ralph, M. R., Foster, R. G., Davis, F. C., and Menaker, M. (1990). Transplanted suprachiasmatic nucleus determines circadian period. *Science, 247,* 975–978.

Ralph, M. R., and Menaker, M. (1988). A mutation of the circadian system in golden hamsters. *Science, 241,* 1225–1227.

Ramachandran, V. S., and Hubbard, E. M. (2001). Psychophysical investigations into the neural basis of synthaesthesia. *Proceedings of the Royal Society of London. Series B: Biological Sciences, 268,* 979–983.

Ramachandran, V. S., and Rogers-Ramachandran, D. (2000). Phantom limbs and neural plasticity. *Archives of Neurology, 57,* 317–320.

Rampon, C., Tang, Y. P., Goodhouse, J., Shimizu, E., et al. (2000). Enrichment induces structural changes and recovery from nonspatial memory deficits in CA1 NMDAR1-knockout mice. *Nature Neuroscience, 3,* 238–244.

Rand, M. N., and Breedlove, S. M. (1987). Ontogeny of functional innervation of bulbocavernosus muscles in male and female rats. *Brain Research, 430,* 150–152.

Rao, P. D. P., and Finger, T. E. (1984). Asymmetry of the olfactory system in the brain of the winter flounder *Pseudopleuronectes americanus. Journal of Comparative Neurology, 225,* 492–510.

Rapoport, J. L. (1989). The biology of obsessions and compulsions. *Scientific American, 260*(6), 82–89.

Rasmussen, L. E., and Greenwood, D. R. (2003). Frontalin: A chemical message of musth in Asian elephants (*Elephas maximus*). *Chemical Senses, 28,* 433–446.

Rasmussen, L. E., Riddle, H. S., and Krishnamurthy, V. (2002). Chemical communication: Mellifluous matures to malodorous in musth. *Nature, 415,* 975–976.

Rasmussen, S., Greenberg, B., Mindus, P., Friehs, G., et al. (2000). Neurosurgical approaches to intractable obsessive-compulsive disorder. *CNS Spectrums, 5*(11), 23–34.

Rathelot, J. A., and Strick, P. L. (2006). Muscle representation in the macaque motor cortex: An anatomical perspective. *Proceedings of the National Academy of Sciences, USA, 103,* 8257–8262.

Rattenborg, N. C. (2006). Do birds sleep in flight? *Naturwissenschaften, 93,* 413–425.

Rauch, S. L., Shin, L. M., and Wright, C. I. (2003). Neuroimaging studies of amygdala function in anxiety disorders. *Annals of the New York Academy of Sciences, 985,* 389–410.

Recanzone, G. H., Schreiner, D. E., and Merzenich, M. M. (1993). Plasticity in the frequency representation of primary auditory cortex following discrimination training in adult owl monkeys. *Journal of Neuroscience, 13,* 87–103.

Rechtschaffen, A., and Bergmann, B. M. (1995). Sleep deprivation in the rat by the disk-over-water method. *Behavioural Brain Research, 69,* 55–63.

Rechtschaffen, A., and Kales, A. (1968). *A manual of standardized terminology, techniques and scoring system for sleep stages of human subjects.* Bethesda, MD: U.S. National Institute of Neurological Diseases and Blindness, Neurological Information Network.

Redican, W. K. (1982). An evolutionary perspective on human facial displays. In P. Ekman (Ed.), *Emotion in the human face* (2nd ed., pp. 212–280). Cambridge, England: Cambridge University Press.

Rehkamper, G., Haase, E., and Frahm, H. D. (1988). Allometric comparison of brain weight and brain structure volumes in different breeds of the domestic pigeon, *Columba livia* f. d. (fantails, homing pigeons, strassers). *Brain, Behavior and Evolution, 31,* 141–149.

Reiner, W. G., and Gearhart, J. P. (2004). Discordant sexual identity in some genetic males with cloacal exstrophy assigned to female sex at birth. *New England Journal of Medicine, 350,* 333–341.

Reisberg, D., and Heuer, F. (1995). Emotion's multiple effects on memory. In J. L. McGaugh, N. M. Weinberger, and G. Lynch (Eds.), *Brain and memory: Modulation and mediation of neuroplasticity* (pp. 84–92). New York, NY: Oxford University Press.

Rempel-Clower, N. L., Zola, S. M., Squire, L. R., and Amaral, D. G. (1996). Three cases of enduring memory impairment after bilateral damage limited to the hippocampal formation. *Journal of Neuroscience, 16,* 5233–5255.

Renner, M. J., and Rosenzweig, M. R. (1987). *Enriched and impoverished environments: Effects on brain and behavior.* New York, NY: Springer-Verlag.

Reppert, S. M., Perlow, M. J., Tamarkin, L., and Klein, D. C. (1979). A diurnal melatonin rhythm in primate cerebrospinal fluid. *Endocrinology, 104,* 295–301.

Reppert, S. M., and Weaver, D. R. (2002). Coordination of circadian timing in mammals. *Nature, 418,* 935–941.

Ressler, K. J., Sullivan, S. L., and Buck, L. B. (1994). A molecular dissection of spatial patterning in the olfactory system. *Current Opinion in Neurobiology, 4,* 588–596.

Reuter, J., Raedler, T., Rose, M., Hand, I., et al. (2005). Pathological gambling is linked to reduced activation of the mesolimbic reward system. *Nature Neuroscience, 8,* 147–148.

Richman, D. P., and Agius, M. A. (2003). Treatment of autoimmune myasthenia gravis. *Neurology, 61,* 1652–1661.

Richter, C. (1967). Sleep and activity: Their relation to the 24-hour clock. *Proceedings of the Association for Research in Nervous and Mental Diseases, 45,* 8–27.

Risch, N., Herrell, R., Lehner, T., Liang, K. Y., et al. (2009). Interaction between the serotonin transporter gene (5-HTTLPR), stressful life events, and risk of depression: A meta-analysis. *JAMA, 301,* 2462–2471.

Rizzolatti, G., and Craighero, L. (2004). The mirror-neuron system. *Annual Review of Neuroscience, 27,* 169–192.

Robins, L. N., and Regier, D. A. (1991). *Psychiatric disorders in America: The epidemiologic catchment area study.* New York, NY: Free Press.

Robinson, D. L., and Petersen, S. E. (1992). The pulvinar and visual salience. *Trends in Neuroscience, 15,* 127–132.

Robinson, R. (2009). Intractable depression responds to deep brain stimulation. *Neurology Today, 9,* 7–10.

Rocca, W. A., Hofman, A., Brayne, C., Breteler, M. M., et al. (1991). The prevalence of vascular dementia in Europe: Facts and fragments from 1980–1990 studies. *Annals of Neurology, 30,* 817–824.

Roenneberg, T., Kuehnle, T., Pramstaller, P. P., Ricken, J., et al. (2004). A marker for the end of adolescence. *Current Biology, 14,* R1038–R1039.

Roffwarg, H. P., Muzio, J. N., and Dement, W. C. (1966). Ontogenetic development of the human sleep-dream cycle. *Science, 152,* 604–619.

Rogan, M. T., Staubli, U. V., and LeDoux, J. E. (1997). Fear conditioning induces associative long-term potentiation in the amygdala. *Nature, 390,* 604–607.

Rogawski, M. A., and Löscher, W. (2004). The neurobiology of antiepileptic drugs. *Nature Reviews Neuroscience, 5,* 553–564.

Roland, P. E. (1993). *Brain activation.* New York, NY: Wiley-Liss.

Rook, J. M., Xiang, Z., Lv, X., Ghoshal, A., et al. (2015). Biased mGlu5-positive allosteric modulators provide in vivo efficacy without potentiating mGlu5 modulation of NMDAR currents. *Neuron, 86,* 1029–1040.

Rose, K. A., Morgan, I. G., Smith, W., Burlutsky, G., et al. (2008). Myopia, lifestyle, and schooling in students of Chinese ethnicity in Singapore and Sydney. *Archives of Ophthalmology, 126,* 527–530.

Roseboom, T. J., Painter, R. C., van Abeelen, A. F., Veenendaal, M. V., et al. (2011). Hungry in the womb: What are the consequences? Lessons from the Dutch famine. *Maturitas, 70,* 141–145.

Roselli, C. E., Larkin, K., Resko, J. A., Stellflug, J. N., et al. (2004). The volume of a sexually dimorphic nucleus in the ovine medial preoptic area/anterior hypothalamus varies with sexual partner preference. *Endocrinology, 145,* 475–477.

Roselli, C. E., and Stormshak, F. (2009). The neurobiology of sexual partner preferences in rams. *Hormones and Behavior, 55,* 611–620.

Rosenbaum, R. S., Köhler, S., Schacter, D. L., Moscovitch, M., et al. (2005). The case of K.C.: Contributions of a memory-impaired person to memory theory. *Neuropsychologia, 43,* 989–1021.

Rosenfield, P. J., Kleinhaus, K., Opler, M., Perrin, M., et al. (2010). Later paternal age and sex differences in schizophrenia symptoms. *Schizophrenia Research, 116,* 191–195.

Rosenkranz, M. A., Jackson, D. C., Dalton, K. M., Dolski, I., et al. (2003). Affective style and in vivo immune response: Neurobehavioral mechanisms. *Proceedings of the National Academy of Sciences, USA, 100,* 11148–11152.

Rosenzweig, M. R. (1946). Discrimination of auditory intensities in the cat. *American Journal of Psychology, 59,* 127–136.

Rosenzweig, M. R., Krech, D., and Bennett, E. L. (1961). Heredity, environment, brain biochemistry, and learning. In *Current trends in psychological theory* (pp. 87–110). Pittsburgh, PA: University of Pittsburgh Press.

Roses, A. D. (1995). On the metabolism of apolipoprotein E and the Alzheimer diseases. *Experimental Neurology, 132,* 149–156.

Ross, C. A., and Akimov, S. S. (2014). Human-induced pluripotent stem cells: Potential for neurodegenerative diseases. *Human Molecular Genetics, 23,* R17–26.

Rossi, D. J., Oshima, T., and Attwell, D. (2000). Glutamate release in severe brain ischaemia is mainly by reversed uptake. *Nature, 403,* 316–321.

Rotarska-Jagiela, A., Schönmeyer, R., Oertel, V., Haenschel, C., et al. (2008). The corpus callosum in schizophrenia—volume and connectivity changes affect specific regions. *NeuroImage, 39,* 1522–1532.

Rothschild, A. J. (1992). Disinhibition, amnestic reactions, and other adverse reactions secondary to triazolam: A review of the literature. *Journal of Clinical Psychiatry, 53,* 69–79.

Rouw, R., and Scholter, H. S. (2007). Increased structural connectivity in grapheme-color synesthesia. *Nature Neuroscience, 10,* 792–797.

Roy, A. (1992). Hypothalamic-pituitary-adrenal axis function and suicidal behavior in depression. *Biological Psychiatry, 32,* 812–816.

Rumbaugh, D. M. (1977). *Language learning by a chimpanzee: The LANA project.* New York, NY: Academic Press.

Rupnick, M. A., Panigrahy, D., Zhang, C. Y., Dallabrida, S. M., et al. (2002). Adipose tissue mass can be regulated through the vasculature. *Proceedings of the National Academy of Sciences, USA, 99,* 10730–10735.

Rupprecht, R., Rammes, G., Eser, D., Baghai, T. C., et al. (2009). Translocator protein (18 kD) as target for anxiolytics without benzodiazepine-like side effects. *Science, 325,* 490–493.

Rusak, B., and Zucker, I. (1979). Neural regulation of circadian rhythms. *Physiological Reviews, 59,* 449–526.

Russell, J. A. (1994). Is there universal recognition of emotion from facial expressions? A review of the cross-cultural studies. *Psychological Bulletin, 115,* 102–141.

Rustichini, A., and Padoa-Schioppa, C. (2015, June 10). A neuro-computational model of economic decisions. *Journal of Neurophysiology.* doi: 10.1152/jn.00184.2015. [Epub ahead of print]

Rymer, R. (1993). *Genie: An abused child's flight from silence.* New York, NY: HarperCollins.

S

Sack, R. L., Brandes, R. W., Kendall, A. R., and Lewy, A. J. (2000). Entrainment of free-running circadian rhythms by melatonin in blind people. *New England Journal of Medicine, 343,* 1070–1077.

Sack, R. L., Lewy, A. J., Blood, M. L., Keith, L. D., et al. (1992). Circadian rhythm abnormalities in totally blind people: Incidence and clinical significance. *Journal of Clinical Endocrinology and Metabolism, 75,* 127–134.

Sahay, A., and Hen, R. (2007). Adult hippocampal neurogenesis in depression. *Nature Neuroscience, 10,* 1110–1114.

Sakurai, T., Amemiya, A., Ishii, M., Matsuzaki, I., et al. (1998). Orexins and orexin receptors: A family of hypothalamic neuropeptides and G protein-coupled receptors that regulate feeding behavior. *Cell, 92,* 573–585.

Salazar, H., Llorente, I., Jara-Oseguera, A., García-Villegas, R., et al. (2008). A single N-terminal cysteine in TRPV1 determines activation by pungent compounds from onion and garlic. *Nature Neuroscience, 11,* 255–260.

Samad, T. A., Moore, K. A., Sapirstein, A., Billet, S., et al. (2001). Interleukin-1β–mediated induction of Cox-2 in the CNS contributes to inflammatory pain hypersensitivity. *Nature, 410,* 471–475.

Samaha, F. F., Iqbal, N., Seshadri, P., Chicano, K. L., et al. (2003). A low-carbohydrate as compared with a low-fat diet in severe obesity. *New England Journal of Medicine, 348,* 2074–2081.

Samson, S., and Zatorre, R. J. (1991). Recognition memory for text and melody of songs after unilateral temporal lobe lesion: Evidence for dual encoding. *Journal of Experimental Psychology. Learning, Memory, and Cognition, 17,* 793–804.

Samson, S., and Zatorre, R. J. (1994). Contribution of the right temporal lobe to musical timbre discrimination. *Neuropsychologia, 32,* 231–240.

Sanderson, D. J., Good, M. A., Seeburg, P. H., Sprengel, R., et al. (2008). The role of the GluR-A (GluR1) AMPA receptor subunit in learning and memory. *Progress in Brain Research, 169,* 159–178.

Saper, C. B., Fuller, P. M., Pedersen, N. P., Lu, J., et al. (2010). Sleep state switching. *Neuron, 68,* 1023–1042.

Sapir, A., Soroker, N., Berger, A., and Henik, A. (1999). Inhibition of return in spatial attention: Direct evidence for collicular generation. *Nature Neuroscience, 2,* 1053–1054.

Sapolsky, R. M. (1992). Neuroendocrinology of the stress-response. In J. B. Becker, S. M. Breedlove, and D. Crews (Eds.), *Behavioral endocrinology* (pp. 287–324). Cambridge, MA: MIT Press.

Sapolsky, R. M. (2001). *A primate's memoir.* New York, NY: Scribner.

Sapolsky, R. M. (2004). *Why zebras don't get ulcers* (3rd ed.). New York, NY: Holt.

Sapolsky, R. (2008, September 23). Stress: Portrait of a killer [National Geographic Special]. [Washington, DC]: National Geographic Television.

Sarver, D. E., Rapport, M. D., Kofler, M. J., Raiker, J. S., et al. (2015). Hyperactivity in attention-deficit/hyperactivity disorder (ADHD): Impairing deficit or compensatory behavior? *Journal of Abnormal Child Psychology,* doi: 10.1007/s10802-015-0011-1

Satinoff, E. (1978). Neural organization and evolution of thermal regulation in mammals. *Science, 201,* 16–22.

Satinoff, E., and Rutstein, J. (1970). Behavioral thermoregulation in rats with anterior hypothalamic lesions. *Journal of Comparative and Physiological Psychology, 71,* 77–82.

Satinoff, E., and Shan, S. Y. (1971). Loss of behavioral thermoregulation after lateral hypothalamic lesions in rats. *Journal of Comparative and Physiological Psychology, 77,* 302–312.

Sato, J. R., Salum, G. A., Gadelha, A., Crossley, N., et al. (2015). Default mode network maturation and psychopathology in children and adolescents. *Journal of Child Psychology and Psychiatry.* doi:10.1111/jcpp.12444

Saul, S. (2006, March 8). Some sleeping pill users range far beyond bed. *The New York Times.*

Savage-Rumbaugh, E. S. (1993). *Language comprehension in ape and child.* Chicago, IL: University of Chicago Press.

Savage-Rumbaugh, [E.] S., and Lewin, R. (1994). *Kanzi: The ape at the brink of the human mind.* New York, NY: Wiley.

Saxe, M. D., Battaglia, F., Wang, J. W., Malleret, G., et al. (2006). Ablation of hippocampal neurogenesis impairs contextual fear conditioning and synaptic plasticity in the dentate gyrus. *Proceedings of the National Academy of Sciences, USA, 103,* 17501–17506.

Saxena, S., Brody, A. L., Ho, M. L., Alborzian, S., et al. (2001). Cerebral metabolism in major depression and obsessive-compulsive disorder occurring separately and concurrently. *Biological Psychiatry, 50,* 159–170.

Saxena, S., and Rauch, S. L. (2000). Functional neuroimaging and the neuroanatomy of obsessive-compulsive disorder. *Psychiatric Clinics of North America, 23,* 563–586.

Scalaidhe, S. P. O., Wilson, F. A. W., and Goldman-Rakic, P. S. (1997). Areal segregation of face-processing neurons in prefrontal cortex. *Science, 278,* 1135–1138.

SCENIHR (Scientific Committee on Emerging and Newly Identified Health Risks). (2008, September 23). *Potential health risks of exposure to noise from personal music players and mobile phones including a music playing function.* Brussels, Belgium: European Commission, Directorate General for Health & Consumers Protection (http://ec.europa.eu/health/ph_risk/committees/04_scenihr/docs/scenihr_o_018.pdf).

Schachter, S. (1975). Cognition and peripheralist-centralist controversies in motivation and emotion. In M. S. Gazzaniga and C. Blakemore (Eds.), *Handbook of psychobiology* (pp. 529–564). New York, NY: Academic Press.

Schachter, S., and Singer, J. (1962). Cognitive, social, and physiological determinants of emotional state. *Psychological Review, 69,* 379–399.

Schacter, D. L., Alpert, N. M., Savage, C. R., Rauch, S. L., et al. (1996). Conscious recollection and the human hippocampal formation: Evidence from positron emission tomography. *Proceedings of the National Academy of Sciences, USA, 93,* 321–325.

Scharff, C., Kirn, J. R., Grossman, M., Macklis, J. D., et al. (2000). Targeted neuronal death affects neuronal replacement and vocal behavior in adult songbirds. *Neuron, 25,* 481–492.

Scheibel, A. B., and Conrad, A. S. (1993). Hippocampal dysgenesis in mutant mouse and schizophrenic man: Is there a relationship? *Schizophrenia Bulletin, 19,* 21–33.

Schein, S. J., and Desimone, R. (1990). Spectral properties of V4 neurons in the macaque. *Journal of Neuroscience, 10,* 3369–3389.

Schenck, C. H., and Mahowald, M. W. (2002). REM sleep behavior disorder: Clinical, developmental, and neuroscience perspectives 16 years after its formal identification in *Sleep. Sleep, 25,* 120–138.

Schieber, M. H, and Hibbard, L. S. (1993). How somatotopic is the motor cortex hand area? *Science, 261,* 489–492.

Schiffman, S. S., Simon, S. A., Gill, J. M., and Beeker, T. G. (1986). Bretylium tosylate enhances salt taste. *Physiology & Behavior, 36,* 1129–1137.

Schildkraut, J. J., and Kety, S. S. (1967). Biogenic amines and emotion. *Science, 156,* 21–30.

Schlaug, G., Jancke, L., Huang, Y., and Steinmetz, H. (1995). In vivo evidence of structural brain asymmetry in musicians. *Science, 267,* 699–701.

Schmidt-Nielsen, K. (1960). *Animal physiology.* Englewood Cliffs, NJ: Prentice-Hall.

Schnapf, J. L., and Baylor, D. A. (1987). How photoreceptor cells respond to light. *Scientific American, 256*(4), 40–47.

Schneider, K. (1959). *Clinical psychopathology* New York, NY: Grune & Stratton.

Schneider, P., Scherg, M., Dosch, H. G., Specht, H. J., et al. (2002). Morphology of Heschl's gyrus reflects enhanced activation in the auditory cortex of musicians. *Nature Neuroscience, 5,* 688–694.

Schneps, M. H., Brockmole, J. R., Sonnert, G., and Pomplun, M. (2012). History of reading struggles linked to enhanced learning in low spatial frequency scenes. *PLOS ONE, 7,* e35724.

Schoenbaum, G., Roesch, M. R., Stalnaker, T. A., and Takahashi, Y. K. (2009). A new perspective on the role of the orbitofrontal cortex in adaptive behaviour. *Nature Reviews Neuroscience, 10,* 885–892.

Schumann, C. M., and Amaral, D. G. (2006). Stereological analysis of amygdala neuron number in autism. *Journal of Neuroscience, 26,* 7674–7679.

Schummers, J., Yu, H., and Sur, M. (2008). Tuned responses of astrocytes and their influence on hemodynamic signals in the visual cortex. *Science, 320,* 1638–1643.

Schuster, C. R. (1970). Psychological approaches to opiate dependence and self-administration by laboratory animals. *Federation Proceedings, 29,* 1–5.

Schwartz, S., and Correll, C. U. (2014). Efficacy and safety of atomoxetine in children and adolescents with attention-deficit/hyperactivity disorder: Results from a comprehensive meta-analysis and metaregression. *Journal of the American Academy of Child and Adolescent Psychiatry, 53,* 174–187.

Schwartz, W. J., Smith, C. B., Davidsen, L., Savaki, H., et al. (1979). Metabolic mapping of functional activity in the hypothalamo-neurohypophysial system of the rat. *Science, 205,* 723–725.

Schwartzer, R., and Gutiérrez-Doña, B. (2000). Health psychology. In K. Pawlik and M. R. Rosenzweig (Eds.), *International handbook of psychology* (pp. 452–465). London, England: Sage.

Schwartzkroin, P. A., and Wester, K. (1975). Long-lasting facilitation of a synaptic potential following tetanization in the in vitro hippocampal slice. *Brain Research, 89,* 107–119.

Scott, D. J., Stohler, C. S., Egnatuk, C. M., Wang, H., et al. (2008). Placebo and nocebo effects are defined by opposite opioid and dopaminergic responses. *Archives of General Psychiatry, 65,* 220–231.

Scoville, W. B., and Milner, B. (1957). Loss of recent memory after bilateral hippocampal lesions. *Journal of Neurology, Neurosurgery and Psychiatry, 20,* 11–21.

Seavey, C., Katz, P., and Zalk, S. R. (1975). Baby X: The effects of gender labels on adult responses to infants. *Sex Roles, 2,* 103–109.

Seeman, P. (1990). Atypical neuroleptics: Role of multiple receptors, endogenous dopamine, and receptor linkage. *Acta Psychiatrica Scandinavica. Supplementum, 358,* 14–20.

Seeman, P., and Tallerico, T. (1998). Antipsychotic drugs which elicit little or no parkinsonism bind more loosely than dopamine to brain D2 receptors, yet occupy high levels of these receptors. *Molecular Psychiatry, 3,* 123–134.

Seiden, R. H. (1978). Where are they now? A follow-up study of suicide attempters from the Golden Gate Bridge. *Suicide and Life Threatening Behavior, 8,* 203–216.

Selkoe, D. J. (1991). Amyloid protein and Alzheimer's disease. *Scientific American, 265*(5), 68–71.

Selye, H. (1956). *The stress of life.* New York, NY: McGraw-Hill.

Semendeferi, K., Lu, A., Schenker, N., and Damasio, H. (2002). Humans and great apes share a large frontal cortex. *Nature Neuroscience, 5,* 272–276.

Senju, A., Southgate, V., White, S., and Frith, W. (2009). Mindblind eyes: An absence of spontaneous theory of mind in Asperger syndrome. *Science, 325,* 883–885.

Serviere, J., Webster, W. R., and Calford, M. B. (1984). Isofrequency labelling revealed by a combined [14C]-2-deoxyglucose, electrophysiological, and horseradish peroxidase study of the inferior colliculus of the cat. *Journal of Comparative Neurology, 228,* 463–477.

Seuss, Dr. (1987). *The tough coughs as he ploughs the dough: Early writings and cartoons by Dr. Seuss.* New York, NY: Morrow.

Sexton, C. E., Mackay, C. E., and Ebmeier, K. P. (2013). A systematic review and meta-analysis of magnetic resonance imaging studies in late-life depression. *American Journal of Geriatric Psychiatry, 21,* 184–195.

Seyfarth, R. M., and Cheney, D. L. (2003). Meaning and emotion in animal vocalizations. *Annals of the New York Academy of Sciences, 1000,* 32–55.

Shackman, A. J., Fox, A. S., Oler, J. A., Shelton, S. E., et al. (2013). Neural mechanisms underlying heterogeneity in the presentation of anxious temperament. *Proceedings of the National Academy of Sciences, USA, 110,* 6145–6150.

Shah, D. B., Pesiridou, A., Baltuch, G. H., Malone, D. A., et al. (2008). Functional neurosurgery in the treatment of severe obsessive compulsive disorder and major depression: Overview of disease circuits and therapeutic targeting for the clinician. *Psychiatry (Edgmont), 5*(9), 24–33.

Shallice, T., and Burgess, P. W. (1991). Deficits in strategy application following frontal lobe damage in man. *Brain, 114,* 727–741.

Shammi, P., and Stuss, D. T. (1999). Humour appreciation: A role of the right frontal lobe. *Brain, 122,* 657–666.

Shapiro, R. M. (1993). Regional neuropathology in schizophrenia: Where are we? Where are we going? *Schizophrenia Research, 10,* 187–239.

Shaw, P., Eckstrand, K., Sharp, W., Blumenthal, J., et al. (2007). Attention-deficit/hyperactivity disorder is characterized by a delay in cortical maturation. *Proceedings of the National Academy of Sciences, USA, 104,* 19649–19654.

Shaw, P., Greenstein, D., Lerch, J., Clasen, L., et al. (2006). Intellectual ability and cortical development in children and adolescents. *Nature, 440,* 676–679.

Shaw, P. J., Tononi, G., Greenspan, R. J., and Robinson, D. F. (2002). Stress response genes protect against lethal effects of sleep deprivation in *Drosophila. Nature, 417,* 287–291.

Shaywitz, S. E., Shaywitz, B. A., Pugh, K. R., Fulbright, R. K., et al. (1998). Functional disruption in the organization of the brain for reading in dyslexia. *Proceedings of the National Academy of Sciences, USA, 95,* 2636–2641.

Sherrington, C. S. (1897). *A textbook of physiology. Part III. The central nervous system* (7th ed.), M. Foster (Ed.). London, England: Macmillan.

Sherrington, C. S. (1898). Experiments in examination of the peripheral distribution of the fibres of the posterior roots of some spinal nerves. *Philosophical Transactions, 190,* 45–186.

Sherry, D. F. (1992). Memory, the hippocampus, and natural selection: Studies of food-storing birds. In L. R. Squire and N. Butters (Eds.), *Neuropsychology of memory* (2nd ed., pp. 521–532). New York, NY: Guilford Press.

Sherry, D. F., and Vaccarino, A. L. (1989). Hippocampus and memory for food caches in black-capped chickadees. *Behavioral Neuroscience, 103,* 308–318.

Sherry, D. F., Vaccarino, A. L., Buckenham, K., and Herz, R. S. (1989). The hippocampal complex of food-storing birds. *Brain Behavior and Evolution, 34,* 308–317.

Sherwin, B. B. (1998). Use of combined estrogen-androgen preparations in the postmenopause: Evidence from clinical studies. *International Journal of Fertility and Women's Medicine, 43*(2), 98–103.

Sherwin, B. B. (2002). Randomized clinical trials of combined estrogen-androgen preparations: Effects on sexual functioning. *Fertility and Sterility, 77*(Suppl. 4), 49–54.

Sherwin, B. B. (2009). Estrogen therapy: Is time of initiation critical for neuroprotection? *Nature Reviews Endocrinology, 5,* 620–627.

Shih, R. A., Belmonte, P. L., and Zandi, P. P. (2004). A review of the evidence from family, twin and adoption studies for a genetic contribution to adult psychiatric disorders. *International Review of Psychiatry, 16,* 260–283.

Shimura, H., Schlossmacher, M. G., Hattori, N., Frosch, M. P., et al. (2001). Ubiquitination of a new form of α-synuclein by parkin from human brain: Implications for Parkinson's disease. *Science, 293,* 263–269.

Shipton, O. A., El-Gaby, M., Apergis-Schoute, J., Deisseroth, K., et al. (2014). Left-right dissociation of hippocampal memory processes in mice. *Proceedings of the National Academy of Sciences, USA, 111,* 15238–15243.

Shu, W., Cho, J. Y., Jiang, Y., Zhang, M., et al. (2005). Altered ultrasonic vocalization in mice with a disruption in the Foxp2 gene. *Proceedings of the National Academy of Sciences, USA, 102,* 9643–9648.

Siegel, J. M. (1994). Brainstem mechanisms generating REM sleep. In M. H. Kryger, T. Roth, and W. C. Dement (Eds.), *Principles and practice of sleep medicine* (2nd ed., pp. 125–144). Philadelphia, PA: Saunders.

Siegel, J. M. (2005). Clues to the function of mammalian sleep. *Nature, 437,* 1264–1271.

Siegel, J. M., Manger, P. R., Nienhuis, R., Fahringer, H. M., et al. (1999). Sleep in the platypus. *Neuroscience, 91,* 391–400.

Siegel, J. M., Nienhuis, R., Gulyani, S., Ouyang, S., et al. (1999). Neuronal degeneration in canine narcolepsy. *Journal of Neuroscience, 19,* 248–257.

Sierakowiak, A., Monnot, C., Aski, S. N., Uppman, M., et al. (2015). Default mode network, motor network, dorsal and ventral basal ganglia networks in the rat brain: Comparison to human networks using resting state-fMRI. *PLOS ONE,* 10, e0120345.

Sikich, L, Frazier, J. A., McClellan, J., Findling, R. L., et al. (2008). Double-blind comparison of first- and second-generation antipsychotics in early-onset schizophrenia and schizo-affective disorder: Findings from the treatment of early-onset schizophrenia spectrum disorders (TEOSS) study. *American Journal of Psychiatry, 165,* 1420–1431.

Silva, D. A., and Satz, P. (1979). Pathological left-handedness. Evaluation of a model. *Brain and Language, 7,* 8–16.

Simons, D. J., and Chabris, C. F. (1999). Gorillas in our midst: Sustained inattentional blindness for dynamic events. *Perception, 28,* 1059–1074.

Simons, D. J., and Jensen, M. S. (2009). The effects of individual differences and task difficulty on inattentional blindness. *Psychonomic Bulletin & Review, 16,* 398–403.

Singer, O., Marr, R. A., Rockenstein, E., Crews, L., et al. (2005). Targeting BACE1 with siRNAs ameliorates Alzheimer disease neuropathology in a transgenic model. *Nature Neuroscience, 8,* 1343–1349.

Singer, P. (1975). *Animal liberation: A new ethics for our treatment of animals.* New York, NY: New York Review.

Singer, T., Seymour, B., O'Doherty, J., Kaube, H., et al. (2004). Empathy for pain involves the affective but not sensory components of pain. *Science, 303,* 1157–1162.

Sirotnak, A. P., Grigsby, T., and Krugman, R. D. (2004). Physical abuse of children. *Pediatrics in Review, 25,* 264–277.

Slee, S. J., and Young, E. D. (2014). Alignment of sound localization cues in the nucleus of the brachium of the inferior colliculus. *Journal of Neurophysiology, 111,* 2624–2633.

Smale, L. (1988). Influence of male gonadal hormones and familiarity on pregnancy interruption in prairie voles. *Biology of Reproduction, 39,* 28–31.

Smale, L., Holekamp, K. E., and White, P. A. (1999). Siblicide revisited in the spotted hyaena: Does it conform to obligate or facultative models? *Animal Behaviour, 58,* 545–551.

Smith, C. (1995). Sleep states and memory processes. *Behavioural Brain Research, 69,* 137–145.

Smith, M. A., Brandt, J., and Shadmehr, R. (2000). Motor disorder in Huntington's disease begins as a dysfunction in error feedback control. *Nature, 403,* 544–549.

Smith, M. J., Cobia, D. J., Wang, L., Alpert, K. I., et al. (2014). Cannabis-related working memory deficits and associated subcortical morphological differences in healthy individuals and schizophrenia subjects. *Schizophrenia Bulletin, 40,* 287–299.

Smith, S. (1997, September 2). Dreaming awake part 1: Living with narcolepsy. *Minnesota Public Radio News* (http://news.minnesota.publicradio.org/features/199709/02_smiths_narcolepsy/narco_1.shtml).

Smoller, J. W., and Finn, C. T. (2003). Family, twin, and adoption studies of bipolar disorder. *American Journal of Medical Genetics. Part C, Seminars in Medical Genetics, 123,* 48–58.

Smyth, K. A., Pritsch, T., Cook, T. B., McClendon, M. J., et al. (2004). Worker functions and traits associated with occupations and the development of AD. *Neurology, 63,* 498–503.

Snyder, J. S., Soumier, A., Brewer, M., Pickel, J., et al. (2011). Adult hippocampal neurogenesis buffers stress responses and depressive behaviour. *Nature, 476,* 458–461.

Solomon, A. (2001). *The noonday demon: An atlas of depression.* New York, NY: Scribner.

Soltis, J., King, L. E., Douglas-Hamilton, I., Vollrath, F., et al. (2014). African elephant alarm calls distinguish between threats from humans and bees. *PLOS ONE, 9,* e89403.

Somerville, M. J., Mervis, C. B., Young, E. J., Seo, E. J., et al. (2005). Severe expressive-language delay related to duplication of the Williams-Beuren locus. *New England Journal of Medicine, 353,* 1694–1701.

Soon, C. S., Brass, M., Heinze, H. J., and Haynes, J. D. (2008). Unconscious determinants of free decisions in the human brain. *Nature Neuroscience, 11,* 543–545.

Sorensen, P. W., and Goetz, F. W. (1993). Pheromonal and reproductive function of F prostaglandins and their metabolites in teleost fish. *Journal of Lipid Mediators, 6,* 385–393.

Sowell, E. R., Kan, E., Yoshii, J., Thompson, P. M., et al. (2008). Thinning of sensorimotor cortices in children with Tourette syndrome. *Nature Neuroscience, 11,* 637–639.

Spiegler, B. J., and Mishkin, M. (1981). Evidence for the sequential participation of inferior temporal cortex and amygdala in the acquisition of stimulus-reward associations. *Behavioural Brain Research, 3,* 303–317.

Spiegler, B. J., and Yeni-Komshian, G. H. (1983). Incidence of left-handed writing in a college population with reference to family patterns of hand preference. *Neuropsychologia, 21,* 651–659.

Spitzer, R. L. (2012). Spitzer reassesses his 2003 study of reparative therapy of homosexuality. *Archives of Sexual Behavior, 41,* 757.

Squire, L. R., Amaral, D. G., Zola-Morgan, S., and Kritchevsky, M. P. G. (1989). Description of brain injury in the amnesic patient N.A. based on magnetic resonance imaging. *Experimental Neurology, 105,* 23–35.

Squire, L. R., and Moore, R. Y. (1979). Dorsal thalamic lesion in a noted case of chronic memory dysfunction. *Annals of Neurology, 6,* 503–506.

Standing, L. G. (1973). Learning 10,000 pictures. *Quarterly Journal of Experimental Psychology, 25,* 207–222.

Stanton, S. J., Beehner, J. C., Saini, E. K., Kuhn, C. M., et al. (2009). Dominance, politics, and physiology: Voters' testosterone changes on the night of the 2008 United States presidential election. *PLOS ONE, 4,* e7543.

Starkstein, S. E., and Robinson, R. G. (1994). Neuropsychiatric aspects of stroke. In C. E. Coffey, J. L. Cummings, M. R. Lovell, and G. D. Pearlson (Eds.), *The American Psychiatric Press textbook of geriatric neuropsychiatry* (pp. 457–477). Washington, DC: American Psychiatric Press.

Staubli, U. V. (1995). Parallel properties of long-term potentiation and memory. In J. L. McGaugh, N. M. Weinberger, and G. Lynch (Eds.), *Brain and memory: Modulation and mediation of neuroplasticity* (pp. 303–318). New York, NY: Oxford University Press.

Stauffer, V. L., Millen, B. A., Andersen, S., Kinon, B. J., et al. (2013). Pomaglumetad methionil: No significant difference as an adjunctive treatment for patients with prominent negative symptoms of schizophrenia compared to placebo. *Schizophrenia Research, 150,* 434–441.

Stefansson, H., Ophoff, R. A., Steinberg, S. Andreassen, O. A., et al. (2009). Common variants conferring risk of schizophrenia. *Nature, 460,* 744–747.

Stein, B. E., and Stanford, T. R. (2008). Multisensory integration: Current issues from the perspective of the single neuron. *Nature Reviews Neuroscience, 9,* 255–266.

Stein, M., and Miller, A. H. (1993). Stress, the hypothalamic-pituitary-adrenal axis, and immune function. *Advances in Experimental Medicine and Biology, 335,* 1–5.

Stein, M., Miller, A. H., and Trestman, R. L. (1991). Depression, the immune system, and health and illness. Findings in search of meaning. *Archives of General Psychiatry, 48,* 171–177.

Stephan, F. K., and Zucker, I. (1972). Circadian rhythms in drinking behavior and locomotor activity of rats are eliminated by hypothalamic lesions. *Proceedings of the National Academy of Sciences, USA, 69,* 1583–1586.

Stern, K., and McClintock, M. (1998). Regulation of ovulation by human pheromones. *Nature, 392,* 177–179.

Stoeckel, C., Gough, P. M., Watkins, K. E., and Devlin, J. T. (2009). Supramarginal gyrus involvement in visual word recognition. *Cortex, 45,* 1091–1096.

Stranahan, A. M., Khalil, D., and Gould, E. (2006). Social isolation delays the positive effects of running on adult neurogenesis. *Nature Neuroscience, 9,* 526–533.

Stronks, H. C., and Dagnelie, G. (2014). The functional performance of the Argus II retinal prosthesis. *Expert Review of Medical Devices, 11,* 23–30.

Substance Abuse and Mental Health Services Administration. (2011). *Results from the 2010 National Survey on Drug Use and Health: Summary of national findings* (NSDUH Series H-41, HHS Publication No. [SMA] 11-4658). Rockville, MD: Substance Abuse and Mental Health Services Administration.

Substance Abuse and Mental Health Services Administration. (2013). *Results from the 2012 National Survey on Drug Use and Health: Mental health findings* (NSDUH Series H-47, HHS Publication No. [SMA] 13-4805). Rockville, MD: Substance Abuse and Mental Health Services Administration.

Sumnall, H. R., and Cole, J. C. (2005). Self-reported depressive symptomatology in community samples of polysubstance misusers who report Ecstasy use: A meta-analysis. *Journal of Psychopharmacology, 19,* 84–92.

Sun, T., Patoine, C., Abu-Khalil, A., Visvader, J., et al. (2005). Early asymmetry of gene transcription in embryonic human left and right cerebral cortex. *Science, 308,* 1794–1798.

Sunn, N., Egli, M., Burazin, T. C. D., Burns, P., et al. (2002). Circulating relaxin acts on subfornical organ neurons to stimulate water drinking in the rat. *Proceedings of the National Academy of Sciences, USA, 99,* 1701–1706.

Sunstein, C. R., and Nussbaum, M. C. (Eds.). (2004). *Animal rights: Current debates and new directions.* Oxford, England: Oxford University Press.

Sutcliffe, J. G., and de Lecea, L. (2002). The hypocretins: Setting the arousal threshold. *Nature Reviews Neuroscience, 3,* 339–349.

Suzuki, S., Brown, C. M., and Wise, P. M. (2009). Neuroprotective effects of estrogens following ischemic stroke. *Frontiers in Neuroendocrinology, 30,* 201–211.

Svenningsson, P., Chergui, K., Rachleff, I., Flajolet, M., et al. (2006). Alterations in 5-HT1B receptor function by p11 in depression-like states. *Science, 311,* 77–80.

Swedo, S. E., Rapoport, J. L., Leonard, H., Lenane, M., et al. (1989). Obsessive-compulsive disorder in children and adolescents: Clinical phenomenology of 70 consecutive cases. *Archives of General Psychiatry, 46,* 335–341.

Szarfman, A., Doraiswamy, P. M., Tonning, J. M., and Levine, J. G. (2006). Association between pathologic gambling and parkinsonian therapy as detected in the Food and Drug Administration Adverse Event database. *Archives of Neurology, 62,* 299–300.

Szente, M., Gajda, Z., Said Ali, K., and Hermesz, E. (2002). Involvement of electrical coupling in the in vivo ictal epileptiform activity induced by 4-aminopyridine in the neocortex. *Neuroscience, 115,* 1067–1078.

T

Taglialatela, J. P., Cantalupo, C., and Hopkins, W. D. (2006). Gesture handedness predicts asymmetry in the chimpanzee inferior frontal gyrus. *NeuroReport, 17,* 923–927.

Takahashi, J. S. (1995). Molecular neurobiology and genetics of circadian rhythms in mammals. *Annual Review of Neuroscience, 18,* 531–554.

Takahashi, T., Svoboda, K., and Malinow, R. (2003). Experience strengthening transmission by driving AMPA receptors into synapses. *Science, 299,* 1585–1588.

Takizawa, R., Maughan, B., and Arseneault, L. (2014). Adult health outcomes of childhood bullying victimization: Evidence from a five-decade longitudinal British birth cohort. *American Journal of Psychiatry, 171,* 777–784.

Tam, J., Duda, D. G., Perentes, J. Y., Quadri, R. S., et al. (2009). Blockade of VEGFR2 and not VEGFR1 can limit diet-induced fat tissue expansion: Role of local versus bone marrow-derived endothelial cells. *PLOS ONE, 4,* e4974.

Tamietto, M., and de Gelder, B. (2010). Neural bases of the non-conscious perception of emotional signals. *Nature Reviews Neuroscience, 11,* 697–709.

Tamietto, M., Pullens, P., de Gelder, B., Weiskrantz, L., et al. (2012). Subcortical connections to human amygdala and changes following destruction of the visual cortex. *Current Biology, 22,* 1449–1455.

Tamminga, C. A., and Schulz, S. C. (1991). *Schizophrenia research.* New York, NY: Raven Press.

Tanaka, K. (1993). Neuronal mechanisms of object recognition. *Science, 262,* 685–688.

Tanda, G., Munzar, P., and Goldberg, S. R. (2000). Self-administration behavior is maintained by the psychoactive ingredient of marijuana in squirrel monkeys. *Nature Neuroscience, 3,* 1073–1074.

Tang, N. M., Dong, H. W., Wang, X. M., Tsui, Z. C., et al. (1997). Cholecystokinin antisense RNA increases the analgesic effect induced by electroacupuncture or low dose morphine: Conversion of low responder rats into high responders. *Pain, 71,* 71–80.

Tang, Y. P., Shimizu, E., Dube, G. R., Rampon, C., et al. (1999). Genetic enhancement of learning and memory in mice. *Nature, 401,* 63–69.

Tang, Y. P., Wang, H., Feng, R., Kyin, M., et al. (2001). Differential effects of enrichment on learning and memory function in NR2B transgenic mice. *Neuropharmacology, 41,* 779–790.

Tanigawa, H., Lu, H. D., and Roe, A. W. (2010). Functional organization for color and orientation in macaque V4. *Nature Neuroscience, 13,* 1542–1548.

Tanji, J. (2001). Sequential organization of multiple movements: Involvement of cortical motor areas. *Annual Review of Neuroscience, 24,* 631–651.

Taub, E. (1976). Movement in nonhuman primates deprived of somatosensory feedback. *Exercise and Sport Sciences Reviews, 4,* 335–374.

Taub, E., Uswatte, G., and Elbert, T. (2002). New treatments in neurorehabilitation founded on basic research. *Nature Reviews Neuroscience, 3,* 228–235.

Temple, E., Deutsch, G. K., Poldrack, R. A., Miller, S. L., et al. (2003). Neural deficits in children with dyslexia ameliorated by behavioral remediation: Evidence from functional MRI. *Proceedings of the National Academy of Sciences, USA, 100,* 2860–2865.

Templeton, C. N., Greene, E., and Davis, K. (2005). Allometry of alarm calls: Black-capped chickadees encode information about predator size. *Science, 308,* 1934–1937.

Terkel, J., and Rosenblatt, J. S. (1972). Humoral factors underlying maternal behavior at parturition: Cross transfusion between freely moving rats. *Journal of Comparative and Physiological Psychology, 80,* 365–371.

Terpstra, N. J., Bolhuis, J. J., Riebel, K., van der Burg, J. M., et al. (2006). Localized brain activation specific to auditory memory in a female songbird. *Journal of Comparative Neurology, 494,* 784–791.

Terrace, H. S. (1979). *Nim.* New York, NY: Knopf.

Terzian, H. (1964). Behavioural and EEG effects of intracarotid sodium amytal injection. *Acta Neurochirurgica, 12,* 230–239.

Thannickal, T. C., Moore, R. Y., Nienhuis, R., Ramanathan, L., et al. (2000). Reduced number of hypocretin neurons in human narcolepsy. *Neuron, 27,* 469–474.

Thomas, S. G., Daniel, R. T., Chacko, A. G., Thomas, M., et al. (2010). Cognitive changes following surgery in intractable hemispheric and sub-hemispheric pediatric epilepsy. *Child's Nervous System, 26,* 1067–1073.

Thompson, P. M., Vidal, C., Giedd, J. N., Gochman, P., et al. (2001). Mapping adolescent brain change reveals dynamic wave of accelerated gray matter loss in very early-onset schizophrenia. *Proceedings of the National Academy of Sciences, USA, 98,* 11650–11655.

Thompson, P. M., Vidal, C., Giedd, J. N., Gochman, P., et al. (2001). Mapping adolescent brain change reveals dynamic wave of accelerated gray matter loss in very early-onset schizophrenia. *Proceedings of the National Academy of Sciences, USA, 98,* 11650–11655.

Thompson, R. F. (1990). Neural mechanisms of classical conditioning in mammals. *Philosophical Transactions of the Royal Society of London. Series B: Biological Sciences, 329,* 161–170.

Thompson, R. F., and Krupa, D. J. (1994). Organization of memory traces in the mammalian brain. *Annual Review of Neuroscience, 17,* 519–549.

Thompson, R. F., and Steinmetz, J. E. (2009). The role of the cerebellum in classical conditioning of discrete behavioral responses. *Neuroscience, 162,* 732–755.

Thompson, T., and Schuster, C. R. (1964). Morphine self-administration, food reinforced and avoidance behaviour in rhesus monkeys. *Psychopharmacologia, 5,* 87–94.

Thornton, A. E., Cox, D. N., Whitfield, K., and Fouladi, R. T. (2008). Cumulative concussion exposure in rugby players: Neurocognitive and symptomatic outcomes. *Journal of Clinical and Experimental Neuropsychology, 30,* 398–409.

Thornton-Jones, Z. D., Kennett, G. A., Benwell, K. R., Revell, D. F., et al. (2006). The cannabinoid CB1 receptor inverse agonist, rimonabant, modifies body weight and adiponectin function in diet-induced obese rats as a consequence of reduced food intake. *Pharmacology, Biochemistry, and Behavior, 84,* 353–359.

Thorpe, S. J., and Fabre-Thorpe, M. (2001). Seeking categories in the brain. *Science, 291,* 260–263.

Timmann, D., Drepper, J., Frings, M., Maschke, M., et al. (2010). The human cerebellum contributes to motor, emotional and cognitive associative learning. A review. *Cortex, 46,* 845–857.

Todorov, A., Said, C. P., Engell, A. D., and Oosterhof, N. N. (2008). Understanding evaluation of faces on social dimensions. *Trends in Cognitive Science, 12,* 455–460.

Tolman, E. C. (1949). There is more than one kind of learning. *Psychological Review, 56,* 144–155.

Tolman, E. C., and Honzik, C. H. (1930). Introduction and removal of reward, and maze performance in rats. *University of California Publications in Psychology, 4,* 257–275.

Tom, S. M., Fox, C. R., Trepel, C., and Poldrack, R. A. (2007). The neural basis of loss aversion in decision-making under risk. *Science, 315,* 515–518.

Tootell, R. B. H., Hadjikhani, N. K., Vanduffel, W., Liu, A. K., et al. (1998). Functional analysis of primary visual cortex (V1) in humans. *Proceedings of the National Academy of Sciences, USA, 95,* 811–817.

Tootell, R. B., Silverman, M. S., Hamilton, S. L., De Valois, R. L., et al. (1988). Functional anatomy of macaque striate cortex. III. Color. *Journal of Neuroscience, 8,* 1569–1593.

Tootell, R. B., Silverman, M. S., Switkes, E., and De Valois, R. L. (1982). Deoxyglucose analysis of retinotopic organization in primate striate cortex. *Science, 218,* 902–904.

Tootell, R. B., Tsao, D., and Vanduffel, W. (2003). Neuroimaging weighs in: Humans meet macaques in "primate" visual cortex. *Journal of Neuroscience, 23,* 3981–3989.

Tordoff, M., Rawson, N., and Friedman, M. (1991). 2,5-Anhydro-D-mannitol acts in liver to initiate feeding. *American Journal of Physiology, 261,* R283–R288.

Torrey, E. F., Bowler, A. E., Taylor, E. H., and Gottesman, I. I. (1994). *Schizophrenia and manic depressive disorder.* New York, NY: Basic Books.

Treesukosol, Y., Lyall, V., Heck, G. L., Desimone, J. A., et al. (2007). A psychophysical and electrophysiological analysis of salt taste in *Trpv1* null mice. *American Journal of Physiology, Regulatory, Integrative, and Comparative Physiology, 292,* R1799–R1809.

Treffert, D. A., and Christensen, D. D. (2005). Inside the mind of a savant. *Scientific American, 293*(6), 108–113.

Treisman, A. [M]. (1996). The binding problem. *Current Opinion in Neurobiology, 6,* 171–178.

Treisman, A. M., and Gelade, G. (1980). A feature-integration theory of attention. *Cognitive Psychology, 12,* 97–136.

Treisman, M. (1977). Motion sickness—Evolutionary hypotheses. *Science, 197,* 493–495.

Trimble, M. R. (1991). Interictal psychoses of epilepsy. *Advances in Neurology, 55,* 143–152.

Tronick, R., and Reck, C. (2009). Infants of depressed mothers. *Harvard Review of Psychiatry, 17,* 147–156.

Tsuchiya, N., and Adolphs, R. (2007). Emotion and consciousness. *Trends in Cognitive Sciences, 11,* 158–167.

Tuller, D. (2002, January 8). A quiet revolution for those prone to nodding off. *The New York Times* (http://query.nytimes.com/gst/fullpage.html?sec=health&res=980DE5DD1439F93BA35752C0A9649C8B63).

Tulving, E. (1972). Episodic and semantic memory. In E. Tulving and W. Donaldson (Eds.), *Organization of memory* (pp. 381–403). New York, NY: Academic Press.

Tulving, E. (1989). Memory: Performance, knowledge, and experience. *European Journal of Cognitive Psychology, 1,* 3–26.

Tulving, E. (2002). Episodic memory: From mind to brain. *Annual Review of Psychology, 53,* 1–25.

Tulving, E., Hayman, C. A., and Macdonald, C. A. (1991). Long-lasting perceptual priming and semantic learning in amnesia: A case experiment. *Journal of Experimental Psychology: Learning, Memory, and Cognition, 17,* 595–617.

Turgeon, J. L., McDonnell, D. P, Martin, K. A, and Wise, P. M. (2004). Hormone therapy: Physiological complexity belies therapeutic simplicity. *Science, 304,* 1269–1273.

Turner, E. H., Matthews, A. M., Linardatos, E., Tell, R. A., et al. (2008). Selective publication of antidepressant trials and its influence on apparent efficacy. *New England Journal of Medicine, 358,* 252–260.

Tyack, P. L. (2003). Dolphins communicate about individual-specific social relationships. In F. de Waal and P. L. Tyack (Eds.), *Animal social complexity: Intelligence, culture, and individualized societies* (pp. 342–361). Cambridge, MA: Harvard University Press.

U

Umilta, M. A., Kohler, E., Galiese, V., ogassi, L., et al. (2001). I know what you are doing: A neurophysiological study. *Neuron,* 1, 155–165.

Ungerleider, L. G., Courtney, S. M., and Haxby, J. V. (1998). A neural system for human visual working memory. *Proceedings of the National Academy of Sciences, USA, 95,* 883–890.

Ursin, H., Baade, E., and Levine, S. (1978). *Psychobiology of stress: A study of coping men.* New York, NY: Academic Press.

V

van Anders, S. M., and Watson, N. V. (2006). Social neuroendocrinology: Effects of social contexts and behaviors on sex steroids in humans. *Human Nature, 17,* 212–237.

Vance, C., Rogelj, B., Hortobágyi, T., De Vos, K. J., et al. (2009). Mutations in FUS, an RNA processing protein, cause familial amyotrophic lateral sclerosis type 6. *Science, 323,* 1208–1211.

Vance, C. G., Dailey, D. L., Rakel, B. A., and Sluka, K. A. (2014). Using TENS for pain control: The state of the evidence. *Pain Management, 4,* 197–209.

Vandermosten, M., Boets, B., Poelmans, H., Sunaert, S., et al. (2012). A tractography study in dyslexia: Neuroanatomic correlates of orthographic, phonological and speech processing. *Brain, 135*(Pt. 3), 935–948.

Van Dongen, H. P., Maislin, G., Mullington, J. M., and Dinges, D. F. (2003). The cumulative cost of additional wakefulness: Dose-response effects on neurobehavioral functions and sleep physiology from chronic sleep restriction and total sleep deprivation. *Sleep, 26,* 117–126.

Van Essen, D. C., and Drury, H. A. (1997). Structural and functional analyses of human cerebral cortex using a surface-based atlas. *Journal of Neuroscience, 17,* 7079–7102.

Van Gaal, L. F., Rissanen, A. M., Scheen, A. J., Ziegler, O., et al. (2005). Effects of the cannabinoid-1 receptor blocker rimonabant on weight reduction and cardiovascular risk factors in overweight patients: 1-year experience from the RIO-Europe study. *Lancet, 365,* 1389–1397.

Vann, S. D., and Aggleton, J. P. (2004). The mammillary bodies: Two memory systems in one? *Nature Reviews Neuroscience, 5,* 35–44.

Van Os, J., and Kapur, S. (2009). Schizophrenia. *Lancet, 374,* 635–645.

Van Os, J., Kenis, G., and Rutten, B. P. (2010). The environment and schizophrenia. *Nature, 468,* 203–212.

Vanston, C. M., and Watson, N. V. (2005). Selective and persistent effect of foetal sex on cognition in pregnant women. *NeuroReport, 16,* 779–782.

Van Tol, M. J., van der Wee, N. J., van den Heuvel, O. A., Nielen, M. M., et al. (2010). Regional brain volume in depression and anxiety disorders. *Archives of General Psychiatry, 67,* 1002–1011.

Van Zoeren, J. G., and Stricker, E. M. (1977). Effects of preoptic, lateral hypothalamic, or dopamine-depleting lesions on behavioral thermoregulation in rats exposed to the cold. *Journal of Comparative and Physiological Psychology, 91,* 989–999.

Vargas, C. D., Aballéa, A., Rodrigues, E. C., Reilly, K. T., et al. (2009). Re-emergence of hand-muscle representations in human motor cortex after hand allograft. *Proceedings of the National Academy of Sciences, USA, 106,* 7197–7202.

Vasey, P. L. (1995). Homosexual behaviour in primates: A review of evidence and theory. *International Journal of Primatology, 16,* 173–204.

Vassar, R., Ngai, J., and Axel, R. (1993). Spatial segregation of odorant receptor expression in the mammalian olfactory epithelium. *Cell, 74,* 309–318.

Vaughan, W., and Greene, S. L. (1984). Pigeon visual memory capacity. *Journal of Experimental Psychology: Animal Behavior Processes, 10,* 256–271.

Veraa, R. P., and Grafstein, B. (1981). Cellular mechanisms for recovery from nervous system injury: A conference report. *Experimental Neurology, 71,* 6–75.

Vernes, S. C., Oliver, P. L., Spiteri, E., Lockstone, H. E., et al. (2011). Foxp2 regulates gene networks implicated in neurite outgrowth in the developing brain. *PLOS Genetics, 7,* e1002145.

Villalobos, M. E., Mizuno, A., Dahl, B. C., Kemmotsu, N., et al. (2005). Reduced functional connectivity between V1 and inferior frontal cortex associated with visuomotor performance in autism. *NeuroImage, 25,* 916–925.

Vincus, A. A., Ringwalt, C., Harris, M. S., and Shamblen, S. R. (2010). A short-term, quasi-experimental evaluation of D.A.R.E.'s revised elementary school curriculum. *Journal of Drug Education, 40,* 37–49.

Vita, A., Dieci, M., Silenzi, C., Tenconi, F., et al. (2000). Cerebral ventricular enlargement as a generalized feature of schizophrenia: A distribution analysis on 502 subjects. *Schizophrenia Research, 44,* 25–34.

Vogt, B. A. (2005). Pain and emotion interactions in subregions of the cingulate gyrus. *Nature Reviews Neuroscience, 6,* 533–544.

Volkow, N. D., Wang, G. J., Kollins, S. H., Wigal, T. L., et al. (2009). Evaluating dopamine reward pathway in ADHD: Clinical implications. *JAMA, 302,* 1084–1091.

Volkow, N. D., and Wise, R.A. (2005). How can drug addiction help us understand obesity? *Nature Neuroscience, 8,* 555–560.

Vythilingam, M., Anderson, E. R., Goddard, A., Woods, S. W., et al. (2000). Temporal lobe volume in panic disorder—A quantitative magnetic resonance imaging study. *Psychiatry Research, 99,* 75–82.

W

Wada, J. A., Clarke, R., and Hamm, A. (1975). Cerebral hemispheric asymmetry in humans. Cortical speech zones in 100 adults and 100 infant brains. *Archives of Neurology, 32,* 239–246.

Wada, J. A., and Rasmussen, T. (1960). Intracarotid injection of sodium amytal for the lateralization of cerebral speech dominance: Experimental and clinical observations. *Journal of Neurosurgery, 17,* 266–282.

Waddell, J., and Shors, T. J. (2008). Neurogenesis, learning and associative strength. *European Journal of Neuroscience, 27,* 3020–3028.

Wager, T. D., Scott, D. J., and Zubieta, J. K. (2007). Placebo effects on human μ-opioid activity during pain. *Proceedings of the National Academy of Sciences, USA, 104,* 11056–11061.

Wagner, A. D., Desmond, J. E., Demb, J. B., Glover, G. H., et al. (1997). Semantic repetition priming for verbal and pictorial knowledge: A functional MRI study of left inferior prefrontal cortex. *Journal of Cognitive Neuroscience, 9,* 714–726.

Wagner, G. C., Beuving, L. J., and Hutchinson, R. R. (1980). The effects of gonadal hormone manipulations on aggressive target-biting in mice. *Aggressive Behavior, 6,* 1–7.

Wagner, U., Gais, S., Haider, H., Verleger, R., et al. (2004). Sleep inspires insight. *Nature, 427,* 352–355.

Wahl, O. F. (1976). Monozygotic twins discordant for schizophrenia: A review. *Psychological Bulletin, 83,* 91–106.

Wahlstrom, J. (2002). Changing times: Findings from the first longitudinal study of later high school start times. *National Association of Secondary School Principals Bulletin, 86,* 3–21.

Wallis, J. D. (2007). Orbitofrontal cortex and its contribution to decision-making. *Annual Review of Neuroscience, 30,* 31–56.

Walters, R. J., Hadley, S. H., Morris, K. D. W., and Amin, J. (2000). Benzodiazepines act on GABAA receptors via two distinct and separable mechanisms. *Nature Neuroscience, 3,* 1273–1280.

Walther, S., Goya-Maldonado, R., Stippich, C., Weisbrod, M., et al. (2010). A supra-modal network for response inhibition. *NeuroReport, 21,* 191–195.

Wang, H., Yu, M., Ochani, M., Amella, C. A., et al. (2003). Nicotinic acetylcholine receptor α7 subunit is an essential regulator of inflammation. *Nature, 421,* 384–388.

Watanabe, M., Cheng, K., Murayama, Y., Ueno, K., et al. (2011). Attention but not awareness modulates the BOLD signal in the human V1 during binocular suppression. *Science, 334,* 829–831.

Watson, N. V. (2001). Sex differences in throwing: Monkeys having a fling. *Trends in Cognitive Sciences, 5,* 98–99.

Watson, N. V., Freeman, L. M., and Breedlove, S. M. (2001). Neuronal size in the spinal nucleus of the bulbocavernosus: Direct modulation by androgen in rats with mosaic androgen insensitivity. *Journal of Neuroscience, 21,* 1062–1066.

Webb, W. B. (1992). *Sleep, the gentle tyrant* (2nd ed.). Bolton, MA: Anker.

Wehr, T. A., Sack, D. A., Duncan, W. C., Mendelson, W. B., et al. (1985). Sleep and circadian rhythms in affective patients isolated from external time cues. *Psychiatry Research, 15,* 327–339.

Wei, F., Wang, G. D., Kerchner, G. A., Kim, S. J., et al. (2001). Genetic enhancement of inflammatory pain by forebrain NR2B overexpression. *Nature Neuroscience, 4,* 164–169.

Weinberger, D. R., Aloia, M. S., Goldberg, T. E., and Berman, K. F. (1994). The frontal lobes and schizophrenia. *Journal of Neuropsychiatry and Clinical Neurosciences, 6,* 419–427.

Weinberger, N. M. (1998). Physiological memory in primary auditory cortex: Characteristics and mechanisms. *Neurobiology of Learning and Memory, 70,* 226–251.

Weiner, R. D. (1994). Treatment optimization with ECT. *Psychopharmacology Bulletin, 30,* 313–320.

Weiss, L. A., Arking, D. E., and The Gene Discovery Project of Johns Hopkins & the Autism Consortium. (2009). A genome-wide linkage and association scan reveals novel loci for autism. *Nature, 461,* 802–808.

Weitzman, E. D. (1981). Sleep and its disorders. *Annual Review of Neurosciences, 4,* 381–417.

Weitzman, E. D., Czeisler, C. A., Zimmerman, J. C., and Moore-Ede, M. C. (1981). Biological rhythms in man: Relationship of sleep-wake, cortisol, growth hormone, and temperature during temporal isolation. In J. B. Martin, S. Reichlin, and K. L. Bick (Eds.), *Neurosecretion and brain peptides* (pp. 475–499). New York, NY: Raven Press.

Wesensten, N. J., Belenky, G., Kautz, M. A., Thorne, D. R., et al. (2002). Maintaining alertness and performance during sleep deprivation: Modafinil versus caffeine. *Psychopharmacology (Berlin), 159,* 238–247.

West, S. L., and O'Neal, K. K. (2004). Project D.A.R.E. outcome effectiveness revisited. *American Journal of Public Health, 94,* 1027–1029.

Westerberg, C. E., Mander, B. A., Florczak, S. M., Weintraub, S., et al. (2012). Concurrent impairments in sleep and memory in amnestic mild cognitive impairment. *Journal of the International Neuropsychological Society, 18,* 490–500.

Wever, R. A. (1979). Influence of physical workload on freerunning circadian rhythms of man. *Pflügers Archiv European Journal of Physiology, 381,* 119–126.

Wexler, N. S., Rose, E. A., and Housman, D. E. (1991). Molecular approaches to hereditary diseases of the nervous system: Huntington's disease as a paradigm. *Annual Review of Neuroscience, 14,* 503–529.

White, N. M., and Milner, P. M. (1992). The psychobiology of reinforcers. *Annual Review of Psychology, 43,* 443–471.

Whitfield-Gabrieli, S., and Ford, J. M. (2012). Default mode network activity and connectivity in psychopathology. *Annual Review of Clinical Psychology, 8,* 49–76.

Whitlock, J. R., Heynen, A. J., Shuler, M. G., and Bear, M. F. (2006). Learning induces long-term potentiation in the hippocampus. *Science, 313,* 1093–1097.

Wible, C. G., Shenton, M. E., Hokama, H., Kikinis, R., et al. (1995). Prefrontal cortex and schizophrenia. A quantitative magnetic resonance imaging study. *Archives of General Psychiatry, 52,* 279–288.

Wiesel, T. N., and Hubel, D. H. (1965). Extent of recovery from the effects of visual deprivation in kittens. *Journal of Neurophysiology, 28,* 1060–1072.

Wilens, T. E., Prince, J. B., Spencer, T. J., and Biederman, J. (2006). Stimulants and sudden death: What is a physician to do? *Pediatrics, 118,* 1215–1219.

Will, B., Galani, R., Kelche, C., and Rosenzweig, M. R. (2004). Recovery from brain injury in animals: Relative efficacy of environmental enrichment, physical exercise

or formal training (1990–2002). *Progress in Neurobiology, 72,* 167–182.

Williams, D. (1969). Neural factors related to habitual aggression. *Brain, 92,* 503–520.

Williams, J. H., Waiter, G. D., Gilchrist, A., Perrett, D. I., et al. (2006). Neural mechanisms of imitation and "mirror neuron" functioning in autistic spectrum disorder. *Neuropsychologia, 44,* 610–621.

Williams, T. J., Pepitone, M. E., Christensen, S. E., Cooke, B. M., et al. (2000). Finger-length ratios and sexual orientation. *Nature, 404,* 455–456.

Willis, S. L., Tennstedt, S. L., Marsiske, M., Ball, K., et al. (2006). Long-term effects of cognitive training on everyday functional outcomes in older adults. *JAMA, 296,* 2805–2814.

Wingfield, J. C., Ball, G. F., Dufty, A. M., Hegner, R. E., et al. (1987). Testosterone and aggression in birds. *American Scientist, 75,* 602–608.

Winkowski, D. E., and Knudsen, E. I. (2006). Top-down gain control of the auditory space map by gaze control circuitry in the barn owl. *Nature, 439,* 336–339.

Winocur, G., Wojtowicz, J. M., Sekeres, M., Snyder, J. S., et al. (2006). Inhibition of neurogenesis interferes with hippocampus-dependent memory function. *Hippocampus, 16,* 296–304.

Winslow, J. T., and Insel, T. R. (2002). The social deficits of the oxytocin knockout mouse. *Neuropeptides, 36,* 221–229.

Wisdom, A. J., Cao, Y., Itoh, N., Spence, R. D., et al. (2013). Estrogen receptor-β ligand treatment after disease onset is neuroprotective in the multiple sclerosis model. *Journal of Neuroscience Research, 91,* 901–908.

Woldorff, M. G., and Hillyard, S. A. (1991). Modulation of early auditory processing during selective listening to rapidly presented tones. *Electroencephalography and Clinical Neurophysiology, 79,* 170–191.

Wolf, S. S., Jones, D. W., Knable, M. B., Gorey, J. G., et al. (1996). Tourette syndrome: Prediction of phenotypic variation in monozygotic twins by caudate nucleus D2 receptor binding. *Science, 273,* 1225–1227.

Wolfe, J. M. (1994). Guided search 2.0: A revised model of visual search. *Psychonomic Bulletin & Review, 1,* 202–238.

Wolfe, J. M., Horowitz, T. S., and Kenner, N. M. (2005). Cognitive psychology: Rare items often missed in visual searches. *Nature, 435,* 439–440.

Wolk, D. A., Price, J. C., Saxton, J. A., Snitz, B. E., et al. (2009). Amyloid imaging in mild cognitive impairment subtypes. *Annals of Neurology, 65,* 557–568.

Wollan, M. (2015, April 10). How to beat a polygraph test. *New York Times Magazine,* p. MM25 (www.nytimes.com/2015/04/12/magazine/how-to-beat-a-polygraph-test.html).

Womelsdorf, T., Anton-Erxleben, K., and Treue, S. (2008). Receptive field shift and shrinkage in macaque middle temporal area through attentional gain modulation. *Journal of Neuroscience, 28,* 8934–8944.

Wood, J. M., Bootzin, R. R., Kihlstrom, J. F., and Schacter, D. L. (1992). Implicit and explicit memory for verbal information presented during sleep. *Psychological Science, 3,* 236–239.

Wood, N., and Cowan, N. (1995). The cocktail party phenomenon revisited: How frequent are attention shifts to one's name in an irrelevant auditory channel? *Journal of Experimental Psychology. Learning, Memory, and Cognition, 21,* 255–260.

Woolf, C. J., and Salter, M. W. (2000). Neuronal plasticity: Increasing the gain in pain. *Science, 288,* 1765–1769.

World Health Organization. (2001). *The world health report.* Geneva, Switzerland: World Health Organization.

Wren, A. M., Seal, L. J., Cohen, M. A., Brynes, A. E., et al. (2001). Ghrelin enhances appetite and increases food intake in humans. *Journal of Clinical Endocrinology and Metabolism, 86,* 5992–5995.

Wren, A. M., Small, C. J., Ward, H. L., Murphy, K. G., et al. (2000). The novel hypothalamic peptide ghrelin stimulates food intake and growth hormone secretion. *Endocrinology, 141,* 4325–4328.

Wright, A. A., Santiago, H. C., Sands, S. F., Kendrick, D. F., et al. (1985). Memory processing of serial lists by pigeons, monkeys, and people. *Science, 229,* 287–289.

Wright, R. D., and Ward, L. M. (2008). *Orienting of attention.* New York, NY: Oxford University Press.

Wuethrich, B. (2000). Learning the world's languages—before they vanish. *Science, 288,* 1156–1159.

Wurtz, R. H., Goldberg, M. E., and Robinson, D. L. (1982). Brain mechanisms of visual attention. *Scientific American, 246*(6), 124–135.

X

Xerri, C., Stern, J. M., and Merzenich, M. M. (1994). Alterations of the cortical representation of the rat ventrum induced by nursing behavior. *Journal of Neuroscience, 14,* 1710–1721.

Xie, L., Kang, H., Xu, Q., Chen, M. J., et al. (2013). Sleep drives metabolite clearance from the adult brain. *Science, 342,* 373–377.

Y

Yaffe, K., Lui, L. Y., Zmuda, J., and Cauley, J. (2002). Sex hormones and cognitive function in older men. *Journal of the American Geriatrics Society, 50,* 707–712.

Yamazaki, S., Numano, R., Abe, M., Hida, A., et al. (2000). Resetting central and peripheral circadian oscillators in transgenic rats. *Science, 288,* 682–685.

Yang, S. H., Cheng, P. H., Banta, H., Piotrowska-Nitsche, K., et al. (2008). Towards a transgenic model of Huntington's disease in a non-human primate. *Nature, 453,* 921–924.

Yang, T. T., Gallen, C. C., Ramachandran, V. S., Cobb, S., et al. (1994). Noninvasive detection of cerebral plasticity in adult human somatosensory cortex. *NeuroReport, 5,* 701–704.

Yehuda, R. (2002). Post-traumatic stress disorder. *New England Journal of Medicine, 346,* 108–114.

Yin, J. C., Del Vecchio, M., Zhou, H., and Tully, T. (1995). CREB as a memory modulator: Induced expression of a dCREB2 activator isoform enhances long-term memory in *Drosophila. Cell, 81,* 107–115.

Yin, L., Wang, J., Klein, P. S., and Lazar, M. A. (2006). Nuclear receptor Rev-erbα is a critical lithium-sensitive component of the circadian clock. *Science, 311,* 1002–1004.

Young, A. B. (1993). Role of excitotoxins in heredito-degenerative neurologic diseases. *Research Publications—Association for Research in Nervous and Mental Disease, 71,* 175–189.

Young, K. D., Erickson, K., Nugent, A. C., Fromm, S. J., et al. (2012). Functional anatomy of autobiographical memory recall deficits in depression. *Psychological Medicine, 42,* 345–357.

Yu, Y. J., Atwal, J. K., Zhang, Y., Tong, R. K., et al. (2015). Therapeutic bispecific antibodies cross the blood-brain barrier in nonhuman primates. *Science Translational Medicine, 6,* 261ra154.

Z

Zaidel, E. (1976). Auditory vocabulary of the right hemisphere following brain bisection or hemidecortication. *Cortex, 12,* 191–211.

Zatorre, R. J., Evans, A. C., and Meyer, E. (1994). Neural mechanisms underlying melodic perception and memory for pitch. *Journal of Neuroscience, 14,* 1908–1919.

Zeki, S., Watson, J. D., Lueck, C. J., Friston, K. J., et al. (1991). A direct demonstration of functional specialization in human visual cortex. *Journal of Neuroscience, 11,* 641–649.

Zeman, A. (2002). *Consciousness: A user's guide.* New Haven, CT: Yale University Press.

Zendel, B. R., and Alain, C. (2009). Concurrent sound segregation is enhanced in musicians. *Journal of Cognitive Neuroscience, 21,* 1488–1498.

Zendel, B. R., Lagrois, M. É., Robitaille, N., and Peretz, I. (2015). Attending to pitch information inhibits processing of pitch information: The curious case of amusia. *Journal of Neuroscience, 35,* 3815–3824.

Zhang, C. L., Zou, Y., He, W., Gage, F. H., et al. (2008). A role for adult TLX-positive neural stem cells in learning and behaviour. *Nature, 451,* 1004–1007.

Zhang, T. Y., and Meaney, M. J. (2010). Epigenetics and the environmental regulation of the genome and its function. *Annual Review of Psychology, 61,* C1–C3.

Zhang, Y., Proenca, R., Maffei, M., Barone, M., et al. (1994). Positional cloning of the mouse obese gene and its human homologue. *Nature, 372,* 425–432.

Zhao, G. Q., Zhang, Y., Hoon, M. A., Chandrashekar, J., et al. (2003). The receptors for mammalian sweet and umami taste. *Cell, 115,* 255–266.

Zheng, J., Shen, W., He, D. Z., Long, K. B., et al. (2000). Prestin is the motor protein of cochlear outer hair cells. *Nature, 405,* 149–155.

Zhong, Z., Deane, R., Ali, Z., Parisi, M., et al. (2008). ALS-causing SOD1 mutants generate vascular changes prior to motor neuron degeneration. *Nature Neuroscience, 11,* 420–422.

Zihl, J., von Cramon, D., and Mai, N. (1983). Selective disturbance of movement vision after bilateral brain damage. *Brain, 106,* 313–340.

Zimmer, C. (2004). *The soul made flesh: The discovery of the brain—and how it changed the world.* New York, NY: Basic Books.

Zola-Morgan, S., and Squire, L. R. (1986). Memory impairment in monkeys following lesions of the hippocampus. *Behavioral Neuroscience, 100,* 155–160.

Zola-Morgan, S., Squire, L. R., and Ramus, S. J. (1994). Severity of memory impairment in monkeys as a function of locus and extent of damage within the medial temporal lobe memory system. *Hippocampus, 4,* 483–495.

Zucker, I. (1976). Light, behavior, and biologic rhythms. *Hospital Practice, 11,* 83–91.

Zucker, L. M., and Zucker, T. F. (1961). "Fatty," a mutation in the rat. *Journal of Heredity, 52,* 275–278.

Author Index

Subject Index

About the book

Editor: Sydney Carroll

Production Editor: Kathleen Emerson

Copyeditor: Lou Doucette

Indexer: Grant Hackett

Production Manager: Christopher Small

Book Design and Production: Joanne Delphia

Typeface: Palatino LT Std Light 10/12

Cover Design: Joanne Delphia

Illustration Program: Dragonfly Media Group

Book and Cover Manufacturer: LSC Communications